Exactly Solved Models: A Journey in Statistical Mechanics

Selected Papers with Commentaries (1963 - 2008)

F. Y. Wu speaking at reception of
Northeastern University Delegation
at the Shanghai University of Science and Technology
(1980)

Exactly Solved Models: A Journey in Statistical Mechanics

Selected Papers with Commentaries (1963 - 2008)

Fa Yueh Wu

Northeastern University, USA

with an introduction by

C N Yang

World Scientific

NEW JERSEY · LONDON · SINGAPORE · BEIJING · SHANGHAI · HONG KONG · TAIPEI · CHENNAI

Published by

World Scientific Publishing Co. Pte. Ltd.
5 Toh Tuck Link, Singapore 596224
USA office: 27 Warren Street, Suite 401-402, Hackensack, NJ 07601
UK office: 57 Shelton Street, Covent Garden, London WC2H 9HE

British Library Cataloguing-in-Publication Data
A catalogue record for this book is available from the British Library.

Painting on the front cover: *Journey to the Mountain*, Painting (circa 1969) by Hsu Yun (1919–1981).

The author and publisher would like to thank the following publishers of the various journals for their assistance and permission to include the selected reprints found in this volume: Elsevier Limited (*Physica A, Physics Letters A, Applied Mathematics Letters*); American Physical Society (*Physical Review Letters, Physical Review, Physical Review B, Physical Review E, Reviews of Modern Physics*); Institute of Physics (*Journal of Physics A, Journal of Physics C*); American Institute of Physics (*Journal of Mathematical Physics, Journal of Applied Physics*); Fourier Institute (*Annales de l'Institut Fourier*); Springer-Verlag (*Journal of Statistical Physics, Letters in Mathematical Physics*); Chinese Physical Society (*Chinese Journal of Physics*).

EXACTLY SOLVED MODELS: A JOURNEY IN STATISTICAL MECHANICS
Selected Papers with Commentaries (1963–2008)

ISBN-13 978-981-281-388-6
ISBN-10 981-281-388-8

Printed by Fulsland Offset Printing (S) Pte Ltd, Singapore

Introduction by C.N. Yang

There is an intriguing difference between mathematics and theoretical physics in their respective long-range value judgment of research work: In mathematics, many 19th century works are still remembered, admired and studied today; while in theoretical physics, very few 19th century works are still remembered, and very very few still studied. If you ask a good graduate student in physics to name ten prominent theorists of the 19th century, most will fail the test, I believe. But for a math graduate student, most will easily pass the test. One might say there is a tremendous "erosion" factor in long-range value judgment in physics.

F.Y. Wu chose to work in an area of theoretical physics which is closely related both in spirit and in method to mathematics. That is the reason why this very impressive book of ten chapters will be remembered, admired and studied for many years to come.

In my generation, many physicists have worked on various areas of research covered in this book. But no one, beyond F.Y. Wu, has made important contributions to all of these areas. I have known F.Y. For over forty years, and have always admired his steady untiring devotion to these important and fascinating areas. Now I shall have the chance to study in detail, from this book, many beautiful developments which I had always wanted to learn about.

C.N. Yang
March 2008

Fa Yueh Wu, Matthews Distinguished University Professor, Northeastern University (1992).

Preface

My journey in statistical mechanics began over four decades ago. The 1960s was a golden time in statistical mechanics, when the modern theory of critical phenomena was taking shape. New models of phase transitions were formulated, analyzed, and solved almost daily, and new knowledge of critical phenomena accumulated at lightning speed. I was fortunate to get involved in these exciting developments at an early stage of my career.

In the last two decades, statistical mechanics has further branched out to other areas of physics and science such as pure mathematics. With statistical mechanics reaching far and beyond, it has become more difficult and formidable to those who wish to break into the field.

This book is aimed at alleviating this situation. Over the years I have established my own way of appreciating and understanding statistical mechanics, particularly with respect to exactly solvable models and their connections and applications outside the traditional realm. I have written expository and pedagogical articles on many new developments, and it has been suggested to me by friends and colleagues that I synthesize my works into a monograph. This book is an attempt in that direction.

The goal of this book is two-fold. First, it introduces students and researchers to the many facets of exactly solvable models and the role they play outside traditional statistical mechanics. Second, the book is a reprint volume of selected papers of mine that are pedagogical in nature. To achieve the goals, I have arranged the reprinted papers into chapters according to topics, and written commentaries introducing each chapter topic starting from scratch. Commentaries are sufficiently self-contained to be useful and resourceful. A diligent student, for example, should be able to work out the Jones polynomial of knots from the Potts model after reading the chapter on knot invariants. The commentaries also describe my work and the background of some of the research papers, thereby giving students a flavor of how research ideas are formed. Lastly, the book is a faithful travelog of the arduous yet thoroughly fulfilling journey I have taken in the past four and half decades.

I wish to thank my colleagues and the numerous collaborators who have contributed either directly or indirectly to the contents of this book. I would like to thank Jacques Perk for a critical reading of the manuscript and for his insightful comments, to my daughters Yvonne, Yolanda and Yelena, and to Corinne McKamey Chang for careful readings and helpful suggestions. The uninterrupted support of my research by the National Science Foundation is acknowledged.

I am especially grateful to my good friend Jean-Marie Maillard who, on the eve of my 70th birthday, took the tedious trouble to write a 52-page review of my whole life's research. His article is included in the Appendix as snapshots from my journey in statistical mechanics. Finally, and most importantly, I thank my wife Jane Ching-Tsu for always being by my side throughout the 45 years of our marriage. Without her unwavering support and understanding, I would not have been able to complete the journey.

Fa Yueh Wu
Boston
May, 2008

Contents

REPRINTED PAPERS

COMMENTARIES

1. Dimer Statistics

Introduction

The dimer problem can be regarded as a realization of adsorption of diatomic molecules. In this picture, dimers are rod-like objects placed along edges of a lattice or a graph, with each dimer covering two neighboring lattice sites and each site accommodating exactly one dimer. Thus a lattice of N sites, N being an even integer, can accommodate $N/2$ dimers. When this happens we have a close-packed dimer configuration. Dimer statistics concerns the enumeration of dimer configurations and its ramifications. For example, one finds that there are 12,988,816 ways that an 8×8 checkerboard can be covered by 32 dominoes (Fisher, 1961).

The dimer problem was first studied within the context of adsorption entropy (Fowler and Rushbrooke, 1937). The approach to the problem was mostly numerical in the early years. A milestone in the development of the dimer problem was the exact solution for the square lattice obtained by Kasteleyn (1961), Temperley and Fisher (1961), and Fisher (1961). In recent years the richness and the combinatorial aspect of the dimer problem have attracted intense interests of mathematicians. In mathematical literature, the close-packed dimer problem is known as the problem of *perfect matchings*.

My interest in the dimer problem began in 1966 when I learned about the subject matter from Elliot Montroll in his lectures at the Brandeis Summer School in Theoretical Physics. That was a golden time in physics as there was plenty of federal funding for summer schools. The Brandeis School, together with the Boulder School (see page 28), was one of the two prominent places where young physicists learned about new advances in theoretical physics.

In February 1967, Elliott Lieb visited the Virginia Polytechnic Institute, where I was teaching, and gave a talk on his new solution of the 6-vertex ice-rule model (Lieb, 1967). Since vertex models and dimer problems are intimately connected (Chapter 2), I naturally tried to apply the method of dimers to the 6-vertex model, but ended up solving only a 5-vertex version (Wu, 1967, **P1**). My interest in dimers continued throughout the years and was particularly strengthened when I spent a sabbatical in the fall of 1980 at the Lorentz Instituut and the Delft Laboratorium voor Technische Natuurkunde in the Netherlands as the guest of Piet Kasteleyn and Hans van Leeuwen. This chapter contains an elementary exposition of the Kasteleyn approach to the dimer problem for readers new to the topic, and a brief description of my own contributions.

The Pfaffian Approach

The pfaffian approach to the dimer problem is based on two mathematical facts: i) An antisymmetric determinant can be written as the complete square of an algebraic quantity known as the *pfaffian*, and ii) there exists, ignoring signs, a bijection between terms in the pfaffian and those in a dimer generating function. Kasteleyn (1961) found a way to fix the signs to make the two entities identical. This reduces the dimer problem to an evaluation of a determinant, a task which can often be carried out by standard means.

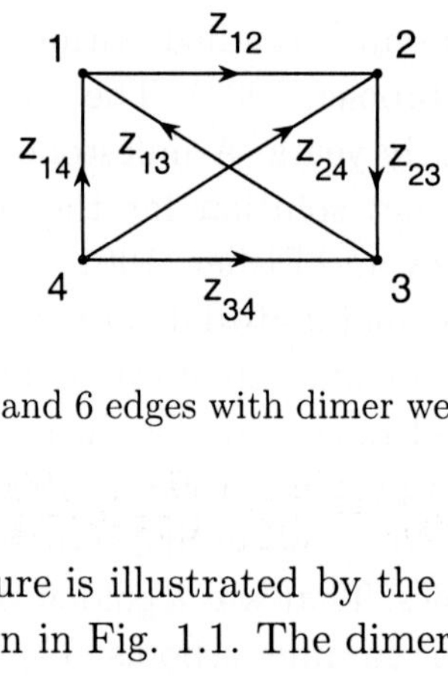

Fig. 1.1. A graph of 4 sites and 6 edges with dimer weights $z_{ij} \geq 0$. Arrows denote a Kasteleyn orientation.

The Kasteleyn procedure is illustrated by the simple example of a (planar) graph of 4 sites shown in Fig. 1.1. The dimer generating function is

$$G(\{z_{ij}\}) = z_{12}z_{34} + z_{13}z_{24} + z_{14}z_{23}, \tag{1.1}$$

where each term corresponds to a close-packed dimer covering. If one naively constructs an antisymmetric matrix M with elements $M_{ij} = z_{ij}, i < j$, or

$$M = \begin{pmatrix} 0 & z_{12} & z_{13} & z_{14} \\ -z_{12} & 0 & z_{23} & z_{24} \\ -z_{13} & -z_{23} & 0 & z_{34} \\ -z_{14} & -z_{24} & -z_{34} & 0 \end{pmatrix} \tag{1.2}$$

and evaluates its determinant, it gives $\det|M| = [\mathrm{Pf}(M)]^2$, where

$$\mathrm{Pf}(M) = z_{12}z_{34} - z_{13}z_{24} + z_{14}z_{23} \tag{1.3}$$

is the pfaffian of the matrix M. While (1.3) and (1.1) are identical term by term, the sign of the second term in (1.3) is wrong. Kasteleyn utilized the fact that one has the additional freedom of assigning a + or − sign to matrix elements. If signs of matrix elements are chosen such that all terms in the pfaffian carry the same sign, then the job is done.

Kasteleyn accomplished this for *any* planar graph, and the rule formulated by him is surprisingly simple. The superposition of two dimer configurations produces transition cycles when traced from dimer to dimer. Kasteleyn (1961) showed that if edges are oriented such that the number of arrows in one direction along any transition cycle is odd, a K-orientation, then the pfaffian of the matrix K, with elements $K_{ij} = z_{ij}$ for edges ij oriented $i \to j$, gives the desired dimer generating function.

In the example of Fig. 1.1, the orientation is a K-orientation and has the K-matrix

$$K = \begin{pmatrix} 0 & z_{12} & -z_{13} & -z_{14} \\ -z_{12} & 0 & z_{23} & -z_{24} \\ z_{13} & -z_{23} & 0 & -z_{34} \\ z_{14} & z_{24} & z_{34} & 0 \end{pmatrix}. \tag{1.4}$$

The pfaffian $\mathrm{Pf}(K) = \sqrt{\det|K|} = z_{12}z_{34} + z_{13}z_{24} + z_{14}z_{23}$ is the desired generating function.

In physics, one often considers a lattice of N sites with N large, and is interested in the closed-form evaluation of the bulk per-dimer "free energy"

$$f(\{z_{ij}\}) = \lim_{N\to\infty} \frac{1}{N/2} \ln G(\{z_{ij}\}), \tag{1.5}$$

and its associated phase transition, if any. The dimer free energy for regular two-dimensional lattices has been evaluated by numerous authors since the early 1960s, with results scattered in the literature. To put things together, I organized the results into a review (Wu, 2006a, **P2**) with particular attention paid to the occurrence of phase transitions.

The Rectangular Lattice

To illustrate what can be done, consider again the example of an 8 × 8 checkerboard. There can be different boundary conditions as shown in Fig. 1.2, where the Möbius strip and Klein bottle are examples of nonorientable surfaces. (The Klein bottle is a Möbius strip with a periodic boundary condition imposed in the other direction.) The enumeration gives the following numbers of coverings

$$
\begin{aligned}
G(\{1\}) &= 12{,}988{,}816 \quad \text{open boundaries} \\
&= 46{,}029{,}729 \quad \text{Mobius strip} \\
&= 71{,}385{,}601 \quad \text{cylinder} \\
&= 220{,}581{,}904 \quad \text{Klein bottle} \\
&= 331{,}853{,}312 \quad \text{torus.}
\end{aligned} \tag{1.6}
$$

For an $M \times N$ lattice with open boundaries and dimer weights x, y, the

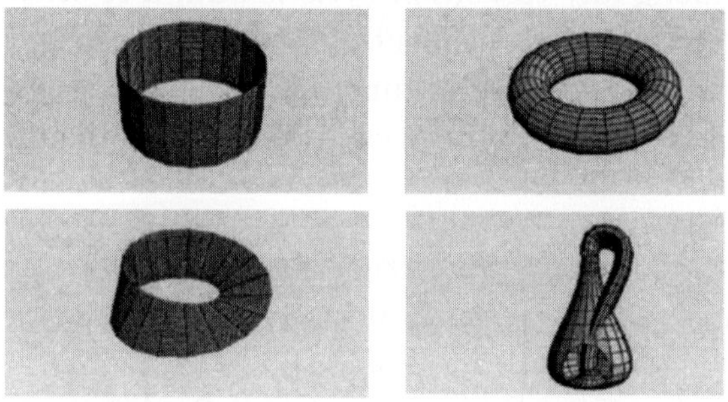

Fig. 1.2. Four different boundary conditions for a rectangular lattice. Cylindrical and toroidal (top row), Möbius strip and Klein bottle (bottom row).

Kasteleyn matrix is

$$K(x,y) = ix\big(F_N - F_N^T\big) \otimes I_M + yI_N \otimes \big(F_M - F_M^T\big) \tag{1.7}$$

where $\otimes$ denotes direct products and T denotes the transpose,

$$F_N = \begin{pmatrix} 0\,1\,0 \dots 0\,0 \\ 0\,0\,1 \dots 0\,0 \\ \vdots\;\vdots\;\vdots\;\ddots\;\vdots\;\vdots \\ 0\,0\,0 \dots 0\,1 \\ 0\,0\,0 \dots 0\,0 \end{pmatrix}, \quad I_N = \begin{pmatrix} 1\,0\,0 \dots 0\,0 \\ 0\,1\,0 \dots 0\,0 \\ \vdots\;\vdots\;\vdots\;\ddots\;\vdots\;\vdots \\ 0\,0\,0 \dots 1\,0 \\ 0\,0\,0 \dots 0\,1 \end{pmatrix}, \tag{1.8}$$

where a factor i is associated with dimers in one spatial dimension adopting a sign convention of T. T. Wu (1962).

The determinant of a matrix is equal to the product of its eigenvalues. Since eigenvalues of $F_N - F_N^T$ are $2i\cos[n\pi/(N+1)]$, $n = 1, 2, ..., N$, one obtains the desired generating function $G(x,y) = \sqrt{\det|K(x,y)|}$ where

$$\det|K(x,y)| = \prod_{m=1}^{M}\prod_{n=1}^{N}\left[-2x\cos\frac{n\pi}{N+1} + 2iy\cos\frac{m\pi}{M+1}\right]. \tag{1.9}$$

This is the result obtained by Kasteleyn (1961).

For both M, N odd, the eigenvalue $m = (M+1)/2, n = (N+1)/2$ in the product (1.9) and the determinant vanish identically, indicating correctly there is no close-packed dimer covering. This observation plays an important role when there is a boundary defect (see next Section).

In 1999, my student Wentao Lu finished his work on the Potts correlation function (Chapter 3) and was in search of a new research topic. It occurred to me that dimers on nonorientable surfaces might be a fruitful problem. The crux of the problem was to find a proper K-orientation of the lattice. It turned out that this can be quite simply done for both M, N even, and we soon obtained the solution for the Möbius strip and Klein bottle (Lu and Wu, 1999). The situation of either M or N odd is more complicated. Three years later, returning from a detour into the exact solution of the Ising model on nonorientable surfaces (see page 30), Lu and I extended the dimer solution to general M, N (Lu and Wu, 2002, **P3**) and obtained the solution in the form of a linear combination of 4 pfaffians. In all cases, the per-dimer free energy (1.5) in the bulk limit is

$$f(x,y) = \frac{1}{4\pi^2}\int_0^{2\pi} d\theta \int_0^{2\pi} d\phi \ln(4x^2\cos^2\theta + 4y^2\cos^2\phi), \tag{1.10}$$

which is independent of the boundary condition.

Rectangular Lattice with a Boundary Vacancy

Dimers on an $M \times N$ rectangular lattice, $MN = \text{odd}$, with open boundaries and a vacant site on the boundary can also be enumerated. Remarkably, the solution is given by the square root of the same product (1.9), only with the zero eigenvalue factor deleted.

During my visit to the National Center for Theoretical Sciences in Taiwan in 2002, I got together with Wen-Jer Tzeng with whom I had collaborated previously on the problem of enumerating spanning trees (Chapter 8). We decided to look into the dimer problem with a boundary vacancy. Very few exact results were known at the time about vacancies. Temperley (1974), however, had put forth an ingenious argument involving a clever bijection between dimer configurations and spanning trees. Manipulating the bijection further, Tzeng and I solved the vacancy problem and arrived at the solution (1.9) without the zero factor (Tzeng and Wu, 2003).

Two years later, I recognized further that the solution is simply a cofactor of the matrix (1.7), a fact which can be seen by introducing a "ghost site" to the lattice (Wu, 2006b, **P4**). The trick of using a ghost site also solves other vacancy problems for which the Temperley bijection cannot be used, such as a cylindrical lattice with a boundary vacancy (Wu, Tzeng and Izmailian, 2009).

The Honeycomb and Kagome Lattices

The dimer problem on the honeycomb lattice is completely equivalent to a 5-vertex model (Wu, 1968, **P5**), a bijection "re-discovered" repeatedly by numerous authors throughout the years. The general 5-vertex model maps into a dimer problem with a nonzero dimer-dimer interaction. This model was solved by myself in collaboration with my student Hsin-Yi Huang and others (Huang, *et al.*, 1996). It is the only soluble dimer system with a nonzero dimer-dimer interaction (see page 14).

The kagome lattice is of special interest in lattice statistics as it often exhibits unique features. In the writing of the aforementioned review **P2**, I found there had been no known results on weighted dimers on the kagome lattice. By mapping the problem into a vertex model, I arrived at the surprisingly simple expression for its free energy,

$$f(x,y,z) = \frac{1}{3}\ln(4xyz)\,, \tag{1.11}$$

where x, y, z are dimer activities in the three principal axes. Details of the derivation were reported in the two papers (Wang and Wu, 2007, **P6**; 2008). Fa Wang, a graduate student at UC Berkeley, and I also showed that the dimer correlation function vanishes identically beyond a short distance, again a feature unique to the kagome lattice. In a further paper (Wu and Wang, 2008) we extended the enumeration to finite lattices with a boundary and formulated a simple derivation using a direct spin variable bijection.

Solution of a Three-Dimensional Dimer Model

The Kasteleyn pfaffian approach invariably fails for lattices in higher-than-two dimensions. In 1996 Vladimir Popkov of the Seoul National University visited Northeastern for one month. During his visit we worked on the Bethe ansatz solution of a 3-dimensional lattice model involving layers of honeycomb dimer lattices interacting with a nonzero inter-layer interaction. Together with Hsin-Yi Huang we obtained its solution. This was the first time that a realistic 3-dimensional dimer system was solved. The resulting phase transition depends crucially on the strength of the inter-layer interaction, and the transition is of either first- or second-order, depending on the nature of the ordered phases (Huang, Popkov and Wu, 1997, **P7**).

Remarks

The dimer problem is related to a host of other outstanding problems in physics and mathematics. Topics discussed in this chapter touch upon only a fraction of current research interests. I have not included any discussion of the height function (Kenyon, 1997), an integer [a] associated with faces of a bipartite lattice which literally elevates the dimer problem into another dimension. Another topic is the equivalence with the sandpile model (Majumdar and Dhar, 1992). Dimers on Aztec diamonds are also of interest, as the problem is related to numerous other outstanding problems, including alternate-sign matrices, the 6-vertex model, and plane partitions of an integer (Propp, 2001). The single monomer problem is also of interest. While the general monomer-dimer problem is known to be intractable, the numerical work by Kong (2006) suggests the single monomer problem might yet be amenable. Particularly, the mobility properties of a single monomer have been studied by Poghosyan, Priezzhev and Ruelle (2008).

Finally, it should be mentioned that the method of dimers plays an important role in solving a particular class of vertex models, the free-fermion model. This connection is discussed in Chapter 2.

References for Chapter 1

Fisher, M. E. (1961), The statistical mechanics of dimers on a planar lattice, *Phys. Rev.* 124, 1664-1672.

Fowler, R. H. and G. S. Rushbrooke (1937), An attempt to extend the statistical theory to perfect solutions, *Trans. Faraday Soc.* 33, 1272-1294.

Huang, H. Y., V. Popkov and F. Y. Wu (1997), **P7**, Exact solution of a three-dimensional dimer system, *Phys. Rev. Lett.* 78, 409-413.

Huang, H. Y., F. Y. Wu, H. Kunz and D. Kim, (1996), Interacting dimers on the honeycomb lattice: Exact solution of the 5-vertex model, *Physica A* 228, 1-32.

Kasteleyn, P. W. (1961), The statistics of dimers on a lattice, *Physica* 27, 1209-1225.

Kenyon, R. (1997), Local statistics of lattice dimers, *Ann. Inst. H. Poincare, Prob. et Stat.* 33, 591-618.

Kong, Y. (2006), Packing dimers on $(2p+1) \times (2q+1)$ lattices, *Phys. Rev. E* 73, 016101.

Lieb, E. H. (1967), Exact solution of the problem of the entropy of two-dimensional ice, *Phys. Rev. Lett.* 18, 692-694.

[a]The height function increases its value by 1 when going around a B (A) site clockwise (counter-clockwise), except that it decreases by $\kappa - 1$ when crossing a dimer, where κ is the degree of the site.

Lu W. T. and F. Y. Wu (1999), Dimer statistics on a Möbius strip and the Klein bottle, *Phys. Lett. A* 259, 108-114.

Lu, W. T. and F.Y. Wu (2002), **P3**, Close-packed dimers on nonorientable surfaces, *Phys. Lett.* A 293, 235-246; Erratum, *Phys. Lett.* A 298, 293.

Majumdar, S. N. and D. Dhar (1992), Equivalence between the abelian sandpile model and the $q \to 0$ limit of the Potts model, *Physica A* 185, 129-145.

Poghosyan, V. S., V. B. Priezzhev and P. Ruelle (2008), Jamming probabilities for a vacancy in the dimer model, *Phys. Rev. E* 77, 041130.

Propp, J. (2001), The many faces of the alternating sign matrices, in *Discrete Models: Combinatorics, Computation, and Geometry*, a special issue of *Discr. Math. and Comput. Science.*

Temperley, H. N. V. (1974), in *Combinatorics: Proceedings of the British Combinatorial Conference* (London Math. Soc. Lecture Notes Series #13). 202-204.

Temperley H. N. V. and Fisher M. E. (1961), Dimer problem in statistical mechanics - An exact result, *Phil. Mag.* 6, 1061-1063.

Tzeng W. J. and F.Y. Wu (2003), Dimers on a simple-quadratic net with a vacancy, *J. Stat. Phys.* 110 671-689.

Wang, F. and F. Y. Wu (2007), **P6**, Close-packed dimers on the kagome lattice, *Phys. Rev. E* 57, 040105(R).

Wang, F. and F. Y. Wu (2008), Dimers on the kagome lattice II: Correlations and the Grassmannian approach, *Physica A* 387, 4157-4162.

Wu, F. Y. (1967), **P1**, Exactly solvable model of the ferroelectric phase transition in two dimensions, *Phys. Rev. Lett.* 18, 605-607.

Wu, F. Y. (1968), **P5**, Remarks on the modified potassium dihydrogen phosphate model of a ferroelectric, *Phys. Rev.* 168, 539-543.

Wu, F. Y. (2006a), **P2**, Dimers on two-dimensional lattices, *Int. J. Mod. Phys. B* 20, 5357-5371.

Wu, F. Y. (2006b), **P4**, The Pfaffian solution of a dimer-monomer problem: Single monomer on the boundary, *Phys. Rev. E* 74, 020104 (R); Erratum: *ibid.* 74, 039907.

Wu, F. Y., W. J. Tzeng and N. S. Izmailian (2009), Close-packed dimers on a self-dual lattice, preprint.

Wu, F. Y. and F. Wang (2008), Dimers on the kagome lattice I: Finite lattices, *Physica A* 387m 4148-4156.

Wu T. T. (1962), Dimers on rectangular lattices, *J. Math. Phys.* 3, 1265-1266.

2. The Vertex Model

Introduction

The notion of vertex model in statistical mechanics arose in the modeling of ferroelectrics (Onsager, 1939). In the structure of hydrogen-bonded ferroelectrics such as KH_2PO_4, there is one hydrogen ion on each bond with the ion located near one or the other end of the bond. By associating energies, or weights, to sites (vertices) according to nearby hydrogen ion configurations, one is led to the notion of a *vertex model.*

A major breakthrough in statistical-mechanical models since Onsager's solution of the Ising model was the exact evaluation of the residual entropy of the two-dimensional ice (Lieb, 1967), the simplest form of a vertex model. The method of the Bethe ansatz used by Lieb in solving ice-rule models opened a new avenue to other lattice models, and the study of vertex models quickly mushroomed into a fast-growing field.

I was fortunate to have met Elliott Lieb in 1967 and to have learned about his solution of the ice problem prior to its publication. I immediately looked into its possible generalizations using the method of dimers I newly learned from Elliot Montroll (see page 3), but was able to solve only a 5-vertex version. In the same year I moved from the Virginia Polytechnic Institute to Northeastern University where Lieb was building a research group. In 1967-68 my first year at Northeastern, Lieb and I collaborated mostly on the Hubbard model (see page 67). In the fall of 1968 at the invitation of Cyril Domb, we began the writing of an expository monograph on vertex models (Lieb and Wu, 1972). The monograph summarized the progress at the time. Further progress up to 1982 can be found in the definitive book by Baxter (1982).

Although Lieb moved to MIT in 1968, our collaboration continued until the completion of the expository monograph one year later. In the ensuing years I continued to work on vertex models. This chapter describes elements of the vertex model and a selection of my work in areas that are of general interest.

The Eight-Vertex Model (Rectangular Lattice)

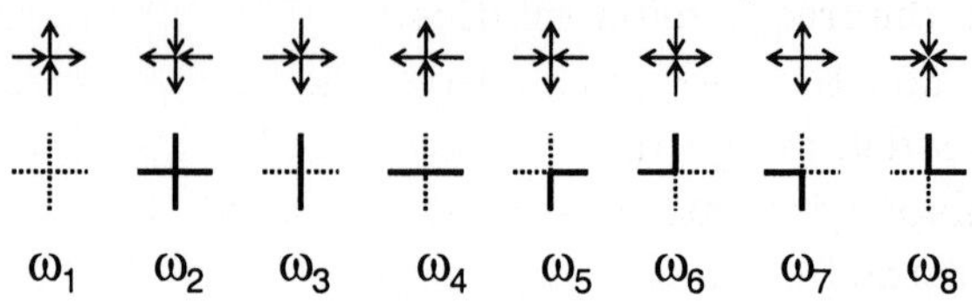

Fig. 2.1. Configurations of the 8-vertex model and the associated weights.

The 8-vertex model is an immediate extension of the ice-rule 6-vertex models. Draw arrows along edges of a square lattice with the restriction that there is always an even number, namely 0, 2, or 4, of arrows pointing into each vertex. There are 8 possible ways to do this as shown in Fig. 2.1, and this defines the *8-vertex model.* Alternately, arrow configurations can be mapped to bond configurations by attaching bonds along edges. One such mapping is shown in Fig. 2.1. The 8-vertex model is equivalent to an Ising model with multi-spin interactions (see page 27).

Denote the vertex weight at the ith site by W_i. The goal is to evaluate the partition function

$$Z_N(W) = \sum_{\text{config.}} \prod_{i=1}^{N} W_i \tag{2.1}$$

where N is the number of sites and the summation is over all arrow configurations on the lattice. The free energy in the bulk limit is

$$f(W) \equiv \lim_{N\to\infty} \frac{1}{N} \ln Z_N(W), \tag{2.2}$$

and one is interested in studying the analytical properties of $f(W)$.

I spent the summer of 1968 at Stony Brook as a guest of Chen Ning Yang. It was an exciting time in statistical mechanics when many lattice models were studied and solved. Yang and his students, Bill Sutherland and Chungpeng Fan, were working on lattice models. While at Stony Brook I collaborated with Fan in a study of the Ising model with first- and second-neighbor interactions (Fan and Wu, 1969). But we spent most of the summer in analyzing the 8-vertex model using a field-theoretical approach involving fermions. We found the 8-vertex model soluble under the condition

$$\omega_1\omega_2 + \omega_3\omega_4 = \omega_5\omega_6 + \omega_7\omega_8 \tag{2.3}$$

which is also the condition of vanishing fermion-fermion interactions. We coined the condition the *free fermion condition*, and the model the *free-*

fermion model (Fan and Wu, 1970, **P8**). These two terms are now widely used and have become part of the lexicon. We also used the method of dimers to derive the free-fermion solution.

The free-fermion 8-vertex model gives rise to a logarithmic singularity as in the Ising model, except in the case of the 5-vertex model which yields a Pokrovsky-Talapov transition (see next Section). Our study of the free-fermion model turned out to be a precursor of the Baxter solution of the 8-vertex model. The Baxter (1971, 1982) solution of the symmetric model ($\omega_1 = \omega_2 = a$, $\omega_3 = \omega_4 = b$, $\omega_5 = \omega_6 = c$, $\omega_7 = \omega_8 = d$) exhibits varied critical behavior. It is an important milestone in the history of exactly solvable models.[b]

In 1974, I visited the National Tsing Hua University in Taiwan, my alma mater, where I collaborated with K. Y. Lin and C. S. Hsue. We extended the free-fermion solution to staggered models (Wu and Lin, 1975, **P9**; Hsue, Lin and Wu, **P10**, 1975). In staggered models, vertex weights are sublattice-dependent thus breaking the translational invariance. We found the model soluble if the free-fermion condition holds locally. The algebraic manipulation was non-trivial at a time without the help of computer algebra and *Mathematica*. The result turned out to be useful 30 years later in the analysis of a different vertex model, the odd 8-vertex model.

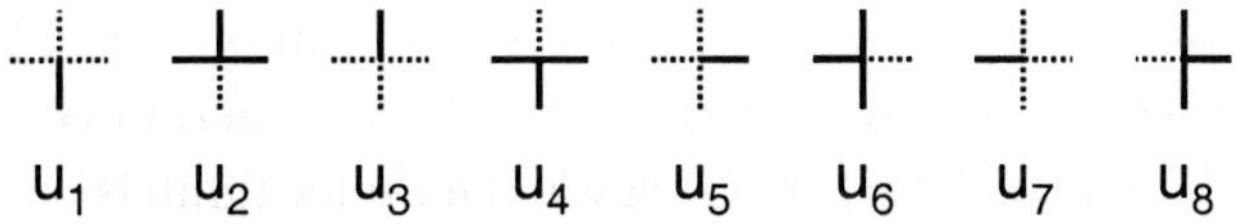

Fig. 2.2. Configurations of the odd 8-vertex model and the associated weights.

In 2004, Hervé Kunz and I introduced the *odd* 8-vertex model, which is another version of the 8-vertex model when the number of bonds incident at a vertex is always odd, namely 1 or 3 (Wu and Kunz, 2004, **P11**). The 8 possible bond configurations are shown in Fig. 2.2. We found the odd 8-vertex model again soluble under a free-fermion condition

$$u_1u_2 + u_3u_4 = u_5u_6 + u_7u_8 \ . \tag{2.4}$$

[b]Rodney Baxter carried out this seminal work aboard a cruise ship on his way from England to Australia in 1971. Years later in Canberra, Rodney showed me the pile of notes inches thick he wrote aboard that storied cruise, demonstrating the hard work behind the discovery.

But the phase transition occurs outside the physical regime of real vertex weights, which was perhaps the reason that the model had not been considered earlier. We dedicated the paper **P11** to Elliott Lieb in a special issue of *Journal of Statistical Physics* in honor of his 70th birthday.

The Five-Vertex Model (Rectangular Lattice)

The 5-vertex model has vertex configurations and weights shown in Fig. 2.3. The problem is equivalent to interacting dimers on the honeycomb lattice.

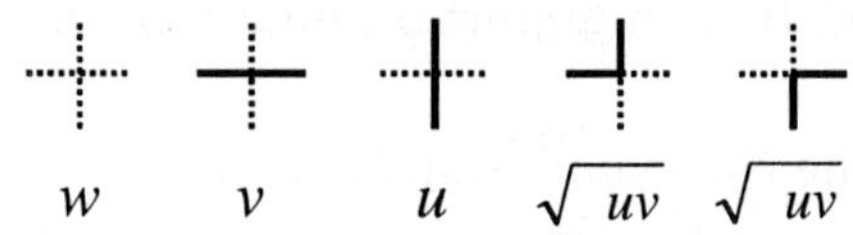

Fig. 2.3. Configurations of the 5-vertex model and the associated vertex weights.

For $\lambda = 1$, the dimers are non-interacting and the 5-vertex model is a free-fermion model soluble by pfaffians. The occurrence of a phase transition in this model was known to Kasteleyn (1963) within the honeycomb dimer model, but its explicit solution was given by myself a few years later in (Wu, 1967, **P1**; 1968, **P5**). The specific heat was found to exhibit the singularity

$$c \sim (T - T_c)^{-1/2}, \quad T \geq T_c \tag{2.5}$$

with the system frozen below T_c at $u + v + w = 2\max\{u, v, w\}$.

The transition (2.5) was "re-discovered" by Pokrovsky and Talapov (1979) some 12 years later, who used the 5-vertex model to model a commensurate-incommensurate transition in line configurations. The singularity (2.5) is henceforth more commonly known as the Pokrovsky-Talapov (PT) transition, and the paper **P1** in which the transition was first studied was largely forgotten. For historical reasons, however, the PT transition should be more appropriately named the Kasteleyn transition.

For $\lambda \neq 1$, dimers interact with *nonzero* interactions and the solution needs to be worked out anew. While Sutherland, Yang and Yang (1967) had described a solution of the general 6-vertex model which includes the 5-vertex model in a delicate limit, details of their solution have remained unpublished. It is more fruitful to consider the 5-vertex model starting from scratch. I put my student Hsin-Yi Huang on this task.

The task involved the carrying out of a convoluted Bethe ansatz analysis in the complex domain. The full analysis was published in the joint paper (Huang, *et al.*, 1996). In addition to the transition (2.5), a new ordered phase with a second-order transition emerges in the $\lambda > 1$ regime.

The history of the joint paper (Huang, *et al.*, 1996) is worthy noting as it typifies many of my publications. Hervé Kunz and I had been interested in the interacting dimer problem since the early 1990s when we found it linked to a directed percolation in 3 dimensions (see page 48). I put Huang to work on the problem. In the meantime in 1993, Doochul Kim of the Seoul National University inquired to me whether anything was known about the 5-vertex model. He and his student J. D. Noh were working on the model from the perspective of the PT transition. This eventually led to the joint paper co-authored with Huang, Kunz and Kim. We dedicated the paper to Hans W. Capel in a special issue of *Physica A* in honor of Capel' 60th birthday.

The Eight-Vertex Model (Honeycomb Lattice)

The 8-vertex model on the honeycomb lattice (Wu, 1974a, **P12**) is of special interest. The model, shown in Fig. 2.4, includes *all* 8 possible bond incident configurations. I became interested in this model since my visit to the Australian National University in 1973.

In the subspace $ad = bc$ and $b^2 = ac$ the model is known to be the Ising model in an external magnetic field. While playing with a weak-graph transformation of the partition function (page 22), I noted the possibility of including a free parameter in the transformation. By choosing the parameter appropriately, one can make the transformed weights to lie on the Ising subspace. This led to a complete equivalence of the 8-vertex model with an Ising model in an external field H. Critical properties of the 8-vertex model could then be determined.

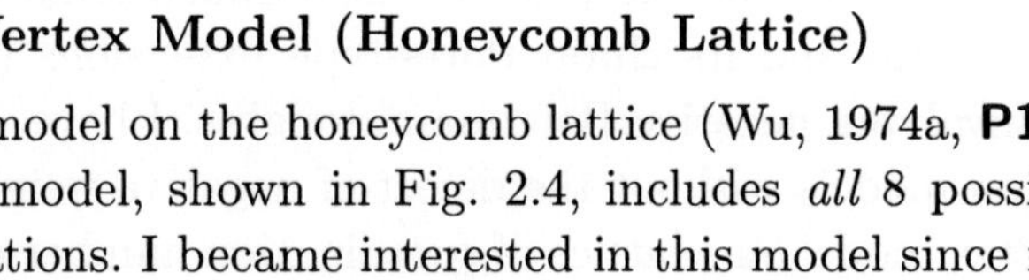

Fig. 2.4. Configurations and vertex weights of the honeycomb 8-vertex model.

The equivalent Ising interaction turned out to be either ferromagnetic with a real field, or anti-ferromagnetic with a pure imaginary field. In the former case the system is critical along a first-order phase boundary at $H = 0$, or

$$P_1 \equiv a(b^3 + d^3) - d(a^3 + c^3) + 3(ab + bc + cd)(c^2 - bd - b^2 + ac) = 0, \quad (2.6)$$

and in the latter case the system exhibits no phase transition.

The formulation was re-visited 15 years later. Xuening Wu began her Ph. D. study on the honeycomb Blume-Emery-Griffiths model (see pp. 40-41) and in the process we took another look at the 8-vertex model. This led to a more elegant analysis (Wu and Wu, 1988). Particularly, we identified the boundary (2.6) as the self-dual trajectory of the weak-graph transformation. One year later, I further reduced the bijection of the 8-vertex and Ising models to a simple spin transformation (Wu, 1990, **P13**).

Vertex Models in Higher Dimensions

There exist essentially no realistic soluble vertex models in higher-than-two dimensions, with possibly the exception of the 3-dimensional Zamolodchikov model (Baxter, 1986) and its chiral Potts-type extensions (Bazhanov and Baxter, 1992) involving negative Boltzmann weights. I have, however, made some progress on models without the defect of negative weights.

One is a vertex model on a lattice of coordination number κ in any spatial dimension, with weights

$$\begin{aligned} \omega_n &= u^{-n}, \qquad n = 0, 1, ..., \kappa - 1 \\ &= \omega, \qquad n = \kappa \end{aligned} \tag{2.7}$$

for vertices with n incident bonds.

The model (2.7) is again equivalent to an Ising model in a nonzero field. This is seen by effecting a mapping akin to a bar-molecule model of Mermin (1971). The model possesses a first-order line

$$(1 + u^2)^{\kappa/2} = w\, u^{\kappa} - 1, \qquad \text{for } 1 + u^2 > e^{4K_c} \tag{2.8}$$

where $K_c > 0$ is the Ising critical point on the same lattice. This remarkable finding was reported in (Wu, 1974b, **P14**).

I have also succeeded in solving other vertex models in d dimensions for general d, including a $(d^2 + 1)$-vertex model (Wu and Huang, 1993, **P15**). In this model each vertex is either empty or having a single line running through in a preferred direction on the lattice with periodic boundary conditions. Line segments carry weights z and loops of lines around the lattice carry factors $y = -1$.

The model is solved by mapping into a dimer problem. Curiously, the loop factors $y = -1$ are taken care of naturally in the Kasteleyn approach. This is a rare case that one can keep track, and make good use, of wrong signs in the Kasteleyn pfaffian. The system is frozen in a state of empty lines for $z < z_c = 1/d$. Above z_c, the specific heat exhibits a singularity

$$c \sim (z - z_c)^{(d-3)/2}, \quad z \to z_c + . \tag{2.9}$$

For $d = 2$ the critical behavior is the same as that of the $(y = 1)$ 5-vertex model as if the sign of y does not matter. In $d = 3$, the model mimics the experimentally observed transition in type II superconductors by regarding lines as magnetic flux (Huang and Wu, 1994).

Huang and I further extended the model to allow lines to intersect. This resulted in a $\binom{2d}{d}$-vertex model. We solved a free-fermion version of the model (Wu and Huang, 1995). The model again exhibits the critical behavior (2.9).

Remarks

Vertex models can be generalized in a number of ways.

The vertex configurations in Figs. 2.1 - 2.4 describe models in which each lattice edge can be in two states. This can be generalized to Q states with edges bearing, say, Q different colors (Perk and Schultz, 1981). One such model is the nonintersecting string model (9.6) considered by Perk and myself. Another is a 21-vertex model solved by Zamolodchikov (1980), which is a $Q = 3$ model allowing either 2 or 4 edges of the same color at a vertex.

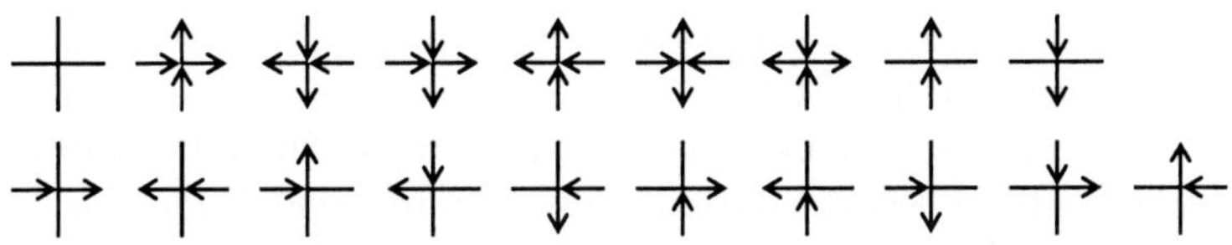

Fig. 2.5. Configurations of a 19-vertex model.

Alternately, lattice edges can be either oriented or un-oriented. For oriented edges this leads to a 19-vertex model if one allows only the same number, namely, 0, 1, or 2, of in- and out-arrows at each vertex. The allowed vertex configurations are shown in Fig. 2.5. I have used a soluble version of the 19-vertex model (Izergin and Korepin, 1981) to construct a new knot invariant (see page 59).

Vertex models can be considered for other lattices. For the triangular lattice it is most natural to consider an "ice-rule" model by restricting to configurations with 3 arrows in and 3 arrows out. This leads to a 20-vertex model (Kelland, 1974a, b) which has played a crucial role in the Baxter-Temperley-Ashley solution of the critical triangular Potts model (see page 21). One can also restrict to vertex configurations with an even number, i.e., 0, 2, 4, or 6, of in-arrows. This leads to the 32-vertex model studied by my student Joseph Sacco and myself (Sacco and Wu, 1975). We found two soluble cases.

Finally, ice-rule and 8-vertex models can be extended to the kagome lattice. The kagome 8-vertex model has been discussed by Baxter (1982), but very little is known about the kagome ice-rule model.

References for Chapter 2

Bazhanov V. V. and R. J. Baxter (1992), New solvable models in 3 dimensions, *J. Stat. Phys.* 69, 453-588.

Baxter, R. J. (1971), Eight-vertex model in lattice statistics, *Phys. Rev. Lett.* 26, 832-833.

Baxter, R. J. (1982), *Exactly Solved Models in Statistical Mechanics* (Academic, New York).

Baxter R. J. (1986), The Yang-Baxter equation and the Zamolodchikov model, *Physica D* 18, 321-347.

Fan, C. and F. Y. Wu (1969), Ising model with second-neighbor interactions. I. Some exact results and an approximate solution, *Phys. Rev.* 179, 560-570.

Fan, C. and F. Y. Wu (1970), **P8**, General lattice model of phase transitions, *Phys. Rev. B* 2, 723-733.

Huang, H. Y. and F. Y. Wu (1994), Exact solution of a model of type-II superconductors, *Physica A* 205, 31-40.

Huang, H. Y., F. Y. Wu, H. Kunz, and D. Kim, (1996), Interacting dimers on the honeycomb lattice: Exact solution of the 5-vertex model, *Physica A* 228, 1-32.

Hsue, C. S., K. Y. Lin and F. Y. Wu (1975), **P10**, Staggered eight-vertex model, *Phys. Rev. B* 12, 429-437.

Izergin A. G. and V. E. Korepin (1981), The inverse scattering method approach to the quantum Shabat-Mikhailov model, *Commun. Math. Phys.* 79, 303-316.

Kasteleyn P. W. (1963), Dimer Statistics and Phase Transitions, *J. Math. Phys.* 4, 287-293.

Kelland, S. B. (1974a), Twenty-vertex model on a triangular lattice, *Aust. J. Phys.* 27, 813-829.

Kelland, S. B. (1974b), Ferroelectric ice model on a triangular lattice, *J. Phys. A* 7, 1907-1912.

Lieb, E. H. (1967), Exact solution of the problem of the entropy of two-dimensional ice, *Phys. Rev. Lett.* 18, 692-694.

Lieb, E. H. and F. Y. Wu (1972), Two-dimensional ferroelectric models, in *Phase Transitions and Critical Phenomena, Vol. 1*, Eds. C. Domb and M. S, Green (Academic, New York), 331-490.

Mermin, N. D. (1971), Solvable model of a vapor-liquid transition with a singular coexistence-curve diameter, *Phys. Rev. Lett.* 26, 169-172.

Onsager, L. (1939), Discussion at a conference, New York Academy of Sciences.

Perk, J. H. H. and C. L. Schultz (1981), New families of commuting transfer matrices in q-state vertex models, *Phys. Lett. A* 84, 407-410.

Pokrovsky, V. I. and A. L. Talapov (1979), Ground state, spectrum, and phase diagram of 2D incommensurate crystals, *Phys. Rev. Lett.* 42, 65-67.

Sacco, J. E. and F. Y. Wu (1975), Thirty-two vertex model on a triangular lattice, *J. Phys. A* 8, 1780-1787; Erratum, *ibid* 10, 1259 (1977).

Sutherland, B., C. N. Yang and C. P. Yang (1967), Exact solution of a model of two-dimensional ferroelectrics in an arbitrary external electric field, *Phys. Rev. Lett.* 19, 588-591.

Wu, F. Y. (1967), **P1**, Exactly solvable model of the ferroelectric phase transition in two dimensions, *Phys. Rev. Lett.* 18, 605-607.

Wu, F. Y. (1968), **P5**, Remarks on the modified potassium dihydrogen phosphate model of a ferroelectric, *Phys. Rev.* 168, 539-543.

Wu, F. Y. (1974a), **P12**, Eight-vertex model on the honeycomb lattice, *J. Math. Phys.* 6, 687-691.

Wu, F. Y. (1974b), **P14**, Phase transition in a vertex model in three dimensions, *Phys. Rev. Lett.* 32, 460-463.

Wu, F. Y. (1990), **P13**, Eight-vertex model and Ising model in a non-zero magnetic field: honeycomb lattice, *J. Phys. A* 23, 375-378.

Wu, F. Y. and H. Y. Huang (1993), **P15**, Exact solution of a vertex model in d dimensions, *Lett. Math. Phys.* 29, 205-213.

Wu, F. Y. and H. Y. Huang (1995), Soluble free-fermion model in d dimensions, *Phys. Rev. E* 51, 889-895.

Wu, F. Y. and H. Kunz (2004), **P11**, The odd 8-vertex model, *J. Stat. Phys.* 116, 67-78.

Wu, F. Y. and K. Y. Lin (1975), **P9**, Staggered ice-rule model: The pfaffian solution, *Phys. Rev. B* 12, 419-428.

Wu, X. N. and F. Y. Wu (1988), Blume-Emery-Griffiths model on the honeycomb lattice, *J. Stat. Phys.* 50, 41-50.

Zamolodchikov, A. B. (1980), Tetrahedra equations and integrable systems in three-dimensional space, *JEPT* 52, 325-336.

3. Duality and Gauge Transformations

Introduction

Duality and gauge transformations play an important role in the study of field theory and statistical systems (Savit, 1980). In lattice statistics, they are relations connecting a lattice model with its dual or gauge-transformed model, which are often used to derive useful information otherwise difficult to obtain. A well-known example is Kramers and Wannier's (1941) determination of the critical temperature of the Ising model from a duality consideration before the exact solution was known. For spin systems on planar lattices, the dual and gauge models are defined on the dual lattice. For vertex models (in *any* spatial dimension), the gauge transformation relates models on the same lattice. Duality relations for spin correlation functions are also important, and have been used in the determination of the equilibrium crystal shape (Rottman and Wortis, 1981).

Duality considerations can be extended to spin systems with chiral interactions. I was fortunate to have recognized this fact, and wrote down the duality relation for the chiral Potts model (Wu and Wang, 1976, **P16**) some 10 years before the model was analyzed (Au-Yang *et al.*, 1987). In the late 1980s, I concentrated in studying gauge transformations of vertex models for the purpose of determining critical frontiers (see Chapter 6). My focus in this direction shifted in the late 1990s to duality relations of Potts correlation functions. These and other topics are described in this chapter.

Duality Relation for the Potts and the Chiral Potts Models

Consider a Potts model (Chapter 5) on a planar lattice, or more generally on a planar graph, G. Let $x_i = 1, 2, ..., q$ denote the spin state at the ith site, and $u(x_i, x_j)$ denote the Boltzmann factor of the edge ij connecting sites i and j. The interaction is *chiral* if $u(x_i, x_j) \neq u(x_j, x_i)$, namely, if the interaction is directionally-dependent. The chiral property is denoted by directing the edge ij from i to j, say. However, one usually considers a cyclic model for which $u(x_i, x_j) = u(x_i - x_j)$, mod q. This is the chiral Potts model.

To a planar G of N sites and E edges, construct a dual graph D whose

N^* sites reside in faces of G and whose E edges intersect those of G. The situation around a typical edge is shown in Fig. 3.1. Denote the respective partition functions by $Z_G(u)$ and $Z_D(u^*)$. Then the duality relation reads

$$q^{-N/2} Z_G(u) = q^{-N^*/2} Z_D(u^*) \tag{3.1}$$

$$u^*(y) = \frac{1}{\sqrt{q}} \sum_{i=1}^{q} e^{2\pi i x y/q}\, u(x). \tag{3.2}$$

Equations (3.1) and (3.2) describe the duality relation of the chiral Potts model (Wu and Wang, 1976, **P16**).

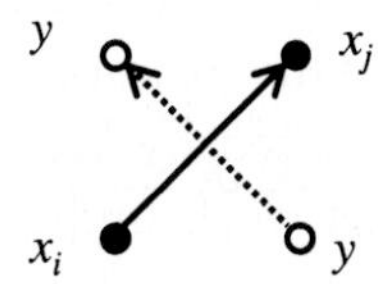

Fig. 3.1. An edge in G (solid line) and the dual edge in D (dotted line).

The dual property shown in Fig. 3.1 is local so the duality identity (3.1) holds quite generally for any planar graph with Boltzmann factors u which may vary from edge to edge. For the standard Potts model with the Boltzmann factor

$$u(x) = 1 + (e^K - 1)\delta_{\mathrm{Kr}}(x, 0), \tag{3.3}$$

where δ_{Kr} is the Kronecker delta, the duality (3.1) and (3.2) read

$$Z_G(e^K) = q^{1-N^*}(e^K - 1)^E Z_D(e^{K^*}), \tag{3.4}$$

$$e^{K^*} = (e^K + q - 1)/(e^K - 1). \tag{3.5}$$

The self-dual point of (3.5) $K = K^* = K_c$ or, explicitly,

$$e^{K_c} = 1 + \sqrt{q}, \tag{3.6}$$

determines the critical point of the square lattice Potts model (Potts, 1952).

The Potts model on the triangular lattice with 2- and 3-site interactions K_2, K_3 in alternate triangles also possesses a duality relation which Baxter, Temperley and Ashley (1978) derived using a Bethe ansatz result on a 20-vertex model due to Kelland (1974a, b). They established that the model possesses the self-dual trajectory

$$e^{3K_2 + K_3} = 3\, e^{K_2} + q - 2. \tag{3.7}$$

Two years later in 1980, Keh-Yin Lin and I re-derived the duality relation using a graphical analysis (see page 55).

Gauge Transformation for the Vertex Model and Syzygies

Consider a vertex model for which each lattice edge can be in q different states. Examples for $q = 2$ are the 8-vertex model of Fig. 2.1 for the square lattice and the 8-vertex model of Fig. 2.4 for the honeycomb lattice. Denote vertex weights at a vertex of degree κ by $W(q_1, q_2, ..., q_\kappa)$, where q_i specifies the respective states of the κ incident edges. The transformation

$$W^*(p_1, ..., p_\kappa) = \sum_{q_1=1}^{q} \cdots \sum_{q_\kappa=1}^{q} R(p_1, q_1) \cdots R(p_\kappa, q_\kappa)\, W(q_1, ..., q_\kappa), \qquad (3.8)$$

which defines a set of new weights W^*, is a gauge transformation if it leaves the partition function invariant. One example is the weak-graph transformation of the 8-vertex model formulated by Wegner (1973).

Xuening Wu began her thesis work under me in 1986 in a study of the Blume-Emery-Griffiths model on the honeycomb lattice. The study very soon led us to examine duality properties of the transformation (3.8). Eventually, we were able to establish that the self-dual manifold for the honeycomb lattice indeed yields the exact criticality (2.6) known from its Ising equivalence (Wu and Wu, 1989, **P17**).

More generally, the determination of invariant manifolds of the gauge transformation has been an outstanding problem of algebraic invariants in mathematics since Hilbert's time. The $q \times q$ matrix $\mathbf{R}$ in (3.8) provides a representation of the orthogonal group O(q). Hilbert (1890) established that the O(q) invariant manifold is of the form of a homogeneous polynomial in the weights W, and that all such polynomials can be expressed in terms of a minimal set of fundamental ones via the use of syzygies. But Hilbert did not pursue the determination of the fundamental invariants.

In the fall of 1989, I visited Jacques Perk at Oklahoma State University, and we looked into the O(2) transformation for symmetric vertex weights a, b, c, d as in Fig. 2.4. We found three fundamental invariants (Perk, Wu and Wu, 1990; **P18**). One of the invariants is precisely the polynomial P_1 given by (2.6); the other two invariants are the polynomials

$$I_1 = (a - 3c)^2 + (3b - d)^2, \quad I_2 = (a + c)^2 + (b + d)^2. \qquad (3.9)$$

The consideration of the O(3) is complicated but is facilitated by the mapping of O(3) to SL(2). In collaboration with Leh-Hun Gwa who visited Northeastern in 1990, we deduced that there are 5 fundamental invariants and worked out their explicit expressions (Gwa and Wu, 1991, **P19**). Physical applications of the O(2) and O(3) invariants are described in Chapter 6 (pp. 39-41).

Duality Relation for Frustrated Spin Systems

In the winter of 2002 I spent a 3-month sabbatical at Berkeley. During that time I frequently had lunch with my host Dung-Hai Lee, and our lunch hour conversations invariably focused on topics of common interest. One of the topics we talked about was the gauge transformation of frustrated systems.

In an Ising model with random $\pm J$ interactions, each face (plaquette) of the lattice is characterized by a parity, which is -1 if the plaquette is frustrated and $+1$ otherwise. Playing with the duality transformation, we found the dual model to be precisely an Ising model in a random external magnetic field, which is 0 for $+1$ plaquettes and $i\pi/2$ for -1 plaquettes. In particular, this identifies the Ising model in the external field $i\pi/2$, whose solution was first obtained by Lee and Yang (1952), as a completely frustrated Ising model on the dual lattice (see also Au-Yang and Perk, 1984).

The consideration can be extended to the Potts model. In a frustrated chiral Potts model the nearest-neighbor Boltzmann factor assumes q different values in the form of $u(x_i - x_j + \lambda_{ij})$ with $\lambda_{ij} = 0, 1, ..., q-1$. Lee and I found the dual model to be a chiral Potts model in a pure imaginary field $h = i\pi\nu\lambda/q$, $\nu = 0, 1, ..., q-1 \,(\mathrm{mod}\, q)$, where λ is the summation of the λ_{ij} factors around the plaquette.

These results were reported in (Lee and Wu, 2003, **P20**).

Duality Relation for Potts Correlation Functions

It had long been known that the correlation function between two boundary Ising spins is related to the interfacial tension of the dual model, which is in turn a correlation function in the dual space (Watson, 1968). In the summer of 1996, Royce Zia, while passing through Boston on his way to a conference, asked me whether such a relation existed for the Potts model. The usual derivation of the Ising relation is based on a high-temperature expansion not readily adapted to the Potts model. Several years earlier the problem had occurred in a joint paper of mine with de Magalhães and Essam (1990). But discussions there were buried underneath the use of a certain percolation average. Obviously, a better and more intuitive formulation was needed.

I came up with the idea of constructing auxiliary lattices by connecting the correlating sites on the boundary in the same spin states. The desired correlation duality can then be deduced from the duality relations (3.4) of the auxiliary lattices. In the case of the Ising correlation between 2 boundary spins, the procedure immediately yields the known duality.

Extending this method to the Potts model, I was able to derive the duality relation for correlation functions of 3 boundary spins (Wu, 1997, **P21**). However, I hastened to state in the paper that the method can be extended "in a straightforward fashion" to higher correlation functions. This claim was soon checked by Jesper Jacobsen, a student of John Cardy, who found the method inadequate for 4-spin correlations (Jacobsen, 1997).

The crux of the matter is that for general n-spin correlations there are b_n unknown correlation functions and c_n auxiliary lattices, and the method requires the determination of the b_n unknowns from c_n equations. For $n = 2, 3$, One has $c_2 = b_2 = 2$, $c_3 = b_3 = 5$, so there is no problem. For $n = 4$, however, one has $b_4 = 15, c_4 = 14$, and the number of relations is insufficient. This is the problem discovered by Jacobsen.

The underlying root of the problem is combinatorial in nature and is related to what is known as crossing partitions. The b_n correlation functions (among n boundary spins) can be characterized by partitioning the n sites into groups of different spin states. For $n = 4$, one can label partitions by indices $ijk\ell$, and there are 15 partitions $\{1111, 2111, 1211, 1121, 1112, 1123, 2113, 2311, 1231, 1213, 2131, 1122, 1221, 1234, 1212\}$. In constructing auxiliary lattices from these partitions, all lines can be made nonintersecting, except in the partition 1212 for which lines connecting 1 to 1 and 2 to 2 always intersect making the auxiliary lattice non-planar. The partition 1212 is the simplest example of a crossing, or non-planar, partition. As non-planar lattices do not possess the duality relation (3.4), one has only 14 equations.[c]

Hsin-Yi Huang and I looked into this problem closely and discovered that all non-planar correlations can actually be expressed in terms of planar ones. For $n = 4$, for example, the relation is

$$P(1212) = P(1213) + P(2131) - P(1234) \tag{3.10}$$

where $P(ijk\ell)$ are the correlation functions. Thus, there are only 14 correlation functions to be determined and all is well. We published the analysis in (Wu and Huang, 1997, **P22**), in which we also conjectured on the expressions of duality relations for general n.

In the meantime, Huang graduated and Wentao Lu began his Ph.D. work with me in 1997. The conjecture became his first assignment. We subsequently established the conjecture in (Lu and Wu, 1998) using a clumsy algebraic approach. Two years later in collaboration with Chris King, a colleague from the Mathematical Department, the conjecture was re-derived as

[c]The 14 planar auxilary lattices are shown in Fig. 3 of **P22**.

the consequence of a graph-theoretical property of the rooted Tutte polynomial (see page 52).

The duality relations for general n are most elegantly stated in terms of partially ordered sets and their Möbius inversions.

Denote correlation functions by $P(X)$ where $X = \{x_1, ..., x_n\}$ specifies the states of the n spins. The n spins divide the dual spins around the lattice boundary into sections of spin states Y. Similarly, denote the dual correlation functions by $P^*(Y)$. Write $P(X)$ and $P^*(Y)$ as summations of partially ordered sets, namely,

$$P(X) = \sum_{X' \preceq X} A(X'), \tag{3.11}$$

$$P^*(Y) = \sum_{Y' \preceq Y} A^*(Y'), \tag{3.12}$$

where X' and Y' are refinement sets within which all states are the same. Then one has the extremely elegant expression

$$A(X) = q^{-|X|-1} A^*(Y), \quad X \text{ planar} \tag{3.13}$$

$$= 0, \quad X \text{ nonplanar}, \tag{3.14}$$

where $|X|$ is the number of blocks in X.

The desired correlation duality relating $P(X)$ and $P^*(Y)$ is obtained by substituting (3.13) into (3.11) with $A^*(Y')$ given by the Möbius inverse of (3.12). Similarly, nonplanar correlations are expressed in terms of planar ones by (3.14) with $A(X)$ given by the Möbius inverse of (3.11).

Several years later, King and I further extended the Potts correlation duality to spins residing in the interior of a lattice (King and Wu, 2002), thus concluding a series of investigations of this fascinating subject.

References for Chapter 3

Au-Yang, H., B. M. McCoy, J. H. H. Perk, S. Tang and M. L. Yan (1987), Commuting transfer matrices in the chiral Potts models: Solution of star-triangle equation with genus > 1, *Phys. Lett. A* 123, 219-223.

Au-Yang, H. and J. H. H. Perk (1984), Ising correlations at the critical temperature, *Phys. Lett.* A 104, 131-134.

Baxter, R. J., H. N. V. Temperley and S. E. Ashley (1978), Triangular Potts model at its transition temperature, and related models, *Proc. R. Soc. Lond. A* 358, 535-559.

de Magalhães, A. C. N., J. W. Essam and F. Y. Wu (1990), Duality relation for Potts multi-spin correlation functions *J. Phys. A* 23, 2651-2669.

Gwa, L. H. and F. Y. Wu (1991), **P19**, The O(3) gauge transformation and 3-state vertex models, *J. Phys. A* 24, L503-L507.

Hilbert, D. (1890), *Math. Ann.* 36, 473.

Jacobsen, J. L. (1997), Comment on "Duality relation for Potts correlation functions" by Wu, *Phys. Lett. A* 223, 489-492.

Kelland, S. B. (1974a), Twenty-vertex model on a triangular lattice, *Aust. J. Phys.* 27, 813-829.

Kelland, S. B. (1974b), Ferroelectric ice model on a triangular lattice, *J. Phys. A* 7, 1907-1912.

King, C. and F. Y. Wu (2002), New correlation duality relations for the planar Potts model, *J. Stat. Phys.* 107, 919-940.

Kramers, H. A. and G. H. Wannier (1941), Statistics of the two-dimensional ferromagnet. Part I, *Phys. Rev. 60*, 252-262.

Lee, D. H. and F. Y. Wu (2003), **P20**, Duality relation for frustrated spin models, *Phys. Rev. E* 67, 026111.

Lee, T. D. and C. N. Yang (1952), Statistical theory of equations of state and phase transitions. II. Lattice gas and Ising model, *Phys. Rev.* 87, 410-419.

Lu, W. T. and F. Y. Wu (1998), On the duality relation for correlation functions of the Potts model, *J. Phys. A* 31, 2823-2836.

Perk, J. H. H., F. Y. Wu and X. N. Wu (1990), **P18**, Algebraic invariants of the symmetric gauge transformation, *J. Phys. A* 23, L131-L135.

Potts, R. B. (1952), Some generalized order-disorder transformations, *Proc. Camb. Phil. Soc.* 48, 106-109.

Rottman, C. and M. Wortis (1981), Exact equilibrium crystal shapes at nonzero temperature in two dimensions, *Phys. Rev. B* 24, 6274-6277.

Savit, R. (1980), Duality relations in field theory and statistical systems, *Rev. Mod. Phys.* 52, 453-487.

Watson, P. G. (1968), Impurities and defects in Ising lattices, *J. Phys. C* 1, 575-581.

Wegner, F. (1973), A transformation including the weak-graph theorem and the duality transformation, *Physica* 68, 570-578.

Wu, F. Y. (1997), **P21**, Duality relation for Potts correlation functions, *Phys. Lett. A* 228, 43-47.

Wu, F. Y. and H. Y. Huang (1997), **P22**, Sum rule identities and the duality relation for the Potts n-point correlation function, *Phys. Rev. Lett.* 79, 4954-4957.

Wu, F. Y. and Y. K. Wang (1976), **P16**, Duality transformation in a many-component spin model, *J. Math. Phys.* 17, 439-440.

Wu, X. N. and F. Y. Wu (1989), **P17**, Duality properties of a general vertex model, *J. Phys. A* **22**, L55-69.

4. The Ising Model

Introduction

My first encounter with the Ising model was in a course on statistical mechanics given by Henry Primakoff in the early 1960s at Washington University. Although Primakoff was a high-energy theorist, his course was fascinating and I soon acquired the "Ising disease" well-known to many avid students. Over the years I made various contributions to the topic, including the writing of an article *The Ising Model* for the *Encyclopedia of Physics* (Wu, 2005) and numerous research papers on various aspects of the Ising model.

The Ising model is a model of ferromagnetism. Place spins σ at sites of a lattice, which can assume two values $\sigma = \pm 1$. Let spins σ_i and σ_j at sites i and j interact with a (reduced) energy $-K\sigma_i\sigma_i$. The goal, which is deceptionally simple, is the closed-form evaluation of the partition function

$$Z = \sum_{\{\text{spin configurations}\}} \prod_{<i,j>} e^{K\sigma_i\sigma_i}. \tag{4.1}$$

The model mimics ferromagnetism since for $K > 0$ the interaction favors spins having the same sign. An excellent description of the history of the Ising model can be found in (Brush, 1967). This chapter describes some of my contributions in areas of general interest.

Ising Representation of the Eight-Vertex Model

A milestone in lattice statistics is the Onsager (1944) solution of the 2-dimensional Ising model. Another milestone is the Baxter solution of the symmetric 8-vertex model (Baxter, 1971) which exhibits a varying critical behavior not seen in the Onsager solution. Since the consideration of vertex models was new and novel at the time, it was desirable to understand the new findings in the context of the more familiar Ising setting. With this purpose in mind, in (Wu, 1971, **P23**) I described the now well-known equivalence of the Baxter 8-vertex model with two Ising models coupled together by 4-spin interactions, an equivalence which was also pointed out by Kadanoff and Wegner (1971) independently. This resolved some of the mystery of the Baxter solution.

The Baxter-Wu Model

I visited the Australian National University in the fall of 1972 on a Fulbright grant and as a guest of Rodney Baxter, at a time not long after he had solved the 8-vertex model (see page 13). That was an exciting time in model solvings, as innovative approaches and methods were being discovered and used to unravel new solutions. After my arrival, Baxter and I focused our attention to the Ising model on the triangular lattice with 3-spin interactions

$$-K\sigma_i\sigma_j\sigma_k \tag{4.2}$$

around every triangular face. The model is self-dual under the transformation $(\tanh K) \leftrightarrow (\tanh K)^{-1}$, leading to a determination of its critical point at the self-dual point $\tanh K_c = 1$. This suggested to us that the model might be soluble. Indeed, we soon obtained its solution using the method of the Bethe ansatz after first converting the problem into an Ashkin-Teller model. It was an ideal collaboration as Baxter has always been the master of algebraic analysis and I contributed largely to the graphical interpretation. We wrote up our results in two papers: a *Letter* communication (Baxter and Wu, 1973, **P24**) and a full paper describing details of the account (Baxter and Wu, 1974).

The free energy of the solution was found to assume an algebraic form, and the specific heat exhibits the critical exponents

$$\alpha = \alpha' = 2/3. \tag{4.3}$$

Both findings were novel at the time. Because of the special symmetry of the interaction (4.2), the 3-spin interaction Ising model has since become a testing ground for simulations and analytic approaches, and acquired the name *Baxter-Wu model.* Three years after the publication of **P24**, using a clever spin transformation Baxter and Enting (1976) showed that the Baxter-Wu model is reducible to a special case of the symmetric 8-vertex model.

Density of Fisher Zeroes

The summer of 1964 marked the end of my first-year teaching at the Virginia Polytechnic Institute. In that summer I participated at the Boulder Summer School of Theoretical Physics. As mentioned in Chapter 1, together with the Brandeis School, the Boulder School was a prominent place to learn about latest developments in theoretical physics. While I was attracted to the School by lectures on liquid heliums given by my thesis advisor Eugene Feenberg, I also attended lectures by Michael Fisher on critical point phenomena.

In his lectures, Fisher (1965) introduced the notion of Fisher zeroes. The remarkable Lee-Yang circle theorem (Lee and Yang, 1952) of the Ising model was already well-known. It dictates that zeroes of the partition function of a ferromagnetic Ising model are located on the unit circle

$$|e^{2L}| = 1 \tag{4.4}$$

for complex magnetic field $H = kTL$. In his 1964 lectures, Fisher proposed that it is also meaningful to consider zeroes of the zero-field Ising partition function for complex temperatures K, and asserted without elaboration that for the square lattice the zeroes are located on the unit circle

$$|\sinh 2K| = 1. \tag{4.5}$$

The Fisher consideration was subject to a number of limitations. Unlike the Lee-Yang circle theorem which holds for all finite (and infinite) lattices, the Fisher circle applies only in the bulk limit. The locus (4.5) was determined essentially by setting the argument of the logarithm in the Onsager (1944) free energy expression equal to zero, a procedure which is clumsy and lacks the rigor usually required in statistical mechanics. The process has been regarded as "hand-waving" by Stephenson and Couzens (1984). Furthermore, Fisher did not compute the density of zeroes along the locus.

These limitations were resolved by Wentao Lu and myself in 2001. The key of our analysis lay in finding an Ising model which permits a direct determination of the zeroes for *finite* lattices. The Onsager solution was of no help since for finite lattices the zeroes can only be determined numerically. Lu and I had earlier in 1988 solved an Ising model on a self-dual $M \times N$ lattice whose zeroes are located on a circle when one spatial dimension $N \to \infty$ (Lu and Wu, 1988). But this did not completely suit our purposes. Then we unearthed a beautiful work of Brascamp and Kunz (1974) which provided precisely the needed link.

For an $M \times 2N$ lattice wound on a cylinder, Brascamp and Kunz showed that its partition function under the application of a special boundary field assumes the simple form

$$Z_{BK} = 2^{2MN} \prod_{1\le i\le N} \prod_{1\le j\le M} \left[1 + (\sinh 2K)^2 - \sinh 2K(\cos\theta_i + \cos\phi_j)\right] \tag{4.6}$$

where $\theta_i = (2i-1)\pi/2N$, $\phi_j = j\pi/(M+1)$. Since zeroes of Z_{BK} are on the unit circle (4.5) for all M, N, one can take the bulk limit with confidence. This firmly establishes the validity of the Fisher circle theorem since the bulk limit is independent of the boundary condition. Furthermore, the partition function (4.6) permits a rigorous determination of the density of zeroes along the circle (4.5). Write $\sinh 2K = e^{i\alpha}$ and let $g(\alpha)$ denote the density of zeroes on the circle. By manipulating (4.6), Lu and I obtained the density function

$$g(\alpha) = \frac{|\sin\alpha|}{\pi^2} K(\sin\alpha) \tag{4.7}$$

where $K(k)$ is the complete elliptical integral of the first kind. These and other results are reported in (Lu and Wu, 2001a, **P25**).

Solution of the Ising Model on Nonorientable Surfaces

The Onsager solution of the Ising model had been deduced under cylindrical and toroidal boundary conditions. For completeness as well as curiosity, I had always wanted to extend the solution to finite lattices under other boundary conditions. One such effort was the solution of the aforementioned self-dual lattices (Lu and Wu, 1988).

In 1986, Blöte, Cardy and Nightingale (1986) found a remarkable connection between the conformal field theory and the boundary free energy of a finite two-dimensional lattice model. Although their theory had been checked against known solutions of lattice models, a lingering question remained regarding its predictions on nonorientable surfaces whose boundaries are somewhat ambiguous. This made the solution of the Ising model on these surfaces more interesting and urgent. Building on our experience with dimers (see Chapter 2), Wentao and I took on the challenge, and proceeded to consider the Ising model on the Möbius strip and the Klein bottle.

We used the Kasteleyn approach (see page 4) in a dimer formulation of the Ising model, and the problem was trickier than that of the simple dimer system. The difficulty lay in keeping track of the number of clockwise arrows in transition cycles looping around the lattice in the Möbius direction. The counting can be done in a straightforward fashion if the strip width has an even number of rows. But for topological reasons the counting is messy if the number of rows is odd. We resolved the problem by devising a trick: For odd numbers of rows we considered the lattice as being obtained by fusing together two central rows in a lattice of even number of rows.

After elaborate algebraic manipulations, we obtained closed-form expressions of the Ising partition function for an $M \times N$ lattice embedded on a Möbius strip and on the Klein bottle (Lu and Wu, 2001b, **P26**). While both expressions have the same bulk limit as the Onsager solution, for finite M and N they differ appreciably depending on whether M and N being even or odd. Finite-size analyses of the solution, however, revealed a simpler picture: The conformal field prediction of the central charge of $c = 1/2$ holds if the Möbius boundary is regarded as a free boundary.

Remarks

The list of topics on the Ising model is endless. Two well-known unsolved problems are the two-dimensional model in a field and the three-dimensional model. While we now know these two problems are intractable, from time to time there are fruitless attempts at their solutions.[d]

I did succeed in making some progress on the finite magnetic field problem. My work included the exact disorder solution of the triangular Ising model in a magnetic field (Wu, 1985), the solution of a special kagome lattice model in a field (Lu and Wu, 2005), and the evaluation of the magnetizations of the checkerboard model in a field $i\pi/2$ (Lin and Wu, 1988).

To completely satisfy my curiosity, I would like very much to see the closed-form solution of the two-dimensional model on a lattice with *free* boundaries. To be sure, the Kasteleyn matrix (page 5) can be written down. For an $M \times N$ lattice with interactions K, the partition function Z is the pfaffian of the $4MN \times 4MN$ antisymmetric matrix

$$K(z) = a_{0,0} \otimes I_M \otimes I_N + a_{1,0} \otimes I_M \otimes F_N - a_{1,0}^T \otimes I_M \otimes F_N^T \\ + a_{0,1} \otimes F_M \otimes I_N - a_{0,1}^T \otimes F_M^T \otimes I_N, \tag{4.8}$$

where the 4×4 matrices F and I are those given in (1.8), and

$$a_{0,0} = \begin{pmatrix} 0 & 1 & -1 & -1 \\ -1 & 0 & 1 & -1 \\ 1 & -1 & 0 & 1 \\ 1 & 1 & -1 & 0 \end{pmatrix}, \quad a_{1,0} = \begin{pmatrix} 0 & z & 0 & 0 \\ 0 & 0 & 0 & 0 \\ 0 & 0 & 0 & 0 \\ 0 & 0 & 0 & 0 \end{pmatrix}, \quad a_{0,1} = \begin{pmatrix} 0 & 0 & 0 & 0 \\ 0 & 0 & 0 & 0 \\ 0 & 0 & 0 & z \\ 0 & 0 & 0 & 0 \end{pmatrix}, \tag{4.9}$$

with $z = \tanh K$. However, the closed-form evaluation of the determinant $\det|K(z)|$ has proven to be elusive.[e]

References for Chapter 4

Baxter, R. J. (1971), Eight-vertex model in lattice statistics, *Phys. Rev. Lett.* 26, 832-833.

Baxter, R. J. and I. Enting (1976), The three-spin Ising model as an 8-vertex model, *J Phys. A* 9, L149-L152.

Baxter, R. J. and F. Y. Wu (1973), **P24**, Exact solution of an Ising model with 3-spin interactions, *Phys. Rev. Lett.* 31, 1294-1297.

[d] See, for example, a critique on one such purported solution (Wu, McCoy, Fisher and Chayes, 2008).

[e] The underlying technical reason making the diagonalization of the matrix $K(z)$ difficult lies in the fact that positions of the nonzero element z are shifted when $a_{1,0}$ and $a_{1,0}$ are transposed.

Baxter, R. J. and F. Y. Wu (1974), Ising model on a triangular lattice with three-spin interactions. I. The eigenvalue equations. *Aust. J. Phys.* 27, 357-367.

Blöte, H. W. J., J. L. Cardy and M. P. Nightingale (1986), *Phys. Rev. Lett.* 56, 742-.

Brascamp, H. J. and H. Kunz (1974), Zeroes of the partition function for the Ising model in the complex temperature plane, *J. Math. Phys.* 15, 65-66.

Brush, G. (1967), History of the Lenz-Ising model, *Rev. Mod. Phys.* 39,883-.

Fisher, M. E. (1965), The nature of critical points, in *Lecture Notes in Theoretical Physics*, Vol. 7c, W. E. Brittin, ed. (University of Colorado Press, Boulder, 1965), 1-159.

Kadanoff, L. P. and F. J. Wegner (1971), Some critical properties of the 8-vertex model, *Phys. Rev. B* 4, 3989-3993.

Lee, T. D. and C. N. Yang (1952), Statistical theory of equation of state and phase transitions.II. Lattice gas and Ising model, *Phys. Rev.* 87, 410-419.

Lin, K. Y. and F. Y. Wu (1988), Magnetization of the Ising model on the generalized checkerboard lattice, *J. Stat. Phys.* 52, 669-677.

Lu, W. T. and F. Y. Wu (1988), Partition function zeroes of a self-dual Ising model, *Physica A* 258, 157-170.

Lu W. T. and F. Y. Wu (1999), Dimer statistics on a Möbius strip and the Klein bottle, *Phys. Lett. A* 259, 108-114.

Lu, W. T. and F. Y. Wu (2001a), **P25**, Density of the Fisher zeroes for the Ising model, *J. Stat. Phys.* 102, 953-970.

Lu, W. T. and F. Y. Wu (2001b), **P26**, Ising model on nonorientable surfaces: Exact solution for the Möbius strip and the Klein bottle, *Phys. Rev. E* 63, 026107.

Lu, W. T. and F. Y. Wu (2005), Soluble kagome Ising model in a magnetic field, *Phys. Rev. E* 71, 042160.

Onsager, L. (1944), Crystal statistics. I. A two-dimensional model with an order-disorder transition, *Phys. Rev.* 65, 117-149.

Stephenson, J. and R. Couzens (1984), Partition function zeros for the two-dimensional Ising model, *Physica A* 129, 201-210.

Wu, F. Y. (1971), **P23**, Ising model with 4-spin interactions, *Phys. Rev. B* 4, 2312-2314.

Wu, F. Y. (1985), Exact solution of the triangular Ising model in a nonzero magnetic field, *J. Phys. A* 40, 613-620.

Wu, F. Y. (2005), The Ising Model, in *Encyclopedia of Physics*, Vol. 1, Eds. R. G. Lerner and G.L. Trigg (Wiley-Vch Verlag, Weinheim), 1145-1146.

Wu, F. Y., B. M. McCoy, M. E. Fisher and L. Chayes (2008), On a recent conjectured solution of the three-dimensional Ising model, *Phil. Mag.* **88**, 3093-3095.

5. The Potts Model

Introduction

After the Ising model, the Potts model (Potts, 1952) is perhaps the most prominent model of phase transitions in statistical mechanics. It generalizes the Ising model (4.1) to more-than-two components by allowing spins to assume q different values $\sigma = 1, 2, ..., q$. In the standard Potts model, two spins interact with a (reduced) energy $-K\delta_{\sigma_i,\sigma_j}$ with the Boltzmann factor (3.3), mimicking a multi-component ferromagnetism. The model finds applications in diverse disciplines.

The Potts model was originally proposed by Cyril Domb as a thesis topic to his student Renfrey Potts.[f] In the early years the model was mostly a curiosity, but its popularity grew rapidly in the 1970's after Rodney Baxter's discovery (Baxter, 1973) of its first-order transition when the number of components is greater than four.

I was visiting the Australian National University at the time when Baxter made the seminal discovery. As Baxter's analysis was based on an equivalence with a 6-vertex model, which is something close to my heart, my interest was also drawn to the Potts model.

In the fall of 1980, I gave a series of lectures on the Potts model during my sabbatical at Delft and, on the basis of the lectures, wrote a tutorial review on the Potts model (Wu, 1982, **P28**). The review appeared at an opportune time when analytic results on the Potts model were converging and interest in the Potts model mounting. Being the first review article on the Potts model, the paper has been widely quoted with more than 50 citations every year since its publication. On several occasions in conferences, people would approach me to express their thanks, telling me they have benefited from the paper. Such expressions are most rewarding to an author.

Rigorous results on the Potts model such as critical exponents continued to accumulate in the two years after the publication of **P28**. I followed up

[f]The thesis by Potts was on a spin model with first- and second-neighbor interactions. It was at the very end of his thesis work that Potts tackled the what is now known as the Potts model (Potts, 1982).

with a sequel (Wu, 1984, **P29**) summarizing the later developments. This short review was written while I was assuming an 18-month "rotator" position as the Director of the Condensed Matter Theory Program at the National Science Foundation.[g]

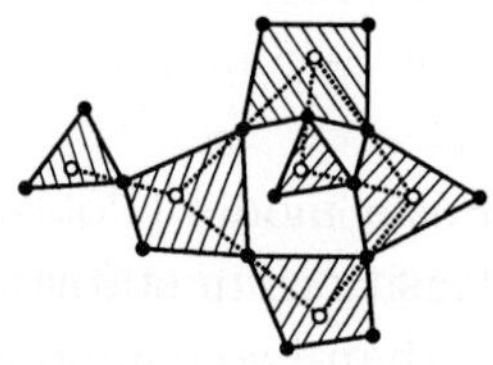

Fig. 5.1. The covering graph of a graph G. Open circles in shaded regions denote sites of G. Solid circles denote sites of the covering graph residing on edges of G. Shaded regions are lands (see text).

Graphical Formulation of the Potts Model

An indispensable tool of studying lattice models is graphical analysis. One example is the high-temperature expansion of the Ising model which forms the basis of the combinatorial approach. For the q-state Potts model (3.3) on a graph G with interaction K, the high-temperature expansion of the partition function (Baxter, Kelland and Wu, 1976, **P27**)

$$Z(q,v) = \sum_{g \subseteq G} q^{n(g)} v^{e(g)}, \qquad v = e^K - 1, \tag{5.1}$$

where the summation is over subgraphs g having $n(g)$ clusters and $e(g)$ edges, forms the basis of almost all analytic and graphical approaches to the Potts model. The polynomial (5.1) is equally prominent in graph theory where it is known as the dichromatic polynomial of Tutte (see pp. 51-52).

The expansion (5.1) can be formulated as an ice-type vertex model through the use of covering graphs. The example of a covering graph is shown in Fig. 5.1. In this picture the plane is divided into "lands" containing sites of G and "seas." The Potts model is transformed to a 6-vertex model on the covering graph. The vertex model formulation was first introduced by Temperley and Lieb (1972) for the square lattice, and extended

[g]Writing in the capacity of a government official, I was obliged to include in the paper a disclaimer: *Any of the opinions expressed herein are those of the author and do not necessarily reflect the views of the NSF.* This presumably says that the NSF does not hold any view on the Potts model, even though it supports its research.

to arbitrary graphs a few years later by Baxter, Kelland and myself in **P27** where we explicitly introduced the scheme of lands and seas. The scheme of lands and seas proves to also play an essential role in graph theory and in knot theory. We were happy at the time to break away from regular lattices, not knowing that doing statistical mechanics on graphs would ten years later lead to far-reaching implications in pure mathematics (Chapter 9).

The partition function (5.1) can also be regarded as that of a random cluster model (Fortuin and Kasteleyn, 1972) by associating a probabilistic measure to the summand. This consideration extends the Potts model to nonintegral values of q (see page 46).

Rigorous Determination of the Potts Critical Point

Potts (1952) determined the critical point (3.6) of the Potts model for the square lattice using a duality argument. The critical point for the triangular and honeycomb lattices

$$\begin{aligned} v &= \sqrt{q}, && \text{square} \\ v^3 + 3\,v^2 &= q, && \text{triangular} \\ q^2 + 3\,q\,v &= v^3, && \text{honeycomb} \end{aligned} \tag{5.2}$$

can be similarly determined. Technically, these determinations are conjectures since they require the additional assumption of a unique transition in the duality argument.

In the summer of 1975, I visited the École Polytechnique Fédéral de Lausanne for the first time, and looked into the conjectured critical point in collaboration with Alfons Hinterman and Hervé Kunz. Two years earlier at the Institut des Hautes Études Scientifiques in Paris, Kunz and I had deduced a number of rigorous results on vertex models (Brascamp, Kunz and Wu, 1973). It was therefore natural that we focused on the vertex model formulation of the Potts model.

The analysis was a good exercise of a function of two complex variables. After twists and turns, we succeeded in rigorously establishing that the conjectured critical points (5.2) are indeed the exact critical points. But we could do this only for $q \geq 4$ (Hinterman, Kunz and Wu, 1978, **P30**).

In the fall of 1979, I returned to the Virginia Polytechnic Institute, where I had taught from 1963 to 1967, for a 3-month visit. Royce Zia, who had since joined the faculty, and I examined the self-dual point (3.7) of the triangular Potts model with 2- and 3-site interactions. Using continuity and uniqueness arguments, we established rigorously that (3.7) is indeed the exact critical point in the ferromagnetic regime for all q (Wu and Zia, 1981).

I have also spent time in locating Potts critical points for other lattices. These efforts will be described in Chapter 6.

Potts Partition Function Zeroes

One way to gain insight into lattice models is to look into the partition function zeroes (see page 29). In the case of the Potts model, numerous authors have studied its partition function zeroes since the 1980s, but failed to draw definite conclusions. I was visiting the Institute of Physics at the Academia Sinica in Taipei in 1994 where Chi-Ning Chen and Chin-Kun Hu had developed a fast algorithm for simulating the Potts model (Chen and Hu, 1991). We got together and used the new algorithm to compute Potts partition function zeroes.

It was by a stroke of luck that we chose to work with *self-dual* lattices. To our surprise, we found that for self-dual lattices all zeroes in the half plane $\mathrm{Re}(x) > 0$ lie precisely on the unit circle

$$|x| = 1\,, \qquad x = (e^K - 1)/\sqrt{q}. \tag{5.3}$$

Earlier studies had used periodic and other boundary conditions, yielding zeroes scattered around the circle. Apparently, it was the self-dual property of the partition function that forced the zeroes to lie precisely on the circle. This is a bit like the Brascamp-Kunz boundary condition of the Ising model which fixes partition function zeroes on a circle (page 29). Our results led us to conjecture that Potts partition function zeroes in the regime $\mathrm{Re}(x) > 0$ lie on the unit circle $|x| = 1$ for self-dual lattices regardless of the lattice size (Chen, Hu and Wu, 1996, **P31**). The conjecture also applies to the bulk limit of any other lattice, since in the bulk limit the zero distribution is independent of the boundary condition.

Remarks

Much remains to be learned about the Potts model. Particularly, the antiferromagnetic (AF) model ($K < 0$) is essentially wide open. To be sure, the critical point of the AF square lattice model has been located by Baxter (1982) at $e^K = \sqrt{4-q} - 1$ for $q \leq 3$. The AF model is harder to analyze because its $T = 0$ partition function (8.3), the chromatic polynomial, is already a convoluted topic in graph theory. Results on the AF model are mostly obtained from numerical investigations. I have contributed to one such endeavor for a 3-dimensional model with competing interactions. Using simulations, Jay Banavar and I identified the nature of its ordered states and the associated phase diagram (Banavar and Wu, 1984).

Finally, I have not touched upon the tremendous past and on-going progress on the integrable chiral Potts model, culminating in the exact determination of its order parameter by Baxter (2005).

References for Chapter 5

Banavar, J. R. and F. Y. Wu (1984), Antiferromagnetic Potts model with competing interactions, *Phys. Rev. B* 29, 1511-1513.

Baxter, R. J. (1982), Critical antiferromagnetic square-lattice Potts model, *Proc. Roy. Soc. A* 388, 43-53.

Baxter, R. J. (2005), Derivation of the order parameter of the chiral Potts model, *Phys. Rev. Lett.* 94, 130602.

Baxter, R. J., S. B. Kelland and F. Y. Wu (1976), **P27**, Equivalence of the Potts model or Whitney polynomial with an ice-type model: A new derivation, *J. Phys. A* 9, 397-406.

Brascamp, H. J., H. Kunz and F. Y. Wu (1973), some rigorous results for the vertex model in statistical mechanics, *J. Math. Phys.* 14, 1927-1932.

Chen, C. N. and C. K. Hu (1991) Fast algorithm to calculate exact geometrical factors for the q-state Potts model, *Phys. Rev. B* 43, 11519-11522.

Chen, C. N., C. K. Hu and F. Y. Wu (1996), **P31**, Partition function zeroes of the square lattice Potts model, *Phys. Rev. Lett.* 76, 173-176.

Fortuin C. M. and P. W. Kasteleyn (1972), On the random-cluster model. I. Introduction and relation to other models, *Physica* 57, 536-564.

Hinterman, A, H. Kunz and F. Y. Wu, **P30**, (1978), Exact results for the Potts model in two dimensions, *J. Stat. Phys.* 19, 623-632.

Potts, R. B. (1952), some generalized order-disorder transformations, *Proc. Camb. Phil.Soc.* 48, 106-109.

Potts, R. B. (1982), Private communication to the author.

Temperley, H. N. V. and E. H. Lieb (1971), Relation between the 'percolation' and 'colouring' problem and other graph-theoretical problems associated with regular planar lattice: some exact results for the 'percolation' problem, *Proc. Roy. Soc.* A 322, 251-280.

Wu, F. Y. (1982), **P28**, The Potts model, *Rev. Mod. Phys.* 54, 235-268.

Wu, F. Y. (1984), **P29**, Potts model of magnetism, *J. Appl. Phys.* 55, 2421-2425.

Wu, F. Y. and R. K. P. Zia, (1981), Critical point of the triangular Potts model with two-and three-site interactions, *J. Phys. A* 14, 721-727.

6. Critical Frontiers

Introduction

An important aspect of understanding lattice models of phase transitions is the determination of the transition point, or the critical frontier, which is a trajectory in the parameter space along which the system is critical. Mathematically, the critical frontier is the trajectory of non-analyticity of the free energy. Since the free energy can be explicitly evaluated for only a limited number of models, one often settles with a less ambitious goal: the determination of the critical frontier of a statistical mechanical model in the absence of an explicit solution.

I have spent very much time in this endeavor using a variety of methods and approaches including, among others, duality considerations, mappings and transformations, algebraic invariants, and finite-size analysis. Some findings on the Potts model have been described in the preceding chapter. This chapter describes further results on critical frontiers.

The Potts Model

By 1978, the critical frontier (5.2) of the Potts model have been determined from duality for the square, honeycomb and triangular lattices. Based on the history of the Ising model, it was expected that the critical point would soon be found for other regular lattices. However, it was a good time to pause and take stock. In the paper (Wu, 1979, **P32**) I wrote in 1979, I summarized everything known at the time on the Potts critical point and proposed several conjectures. It was not anticipated that, 30 years later, we still would not know much more than we did then.

In addition to the critical point (5.2) for models with pure nearest-neighbor interactions, the critical frontier of the triangular model with 2-site interactions K_2 and 3-site interactions $\{0, K_3\}$ in alternate triangles had been found to be (3.7). The critical point (3.7) was determined by Baxter, Temperley and Ashley (1978) algebraically, and rederived later by Lin and myself using a graphical analysis (see page 55).

A quick glance of (3.7) suggests that for alternate 3-site interactions

$\{K_3, K_3'\}$ the critical frontier would likely to assume the form

$$e^{3K_2+K_3+K_3'} = 3\,e^{K_2} + q - 2. \tag{6.1}$$

After all, the expression (6.1) is the most natural extension of (3.7) if the critical frontier remains algebraic. I made this conjecture in **P32**. Once accepting the conjecture, the critical frontier can be worked out for a number of other regular lattices including the 3-12 and the kagome lattices.

The critical frontier of the kagome lattice has been the subject of numerous studies with several competing conjectures (see Chen, Hu and Wu, 1980). Particularly, one deduces from (6.1) the critical frontier[h]

$$v^4(v+3)^2 = q^3 + 6q^2v + 2qv^2(v+6), \quad v = e^K - 1. \tag{6.2}$$

This and other conjectures have been tested numerically by many authors. Numerical data to date suggest that (6.2) is extremely accurate, and may very well be the exact expression (Chen, Hu and Wu, 1980; Scullard and Ziff, 2006).

On the other hand, the critical frontier (3.7) is exact and leads to the determination of the critical frontier of several new lattices. In a renewed effort of locating percolation thresholds using duality, Scullard and Ziff (2006), Ziff (2006), and Ziff and Scullard (2008) introduced several new lattices. Subsequently, I worked out the corresponding Potts critical frontiers, thus greatly enhancing the repertoire of known Potts critical points (Wu, 2006, **P33**). The Scullard-Ziff percolation thresholds are recovered by simply setting $q = 1$ in the new Potts critical frontiers.

The Antiferromagnetic Ising Model in a Magnetic Field

The Ising model in a nonzero magnetic field H is a well-known unsolved problem. It is known that there exists a field-induced phase transition in the antiferromagnetic model, but the determination of the exact critical point has been elusive.

One approach to the nonzero-field problem is to consider the high-temperature expansion. The high-temperature expansion of the Ising partition function with a field can be formulated as a vertex model, and for the honeycomb lattice it is the 8-vertex model of Fig. 2.4. Having been interested in the honeycomb 8-vertex model for many years, I naturally approached the Ising problem from this formulation.

[h]Equation (4) in (Chen, Hu and Wu, 1998) contains a typo missing a factor (w+6) in the fourth term.

The honeycomb 8-vertex model possesses an O(2) gauge transformation. As described in Chapter 3, Xuening Wu and I studied the algebraic invariant of O(2), and found fundamental invariants P_1 given in (2.6) and I_1, I_2 given in (3.8). Since the invariant $P_1 = 0$ coincides precisely with the known critical frontier $H = 0$ of the ferromagnetic Ising model, it was reasonable for us to assume that the critical frontier of the antiferromagnetic Ising model would also be related to the 3 invariant manifolds. However, we did not know how to extract the precise relation.

At this point we sought help from Henk Blöte whom I had known since my 1980 sabbatical days at Delft. Blöte is an expert on finite-size analysis and at that time was visiting the University of Rhode Island nearby. He came to Boston to teach us the art of finite-size analysis, and together we carried out an elaborate and highly-accurate numerical determination of the nonzero field honeycomb Ising critical point. By effecting a fit with 4 free parameters, we obtained a closed-form expression of the critical frontier in terms of the three invariants P_1, I_1, I_2, reproducing numerical data within 7 decimal places (Wu, Wu and Blöte, 1989, **P34**; Blöte, Wu and Wu, 1990). It is a remarkable fit. The analysis was later extended to the square lattice (Wu and Wu, 1990).

The Blume-Emery-Griffiths Model

The Blume-Emery-Griffiths (BEG) model (1971) is a spin-1 Ising system with the Hamiltonian

$$-\mathcal{H} = J \sum_{\langle i,j \rangle} S_i S_j + K \sum_{\langle i,j \rangle} S_i^2 S_j^2 - \Delta \sum_i S_i^2 + H \sum_i S_i \qquad (6.3)$$

where $S_i = 0, \pm 1$. The Hamiltonian was originally devised to model the λ transition in liquid helium mixtures. But it is also of interest in its own right as a lattice model.

When $J = 0$ the model reduces to a spin-$\frac{1}{2}$ Ising model. Using this mapping the exact critical frontier can be worked out for all K, Δ, and H. This analysis was published in a special issue of the *Chinese Journal of Physics* (Wu, 1978) on the occasion of the 70th birthday of the great theoretical physicist and science administrator Ta-You Wu (see also page 68).

The BEG model has a vertex model representation. For the honeycomb lattice the vertex model is a 27-vertex model where each lattice edge can be in 3 states (Gwa and Wu, 1991b, **P35**). In the subspace

$$K = -\ln \cosh J, \qquad (6.4)$$

the 27-vertex model reduces to the 8-vertex model of Fig. 2.4, making it possible to study its critical frontier as discussed in Chapter 2. The study led to a complete determination of the exact critical frontier in this subspace (Wu and Wu, 1988).

For the general BEG Hamiltonian (6.3), one needs to know the invariants of the O(3) gauge transformation. This investigation was carried out in collaboration with Leh-Hun Gwa of Rutgers University. To simplify the algebra we focused on the subspace

$$H = 0 \tag{6.5}$$

for which the 27-vertex model reduces to a more manageable 14-vertex model.

After some hard work of deciphering century-old mathematical literature on algebraic invariants, we found 5 fundamental O(3) invariants (Gwa and Wu, 1991a, **P19**). We also carried out finite-size analysis on the BEG model. Using a fit of only 6 adjustable parameters, Gwa and I obtained a closed-form expression of the critical frontier of the honeycomb BEG model which fits numerical data for *all* J, K, and Δ extremely well (Gwa and Wu, 1991b, **P35**). It was a remarkable piece of work combining hard mathematics with numerical analysis.

The Ashkin-Teller Model

The Ashkin-Teller model (1943) is a generalization of the Ising model to 4 components. The model possesses a duality relation which unfortunately does not determine its critical point except in a degenerate case. In my sabbatical at the National Tsing Hua University in Taiwan in 1974, I collaborated with Keh-Yin Lin on several problems. Besides the staggered vertex models described in Chapter 2, we also studied the Ashkin-Teller model.

The Ashkin-Teller model can be formulated as two Ising models with respective interactions K_1 and K_2 coupled together with a 4-spin interaction K_3. The model can be mapped into a staggered 8-vertex model by effecting a duality transformation to one of the two Ising models. When $K_3 = 0$, the two Ising models are decoupled and there are two distinct transitions. What was not clear was what happens when K_3 is turned on. By piecing together exact information known at the time and using essentially a continuity argument, Lin and I established with certainty that there are two phase transitions in the Ashkin-Teller model (Wu and Lin, 1974, **P36**). We also produced a sketch of the shape of the phase diagram in the parameter space. The paper **P36** has since become a corner stone in the manual of lattice model solutions.

The O(n) Model

One of my favorite lattice models is the O(n) model. The O(n) model can be formulated either as an n-component corner-cubic spin model, or as a graph-theoretical "loop" model which is conceptually simpler. In the graph-theoretical language the partition function reads

$$Z = \sum_{\text{loops}} n^{\ell} x^{e}, \tag{6.6}$$

where the summation is over all loop diagrams on the lattice, ℓ is the number of loops, e the number of edges in loops, and x the weight of each loop edge. The loop model is particularly simple on the honeycomb lattice since loops do not intersect.

For the honeycomb lattice, Nienhuis (1982) has determined the critical point $x_c = 1/\sqrt{2+\sqrt{2-n}}$ in $1 \leq n \leq 2$, and it was generally believed that the model does not have a transition for $n > 2$. To examine whether the belief holds, Hervé Kunz and I studied zeroes of the partition function (6.6), and established the absence of transitions for sufficiently large n (Kunz and Wu, 1988).

The question of criticality for $n > 2$ was taken up again 10 years later by Wenan Guo and Henk Blöte. Through a careful numerical analysis of the honeycomb corner-cubic O(n) model, they located a critical frontier in the $n > 2$ regime, and found it intrude into a region proclaimed to be free of transitions in (Kunz and Wu, 1988). At the time of their work I happened to be at the Beijing Normal University where the numerical analysis was done. Working together with Guo and Blöte, I re-examined (Kunz and Wu, 1988) and discovered a typo (of a factor of 2) in one of the equations. The correction, together with the numerical determination of the critical point for $n > 2$, were reported in the joint paper (Guo, Blöte and Wu, 2000).

This work was followed a few years later by a study of the O(n) model on the square lattice. While the critical point of the O(n) model was known for the honeycomb model for $1 \leq n \leq 2$, there had been no definitive study for the square lattice. I had earlier worked out a duality relation yielding a self-dual trajectory $x = (\sqrt{2n} - 1)/n$ which gives the known critical point at $n = 1$ and 2. To examine whether it also gives the actual critical point in $1 < n < 2$, Guo, Blöte and Xiafeng Qian carried out finite-size analysis on the transfer matrix. The result showed that actual critical points definitely do not lie on the self-dual trajectory in the regime $1 < n < 2$. This is another example that self-dual arguments should be used with care. The findings are reported in the joint paper (Guo *et al.*, 2006).

Remarks

There is a myriad of lattice models whose critical frontiers are unknown. Here I mention just one.

A model akin to the Potts model is the $Z(N)$ model. It describes spins confined in a plane pointing to one of the N equally spaced directions. The Hamiltonian is

$$-\mathcal{H} = \sum_{<i,j>} J(\theta_i - \theta_j), \tag{6.7}$$

where the function $J(\theta)$ is 2π-periodic. The special case of $J(\theta) = \epsilon \cos\theta$ is the vector Potts, or the clock, model to which I have made some progress for $N = 5$. Based on a consideration of its thermodynamic path, the trajectory traced by a thermodynamic system in the parameter space when the temperature is raised from 0 to ∞, the $Z(5)$ model is found to have two phase transitions (Wu, 1979a).

The $Z(N)$ model of general N remains open.

References for Chapter 6

Ashkin, J. and E. Teller (1943), Statistics of two-dimensional lattices with four components, *Phys. Rev.* 64, 178-184.

Baxter, R. J., H. N. V. Temperley and S. E. Ashley (1978), Triangular Potts model at its transition temperature, and related models, *Proc. R. Roc. Lond. A* 358, 535-559.

Blöte, H. W. J., F. Y. Wu and X. N. Wu (1990), Critical point of the honeycomb antiferromagnetic Ising model in a nonzero magnetic field: Finite-size analysis, *Int. J. Mod. Phys. B* 4, 619-629.

Blume, M., V. J. Emery and R. B. Griffiths (1971), Ising model for the λ transition and phase separation in He^3-He^4-mixtures. *Phys. Rev. A* 4, 1071-1077.

Chen, J. A., C. H. Hu and F. Y. Wu (1998), Critical point of the kagome lattice Potts model: a Monte Carlo renormalization group and scaling determination, *J. Phys. A* 31, 7855-7864.

Guo, W., H. W. J. Blöte and F. Y. Wu (2000), Phase transition in the $n > 2$ honeycomb $O(n)$ model, *Phys. Rev. Lett.* 85, 3874-3877.

Guo, W., X. Qian, H. W. J. Blöte and F. Y. Wu (2006), Critical line of an n-component cubic model, *Phys. Rev. E* 73, 026104.

Gwa, L. H. and F. Y. Wu (1991a), **P19**, The O(3) gauge transformation and 3-state vertex models, *J. Phys.. A* 24, L503-L507.

Gwa, L. H. and F. Y. Wu (1991b), **P35**, Critical surface of the Blume-Emery-Griffiths model on the honeycomb lattice, *Phys. Rev. B* 43, 13755-13757.

Kunz, H. and F. Y. Wu (1988), Exact results for an $O(n)$ model in two dimensions, *J. Phys. A* 21, L1141-L1144.

Nienhuis, B. (1982), Exact critical point and critical exponents of $O(n)$ models in two dimensions, *Phys. Rev. Lett.* 49, 1062-1065.

Scullard, C. R. and R. M. Ziff (2006), Predictions of bond percolation thresholds for the kagome and Archimedean (3.12^2) lattices, *Phys. Rev. E* 73, 045102.

Wu, F. Y. (1978), Phase diagram of a spin-one Ising system, *Ch. J. Phys.* 16, 153-156.

Wu, F. Y. (1979), **P32**, Critical point of planar Potts models, *J. Phys. C* 12, L645-L650.

Wu, F. Y. (1979a), Phase diagram of a 5-state spin system, *J. Phys. A* 19, L317-L320.

Wu, F. Y. (2006), **P33**, New critical frontiers for the Potts and percolation models, *Phys. Rev. Lett.* 96, 090602.

Wu, F. Y. and K. Y. Lin (1974), **P36**, Two phase transitions in the Ashkin-Teller model, *J. Phys. C* 7, L181-L184.

Wu, X. N. and F. Y. Wu (1988), Blume-Emery-Griffiths model on the honeycomb lattice, *J. Stat. Phys.* 50, 41-55.

Wu, X. N. and F. Y. Wu (1990), Critical line of the square-lattice antiferromagnetic Ising model in a magnetic field, *Phys. Lett. A* 144, 123-126.

Wu, F. Y., X. N. Wu and H. W. J. Blöte (1989), **P34**, Critical frontier of the antiferromagnetic Ising model in a nonzero magnetic field: The honeycomb lattice, *Phys. Rev. Lett.* 62, 2773-2776.

Ziff, R. M. (2006), Generalized cell-dual-cell transformation and exact thresholds for percolation, *Phys. Rev. E* 73, 016134.

Ziff, R. M. and C. R. Scullard (2008), Exact bond percolation thresholds in two dimensions, *J. Phys. A* 39, 15083-15090.

7. Percolation

Introduction

Percolation describes the process of a fluid flowing through a random porous media (Shante and Kirkpatrick, 1971; Kesten, 2006). To formulate the process mathematically, one considers a lattice, or a graph, in which each site (or bond) is independently "occupied" with a probability p to allow fluid to flow through. This describes a site (bond) percolation. Thus, the probability measure for a configuration g with $o(g)$ occupied sites (bonds) to occur is

$$\pi(g) = p^{o(g)}(1-p)^{N-o(g)}, \tag{7.1}$$

where N is the total number of sites (bonds). The expectation, or the mean, value of a variable $A(g)$ is given by

$$\langle A \rangle = \sum_g \pi(g)A(g). \tag{7.2}$$

Many questions can be asked about the percolation process. Two occupied sites (bonds) belong to the same cluster if they are connected by consecutive occupied sites (bonds). The simplest question one can ask is the behavior of the *percolation probability*

$$P(p) = \lim_{N\to\infty} P_N(p), \tag{7.3}$$

where $P_N(p)$ is the probability that a given site (bond) is in a cluster of size of at least N. The expression (7.3) gives the probability that the media is "percolating."

It is clear that $P(0) = 0$ and $P(1) = 1$. It perhaps is also clear that $P(p)$ would remain zero for very small p. At some critical p_c, the percolation threshold, the percolation probability must arises from 0, and we write

$$P(p) \sim (p-p_c)^{\beta}, \quad p = p_c + . \tag{7.4}$$

This describes a critical behavior characterized by a critical exponent β. Other critical behavior and critical exponents can be similarly defined (Essam, 1972, 1980).

My interest in percolation stemmed from its connection with the Potts model. In the early years the percolation problem was studied from the point of view of a probabilistic process. A breakthrough which permitted a conventional statistical mechanical approach is its connection with the Potts model via the random-cluster model put forth by Fortuin and Kasteleyn.

However, Fortuin and Kasteleyn (1972) focused their attention to graph-theoretical aspects of the random-cluster model, and did not fully explore its connection with percolation. To elucidate the matter, I followed up with an expository article (Wu, 1978, **P37**) describing the precise formulation of the bond percolation in terms of a Potts model. In a subsequent work, Hervé Kunz and I extended the formulation to site percolation (Kunz and Wu, 1978). With these new tools, the Potts model provides a convenient Hamiltonian approach to the percolation problem.

The Random-Cluster model and the Potts Model

The random-cluster model (Fortuin and Kasteleyn, 1972) describes a probabilistic process over a graph in which edges are randomly broken so vertices are broken into random clusters. Identifying connecting edges as the occupied edges in a bond percolation, the random cluster model describes a bond percolation.

The random cluster model is also a Potts model. In the high-temperature expansion (5.1) of the Potts partition function, write $v = p/(1-p)$. Then (5.1) assumes the form

$$Z(q,v) = e^{NK}\langle q^n\rangle \tag{7.5}$$

of the mean value of q^n, where n is the number of clusters with

$$p = 1 - e^{-K}. \tag{7.6}$$

The expression (7.5) links the Potts model to percolation. It also extends the Potts model to non-integral values of q permitting, for example, the taking of derivatives to obtain

$$\langle n\rangle = \left[\frac{\partial}{\partial q}\ln Z(q,v)\right]_{q=1}. \tag{7.7}$$

Using (7.7), one finds the mean number of clusters at p_c to be $\langle n\rangle_c = 0.09807\,N$ for the square lattice (Temperley and Lieb, 1972) and $0.01118\,N$ for the triangular lattice (Baxter, Temperley and Ashley, 1978).

Furthermore, by introducing an external field L to one of the q Potts spin states and denoting the partition function by $Z(q,v,L)$, one has the further relation

$$P(p) = 1 + \frac{1}{N}\left[\frac{\partial^2}{\partial L\partial q}\ln Z(q,v,L)\right]_{q=1,\,L=0+}. \tag{7.8}$$

These and other similar expressions relate percolation averages to derivatives of the Potts model partition function (Wu, 1978, **P37**; 1982, **P27**). Particularly, percolation thresholds are given by the Potts critical points at $q = 1$. This gives rise to the folklore that percolation is the $q = 1$ Potts model.

Percolation Thresholds

The threshold probability p_c of a percolation process is traditionally determined from duality arguments (Essam, 1972). They are given by

$$\begin{aligned} p_c &= 1/2, && \text{square} \\ 1 - 3p_c + p_c^3 &= 0, && \text{triangular} \\ 1 - 3p_c^2 + p_c^3 &= 0, && \text{honeycomb} \end{aligned} \tag{7.9}$$

for the 3 lattices as indicated. These thresholds are also the Potts critical points (5.2) at $q = 1$. Similarly, the percolation thresholds for several new lattices (Scullard and Ziff, 2006; Ziff, 2006; Ziff and Scullard, 2008) are obtained by setting $q = 1$ in the corresponding Potts critical frontiers (see page 39).

An Infinite-Range Bond Percolation

The Potts model formulation can be fully carried out in the case of an infinite-range bond percolation. Consider a bond percolation process on a complete graph of N vertices, in which every vertex is connected to every other with a probability p/N. This describes an infinite-range bond percolation in the limit of $N \to \infty$. It is also the random graph problem introduced by Erdös and Rényi (1960).

In the Potts model formulation of percolation, one obtains from (7.6) the Potts interaction K/N (for N large). The Potts model on a complete graph with this interaction was solved in (Wu, 1982, **P28**) as the mean-field solution. Using the mean-field results, one obtains directly the Erdös-Rényi threshold probability

$$p_c = 1/2 \tag{7.10}$$

and the critical exponent $\beta = 1$. The Potts model formulation therefore provides a heuristic approach to the random graph problem (see page 53). These results are reported in (Wu, 1982a, **P38**).

Directed Percolation

Directed percolation is a Markovian process in which edges in a graph or lattice are directed to allow a fluid to "percolate" in only one direction. Very few exact results are known about directed percolation. In 1981, Domany and Kinzel solved one version of a directed percolation where the bond occupation probability is fixed at unity in one spatial direction of a rectangular lattice (Domany and Kinzel, 1981). The analysis made use of the Stirling approximation.

Returning from a yearlong sabbatical in the Netherlands and Germany in 1981, I spent the summer months at Boston University as a guest of Gene Stanley. After reading the Domany-Kinzel paper, I reckoned that their directed percolation model is actually a random walk problem, which can be more precisely formulated and analyzed. Stanley and I looked into this possibility. This resulted in the paper (Wu and Stanley, 1982, **P39**) in which the problem was re-formulated as a random walk. The rigorous analysis gives a firm footing to the Domany-Kinzel analysis.

However, the Domany-Kinzel model is essentially one-dimensional in nature. To uncover the genuine nature of a two-dimensional directed percolation, one needs to relax the restriction of full occupation of edges in one spatial direction.

In the fall of 2004, Lun-Chi Chen of the Institute of Mathematics of Academia Sinica (Taipei) visited me for 3 months. Chen and I re-examined the Domany-Kinzel model by relaxing the restriction of unity occupation probability to every other row of vertical edges. We succeeded in solving the model exactly. While the model retains the same critical exponent of the Domany-Kinzel model, it represents a step forward toward solving a true two-dimensional problem. We published the analysis (Chen and Wu, 2006) in a special issue of *Nankai Tracts in Mathematics* dedicated to the great mathematician S. S. Chern, who passed away in December 2004, at the time when Chen and I were working on the problem.

Remarks

The Domany-Kinzel model of directed percolation can be extended to three dimensions. One way to do this is to assign full occupation probability to edges within a plane, instead of within a row as in the Domany-Kinzel model. Hervé Kunz and I have pondered over this extension. We found a bijection mapping the model to a counting problem of walks on a 3-dimensional terrace, which in turn is mapped to an *interacting* dimer problem on an honeycomb lattice with a boundary. However, since the solution of the latter problem depends crucially on aspect ratios of the lattice boundary (Elser, 1984), which is yet unresolved, the full solution still awaits any further progress on the dimer problem.

References for Chapter 7

Baxter, R. J., H. N. V. Temperley and S. E. Ashley (1978), Triangular Potts model at its transition temperature, and related models, *Proc. Roy. Soc.* A 358, 535-559.

Chen, L. C. and F. Y. Wu (2006), Directed Percolation in Two Dimensions: An Exact Solution, in *Differential Geometry and Physics, Nankai Tracts in Mathematics*, Vol. 10, Eds. M. L. Ge and W. Zhang (World Scientific) 160-168.

Domany, E. and W. Kinzel (1981), Directed percolation in two dimensions: Numerical analysis and an exact solution, *Phys. Rev. Lett.* 47, 5-8.

Elser, V. (1984), Solution of the dimer problem on a hexagonal lattice with boundary *J. Phys. A* 17 1509-1514.

Erdös, P. and A. Rényi (1960), On the evolution of random graphs, *Publ. Math. Inst. Hung. Acad. Sci.* 5, 17-60.

Essam, J. W.(1972), Percolation and cluster size, in *Phase Transitions and Critical Phenomena, Vol. 2*, Eds. C. Domb and M. S, Green (Academic, New York), 197-270.

Essam, J. W.(1980), Percolation theory, *Rep. Prog. Phys.* 43, 833-912.

Fortuin C. M. and P. W. Kasteleyn (1972), On the random-cluster model. I. Introduction and relation to other models, *Physica* 57, 536-564.

Kesten, H. (2006), What is percolation, *Notices of the American Mathematical Society*, May, 572-573.

Kunz, H and F. Y. Wu (1978), Site percolation as a Potts model, *J. Phys. C* 11, L1-L4; Erratum, *ibid.* 11, 359.

Scullard, C. R. and R. M. Ziff (2006), Predictions of bond percolation thresholds for the kagome and Archimedean (3.12^2) lattices, *Phys. Rev. E* 73, 045102.

Shante, V. K. S. and S. Kirkpatrick (1971), An introduction to percolation theory, *Adv. Phys.* 20, 325-327.

Temperley, H. N. V. and E. H. Lieb (1971), Relation between the 'percolation' and 'colouring' problem and other graph-theoretical problems associated with regular planar lattice: Some exact results for the 'percolation' problem, *Proc. Roy. Soc.* A 322, 251-280.

Wu, F. Y. (1978), **P37**, Percolation and the Potts model, *J. Stat. Phys.* 18, 115-123.

Wu, F. Y. (1982), **P28**, The Potts model, *Rev. Mod. Phys.* 54, 235-268.

Wu, F. Y. (1982a), **P38**, An infinite-range bond percolation, *J. Appl. Phys.* 53, 7977.

Wu, F. Y. and H. E. Stanley (1982), **P39**, Domany-Kinzel model of directed percolation: Formulation as a random-walk problem and some exact results, *Phys. Rev. Lett.* 48, 775-777.

Ziff, R. M. (2006), Generalized cell-dual-cell transformation and exact thresholds for percolation, *Phys. Rev. E* 73, 016134.

Ziff, R. M. and C. R. Scullard (2008), Exact bond percolation thresholds in two dimensions, *J. Phys. A* 39, 15083-15090.

8. Graph Theory

Introduction

Studies in statistical mechanics are often facilitated by the use of graphical terms and graphical analysis (Kasteleyn, 1967; Essam, 1971; Wu, 1978). Examples include the Mayer cluster expansion of an imperfect gas (Mayer, 1937), the high- and low-temperature expansions of the Ising partition function, and the vertex model considered as a graph-theoretical problem. I have always been fond of graphical analysis, and graph-theoretical studies have been my favorite subject. The review article on graph theory and statistical physics (Wu, 1978) is the outgrowth of a talk I gave at the 1975 Recontre on Combinatorial Mathematics and Applications held in Aberdeen, Scotland.

Graphical Analyses in Many-Body Theory

My interest in graph method began early in my graduate student days.

I did my Ph.D. work in many-body theory under the late Eugene Feenberg at Washington University. Feenberg was a prominent nuclear theorist who had shifted his interest to many-body theory in the late 1950s. I entered Washington University in the fall of 1959 and was one of his first students to work in this new direction. My classmates at the time included Walter Massey, who later became the director of the National Science Foundation, and Chia-Wei Woo, who 20 years later would be the founder and the first president of the Hong Kong University of Science and Technology. It was an exciting time in the Feenberg school.

Feenberg pioneered in the correlated-basis function approach to the many-body problem using Jastrow-type wavefunctions, culminating in the publication of the definitive book (Feenberg, 1969) summarizing his accomplishments. Under his guidance, I completed two papers on applications of correlated-basis functions, one on liquid helium II (Wu and Feenberg, 1961) and one on Fermi liquids (Wu and Feenberg, 1962). But I felt more at home in developing the methodology of the formulation.

Feenberg and I used a cluster expansion of the Jastrow wavefunction to study Fermi liquids (Feenberg and Wu, 1962). While only the first 2 terms in the expansion were needed in the numerical calculation, I became interested in the expansion itself, and undertook the task of extending it to all orders. I developed a graphical method which eventually led to an elegant linked-cluster type expansion (Wu, 1963, **P40**). The final result generalizes the well-known Ursell-Mayer expansion of a partition function to functions with indices such as those associated with Fermi systems.

That was my first taste of graphical analysis.

The correlated-basis function approach involves the use of the n-particle distribution function g_n, which is unknown for $n \geq 3$. The usual way to get around for $n = 3$ is to use the Kirkwood superposition approximation

$$g_3(1,2,3) \cong g_2(1,2)\, g_2(2,3)\, g_2(3,1). \tag{8.1}$$

The approximation (8.1), while convenient to use, fails to satisfy a sequential requirement on the distribution function

$$\rho \int g_n(1,2,\cdots,n) d\mathbf{r}_n = (N-n+1)\, g_{n-1}(1,2,\cdots,n), \tag{8.2}$$

where N is the number of particles and ρ is the particle density.

In a concurrent study of liquid helium, H. Woody Jackson, another classmate of mine, and Feenberg (1962) introduced a convolution approximation for g_3 which obeys (8.2) exactly. Several years later at the Virginia Polytechnic Institute, my student Ming-Kang Chien and I extended the convolution form to all orders (Wu and Chien, 1970). The final expression satisfies (8.2) exactly for all n, and is expressed graphically in terms of rooted Cayley trees. This exposure to rooted graphs eventually led to the notion of the rooted Tutte polynomial introduced by me more than 30 years later (see next Section).

The Potts Model and the Tutte Polynomial

The Potts partition function (5.1) is the starting point of its graphical analysis (Baxter, Kelland and Wu, 1976, **P27**). Consider, for example, the antiferromagnetic Potts model at $T = 0$. In the ground state all neighboring sites are in distinct states so one has $K = -\infty$ and $v = -1$. The partition function (5.1) then immediately gives the number of q-colorings of G in the form of a polynomial in q, or

$$Z_G(q,-1) = \sum_{g \subseteq G} (-1)^{e(g)} q^{n(g)}. \tag{8.3}$$

This is the celebrated Birkhoff (1912) formula of the chromatic function, which Birkhoff deduced using an inclusion-exclusion argument.

The Potts model partition function (5.1) is known as the dichromatic polynomial of Tutte (1954) in graph theory. The dichromatic polynomial was introduced by Tutte in 1954, two years after the Potts (1952) introduction of the Potts model. It is of more than passing interest to note that an important entity arose in two disciplines independently at about the same time. This occurs often to the emergence of great ideas (see also pp. 60-61).

For a (connected) graph G of N sites, the Tutte dichromatic polynomial is related to the Potts partition function by

$$T_G(x,y) \equiv (x-1)^{-1}(y-1)^{-N}\, Z_G(q,v)\Big|_{q=(x-1)(y-1),\, v=y-1}, \tag{8.4}$$

$$x = 1 + q/v, \qquad y = 1 + v. \tag{8.5}$$

Furthermore, if one expands (5.1) as a power series in e^{-K^*}, where K^* is given by (3.5), the expansion coefficients are the flow polynomial in grapy theory (Tutte, 1954, 1984).

The connection of the Potts model with graph theory was known to physicists beginning in the 1960s, but discussions were scattered in the literature. In the fall of 1988, I visited the University of Washington as a guest of Michael Schick. The visit gave me the time to pause, and I wrote an expository paper (Wu, 1988, **P41**) elucidating graph-theoretical aspects of the Potts model. The work also revealed some curious results, among them a sum rule of the Potts partition function for any graph G,

$$\sum_{g\subseteq G}(-1)^{e(g)} Z_g(q,v) = (-1)^{e(G)} q\,. \tag{8.6}$$

I came upon with the notion of rooted Tutte polynomial some 10 years later. In the study of Potts duality relations (Chapter 3), partition functions (sums) in the form of (5.1) arose which disregard components containing correlating sites in the counting of $n(g)$. I named these reduced partition functions the *rooted Tutte polynomial.* The correlation duality relation (3.13) then emerged as a simple consequence in a graphical analysis of the rooted Tutte polynomial. This analysis was carried out in collaboration with Chris King and Wentao Lu and published in (Wu, King and Lu, 1999, **P42**).

The work on the rooted Tutte polynomial was reported at a symposium in memory of the late knot theorist François Jaeger held in Grenoble in 1998. I had the good fortune of meeting William Tutte at the symposium, and presenting to Tutte the association of roots to a polynomial that was famously named after him. Tutte later reviewed **P42** (Tutte, 2000), commenting that the theory of rooted Tutte polynomials is reminiscent to the Birkhoff-Lewis equations (Tutte, 1993).

Random Graphs

A topic of fundamental importance in graph theory is the problem of random graphs (Erdös and Rényi, 1960; Bollobás, 2001). The random graph problem concerns with a complete graph whose edges are randomly broken. Erdös and Rényi studied clustering properties of the graph from a probabilistic approach.

While writing the Potts model review (Wu, 1982, **P28**), I was struck by the similarity between random graphs and the random cluster model of Kasteleyn and Fortuin (1969). This suggested to me the possibility of formulating random graphs as a Potts model. The formulation resulted in the infinite range bond percolation problem described in the preceding chapter. It can also be interpreted as a problem of network communication.

Consider a network of N communication stations, where each station is connected to every other with a probability p/N. For $p = 0$ there is no communication between all stations. For p large stations can communicate to each other, via a third station if necessary. One asks how does the connectivity change as a function of p.

By formulating the network as a Potts model, I was able to deduce clustering properties of the network. This led to the determination of a critical $p_c = 1/2$ in the limit of $N \to \infty$, with the network breaking into isolated connected clusters when $p < p_c$. This reproduces the result of Erdös and Rényi on random graphs, now deduced from a new viewpoint. Other properties including the mean cluster size can also be computed (Wu, 1982b, **P43**).

This heuristic approach to random graphs appeared to have drawn some interest from the graph theory community, and I was invited to a conference on random graphs held in Poland in 1983.

Spanning Trees

The spanning tree of a graph G of N sites is a connected subgraph (of $N-1$ edges) which covers all sites and has no circuits. The example of a spanning tree on a 4×4 lattice is shown in Fig. 8.1.

Consider a symmetric Laplacian matrix $\mathbf{L}$ with elements

$$\begin{aligned} L_{ij} &= -x_{ij}, & i &\neq j \\ &= \sum_{k\neq i} x_{ik} & i &= j, \end{aligned} \tag{8.7}$$

where x_{ij} is the weight of edge ij in G. Since the sum of every row and column is equal to zero, there is one zero eigenvalue, $\lambda_1 = 0$. The matrix

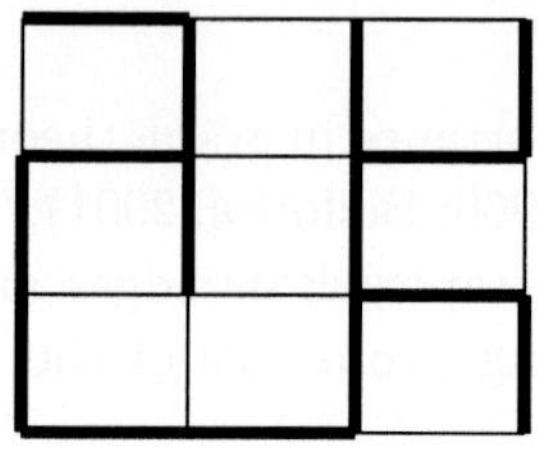

Fig. 8.1. A spanning tree (heavy lines) on a 4 × 4 lattice.

L plays an important role in graph theory (Biggs, 1993). Particularly, all cofactors of the Laplacian matrix are equal and equal to the spanning tree generating function

$$Z_G(\{x_{ij}\}) \equiv \sum_{\text{spanning trees}} \prod_{\text{edges}} x_{ij} = \frac{1}{N}\prod_{i=2}^{N} \lambda_i. \tag{8.8}$$

The Laplacian matrix L also plays a central role in electric circuit theory (see page 63).

The expression (8.8) gives the number of spanning trees N_{ST} on G by setting $x_{ij} = 1$. This is one of the rare cases in lattice statistics that an enumeration problem can be done with complete rigor for lattices in any spatial dimension. Spanning trees can also be enumerated by the Tutte polynomial (8.4) as $N_{ST} = T_G(1,1)$. It is also enumerated by the Potts partition function $Z_G(q,v)$ in the limit of $q = v \to 0$. Using the Potts model realization, I was able to enumerate N_{ST} for several two-dimensional lattices in (Wu, 1977).

I organized a summer school in mathematical physics in 1999 in Taiwan. During the School I worked with Wen-Jer Tzeng on a project of enumerating $Z(\{x_{ij}\})$ for nonorientable surfaces and hypercubic lattices. While a straightforward task of diagonalizing the Laplacian matrix associated with each lattice, it was surprising that the work had not been done before. Bernard Nienhuis later told me in Amsterdam that the project was also on his next-to-do list. Tzeng and I wrote up our results in (Tzeng and Wu, 2000; **P44**).

In another project initiated at the School, Robert Shrock and I completed an evaluation of the entropy of spanning trees for all regular two-dimensional lattices (Shrock and Wu, 2000).

Graphical Analysis of Lattice Models

Graphical analysis offers direct and often intuitive derivation of physical results. A good example is the triangular Potts model with two- and alternate three-site interactions. The model possesses the duality relation (3.7) obtained algebraically by Baxter, Temperley and Ashley (1978). Using a graphical approach, Keh-Yin Lin and I re-derived the duality relation as an instance of a symmetry relation of a vertex model (Wu and Lin, 1980, **P45**).

Another example of a graphical analysis is the nonintersecting string model (9.6) studied by Jacques Perk and myself. The nonintersecting string model generalizes the 6-vertex model to allow Q different lattice edge states. The model had been studied earlier by Perk and Schultz (1983) and by Truong (1986) using algebraic methods. Using graphical analysis, Perk and I deduced further results on the model (9.6) including a heuristic derivation of its inversion relation (Perk and Wu, 1986b). We also established its equivalence with a Q^2-state Potts model (Perk and Wu, 1986a, **P46**). This nonintersecting model emerged one year later prominently in knot theory as the state model of Kauffman (1987) (pp. 60-61).

References for Chapter 8

Baxter, R. J., S. B. Kelland and F. Y. Wu (1976), **P27**, Equivalence of the Potts model or Whitney polynomial with an ice-type model: A new derivation, *J. Phys. A* 9, 439-440.

Biggs, N. L. (1993), *Algebraic Graph Theory, 2nd Ed.* (Cambridge University Press, Cambridge).

Birkhoff, G. D. (1912), *Ann. Math.* 14. 42-44.

Bollobás, B. (2001), *Random Graphs* (Cambridge University Press, New York).

Essam, J. W. (1971), Graph theory and statistical physics, *Discrete Math.* 1, 83-112.

Erdös, P. and A. Rényi (1960), On the evolution of random graphs, *Publ. Math. Inst. Hung. Acad. Sci.* 5, 17-60.

Feenberg, E. (1969), *Theory of Quantum Fluids* (Academic Press, New York).

Jackson, H. W. and E. Feenberg, (1962), Energy spectrum of elementary excitations in helium II, *Rev. Mod. Phys.* 34, 686-693.

Kasteleyn, P. W. (1967), Graph theory and crystal physics, in *Graph Theory and Theoretical Physics*, Ed. F. Harary (Academic Press, London), 43-110.

Kauffman, L. H. (1987), State models and the Jones polynomial, *Topology* 26, 396-407.

Lu, W. T. and F. Y. Wu (1998), On the duality relation for correlation functions of the Potts model, *J. Phys. A* 31, 2823-2836.

Mayer, J. (1937), The statistical mechanics of condensed systems: I, *Proc. Roy. Soc.* 5, 67-73.

Perk, J. H. H. and F. Y. Wu (1986a), **P46**, Nonintersecting string model and graphical approach: Equivalence with a Potts model, *J. Stat. Phys.* 42, 727-742.

Perk, J. H. H. and F. Y. Wu (1986b), Graphical approach to the nonintersecting string model: Star-triangle relation, inversion relation and exact solution, *Physica A* 138, 100-124.

Potts, R. B. (1952), some generalized order-disorder transformations, *Proc. Camb. Phil.Soc.* 48, 106-109.

Shrock, R. and F. Y. Wu (2000), spanning trees on graphs and lattices in d dimensions. *J. Phys. A* 33, 3881-3902.

Truong, T. T. (1986), Structure properties of a $Z(N^2)$-vertex spin model and its equivalent $Z(N)$-vertex model, *J. Stat. Phys.* 42, 349-379.

Tutte, W. T. (1954), A contribution to the theory of chromatic polynomials, *Can. J. Math.* 6, 80-91.

Tutte, W. T. (1984), *Encyclopedia of Mathematics and Its Applications*, Vol. 21, *Graph Theory* (Addison-Wesley, Reading, Massachusetts), Chapter 9.

Tutte, W. T. (1993), The matrix of chromatic joins, *J. Comb. Theory B* 57, 269-288.

Tutte, W. T. (2000), MathSciNet, MR1703441.

Tzeng, W.-J. and F. Y. Wu (2000), **P44** Spanning trees on hypercubic lattices and nonorientable surfaces, *Appl. Math. Lett.* 13:6, 19-25.

Wu, F. Y. (1963), **P40**, Cluster development in an N-body problem, *J. Math. Phys.* 4, 1438-1443.

Wu, F. Y. and M. K. Chien (1970), Convolution approximation for the n-particle distribution function, *J. Math. Phys.* 11, 1912-1916.

Wu, F. Y. (1977), Number of spanning trees on a lattice, *J. Phys. A* 10, L113-L115.

Wu, F. Y. (1978), Graph theory in statistical physics, in *Studies in Foundations of Combinatorics, Adv. in Math.: Supp. V.1*, Ed. G.-C. Rota, 151-166, (Academic Press, New York 1978).

Wu, F. Y. (1982), **P28**, The Potts model, *Rev. Mod. Phys.* 54, 235-268.

Wu, F. Y. (1982b), **P43**, Random graphs and network communication, *J. Phys. A* 15, L395-L398.

Wu, F. Y. (1988), **P41**, Potts model and graph theory, *J. Stat. Phys.* 52, 99-112.

Wu, F. Y. and E. Feenberg, (1961), Ground state of liquid helium (Mass 4), *Phys. Rev.* 122, 739-742.

Wu, F. Y. and E. Feenberg, (1962), Theory of the Fermion liquid, *Phys. Rev.* 128, 943-955.

Wu, F. Y., C. King and W. T. Lu, (1999), **P42**, On the rooted Tutte Polynomial, *Ann. Inst. Fourier, Grenoble*, 49:3, 1103-1114.

Wu, F. Y. and K. Y. Lin (1980), **P45**, On the triangular Potts model with two- and three-site interactions, *J. Phys. A* 14, 629-635.

9. Knot Invariants

Introduction

A knot is the embedding of a circle in the three-dimensional space, and a link is the embedding of two or more circles. For visualization purposes, knots and links are described by their projections onto a plane as shown in Fig. 53 in (Wu, 1992, **P47**). For brevity, these projections are called knots.

Two knots are equivalent if they can be transformed into each other by a continuous deformation of the circle(s) in the 3-dimensional space. A central problem in the mathematical theory of knots has been the finding of a way to determine from their projections whether two knots are equivalent. This has led to the notion of knot invariants.

Knot invariants are algebraic entities associated with knots which remain unchanged when knot lines are deformed. Very few knot invariants were known before the 1980s. The situation changed abruptly with the discovery of the Jones polynomial (Jones, 1985) and the subsequent revelation of its connection with statistical mechanics (Kauffman, 1987; Jones, 1989). Many new invariants have since been discovered by making use of this connection.

In the years following Jones' discovery, papers on this new development were written mostly for the consumption by mathematicians. In 1991, motivated by the reading of an article by Jones (1990) in *Scientific American* aimed to a general readership, I undertook the task of writing an expository article on knot invariants for the benefit of physicists. This resulted in the review paper (Wu, 1992, **P47**), the second review article (after the Potts model review **P28**) I wrote for the *Reviews of Modern Physics.*

Lattice Models and Knot Invariants

The idea of the statistical mechanical approach to knot invariants is extremely simple once it is understood. Briefly stated, it is the construction of lattice models whose partition functions are taken to be knot invariants.

Specifically, starting from a given knot projection, one constructs a lattice and a lattice model on this lattice. If model parameters are chosen such that the partition function of the model remains invariant when the knot (and the lattice) is deformed, the partition function is a knot invariant by definition.

To accomplish this goal, it is necessary to understand the effect of knot deformations in 3 dimensions on knot projections. It was established by Reidemeister (1948) that all deformations of knots in the 3-dimensional space induce sequences of three basic types of line moves in the projection, the Reidemeister moves I, II, III. These three moves are shown in Fig. 9.1. It follows that one needs only to require the invariance of the partition function under these three Reidemeister moves.

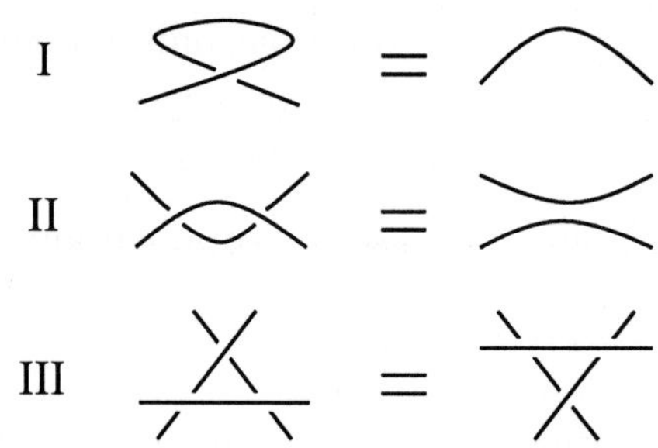

Fig. 9.1. Reidemeister moves for unoriented knots.

Lattice models constructed from knots can be either vertex or spin models. It is usually straightforward to fix model parameters to satisfy type I and II requirements. The requirement of the invariance under type III Reidemeister moves is more subtle, and turns out to coincide with the Yang-Baxter equation (Yang, 1967; Baxter, 1978) of the lattice model. It follows that the Yang-Baxter equation of any lattice model can be used to construct a knot invariant, regardless of the solubility of the model. Jones (1989) showed that all knot invariants known at the time, including the two-variable homfly polynomial (Freyd *et al.*, 1985), can be constructed in this fashion.

Two New Knot Invariants

I have contributed to the construction of two new knot invariants using the Yang-Baxter approach.

One invariant is a set of scalars derived from the chiral Potts model. The investigation of the chiral Potts model was initiated by Au-Yang *et al.* (1987), and the full solution of the Yang-Baxter equation was obtained by Baxter, Au-Yang and Perk (1988). In 1992, Au-Yang remarked to me that the Yang-Baxter solution of the chiral Potts model bears much in common with Reidemeister moves. This prompted me to look into the resemblance.

Together with my student Predeep Pant and colleague Chris King, I derived a new knot invariant in the form of a set of scalars. The derivation requires the use of a generalized Gaussian summation identity which we worked out (and learned later that it had been known since 1960). For the trefoil knot, for example, the chiral Potts invariant is the set of scalars

$$I_{\text{trefoil}}(N) = -i(1 + 2e^{2\pi i N/3})/\sqrt{3}, \quad N = 2, 3, \cdots . \tag{9.1}$$

This finding was reported as a *Letter* in (Wu, Pant, and King, 1994, **P48**) and later in a full paper (Wu, Pant, and King, 1995). Invariants for knots and links up to 8 crossings were computed.

The second new invariant is a polynomial deduced from the soluble 19-vertex model of Izergin and Korepin (1981) shown in Fig. 2.5. The invariant turned out to be identical to one obtained by Akutsu, Deguchi, and Wadati (1987) who considered only the case of 3-braided knots. For the trefoil, for example, the invariant is the polynomial

$$I_{\text{trefoil}}(t) = t^2(1 + t^3 - t^5 + t^6 - t^7 - t^8 + t^9). \tag{9.2}$$

This new invariant was reported in (Wu and Pant, 1997).

Work on the two new knot invariants formed part of the Ph.D. dissertation of Predeep Pant who graduated in 1997.

Knot Invariants from Spin Models

Knot invariants can be constructed from lattice models in a variety of ways (see **P47**). It is the simplest to use spin models with edge interactions.

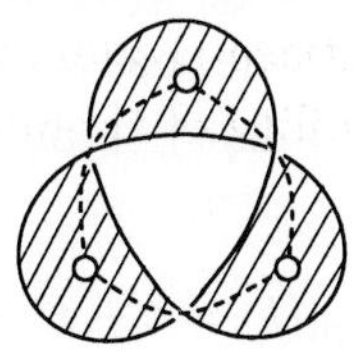

Fig. 9.2. A lattice (broken lines) constructed from a trefoil (solid lines).

Starting from a given knot, shade every *other* face of the knot diagram and place spins inside shaded faces with interactions crossing the knot line crossings. This yields a spin lattice whose covering graph is the knot itself (see Fig. 5.1). The example of a trefoil knot leading to a lattice of three spins is shown in Fig. 9.2.

The scheme of the face shading is the same as that devised by Baxter, Kelland and myself in the graphical analysis of the Potts model in **P27**, which is also the scheme used by Perk and myself in analyzing the nonintersecting string model in **P46**.

There are two kinds of topologically different line crossings, + and −, shown in Figs. 9.3(b) and (c), thus requiring 2 kinds of spin interactions. In the case of the Potts model denote the two kinds of interactions by K_+

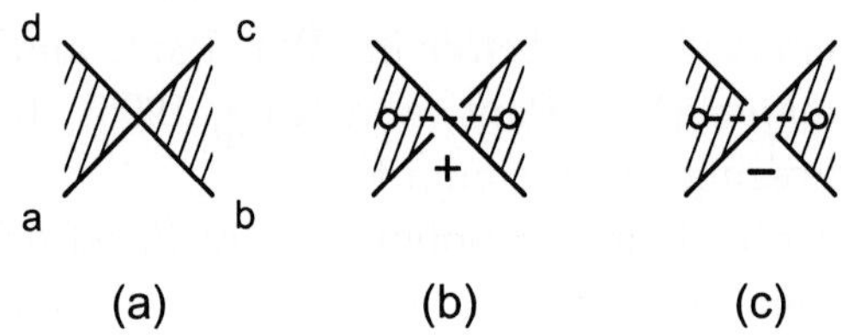

Fig. 9.3. Configurations of line crossings. (a) Weight (9.6). (b),(c) Two types of line crossings.

and K_-. It can be verified (see Section VII.B of **P47**) that the Potts partition function $Z_{\text{Potts}}(q, v_+, v_-)$ given by (5.1) is invariant under all three Reidemeister moves of knot lines, provided that one has

$$v_\pm = e^{K_\pm} - 1 = -(1 + t^{\mp 1}), \quad q = t + 2 + t^{-1}. \tag{9.3}$$

The Jones polynomial of oriented knots is then given by

$$V(t) = q^{-(N+1)/2}(-t)^{n_+ - n_-} Z_{\text{Potts}}(q, v_+, v_-), \tag{9.4}$$

where N is the number of crossings and $n_\pm$ is the number of $\pm$ crossings in the (oriented) knot as defined in Fig. 2(b) in **P47**.

For the trefoil of Fig. 9.2, one has $N = 3$, $n_+ = 3, n_- = 0$ and $Z_{\text{Potts}}(q, v_+, v_-) = q^3 + 3q^2 v_+ + 3qv_+^2 + 3qv_+^3$. Then (9.4) gives the Jones polynomial

$$V_{\text{trefoil}}(t) = t + t^3 - t^4. \tag{9.5}$$

Alternately, the Jones polynomial can also be constructed from a vertex model (Wu, 1992a).

Bracket Polynomial and the Nonintersecting String Model

Soon after Jones' discovery of the Jones polynomial, Kauffman (1987) showed that the Jones polynomial can be obtained from the "bracket polynomial" of a state model. The state model turned out to be the nonintersecting string model studied by Jacques Perk and myself one year earlier (Perk and Wu, 1986a, **P46**), and the bracket polynomial is its partition function. This is another case that an important entity arose independently in two seemingly unrelated disciplines at about the same time.

Perk and I studied a Q-state nonintersecting string model whose edges can assume Q different colors. The model has site-dependent vertex weight

$$W_i = A_i\,\delta_{ab}\,\delta_{cd} + B_i\,\delta_{ad}\,\delta_{bc}, \quad i = 1, 2, \cdots, N \tag{9.6}$$

where a, b, c, d specify the colors of the 4 lines incident to the site in the order shown in Fig. 9.3(a).

Connect lines of the same color as dictated by the Kronecker deltas in (9.6) (there are 2 non-crossing decompostions of the 4 lines in the case they bear the same color). The line graph on the lattice is then decomposed into polygonal configurations P consisting of nonintersecting polygons. The partition function assumes the form

$$Z_{\mathrm{NIS}}(A_i, B_i) = \sum_P Q^{p(P)} \prod_i w_i(P)\,, \tag{9.7}$$

where $p(P)$ is the number of polygons in P, $w_i(P)$ is either A_i or B_i, depending on the decomposition of the lines at the ith vertex.

Perk and I established the equivalence of the model (9.6) with a $q = Q^2$ state Potts model with interactions K_i given by

$$e^{K_i} = 1 + QA_i/B_i\ . \tag{9.8}$$

For uniform weights $A_i = A$, $B_i = B$, the model (9.6) becomes the Kauffman state model, and its partition function $Z_{\mathrm{NIS}}(A, B)$ is the bracket polynomial.

Kauffman (1987) considered the state model on knot graphs, and showed that the bracket polynomial gives the Jones polynomial by taking

$$A = B^{-1} = t^{-1/4}, \quad Q = -(\sqrt{t} + 1/\sqrt{t}). \tag{9.9}$$

It can be verified that (9.8) and (9.9) are the same as (9.3) with $A_+ = B_- = A$ and $A_- = B_+ = B$.

Remarks

It is clear that two equivalent knots always have the same invariant. But the inverse is not necessarily true. Namely, having the same invariant does not necessarily imply two knots are equivalent. An outstanding open question in knot theory is to find invariants which can distinguish all knots. In fact, it is not even known whether the Jones polynomial fulfills this requirement.

References for Chapter 9

Akutsu, Y., T. Deguchi and M. Wadati (1987), Exactly solvable models and new link polynomials. II. Link polynomials for closed 3-braids. *J. Phys. Soc., Japan* 56, 3464-3470.

Au-Yang, H, B. M. McCoy, J. H. H. Perk, S. Tang and M.-L. Yan (1987), Commuting transfer matrices in the chiral Potts models: Solutions of the star-triangle equations with genus > 1, *Phys. Lett. A* 123, 219-223.

Baxter, R. J., S. B. Kelland and F. Y. Wu (1976), **P27**, Equivalence of the Potts model or Whitney polynomial with an ice-type model: A new derivation, *J. Phys. A* 9, 439-440.

Baxter, R. J., J. H. H. Perk and H. Au-Yang (1988), New solutions of the star-triangle relations for the chiral Potts model, *Phys. Lett. A* 128, 138-142.

Freyd, D., P. Yetter, J. Hoste, W. B. R. Lickorish, K. C. Millet and A. Oceanau (1985), A new polynomial invariants of knots and links, *Bull. Am. Math. Soc.* 12, 239-246.

Izergin A. G. and V. E. Korepin (1981), The inverse scattering method approach to the quantum Shabat-Mikhailov model, *Commun. Math. Phys.* 79, 303-316.

Jones, V. F. R. (1985), A polynomial invariant for links via von Neumann algebras, *Bull. Am. Math. Soc.* 12, 103-112.

Jones, V. F. R. (1989), On knot invariants related to some statistical mechanical models, *Pacific J. Math.* 137, 311-334.

Jones, V. F. R. (1990), Knot theory and statistical mechanics, *Sci. Am.* November, 98-103.

Kauffman, L. H. (1987), State models and the Jones polynomial, *Topology* 26, 396-407.

Pant, P. and F. Y. Wu (1997), Link invariant of the Izergin-Korepin model, *J. Phys. A* 30, 7775-7782.

Perk, J. H. H. and F. Y. Wu (1986a), **P46**, Nonintersecting string model and graphical approach: Equivalence with a Potts model, *J. Stat. Phys.* 42, 727-742.

Reidemeister, K. (1948), *Knotentheorie* (Chelsea, New York).

Wu, F. Y. (1992), **P47**, Knot theory and statistical mechanics, *Rev. Mod. Phys.* 64, 1099-1131.

Wu, F. Y. (1992a), Jones polynomial as a Potts model partition function, *J. Knot Theory and Its Ramifications*, 1, 47-57.

Wu, F. Y., P. Pant and C. King (1994), **P48**, New link invariant from the chiral Potts model, *Phys. Rev. Lett.* 72, 3937-3940.

Wu, F. Y., P. Pant and C. King (1995), Knot invariant and the chiral Potts model, *J. Stat. Phys.* 78, 1253-1276.

Yang, C. N. (1967), Some exact results for the many-body problem in one dimension with repulsive delta function interaction, *Phys. Rev. Lett.* 67, 1312-1315.

10. Other Topics

This chapter describes selected works outside the scope of the preceding chapters. Topics range from a perplexing problem in the theory of electric circuits that had lingered since Kirchhoff's time, to the solution of the one-dimensional Hubbard model, which has played a very important role in the theory of high-T_c superconductivity.

Theory of Electric Circuits

The theory of electric networks was formulated by Kirchhoff (1847) more than 160 years ago. One central problem in network theory is the computation of two-point resistances.

Consider a resistor network consisting of N nodes with nodes i and j connected by a resistance r_{ij}. Let V_i and I_i denote, respectively, the electric potential and current flowing into the network at node i. To compute the effective resistance $R_{\alpha\beta}$ between nodes α and β, one injects into the network a current I at node α and collects an exiting current I at node β, namely, setting $I_i = I(\delta_{i,\alpha} - \delta_{i,\beta})$. Then, by Ohm's law, the effective resistance is computed as $R_{\alpha\beta} = (V_\alpha - V_\beta)/I$.

The potential $\mathbf{V} = (V_1, V_2, ..., V_N)$ and current $\mathbf{I} = (I_1, I_2, ..., I_N)$ are related by the Kirchhoff equation

$$\mathbf{L}\,\mathbf{V} = \mathbf{I}, \tag{10.1}$$

where $\mathbf{L}$ is the Laplacian matrix (8.7) with elements $x_{ij} = 1/r_{ij}$. The crux of the matter is to solve $\mathbf{V}$ for the given $\mathbf{I}$.

Equation (10.1) cannot be solved directly by inversion since $\mathbf{L}$ is singular. Kirchhoff recognized this fact, and instead made use of graph-theoretical aspects of the Laplacian matrix to formulate $R_{\alpha\beta}$ as the ratio of 2-rooted spanning forests and spanning trees. But the formulation, while elegant, is not very useful in practice, since it is not straightforward to enumerate rooted spanning forests. The evaluation of two-point resistances has therefore remained a topic of continuing interest over many years (see, for example, van der Pol, 1959; Doyle and Snell, 1984). Furthermore, past studies have focused on infinite networks (Cserti, 2002), with little attention paid to finite networks, even though the latter are those occurring in real life.

During my sabbatical in Berkeley in 2002, Dung-Hai Lee and I often chatted about elastic properties of a solid. The potential energy of a solid is governed by the same Laplacian matrix (8.7), and the intriguing resemblance prompted me to ponder over the resistor problem. This led to the paper (Wu, 2004, **P49**) in which I obtained a concise formulation for the two-point resistance.

For resistor networks the Laplacian matrix $\mathbf{L}$ is Hermitian. As a Hermitian matrix, $\mathbf{L}$ has orthonormal eigenvectors $\Psi_i = (\psi_{i1}, \psi_{i2}, \cdots, \psi_{iN})$ and eigenvalues λ_i, $i = 1, 2, 3, \cdots, N$, with $\lambda_1 = 0$ (see page 53). In **P49**, I deduced the formula

$$R_{\alpha\beta} = \sum_{i=2}^{N} \frac{1}{\lambda_i} \left|\psi_{i\alpha} - \psi_{i\beta}\right|^2 \tag{10.2}$$

for two-point resistances, and applied it to various resistor networks. Since a network is completely described by its Laplacian matrix, it is only natural that information regarding the network should be given in terms of the eigenvalues and eigenvectors of the matrix.

The formulation (10.2) solves the classical problem of corner-to-corner resistance of a rectangular resistor network. But the application of (10.2) yields an expression in the form of a double summation whose physical and mathematical contents are not immediately clear. In the Spring of 2008 I was at the Issac Newton Institute for Mathematical Sciences in Cambridge participating in a Program on combinatorics. I took the opportunity to collaborate with John Essam of the Royal Holloway College of the University of London. After some hard work, we deduced the asymptotic expansion of the resistance with close-form expressions to all orders (Essam and Wu, 2009). For an $N \times N$ square lattice net with unit link resistors, the result gives

$$R_{N\times N} = \frac{4}{\pi}\log N + 0.077\,318 + \frac{0.266\,070}{N^2} - \frac{0.534\,779}{N^4} + O(\frac{1}{N^6}). \tag{10.3}$$

For impedance networks containing capacitances and inductances, one must use the *phaser* language of electric circuits for which impedances are complex entities. This makes the Laplacian matrix non-Hermitian and (10.2) no long holds. Three years later in 2005, Wen-Jer Tzeng and I circumvented the difficulty by considering instead the matrix $\mathbf{L}^\dagger\mathbf{L}$ which is Hermitian. We obtained an expression for the 2-point impedance similar to (10.2) in terms of eigenvalues and eigenfunctions of $\mathbf{L}^\dagger\mathbf{L}$ (Tzeng and Wu, 2006). An interesting ramification of the theory is the prediction of multiple resonances in a network of capacitors and inductances.

Quantization of the orbital angular momentum

It is well-known in elementary quantum mechanics that eigenvalues of the orbital angular momentum operator

$$L_z = xp_y - yp_x \tag{10.4}$$

are $n\hbar$, where $n = 0, \pm1, \pm2, \cdots$. The standard textbook derivation of the eigenvalues consists of two steps. First, one makes use of the commutation relations among L_x, L_y, L_z, which hold for any angular momentum, to conclude that eigenvalues of L_z are 0, $\pm1/2\hbar$, $\pm\hbar$, $\pm3/2\hbar$, $\cdots$. One then uses a single-valuedness argument of its eigenfunction in the Schrödinger representation to rule out the half-odd integral values. But the need of using a specific representation to render a physical conclusion had been regarded by some as unsettling.

In the 1960s I taught a graduate course of quantum mechanics at the Virginia Polytechnic Institute. After encountering this unsettling point and thinking it over, it became clear that the particular form of L_z given in (10.4) had not been fully used in the textbook derivation. My colleague David Kaplan and I pursued along this line, and succeeded to show that integral eigenvalues follow directly from an operator identity without the need of using specific operator representation. By writing L_z as the difference of two commuting operators (Eq. (2) of **P50**), each having non-negative integral eigenvalues, the desired integral eigenvalues of L_z follow immediately as a consequence.

But we had a hard time in getting our result published. The referee insisted that the proof was not new, and the fact that the authors were relatively unknown did not help either. Eventually, we settled in having our short paper published overseas (Kaplan and Wu, 1971, **P50**).

To this date, I continue to regard this little observation on orbital angular momentum as a *gem* which should be taught in every class of elementary quantum mechanics.

The Vicious Neighbor Problem

While on a plane to Texas in the summer of 1986, I grabbed a copy of *Omni* magazine from the magazine rack. What caught my eye next was the posting of a cash prize in the *Games* section, offering to anyone who could solve a "riflemen puzzle" (Morris, 1986).

The puzzle asks for the survival probability, the probability that a person survives amongst a shootout in a large crowd where everyone shoots, and kills, his/her nearest neighbor. This problem was originally posted by the

Brandeis mathematicians Abilock and Goldberg (1967) in *American Mathematical Monthly*; it had remained unsolved for almost 2 decades.

The puzzle caught my attention because it appeared to be a problem of statistical mechanics. After returning to Boston I got my colleague Rongjia Tao interested in the problem, and we proceeded to work on its solution. We considered the more general version of a d-dimensional domain where everyone kills the nearest neighbor with a probability p. The Abilock-Goldberg version corresponds to $d = 2, p = 1$.

The key to the solution lies in the fact that a person can receive only a limited number of bullets. In two dimensions, for example, a person can receive at most 5 bullets. Then the survival probability is obtained by subtracting from 1 the respective probabilities of receiving 1, 2, 3, 4, or 5 bullets. The problem became an exercise of computing the overlapping area of circles. After a combination of analytical and numerical chore, our analysis yielded the survival probability 0.284 051 for $d = 2$ and $p = 1$.

We claimed the *Omni* prize with this answer (Morris, 1987), and published the analysis in (Tao and Wu, 1987, **P51**). We also coined the name *vicious neighbor problem* to the puzzle, adopting the word *vicious* from Michael Fisher's Boltzmann medalist address on Vicious Walkers (Fisher, 1984). The vicious neighbor problem has since drawn the attention of other researchers, extending it to the more general framework of nearest-neighbor graphs (Finch, 2008).

Restricted Partitions of an Integer

A classic problem in combinatorics is the partition of an integer (MacMahon, 1916), the computation of the number of ways, P_n, that an integer n can be written as a summation of integers. For example, $P_2 = 2$ since one has $2 = 0 + 2 = 1 + 1$, and $P_3 = 3$ since $3 = 0 + 3 = 1 + 2 = 1 + 1 + 1$, etc.

The partition of an integer can also be studied with restrictions, and this leads to *restricted partitions*. Restricted partitions are most conveniently described by graphs, giving rise to plane partitions if planar graphs are used, solid partitions if using solids, etc. The plane partition problem has been solved since Gauss' time, but the solid partition has remained an outstanding unsolved problem in combinatorics for almost one century.

I became interested in restricted partitions due to its connection with the Potts model. In (Wu, 1997) I established that the restricted partition is equivalent to an infinite-state Potts model. In collaboration with Jean-Marie Maillard and others, I established the further connection of restricted partitions to a lattice animal problem (Wu, *et al.*, 1996, **P52**).

The equivalence of restricted partitions with a Potts model is intriguing. Since zeroes of the infinite-state Potts partition function tend to a unit circle in large lattices, it suggests that one should look at zeroes of the generating functions. My student Hsin-Yi Huang and I carried out this study for solid partitions, and found indeed that zeroes of the generating function approach the unit circle as the domain size increases.

Based on our work, we conjectured that all zeroes of the generating function of solid partitions lie on the unit circle in the limit when any linear dimension of the solid goes to infinity (Huang and Wu, 1997, **P53**). The paper **P53** was reviewed by George Andrews (1998) who remarked that "the observations are truly intriguing and point to the possibility that there are some deep and beautiful truths lurking in general d-dimensional partitions".

The Hubbard Model

I joined the faculty at Northeastern University in 1967 where Elliott Lieb was building a research group. In December that year, C. N. Yang (1967) published his renowned solution to the one-dimensional delta-function Fermi gas. Lieb recognized the implication of Yang's solution on the Hubbard model in condensed matter physics. This prompted us to investigate the latter problem using the method of nested Bethe ansatz devised by Yang.

The Hubbard model describes a system of electrons on a lattice. The Hamiltonian

$$\mathcal{H} = T \sum_{<ij>} \sum_{\sigma} c^\dagger_{i\,\sigma}\, c_{j\,\sigma} + U \sum_{i} n_{i\,+}\, n_{i\,-} \tag{10.5}$$

describes itinerant electrons hopping between the Wannier states of neighboring lattice sites with an on-site overlapping interaction U. Here $c^\dagger_{i\,\sigma}$ and $c_{i\,\sigma}$ are, respectively, the creation and annihilation operators for an electron of spin σ in the Wannier state at the i-th lattice site and $n_{i\,\sigma} = c^\dagger_{i\,\sigma}\, c_{i\,\sigma}$. The summation $< ij >$ is over nearest neighbors. It was believed that, as a function of U, the model describes a Mott transition between insulating and conducting states. Lieb and I considered the one-dimensional version.

We found the Bethe ansatz equations of the Hubbard model to be the same as those of the delta-function gas, except with the replacement of momentum k by $\sin k$ in the integral equations. The non-monotonicity of $\sin k$ in k caused a slight problem, but after the problem was resolved, we obtained a nice and concise expression for the ground state energy. It shows there is no Mott transition in one dimension. We reported the finding in a *Letter* (Lieb and Wu, 1968, **P54**).

Lieb moved across the Charles River from Northeastern to MIT soon thereafter. In 1968-69, Lieb and I continued to work on the expository monograph on vertex models (see page 11). The Hubbard model was put aside and was soon forgotten as each of us moved onto other directions. We did not get back to write the full paper on the Hubbard model until 34 years later.

With the advent of the high-T_c superconductivity in the 1980s, the Hubbard model suddenly became prominent in condensed matter physics. Several books have appeared on the subject matter of the one-dimensional model alone (Essler *et al.*, 2005). While we had often been urged to follow up with the full account of our work, it was not until 2001 when Lieb and I finally settled down to write the full paper (Lieb and Wu, 2002, **P55**). [i]

Review of a Book by Ta-You Wu

Ta-You Wu (1907-2000), a prominent teacher, educator, science administrator, and an outstanding theoretical physicist, was the father of scientific developments in Taiwan. At the invitation of President Chiang Kai-Shek, in the 1960s he single-handedly established a long-term scientific development plan, which was largely responsible for creating the scientific and industry complex that makes up Taiwan today. He had served as the President of the Academia Sinica and his many illustrious students included T. D. Lee and C. N. Yang. He was teaching at the National Tsing Hua University in Taiwan when I entered its graduate school in 1957.

Ta-You Wu has published several books in physics at the graduate level. The book *Lectures on the kinetic theory of gases, nonequilibrium thermodynamics and statistical theories* records expanded lectures delivered by him at the ripe age of 87 at the National Tsing Hua University. In this book, Ta-You Wu presented his unique view on the largely uncharted territory of nonequilibrium statistical mechanics. I reviewed the book in collaboration with my Northeastern colleague Allan Widom (Widom and Wu, 2005, **P56**).

Eulogy on Shang-Keng Ma

Shang-Keng Ma (1940-1983) was a professor of theoretical physics at the University of California, San Diego when he passed away at the young age of 43. He was a brilliant physicist who began his career in many-body theory. He visited Cornell University in 1972 where he became involved in the

[i]The full paper ***P55*** was published in the Proceedings of a Symposium held in Taiwan in 2002 on the occasion of my 70th birthday.

development of the renormalization theory of critical phenomena. Gradually, his interest shifted to statistical physics.

In 1981-82 Shang-Keng visited the National Tsing Hua University in Taiwan, where he authored the book *Statistical Mechanics* written in Chinese. In the book, which was translated into English after his death (Ma, 1985), he presented his viewpoint on statistical physics based on dynamical origins. It was during this visit that he became ill. Ma was modest and unassuming. After learning that he had cancer, he wrote me simply to say that "it is a new experience." He passed away on Thanksgiving day, November 24, 1983.

At the time I was the Director of the Condensed Matter Theory Program at the National Science Foundation administering his research grant. I learned of the sad news the day after his death when his colleagues called regarding his grant. The 50th Statistical Mechanics Meeting held 3 weeks later at Rutgers University, and I gave a eulogy on Shang-Keng at the meeting. The paper (Wu, 1983, **P57**) is a slightly edited version (by Harvey Gould) of the writeup of the talk.

Professor C. N. Yang and Statistical Mechanics

In 2007, Dr. K. K. Phua of the Nanyang Technological University invited me to give a talk at a symposium held in November of that year in honor of Professor C. N. Yang's 85th birthday. Chen Ning Yang is known throughout the world for his 1957 Nobel prize work in particle physics. But he has made equally important and seminal contributions in statistical physics which are less-widely known. I therefore chose to talk about Professor Yang's works in statistical mechanics and, based on my talk, wrote (Wu, 2008, **P58**).

References for Chapter 10

Abilock, R. and M. Goldberg (1967), N riflemen, *Amer. Math. Monthly* 74, 720.

Andrews, G. E. (1998), MathSciNet, MR1435061.

Cserti, J. (2002), Application of the lattice Green's function for calculating the resistance of infinite networks of resistors, *Am. J. Phys.* 68, 896-906.

Doyle, P. G. and J. L. Snell (1984), *Random walks and electric networks,* The Carus Mathematical Monograph, Series 22 (The Mathematical Association of America), 83-149; arXiv:math.PR/0001057.

Essam, J. W. and F. Y. Wu (2009), The exact corner-to-corner resistance of an $M \times N$ resisitor network: Asymptotic expansion, *J. Phys. A* 42, 025205.

Essler, F. H. L., F. H. L., H. Frahm, F. Göhmann, A. Klümper and V. E. Korepin, *The One-dimensional Hubbard Model,* (Cambridge University Press, 2005).

Finch, S. (2008), Nearest-neighbor graphs, http://algo.inria.fr/bsolve. I am grateful to Steven Finch for this reference.

Fisher, M. E. (1984), Walks, walls, wetting, and melting, *J. Stat. Phys.* 34, 667-729.

Huang, H. Y. and F. Y. Wu (1997), **P53**, The infinite-state Potts model and solid partitions of an integer, *Int. J. Mod. Phys.* 11, 121-126.

Kaplan, D. M. and F. Y. Wu (1971), **P50**, On the eigenvalues of orbital angular momentum, *Ch. J. Physics*, 9, 31-33.

Kirchhoff, G. (1847), Über die Auflösung der Gleichungen, auf welche man bei der Untersuchung der linearen Verteilung galvanischer Ströme geführt wird, Ann. Phys. und Chemie 72, 497-508.

Lieb, E. H. and F. Y. Wu (1967), **P54**, Absence of Mott transition in an exact solution of the short-range one-band model in one dimension, *Phys. Rev. Lett.* 20, 1445-1448.

Lieb, E. H. and F. Y. Wu (2003), **P55**, The one-dimensional Hubbard model: A reminiscence, *Physica A* 321, 1-27.

Ma, S. K. (1985), *Statistical Mechanics*, (World Scientific, Singapore, 1985).

MacMahon, P. A. (1916), *Combinatory Analysis* (Cambridge University Press, Cambridge, U.K.), Vol. 2.

Morris, S. (1986), Riffle puzzle, *Omni* Vol. 8, #4, 113.

Morris, S. (1987), Contest winners, *Omni* Vol. 9, #7, 141.

Tao, R. and F. Y. Wu (1987), **P51**, The vicious neighbour problem, *J. Phys. A* 20, L299-L306.

Tzeng, W. J. and F. Y. Wu (2006), Theory of impedance networks: the two-point impedance and *LC* resonances, *J. Phys. A* 39, 8579-8591.

van der Pol, B. (1959), The finite-difference analogy of the periodic wave equation and the potential equation, in *Probability and Related Topics in Physical Sciences*, Lectures in Applied Mathematics, Vol. 1, Ed. M. Kac (Interscience Publ. London) 237-257.

Widom, A. and F. Y. Wu (2005), **P56**, Book Review: Lectures on the Kinetic theory of Gases, Nonequilibrium Thermodynamics and Statistical Theories, *J. Stat. Phys.* 119, 945-948.

Wu, F. Y. (1983), **P57**, In Memorial of Shang-Keng Ma, unpublished.

Wu, F. Y. (1997), The infinite-state Potts model and restricted multi-dimensional partitions of an integer, *Math. and Comp. Modeling* 26, 269-274.

Wu, F. Y. (2004), **P49**, Theory of resistor networks: The two-point resistance, *J. Phys. A: Math. Gen.* 38, 6653-6673.

Wu, F. Y. (2008), **P58**, Professor C. N. Yang and Statistical Mechanics, *Int. J. Mod. Phys. B* 22, 1899-1909.

Wu, F. Y., G. Rollet, H. Y. Huang, J. M. Maillard, C. K. Hu and C. N. Chen (1996), **P52**, Directed lattice animals, restricted partitions of numbers, and the infinite-state Potts model, *Phys. Rev. Lett.* 76, 173-176.

Yang, C. N. (1967), Some exact results for the many-body problem in one dimension with repulsive delta-function interaction, *Phys. Rev. Lett.* 19, 1312-1315.

PHOTOGRAPHS

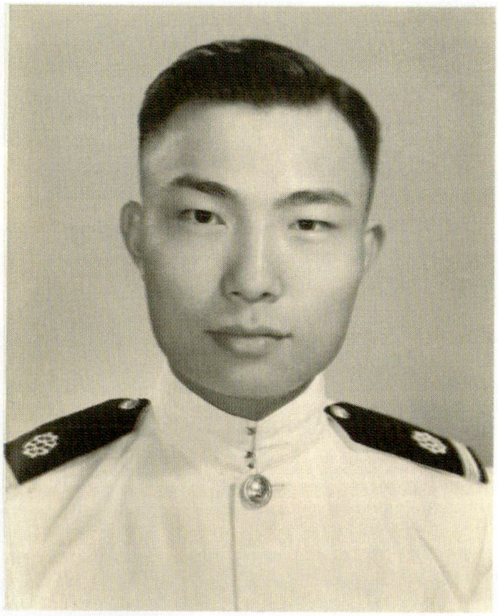

Ensign Wu, Chinese Navy, Taiwan (1954).

Class of U.S. Naval Instructors School (1956).

With Hao Bailin and Yu Lu (1979).

Class of 1959 Graduate School, National Tsing Hua University.
Left to Right: Front row: T.S. Kuo, Y. Liu (professor), T. Lang, F.Y. Wu, W. Yeh, C.T. Chen-Tsai. **Second row:** T.S. Chao, K.H. Lin, S.Y. Wang, Y. Shan, T.S. Yi. **Rear row:** S.S. Tsai, C.H. Chen, T.C. Ho, W. Mo. C.T. Chang, L.H. Tang (1958).

With J.S. (Zhuxi) Wang, Beijing (1979).

Nankai Conference in honor of F.Y. Wu's 70th birthday.
Left to Right: Sitting, S.S. Chern, F.Y. Wu, C.N. Yang. **Standing:** Y.C. Ho, C.W. Woo, M.L. Ge, Z.S. Hou, B.Y. Hou (2001).

With Rodney and Elizabeth Baxter in Canberra (1990).

With Catherine Kunz, Jane Wu and Herve Kunz in Swiss Alps (1991).

With J.-M. Maillard in Taiwan (1991).

With A.J. Guttmann, Marseille (1995).

With H.N.V. Temperley at Berkeley (1998).

With C.N. Yang and Mrs. Yang on the occasion of Professor Yang's retirement, Stony Brook (1999). **In the background:** L.D. Faddeev (center) and B. Sutherland (right).

With H.E. Stanley at Wu Fest (2003).

Wu Fest in honor of F.Y. Wu, Northeastern University (2003).

Left to Right: R. Shrock, P. Nath, J.R. Stellar, J.-M. Maillard, F.Y. Wu, S. Recucroft, E.H. Lieb, D.H. Lee, B.M. McCoy, J.H.H. Perk, A. Bansil, H. Au-Yang, J. Jose, M.E. Fisher, X.K. Wen (2003).

With E.H. Lieb (2003).

With Chia-Wei Woo (2001).

With S.S. Chern at Professor Chern's home (2002).

With C.N. Yang and Mrs. Yang
on the occasion of Professor Yang's 85th birthday, Singapore (2007).

REPRINTS OF PAPERS

1. Dimer Statistics

Volume 18, Number 15 PHYSICAL REVIEW LETTERS 10 April 1967

EXACTLY SOLUBLE MODEL OF THE FERROELECTRIC PHASE TRANSITION IN TWO DIMENSIONS

F. Y. Wu
Department of Physics, Virginia Polytechnic Institute, Blacksburg, Virginia
(Received 17 March 1967)

The Slater model of the two-dimensional potassium dihydrogen phosphate crystal is solved exactly under the additional assumption that the dipoles are excluded from pointing along one direction of the crystal axis. The Curie temperature T_c is not affected by this additional assumption but the phase change becomes a second-order transition. Complete polarization occurs below T_c with specific heat $\sim(T-T_c)^{-1/2}$ near and above the Curie point.

This Letter reports a model of the ferroelectric phase transition which is exactly soluble in the two-dimensional case. We fix our attention on the potassium dihydrogen phosphate (KDP) crystal, KH_2PO_4, which undergoes a second-order phase transition at 123°K. Slater[1] was the first to point out the important role played by the hydrogen atoms in the mechanism of this phase transition. The detailed structure of the KDP crystal proposed by him allows six possible configurations for the four hydrogen atoms attached to each PO_4 group. This simplified picture permits one to construct a well-defined mathematical model for the KDP crystal by associating arrows to the lattice bonds and energies to the lattice sites.[2] However, in spite of the simplicity of the statement of this problem, rigorous approaches to the solution have been lacking. Most of the previous treatments based on the Slater model and its modifications have been essentially mean-field methods yielding a first-order phase transition,[2-5] while the experimentally observed transition is a second-order one. The best statistical mechanical treatment to date has been given by Nagle,[5] who obtained both the high and low temperature expansions of the partition function and located the Curie point. These expansions, however, yield no information about the behavior of the specific heat, which is of considerable theoretical interest. It therefore seems desirable to have an exactly soluble model which can exhibit the character of the discontinuity, while serving as a model for testing the validity of other approximation procedures.

We first describe the Slater KDP model.[1,2] Consider a diamond-type lattice (four nearest

VOLUME 18, NUMBER 15 PHYSICAL REVIEW LETTERS 10 APRIL 1967

neighbors to each site) with directed arrows attached to all the lattice bonds. The rule is that there are always two arrows pointing toward and two arrows pointing away from a given lattice site. Then there are altogether six possible arrow configurations that can be associated to a site. A zero site energy is associated with two of the six configurations and an energy $\epsilon > 0$ with the remaining four (see Fig. 1). Each distinct way of associating arrows to the lattice as a whole will be called a state of the lattice. The energy of a given state is simply $n(\epsilon)\epsilon$, where $n(\epsilon)$ is the number of sites with energy ϵ. The partition function is now given by

$$Z = \sum_{\text{all states}} e^{-n(\epsilon)\epsilon/kT}. \tag{1}$$

The model we propose is the same as above except that we impose the further restriction that only one of the zero-energy configurations is allowed. Physically this corresponds to the assumption that the dipoles are excluded from pointing along one direction of the crystal axis.[1] Nevertheless, this provides us with a model which can now be treated with mathematical rigor.[6]

In this note we shall only write down the final results and give a brief description of the intermediate steps, while reserving the details, together with some interesting observations on the dimer method, for another communication. The partition function of our model is still given by (1), although the summation is now taken over a more restricted set of states. For an infinite rectangular lattice wrapped around a torus, the logarithm of the partition function, the energy, and the specific heat are given, respectively, by (N = number of lattice sites)

$$\ln Z = \frac{N}{8\pi^2}\int_0^{2\pi} d\theta \int_0^{2\pi} d\varphi \ln[1 + 2e^{-2\epsilon/kT} + 2e^{-2\epsilon/kT}\cos(\theta-\varphi) - 2e^{-\epsilon/kT}(\cos\theta + \cos\varphi)]; \tag{2}$$

$$E = 0, \qquad T < T_c = \epsilon/k\ln 2;$$
$$= N(2\epsilon/\pi)\cos^{-1}(\tfrac{1}{2}e^{\epsilon/kT}), \quad T > T_c;$$
$$C = 0, \qquad T < T_c;$$
$$= (N\epsilon^2/\pi kT^2)(e^{-2\epsilon/kT} - \tfrac{1}{4})^{-1/2}, \quad T > T_c.$$

Clearly a phase change occurs at $T_c = \epsilon/k\ln 2$ with $c \sim (T-T_c)^{-1/2}$ near and above the Curie point. The location of the Curie temperature T_c is in agreement with previous results on the Slater KDP model.[1-5] There is also total polarization below T_c: All sites take the zero-energy configuration to form an ordered state in which all dipoles point in the same direction. It is interesting to note that E is continuous at $T = T_c$ as is the case for the three-dimensional KDP crystal (previous theoretical discussions have always led to a first-order phase transtion).

Our method of attacking the problem consists of first converting the problem of counting the arrow configurations into a problem of drawing restricted closed polygons on the crystal lattice. The latter is equivalent to a dimer problem[7] and hence the evaluation of a Pfaffian.[8]

Let us take any state of the crystal lattice L as a standard state. Then, as compared with this standard one, an arbitrary state of L has either 0, 2, or 4 arrows reversed at each site. If we denote each of these reversed arrows by a bond connecting the two corresponding sites, then an arbitrary state of L is now transformed into a bond diagram made of closed polygons. The original restrictions on the ar-

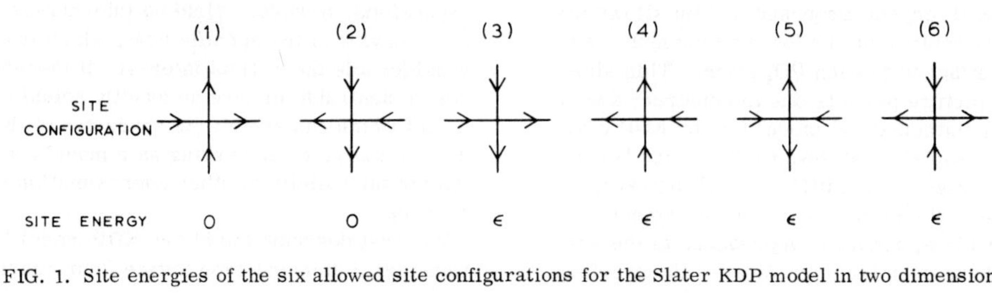

FIG. 1. Site energies of the six allowed site configurations for the Slater KDP model in two dimensions.

row configurations at each site now impose the restriction that there are only five possible bond configurations at each site.

Next we expand the lattice L into a terminal (dimer) lattice L^Δ by replacing each site of L by a "city" of internally connected points.[9] The one-to-one correspondence between the configurations of closed polygons on L and the dimer configurations on L^Δ (provided that each city contains an even number of points) is well known.[7,10] We need only to choose the cities and the weights associated with their internal bonds properly, to take care of the restrictions on the bond configurations.

It is found that this trick can be accomplished by the dimer city shown in Fig. 2, where the standard state has been taken to have the configuration (1) of Fig. 1 at all sites.

Once the dimer cities are properly drawn

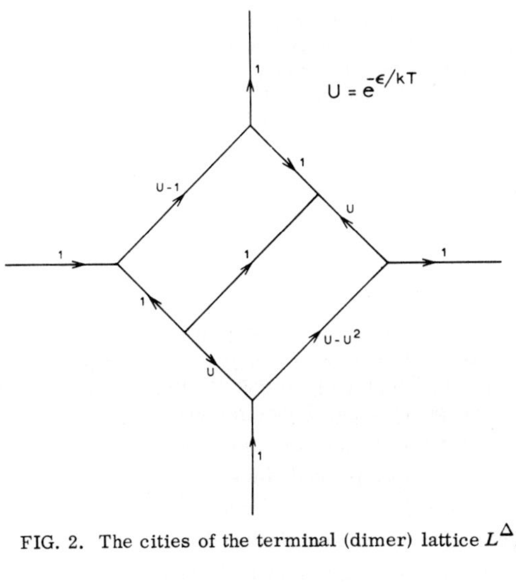

FIG. 2. The cities of the terminal (dimer) lattice L^Δ.

and the arrows on the terminal lattice L^Δ properly attached,[10,11] the evaluation of the partition function (1) is straightforward[7] and yields

$$\lim_{N\to\infty} \ln Z = \frac{N}{8\pi^2}\int_0^{2\pi} d\theta \int_0^{2\pi} d\varphi \ln D,$$

where D is the determinant given by

$$D = \begin{vmatrix} 0 & 0 & 0 & U-1 & -e^{-i\theta} & -1 \\ 0 & 0 & 0 & -1 & -U & -1 \\ 0 & 0 & 0 & -e^{-i\varphi} & U-U^2 & -U \\ 1-U & 1 & e^{i\varphi} & 0 & 0 & 0 \\ e^{i\theta} & U & U^2-U & 0 & 0 & 0 \\ 1 & 1 & U & 0 & 0 & 0 \end{vmatrix}$$

with $U = e^{-\epsilon/kT}$. Substitution now leads to Eq. (2).

The author is indebted to Professor Elliott Lieb for a valuable conversation.

---

[1]J. C. Slater, J. Chem. Phys. 9, 16 (1941).

[2]H. Takahashi, Proc. Phys.-Math. Soc. Japan 23, 1069 (1941).

[3]S. Yomosa and T. Nagamiya, Progr. Theoret. Phys. (Kyoto) 4, 263 (1949).

[4]Y. Takagi, J. Phys. Soc. Japan 3, 271 (1948).

[5]J. F. Nagle, J. Math. Phys. 7, 1492 (1966).

[6]We mention in passing that another model, in which the configuration of four arrows pointing away from a site is allowed in [illegible] to those of the present model, can also be solved exactly. The conclusions are identical with ours.

[7]C. A. Hurst and H. S. Green, J. Chem. Phys. 33, 1059 (1960).

[8]M. E. Fisher, Phys. Rev. 124, 1664 (1961).

[9]M. E. Fisher, J. Math. Phys. 7, 1776 (1966).

[10]P. W. Kasteleyn, J. Math. Phys. 4, 287 (1963).

[11]P. W. Kasteleyn, Physica 27, 1209 (1961).

Review

International Journal of Modern Physics B
Vol. 20, No. 32 (2006) 5357–5371

World Scientific
www.worldscientific.com

DIMERS ON TWO-DIMENSIONAL LATTICES

F. Y. WU

Department of Physics, Northeastern University, Boston, Massachusetts 02115, USA

Received 4 December 2006

We consider close-packed dimers, or perfect matchings, on two-dimensional regular lattices. We review known results and derive new expressions for the free energy, entropy, and the molecular freedom of dimers for a number of lattices including the simple-quartic, honeycomb, triangular, kagome, 3-12 and its dual, and 4-8 and its dual Union Jack lattices. The occurrence and nature of phase transitions are also elucidated and discussed in each case.

Keywords: Close-packed dimers; two-dimensional lattices; exact results; phase transitions.

PACS numbers: 05.50.+q, 04.20.Jb, 02.10.Ox

1. Introduction

A central problem in statistical physics and combinatorial mathematics is the enumeration of close-packed dimers, often referred to as perfect matchings in mathematical literature, on lattices which mimic the adsorption of diatomic molecules on a surface.[1] A folklore in lattice statistics states that close-packed dimers can always be enumerated for two-dimensional lattices. Indeed, the seminal works of Kasteleyn[2] and Fisher and Temperley[3,4] on the simple-quartic lattice produced the first exact solution. However, a search of the literature indicates that very little else has appeared in print on other lattices. This paper is an attempt to fill the gap. Here we review known exact results and in some cases derive new expressions for the free energy, entropy, and molecular freedom for dimers on various two-dimensional lattices. We also discuss and elucidate analytic properties of the free energy.

We consider a regular two-dimensional array of N (= even) lattice points which can be covered by $N/2$ dimers and in the large N limit. We denote the dimer weights by $\{z_i\}$ and define the generating function

$$Z(\{z_i\}) = \sum_{\text{dimer coverings}} \prod_i z_i^{n_i} \tag{1}$$

where the summation is over all dimer coverings and n_i is the number of dimers with weight z_i. For a large lattice $Z(\{z_i\})$ is expected to grow exponentially in N.

Our goal is to compute the "free energy" per dimer

$$f(\{z_i\}) = \lim_{N\to\infty} \frac{1}{N/2} \ln Z(\{z_i\}) . \tag{2}$$

Setting $z_i = 1$, the numerical value

$$S = f(\{1\}) \tag{3}$$

is the entropy of adsoptions of diatomic molecules and

$$W = \lim_{N\to\infty} [Z_N(\{1\})]^{2/N} = e^S \tag{4}$$

is often known as the per-dimer molecular freedom. These are quantities of primary interest in mathematics, physics, and chemistry.[a]

The following two integration formulas are useful for our purposes:

$$\frac{1}{2\pi}\int_0^{2\pi} d\theta \ln(2A + 2B\cos\theta + 2C\sin\theta) = \ln[A + \sqrt{A^2 - B^2 - C^2}] \tag{5}$$

for real A, B ,C, and

$$\frac{1}{2\pi}\int_0^{2\pi} d\theta \ln|A + Be^{i\theta}| = \ln\max\{|A|, |B|\} \tag{6}$$

for real or complex A and B.

We also have the following result (used in Sec. 9) which holds generally for any graph, whether planar or nonplanar:

Proposition. *Let G be a bipartite graph consisting of two sets of equal number of vertices A and B, with vertices in A connected only to vertices in B, and vice versa. Let G^* be a graph generated from G by adding edges connecting vertices within one set. Let Z_G and Z_{G^*} be the respective dimer generating functions. Then we have the identity*

$$Z_{G^*} = Z_G . \tag{7}$$

Namely, the addition of edges connecting vertices within one set of vertices in a bipartite graph does not alter the dimer generating function.

Proof. Let the inserted edges connect A vertices. In any dimer configuration in G^*, every B vertex must be covered by a dimer and this dimer must end at an A vertex. Since the numbers of A and B sites are equal, these AB-dimers cover all vertices. Thus, the inserted edges do not enter the picture. Q.E.D.

We consider individual lattices in ensuing sections.

[a] In some earlier papers relevant quantities are defined on a per lattice site basis with $W_{\text{per site}} = \sqrt{W}$, etc.

2. The Simple-Quartic Lattice

For a simple-quartic lattice with uniform dimer weights z_1 and z_2 in the two (horizontal and vertical) directions, we have[2–4] (see also Sec. 4)

$$f_{\rm SQ}(z_1,z_2) = \frac{1}{8\pi^2}\int_0^{2\pi} d\theta \int_0^{2\pi} d\phi \, \ln[2(z_1^2+z_2^2 - z_1^2\cos(\theta+\phi) - z_2^2\cos(\theta-\phi)] \,. \tag{8}$$

The free energy $f_{\rm SQ}(z_1,z_2)$ is regular in z_1 and z_2.

Setting $z_1 = z_2 = 1$ and making use of the integration identity (5), we obtain

$$\begin{aligned} S_{\rm SQ} &= \frac{1}{8\pi^2}\int_0^{2\pi} d\theta \int_0^{2\pi} d\phi \, \ln[4(1-\cos\theta\cos\phi)] \\ &= \frac{1}{\pi}\int_0^{\pi/2} d\theta \, \ln[2(1+\sin\,\theta)] \\ &= \frac{4}{\pi}\int_0^{\pi/4} d\theta \, \ln(2\,\cos\,\theta) \\ &= \frac{2}{\pi} G \end{aligned} \tag{9}$$

where the last step follows from the identity (4.224.5) of Ref. 6 and

$$G = 1 - 3^{-2} + 5^{-2} - 7^{-2} + \cdots = 0.915\ 965\ 594\ldots$$

is the Catalan constant. It follows that we have

$$\begin{aligned} S_{\rm SQ} &= 0.583\ 121\ 808\ldots \\ W_{\rm SQ} &= e^{S_{\rm SQ}} = 1.791\ 622\ 812\ldots. \end{aligned} \tag{10}$$

3. The Honeycomb Lattice

Kasteleyn[5] has pointed out that phase transitions can occur in dimer systems with anisotropic weights and cited the honeycomb lattice as an example. The honeycomb dimer problem describes a modified KDP model whose statistical property was later analyzed in details.[7,8] Here is a summary of the findings.

Let z_1, z_2, z_3 be the dimer weights along the three edge directions of the honeycomb lattice. We have[7] (see also Sec. 4)

$$f_{\rm HC}(z_1,z_2,z_3) = \frac{1}{8\pi^2}\int_0^{2\pi} d\theta \int_0^{2\pi} d\phi \, \ln[z_1^2+z_2^2+z_3^2+2z_1z_2\cos\theta + 2z_2z_3\cos\,\phi + 2z_3z_1\cos(\theta-\phi)] \,. \tag{11}$$

When one of the dimer weight dominates so that z_1, z_2, z_3 do not form a triangle, the free energy (11) is frozen and one has

$$f_{\rm HC}(z_1,z_2,z_3) = \ln\, z_i \,, \quad z_i \geq z_j + z_k \,, \quad i,j,k \text{ distinct.} \tag{12}$$

The second derivatives of the free energy in the $\{z_1, z_2, z_3\}$-space exhibit an inverse square-root singularity near the phase boundaries $z_i = z_j + z_k$. We shall refer to this singular behavior as the KDP-type transition.

Setting $z_1 = z_2 = z_3 = 1$ and carrying out the ϕ-integration, we obtain

$$\begin{aligned} S_{\rm HC} &= \frac{1}{8\pi^2}\int_0^{2\pi} d\theta \int_0^{2\pi} d\phi \ln[3 + 2\cos\theta + 2\cos\phi + 2\cos(\theta-\phi)] \\ &= \frac{1}{4\pi}\int_{-\pi}^{\pi} d\theta \ln\frac{1}{2}[3 + 2\cos\theta + |1 + 2\cos\theta|] \\ &= \frac{1}{4\pi}\int_{-2\pi/3}^{2\pi/3} d\theta \ln(2 + 2\cos\theta) \\ &= \frac{2}{\pi}\int_0^{\pi/3} d\theta \ln(2\cos\theta) \\ &= 0.323\ 065\ 947\ldots \end{aligned} \tag{13}$$

$$W_{\rm HC} = e^{S_{\rm HC}} = 1.381\ 356\ 444\ldots. \tag{14}$$

The resemblance of the last integral in (13) with that in (9) is striking. We also remark that the entropy $S_{\rm HC}$ is the same as that of the ground state of an isotropic antiferromagnetic triangular Ising model.[9] The integral in Eq. (13) was first obtained and evaluated by Wannier in a study of the latter problem more than half century ago, but his numerical evaluation of the integral is off by 5%.[10]

Finally, we point out that the honeycomb free energy can also be evaluated when there exists a dimer-dimer interaction.[11] Analyses of the phase diagram and the associated critical behavior make use of the method of the Bethe ansatz and are fairly involved. Readers are referred to Ref. 11 for details.

4. The Checkerboard Square Lattice

The checkerboard lattice is a simple-quartic lattice with anisotropic dimer weights z_1, z_2, z_3 and z_4 as shown in Fig. 1. Again, the solution of this problem was certainly known to Kasteleyn[5] who cited that the model exhibits phase transitions. The solution has also been mentioned by Montroll,[12] and studied recently in some detail by Cohn, Kenyon and Popp.[13] Here we provide a concise analysis using previously known results on vertex models.

Orient lattice edges as shown for which it is known[14] that the Kasteleyn clockwise odd sign rule[2] can be realized by setting $z_2 \to iz_2$, $z_4 \to iz_4$ in the evaluation of the Pfaffian. This permits us to take unit cells of two lattice sites as shown in Fig. 1. Since the cells form a rectangular array, following the standard procedure[12] one obtains its free energy given by the generally valid expression

$$f(\{z_i\}) = \frac{1}{8\pi^2}\int_0^{2\pi} d\theta \int_0^{2\pi} d\phi \ln \det F(\theta,\phi), \tag{15}$$

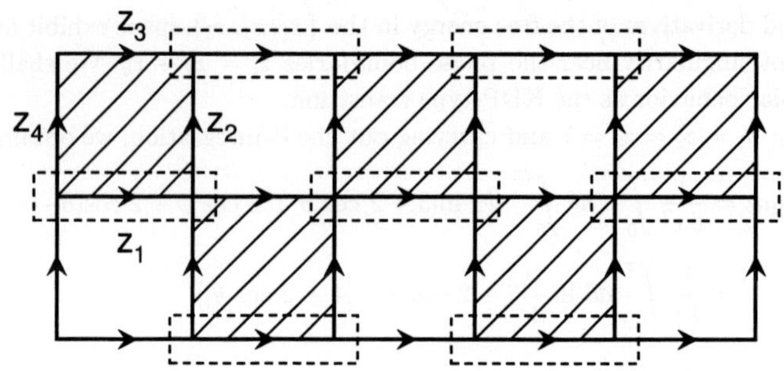

Fig. 1. Edge orientation and unit cells of the checkerboard lattice. Shaded squares are repeated. Weights z_2 and z_4 are replaced by iz_2 and iz_4 in the evaluation of the Pfaffian (see text).

where

$$F(\theta,\phi) = M_{(0,0)} + M_{(1,0)}e^{i\theta} + M_{(-1,0)}e^{-i\theta} + M_{(0,1)}e^{i\phi}$$
$$+ M_{(0,-1)}e^{-i\phi} + M_{(1,1)}e^{i(\theta+\phi)} + M_{(-1,-1)}e^{-i(\theta+\phi)}\,, \tag{16}$$

and the M's are matrices reflecting the orientation and connectivity of the edges.

For the checkerboard lattice we have[b]

$$M_{(0,0)} = \begin{pmatrix} 0 & z_1 \\ -z_1 & 0 \end{pmatrix}, \quad M_{(1,0)} = \begin{pmatrix} 0 & 0 \\ -iz_4 & 0 \end{pmatrix}, \quad M_{(-1,0)} = \begin{pmatrix} 0 & iz_4 \\ 0 & 0 \end{pmatrix},$$
$$M_{(0,1)} = \begin{pmatrix} 0 & 0 \\ iz_2 & 0 \end{pmatrix}, \quad M_{(0,-1)} = \begin{pmatrix} 0 & -iz_2 \\ 0 & 0 \end{pmatrix}, \tag{17}$$
$$M_{(1,1)} = \begin{pmatrix} 0 & 0 \\ z_3 & 0 \end{pmatrix}, \quad M_{(-1,-1)} = \begin{pmatrix} 0 & -z_3 \\ 0 & 0 \end{pmatrix}.$$

Explicitly, this leads to

$$f_{\text{CKB}}(z_1,z_2,z_3,z_4) = \frac{1}{8\pi^2}\int_0^{2\pi} d\theta \int_0^{2\pi} d\phi \,\ln|z_1 - iz_4e^{i\theta} + iz_2e^{i\phi} - z_3e^{i(\theta+\phi)}|^2$$
$$= \frac{1}{8\pi^2}\int_0^{2\pi} d\theta \int_0^{2\pi} d\phi \,\ln[z_1^2 + z_2^2 + z_3^2 + z_4^2$$
$$+ 2(z_1z_4 - z_2z_3)\sin\theta - 2(z_1z_2 - z_3z_4)\sin\phi$$
$$- 2z_1z_3\cos(\theta+\phi) - 2z_2z_4\cos(\theta-\phi)]\,. \tag{18}$$

[b]The convention used here in writing down (17) is such that the direction to the right in Fig. 1 is the $(1,1)$-direction.

For $z_1 = z_3$, $z_2 = z_4$ the solution reduces to the integral (8) for the simple-quartic lattice with uniform weights, and for $z_4 = 0$ the solution reduces to the integral (11) for the honeycomb lattice. Comparing the second line of the integral (18) (after changing $\theta \to \pi/2 - \theta, \phi \to \pi/2 - \phi$) with Eq. (16) of Ref. 15, we see that the checkerboard dimer model is completely equivalent to a free-fermion 8-vertex model with weights

$$\begin{gathered} \omega_1 = z_4\,, \quad \omega_2 = z_2\,, \quad \omega_3 = z_1\,, \quad \omega_4 = z_3\,, \\ \omega_5\omega_6 = 0\,, \quad \omega_7\omega_8 = z_1 z_3 + z_2 z_4\,. \end{gathered} \tag{19}$$

To analyze the free energy it is most convenient to apply the integration formula (6) to the first line of the integral (18). This gives

$$f_{\mathrm{CKB}}(z_1, z_2, z_3, z_4) = \frac{1}{2\pi}\int_0^{2\pi} d\theta \ln \max\{|z_1 + iz_4 e^{-i\theta}|, |iz_2 + z_3 e^{-i\theta}|\}\,. \tag{20}$$

It is then seen that the free energy f_{CKB} is analytical in z_i, except when one of the weights dominates so that z_1, z_2, z_3, z_4 do not form a quadrilateral, namely,

$$2z_i \geq z_1 + z_2 + z_3 + z_4\,, \quad i = 1, 2, 3, 4\,. \tag{21}$$

When this happens the free energy is frozen and

$$f_{\mathrm{CKB}}(z_1, z_2, z_3, z_4) = \ln z_i\,. \tag{22}$$

The free-fermion model with free energy (20) has been studied in details by Wu and Lin.[16,c] It is found that, provided that either $z_1 \neq z_3$ or $z_2 \neq z_4$ (or both), namely, it is not the uniform simple-quartic model discussed in Sec. 2, the second derivatives of f_{CKB} in the $\{z_1, z_2, z_3, z_4\}$-space exhibit an inverse square-root singularity of the KDP-type transition near the phase boundary (21). This shows that the uniform model of Sec. 2 is a unique degenerate case for which the free energy is analytic and the model exhibits no phase transition.

5. The Triangular Lattice

The study of dimers on the triangular lattice has been of interest for many years (see, for example, Refs. 17–20). Although an edge orientation of the triangular lattice satisfying the Kasteleyn clockwise-odd sign rule[2] for a Pfaffian evaluation has been given by Montroll,[12] it seems that the closed-form expression of the solution appeared in print only very recently.[19,20] Here, we provide a derivation of the solution using the Montroll edge orientation and analyze its physical properties.

[c] The model (19) is transformed into one discussed in Ref. 16 after reversing all horizontal arrows in the vertex configurations. As found in Ref. 16, results reported therein apply to uniform as well as staggered models.

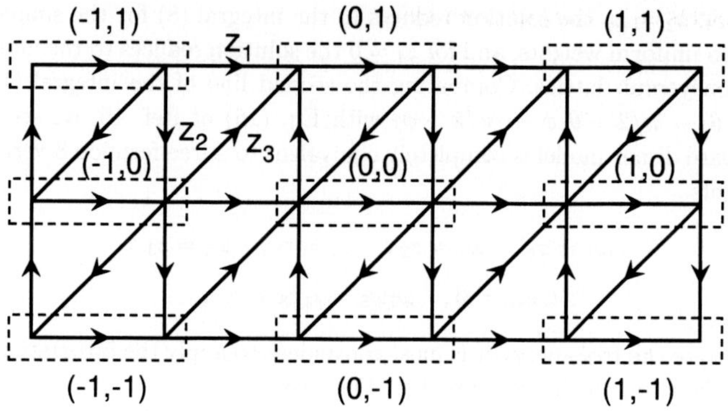

Fig. 2. Edge orientation and unit cells of the triangular lattice.

Divide the lattice into unit cells containing two sites as shown in Fig. 2. Then, the free energy is given by the general expression (15) with

$$M_{(0,0)} = \begin{pmatrix} 0 & z_1 \\ -z_1 & 0 \end{pmatrix}, \quad M_{(1,0)} = \begin{pmatrix} 0 & 0 \\ z_1 & 0 \end{pmatrix}, \quad M_{(-1,0)} = \begin{pmatrix} 0 & -z_1 \\ 0 & 0 \end{pmatrix},$$
$$M_{(0,1)} = \begin{pmatrix} z_2 & -z_3 \\ 0 & -z_2 \end{pmatrix}, \quad M_{(0,-1)} = \begin{pmatrix} -z_2 & 0 \\ z_3 & z_2 \end{pmatrix}, \tag{23}$$
$$M_{(1,1)} = \begin{pmatrix} 0 & 0 \\ z_3 & 0 \end{pmatrix}, \quad M_{(-1,-1)} = \begin{pmatrix} 0 & -z_3 \\ 0 & 0 \end{pmatrix}.$$

This leads to

$$\begin{aligned} f_{\mathrm{TRI}}(z_1, z_2, z_3) &= \frac{1}{8\pi^2}\int_0^{2\pi} d\theta \int_0^{2\pi} d\phi \ln 2[z_1^2 + z_2^2 + z_3^2 \\ &\quad - z_1^2 \cos\theta - z_2^2 \cos 2\phi + z_3^2 \cos(\theta + 2\phi)] \\ &= \frac{1}{8\pi^2}\int_0^{2\pi} d\alpha \int_0^{2\pi} d\beta \ln 2[z_1^2 + z_2^2 + z_3^2 \\ &\quad + z_1^2 \cos\alpha + z_2^2 \cos\beta + z_3^2 \cos(\alpha + \beta)], \end{aligned} \tag{24}$$

where in the last step we have changed variables to $\alpha = \pi - \theta$, $\beta = \pi - 2\phi$. The free energy (24) is again of the form of that of a free-fermion 8-vertex model.[15] Comparing the integral (24) with Eq. (16) of Ref. 15, one obtains the free-fermion vertex weights $\omega_1, \ldots, \omega_8$ as given by

$$\begin{aligned} \omega_1^2+\omega_2^2+\omega_3^2+\omega_4^2 &= 2(z_1^2+z_2^2+z_3^2) \\ \omega_1\omega_3-\omega_2\omega_4 &= z_1^2 \\ \omega_1\omega_4-\omega_2\omega_3 &= z_2^2 \\ \omega_3\omega_4-\omega_1\omega_2 &= z_3^2 \\ \omega_5\omega_6 &= \omega_1\omega_2 \\ \omega_7\omega_8 &= \omega_3\omega_4\,. \end{aligned} \tag{25}$$

It was found[15] that the model exhibits a transition at the critical point

$$\begin{aligned} &(-\omega_1+\omega_2+\omega_3+\omega_4)(\omega_1-\omega_2+\omega_3+\omega_4)(\omega_1+\omega_2-\omega_3+\omega_4) \\ &\quad\times(\omega_1+\omega_2+\omega_3-\omega_4)=0\,, \end{aligned} \tag{26}$$

and that the transition is of a KDP-type transition with an inverse square-root singularity in the second derivative of the free energy if $\omega_1\omega_2\omega_3\omega_4 = 0$, and of an Ising-type transition with a logarithmic singularity in the second derivative if $\omega_1\omega_2\omega_3\omega_4 \neq 0$.

We can solve for the ω's by forming linear combinations of the equalities in the vertex weights (25) to obtain

$$\begin{aligned} (-\omega_1+\omega_2+\omega_3+\omega_4)^2 &= 4z_3^2 \\ (\omega_1-\omega_2+\omega_3+\omega_4)^2 &= 4(z_1^2+z_2^2+z_3^2) \\ (\omega_1+\omega_2-\omega_3+\omega_4)^2 &= 4z_2^2 \\ (\omega_1+\omega_2+\omega_3-\omega_4)^2 &= 4z_1^2\,. \end{aligned} \tag{27}$$

This leads to the explicit solution

$$\begin{aligned} \omega_1 &= \frac{1}{2}\Big[z_1+z_2-z_3+\sqrt{z_1^2+z_2^2+z_3^2}\Big] \\ \omega_2 &= \frac{1}{2}\Big[z_1+z_2+z_3-\sqrt{z_1^2+z_2^2+z_3^2}\Big] \\ \omega_3 &= \frac{1}{2}\Big[z_1-z_2+z_3+\sqrt{z_1^2+z_2^2+z_3^2}\Big] \\ \omega_4 &= \frac{1}{2}\Big[-z_1+z_2+z_3+\sqrt{z_1^2+z_2^2+z_3^2}\Big] \end{aligned} \tag{28}$$

using which one verifies that we have $\omega_1\omega_2\omega_3\omega_4 \neq 0$. This implies that the triangular dimer model exhibits Ising-type transitions at

$$z_i = 0\,, \quad i = 1, 2, 3\,. \tag{29}$$

Namely, the uniform simple-quartic model discussed in Sec. 2 is precisely the critical manifold of the triangular model.

Setting $z_1 = z_2 = z_3 = 1$, we obtain

$$\begin{aligned} S_{\text{TRI}} &= \frac{1}{8\pi^2}\int_0^{2\pi} d\alpha \int_0^{2\pi} d\beta \ln[6 + 2\cos\alpha + 2\cos\beta + 2\cos(\alpha+\beta)] \\ &= \frac{1}{2\pi}\int_0^{\pi} d\alpha \ln\left[3 + \cos\alpha + \sqrt{7 + 4\cos\alpha + \cos^2\alpha}\right] \\ &= 0.857\,189\,074\ldots \end{aligned} \tag{30}$$

$$W_{\text{TRI}} = e^{S_{\text{TRI}}} = 2.356\,527\,353\ldots.$$

It is of interest to note that an integral similar to the integral (30),

$$\begin{aligned} S_{\text{SPT}} &= \frac{1}{4\pi^2}\int_0^{2\pi} d\alpha \int_0^{2\pi} d\beta \ln[6 - 2\cos\alpha - 2\cos\beta - 2\cos(\alpha+\beta)] \\ &= \frac{1}{\pi}\int_0^{\pi} d\alpha \ln\left[3 - \cos\alpha + \sqrt{7 - 8\cos\alpha + \cos^2\alpha}\right] \\ &= \frac{3\sqrt{3}}{\pi}(1 - 5^{-2} + 7^{-2} - 11^{-2} + 13^{-2} - \cdots) \\ &= 1.615\,329\,736\ldots, \end{aligned} \tag{31}$$

gives the per-site free energy of spanning trees on the triangular lattice[21,22] which is reducible to a simple numerical series akin to the Catalan constant. The reduction of the integral (31) into the form of a series was first reported in Ref. 21 without providing details. Details of the derivation were published only recently,[24] where the series was deduced from the integral in two different ways: By mapping the spanning tree problem to a Potts model and in turn to an F model on the triangular lattice,[23] and by a direct algebraic manipulation.

6. The Kagome Lattice

The kagome lattice is shown in Fig. 3 with dimer weights z_1, z_2, z_3 along the 3 edge directions. The study of the molecular freedom for the kagome lattice has been a subject matter of interest for many years (see, for example, Refs. 25 and 26), but most of the studies have been numerical or approximate.

The exact kagome free energy has been obtained recently.[27] The solution is found to be surprisingly simple and is given by

$$f_{\text{KG}}(z_1, z_2, z_3) = \frac{1}{3}\ln(4z_1z_2z_3)\,. \tag{32}$$

The solution (32) differs fundamentally from those of other lattices as it does not have a series expansion. This explains why most of other approaches, which are invariably based on series expansions, are not very effective.[d]

[d]However, the series expansion scheme is recovered if one introduces further symmetry-breaking weights.[27]

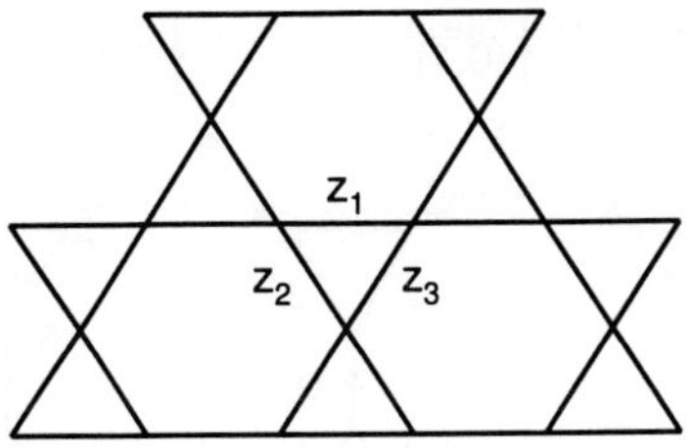

Fig. 3. The kagome lattice.

From the solution (32) we now have

$$\begin{aligned} S_{\mathrm{KG}} &= \frac{2}{3}\ \ln 2 = 0.462\,098\,120\ldots \\ W_{\mathrm{KG}} &= e^{S_{\mathrm{KG}}} = 1.587\,401\,051\ldots. \end{aligned} \tag{33}$$

7. The 3-12 Lattice

The 3-12 lattice is shown in Fig. 4. We consider the case of six different dimer weights x, y, z, u, v, w.

The 3–12 lattice has been used by Fisher[28] in a dimer formulation of the Ising model. Using Fisher's edge orientation which we show in Fig. 4, one finds the free energy given by the general expression (15) with

$$F(\theta,\phi) = \begin{pmatrix} 0 & x & z & 0 & -we^{-i\theta} & 0 \\ -x & 0 & y & 0 & 0 & -ve^{-i\phi} \\ -z & -y & 0 & u & 0 & 0 \\ 0 & 0 & -u & 0 & z & y \\ we^{i\theta} & 0 & 0 & -z & 0 & x \\ 0 & we^{i\phi} & 0 & -y & -x & 0 \end{pmatrix}. \tag{34}$$

Namely,

$$\begin{aligned} f_{3-12}(x,y,z;u,v,w) = \frac{1}{3}\cdot\frac{1}{8\pi^2}\int_0^{2\pi} d\theta \int_0^{2\pi} d\phi\ \ln[\omega_1^2+\omega_2^2+\omega_3^2+\omega_4^2 \\ +2(\omega_1\omega_3-\omega_2\omega_4)\cos\theta + 2(\omega_1\omega_4-\omega_2\omega_3)\cos\phi \\ +2(\omega_3\omega_4-\omega_1\omega_2)\cos(\theta-\phi)] \end{aligned} \tag{35}$$

where the factor 1/3 is due to the fact that there are 3 dimers per unit cell, and

$$\omega_1 = x^2u\,,\quad \omega_2 = uvw\,,\quad \omega_3 = y^2w\,,\quad \omega_4 = z^2v\,. \tag{36}$$

The free energy (36) is again of the form of that of a free-fermion 8-vertex model with vertex weights ω_1, ω_2, ω_3, ω_4, $\omega_5\omega_6 = \omega_3\omega_4$, and $\omega_7\omega_8 = \omega_1\omega_2$ and the critical

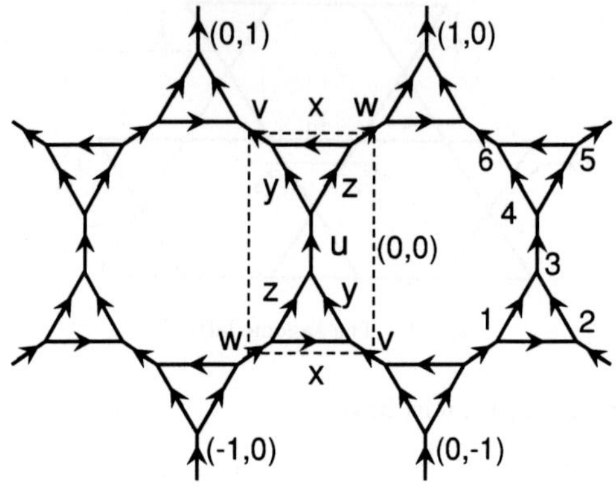

Fig. 4. Edge orientation and a unit cell of the 3-12 lattice.

point (26). For the Ising model with interactions K_1 and K_2,[28] for example, we have $x = y = z = u = 1$, $v = \tanh K_1$, $w = \tanh K_2$. It is then verified that the critical condition $\omega_1 = \omega_2 + \omega_3 + \omega_4$ can be realized and gives the known Ising critical point

$$\sinh 2K_1 \sinh 2K_2 = 1 . \tag{37}$$

Finally, setting $x = y = z = u = v = w = 1$, we obtain

$$\begin{aligned} S_{3-12} &= \frac{1}{3} \ln 2 = 0.231\ 049\ 060 \ldots \\ W_{3-12} &= e^{S_{3-12}} = 1.259\ 921\ 049 \ldots . \end{aligned} \tag{38}$$

8. The 4-8 Lattice

Dimer models on the 4-8 lattice shown in Fig. 5 have been used to describe phase transitions in physical systems.[29–31] By setting $u = v$, for example, the dimer model describes the phase transition in the layered hydrogen-bonded $SnCl_2 \cdot 2H_2O$ crystal.[30]

We orient the lattice as shown, and this leads to the free energy (15) with

$$F(\theta, \phi) = \begin{pmatrix} 0 & xe^{i\theta} & -v & -u \\ -xe^{-i\theta} & 0 & u & -v \\ v & -u & 0 & ye^{i\phi} \\ u & v & -ye^{-i\phi} & 0 \end{pmatrix} . \tag{39}$$

Namely,

$$\begin{aligned} f_{4-8}(x, y; u, v) = \frac{1}{2} \cdot \frac{1}{8\pi^2} \int_0^{2\pi} d\theta \int_0^{2\pi} d\phi \ln[x^2y^2 + (u^2 + v^2)^2 \\ - 2xyu^2 \cos(\theta + \phi) - 2xyv^2 \cos(\theta - \phi)] . \end{aligned} \tag{40}$$

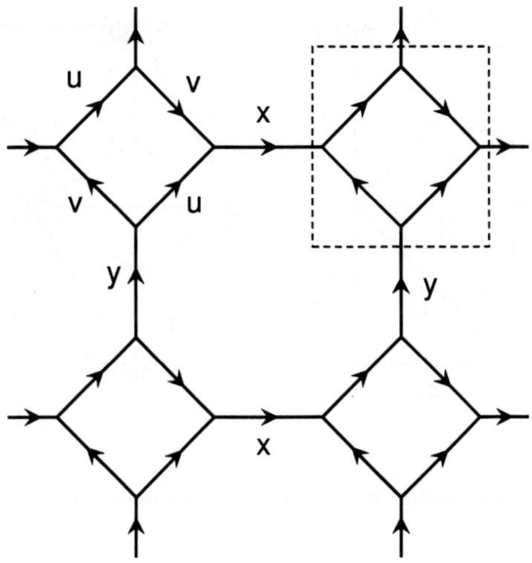

Fig. 5. Edge orientation and a unite cell of the 4-8 lattice.

The free energy (40) is also of the form of the that of a free-fermion 8-vertex model[15] and exhibits an Ising type transition at

$$u^2 + v^2 = xy\,. \tag{41}$$

Setting $x = y = u = v = 1$, one obtains

$$\begin{aligned} S_{4-8} &= \frac{1}{16\pi^2}\int_0^{2\pi} d\theta \int_0^{2\pi} d\phi \ \ln[5 - 2\cos(\theta+\phi) - 2\cos(\theta-\phi) \\ &= \frac{1}{16\pi^2}\int_0^{2\pi} d\theta \int_0^{2\pi} d\phi \ \ln[5 - 4\cos\theta\cos\phi] \\ &= \frac{1}{2\pi}\int_0^{\pi/2} d\theta \ \ln\left[\frac{5+\sqrt{25-16\cos^2\theta}}{2}\right] \\ &= 0.376\,995\,650\ldots \end{aligned} \tag{42}$$

$$W_{4-8} = e^{S_{4-8}} = 1.457\,897\,968\ldots.$$

9. The Union Jack Lattice

The Union Jack lattice shown in Fig. 6(a) is the dual of the 4-8 lattice. It is constructed by inserting diagonal edges with weights u and v to a checkerboard lattice. Since the checkerboard lattice is bipartite and the inserted diagonals connect vertices of one sublattice only, the proposition established in Sec. 1 now implies the

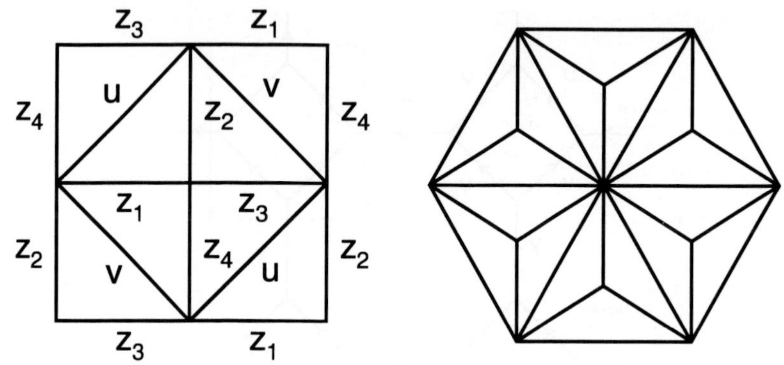

(a) (b)

Fig. 6. (a) The Union Jack lattice is the dual of the 4-8 lattice, and (b) The dual of the 3-12 lattice.

identity

$$Z_{\mathrm{UJ}}(z_1, z_2, z_3, z_4; u, v) = Z_{\mathrm{CKB}}(z_1, z_2, z_3, z_4)\,. \tag{43}$$

That is, the solution for the Union Jack lattice is identical to that of a simple-quartic lattice as if the u, v edges were absent. Particularly, they have identical entropy and molecular freedom.

10. The Dual of the 3-12 Lattice

The lattice shown in Fig. 6(b) is the dual of the 3-12 lattice. It consists of two sets of vertices, set A vertices each of which having coordination number 3 and set B vertices each having coordination number 12. The number of A vertices is twice that of B, and the A vertices are connected to B vertices only.

The dual of the 3-12 lattice does not admit dimer coverings. This follows from the fact that in a proper dimer covering each A vertex must by covered by a dimer and the dimer must end at a B vertex. This means some B vertices will have more than one dimers and this is not possible. Thus, there is no proper dimer coverings and the generating function is identically zero.

11. Summary

We have presented analytic and numerical results on the free energy, entropy, and molecular freedom for close-packed dimers on two-dimensional lattices which have unit cells arranged on a rectangular array. For the anisotropic checkerboard lattice the free energy is found to exhibit a KDP-type singularity, except in the degenerate case of uniform dimer weights the free energy is analytic with no transitions. For the triangular lattice the free energy is analytic for nonzero z_i and is critical at

Table 1. Summary of numerical results. Phase transition occurs when dimer weights are anisotropic (see text).

Lattice	Entropy S	Molecular freedom $W = e^S$	Phase transition
Honeycomb	0.323 065 947	1.381 356 444	KDP
Checkerboard	0.583 121 808	1.791 622 812	KDP
Triangular	0.857 189 074	2.356 527 353	Ising
Kagome	0.462 098 120	1.587 401 051	None
3-12	0.231 049 060	1.259 921 049	Ising
4-8	0.376 995 650	1.457 897 968	Ising
Union Jack	0.583 121 808	1.791 622 812	Ising
Dual of 3-12 lattice	No dimer coverings	No dimer coverings	None

the manifold $z_1 = 0$. For 4-8 and 3-12 lattices the dimer models exhibit Ising-type transitions.

Numerical results on the entropy and molecular freedom are summarized in Table 1. We observe that the entropy is not necessarily a monotonic function of the coordination number of the lattice.

Acknowledgments

The author would like to thank R. Shrock for a critical reading of the manuscript, J. Propp for calling his attention to Ref. 13, and F. Hucht for pointing a numerical error in Eq. (31) in an earlier version of the manuscript. He is grateful W. T. Lu for help in preparing the manuscript.

References

1. R. H. Fowler and G. S. Rushbrooke, An attempt to extend the statistical theory to perfect solutions, *Trans. Faraday Soc.* **33**, 1272 (1937).
2. P. W. Kasteleyn, The statistics of dimers on a lattice, *Physica* **27**, 1209 (1961).
3. H. N. V. Temperley and M. E. Fisher, Dimer problem in statistical mechanics — An exact result, *Phil. Mag.* **6**, 1061 (1961).
4. M. E. Fisher, Statistical mechanics of dimers on a plane lattice, *Phys. Rev.* **124**, 1664 (1961).
5. P. W. Kasteleyn, Dimer statistics and phase transitions, *J. Math. Phys.* **4**, 287 (1963).
6. I. S. Gradshteyn and I. M. Ryzhik, *Tables of Integrals, Series and Products* (Academic Press, San Diego, 2000).
7. F. Y. Wu, Remarks on the modified potassium dihydrogen phosphate model of a ferroelectric, *Phys. Rev.* **168**, 539 (1968).
8. F. Y. Wu, Exactly soluble model of the ferroelectric phase transition in two dimensions, *Phys. Rev. Lett.* **18**, 605 (1967).
9. See, for example, H. W. J. Blöte and H. J. Hilhorst, Roughening transitions and the zero-temperature triangular Ising antiferromagnet, *J. Phys. A* **15**, L631 (1982).
10. G. H. Wannier, Antiferromagnetism. The triangular Ising net, *Phys. Rev.* **79**, 357 (1950).
11. H. Y. Huang, F. Y. Wu, H. Kunz and D. Kim, Interacting dimers on the honeycomb lattice: An exact solution of the 5-vertex model, *Physica A* **228**, 1 (1966).

12. E. W. Montroll, *Lattice Statistics, in Applied Combinatorial Mathematics*, ed. E. F. Beckenbach (John Wiley, New York, 1964).
13. H. Cohn, R. Kenyon and J. Propp, A variational principle for domino tilings, *J. AMS* **14**, 297 (2001).
14. T. T. Wu, Dimers on rectangular lattices, *J. Math. Phys.* **3**, 1265 (1962).
15. C. Fan and F. Y. Wu, General lattice model of phase transitions, *Phys. Rev. B* **2**, 723 (1970).
16. F. Y. Wu and K. Y. Lin, Staggered ice-rule vertex model — The Pfaffian solution, *Phys. Rev. B* **12**, 419 (1975).
17. J. F. Nagle, New series-expansion method for the dimer problem, *Phys. Rev.* **152**, 190 (1966).
18. A. J. Phares and F. J. Wunderlich, Thermodynamics and molecular freedom of dimers on plane triangular lattices, *J. Math. Phys.* **27**, 1099 (1985).
19. R. Kenyon, The planar dimer model with boundary: a survey, Directions in Mathematical Quasicrystals, *CRM Monogr. Ser. Amer. Math. Soc.* **13**, 307 (2000).
20. P. Fendley, R. Moessner and S. L. Sondhi, Classical dimer model on the triangular lattice, *Phys. Rev. B* **66**, 214513 (2002).
21. F. Y. Wu, Number of spanning trees on a lattice, *J. Phys. A* **10**, L113 (1977).
22. R. Shrock and F. Y. Wu, Spanning trees on graphs and lattices in d dimensions, *J. Phys. A* **33**, 3881 (2000).
23. R. J. Baxter, F model on a triangular lattice, *J. Math. Phys.* **10**, 1211 (1969).
24. M. L. Glasser and F. Y. Wu, On the entropy of spanning trees on a large triangular lattice, *Ramanujian Journal* **10**, 205 (2005).
25. A. J. Phares and F. J. Wunderlich, Thermodynamics and molecular freedom of dimers on plane honeycomb and kagomé lattices, *Il Nuovo Cimento B* **101**, 653 (1988).
26. V. Elser, Nuclear antiferromagnetism in a registered ^{3}He solid, *Phys. Rev. Lett.* **62**, 2405 (1989).
27. F. Wang and F. Y. Wu, Exact solution of close-packed dimers on the kagome lattice, preprint.
28. M. E. Fisher, On the dimer solution of planar Ising models, *J. Math. Phys.* **7**, 1776 (1966).
29. G. R. Allen, Dimer models for the antiferroelctric transition in copper formate tetrahydrate, *J. Chem. Phys.* **60**, 3299 (1974).
30. S. R. Salinas and J. F. Nagle, Theory of the phase transition in the layered hydrogen-bonded $SnCl_2 \cdot 2H_2O$ crystal, *Phys. Rev. B* **9**, 4920 (1974).
31. J. F. Nagle, C. S. O. Yokoi and S. M. Bhattacharjee, Dimer models on anisotropic lattices, in *Phase Transitions and Critical Phenomena*, Vol. 13, eds. C. Domb and J. L. Lebowitz (Academic Press, New York, 1989).

ELSEVIER

4 February 2002

PHYSICS LETTERS A

Physics Letters A 293 (2002) 235–246

www.elsevier.com/locate/pla

Close-packed dimers on nonorientable surfaces

Wentao T. Lu *, F.Y. Wu

Department of Physics, Northeastern University, Boston, MA 02115, USA

Received 19 September 2001; received in revised form 19 December 2001; accepted 19 December 2001

Communicated by C.R. Doering

Abstract

The problem of enumerating dimers on an $\mathcal{M} \times \mathcal{N}$ net embedded on nonorientable surfaces is considered. We solve *both* the Möbius strip and Klein bottle problems for all $\mathcal{M}$ and $\mathcal{N}$ with the aid of imaginary dimer weights. The use of imaginary weights simplifies the analysis, and as a result we obtain new compact solutions in the form of double products. The compact expressions also permit us to establish a general reciprocity theorem. © 2002 Elsevier Science B.V. All rights reserved.

Keywords: Close-packed dimers; Nonorientable surfaces; Reciprocity theorem

1. Introduction

A seminal development in modern lattice statistics is the solution of enumerating close-packed dimers, or perfect matchings, on a finite $\mathcal{M} \times \mathcal{N}$ simple-quartic net obtained by Kasteleyn [1] and by Temperley and Fisher [2,3] more than 40 years ago. In their solutions the simple-quartic net is assumed to possess free or periodic boundary conditions [1]. In view of the connection with the conformal field theory [4], where the boundary conditions play a crucial role, there has been considerable renewed interest to consider lattice models on nonorientable surfaces [5–9]. For close-packed dimers the present authors [6] have obtained the generating function for the Möbius strip and the Klein bottle for even $\mathcal{M}$ and $\mathcal{N}$. Independently, Tesler [10] has solved the problem of perfect matchings on Möbius strips and deduced solutions in terms of a q-analogue of the Fibonacci numbers for all $\{\mathcal{M}, \mathcal{N}\}$. It turns out that the explicit expression of the generating function depends crucially on whether $\mathcal{M}$ and $\mathcal{N}$ being even or odd, and the analysis differs considerably when either $\mathcal{M}$ or $\mathcal{N}$ is odd. In this Letter we consider the general $\{\mathcal{M}, \mathcal{N}\}$ problems for *both* the Möbius strip and the Klein bottle by introducing imaginary dimer weights. The use of imaginary weights simplifies the analysis and as a result we obtain compact expressions of the solutions without the recourse of Fibonacci numbers. The compact expressions of the solutions also permit us to establish a reciprocity theorem on the enumeration of dimers.

* Corresponding author.
E-mail address: wtlu@neu.edu (W.T. Lu).

PII: S0375-9601(02)00019-1

2. Summary of results

For the convenience of references, we first summarize our main results. Details of derivation will be presented in subsequent sections.

Consider an $\mathcal{M} \times \mathcal{N}$ simple-quartic net consisting $\mathcal{MN}$ sites arranged in an array of $\mathcal{M}$ rows and $\mathcal{N}$ columns. The net forms a Möbius strip if there is a twisted boundary condition in the horizontal direction as shown in Fig. 1, and a Klein bottle if, in addition to the twisted boundary condition, there is also a periodic boundary condition in the vertical direction. Let the dimer weights be z_h and z_v, respectively, in the horizontal and vertical directions. We are interested in the close-form evaluation of the dimer generating function

$$Z_{\mathcal{M},\mathcal{N}}(z_h, z_v) = \sum z_h^{n_h} z_v^{n_v}, \tag{1}$$

where n_h, n_v are, respectively, the number of horizontal and vertical dimers, and the summation is taken over all close-packed dimer coverings of the net.

Our results are as follows: For both $\mathcal{M}$ and $\mathcal{N}$ even, we have [6]

$$Z^{\mathrm{Mob}}_{\mathcal{M},\mathcal{N}}(z_h, z_v) = \prod_{m=1}^{\mathcal{M}/2} \prod_{n=1}^{\mathcal{N}/2} \left[4z_h^2 \sin^2 \frac{(4n-1)\pi}{2\mathcal{N}} + 4z_v^2 \cos^2 \frac{m\pi}{\mathcal{M}+1} \right], \tag{2}$$

$$Z^{\mathrm{Kln}}_{\mathcal{M},\mathcal{N}}(z_h, z_v) = \prod_{m=1}^{\mathcal{M}/2} \prod_{n=1}^{\mathcal{N}/2} \left[4z_h^2 \sin^2 \frac{(4n-1)\pi}{2\mathcal{N}} + 4z_v^2 \sin^2 \frac{(2m-1)\pi}{\mathcal{M}} \right], \tag{3}$$

where the superscripts refer to the type of the nonorientable surface under consideration.

For $\mathcal{M}$ even and $\mathcal{N}$ odd, we have

$$Z^{\mathrm{Mob}}_{\mathcal{M},\mathcal{N}}(z_h, z_v) = \mathrm{Re}\left[(1-i) \prod_{m=1}^{\mathcal{M}/2} \prod_{n=1}^{\mathcal{N}} \left(2i(-1)^{\mathcal{M}/2+m+1} z_h \sin \frac{(4n-1)\pi}{2\mathcal{N}} + 2z_v \cos \frac{m\pi}{\mathcal{M}+1} \right) \right], \tag{4}$$

$$Z^{\mathrm{Kln}}_{\mathcal{M},\mathcal{N}}(z_h, z_v) = \mathrm{Re}\left[(1-i) \prod_{m=1}^{\mathcal{M}/2} \prod_{n=1}^{\mathcal{N}} \left(2i(-1)^{\mathcal{M}/2+m+1} z_h \sin \frac{(4n-1)\pi}{2\mathcal{N}} + 2z_v \sin \frac{(2m-1)\pi}{\mathcal{M}} \right) \right], \tag{5}$$

and for $\mathcal{M}$ odd and $\mathcal{N}$ even, we have

$$Z^{\mathrm{Mob}}_{\mathcal{M},\mathcal{N}}(z_h, z_v) = z_h^{-\mathcal{N}/2} \prod_{n=1}^{\mathcal{N}/2} \prod_{m=1}^{(\mathcal{M}+1)/2} \left[4z_h^2 \sin^2 \frac{(4n-1)\pi}{2\mathcal{N}} + 4z_v^2 \cos^2 \frac{m\pi}{\mathcal{M}+1} \right], \tag{6}$$

$$Z^{\mathrm{Kln}}_{\mathcal{M},\mathcal{N}}(z_h, z_v) = z_h^{-\mathcal{N}/2} \prod_{n=1}^{\mathcal{N}/2} \prod_{m=1}^{(\mathcal{M}+1)/2} \left[4z_h^2 \sin^2 \frac{(4n-1)\pi}{2\mathcal{N}} + 4z_v^2 \sin^2 \frac{(2m-1)\pi}{\mathcal{M}} \right]. \tag{7}$$

For $\mathcal{M}$ and $\mathcal{N}$ both odd, the generating function is zero.

3. The Möbius strip

It is well-known that there is a one-one correspondence between dimer coverings and terms in a Pfaffian defined by the dimer weights. However, since terms in the Pfaffian generally do not possess the same sign, the evaluation of the Pfaffian does not necessarily produce the desired dimer generating function. The crux of matter is to attach signs, or more generally factors, to the dimer weights so that all terms in the Pfaffian have the same sign, and the task is reduced to that of evaluating a Pfaffian.

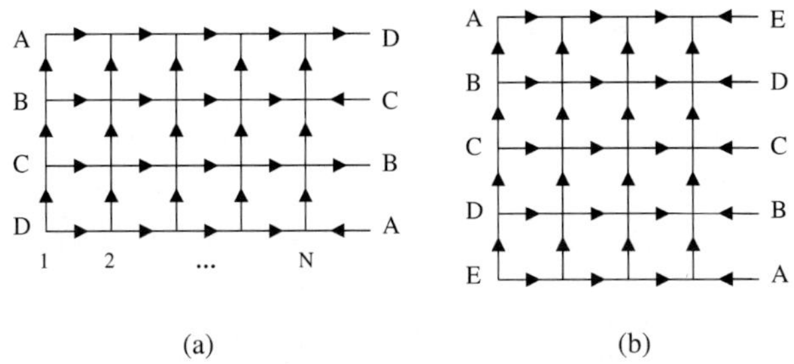

Fig. 1. Edge orientations of Möbius strips with twisted boundary conditions in the horizontal direction. A, B, C, D, E denote repeated sites. (a) $\mathcal{M}=4, \mathcal{N}=5$. (b) $\mathcal{M}=5, \mathcal{N}=4$.

These tasks were first achieved by Kasteleyn [1] for the simple-quartic dimer lattice with free and periodic boundary conditions, who showed how to attach signs to dimer weights and how to evaluate the Pfaffian. Soon after the publication of [1], Wu [11] pointed out that the structure of the Pfaffian, and hence its evaluation, can be simplified if a factor i is associated to dimer weights in one spatial direction. Indeed, the Wu prescription requires only uniformly directed lattice edges with the association of a factor i to dimer weights in the direction in which the number of lattice sites is odd. (If the number of lattice sites is even in both directions, then the factor i can be associated to dimers in either direction.) For $\{\mathcal{M},\mathcal{N}\}=\{\text{even, odd}\}$, for example, one replaces z_h by iz_h.

To see that the Wu procedure is correct, one considers a standard dimer covering C_0 in which the lattice is covered only by parallel dimers with real weights. Then, the two terms in the Pfaffian corresponding to C_0 and any other dimer covering C_1 have the same sign, since the superposition polygon produced by C_0 and C_1 contains an even number of arrows pointing in one direction as well as a factor $i^{4n+2}=-1$, where n is a nonnegative integer. Namely, the superposition polygon is "clockwise-odd" [11]. This use of imaginary dimer weights is the starting point of our analysis.

3.1. $\mathcal{M}=$ even Möbius strip

For $\mathcal{M}$ even and $\mathcal{N}=$ even or odd, we write for definiteness $\mathcal{M}=2M$, $\mathcal{N}=N$, where M, N are positive integers. We orient lattice edges as shown in Fig. 1(a). For the time being consider more generally that the horizontal dimers connecting the first and $\mathcal{N}$th columns have weights z and the associated generating function

$$Z^{\text{Mob}}_{\mathcal{M},\mathcal{N}}(z_h,z_v;z)=\sum_{m=0}^{2M} z^m T_m, \tag{8}$$

where $T_m\equiv T_m(z_h,z_v)$ is a multinomial in z_h and z_v with strictly positive coefficients. The desired result is then obtained by setting $z=z_h$. Note that T_0 is precisely the generating function with free boundary conditions.

Attach a factor i to all (horizontal) dimer weights z_h. This leads us to consider the antisymmetric matrix

$$A(z)=iz_h\big(F_N-F_N^{\mathrm{T}}\big)\otimes I_{2M}+z_v I_N\otimes\big(F_{2M}-F_{2M}^{\mathrm{T}}\big)+z\big(K_N+K_N^{\mathrm{T}}\big)\otimes J_{2M}, \tag{9}$$

where I_N is the $N\times N$ identity matrix, F^{T} is the transpose of F, and F_{2M}, K_N and J_{2M} are matrices of the order given by the subscripts:

$$F_N=\begin{pmatrix}0&1&0&\dots&0\\0&0&1&\dots&0\\ \vdots&\vdots&\vdots&\ddots&\vdots\\0&0&0&\dots&1\\0&0&0&\dots&0\end{pmatrix},\qquad K_N=\begin{pmatrix}0&0&\dots&0\\ \vdots&\vdots&\ddots&\vdots\\0&0&\dots&0\\1&0&\dots&0\end{pmatrix},\qquad J_{2M}=\begin{pmatrix}&&&&1\\&&&-1&\\&&\dots&&\\&1&&&\\-1&&&&\end{pmatrix}. \tag{10}$$

Now, the Pfaffian of $A(z)$ gives the correct generating function T_0 in the case of $z = 0$ [11]. For general z we have the following result:

Theorem. *The dimer generating function for the simple-quartic net with a twisted boundary condition in the horizontal direction is*

$$Z^{\mathrm{Mob}}_{\mathcal{M},\mathcal{N}}(z_h, z_v; z_h) = \frac{1}{2}\big[(1-i)\,\mathrm{Pf}\,A(iz_h) + (1+i)\,\mathrm{Pf}\,A(-iz_h)\big] = \mathrm{Re}\big[(1-i)\,\mathrm{Pf}\,A(iz_h)\big]. \tag{11}$$

Proof. It is clear that the term in (8) corresponding to the configuration C_0 ($m = 0$) has the correct sign. For any other dimer configuration C_1, the superposition of C_0 and C_1 forms superposition polygons containing z edges. We have the following facts which can be readily verified:

(i) The sign of a superposition polygon remains unchanged under deformation of *its* border which leaves n_z, the number of z edges it contains, invariant.
(ii) Deformations of the border of a superposition polygon can change n_z only by multiples of 2, and the sign of the superposition polygon reverses whenever n_z changes by 2.
(iii) Superposition polygons having 0 or 1 z edges have the sign $+$.

As a result, we obtain

$$\mathrm{Pf}\,A(z) \equiv \sqrt{|A(z)|} = X_0 + zX_1 - z^2X_2 - z^3X_3, \tag{12}$$

where $|\cdot|$ denotes the determinant of $\cdot$ and

$$X_\alpha = T_\alpha + z^4T_{\alpha+4} + z^8T_{\alpha+8} + \cdots, \quad \alpha = 0, 1, 2, 3. \tag{13}$$

The theorem is now a consequence of the fact $Z^{\mathrm{Mob}}_{\mathcal{M},\mathcal{N}}(z_h, z_v; z_h) = X_0 + z_hX_1 + z_h^2X_2 + z_h^3X_3$. □

Remarks. The theorem holds also for the Klein bottle which, in addition to a twisted boundary condition in the horizontal direction, has a periodic boundary condition in the vertical direction (see below). For the Möbius strip we have $X_1 = X_3 = 0$ when $N =$ even.

It now remains to evaluate $\mathrm{Pf}\,A(\pm iz_h)$. To evaluate $\mathrm{Pf}\,A(\pm iz_h) = \sqrt{|A(\pm iz_h)|}$ we make use of the fact that, since $F_{2M} - F_{2M}^{\mathrm{T}}$ commutes with J_{2M}, the $2MN \times 2MN$ matrix $A(z)$ can be diagonalized in the $2M$-subspace [6]. Introducing the $2M \times 2M$ matrix U whose elements are

$$\begin{aligned} U_{m,m'} &= i^m\sqrt{\frac{2}{2M+1}}\sin\left(\frac{mm'\pi}{2M+1}\right), \\ \left(U^{-1}\right)_{m,m'} &= (-i)^{m'}\sqrt{\frac{2}{2M+1}}\sin\left(\frac{mm'\pi}{2M+1}\right), \quad m, m' = 1, 2, \ldots, 2M, \end{aligned} \tag{14}$$

we find

$$\begin{aligned} \left(U\left(F_{2M} - F_{2M}^{\mathrm{T}}\right)U^{-1}\right)_{m,m'} &= (2i\cos\phi_m)\delta_{m,m'}, \\ \left(U^{-1}J_{2M}U\right)_{m,m'} &= i(-1)^{M+m}\delta_{m,m'}, \quad m, m' = 1, 2, \ldots, 2M, \end{aligned} \tag{15}$$

where $\phi_m = m\pi/(2M+1)$. Thus, we can replace the $2M \times 2M$ matrices in (9) by their respective eigenvalues, and express $|A(z)|$ as a product of the replaced determinants, namely,

$$\left|A(z)\right| = i^{2MN}\prod_{m=1}^{2M}\left|A_N^{(m)}(z)\right|, \tag{16}$$

where we have taken out a common factor i from each element of the $N \times N$ matrix

$$A_N^{(m)}(z) = 2z_v \cos\phi_m I_N + z_h\big(F_N - F_N^{\mathrm{T}}\big) + (-1)^{M+m} z\big(K_N + K_N^{\mathrm{T}}\big). \tag{17}$$

The matrix $A_N^{(m)}(z)$ can be evaluated for general z in terms of a q-analogue of Fibonacci numbers, but for our purposes when $z = \pm i z_h$, the matrix can be diagonalized directly.

Define the $N \times N$ matrix

$$T_N = F_N + i(-1)^{M+m} K_N = \begin{pmatrix} 0 & 1 & 0 & \dots & 0 \\ 0 & 0 & 1 & \dots & 0 \\ \vdots & \vdots & \vdots & \ddots & \vdots \\ 0 & 0 & 0 & \dots & 1 \\ i(-1)^{M+m} & 0 & 0 & \dots & 0 \end{pmatrix}, \tag{18}$$

we can rewrite (17) when $z = i z_h$ as

$$A_N^{(m)}(i z_h) = 2z_v \cos\phi_m I_N + z_h\big(T_N - T_N^{\dagger}\big). \tag{19}$$

Now T_N and $T_N^{\dagger}$ commute so they can be diagonalized simultaneously with respective eigenvalues $e^{i\theta_n}$ and $e^{-i\theta_n}$, where[1]

$$\theta_n = (-1)^{M+m+1}(4n-1)\pi/2N. \tag{20}$$

Thus, we obtain

$$\big|A_N^{(m)}(i z_h)\big| = \prod_{n=1}^{N}\left[2z_v \cos\frac{m\pi}{2M+1} + 2i(-1)^{M+m+1} z_h \sin\frac{(4n-1)\pi}{2N}\right] \tag{21}$$

and, as a result,

$$\big|A(i z_h)\big| = \prod_{m=1}^{M}\prod_{n=1}^{N}\left[2z_v \cos\frac{m\pi}{2M+1} + 2i(-1)^{M+m+1} z_h \sin\frac{(4n-1)\pi}{2N}\right]^2, \tag{22}$$

where we have made use of the fact that $\cos\phi_{2M+1-m} = -\cos\phi_m$, $(-1)^{2M+1-m} = -(-1)^m$, and $i^{2MN} = (-1)^{MN}$. We thus obtain after taking the square root of (22):

$$\mathrm{Pf}\, A(i z_h) = \prod_{m=1}^{M}\prod_{n=1}^{N}\left[2z_v \cos\frac{m\pi}{2M+1} + 2i(-1)^{M+m+1} z_h \sin\frac{(4n-1)\pi}{2N}\right]. \tag{23}$$

The substitution of (23) into (11) now yields (4). For $N =$ even the Pfaffian (23) is real and (4) reduces to (2). There is no such simplification for $N =$ odd.

3.2. $\mathcal{M} =$ odd Möbius strip

For $\mathcal{M} =$ odd and $\mathcal{N} =$ even or odd, we write for definiteness $\mathcal{M} = 2M+1$, $\mathcal{N} = N$. Since the number of rows $\mathcal{M}$ is odd, we now attach a factor i to dimers in the vertical direction. Again, we assign weights z to horizontal dimers connecting the first and Nth columns and consider the generating function

$$Z^{\mathrm{Mob}}_{\mathcal{M},\mathcal{N}}(z_h, z_v; z) = \sum_{m=0}^{2M+1} z^m T_m \tag{24}$$

[1] A similar expression of θ_n given in (17) of Ref. [6] contains a typo where $M+m+1$ in the exponent should read $M+m$. This does not alter the results of Ref. [6].

defined similar to (8). It is readily verified that, with lattice edge orientations shown in Fig. 1(b), all terms in T_0 have the same sign. It follows that we can use the theorem of the preceding subsection where Pf A is the Pfaffian of the antisymmetric matrix

$$A(z) = z_h\big(F_N - F_N^{\mathrm{T}}\big) \otimes I_{2M+1} - i z_v I_N \otimes \big(F_{2M+1} - F_{2M+1}^{\mathrm{T}}\big) + zG_N \otimes H_{2M+1}, \tag{25}$$

with $G_N = K_N - K_N^{\mathrm{T}}$ and

$$H_{2M+1} = \begin{pmatrix} & & & 1 \\ & & 1 & \\ & \cdots & & \\ 1 & & & \end{pmatrix}. \tag{26}$$

Apply to (25) the unitary transformation (14) (with $2M$ replaced by $2M+1$). The transformation diagonalizes $(F_{2M+1} - F_{2M+1}^{\mathrm{T}})$ as in (15) and, in addition, produces

$$U^{-1} H_{2M+1} U = (-1)^M J_{2M+1}. \tag{27}$$

Thus, we obtain

$$\bar{A}(z) \equiv \big(I_N \otimes U^{-1}\big) A(z) (I_N \otimes U) = \big[z_h\big(F_N - F_N^{\mathrm{T}}\big) + 2z_v \cos\bar{\phi}_m I_N\big] \otimes I_{2M+1} - zG_N \otimes J_{2M+1}, \tag{28}$$

where $\bar{\phi}_m = m\pi/(2M+2)$.

Writing

$$B_N^{(m)} \equiv z_h\big(F_N - F_N^{\mathrm{T}}\big) + 2z_v \cos\bar{\phi}_m I_N, \tag{29}$$

then the matrix $\bar{A}(z)$ assumes a form shown below in the case of $2M+1=5$:

$$\bar{A}(z) = \begin{pmatrix} B_N^{(1)} & & & & zG_N \\ & B_N^{(2)} & & -zG_N & \\ & & B_N^{(3)} + zG_N & & \\ & -zG_N & & B_N^{(4)} & \\ zG_N & & & & B_N^{(5)} \end{pmatrix}. \tag{30}$$

Interchanging rows and columns, this matrix can be changed into a block-diagonal form having the same determinant:

$$\begin{pmatrix} B_N^{(1)} & zG_N & & & \\ zG_N & B_N^{(5)} & & & \\ & & B_N^{(2)} & -zG_N & \\ & & -zG_N & B_N^{(4)} & \\ & & & & B_N^{(3)} + zG_N \end{pmatrix}. \tag{31}$$

For general M we define the $2N \times 2N$ matrix

$$A_{2N}^{(m)}(z) = \begin{pmatrix} B_N^{(m)} & (-1)^{M+m+1} zG_N \\ (-1)^{M+m+1} zG_N & B_N^{(2M+1-m)} \end{pmatrix}, \tag{32}$$

and use the result

$$\big|B_N^{(M+1)} + zG_N\big| = \frac{1}{2}\big[1 + (-1)^N\big] z_h^{N-2} (z_h + z)^2 \tag{33}$$

to obtain

$$|A(z)| = |\bar{A}(z)| = \frac{1}{2}[1+(-1)^N]z_h^{N-2}(z_h+z)^2 \prod_{m=1}^{M} |A_{2N}^{(m)}(z)|. \tag{34}$$

It therefore remains to evaluate $|A_{2N}^{(m)}(z)|$.

The matrix $A_{2N}^{(m)}(z)$ can be diagonalized for $z = \pm i z_h$. To proceed, it is convenient to multiply from the right by a $2N \times 2N$ matrix (whose determinant is $(-1)^N$) to obtain

$$\begin{aligned}\bar{A}_N^{(m)}(iz_h) &\equiv A_{2N}^{(m)}(iz_h)\left[I_N \otimes \begin{pmatrix}1 & 0\\ 0 & -1\end{pmatrix}\right]\\ &= 2z_v\cos\bar{\phi}_m I_{2N} + z_h\left(F_N - F_N^{\mathrm{T}}\right)\otimes\begin{pmatrix}1 & 0\\ 0 & -1\end{pmatrix} + (-1)^{M+m+1} i z_h G_N \otimes \begin{pmatrix}0 & -1\\ 1 & 0\end{pmatrix}\\ &= 2z_v\cos\bar{\phi}_m I_{2N} + z_h\left(Q_{2N} - Q_{2N}^{\dagger}\right),\end{aligned} \tag{35}$$

where

$$Q_{2N} = \begin{pmatrix} F_N & i(-1)^{M+m}K_N \\ -i(-1)^{M+m}K_N & -F_N \end{pmatrix}. \tag{36}$$

Now Q_{2N} commutes with $Q_{2N}^{\dagger}$ so they can be diagonalized simultaneously with the respective eigenvalues $e^{i\bar{\theta}_n}$ and $e^{-i\bar{\theta}_n}$, where $\bar{\theta}_n = (2n-1)/2N$, $n = 1, 2, \ldots, 2N$. Thus we obtain

$$|A_{2N}^{(m)}(iz_h)| = (-1)^N |\bar{A}_{2N}^{(m)}(iz_h)| = (-1)^N \prod_{n=1}^{N}\left(4z_h^2\sin^2\bar{\theta}_n + 4z_v^2\cos^2\bar{\phi}_m\right), \tag{37}$$

and, taking the square root of (34) with z replaced by iz_h,

$$\mathrm{Pf}\,A(iz_h) = \frac{1}{2}[1+(-1)^N](1+i)z_h^{N/2}\prod_{n=1}^{N/2}\prod_{m=1}^{M}\left[4z_h^2\sin^2\frac{(2n-1)\pi}{2N} + 4z_v^2\cos^2\frac{m\pi}{2M+2}\right]. \tag{38}$$

Therefore the generating function vanishes identically for N = odd. For N = even we substitute (38) into (11) and make use of the fact that the two sets $\sin^2(2n-1)\pi/2N$ and $\sin^2(4n-1)\pi/2N$, $n = 1, 2, \ldots, N/2$, are identical. This leads to result (6).

4. The Klein bottle

A Klein bottle is constructed by inserting $\mathcal{N}$ extra vertical edges with weight z_v to the boundary of the Möbius strips of Figs. 1(a) and (b), so that there is a periodic boundary condition in the vertical direction. The extra edges are oriented upward as the other vertical edges. The consideration of the Klein bottle parallels to that of the Möbius strip. Again, we need to consider the cases of even and odd $\mathcal{M}$ separately.

4.1. $\mathcal{M}$ = even Klein bottle

For a $2M \times N$ Klein bottle with horizontal edges connecting the first and $\mathcal{N}$th row having weights z, we have as in (8) the generating function

$$Z_{\mathcal{M},\mathcal{N}}^{\mathrm{Kln}}(z_h, z_v; z) = \sum_{m=0}^{2M} z^m T_m. \tag{39}$$

The dimer weights now generate the antisymmetric matrix

$$A^{\mathrm{Kln}}(z) = A(z) - z_v I_N \otimes \left(K_{2M} - K_{2M}^{\mathrm{T}}\right), \tag{40}$$

where $A(z)$ is given by (9). Following the same analysis as in the case of the Möbius strip, since (i)–(iii) still hold, we find the desired dimer generating function again given by theorem (11) or, explicitly,

$$Z_{\mathcal{M},\mathcal{N}}^{\mathrm{Kln}}(z_h, z_v; z_h) = \frac{1}{2}\left[(1-i)\,\mathrm{Pf}\,A^{\mathrm{Kln}}(iz_h) + (1+i)\,\mathrm{Pf}\,A^{\mathrm{Kln}}(-iz_h)\right]. \tag{41}$$

To evaluate $\mathrm{Pf}\,A^{\mathrm{Kln}}(z)$, we note that the $2M \times 2M$ matrices $(F_{2M} - K_{2M} - F_{2M}^{\mathrm{T}} + K_{2M}^{\mathrm{T}})$ and J_{2M} commute, and can be diagonalized simultaneously by the $2M \times 2M$ matrix V whose elements are

$$V_{mm'} = \frac{1}{\sqrt{2M}} e^{im(2m'-1)\pi/2M},$$
$$\left(V^{-1}\right)_{mm'} = \frac{1}{\sqrt{2M}} e^{-im'(2m-1)\pi/2M}, \quad m, m' = 1, 2, \ldots, 2M. \tag{42}$$

We find

$$\left[V\left(F_{2M} - K_{2M} - F_{2M}^{\mathrm{T}} + K_{2M}^{\mathrm{T}}\right)V^{-1}\right]_{m,m'} = (2i \sin\alpha_m)\delta_{m,m'},$$
$$\left(U^{-1} J_{2M} U\right)_{m,m'} = i(-1)^{M+m}\delta_{m,m'}, \tag{43}$$

where $\alpha_m = (2m-1)\pi/2M$. Diagonalizing $A(z)$ in the $2M$-subspace, we obtain

$$\left|A^{\mathrm{Kln}}(z)\right| = i^{2MN} \prod_{m=1}^{2M} \left|A_N^{(m)}(z)\right|, \tag{44}$$

where

$$A_N^{(m)}(z) = 2z_v \sin\alpha_m I_N + z_h\left(F_N - F_N^{\mathrm{T}}\right) + (-1)^{M+m} z\left(K_N + K_N^{\mathrm{T}}\right). \tag{45}$$

This expression is the same as (17) for the Möbius strip, except with the substitution of $\cos\phi_m$ by $\sin\alpha_m$. Thus, following the same analysis, we obtain

$$\mathrm{Pf}\,A(iz_h) = \prod_{m=1}^{M}\prod_{n=1}^{N}\left[2z_v \sin\frac{(2m-1)\pi}{2M} + 2i(-1)^{M+m+1} z_h \sin\frac{(4n-1)\pi}{2N}\right]. \tag{46}$$

The substitution of (46) into (41) now gives result (5). For $N =$ even, Pfaffian (46) is real and (5) reduces to (3).

4.2. $\mathcal{M} =$ *odd Klein bottle*

For a $(2M+1) \times N$ Klein bottle, the inserted vertical edges have dimer weights iz_v. The consideration then parallels that of the preceding sections. Particularly, the desired dimer generating function is also given by (41), but now with

$$A^{\mathrm{Kln}}(z) = A(z) + iz_v I_N \otimes \left(K_{2M+1} - K_{2M+1}^{\mathrm{T}}\right). \tag{47}$$

To evaluate $\mathrm{Pf}\,A^{\mathrm{Kln}}(z)$, one again applies in the $(2M+1)$-subspace the unitary transformation (42) with $2M$ replaced by $2M+1$, which diagonalizes $F_{2M+1} - K_{2M+1} - F_{2M+1}^{\mathrm{T}} + K_{2M+1}^{\mathrm{T}}$. Using the result

$$\left(V^{-1} H_{2M+1} V\right)_{mm'} = -e^{-i\bar{\alpha}_m}\delta_{m,M+1-m'}, \tag{48}$$

where $\bar{\alpha}_m = (2m-1)\pi/(2M+1)$, $m = 1, 2, \ldots, 2M+1$, then the matrix $\bar{A}(z) = (I_N \otimes U^{-1})A(z)(I_N \otimes U)$ assumes the form in the case of $2M+1=5$:

$$\bar{A}(z) = \begin{pmatrix} B_N^{(1)} & & & & -ze^{-i\bar{\alpha}_1}G_N \\ & B_N^{(2)} & & -ze^{-i\bar{\alpha}_2}G_N & \\ & & B_N^{(3)} + zG_N & & \\ & -ze^{i\bar{\alpha}_2}G_N & & B_N^{(4)} & \\ -ze^{i\bar{\alpha}_1}G_N & & & & B_N^{(5)} \end{pmatrix}. \tag{49}$$

Here $B_N^{(m)} = z_h Q_N + 2z_v \sin\bar{\alpha}_m$. Again, interchanging rows and columns, one changes $\bar{A}(z)$ into the block-diagonal form

$$\begin{pmatrix} B_N^{(1)} & -ze^{-i\bar{\alpha}_1}G_N & & & \\ -ze^{i\bar{\alpha}_1}G_N & B_N^{(5)} & & & \\ & & B_N^{(2)} & -ze^{-i\bar{\alpha}_2}G_N & \\ & & -ze^{i\bar{\alpha}_2}G_N & B_N^{(4)} & \\ & & & & B_N^{(3)} + zG_N \end{pmatrix}. \tag{50}$$

Explicitly, for general M, the mth block is

$$\begin{aligned} A_{2N}^{(m)}(z) &= \begin{pmatrix} B_N^{(m)} & -ze^{-i\bar{\alpha}_m}G_N \\ -ze^{i\bar{\alpha}_m}G_N & B_N^{(2M+1-m)} \end{pmatrix} \\ &= z_h\bigl(F_N - F_N^{\mathrm{T}}\bigr) \otimes I_2 + 2z_v \sin\bar{\alpha}_m I_N \otimes \begin{pmatrix} 1 & 0 \\ 0 & -1 \end{pmatrix} - zG_N \otimes \begin{pmatrix} 0 & e^{-i\bar{\alpha}_m} \\ e^{i\bar{\alpha}_m} & 0 \end{pmatrix}. \end{aligned} \tag{51}$$

We proceed as in (35) by multiplying a $2N \times 2N$ matrix whose determinant is $(-1)^N$ from the right, and obtain

$$\begin{aligned} \bar{A}_{2N}^{(m)}(iz_h) &\equiv A_{2N}^{(m)}(iz_h)\left[I_N \otimes \begin{pmatrix} 1 & 0 \\ 0 & -1 \end{pmatrix}\right] \\ &= 2z_v \sin\bar{\alpha}_m I_{2N} + z_h\bigl(F_N - F_N^{\mathrm{T}}\bigr) \otimes \begin{pmatrix} 1 & 0 \\ 0 & -1 \end{pmatrix} + iz_h G_N \otimes \begin{pmatrix} 0 & e^{-i\bar{\alpha}_m} \\ -e^{i\bar{\alpha}_m} & 0 \end{pmatrix} \\ &= 2z_v \sin\bar{\alpha}_m I_{2N} + z_h\bigl(Q_{2N} - Q_{2N}^{\dagger}\bigr), \end{aligned} \tag{52}$$

where

$$Q_{2N} = \begin{pmatrix} F_N & ie^{-i\bar{\alpha}_m}K_N \\ -ie^{i\bar{\alpha}_m}K_N & -F_N \end{pmatrix}. \tag{53}$$

Now Q_{2N} commutes with $Q_{2N}^{\dagger}$, and they can be diagonalized simultaneously with the respective eigenvalues $e^{i\bar{\beta}_n}$ and $e^{-i\bar{\beta}_n}$, where $\bar{\beta}_n = (2n-1)/2N$, $n = 1, 2, \ldots, 2N$. Thus we obtain

$$\bigl|\bar{A}_{2N}^{(m)}(iz_h)\bigr| = \prod_{n=1}^{N}\left[4z_h^2 \sin^2\frac{(2n-1)\pi}{2N} + 4z_v^2 \sin^2\frac{(2m-1)\pi}{2M+1}\right]. \tag{54}$$

Using (52) and (33), we get

$$\operatorname{Pf} A(iz_h) = \frac{1}{2}\bigl(1+(-1)^N\bigr)(1+i)z_h^{N/2}\prod_{n=1}^{N/2}\prod_{m=1}^{M}\left[4z_h^2 \sin^2\frac{(2n-1)\pi}{2N} + 4z_v^2 \sin^2\frac{(2m-1)\pi}{2M+1}\right]. \tag{55}$$

Thus, the generating function (41) vanishes identically for $N =$ odd. For $N =$ even, we replace as before $\sin^2(2n-1)\pi/N$ by $\sin^2(4n-1)\pi/N$. The substitution of (55) into (41) now leads to result (7).

5. A reciprocity theorem

Using the explicit expression of dimer enumerations on a simple-quartic lattice with free boundaries, Stanley [12] has shown that the enumeration expression satisfies a certain reciprocity relation, a relation rederived recently by Propp [13] from a combinatorial approach. Here, we show that our solutions of dimer enumerations lead to an extension of the reciprocity relation to enumerations on cylindrical, toroidal, and nonorientable surfaces.

We first consider solutions (2), (3), (6) and (7) for $\mathcal{N}=$ even. Writing

$$T^{\mathrm{Mob}}(\mathcal{M},\mathcal{N})=Z^{\mathrm{Mob}}_{\mathcal{M},\mathcal{N}}(1,1),\qquad T^{\mathrm{Kln}}(\mathcal{M},\mathcal{N})=Z^{\mathrm{Kln}}_{\mathcal{M},\mathcal{N}}(1,1) \tag{56}$$

and using the identity [14]

$$\prod_{k=0}^{n-1}\left[x^2-2x\cos\left(\alpha+\frac{2k\pi}{n}\right)+1\right]=x^{2n}-2x^n\cos(n\alpha)+1 \tag{57}$$

repeatedly, we can rewrite our solutions (for general z_h and z_v) in the form of

$$\begin{aligned} Z^{\mathrm{Mob}}_{\mathcal{M},\mathcal{N}}(z_h,z_v)&=z_h^{\mathcal{M}\mathcal{N}/2}\prod_{m=1}^{[(\mathcal{M}+1)/2]}\left(x_m^{\mathcal{N}}+x_m^{-\mathcal{N}}\right)\\ &=\sqrt{\frac{3-(-1)^{\mathcal{M}}}{2}}\,z_v^{\mathcal{M}\mathcal{N}/2}\prod_{n=1}^{\mathcal{N}/2}\left[\frac{y_n^{\mathcal{M}+1}+(-1)^{\mathcal{M}}y_n^{-\mathcal{M}-1}}{y_n+y_n^{-1}}\right], \end{aligned} \tag{58}$$

$$\begin{aligned} Z^{\mathrm{Kln}}_{\mathcal{M},\mathcal{N}}(z_h,z_v)&=z_h^{\mathcal{M}\mathcal{N}/2}\prod_{m=1}^{[(\mathcal{M}+1)/2]}\left(t_m^{\mathcal{N}}+t_m^{-\mathcal{N}}\right)\\ &=\sqrt{\frac{3-(-1)^{\mathcal{M}}}{2}}\,z_v^{\mathcal{M}\mathcal{N}/2}\prod_{n=1}^{\mathcal{N}/2}\left[y_n^{\mathcal{M}}+(-y_n)^{-\mathcal{M}}+1+(-1)^{\mathcal{M}}\right], \end{aligned} \tag{59}$$

where

$$x_m=G\left(\frac{z_v}{z_h}\cos\frac{m\pi}{\mathcal{M}+1}\right),\qquad y_n=G\left(\frac{z_h}{z_v}\sin\frac{(2n-1)\pi}{2\mathcal{N}}\right),\qquad t_m=G\left(\frac{z_v}{z_h}\sin\frac{(2m-1)\pi}{\mathcal{M}}\right) \tag{60}$$

and $G(y)=y+\sqrt{y^2+1}$. Thus, the following reciprocity relations are obtained by inspection:

$$\begin{aligned} T^{\mathrm{Mob}}(\mathcal{M},\mathcal{N})&=T^{\mathrm{Mob}}(\mathcal{M},-\mathcal{N})=\epsilon_{\mathcal{N}\mathcal{M}}T^{\mathrm{Mob}}(-M-2,N),\\ T^{\mathrm{Kln}}(\mathcal{M},\mathcal{N})&=T^{\mathrm{Kln}}(\mathcal{M},-\mathcal{N})=\epsilon_{\mathcal{N}\mathcal{M}}T^{\mathrm{Kln}}(-\mathcal{M},\mathcal{N}), \end{aligned} \tag{61}$$

where

$$\epsilon_{\mathcal{N}\mathcal{M}}=\begin{cases}(-1)^{\mathcal{M}}, & \mathcal{N}=2\ (\mathrm{mod}\ 4),\\ +1, & \text{otherwise}.\end{cases} \tag{62}$$

There are no reciprocity relations for $\mathcal{N}=$ odd.

We have carried out similar analyses for dimer enumerations on a simple-quartic net embedded on a cylinder and a torus, using the solutions given in [1,15], and have discovered universal rules of associating reciprocity relations to specific boundary conditions. Generally, there are 3 different boundary conditions, or "matchings", that can be imposed between 2 opposite boundaries of an $\mathcal{M}\times\mathcal{N}$ net. The conditions can be *twisted* such as those shown in the horizontal direction in Fig. 1, *periodic* such as on a torus, or *free* which means free standing. To establish the convention we shall refer to the boundary condition between the first and the $\mathcal{N}$th columns as (the boundary condition) in the $\mathcal{N}$ direction, and similar that between the first and the $\mathcal{M}$th rows as in the $\mathcal{M}$ direction. Then, our findings together with those of Ref. [13] lead to the following theorem applicable to all cases:

Reciprocity theorem. *Let $T(\mathcal{M},\mathcal{N})$ be the number of close-packed dimer configurations (perfect matchings) on an $\mathcal{M}\times\mathcal{N}$ simple-quartic lattice with free, periodic, or twisted boundary conditions in either direction. (The case of twisted boundary conditions in both directions is excluded.) If the twisted boundary condition, if occurring, is in the $\mathcal{M}$ ($\mathcal{N}$) direction, we restrict to $\mathcal{M}$ ($\mathcal{N}$) = even. Then, we have*

1. $T(\mathcal{M},\mathcal{N})=\epsilon_{\mathcal{N}\mathcal{M}}T(-2-\mathcal{M},\mathcal{N})$ *if the boundary condition in the $\mathcal{M}$ direction is free.*
2. $T(\mathcal{M},\mathcal{N})=\epsilon_{\mathcal{N}\mathcal{M}}T(-\mathcal{M},\mathcal{N})$ *if the boundary condition in the $\mathcal{M}$ direction is periodic or twisted.*

6. Summary and discussion

We have evaluated the dimer generating function (1) for an $\mathcal{M}\times\mathcal{N}$ simple-quartic net embedded on a Möbius strip and a Klein bottle for all $\mathcal{M},\mathcal{N}$. The results are given by (2)–(7). Our results can also be written in terms of the q-analogue of the Fibonacci numbers $\mathcal{F}_n(q)$ defined by

$$\frac{1}{1-qs-s^2}=\sum_{n=0}^{\infty}\mathcal{F}_n(q)s^n. \tag{63}$$

Using the first line of (58), for example, and the identity

$$\mathcal{F}_n(q)+\mathcal{F}_{n-2}(q)=x^n+(-x)^{-n},\quad q\equiv x-x^{-1}, \tag{64}$$

one can verify that our results (2) and (6) for the Möbius strip are the same as those given by Tesler [10]. Details of the proof which also lead to some new product identities involving the Fibonacci numbers will be given elsewhere [16]. We have also deduced a reciprocity theorem for the enumeration $T(\mathcal{M},\mathcal{N})$ of dimers on an $\mathcal{M}\times\mathcal{N}$ lattice under arbitrary including free, periodic, and twisted boundary conditions.

Finally, we point out that results (2)–(7) can be put in a compact expression valid for all cases as

$$Z_{\mathcal{M},\mathcal{N}}(z_h,z_v)=z_h^{\mathcal{M}\mathcal{N}/2}\,\mathrm{Re}\left[(1-i)\prod_{m=1}^{[(\mathcal{M}+1)/2]}\prod_{n=1}^{\mathcal{N}}\left(2i(-1)^{\mathcal{M}/2+m+1}\sin\frac{(4n-1)\pi}{2\mathcal{N}}+2X_m\right)\right], \tag{65}$$

where $[x]$ is the integral part of x, and

$$X_m=\begin{cases}\left(\frac{z_v}{z_h}\right)\cos\frac{m\pi}{\mathcal{M}+1} & \text{for the Möbius strip,}\\ \left(\frac{z_v}{z_h}\right)\sin\frac{(2m-1)\pi}{\mathcal{M}} & \text{for the Klein bottle.}\end{cases} \tag{66}$$

Acknowledgement

Work has been supported in part by National Science Foundation Grant DMR-9980440.

References

[1] P.W. Kasteleyn, Physica 27 (1961) 1209.
[2] H.N.V. Temperley, M.E. Fisher, Philos. Mag. 6 (1961) 1061.
[3] M.E. Fisher, Phys. Rev. 124 (1961) 1664.
[4] H.W.J. Blöte, J.L. Cardy, M.P. Nightingale, Phys. Rev. Lett. 56 (1986) 742.
[5] R. Shrock, Phys. Lett. A 261 (1999) 57.
[6] W.T. Lu, F.Y. Wu, Phys. Lett. A 259 (1999) 108.

[7] W.-J. Tzeng, F.Y. Wu, Appl. Math. Lett. 13 (2000) 19.
[8] W.T. Lu, F.Y. Wu, Phys. Rev. E 63 (2001) 026107.
[9] K. Kaneda, Y. Okabe, Phys. Rev. Lett. 86 (2001) 2134.
[10] G. Tesler, J. Combin. Theory B 78 (2000) 198.
[11] T.T. Wu, J. Math. Phys. 3 (1962) 1265.
[12] R. Stanley, Discrete Appl. Math. 12 (1985) 81.
[13] J. Propp, math.CO/0104011.
[14] I.S. Gradshteyn, I.M. Ryzhik, Table of Integrals, Series, and Products, Academic Press, New York, 1994, 1.394 or, specifically, 1.396.1, 1.396.2 and 1.396.4.
[15] B.M. McCoy, T.T. Wu, The Two-Dimensional Ising Model, Harvard University Press, Cambridge, 1973.
[16] W.T. Lu, F.Y. Wu, in preparation.

RAPID COMMUNICATIONS

PHYSICAL REVIEW E 74, 020104(R) (2006)

Pfaffian solution of a dimer-monomer problem: Single monomer on the boundary

F. Y. Wu

Department of Physics, Northeastern University, Boston, Massachusetts 02115, USA

(Received 23 June 2006; published 18 August 2006)

We consider the dimer-monomer problem for the rectangular lattice. By mapping the problem into one of close-packed dimers on an extended lattice, we rederive the Tzeng-Wu solution for a single monomer on the boundary by evaluating a Pfaffian. We also clarify the mathematical content of the Tzeng-Wu solution by identifying it as the product of the nonzero eigenvalues of the Kasteleyn matrix.

DOI: 10.1103/PhysRevE.74.020104 PACS number(s): 05.50.+q, 04.20.Jb, 02.10.Ox

I. INTRODUCTION

An outstanding unsolved problem in lattice statistics is the dimer-monomer problem. While it is known [1] that the dimer-monomer system does not exhibit a phase transition, there have been only limited closed-form results. The case of close-packed dimers on planar lattices has been solved by Kasteleyn [2] and by Temperley and Fisher [3,4], and the solution has been extended to nonorientable surfaces [5,6]. But the general dimer-monomer problem has proven to be intractable [7].

In 1974 Temperley [8] pointed out a bijection between configurations of a single monomer on the boundary of a planar lattice and spanning trees on a related lattice. The bijection was used in [8] to explain why enumerations of close-packed dimers and spanning trees on square lattices yield the same Catalan constant. More recently Tzeng and Wu [9] made further use of the Temperley bijection to obtain the closed-form generating function for a single monomer on the boundary. The derivation is, however, indirect since it makes use of the Temperley bijection which obscures the underlining mathematics of the closed-form solution.

Motivated by the Tzeng-Wu result, there has been renewed interest in the general dimer-monomer problem. In a series of papers Kong [10–12] has studied numerical enumerations of such configurations on $m \times n$ rectangular lattices for varying m,n, and extracted finite-size correction terms for the single-monomer [10] and general monomer-dimer [11,12] problems. Of particular interest is the finding [10] that in the case of a single monomer the enumeration exhibits a regular pattern similar to that found in the Kasteleyn solution of close-packed dimers. This suggests that the general single-monomer problem might be soluble.

As a first step toward finding that solution it is necessary to have an alternate and direct derivation of the Tzeng-Wu solution without recourse to the Temperley bijection. Here we present such a derivation. Our approach points the way to a possible extension toward the general single-monomer problem. It also shows that, apart from an overall constant, the Tzeng-Wu solution is given by the square root of the product of the nonzero eigenvalues of the Kasteleyn matrix, and thus clarifies its underlining mathematics.

II. THE SINGLE-MONOMER PROBLEM

Consider a rectangular lattice $\mathcal{L}$ consisting of an array of M rows and N columns, where both M and N are odd. The lattice consists of two sublattices A and B. Since the total number of sites MN is odd, the four corner sites belong to the same sublattice, say, A and there is one more A than B sites. The lattice can therefore be completely covered by dimers if one A site is left open. The open A site can be regarded as a monomer.

Assign non-negative weights x and y, respectively, to horizontal and vertical dimers. When the monomer is on the boundary, Tzeng and Wu [9] obtained the following closed-form expression for the generating function:

$$G(x,y) = x^{(M-1)/2}y^{(N-1)/2} \times \prod_{m=1}^{(M-1)/2} \prod_{n=1}^{(N-1)/2} \left(4x^2 \cos^2\frac{m\pi}{M+1} + 4y^2 \cos^2\frac{n\pi}{N+1}\right). \tag{1}$$

This result is independent of the location of the monomer provided that it is an A site on the boundary.

We rederive the result (1) using a formulation that is applicable to any dimer-monomer problem. We first expand $\mathcal{L}$ into an extended lattice $\mathcal{L}'$ constructed by connecting each site occupied by a monomer to a new added site, and then consider close-packed dimers on $\mathcal{L}'$. Since the newly added sites are all of degree 1, all edges originating from the new sites must be covered by dimers. Consequently, the dimer-monomer problem on $\mathcal{L}$ (with fixed monomer sites) is mapped to a close-packed dimer problem on $\mathcal{L}'$, which can be treated by standard means.

We use the Kasteleyn method [2] to treat the latter problem. Returning to the single-monomer problem let the boundary monomer be at site $s_0=(1,n)$ as demonstrated in Fig. 1(a). The site s_0 is connected to a new site s' by an edge with weight 1 as shown in Fig. 1(b). To enumerate close-packed dimers on $\mathcal{L}'$ using the Kasteleyn approach, we need to orient, and associate phase factors with, edges so that all terms in the resulting Pfaffian yield the same sign.

A convenient choice of orientation and assignment of phase factors is the one suggested by Wu [13]. While Wu considered the case of MN even, the consideration can be extended to the present case. Orient all horizontal (vertical) edges in the same direction and the new edge from s' to s_0, and introduce a phase factor i to all horizontal edges as shown in Fig. 1(b). Then all terms in the Pfaffian assume the same sign. To prove this assertion it suffices to show that a typical term in the Pfaffian associated with a dimer configu-

1539-3755/2006/74(2)/020104(4)

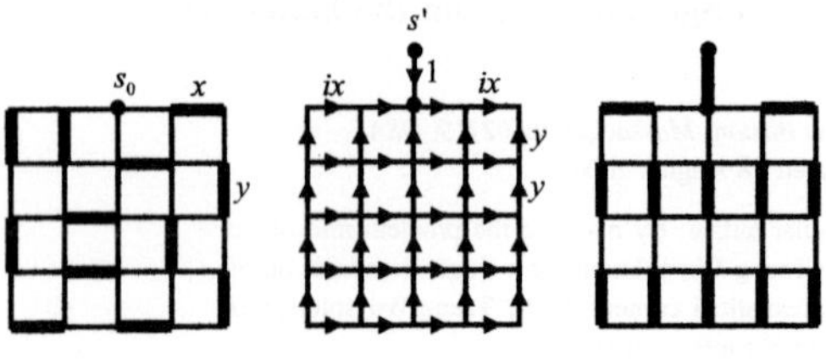

FIG. 1. (a) A dimer-monomer configuration on a 5×5 lattice $\mathcal{L}$ with a single monomer at $s_0=(1,3)$. (b) The extended lattice $\mathcal{L}'$ with edge orientation and a phase factor i to horizontal edges. (c) The reference dimer configuration C_0.

ration C has the same sign as the term associated with a reference configuration C_0. For C_0 we choose the configuration shown in Fig. 1(c), in which horizontal dimers are placed in the first row with vertical dimers covering the rest of the lattice. Then C and C_0 assume the same sign.

The simplest way to verify the last statement is to start from a configuration in which every heavy edge in C_0 shown in Fig. 1(c) is occupied by two dimers, and view each of the doubly occupied dimers as a polygon of two edges. Then the "transposition polygon" (cf. [2]) formed by superimposing any C and C_0 can always be generated by deforming some of the doubly occupied edges into bigger polygons, a process that does not alter the overall sign. It follows that C and C_0 have the same sign for any C. This completes the proof.

Here we have implicitly made use of the fact that the monomer is on the boundary. If the monomer resides in the interior of $\mathcal{L}$, then there exist transposition polygons encircling the monomer site which may not necessarily carry the correct sign. The Pfaffian, while it can still be evaluated, does not yield the dimer-monomer generating function. We shall consider this general single-monomer problem subsequently [14].

With the edge orientation and phase factors in place, the dimer generating function G is obtained by evaluating the Pfaffian

$$G(x,y) = \mathrm{Pf}(A') = \sqrt{\mathrm{Det}\,A'} \tag{2}$$

where A' is the antisymmetric Kasteleyn matrix of dimension $(MN+1)\times(MN+1)$. Explicitly, it reads

$$A' = \begin{pmatrix} 0 & 0 \; \cdots \; 0 \; 1 \; 0 \; \cdots \; 0 \\ 0 & \\ \vdots & \\ 0 & \\ -1 & A \\ 0 & \\ \vdots & \\ 0 & \end{pmatrix}. \tag{3}$$

Here, A is the Kasteleyn matrix of dimension MN for $\mathcal{L}$ given by

$$A = ixT_M \otimes I_N + yI_M \otimes T_N, \tag{4}$$

with I_N the $N\times N$ identity matrix and T_N the $N\times N$ matrix,

$$T_N = \begin{pmatrix} 0 & 1 & 0 & \cdots & 0 & 0 \\ -1 & 0 & 1 & \cdots & 0 & 0 \\ 0 & -1 & 0 & \cdots & 0 & 0 \\ \cdots & \cdots & \cdots & \cdots & \cdots & \cdots \\ 0 & 0 & 0 & \cdots & 0 & 1 \\ 0 & 0 & 0 & \cdots & -1 & 0 \end{pmatrix}. \tag{5}$$

Note that elements of A are labeled by $\{(m,n);(m',n')\}$, where (m,n) is a site index, and the element 1 in the first row of A' is at position $(1,n)$ of A, n=odd.

Expanding (3) along the first row and column, we obtain

$$\mathrm{Det}\,A' = C(A;\{(1,n);(1,n)\}) \tag{6}$$

where $C(A;\{(1,n);(1,n)\})$ is the cofactor of the $\{(1,n);(1,n)\}$th element of Det A.

The cofactor $C(\alpha,\beta)$ of the (α,β)th element of any nonsingular A can be computed using the identity

$$C(A;\alpha,\beta) = A^{-1}(\beta,\alpha)\mathrm{Det}\,A, \tag{7}$$

where $A^{-1}(\beta,\alpha)$ is the (β,α)th element of A. However, the formula is not directly useful in the present case since the matrix A is singular. We shall return to its evaluation in Sec. IV.

III. EIGENVALUES OF THE DETERMINANT A

In this section we enumerate the eigenvalues of A.

The matrix T_N is diagonalized by the similarity transformation

$$U_N^{-1}T_NU_N = \Lambda_N$$

where U_N and U_N^{-1} are $N\times N$ matrices with elements

$$U_N(n_1,n_2) = \sqrt{\frac{2}{N+1}}\,i^{n_1}\sin\frac{n_1n_2\pi}{n+1},$$

$$U_N^{-1}(n_1,n_2) = \sqrt{\frac{2}{N+1}}\,(-i)^{n_2}\sin\frac{n_1n_2\pi}{N+1}, \tag{8}$$

and Λ_N is an $N\times N$ diagonal matrix whose diagonal elements are the eigenvalues of T_N,

$$\lambda_m = 2i\cos\frac{m\pi}{N+1}, \quad m=1,2,\ldots,N. \tag{9}$$

Similarly the $MN\times MN$ matrix A is diagonalized by the similarity transformation generated by $U_{MN}=U_M\otimes U_N$; namely,

$$U_{MN}^{-1}AU_{MN} = \Lambda_{MN}, \tag{10}$$

where Λ_{MN} is a diagonal matrix with eigenvalues

$$\lambda_{mn} = 2i\left(ix\cos\frac{m\pi}{M+1} + y\cos\frac{n\pi}{N+1}\right),$$

$$m=1,2,\ldots,M, \quad n=1,2,\ldots,N, \tag{11}$$

on the diagonal, and elements of U_{MN} and U_{MN}^{-1} are

$$U_{MN}(m_1,n_1;m_2,n_2)=U_M(m_1,m_2)U_N(n_1,n_2),$$

$$U_{MN}^{-1}(m_1,n_1;m_2,n_2)=U_M^{-1}(m_1,m_2)U_N^{-1}(n_1,n_2).$$

Then we have

$$\mathrm{Det}\,A=\prod_{m=1}^{M}\prod_{n=1}^{N}\lambda_{mn}. \tag{12}$$

As in (2), close-packed dimers on $\mathcal{L}$ are enumerated by evaluating $\sqrt{\mathrm{Det}\,A}$. For MN even, this procedure gives precisely the Kasteleyn solution [2]. For MN odd, the case we are considering, the eigenvalue $\lambda_{mn}=0$ for $m=(M+1)/2,n=(N+1)/2$, and hence $\mathrm{Det}\,A=0$, indicating correctly that there is no dimer covering of $\mathcal{L}$. However, it is useful for later purposes to consider the product of the nonzero eigenvalues of A,

$$P\equiv\prod_{m=1}^{M}\prod_{n=1}^{N}{}'\lambda_{mn}, \tag{13}$$

where the prime over the product denotes the restriction $(m,n)\neq((M+1)/2,(N+1)/2)$.

Using the identity

$$\cos\left(\frac{m}{M+1}\right)\pi=-\cos\left(\frac{M-m+1}{M+1}\right)\pi,$$

one can rearrange factors in the product to arrive at

$$P=Q\prod_{m=1}^{(M-1)/2}\prod_{n=1}^{(N-1)/2}\left(4x^2\cos^2\frac{m\pi}{M+1}+4y^2\cos^2\frac{n\pi}{N+1}\right)^2 \tag{14}$$

where the factor Q is the product of factors with either $m=(M+1)/2$ or $n=(N+1)/2$. That is,

$$Q=\left(\prod_{m=1}^{(M-1)/2}4x^2\cos^2\frac{m\pi}{M+1}\right)\left(\prod_{n=1}^{(N-1)/2}4y^2\cos^2\frac{n\pi}{N+1}\right)=\left(\frac{(M+1)(N+1)}{4}\right)x^{M-1}y^{N-1}, \tag{15}$$

where we have made use of the identity

$$\prod_{n=1}^{(N-1)/2}\left(4\cos^2\frac{n\pi}{N+1}\right)=\frac{N+1}{2},\quad N\text{ odd}.$$

The expression (14) for P will be used in the next section.

IV. EVALUATION OF THE COFACTOR

We now return to the evaluation of the cofactor $C(A;\{(1,n);(1,n)\})$. We shall, however, evaluate the cofactor $C(A;\{(m,n);(m',n')\})$ for general m,m',n,n', although only the result of $m=m'=1,n=n'$ is needed here.

To circumvent the problem of using (7) caused by the vanishing of $\mathrm{Det}\,A=0$, we replace A by the matrix

$$A(\epsilon)=A+\epsilon I_{MN},\quad \epsilon\neq0,$$

whose inverse exists, and take the $\epsilon\to0$ limit to rewrite (7) as

$$C(A;\{(m,n);(m',n')\})=\lim_{\epsilon\to0}\{[A^{-1}(\epsilon)](m',n';m,n)\mathrm{Det}\,A(\epsilon)\}. \tag{16}$$

Quantities on the right-hand side of (16) are now well defined and the cofactor can be evaluated accordingly. Consideration of the inverse of a singular matrix along this line is known in mathematics literature as finding the pseudoinverse [15,16]. The method of taking the small-ϵ limit used here has previously been used successfully in analyses of resistance [17] and impedance [18] networks.

The eigenvalues of $A(\epsilon)$ are $\lambda_{mn}(\epsilon)=\lambda_{mn}+\epsilon$ and hence we have

$$\mathrm{Det}\,A(\epsilon)=\prod_{m=1}^{M}\prod_{n=1}^{N}(\lambda_{mn}+\epsilon)=\epsilon P+O(\epsilon^2), \tag{17}$$

where P is the product of nonzero eigenvalues given by (14).

We next evaluate $A^{-1}(\epsilon)(m',n';m,n)$ and retain only terms of the order of $1/\epsilon$. Taking the inverse of (10) with $A(\epsilon)$ in place of A, we obtain

$$A^{-1}(\epsilon)=U_{MN}\Lambda_{MN}^{-1}(\epsilon)U_{MN}^{-1}.$$

Writing out its matrix elements explicitly, we have

$$A^{-1}(\epsilon)(m',n';m,n)=\sum_{m''=1}^{M}\sum_{n''=1}^{N}\times\frac{U_{MN}(m',n';m'',n'')U_{MN}^{-1}(m'',n'';m,n)}{\lambda_{m'',n''}+\epsilon}. \tag{18}$$

For ϵ small the leading term comes from $(m'',n'')=((M+1)/2,(N+1)/2)$ for which $\lambda_{m'',n''}=0$. Using $U_{MN}^{-1}(m,n;m',n')=U_M^{-1}(m,m')U_N^{-1}(n,n')$ and (8), this leads to the expression

$$A^{-1}(\epsilon)(m',n';m,n)=\left(\frac{1}{\epsilon}\right)\left(\frac{4i^{m'+n'}(-i)^{m+n}}{(M+1)(N+1)}\right)\sin\frac{m'\pi}{2}\times\sin\frac{n'\pi}{2}\sin\frac{m\pi}{2}\sin\frac{n\pi}{2}+O(1).$$

Thus, after making use of (16) and (17) we obtain

$$C(A;\{(m,n);(m',n')\})=\sin\frac{m\pi}{2}\sin\frac{n\pi}{2}\sin\frac{m'\pi}{2}\sin\frac{n'\pi}{2}\times\left(\frac{4i^{m'+n'}(-i)^{m+n}P}{(M+1)(N+1)}\right). \tag{19}$$

Finally, specializing to $m=m'=1$, $n=n'$ and combining (2), (6), and (19), we obtain

$$G(x,y) = \sqrt{C(A;\{(1,n);(1,n)\})} = \begin{cases} \sqrt{\dfrac{4P}{(M+1)(N+1)}} & \text{for } n \text{ odd } (A \text{ site}), \\ 0 & \text{for } n \text{ even } (B \text{ site}). \end{cases} \tag{20}$$

This gives the result (1) after introducing (14) for P. It also says that there is no dimer covering if the monomer is on a B site.

The expression (20) clarifies the underlining mathematical content of the Tzeng-Wu solution (1) by identifying it as the product of the *nonzero* eigenvalues of the Kasteleyn matrix. This is compared to the Kasteleyn result [2] that for MN even the dimer generating function is given by the product of *all* eigenvalues.

V. DISCUSSIONS

We have used a direct approach to derive the closed-form expression of the dimer-monomer generating function for the rectangular lattice with a single monomer on the boundary. Our approach is to first convert the problem into one of close-packed dimers without monomers, and consider the latter problem using established means. This approach suggests a possible route toward analyzing the general dimer-monomer problem.

We have also established that the Tzeng-Wu solution (1) is given by the product of the nonzero eigenvalues of the Kasteleyn matrix of the lattice. This is reminiscent of the well-known result in algebraic graph theory [19] that spanning trees on a graph are enumerated by evaluating the product of the nonzero eigenvalues of its tree matrix. The method of evaluating cofactors of a singular matrix as indicated by (16), when applied to the tree matrix of spanning trees, details of which can be easily worked out, offers a simple and direct proof of the fact that all cofactors of a tree matrix are equal and equal to the product of its nonzero eigenvalues. The intriguing similarity of the results suggests that there might be something deeper lurking behind our analysis.

[1] E. H. Lieb and O. J. Heilmann, Phys. Rev. Lett. **24**, 1412 (1970).
[2] P. W. Kasteleyn, Physica (Amsterdam) **27**, 1209 (1961).
[3] H. N. V. Temperley and M. E. Fisher, Philos. Mag. **6**, 1061 (1961).
[4] M. E. Fisher, Phys. Rev. **124**, 1664 (1961).
[5] W. T. Lu and F. Y. Wu, Phys. Lett. A **259**, 108 (1999).
[6] G. Tesler, J. Comb. Theory, Ser. B **78**, 198 (2000).
[7] M. Jerrum, J. Stat. Phys. **48**, 121 (1987); **59**, 1087 (1990).
[8] H. N. V. Temperley, in *Combinatorics: Proceedings of the British Combinatorial Conference*, London Mathematical Society Lecture Notes Series Vol. 13, (Cambridge University Press, Cambridge, U.K., 1974), p. 202.
[9] W.-J. Tzeng and F. Y. Wu, J. Stat. Phys. **10**, 671 (2003).
[10] Y. Kong, Phys. Rev. E **73**, 016106 (2006).
[11] Y. Kong, Phys. Rev. E **74**, 011102 (2006).
[12] Y. Kong (private communication).
[13] T. T. Wu, J. Math. Phys. **3**, 1265 (1962).
[14] W. T. Lu and F. Y. Wu (unpublished).
[15] A. Ben-Israel and T. N. E. Greville, *Generalized Inverses: Theory and Applications*, 2nd ed. (Springer-Verlag, New York, 2003).
[16] S. H. Friedberg, A. J. Insel, and L. E. Spence, *Linear Algebra*, 4th ed. (Prentice-Hall, New York, 2002), Sec. 6.7.
[17] F. Y. Wu, J. Phys. A **37**, 6653 (2004).
[18] W.-J. Tzeng and F. Y. Wu, J. Phys. A **39**, 8579 (2006).
[19] See, for example, N. L. Bigg, *Algebraic Graph Theory*, 2nd ed. (Cambridge University Press, Cambridge, U.K., 1993).

Reprinted from THE PHYSICAL REVIEW, Vol. 168, No. 2, 539–543, 10 April 1968
Printed in U. S. A.

Remarks on the Modified Potassium Dihydrogen Phosphate Model of a Ferroelectric

F. Y. WU
Department of Physics, Northeastern University, Boston, Massachusetts

(Received 4 December 1967)

The modified potassium dihydrogen phosphate (KDP) model of a two-dimensional ferroelectric with the inclusion of an arbitrary electric field $\mathcal{E}$ is considered. The nature of the phase transition is unaffected by the inclusion of an external field, whereas the transition temperature depends on $\mathcal{E}$ with $T_c \to \infty$ as $\mathcal{E} \to \infty$, and $T_c \to 0$ as $\mathcal{E} \to$some special values. We also find that T_c is determined by different relations in different regions in the $\mathcal{E}$ plane, a property not shared by the Slater KDP model. Among the other thermodynamic properties discussed, the critical behavior of the polarizability is shown to be $\chi \sim (T-T_c)^{-1/2}$. It is shown that this model is identical to the problem of close-packed dimers on a hexagonal lattice.

I. INTRODUCTION

RECENTLY, the problem of the two-dimensional hydrogen-bonded crystals as soluble models in statistical mechanics has received considerable attention. After Lieb's celebrated evaluation of the residual entropy of the two-dimensional ice through the use of the method of transfer matrix,[1,2] the solutions have since been extended to include the F model[3] and the Slater potassium dihydrogen phosphate (KDP) model, both with a vertical field,[4,5] and to the general case when the energy parameters are arbitrary.[6,7] In these discussions, the partition function of the crystal is identified as the largest eigenvalue of a certain matrix, while the associated eigenvector proves to be identical to the ground state of a one-dimensional anisotropic Heisenberg chain. Use is then made of the properties of the latter problem which have been discussed extensively.[8–10] While in principle the thermodynamic properties of the hydrogen-bonded crystals follow from the ground-state properties of the linear chain, detailed studies of these properties involve the solution of an integral equation which is analytically soluble only in certain special instances. As a consequence, the partition functions of the F model and the Slater KDP model are given explicitly only when there is no external field.[3,4] With a finite external field, the integral equation cannot be solved in terms of known functions.

In a previous paper[11] (hereafter referred to as I), we have proposed a *modified* KDP model which exhibits the main features of the two-dimensional Slater KDP model, while mathematically it is much easier to deal with. This model is also interesting in that the inclusion of an external field does not present any mathematical problem. Explicit and closed expression can be obtained for the partition function and one has at least one model of a phase transition which is *explicitly* solvable when there is a finite external field. In this paper we shall obtain this solution. First, in Sec. II a generalized version of the modified KDP model is considered and it is shown that this model is equivalent to the problem of close-packed dimers on a hexagonal lattice, namely, the solution to this model is exactly the generating function for the related dimer problem. The modified KDP model in an arbitrary field now appears as a special case and is discussed in Sec. III with the complete thermodynamic properties derived and comparisons with the Slater KDP model given.

II. GENERALIZED MODEL

As is well known,[12] a simple picture of the structure of the KDP crystal (coordination number is 4) allows one hydrogen atom sitting off center on each lattice edge so that the crystal L can be represented by a directed graph. The ice condition (or the condition of local electrical neutrality) requires that there are precisely two arrows pointing into a vertex. Energy values are assigned to the crystal vertices according to the arrow configurations, and we are required to evaluate the partition function

$$Z=\sum_{\text{allowed configurations on } L} \exp\left(-\beta \sum_{\alpha} \epsilon_a\right), \tag{1}$$

where ϵ_a is the energy of the αth lattice vertex and $\beta \equiv 1/kT$.

As pointed out in I, considerations relating to the directed graphs can always be transformed into the language of closed polygons. One simply compares an arbitrary directed graph with a standard one and observes that there is always an even number of arrows reversed at each vertex. Replacing these reversed arrows

---

[1] E. H. Lieb, Phys. Rev. Letters **18**, 692 (1967).
[2] E. H. Lieb, Phys. Rev. **162**, 162 (1967).
[3] E. H. Lieb, Phys. Rev. Letters **18**, 1046 (1967).
[4] E. H. Lieb, Phys. Rev. Letters **19**, 108 (1967).
[5] B. Sutherland, Phys. Rev. Letters **19**, 103 (1967).
[6] C. P. Yang, Phys. Rev. Letters **19**, 586 (1967).
[7] B. Sutherland, C. N. Yang, and C. P. Yang, Phys. Rev. Letters **19**, 588 (1967).
[8] C. N. Yang and C. P. Yang, Phys. Rev. **150**, 321 (1966).
[9] C. N. Yang and C. P. Yang, Phys. Rev. **150**, 327 (1966).
[10] J. des Cloizeaux and M. Gaudin, J. Math. Phys. **7**, 1384 (1966).
[11] F. Y. Wu, Phys. Rev. Letters **18**, 605 (1967).
[12] J. F. Nagle, Phys. Rev. **152**, 190 (1966).

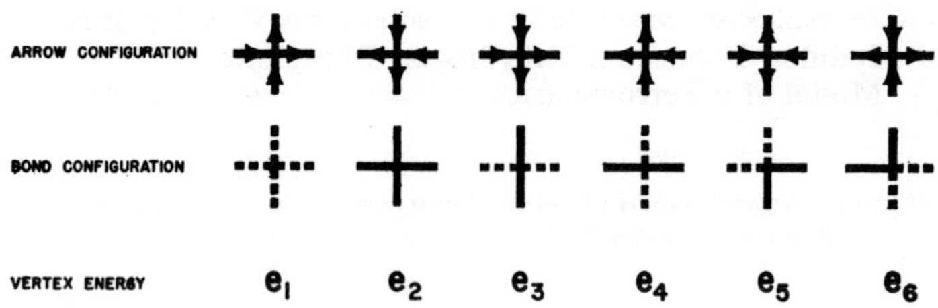

FIG. 1. The six kinds of arrow configurations allowed by the "ice condition" and the associated bond configurations. Configuration (1) is taken as the standard in obtaining the bond configurations.

by bonds, one recovers closed polygons consisting of bonds. These comparisons and correspondences are exhibited in Fig. 1 for the case of a rectangular lattice. Let e_i be the energy of a vertex having the ith ($i=1, 2, \cdots, 6$) configuration. The Slater KDP model corresponds to $e_1=e_2=0$, $e_3=e_4=e_5=e_6=\epsilon>0$; while the model we now proceed to solve has the following restrictions:

$$e_1=\infty \text{ [configuration (1) forbidden]},$$

$$e_3+e_4=e_5+e_6. \tag{2}$$

On taking $e_2=0$ and $e_3=e_4=e_5=e_6=\epsilon>0$, one recovers the model considered in I.

It has been pointed out that the restriction $e_1=\infty$ is equivalent to taking a certain limit in the more general model considered by Sutherland, Yang, and Yang.[7] However, we shall proceed here with the method of Pfaffians because it allows us to see directly the equivalence of this model and the problem of dimers on a hexagonal lattice. The partition function we wish to evaluate is given by (1) with the summation now extending over all allowed bond configurations on L. Since it is well known[13–15] that such a sum can be transformed into a dimer generating function, we shall describe the procedures only briefly. As illustrated in Figs. 2 and 3, we first construct a terminal (dimer) lattice L^Δ by expanding each vertex on L into a city of internally connected points.[16] Next, we cover L^Δ by placing dimers along the edges so that (a) each dimer covers two (neighboring) points on L^Δ, and (b) each point on L^Δ is covered by one and only one dimer (close-packed configuration). For any allowed dimer configuration, we note that there are either two or four dimers leading into a city, corresponding to an allowed bond configuration on L (two or four bonds leading into a vertex). In fact, as shown in Fig. 4, the correspondence between the dimer configurations (within a city on L^Δ) and the allowed bond configurations (at a vertex on L) is actually one-to-one. Therefore to each dimer configuration on L^Δ, there corresponds a bond configuration on L, and vice versa. Now we assign weights (or activities) to the edges on L^Δ (according to Fig. 2) and consider the product of the activities of the covered edges as the configurational weight of a dimer configuration. It is then easy to verify, using the relation $[u_i\equiv\exp(-e_i/kT)]$

$$u_3u_4=u_5u_6, \tag{3}$$

that the dimer configurational weights of L^Δ are just the needed Boltzmann factors for the corresponding bond configurations on L. It follows then the partition function Z is exactly the dimer generating (partition) function Δ defined by

$$Z=\Delta\equiv\sum_{\text{all dimer configurations on } L^\Delta}(\text{configurational weight of } L^\Delta). \tag{4}$$

The dimer lattice L^Δ, as shown in Fig. 3, can still

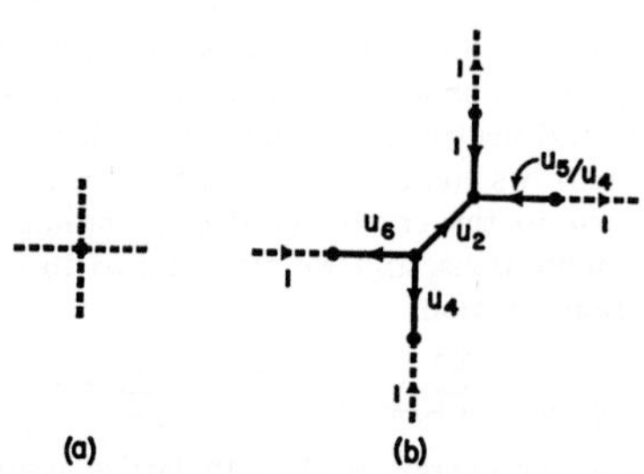

FIG. 2. The expansion of (a) a vertex point on L into (b) a city on L^Δ. The dotted lines denote the lattice edges originally on L and the solid lines denote those gererated by the expansion procedure.

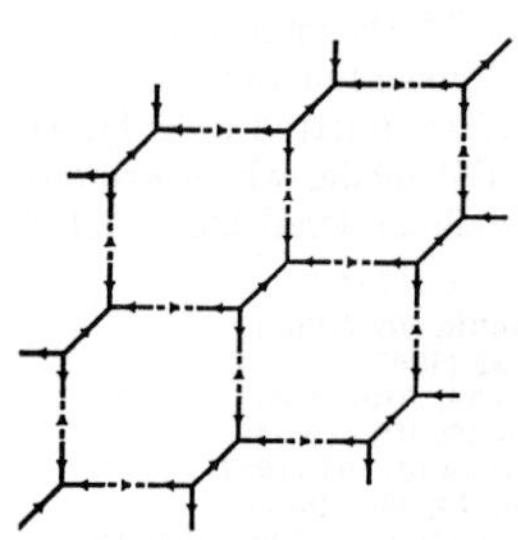

FIG. 3. The dimer lattice L^Δ generated by the expansion procedure shown in Fig. 2. The meaning of the arrows attached to the edges are explained in Ref. 19.

[13] E. W. Montroll, in *Applied Combinatorial Mathematics*, edited by E. F. Beckenbach (John Wiley & Sons, Inc., New York, 1964), Chap. 4.

[14] H. S. Green and C. A. Hurst, in *Order-Disorder Phenomena*, edited by I. Prigogine (Interscience Publishers, Inc., New York, 1964).

[15] M. E. Fisher, J. Math. Phys. **7**, 1776 (1966).

[16] The structure of the city used here, which allows a direct identification with the problem of close-packed dimers on a hexagonal lattice, is simpler than the one adopted in I. This possible simplification is also observed by M. E. Fisher (private communication).

be further simplified. We observe that L^Δ consists of chains of three edges with activities 1, 1, u_4 and u_6, 1, u_5/u_4, respectively. Since the middle edge of these chains always has a unity activity, we may replace each chain by a single edge of activity u_4 or $u_6u_5/u_4=u_3$, respectively. The resulting dimer lattice is therefore a hexagonal (honeycomb) one (see Fig. 5) with activities u_2, u_3, and u_4 respectively along the directions of the three principle axes.[17] Therefore, we have established that the partition function of the generalized (modified) KDP model is identical to the generating (partition) function of the problem of close-packed dimers on a hexagonal lattice, with the dimer lattice containing twice as many vertex points.

The problem of dimers on a hexagonal lattice has been discussed by Kasteleyn.[18] The main feature of the solution is that the partition function is a smoothly varying function of the activities u_2, u_3, and u_4 whenever the activities satisfy the triangle inequalities $u_2+u_3>u_4$, etc.; otherwise the largest activity prevails resulting in a perfect ordering state. Since Kasteleyn did not write the partition function, and it has not been given in the existing literature, we shall supply it here. The derivation is through the use of Pfaffian and is quite straightforward, if one goes back to Fig. 2 and uses the dimer city and the associated activities given there. We refer the readers to Ref. 13 for details and only quote the result.[19] For an infinite lattice (N is the number of vertices on the KDP lattice or $2N$ is the number of vertices on the hexagonal lattice) wrapped around a torus, we find the free energy F per vertex (for the KCP lattice) given by

$$-\beta F=\lim_{N\to\infty} N^{-1}\ln Z$$

$$=\frac{1}{8\pi^2}\int_0^{2\pi}d\theta\int_0^{2\pi}d\phi\ln[u_2^2+u_3^2+u_4^2+2u_2u_3\cos\theta+2u_2u_4\cos\phi+2u_3u_4\cos(\theta-\phi)]. \quad (5)$$

Equation (5) is obviously symmetric in u_2, u_3, and u_4. With the aid of the formula

$$\int_0^{2\pi}\ln[2a+2b\cos\phi+2c\sin\phi]d\phi=2\pi\ln[a+(a^2-b^2-c^2)^{1/2}], \quad (6)$$

one of the integrations can be performed, yielding

$$-\beta F=\frac{1}{4\pi}\int_0^{2\pi}d\theta\ln\max\{u_4^2,\ u_2^2+u_3^2-2u_2u_3\cos\theta\}. \quad (7)$$

Despite its apparent asymmetric appearance, Eq. (7) is still symmetric in u_2, u_3, and u_4. It is then clear that Z is a smoothly varying function in u_2, u_3, and u_4 whenever u_2, u_3, and u_4 satisfy the triangle inequalities $u_2+u_3>u_4$, $u_3+u_4>u_2$ and $u_4+u_2>u_3$. Suppose, on the other hand, $u_4>u_2+u_3$; then one has identically

$$Z=u_4^N,\qquad u_2+u_3<u_4. \quad (8)$$

Kasteleyn has given the reason for this from the point of view of the dimer lattice.[18] It is also easy to see why an ordered state should occur from the considerations of the KDP lattice. If $e_2\neq e_3\neq e_4$, then, at a sufficiently low temperature, the configuration with the lowest energy dominates, thus forming an ordered state.[20]

The energy per vertex E can now be computed for all temperatures. However, it is easier (for $T>T_c$) to start from the expression of the free energy given by Eq. (5). We obtain (assuming $e_2<e_3$, e_4)

$$E=\partial(\beta F)/\partial\beta$$
$$=\tfrac{1}{2}(e_3+e_4)+(4\pi)^{-1}(e_3-e_2)\times\int_0^{2\pi}d\phi\,\mathrm{sgn}(2u_2u_4\cos\phi+u_3^2-u_4^2-u_2^2)+(4\pi)^{-1}(e_4-e_2)\int_0^{2\pi}d\phi\,\mathrm{sgn}(2u_2u_3\cos\phi+u_4^2-u_3^2-u_2^2) \quad (9)$$

$$=e_2\qquad u_3+u_4<u_2$$
$$=e_2+[(e_3-e_2)/\pi]\cos^{-1}[(u_4^2-u_3^2+u_2^2)/2u_2u_4]+[(e_4-e_2)/\pi]\cos^{-1}[(u_3^2-u_4^2+u_2^2)/2u_2u_3],\qquad u_3+u_4>u_2. \quad (10)$$

Here we have used the following identity in obtaining the expression (9):

$$\int_0^{2\pi}\frac{d\theta}{A+B\cos\theta+c\sin\theta}=2\pi/(A^2-B^2-C^2)^{1/2},\qquad \text{for } A^2>B^2+C^2$$
$$=0,\qquad \text{for } A^2<B^2+C^2. \quad (11)$$

As concluded in I, a second-order phase transition (without latent heat) occurs at T_c defined by

$$\exp(-e_3/kT_c)+\exp(-e_4/kT_c)=\exp(-e_2/kT_c). \quad (12)$$

This is in contrast to the result on the Slater KDP model with zero field (phase change with latent heat).[4]

[17] It is easy to see why the result does not depend on e_5 or e_6. This is because the configurations (5) and (6) always occur in pairs with energy $e_5+e_6=e_3+e_4$.

[18] P. W. Kasteleyn, J. Math. Phys. **4**, 287 (1963).

[19] We mention only one important point in the evaluation of the dimer generating function Δ through the use of Pfaffians. In order to transform Δ directly and correctly into a Pfaffian, it suffices to orient the edges of the dimer lattice L^Δ such that if one traces around any closed path (cycle) on L^Δ according to the two rules: (a) The number of edges contained in the cycle is even, and (b) the number of lattice points enclosed by the cycle is also even; one always finds an odd number of edges oriented in the clockwise (or the counterclockwise) direction. This is the key step involved in the evaluation of Δ, and it can be shown that this proper orientation can always be realized for planar lattices (Ref. 15). Interested readers may check that this is indeed the case for the oriented dimer lattice shown in Fig. 3.

[20] Provided that this energy is also lower than $\frac{1}{2}(e_5+e_6)$. Otherwise, we have an antiferroelectric.

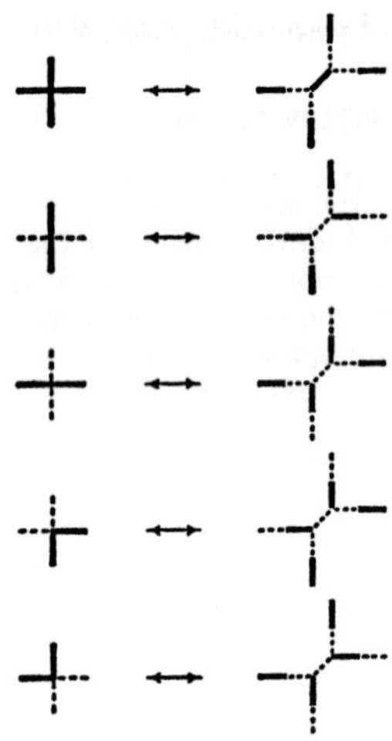

FIG. 4. The one-to-one correspondence between the bond configurations on L and the dimer configurations on L^{Δ}.

It is straightforward to compute the specific heat $c=\partial E/\partial T$. We find $c=0$ below T_c and $c\sim(T-T_c)^{-1/2}$ near and above the Curie temperature, in agreement with the previous conclusions.[4,11]

III. MODIFIED KDP MODEL IN AN ARBITRARY FIELD

We now turn to the modified KDP model in the presence of an arbitrary external field $\boldsymbol{\mathcal{E}}=(\mathcal{E}_x,\mathcal{E}_y)$.[21] In addition to the vertex energies, we now also have bond energies due to the dipole moment d of each arrow. However, one may split the bond energy into two halves and associate one half to each of the two vertices the bond connects. Thus for the modified KDP model in an external field $\boldsymbol{\mathcal{E}}$, one makes the substitutions

$$e_2=(\mathcal{E}_x+\mathcal{E}_y)d, \qquad e_3=\epsilon-(\mathcal{E}_x-\mathcal{E}_y)d,$$
$$e_4=\epsilon+(\mathcal{E}_x-\mathcal{E}_y)d, \qquad e_5=e_6=\epsilon.$$

Since condition (2) is satisfied, the results of the last section apply and the following conclusions are immediate: The transition temperature T_c now depends on the external field and, with the $\boldsymbol{\mathcal{E}}$ plane divided into three regions (Fig. 6), the critical condition takes different forms in each of the three regions. They are $[K\equiv e^{\beta\epsilon},\ X\equiv\exp(2\beta\mathcal{E}_x d),\ Y\equiv\exp(2\beta\mathcal{E}_y d)]$.

$$\begin{aligned}
K&=X+Y, &&(\epsilon>2\mathcal{E}_x d,\ \epsilon>2\mathcal{E}_y d;\ \text{region I})\\
K&=X-Y, &&(\epsilon<2\mathcal{E}_x d,\ \mathcal{E}_x>\mathcal{E}_y;\ \text{region II})\\
K&=Y-X, &&(\epsilon<2\mathcal{E}_y d,\ \mathcal{E}_y>\mathcal{E}_x;\ \text{region III}).
\end{aligned}\tag{13}$$

We observe that $T_c\rightarrow\infty$ as $\mathcal{E}\rightarrow\infty$ and $T_c\rightarrow 0$ as $\boldsymbol{\mathcal{E}}$ approaches the boundaries between the three regions. That T_c may go to zero is a unique property of this model and is also what we expect on physical grounds. On these boundaries two or more of the five allowed vertex configurations have the lowest energy and one no longer has a unique, energetically preferred state. The relation analogous to (13) for the Slater KDP model in a vertical field only is[22]

$$K=1+\exp(-2\beta\,|\,\mathcal{E}_y\,|\,d),\tag{14}$$

which agrees with our expression when $\mathcal{E}_y<0$ ($\mathcal{E}_x=0$). This reflects the fact that our model does not possess the up-down symmetry of the Slater KDP model. It is also straightforward to compute the polarization per vertex $\mathbf{P}=-\partial F/\partial\boldsymbol{\mathcal{E}}$. Again using (11), one finds

$T>T_c$:

$$P_x=\left[\frac{2}{\pi}\cos^{-1}\frac{Y^2-X^2+K^2}{2KY}-1\right]d,$$
$$P_y=\left[\frac{2}{\pi}\cos^{-1}\frac{X^2-Y^2+K^2}{2KX}-1\right]d,\tag{15}$$

$T<T_c$:

$$\begin{aligned}
\mathbf{P}&=(-d,-d), &&\text{(region I)}\\
&=(+d,-d), &&\text{(region II)}\\
&=(-d,+d), &&\text{(region III)}.
\end{aligned}\tag{16}$$

Below T_c, all vertices take the configurations (2), (3),

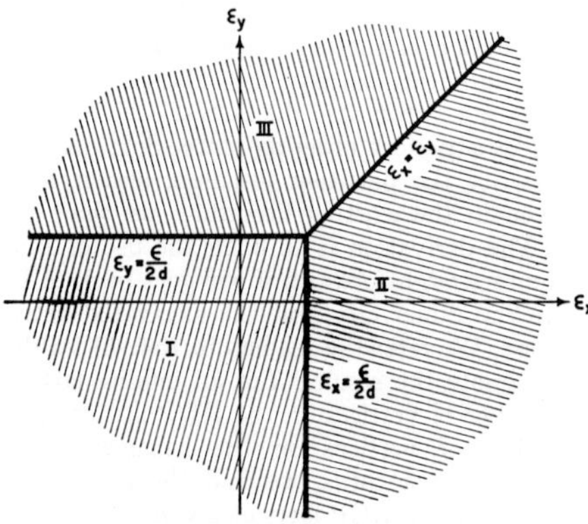

FIG. 6. The three regions in $\boldsymbol{\mathcal{E}}$ the plane and the equations for the boundaries.

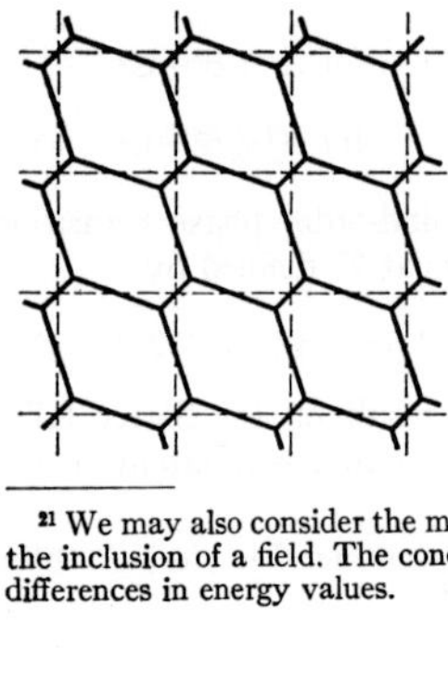

FIG. 5. The hexagonal dimer lattice superimposed on the original lattice L (denoted by the thin lines).

[21] We may also consider the model specified by relation (2) with the inclusion of a field. The conclusions are unchanged except the differences in energy values.

[22] Equation (14) is implicit in Ref. 4.

or (4) for $\mathcal{E}$ in the regions I, II, or III, respectively. It is also interesting to note that there is a finite polarization for $T > T_c$ in the absence of an external field, again a consequence of the lack of symmetry of the model. The polarizability $\chi \equiv \partial \mathbf{P}/\partial \mathcal{E}$ can now be computed exactly. We note only its critical behavior $\chi \sim (T-T_c)^{-1/2}$ as compared to the $(T-T_c)^{-1}$ singularity of the Slater KDP model.[4]

ACKNOWLEDGMENT

I should like to thank Professor Elliott H. Lieb for useful comments.

RAPID COMMUNICATIONS

PHYSICAL REVIEW E 75, 040105(R) (2007)

Exact solution of close-packed dimers on the kagomé lattice

Fa Wang[1,2] and F. Y. Wu[3]
[1]Department of Physics, University of California, Berkeley, California 94720, USA
[2]Material Sciences Division, Lawrence Berkeley National Laboratory, Berkeley, California 94720, USA
[3]Department of Physics, Northeastern University, Boston, Massachusetts 02115, USA
(Received 25 December 2006; published 19 April 2007)

It is well known that exact enumerations of close-packed dimers can be carried out for two-dimensional lattices. While details of results are now known for most lattices, due to the unique nature of the lattice structure, there has been no complete analysis for the kagomé lattice. Here we derive the close-form expression $(1/3)\ln(4xyz)$ for the free energy of close-packed dimers on the kagomé lattice, where x, y, and z are dimer weights. We use two different approaches: the Kasteleyn method of evaluating a Pfaffian and an alternative vertex model formulation. Both methods lead to the same final expression. The correlation function between two dimers at a distance equal or greater than two lattice spacings is found to vanish identically.

DOI: 10.1103/PhysRevE.75.040105 PACS number(s): 05.50.+q, 02.10.Ox

I. INTRODUCTION

A central problem in statistical physics is the enumeration of close-packed dimers on lattices. The origin of the problem has a long history tracing back to the 1937 paper of Fowler and Rushbrooke [1] in their attempt at enumerating the absorption of diatomic molecules on a surface. A milestone in the history of the dimer problem is the exact solution for the square lattice obtained by Kasteleyn [2] and Temperley and Fisher [3] in 1961. Indeed, the method of Kasteleyn is quite general and applicable to all planar lattices [4]. Exact results obtained in this way are summarized in a recent review [5] for a number of two-dimensional lattices.

In the case of the kagomé lattice, however, there has been no complete analysis of the dimer problem other than studies of pure dimer enumerations, most of which are numerical and series expansions (see [6] and references cited therein). In recent years there has been considerable interest in the study of physical phenomena on the kagomé lattice. These range from high-T_c superconductivity [7], Heisenberg antiferromagnets [8–12], and quantum dimers [13] to the occurrence of spin-liquid states [14]. It has also been shown that the consideration of close-packed dimers is related to the ground state of a quantum dimer model [15]. In light of these developments, it is of pertinent interest to take a fresh look at close-packed dimers for the kagomé lattice.

In this Rapid Communication we consider this problem and derive the closed-form expression

$$f_{\text{kagomé}}(x,y,z) = (1/3)\ln(4xyz) \quad (1)$$

for the free energy (for definition of terms see below), a formula quoted in [5]. As exact solutions for other lattices are invariably of the form of a double integral akin to the Onsager solution of the Ising model [5], the very simple expression of the solution (1) and its logarithmic dependence on dimer weights are novel and unique. It points to the special role played by the kagomé lattice (which often makes a problem more amenable) and suggests that caution must be taken in generalizing physical results derived from the kagomé lattice. For example, as we shall see in Sec. IV below, the correlation between two dimers is identically zero at distances equal or greater than two lattice spacings on the kagomé lattice, but this conclusion does not hold for other lattices.

The kagomé lattice is shown in Fig. 1, where x, y, and z are dimer weights along the three principal directions. We denote the lattice by $\mathcal{L}$. Let N (=even) be the number of sites of $\mathcal{L}$, so the lattice can be completely covered by $N/2$ dimers. The dimer generating function is defined to be the summation

$$Z_{\text{kagomé}}(x,y,z) = \sum_{\text{dimer coverings}} x^{n_x}y^{n_y}z^{n_z} \quad (2)$$

over all close-packed dimer configurations of $\mathcal{L}$. Here, n_x is the number of dimers with weight x, etc., subject to $n_x+n_y+n_z=N/2$. Our goal is to evaluate the *per-dimer* free energy

$$f_{\text{kagomé}}(x,y,z) = \lim_{N\to\infty} \frac{1}{N/2} \ln Z_{\text{kagomé}}(x,y,z) \quad (3)$$

in a close form. Past attempts have been confined to enumerations of $f(1,1,1)$. Here we consider the problem for general x, y, and z, a consideration which can find application to anisotropic kagomé systems such as the volborthite antiferromagnet [12].

We derive the solution (1) using two different methods: the Kasteleyn method of evaluating a Pfaffian and alternately a method of a vertex model formulation, which we describe in the next two sections.

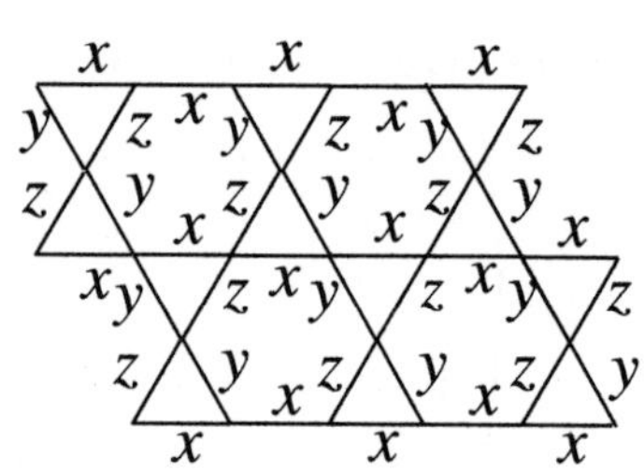

FIG. 1. The kagomé lattice $\mathcal{L}$ with dimer weights x, y, and z in the three principal directions.

1539-3755/2007/75(4)/040105(4)

II. PFAFFIAN APPROACH

The first step of the Pfaffian method is to find a Kasteleyn orientation [2] of lattice edges. A Kasteleyn orientation of a planar lattice is an orientation of edges such that every transition cycle consisting of a loop of edges derived from the superposition of two dimer coverings has an odd number of arrows pointing in the clockwise direction, a property which we term as *clockwise odd*. While Kasteleyn [4] has demonstrated that such an orientation is possible for all planar graphs, the actual orientation of edges for a given lattice, or graph, still needs to be worked out and the crux of the matter of the Kasteleyn method is the finding of the appropriate clockwise-odd orientation.

For the kagomé lattice a Kasteleyn orientation can be taken as that shown in Fig. 2. The kagomé lattice is composed of up-pointing and down-pointing triangles. The orientation in Fig. 2 consists of orienting all up-pointing triangles and every other down-pointing triangles in the counterclockwise direction, with the other half of the down-pointing triangles oriented as shown. In this orientation a unit cell of the lattice consists of the six sites forming two neighboring down-pointing triangles, which are numbered 1,...,6 as shown. Our orientation is different from that used in [6].

To see that the orientation in Fig. 2 is indeed a Kasteleyn orientation, we note that all transition cycles in Fig. 2 are clockwise odd. As all transition cycles on the lattice can be formed by deforming those in Fig. 2 without altering the clockwise odd property, all transition cycles are also clockwise odd so the orientation in Fig. 2 is a good Kasteleyn orientation [16]. This is essentially the argument of Kasteleyn [4].

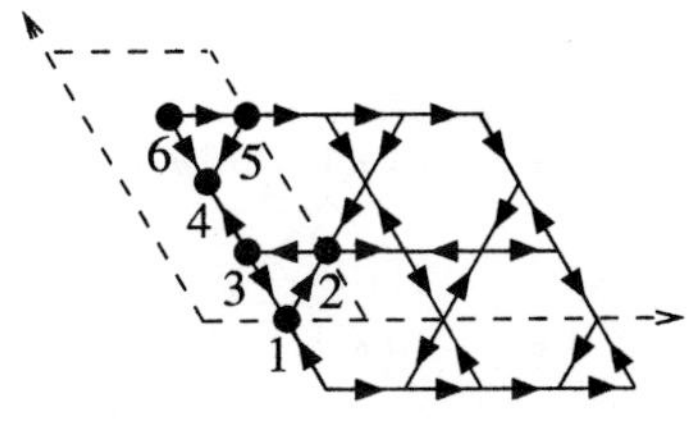

FIG. 2. The Kasteleyn orientation of the kagomé lattice. A unit cell is the region bounded by dashed lines containing the six sites numbered 1,...,6.

After rendering the Kasteleyn orientation, we can write the dimer partition function as a Pfaffian. As the analysis is now standard we follow the standard procedure [see, for example, Eq. (4.34) of [17]] and arrive at the result

$$
\begin{aligned}
f_{\text{kagomé}}(x,y,z) &= \left(\frac{1}{3}\right)\frac{1}{(2\pi)^2}\int_0^{2\pi}\int_0^{2\pi} \ln \mathrm{Pf}[M(\theta,\phi)]d\theta\, d\phi \\
&= \left(\frac{1}{6}\right)\frac{1}{(2\pi)^2}\int_0^{2\pi}\int_0^{2\pi} \ln \det|M(\theta,\phi)|d\theta\, d\phi.
\end{aligned} \tag{4}
$$

Here the factor of $1/3$ in the first line comes from the fact that there are three dimers per unit cell in a close-packed configuration, and $\mathrm{Pf}[M(\theta,\phi)]=\sqrt{\det|M(\theta,\phi)|}$ is the Pfaffian of the 6×6 matrix

$$
\begin{aligned}
M(\theta,\phi) &= a(0,0)+a(1,0)e^{i\theta}+a(-1,0)e^{-i\theta}+a(0,1)e^{i\phi}+a(0,-1)e^{-i\phi}+a(1,1)e^{i(\theta+\phi)}+a(-1,-1)e^{-i(\theta+\phi)} \\
&= \begin{pmatrix}
0 & z & -y & 0 & ze^{-i(\theta+\phi)} & -ye^{-i\phi} \\
-z & 0 & x(1+e^{i\theta}) & -ze^{i\theta} & 0 & 0 \\
y & -x(1+e^{-i\theta}) & 0 & y & 0 & 0 \\
0 & ze^{-i\theta} & -y & 0 & -z & -y \\
-ze^{i(\theta+\phi)} & 0 & 0 & z & 0 & x(-1+e^{i\theta}) \\
ye^{i\phi} & 0 & 0 & y & x(1-e^{-i\theta}) & 0
\end{pmatrix}.
\end{aligned} \tag{5}
$$

The a matrices are read off from Fig. 2 to be

$$
a(0,0)=\begin{pmatrix}
0 & z & -y & 0 & 0 & 0 \\
-z & 0 & x & 0 & 0 & 0 \\
y & -x & 0 & y & 0 & 0 \\
0 & 0 & -y & 0 & -z & -y \\
0 & 0 & 0 & z & 0 & -x \\
0 & 0 & 0 & y & x & 0
\end{pmatrix}, \tag{6a}
$$

$$
a(-1,0)=-a^T(1,0), \tag{6b}
$$

$$
a(0,-1)=-a^T(0,1), \tag{6c}
$$

$$
a(-1,-1)=-a^T(1,1), \tag{6d}
$$

$$
a(1,0)=\begin{pmatrix}
0 & 0 & 0 & 0 & 0 & 0 \\
0 & 0 & x & -z & 0 & 0 \\
0 & 0 & 0 & 0 & 0 & 0 \\
0 & 0 & 0 & 0 & 0 & 0 \\
0 & 0 & 0 & 0 & 0 & x \\
0 & 0 & 0 & 0 & 0 & 0
\end{pmatrix}, \tag{6e}
$$

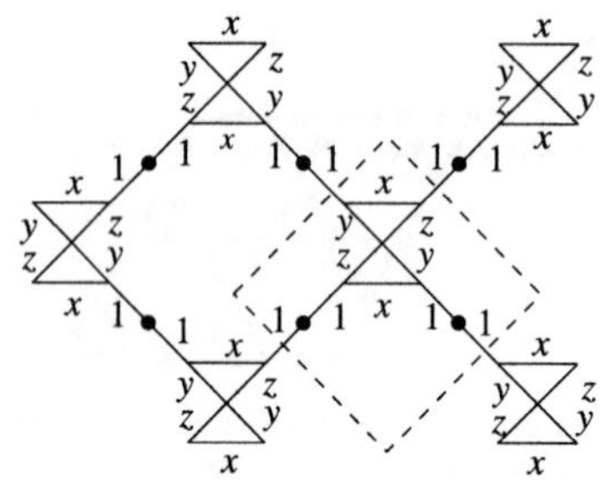

FIG. 3. An extended kagomé lattice $\mathcal{L}'$ constructed by inserting a decorating site attached to two inserted edges of weight 1 as shown. The decorating sites are denoted by solid circles. The unit cell is the region bounded by dashed lines. Repeating unit cells form a square lattice.

$$a(0,1) = \begin{pmatrix} 0 & 0 & 0 & 0 & 0 & 0 \\ 0 & 0 & 0 & 0 & 0 & 0 \\ 0 & 0 & 0 & 0 & 0 & 0 \\ 0 & 0 & 0 & 0 & 0 & 0 \\ 0 & 0 & 0 & 0 & 0 & 0 \\ y & 0 & 0 & 0 & 0 & 0 \end{pmatrix}, \tag{6f}$$

$$a(1,1) = \begin{pmatrix} 0 & 0 & 0 & 0 & 0 & 0 \\ 0 & 0 & 0 & 0 & 0 & 0 \\ 0 & 0 & 0 & 0 & 0 & 0 \\ 0 & 0 & 0 & 0 & 0 & 0 \\ -z & 0 & 0 & 0 & 0 & 0 \\ 0 & 0 & 0 & 0 & 0 & 0 \end{pmatrix}, \tag{6g}$$

where the superscript T denotes the matrix transpose.

The evaluation of the determinant in Eq. (4) gives the surprisingly simple result

$$\det[M(\theta,\phi)] = 16x^2y^2z^2. \tag{7}$$

The substitution of Eq. (7) into Eq. (4) now yields Eq. (1). The expression (7) and result (1) have previously been obtained for $x=y=z=1$ in [6] for pure dimer enumerations.

III. VERTEX-MODEL APPROACH

The kagomé dimer problem can also be solved using a vertex model approach which is conceptually simpler. This involves the mapping of the dimer problem onto a vertex model for which the solution is known.

The first step of the mapping is to introduce the extended kagomé lattice $\mathcal{L}'$ of Fig. 3. The extended lattice $\mathcal{L}'$ is constructed from $\mathcal{L}$ by introducing $4N/3$ extra lattice edges with weight 1 and $2N/3$ new (decorating) sites as shown. By inspection it is clear that a bijection exists between dimer configurations on $\mathcal{L}$ and $\mathcal{L}'$. This permits us to consider instead the dimer problem on $\mathcal{L}'$.

The dimer problem on $\mathcal{L}'$ is next mapped onto a vertex model.

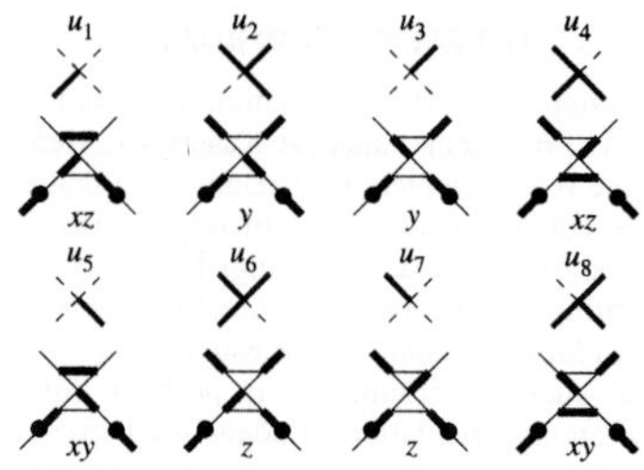

FIG. 4. Mapping between vertex and dimer configurations and the corresponding weights.

The extended lattice $\mathcal{L}'$ consists of $N/3$ unit cells each of which is the region bounded by dashed lines shown in Fig. 3. The unit cells form a square lattice $\mathcal{S}$. We next map dimer configurations on $\mathcal{L}'$ onto vertex configurations on $\mathcal{S}$, by regarding the four edges extending from a unit cell of $\mathcal{L}'$ as the four edges incident to a site on $\mathcal{S}$. To each extending edge on $\mathcal{L}'$ covered by a dimer, draw a *bond* on the corresponding edge on $\mathcal{S}$, and to each extending edge not covered by a dimer, leave the corresponding edge *empty*. Then, as shown in Fig. 4, dimer coverings of a unit cell are mapped onto vertex configurations on $\mathcal{S}$. Since the number of bonds extending from each vertex is either 1 or 3, which is an odd number, we are led to the *odd eight-vertex model* considered in [18].

Vertex weights of the odd eight-vertex model can be read off from Fig. 4 as

$$u_1 = xz, \quad u_2 = y, \quad u_3 = y, \quad u_4 = xz,$$

$$u_5 = xy, \quad u_6 = z, \quad u_7 = z, \quad u_8 = xy. \tag{8}$$

The per-vertex eight-vertex model free energy is then

$$f_{8v}(x,y,z) = \lim_{N\to\infty} \frac{1}{N/3} \ln Z_{\text{kagomé}}(x,y,z). \tag{9}$$

Comparing Eq. (9) with Eq. (3), we obtain the equivalence

$$f_{\text{kagomé}}(x,y,z) = \left(\frac{2}{3}\right) f_{8v}(x,y,z). \tag{10}$$

Now the weights (8) satisfy the free-fermion condition

$$u_1u_2 + u_3u_4 = u_5u_6 + u_7u_8 \tag{11}$$

for which the per-vertex eight-vertex model free energy has been evaluated in [18] as

$$f_{8v} = \frac{1}{16\pi^2}\int_0^{2\pi} d\theta \int_0^{2\pi} d\phi \ln F(\theta,\phi), \tag{12}$$

where

$$F(\theta,\phi) = 2A + 2D\cos(\theta-\phi) + 2E\cos(\theta+\phi) + 4\Delta_1 \sin^2\phi + 4\Delta_2 \sin^2\theta,$$

with

$$A = (u_1u_3 + u_2u_4)^2 + (u_5u_7 + u_6u_8)^2,$$

$$D = (u_5u_7)^2 + (u_6u_8)^2 - 2u_1u_2u_3u_4,$$

$$E = -(u_1u_3)^2 - (u_2u_4)^2 + 2u_5u_6u_7u_8,$$

$$\Delta_1 = (u_1u_2 - u_5u_6)^2,$$

$$\Delta_2 = (u_3u_4 - u_5u_6)^2. \quad (13)$$

The solution (1) is now obtained by substituting Eqs. (13) into Eqs. (12) and (10).

IV. DIMER-DIMER CORRELATION

The dimer-dimer correlation function can be evaluated by either considering a perturbation of the Pfaffian as in [19,20] or by applying the Grassmannian method of [21,22]; details of both approaches will be given elsewhere. Here we sketch steps in the Pfaffian computation.

Define for the lattice edge connecting sites i and j in unit cell at $\mathbf{r}=(r_x,r_y)$ an edge *vacancy* number

$$\begin{aligned} n_{ij,\mathbf{r}} &= 1, \quad \text{if } ij \text{ is empty,} \\ &= 0, \quad \text{if } ij \text{ is occupied,} \end{aligned} \quad (14)$$

where $\langle\cdot\rangle$ denotes the configurational average. Then, the correlation function between two dimers on edges ij in cell $\mathbf{r}_1$ and $k\ell$ in cell $\mathbf{r}_2$ is

$$c(ij,\mathbf{r}_1;k\ell,\mathbf{r}_2) = \langle \bar{n}_{ij,\mathbf{r}_1}\bar{n}_{k\ell,\mathbf{r}_1}\rangle - \langle \bar{n}_{ij;\mathbf{r}_1}\rangle\langle \bar{n}_{k\ell;\mathbf{r}_2}\rangle. \quad (15)$$

To make use of Eq. (15) we need to compute the dimer generating function with specific edge(s) missing. Let A be the antisymmetric matrix derived from the Kasteleyn orientation, and let A' denote the antisymmetric matrix derived from A with edge ij—say, in computing $\langle \bar{n}_{ij}\rangle$—missing. Write

$$Z = \text{Pf}A,$$

$$Z' = \text{Pf}A' = \text{Pf}[A+\Delta], \quad (16)$$

where Δ is the matrix with zero elements everywhere except the ij element is $-A_{ij}$ and the ji element is $-A_{ji}(=A_{ij})$. Then

$$\langle \bar{n}_{ij}\rangle = Z'/Z = \text{Pf}A'/\text{Pf}A \quad (17)$$

and

$$\langle \bar{n}_{ij}\rangle^2 = \frac{\det A'}{\det A} = \frac{\det[A(I+G\Delta)]}{\det A} = \det(I+G\Delta), \quad (18)$$

where $G=A^{-1}$ is the Green's function matrix and I the identity matrix.

In computing Eq. (18) we need only to keep those row(s) and column(s) in Δ and A^{-1} where elements of Δ are nonzero. In addition, in the interior of a large lattice, the correlation depends only on the difference $\mathbf{r}=\mathbf{r}_1-\mathbf{r}_2=\{r_x,r_y\}$, so elements of G are given by

$$G(\mathbf{r}) = \frac{1}{(2\pi)^2}\int_0^{2\pi}\int_0^{2\pi} d\theta\, d\phi\, e^{i(r_x\theta+r_y\phi)}A^{-1}(\theta,\phi). \quad (19)$$

These considerations lead to the explicit evaluations of Eq. (18) and, hence, the correlation (15).

Particularly, due to the fact that elements in $A^{-1}(\theta,\phi)$ contain only a monomial of $e^{\pm i\theta}$ and $e^{\pm i\phi}$, a consequence of the fact that the determinant $\det A$ is given by the simple expression (7), the integral (19) vanishes for $|r_{1x}-r_{2x}|>1$ or $|r_{1y}-r_{2y}|>1$. This leads to the result

$$c(ij,\mathbf{r}_1;k\ell,\mathbf{r}_2) = 0, \quad |\mathbf{r}_1-\mathbf{r}_2| \geqslant 2. \quad (20)$$

The absence of the dimer-dimer correlation beyond a certain distance, which is also found in the Sutherland-Rokhsar-Kivelson state of a quantum dimer model [13], is a property unique to the kagomé lattice. This underscores the special role played by the kagomé lattice in the statistical mechanics and quantum physics of lattice systems.

ACKNOWLEDGMENT

One of us (F.W.) is supported in part by DOE Grant No. LDRD DEA 3664LV.

[1] R. H. Fowler and G. S. Rushbrooke, Trans. Faraday Soc. **33**, 1272 (1937).
[2] P. W. Kasteleyn, Physica (Amsterdam) **27**, 1209 (1961).
[3] H. N. V. Temperley and M. E. Fisher, Philos. Mag. **6**, 1061 (1961).
[4] P. W. Kasteleyn, Physica (Amsterdam) **27**, 1209 (1961).
[5] F. Y. Wu, Int. J. Mod. Phys. B **20**, 5357 (2006).
[6] A. J. Phares and F. J. Wunderlich, Nuovo Cimento Soc. Ital. Fis., B **101**, 653 (1988).
[7] K. Obredors *et al.*, Solid State Commun. **65**, 189 (1988).
[8] V. Elser, Phys. Rev. Lett. **62**, 2405 (1989).
[9] J. S. Helton *et al.*, Phys. Rev. Lett. **98**, 107204 (2007).
[10] O. Ofer *et al.*, e-print cond-mat/0610540.
[11] P. Mendels *et al.*, Phys. Rev. Lett. **98**, 077204 (2007).
[12] F. Bert *et al.*, Phys. Rev. Lett. **95**, 087203 (2005).
[13] G. Misguich, D. Serban, and V. Pasquier, Phys. Rev. Lett. **89**, 137202 (2002).
[14] F. Wang and A. Vishwanath, Phys. Rev. B **74**, 174423 (2006).
[15] D. S. Rokhsar and S. A. Kivelson, Phys. Rev. Lett. **61**, 2376 (1988).
[16] This argument holds as long as the transition cycle does not loop around the lattice.
[17] E. W. Montroll, in *Applied Combinatorial Mathematics*, edited by E. F. Beckenbach (Wiley, New York, 1964).
[18] F. Y. Wu and H. Kunz, J. Stat. Phys. **116**, 67 (2004).
[19] E. W. Montroll, R. B. Potts, and J. C. Ward, J. Math. Phys. **4**, 308 (1963).
[20] M. E. Fisher and J. Stephenson, Phys. Rev. **132**, 1411 (1963).
[21] S. Samuel, J. Math. Phys. **21**, 2815 (1980).
[22] P. Fendley, R. Moessner, and S. L. Sondhi, Phys. Rev. B **66**, 214513 (2002).

VOLUME 78, NUMBER 3 PHYSICAL REVIEW LETTERS 20 JANUARY 1997

Exact Solution of a Three-Dimensional Dimer System

H. Y. Huang,[1] V. Popkov,[2,*] and F. Y. Wu[1]

[1]Department of Physics and Center for Interdisciplinary Research in Complex Systems, Northeastern University, Boston, Massachusetts 02115

[2]Center for Theoretical Physics, Seoul National University, Seoul 151-742, Korea

(Received 16 September 1996)

We present the exact solution of a three-dimensional lattice-statistical model consisting of layers of vertex models coupled with interlayer interactions. For a particular nontrivial interlayer interaction between charge-conserving vertex models and using a transfer matrix approach, we show that the eigenvalues and eigenvectors of the transfer matrix are related to those of the two-dimensional vertex model. As an illustration of the general solution, we analyze the phase transitions in a realistic three-dimensional dimer system and determine its phase diagram and nature of transitions. [S0031-9007(96)02210-7]

PACS numbers: 05.50.+q

An outstanding unsolved problem in the statistical mechanics of critical phenomena is the pursuit of exact solutions for realistic three-dimensional (3D) systems. While a large number of 2D systems have yielded to analyses [1], only a limited number of 3D systems have been solved. They include the 3D Ising model solved by Suzuki [2], the Zamolodchikov model [3] solved by Baxter [4], and its more recent N-state extension by Bazhanov and Baxter [5]. However, these models invariably suffer defects in one way or another: The Suzuki model turns out to be a 2D system in disguise, while the Zamolodchikov model and its extension involve unphysical negative Boltzmann weights. Similarly, a continuous string model in general d dimensions solved by two of us [6] also involves negative weights. The solution of realistic physical 3D models has remained very much an open problem.

One approach toward solving realistic 3D models is to build from 2D systems. Indeed, such an approach has been suggested [7] and applied with some success recently [8,9]. However, in these considerations much attention has been placed to the algebraic structure of the transfer matrix and the associated Yang-Baxter equation, to the extent of masking the elegance of the solution. Here we present a more general formulation of the model and deduce its solution via an alternate and yet much simpler consideration. As an illustration of the general formulation, the result is applied to analyze a realistic layered 3D dimer system.

Consider a simple-cubic lattice $\mathcal{L}$ of size $K \times M \times N$ with periodic boundary conditions. Regard $\mathcal{L}$ as consisting of K copies of square lattices, of M rows and N columns each and stacked together as shown in Fig. 1. For simplicity, we shall speak of the square lattices as "layers" of $\mathcal{L}$. Label sites of $\mathcal{L}$ by indices $\{m, j, k\}$, with $1 \le m \le M$, $1 \le j \le N$, and $1 \le k \le K$. Within each layer of $\mathcal{L}$ define a 2D q-state vertex model whose lattice edges can be in q distinct states. Label the state of the horizontal (vertical) edge incident at the site $\{m, j, k\}$ in the direction of, say, decreasing $\{m, j\}$ by α_{mjk} (β_{mjk}). It is convenient at times to suppress the subscripts m and/or k by adopting the notation

$$\beta_{m+1,j,k} \longrightarrow \beta'_j, \qquad \beta_{m,j,k+1} \longrightarrow \tilde{\beta}_j, \tag{1}$$

and similarly for the α's. Associate vertex weight W_{mjk} to site $\{m, j, k\}$ which is a function of the configuration $C_{m,j,k} \equiv \{\alpha_{mjk}, \beta_{mjk}, \alpha_{m,j+1,k}, \beta_{m+1,j,k}\} \longrightarrow \{\alpha_j, \beta_j, \alpha_{j+1}, \beta'_j\}$ of the four edges incident at the site $\{m, j, k\}$. Let the $\{m, j\}$ sites of two adjacent layers k and $k+1$ interact with a Boltzmann factor B_{mjk} which in the most general case is a function of the configurations $C_{m,j,k}$ and $C_{m,j,k+1}$ of the eight edges incident to the two sites. Then the problem at hand is the evaluation of the partition function

$$Z_{MNK} = \sum_{\alpha_{mjk}} \sum_{\beta_{mjk}} \prod_{k=1}^{K} \prod_{m=1}^{M} \prod_{j=1}^{N} (B_{mjk} W_{mjk}), \tag{2}$$

where the summations are taken over all edge states α_{mjk} and β_{mjk}, and the per-site "free energy" for any K

$$f_K = K^{-1} \lim_{M,N \to \infty} (MN)^{-1} \ln Z_{MNK}. \tag{3}$$

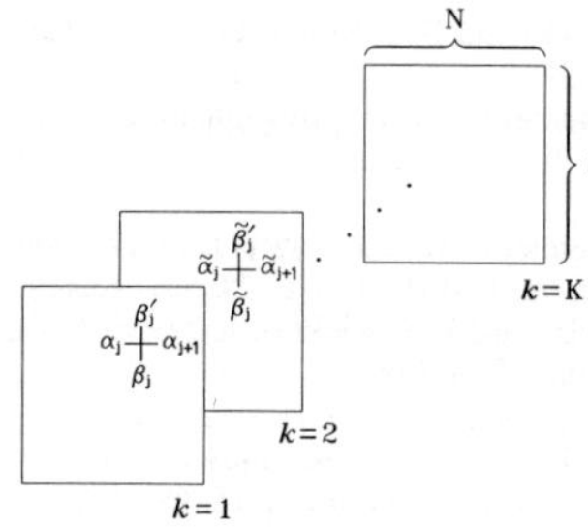

FIG. 1. A three-dimensional lattice model consisting of layered vertex models.

Transfer matrix.—The partition function (2) can be evaluated by applying a transfer matrix in the vertical direction. In a horizontal cross section of $\mathcal{L}$ there are NK vertical edges. Let $\{\beta_m\} = \{\beta_{mjk} | 1 \leq j \leq N, 1 \leq k \leq K\}$, $1 \leq m \leq M$ denote the states of these NK vertical edges, and define a $2^{NK} \times 2^{NK}$ matrix $\mathbf{T}$ with elements

$$T(\{\beta_m\}, \{\beta_{m+1}\}) = \sum_{\alpha_{mjk}} \prod_{k=1}^{K} \prod_{j=1}^{N} (B_{mjk} W_{mjk}), \quad 1 \leq m \leq M. \tag{4}$$

Then one has

$$\begin{aligned} Z_{MNK} &= \sum_{\beta_{mjk}} \prod_{m=1}^{M} T(\{\beta_m\}, \{\beta_{m+1}\}) \\ &= \mathrm{Tr}\, \mathbf{T}^M \\ &\sim \Lambda_{\max}^M, \qquad M \text{ large} \end{aligned} \tag{5}$$

where $\Lambda_{\max}$ is the largest eigenvalue of $\mathbf{T}$.

It is clear that we need to restrict considerations to models which are soluble when the interlayer interaction is absent, or $B_{mjk} = 1$. This leads us to build 3D systems from soluble 2D models. It is also clear that the interlayer interaction B_{mjk} should be such that the overall interlayer factor $\prod_{m,j,k} B_{mjk}$ can be conveniently treated. For this purpose we restrict considerations to 2D charge conserving models.

For definiteness let the labels α_{mjk} and β_{mjk} take on a set I of q integral values. For example, one can take $I = \{+1, -1\}$ for $q = 2$ and $I = \{+1, 0, -1\}$ for $q = 3$. A 2D vertex model is charge conserving if its vertex weights are nonvanishing only when

$$\alpha_j + \beta_j = \alpha_{j+1} + \beta_j' \quad \text{(charge conservation)} \tag{6}$$

holds at all sites. Examples of charge conserving models are the $q = 2$ ice-rule models [10], the q-state string model [11], the $q = 3$ Izergin-Korepin model [12], and others [13].

A direct consequence of the charge conserving rule (6) is deduced by summing (6) from $j = 1$ to $j = N$. This yields

$$y_k \equiv \frac{1}{N} \sum_{j=1}^{N} \beta_j = \frac{1}{N} \sum_{j=1}^{N} \beta_j', \tag{7}$$

showing that the quantity $-1 \leq y_k \leq 1$ is independent of m. (For ice-rule models this fact is well known.)

Next one introduces the interlayer interaction

$$\begin{aligned} B_{mjk} = \exp[&h(\alpha_j \tilde{\beta}_j - \tilde{\alpha}_{j+1} \beta_j') + c_1 h(\alpha_j - \tilde{\alpha}_{j+1}) \\ &+ c_2 h(\tilde{\beta}_j - \beta_j')], \end{aligned} \tag{8}$$

where h, c_1, and c_2 are constants. Since the negation of h corresponds to a reversal of the layer numberings, without loss of generality we can take $h \geq 0$. Note that the terms involving c_1 and c_2 are gauge factors which are canceled in the product in (2) and do not contribute in the free energy (3). So we need only to consider $c_1 = c_2 = 0$ as in [7–9]. But the retaining of c_1 and c_2 in (8) gives us the flexibility of introducing desired local Boltzmann weights in applications (see below).

We now show quite generally that the interlayer interaction (8) (with $c_1 = c_2 = 0$) leads to a considerable simplification of the transfer matrix. Consider first the product

$$\prod_{j=1}^{N} B_{mjk} = \exp\left[h \sum_{j=1}^{N} (\alpha_j \tilde{\beta}_j - \tilde{\alpha}_{j+1} \beta_j') \right]. \tag{9}$$

Summing over (6), or $\alpha_i + \beta_i = \alpha_{i+1} + \beta_i'$, for $i = \{1, j-1\}$ and $i = \{j+1, N\}$ for the layer $k+1$, one obtains, respectively, the identities

$$\begin{aligned} \alpha_j &= \alpha_1 + \sum_{i=1}^{j-1} (\beta_i - \beta_i'), \quad j = 2, 3, \ldots, N, \\ \tilde{\alpha}_{j+1} &= \tilde{\alpha}_1 - \sum_{i=j+1}^{N} (\tilde{\beta}_i - \tilde{\beta}_i'), \quad j = 1, 2, \ldots, N-1, \end{aligned} \tag{10}$$

where we have used $\tilde{\alpha}_{N+1} = \tilde{\alpha}_1$. Substituting (10) into (9) and making use of the identity $\sum_{j=2}^{N} \sum_{i=1}^{j-1} = \sum_{i=1}^{N-1} \sum_{j=i+1}^{N}$ in the first summation in (9), one arrives after a little algebra at

$$\begin{aligned} \prod_{j=1}^{N} B_{mjk} = \exp\{&Nh(\alpha_1 \tilde{y} - \tilde{\alpha}_1 y) \\ &+ Nh[f(\beta, \tilde{\beta}) - f(\beta', \tilde{\beta}')]\}, \end{aligned} \tag{11}$$

where $f(\beta, \tilde{\beta}) \equiv \sum_{j=2}^{N} \sum_{i=1}^{j-1} \beta_i \tilde{\beta}_j$. The numerical factor $f(\beta, \tilde{\beta})$, which is defined for each fixed m, is canceled in the further product

$$\prod_{m=1}^{M} \prod_{j=1}^{N} B_{mjk} = \prod_{m=1}^{M} \exp[Nh(\alpha_{m,1,k} y_{k+1} - \alpha_{m,1,k+1} y_k)]. \tag{12}$$

As a result, only the conserved quantities y_k and the state $\alpha_{m,1,k}$ of the extremities of a row of horizontal edges appear in the product (12). This leads us to rewrite the partition function (2) as

$$Z_{MNK} = \mathrm{Tr}(\mathbf{T}^{\mathrm{eff}})^M, \tag{13}$$

where $\mathbf{T}^{\mathrm{eff}}$ is a matrix with elements

$$\begin{aligned} T^{\mathrm{eff}}(\{\beta_m\}, \{\beta_{m+1}\}) = \sum_{\alpha_{mjk}} \prod_{k=1}^{K} \\ \times \left(e^{Nh\alpha_{m,1,k}(y_{k+1} - y_{k-1})} \prod_{j=1}^{N} W_{mjk} \right). \end{aligned} \tag{14}$$

The problem is now reduced to one of finding the largest eigenvalue of $\mathbf{T}^{\mathrm{eff}}$. In fact, expression (11) shows that $\mathbf{T}^{\mathrm{eff}}$ is related to $\mathbf{T}$ by a similarity transformation $\mathbf{T}^{\mathrm{eff}} = \mathbf{S}\ \mathbf{T}\ \mathbf{S}^{-1}$ where $\mathbf{S}$ is diagonal. It follows that $\mathbf{T}$ and $\mathbf{T}^{\mathrm{eff}}$ have the same eigenvalues, and their eigenvectors are related. The task is now considerably simpler since one needs only to keep track of the 2D system. The

VOLUME 78, NUMBER 3 PHYSICAL REVIEW LETTERS 20 JANUARY 1997

problem is solved if the eigenvalues of the transfer matrix for the 2D vertex model can be evaluated for fixed y_k and $\alpha_{m,1,k}$.

Ice-rule model.—To illustrate the usefulness of this formulation, we now apply it to layers of ice-rule model with vertex weights $\{\omega_1, \omega_2, \ldots, \omega_6\}$ (for standard notations relevant to present discussions, see, for example, [10]). Let $\alpha = +1$ (-1) denote arrows pointing toward right (left), and $\beta = +1$ (-1) arrows pointing up (down). Then one verifies that the charge conserving condition (6) is satisfied with $y_k = 1 - 2n_k/N$, where n_k is the number of down arrows in a row of vertical edges in the kth layer. Introducing next the interlayer interaction (8) (with $c_1 = c_2 = 0$), the eigenvalues of the matrix (14) are obtained by applying a global Bethe ansatz consisting of the usual Bethe ansatz for each layer. The algebra is straightforward and one obtains

$$Z_{MNK} \sim \max_{1\le n_k \le N} \prod_{k=1}^{K} [\Lambda_R(n_k) + \Lambda_L(n_k)]^M, \tag{15}$$

with

$$\Lambda_R(n_k) = e^{-2h(n_{k+1}-n_{k-1})} \omega_1^{N-n_k} \prod_{j=1}^{n_k} \times \left(\frac{\omega_3\omega_4 - \omega_5\omega_6 - \omega_1\omega_3 z_j^{(k)}}{\omega_4 - \omega_1 z_j^{(k)}} \right),$$

$$\Lambda_L(n_k) = e^{2h(n_{k+1}-n_{k-1})} \omega_4^{N-n_k} \prod_{j=1}^{n_k} \times \left(\frac{\omega_1\omega_2 - \omega_5\omega_6 - \omega_2\omega_4/z_j^{(k)}}{\omega_1 - \omega_4/z_j^{(k)}} \right), \tag{16}$$

where Λ_R (Λ_L) refers to the eigenvalue for $\alpha_{m,1,k} = +1$ (-1) [10] and, for each $1 \le k \le K$, the n_k complex numbers $z_j^{(k)}$, $j = 1, 2, \ldots, n_k$ are the solutions of the Bethe ansatz equations

$$e^{4h(n_{k+1}-n_{k-1})}(z_j^{(k)})^N = (-1)^{n_k+1} \prod_{i=1}^{n_k} \left[\frac{B(z_i, z_j)}{B(z_j, z_i)} \right],$$

$$j = 1, 2, \ldots, n_k, \tag{17}$$

where $B(z, z') = \omega_2\omega_4 + \omega_1\omega_3 z z' - (\omega_1\omega_2 + \omega_3\omega_4 - \omega_5\omega_6)z'$. The Bethe ansatz equation (17), which is obtained by imposing the cancellation of unwanted terms in the Bethe ansatz solution, differs from its usual form (see, for example, [14]) in the inclusion of the exponential factor involving h. Various special forms of this solution have been given previously [7–9].

Dimer system with interlayer interactions.—We now consider a 3D lattice model consisting of layers of honeycomb dimer lattices. The dimers, which carry weights u, v, w along the three honeycomb edge directions, are close packed within each layer and, in addition, interact between layers. For two dimers incident at the same $\{m, j\}$ site in adjacent layers, the interaction energy is given in Table I. Note that the interaction is completely symmetric in u, v, w.

The 2D honeycomb dimer system can be formulated as a five-vertex model, namely, an ice-rule model with the weights [14,15]

$$\{\omega_1, \omega_2, \omega_3, \omega_4, \omega_5, \omega_6\} = \{0, w, v, u, \sqrt{uv}, \sqrt{uv}\}. \tag{18}$$

TABLE I. Interaction energy between two dimers incident at the same $\{m, j\}$ site of adjacent layers. For example, a v dimer in the kth layer interacts with a w dimer in the $(k+1)$th layer with an energy $2h/3$.

Layer $k \to k+1$	w	v	u
w	0	$-2h/3$	$2h/3$
v	$2h/3$	0	$-2h/3$
u	$-2h/3$	$2h/3$	0

The five-vertex model is defined on a square lattice of size $M \times N$ mapping to a honeycomb lattice of $2MN$ sites [14,15]. The mapping is such that the edge state $\alpha = +1$ $(\beta = +1)$ corresponds to the presence of a v (u) dimer. It can then be verified that the interlayer interaction given in Table I can be written precisely in the form of (8) with $c_1 = c_2 = \frac{1}{3}$ [16], and therefore we can use the ice-rule model results established in the above.

Substituting (18) into (15), one obtains

$$Z_{MNK} \sim u^{MNK} \max_{1\le n_k\le N} \prod_{k=1}^{K} \prod_{j=1}^{n_k} \left(\frac{w}{u} + \frac{v}{u} z_j^{(k)} \right)^M \tag{19}$$

with the Bethe ansatz solution

$$z_j^{(k)} = e^{i\theta_j} e^{2h(y_{k+1}-y_{k-1})}, \quad j = 1, 2, \cdots, n_k, \tag{20}$$

where $e^{i\theta_j}$ are n_k distinct Nth roots of $(-1)^{n_k+1}$. For a given n_k, the factor inside the parentheses in (19) attends its maximum if the θ_j's lie on an arc crossing the positive real axis and extending from $-\pi(1 - y_k)/2$ to $\pi(1 - y_k)/2$. Using (3) this leads to the per-site free energy

$$f_K = \ln u + \max_{-1\le y_k\le 1} \frac{1}{K} \sum_{k=1}^{K} \frac{1}{2\pi} \times \int_{-\pi(1-y_k)/2}^{\pi(1-y_k)/2} \ln\left(\frac{w}{u} + \frac{v}{u} e^{2h(y_{k+1}-y_{k-1})} e^{i\theta} \right) d\theta. \tag{21}$$

This is our main result.

We have carried out analytic as well as numerical analyses of the free energy (21) for $K = 3 \times$ integer. Here we summarize the findings. For $h = 0$, the layers are decoupled and the property of the system is the same as that of the 2D system [14,15]. For large h, it is readily seen from Table I that the energetically preferred state is one in which each layer is occupied by one kind of dimers, u, v, or w, and the layers are ordered in the sequence of $\{w, v, u, w, v, u, \ldots\}$. It is also clear that for large u, v, or w the system is also frozen with complete ordering of u, v, or w dimers. These orderings are referred to as the H, U, V, and W phases, respectively. The system can also be

VOLUME 78, NUMBER 3 PHYSICAL REVIEW LETTERS 20 JANUARY 1997

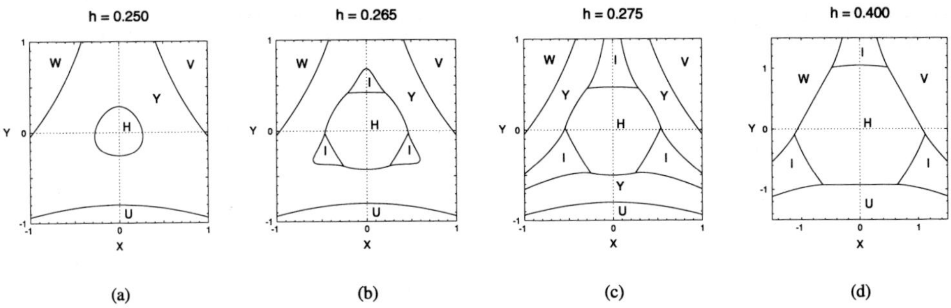

FIG. 2. Phase diagrams of the dimer system. (a) $h_0 < h < h_1$, (b) $h_1 < h < h_2$, (c) $h_2 < h < h_3$, (d) $h > h_3$.

in two other phases. A Y phase in which all layers have the same value of $y_k = y$ determined straightforwardly by maximizing (21), or

$$w^2 + v^2 + 2wv \cos\left[\frac{\pi}{2}(1-y)\right] = u^2, \tag{22}$$

and an I phase which is the H phase with any of the w, v, or u layers replaced by layers with $y_k = y$. If the v layer is replaced by a y layer so that the ordering is $\{w, y, u, w, y, u, \ldots\}$, for example, then y is given by (22) with v replaced by ve^{4h}.

Since the phase diagram is symmetric in w, v, and u, it is convenient to introduce the coordinates

$$X = \ln(v/w) \qquad Y = (\sqrt{3})^{-1}\ln(vw/u^2) \tag{23}$$

so that any interchange of the three variables w, v, and u corresponds to a 120° rotation in the $\{X, Y\}$ plane. The phase diagram for $h < h_0 = 0.2422995\ldots$ is the same as in Fig. 2(a) but without the H regime. Increasing the value of h one finds the H phase appear in $h_0 < h < h_1 = 0.2552479\ldots$ as shown in Fig. 2(a). At $h = h_1$ the I phase appears [Fig. 2(b)], with its region extending to infinity when h reaches $h_2 = (\ln 3)/4 = 0.2746531\ldots$ [Fig. 2(c)]. When h reaches $h_3 = 0.3816955\ldots$ and higher, the Y phase disappears completely as shown in Fig. 2(d). All transitions are found to be of first order except the transitions between the $\{U, V, W\}$ and Y phases, and between the I and H phases, which are found to be of second order with a square-root divergence in the specific heat.

In summary, we have presented the solution of a general 3D lattice model with strictly positive Boltzmann weights, and applied it to solve a realistic 3D dimer system. The analysis can be extended to include dimer-dimer interactions within each layer [14], and further-neighbor interlayer interactions. Details of the present and further analyses will be presented elsewhere.

Work by H. Y. H. and F. Y. W. has been supported in part by NSF Grants No. DMR-9313648 and No. DMR-9614170, and work by V. P. has been supported in part by INTAS Grants No. 93-1324 and No. 93-0633, and the Korea Science and Engineering Foundation through the SRC program. V. P. would like to thank Professor D. Kim for discussions.

*Permanent address: Institute for Low Temperature Physics, Kharkov, Ukraine.

[1] See R. J. Baxter, *Exactly Solved Models in Statistical Mechanics* (Academic, New York, 1982).
[2] M. Suzuki, Phys. Rev. Lett. **28**, 507 (1972).
[3] A. B. Zamolodchikov, JETP **52**, 325 (1980).
[4] R. J. Baxter, Commun. Math. Phys. **88**, 185 (1983).
[5] V. V. Bazhanov and R. J. Baxter, J. Stat. Phys. **69**, 453 (1992).
[6] F. Y. Wu and H. Y. Huang, Lett. Math. Phys. **29**, 105 (1993).
[7] V. Popkov, Phys. Lett. A **192**, 337 (1994).
[8] A. E. Borovick, S. I. Kulinich, V. Popkov, and Yu. M. Strzhemechny, Int. J. Mod. Phys. B **10**, 443 (1996).
[9] V. Popkov and B. Nienhuis, J. Phys. A **30**, 99 (1997).
[10] See, for example, E. H. Lieb and F. Y. Wu, in *Phase Transitions and Critical Phenomena,* edited by C. Domb and M. S. Green (Academic Press, New York, 1972), Vol. I.
[11] J. H. H. Perk and C. L. Schultz, Phys. Lett. A **84**, 407 (1981).
[12] A. G. Izergin and V. E. Korepin, Commun. Math. Phys. **79**, 303 (1981).
[13] See, for example, M. Wadati, T. Deguchi, and Y. Akutsu, Phys. Rep. **180**, 247 (1989).
[14] H. Y. Huang, F. Y. Wu, H. Kunz, and D. Kim, Physica (Amsterdam) **228A**, 1 (1996).
[15] F. Y. Wu, Phys. Rev. **168**, 539 (1968).
[16] To be more precise, (8) leads to Table I with uu and vv interactions $2\epsilon h$ instead of 0, where $\epsilon = +1$ (-1) if the site is in sublattice A (B). But the two assignments of energies are equivalent, since the interacting uu or vv dimers are parallel covering the same A and B sites.

2. The Vertex Models

PHYSICAL REVIEW B VOLUME 2, NUMBER 3 1 AUGUST 1970

General Lattice Model of Phase Transitions

Chungpeng Fan
Department of Physics, Rutgers, The State University, New Brunswick, New Jersey 08903
and
F. Y. Wu*
Department of Physics, Northeastern University, Boston, Massachusetts 02115
(Received 16 December 1969)

A general lattice-statistical model which includes all soluble two-dimensional model of phase transitions is proposed. Besides the well-known Ising and "ice" models, other soluble cases are also considered. After discussing some general symmetry properties of this model, we consider in detail a particular class of the soluble cases, the "free-fermion" model. The explicit expressions for all thermodynamic functions with the inclusion of an external electric field are obtained. It is shown that both the specific heat and the polarizability of the free-fermion model exhibit in general a logarithmic singularity. An inverse-square-root singularity results, however, if the free-fermion model also satisfies the ice condition. The results are illustrated with a specific example.

I. INTRODUCTION

Considerations of the phenomena of phase transitions have been, to a large extent, centered around the study of lattice systems. Besides the intrinsic interest surrounding the lattice systems as models of real physical situations, one is further attracted to their consideration by the possibility of obtaining exact nontrivial solutions. But the soluble problems are very few in number. The Ising model[1,2] of magnetism, first proposed some

40 years ago, still stands at the very frontiers of present knowledge. The only other nontrivial models of phase transitions possessing rigorous solutions are the recently solved models of hydrogen-bonded ferroelectrics and antiferroelectrics.[3–9] It is perhaps not too surprising to find that, while the physical mechanisms responsible for the phase changes associated with the Ising and the ferroelectric models are quite distinct, the mathematical descriptions of these models are not too different. The central mathematical problem involved in all these models is to evaluate a certain generating function in the language of linear graphs. In an effort to search for further soluble problems, we have previously extended these considerations by proposing a general lattice-statistical problem.[9] While on the one hand this problem appears as a general model of ferroelectrics, including all the previously solved models, on the other, it also includes a number of yet unsolved statistical problems. In Ref. 9, we considered an approximate treatment of one aspect of the unsolved problems, namely, the next-neighbor Ising problem. However, there exist other soluble cases of this general problem which do not correspond to any of the known solutions that have been hitherto discussed. In the present paper we return to the study of these situations. The problem under consideration is first defined in Sec. II. Some general symmetry properties of the model are considered in Sec. III where the soluble models are categorized. The thermodynamic properties of one such category, the free-fermion model, are discussed in detail in Sec. IV. The results are illustrated by a specific example in Sec. V.

II. DEFINITION OF PROBLEM

Consider a periodic square lattice composed of N lattice sites (or vertices) and of $2N$ lattice edges. An edge can be either covered by a bond or empty. A definite covering of the lattice edges is called a bond complection G. Clearly there are a total of 2^{2N} distinct bond complections. The number of bonds incident to a vertex in a given bond complection is the degree of the vertex. A bond complection will in general consist of vertices of degrees ranging from 0 to q, where q is the coordination number of the lattice ($q=4$ for square lattice). We shall confine ourselves to considerations of bond complections G' consisting of vertices of *even* degrees only. There are then eight different types of bond configurations which may appear at a vertex. These are numbered from 1 to 8 as shown in Fig. 1. An empty lattice edge will be called a hole. Two bond configurations (or complections) are conjugate to each other if they are related through the interchange of all holes and bonds. Thus, for instance, the vertex types (1) and (2) of Fig. 1 are mutually conjugate. Next, a weight factor $\omega(\xi)$ is associated with each vertex configuration of type $\xi(=1, 2, \ldots, 8)$. The weight W_G of a bond complection G is then taken to be the product of all N vertex weights. The mathematical problem we face is to evaluate the partition sum or the generating function

$$Z=\sum_{G'} W_{G'} = \sum_{G'} \prod_{i=1}^{N} \omega(\xi_i) . \tag{1}$$

Here, the summation is extended over all bond complections G' consisting of vertices of even degrees; the symbol ξ_i refers to the type of configuration at the ith vertex for a given bond complection G'.

For problems of physical interest, energies are assigned to the different vertex configurations and the weights $\omega(\xi)$ are simply the Boltzmann factors

$$\omega(\xi) = \exp(-\beta e_\xi) , \tag{2}$$

where e_ξ is the energy assigned to the ξth type of vertex configuration, $T=(k\beta)^{-1}$ is the temperature, and k is the Boltzmann constant. In such cases, Z is the partition function of the system and the thermodynamic functions can be deduced from the free energy per vertex

$$f=-\beta^{-1} \lim_{N\to\infty} \frac{1}{N} \ln Z. \tag{3}$$

The previous models of phase transitions are recovered for special choices of the vertex energies. For easy reference, a collection of these specializations is included in Table I.

One physical quantity of interest in the consideration of the ferroelectric and antiferroelectric models is the polarization. In these models, each lattice edge is considered to carry a (unit) dipole moment. We now make the correspondence that an edge covered by a bond means a dipole pointing toward the left (for horizontal bonds) or in the upward direction (for vertical bonds). Then, in the presence of an external field $\vec{\mathscr{E}}=(h, v)$, the dipole energies $-\vec{d}\cdot\vec{\mathscr{E}}$ are included by redefining the vertex energies

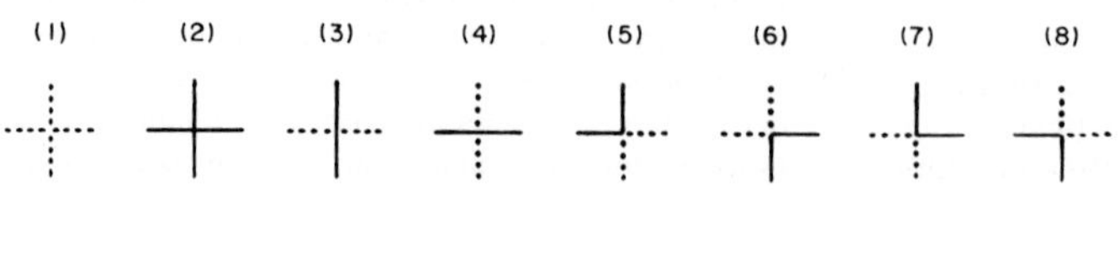

FIG. 1. The eight different kinds of vertex configurations.

TABLE I. Reduction of the general problem to soluble models.

	e_1	e_2	e_3	e_4	e_5	e_6	e_7	e_8	Singularity in specific heat
KDP[a]	0	0	ϵ	ϵ	ϵ	ϵ	∞	∞	$(T-T_c)^{-1/2}$
F[b]	ϵ	ϵ	ϵ	ϵ	0	0	∞	∞	none[c]
Modified KDP[d]	∞	ϵ_2	ϵ_3	ϵ_4	ϵ_5	ϵ_6	∞	∞	$(T-T_c)^{-1/2}$
Modified F[e]	ϵ	ϵ	ϵ	ϵ	0	0	2ϵ	2ϵ	$\ln\lvert T-T_c\rvert$
Rectangular Ising[f]	$-(J_1+J_2)$	J_1+J_2	J_1-J_2	$-J_1+J_2$	0	0	0	0	$\ln\lvert T-T_c\rvert$
Triangular Ising[g]	$-(J_1+J_2+J_3)$	$J_1+J_2-J_3$	$J_1-J_2+J_3$	$-J_1+J_2+J_3$	J_3	J_3	$-J_3$	$-J_3$	$\ln\lvert T-T_c\rvert$
Free-fermion Models[h]			$e_1+e_2=e_7+e_8$,	$e_3+e_4=e_5+e_6$					$\ln\lvert T-T_c\rvert$[h]
Conjugate Models[i]	δ	δ	$-\delta$	$-\delta$	$-\epsilon$	$-\epsilon$	$-\epsilon$	$-\epsilon$	$\ln\lvert T-T_c\rvert$

[a]Reference 3.
[b]Reference 3.
[c]A limiting situation of λ transitions, see Ref. 8.
[d]Reference 6, restricted to $\epsilon_3+\epsilon_4=\epsilon_5+\epsilon_6$.
[e]Reference 7. This is a special case of the conjugate model.
[f]Reference 9.
[g]The vertex energies are obtained by putting $J'=0$ in Eq. (A1) of Ref. 7. The resulting critical condition (34) for the triangular lattice now has a compact form and applies to both ferromagnetic and antiferromagnetic interactions.
[h]This becomes the modified KDP model if $e_1=\infty$.
[i]See description in text. Here we have taken $u_1u_2=u_3u_4=1$.

$$\begin{aligned} \bar{e}_1 &= e_1-(h+v), & \bar{e}_5 &= e_5\,, \\ \bar{e}_2 &= e_2+(h+v)\,, & \bar{e}_6 &= e_6\,, \\ \bar{e}_3 &= e_3-(h-v), & \bar{e}_7 &= e_7\,, \\ \bar{e}_4 &= e_4+(h-v), & \bar{e}_8 &= e_8\,. \end{aligned} \tag{4}$$

Consequently, the polarization $\vec{\mathrm{P}}$ is given by the expressions

$$P_x=-kT\frac{\partial\bar{f}}{\partial h}\,, \qquad P_y=-kT\frac{\partial\bar{f}}{\partial v}\,, \tag{5}$$

where $\bar{f}$ is the free energy (3) evaluated with the energies $\bar{e}_i$ in the place of e_i.

III. GENERAL CONSIDERATIONS

The partition function (1) has not been evaluated in a closed form for its most general expression with arbitrary vertex energies. The general partition function possesses, however, a number of symmetry properties that can be obtained through the following considerations.[10]

First, it is easy to see that pairs of vertices (5) and (6) or (7) and (8) occur together. Therefore, without loss of generality, we may take $\omega(5)=\omega(6)=a$, $\omega(7)=\omega(8)=b$. Consequently, Z is invariant under the change of sign of a or b. The left-right symmetry then leads to the symmetry relation

$$Z(a,b)=Z(b,a)\,. \tag{6}$$

We shall now temporarily disregard the weights a and b and write $Z=Z_{1234}$, where each numerical $\xi(=1,2,3,4)$ stands for the vertex weight $\omega(\xi)$. Since it is immaterial whether to call the bonds holes or bonds, Z is invariant under the interchange of bonds and holes in a given (vertical or horizontal) direction or in both directions simultaneously. We then have the symmetry relation[10]

$$Z=Z_{1234}=Z_{2143}=Z_{4321}=Z_{3412}\,. \tag{7a}$$

Furthermore, a 90° rotation of the lattice interchanges only the indices 3 and 4 (and also the symmetric weights a and b); hence we have

$$Z=Z_{1243}=Z_{2134}=Z_{4312}=Z_{3421}\,. \tag{7b}$$

Relations (6) and (7) tell us that the partition function is invariant with respect to interchanges between conjugate pairs provided that no changes between the vertices 1, 2, 3, 4 and 5, 6, 7, 8 occur.

If the conjugate pairs have the same weights, further symmetry relations exist which permit permutations between the vertices 1, 2, 3, 4 and 5, 6, 7, 8. Let us denote the weights by

$$\omega(1)=\omega(2)=u_1, \qquad \omega(3)=\omega(4)=u_2\,,$$

$$\omega(5)=\omega(6)=u_3, \qquad \omega(7)=\omega(8)=u_4, \tag{8}$$

and write $Z=Z(u_1, u_2; u_3, u_4)$.

Equations (6) and (7) now read

$$Z=Z(u_2, u_1; u_3, u_4)=Z(u_1, u_2; u_4, u_3)\,. \tag{9}$$

We now interchange the bonds and holes along the zigzag paths shown in Fig. 2. Decomposing the lattice into two interconnecting sublattices A and B, it is then easy to see that if on sublattice A we have the following vertex interchanges:

$$\begin{aligned}&(1)\leftrightarrow(6), \quad (2)\leftrightarrow(5),\\&(3)\leftrightarrow(7), \quad (4)\leftrightarrow(8),\end{aligned} \tag{10a}$$

then on sublattice B we have the interchanges

$$\begin{aligned}&(1)\leftrightarrow(5), \quad (2)\leftrightarrow(6),\\&(3)\leftrightarrow(8), \quad (4)\leftrightarrow(7).\end{aligned} \tag{10b}$$

Using (8), we see that (10a) and (10b) are identical and we are led to the further relation

$$Z\equiv Z(u_1, u_2; u_3, u_4)=Z(u_3, u_4; u_1, u_2)\,. \tag{11}$$

In both (9) and (11), it should be remembered that Z is also invariant under the replacement of any u_i by $-u_i$.

Another useful relation can be obtained by applying to the partition function a rearrangement procedure, known as the method of weak-graph expansion.[11] The discussion of the method and its application to the present problem have been previously given by one of us.[7] We write here only the resulting relation[12]

$$\begin{aligned}Z(u_1, u_2; u_3, u_4)&\\&=Z(\tfrac{1}{2}(u_1+u_2+u_3+u_4), \tfrac{1}{2}(u_1+u_2-u_3-u_4);\\&\quad\tfrac{1}{2}(u_1-u_2+u_3-u_4), \tfrac{1}{2}(u_1-u_2-u_3+u_4))\,.\end{aligned} \tag{12}$$

Further iterations of (12) yield no new relations except those specified by (9) and (11) and the replacement of u_i by $-u_i$.

The reduction of this general problem into specific models has been given in Table I. For some of these special cases, a closed form of the free energy can be obtained. We classify these soluble cases into the following categories.

a. The ice models.[3–6] Included in this category are the models with the ice-condition constraint $\omega(7)=\omega(8)=0$ or equivalently $e_7=e_8=\infty$. These include the potassium dihydrogen phosphate KDP model of ferroelectrics and the F model of antiferroelectrics considered by Lieb.[3] The exact solution of the ice models can be obtained by a method which explicitly uses the fact that the vertices (7) and (8) are excluded.[13] The readers are referred to Refs. 3 and 4 for detailed discussions of these solutions. The specific heat exhibits in general a $|T-T_c|^{-1/2}$ singularity in the ice models.[14]

b. The conjugate models. Included in this category are the models specified by (8) but with the further constraint

$$u_1u_2=u_3u_4\,. \tag{13}$$

It has been shown[15] that this model is equivalent to an Ising model of a rectangular lattice with the interactions $J=\frac{1}{2}(e_1-e_3)$ and $J'=\frac{1}{2}(e_2-e_3)$. Hence, the specific heat exhibits a $\ln|T-T_c|$ singularity. The free energy has the following closed expression[12]:

$$\begin{aligned}\beta f=-\frac{1}{8\pi^2}\int_0^{2\pi}d\theta\int_0^{2\pi}d\phi\, \ln[(u_1+u_2)^2+(u_3-u_4)^2&\\+2(u_1-u_3)(u_2-u_4)\cos(\theta-\phi)&\\+2(u_1-u_4)(u_2-u_3)\cos(\theta+\phi)]\,.&\end{aligned} \tag{14}$$

This expression can also be obtained by using (12) to convert the model into the free-fermion model considered in the following. An interesting application is the modified F model of antiferroelectrics considered by one of us.[7]

c. The free-fermion models. Included in this category are the models satisfying the relation[16]

$$\omega(1)\omega(2)+\omega(3)\omega(4)=\omega(5)\omega(6)+\omega(7)\omega(8), \tag{15}$$

which we refer to as the free-fermion condition. This category includes the various planar Ising

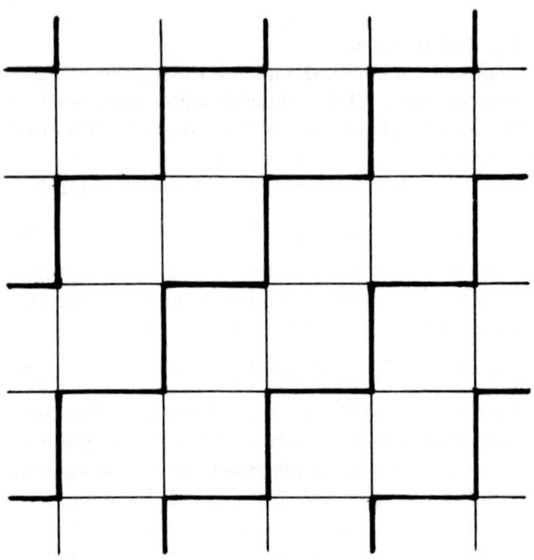

FIG. 2. The zig-zag paths along which the bonds and holes are interchanged to obtain the symmetry relation (11).

models and the modified KDP model of ferroelectrics.[5,6] A closed expression for the free energy, which is valid when (15) holds, has been obtained in Ref. 9 using a method similar to the S-matrix formulation for the many-body problem. It was seen there that the imposition of the condition (15) is equivalent to the consideration of a system of noninteracting fermions. For completeness, we now give below the closed expression for the free energy f and include in the Appendix an alternative derivation using the method of dimers:

$$\beta f = -\frac{1}{8\pi^2}\int_0^{2\pi} d\theta \int_0^{2\pi} d\phi \ln[2a + 2b\cos\theta + 2c\cos\phi + 2d\cos(\theta - \phi) + 2e\cos(\theta + \phi)], \tag{16}$$

where

$$\begin{aligned} 2a &= [\omega(1)]^2 + [\omega(2)]^2 + [\omega(3)]^2 + [\omega(4)]^2, \\ b &= \omega(1)\omega(3) - \omega(2)\omega(4), \\ c &= \omega(1)\omega(4) - \omega(2)\omega(3), \\ d &= \omega(3)\omega(4) - \omega(7)\omega(8), \\ e &= \omega(3)\omega(4) - \omega(5)\omega(6). \end{aligned} \tag{17}$$

It is to be noted that the free-fermion condition (15) remains as an identity when we replace e_i by $\bar{e}_i$ as given by (4). Hence, it is possible to discuss the properties of the free-fermion model under an external electric field. As seen from Table I, special cases of the free-fermion model include the various planar Ising models as well as the modified KDP model whose thermodynamic properties are well known. Besides these special cases, however, discussions of the thermodynamic properties of the free-fermion models in general have not been given.[17] Such a study is especially useful in view of the different critical behaviors exhibited by the Ising and the modified KDP models. For example, it would be illuminating to see how the change of critical behavior comes about as the vertex energies are varied. These discussions will be the subject of Sec. IV.

IV. FREE-FERMION MODEL

In this section we discuss the thermodynamic properties of the free-fermion model defined by (15). For the free-fermion condition (15) to hold at all temperatures, we must have the identities

$$\{e_1 + e_2,\ e_3 + e_4\} = \{e_5 + e_6,\ e_7 + e_8\}. \tag{18}$$

Because of the symmetry relation (6), we may take, without loss of generality,

$$e_1 + e_2 = e_7 + e_8, \quad e_3 + e_4 = e_5 + e_6. \tag{19}$$

Then, the free energy is given only in terms of ω_1, ω_2, ω_3, and ω_4. Writing $\omega_i \equiv \omega(i)$, we have explicitly

$$\begin{aligned} a &= \tfrac{1}{2}(\omega_1^2 + \omega_2^2 + \omega_3^2 + \omega_4^2), \quad b = \omega_1\omega_3 - \omega_2\omega_4, \\ c &= \omega_1\omega_4 - \omega_2\omega_3, \quad d = \omega_3\omega_4 - \omega_1\omega_2, \quad e = 0. \end{aligned} \tag{20}$$

One of the two integrations in (16) can be performed. Here, we use the integration formula

$$\int_0^{2\pi} d\theta \ln(2A + 2B\cos\theta + 2C\sin\theta) = 2\pi \ln[A + (A^2 - B^2 - C^2)^{1/2}], \tag{21}$$

where for our purposes

$$A = a + c\cos\phi, \quad B = b + d\cos\phi, \quad C = -d\sin\phi. \tag{22}$$

Then, expressions (16) and (21) lead to the following expression:

$$\beta f = -\frac{1}{4\pi}\int_0^{2\pi} d\phi \ln\{A + [Q(\phi)]^{1/2}\}. \tag{23}$$

Here

$$\begin{aligned} Q(\phi) &= A^2 - B^2 - C^2 = y^2 + z^2 - x^2 - 2yz\cos\phi + x^2\cos^2\phi \\ &= x^2(\cos\phi - yz/x^2)^2 + (1/x^2)(x^2 - y^2)(z^2 - x^2), \end{aligned} \tag{24}$$

and

$$x = \omega_2\omega_3 - \omega_1\omega_4, \quad y = \tfrac{1}{2}(\omega_1^2 - \omega_2^2 - \omega_3^2 + \omega_4^2), \quad z = \omega_2\omega_3 + \omega_1\omega_4. \tag{25}$$

Clearly, in the discussion of the analytic properties of f, it is important to consider whether $Q(\phi)$ is a complete square. We have the following cases to consider.

Case (1): *$Q(\phi)$ is a complete square.* These are two possibilities.

(a) $x^2 = z^2$: This is equivalent to the condition $\omega_1\omega_2\omega_3\omega_4 = 0$. Without loss of generality, we may take $\omega_1 = 0$. Then, the free-fermion condition (19) also implies $\omega_7\omega_8 = 0$ and the problem reduces to a special case of the ice models, i.e., the modified KDP model considered in Ref. 6. More explicitly, (23) becomes

$$\beta f = -\frac{1}{4\pi}\int_0^{2\pi} d\phi \ln\tfrac{1}{2}(\omega_2^2 + \omega_3^2 + \omega_4^2 - 2\omega_2\omega_3\cos\phi + |\omega_2^2 + \omega_3^2 - \omega_4^2 - 2\omega_2\omega_3\cos\phi|), \tag{26}$$

which is precisely Eq. (7) of Ref. 6. The important thermodynamic properties of this model are summarized as follows. A second-order phase transition occurs at a temperature T_c determined by

$$\omega_2 + \omega_3 + \omega_4 = 2\max\{\omega_2, \omega_3, \omega_4\}. \tag{27}$$

The specific heat c vanishes for $T < T_c$ and behaves as $(T - T_c)^{-1/2}$ near and above T_c. We note that we have obtained here the $(T - T_c)^{-1/2}$ singularity of the ice models.

(b) $x^2=y^2$: This is equivalent to the condition

$$(\omega_1+\omega_2-\omega_3-\omega_4)(\omega_1-\omega_2-\omega_3+\omega_4)\times(\omega_1-\omega_2+\omega_3-\omega_4)=0. \quad (28)$$

We then have two alternatives if (28) is to hold at all temperatures. (i) $\{e_1, e_2\}=\{e_3, e_4\}$. It is easy to see that (23) reduces to

$$\beta f=-\tfrac{1}{2}\ln(2(\omega_1^2+\omega_2^2)). \quad (29)$$

(ii) $\{e_1, e_3\}=\{e_2, e_4\}$. Again we have the simple expression

$$\beta f=-\tfrac{1}{2}\ln(2(\omega_1^2+\omega_3^2+\omega_1\omega_3)). \quad (30)$$

In either case, the model exhibits no phase transition, a fact which can also be seen from the fact that the ground state is macroscopically degenerate.

Case (2): $Q(\phi)$ *is not a complete square*. One observes from (23) that f is analytic in T except at the points given by $Q(\phi)=0$, where some derivatives of f diverge. Now we find $Q(\phi)\geq 0$ by (24). Therefore, the critical condition is obtained by setting the absolute minimum of $Q(\phi)$ equal to zero. Since

$$Q'(\phi)=2x^2\sin\phi\,(yz/x^2-\cos\phi), \quad (31)$$

there are three possible extreme values for $Q(\phi)$,

$$Q(0)=(y-z)^2,\quad Q(\pi)=(y+z)^2,$$
$$Q(\phi_0)=(1/x^2)(x^2-y^2)(z^2-x^2)>0, \quad (32)$$

where $\cos\phi_0=yz/x^2$ for $|yz|\leq x^2$, and $Q(\phi_0)$ is positive because $z^2\geq x^2$, and hence $x^2\geq y^2$. Therefore, the only possibilities are to set $Q(0)$ or $Q(\pi)$ equal to zero. This is equivalent to $z^2=y^2$ and hence to the critical condition

$$(-\omega_1+\omega_2+\omega_3+\omega_4)(\omega_1-\omega_2+\omega_3+\omega_4)(\omega_1+\omega_2-\omega_3+\omega_4)(\omega_1+\omega_2+\omega_3-\omega_4)=0. \quad (33)$$

Equivalently, (33) can be rewritten as

$$\omega_1+\omega_2+\omega_3+\omega_4=2\max\{\omega_1, \omega_2, \omega_3, \omega_4\}. \quad (34)$$

It is now straightforward, although tedious, to compute the energy E through direct differentiation of (23): We find

$$E=\frac{\partial}{\partial\beta}(\beta f)=-\frac{1}{4\pi}\int_0^{2\pi}d\phi\left[\frac{BB'+CC'}{B^2+C^2}+\left(A'-A\frac{BB'+CC'}{B^2+C^2}\right)[Q(\phi)]^{-1/2}\right], \quad (35)$$

where the primes denote derivatives with respect to β. We now introduce (22) and after some reduction we arrive at the expression

$$E=C_0+C_1I_1+C_2I_2+C_3I_3, \quad (36)$$

where

$$C_0=-\frac{1}{4\pi}\int_0^{2\pi}d\phi\left(\frac{BB'+CC'}{B^2+C^2}\right)$$
$$=-\frac{b'}{2b}\quad \text{if } b^2>d^2 \quad\text{or } \operatorname{sgn}(e_1-e_4)=\operatorname{sgn}(e_3-e_2),$$
$$=-\frac{d'}{2d}\quad \text{if } b^2<d^2 \quad\text{or } \operatorname{sgn}(e_1-e_4)=\operatorname{sgn}(e_2-e_3),$$
$$C_1=-a'+\frac{a}{2}\left(\frac{b'}{b}+\frac{d'}{d}\right)+\frac{c}{4bd}\left(\frac{b'}{b}-\frac{d'}{d}\right)(b^2-d^2), \quad (37)$$
$$C_2=-c'+\frac{c}{2}\left(\frac{b'}{b}+\frac{d'}{d}\right),$$
$$C_3=\frac{1}{2}\left(\frac{b'}{b}-\frac{d'}{d}\right)(b^2-d^2)\left(\frac{a}{b^2+d^2}-\frac{c}{2bd}\right),$$

and I_1, I_2, and I_3 are definite integrals given by

$$I_1=(4\pi)^{-1}\int_0^{2\pi}[Q(\phi)]^{-1/2}d\phi,$$
$$I_2=(4\pi)^{-1}\int_0^{2\pi}\cos\phi[Q(\phi)]^{-1/2}d\phi, \quad (38)$$
$$I_3=(4\pi)^{-1}\int_0^{2\pi}(1+\omega\cos\phi)^{-1}[Q(\phi)]^{-1/2}d\phi,$$
$$\omega=2bd/(b^2+d^2).$$

In all these formulas, $Q(\phi)$ is given by (24) and a, b, c, d given by (20). The integrals I_1, I_2, and I_3 can be expressed in terms of the complete elliptical integrals of the first and the third kinds. In evaluating these integrals, it is sufficient to consider positive x, y, and z only. For with negative x, y, or z, we may use the simple relations

$$I_1(x, y, z)=I_1(\pm x, \pm y, \pm z),$$
$$I_2(x, -y, z)=-I_2(x, y, z),$$
$$I_3(x, -y, z, \omega)=I_3(x, y, z, -\omega),\ \text{etc.}$$

For positive x, y, z, the following definitions of a new variable α are useful:

$$z>x>y>0,\quad \cos\alpha=(z\cos\phi-y)/(z-y\cos\phi),$$
$$z>y>x>0,\quad \sin\alpha=(z\cos\phi-y)/(z-y\cos\phi), \quad (39)$$
$$y>z>x>0,\quad \sin\alpha=(y\cos\phi-z)/(y-z\cos\phi).$$

It is then elementary to find

(i) $z>x>y>0$:

$$I_1=\frac{1}{\pi(z^2-y^2)^{1/2}}K(k),$$
$$I_2=\frac{z}{\pi y(z^2-y^2)^{1/2}}[K(k)-\Pi(n, k,)],$$

$$I_3 = \frac{1}{\pi(y+\omega z)(z^2-y^2)^{1/2}} \times\left(yK(k) + \frac{\omega(z+\omega y)}{1-\omega^2}\,\Pi(m,k)\right),$$

where
$$k^2 = (x^2-y^2)/(z^2-y^2),\quad n = y^2/(z^2-y^2),\quad m = (\omega y+z)^2/(1-\omega^2)(z^2-y^2). \tag{40}$$

(ii) $z>y>x>0$:

$$I_1 = \frac{1}{\pi(z^2-x^2)}K(g),$$

$$I_2 = \frac{1}{\pi y(z^2-x^2)^{1/2}}\,[z^2K(g) + (y^2-z^2)\Pi(r,k)],$$

$$I_3 = \frac{1}{\pi(y+\omega z)(z^2-x^2)^{1/2}}\left(gK(g) + \frac{\omega(z^2-y^2)}{z+\omega y}\,\Pi(s,k)\right).$$

where
$$q^2 = (y^2-x^2)/(z^2-x^2),\quad r = -y^2/z^2,\quad s = -(\omega z+y)^2/(z+\omega y)^2. \tag{41}$$

(iii) $y>z>x>0$: Same as case (ii) only with y and z interchanged. In these expressions, K and Π are, respectively, the complete elliptical integrals of the first and the third kinds defined by

$$K(k) = \int_0^{\pi/2}(1-k^2\sin^2\alpha)^{-1/2}\,d\alpha,\qquad \Pi(n,k) = \int_0^{\pi/2}(1+n\sin^2\alpha)^{-1}(1-k^2\sin^2\alpha)^{-1/2}\,d\alpha. \tag{42}$$

The phase transition occurs at $|y|=|z|$ or $q=1$, $r=-1$. While K and Π diverge as $\ln|T-T_c|$ when $q\to 1$, it can be seen from (36) and (37) that

$$E \to C_0 + |T-T_c|\ln|T-T_c|.$$

Hence, E is continuous at T_c and the specific heat diverges logarithmically near T_c.

A special feature is that the energy E has different expressions for $|x|>|y|$ and $|x|<|y|$, although $|x|=|y|$ is not a singular point. Since for the quartic Ising model only one expression is needed for E, it is of interest to investigate the condition under which $|x|=|y|$ may occur. To obtain such a criterion, let us arrange the energy values e_1, e_2, e_3, and e_4 according to their magnitudes such that

$$e_\alpha \ge e_\beta \ge e_\gamma \ge e_\delta .$$

Then, $x^2=y^2$ corresponds to

$$\omega_\alpha + \omega_\delta = \omega_\beta + \omega_\gamma, \tag{43}$$

where $\omega_\alpha = \exp(-e_\alpha/kT)$, etc. Equation (43) can be rewritten as

$$\exp\left\{\frac{1}{2kT}[(e_\beta+e_\gamma)-(e_\alpha+e_\delta)]\right\} = \cosh\left(\frac{e_\alpha-e_\delta}{2kT}\right)\Big/\cosh\left(\frac{e_\beta-e_\gamma}{2kT}\right) > 1.$$

Hence, (43) has a solution, and thus it is possible to have $x^2=y^2$ only for

$$e_\beta + e_\gamma < e_\alpha + e_\delta .$$

It is not difficult to show that, in fact, $|x|>|y|$ at all temperatures if $e_\beta+e_\gamma \ge e_\alpha+e_\delta$. An example is the quartic Ising model for which the equality holds (cf. Table I), and the energy is given by the same single expression both above and below the transition temperature T_c. For a triangular Ising lattice with the corresponding vertex energies given in Table I, it is possible to satisfy (43) whenever the interactions are antiferromagnetic. The temperature thus determined is the disorder temperature[18] at which the spin correlations undergo a remarkable change.[19]

Finally, we consider the inclusion of an external field $\vec{\mathcal{E}}$. Earlier, we have shown that the inclusion of a field amounts simply to redefining the vertex energies. Since the new vertex energies (4) also happen to satisfy the free-fermion condition, we have already solved the problem. In fact, all analyses of this section again go through provided that we use the new vertex energies in all formulas. In particular, the transition temperature T_c now depends on the external field and is determined by

$$\omega_1HV + \omega_2(HV)^{-1} + \omega_3HV^{-1} + \omega_4H^{-1}V = 2\max\{\omega_1HV, \omega_2(HV)^{-1}, \omega_3HV^{-1}, \omega_4H^{-1}V\}, \tag{44}$$

where $H \equiv e^{\beta h}$, $V \equiv e^{\beta v}$.

The polarizations $\vec{P} = (P_x, P_y)$ for the free-fermion model can be obtained from (5) and (23). The expression for P_x is given below in (45) and (46), while the expression for P_y is the same except for the interchange of ω_3 and ω_4:

$$P_x = \frac{2}{\pi}\frac{z^2-y^2}{y(z^2-x^2)^{1/2}}[\Pi(n,k)-K(k)],\qquad T>T_c$$
$$P_x = \frac{2}{\pi}\frac{y^2-z^2}{y(y^2-x^2)^{1/2}}\,\Pi(n,k),\qquad T<T_c. \tag{45}$$

Near T_c, we have asymptotically

$$P_x = \frac{z}{x}\left(1-\frac{2}{\pi}\cos^{-1}\frac{x}{z} + \frac{x(z^2-y^2)}{2\pi z^2(z^2-x^2)^{1/2}}\right) + \frac{y^2-z^2}{\pi z(z^2-x^2)^{1/2}}\ln\frac{4}{k'} + O(k'^4),\qquad T>T_c$$
$$P_x = \frac{z}{x}\left(1-\frac{2}{\pi}\cos^{-1}\frac{x}{y} + \frac{y^2-z^2}{2\pi y^2(y^2-x^2)^{1/2}}\right) \tag{46}$$

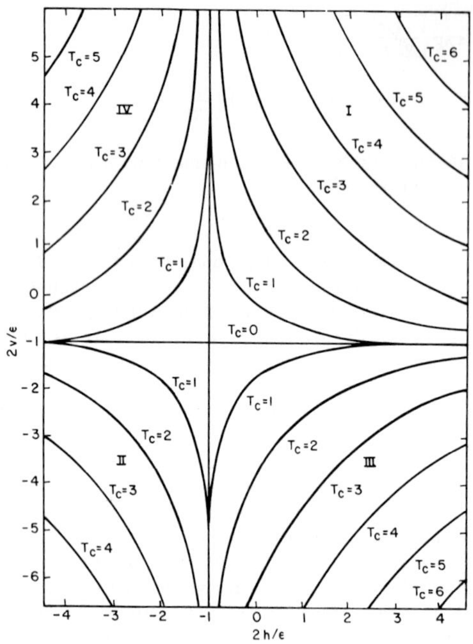

FIG. 3. Constant T_c contours in the hv plane. The transition temperature T_c is measured in units ϵ/k.

$$+\frac{z^2-y^2}{\pi y(y^2-x^2)^{1/2}}\ln\frac{4}{k'}+O(k'^4)\,,\qquad T<T_c$$

where $k'=(1-k^2)^{1/2}$. The polarization is nonvanishing and continuous at all temperatures. It behaves near T_c as $C_0+C_1(T-T_c)\ln(T-T_c)$. Therefore, the polarizability $\partial\vec{P}/\partial\vec{\epsilon}$ exhibits a logarithmic singularity. Thus, we see that in the ferroelectric version of the free-fermion model, the phase transition is not marked by the appearance or vanishing of the order-parameter polarization. It is instead associated with the logarithmic divergence in the specific heat and the polarizability. We also note that none of these thermodynamic functions has a simple power-law dependence on $|T-T_c|$ near T_c.

V. NUMERICAL EXAMPLE

To illustrate the above results, we shall now consider a specific example with vertex energies

$$e_1=0,\qquad e_2=2\epsilon>0\,,$$
$$e_3=e_4=e_5=e_6=e_7=e_8=\epsilon\,. \tag{47}$$

As seen from Table I, this specifies precisely the regular Ising model with an interaction parameter $J_1=J_2=\frac{1}{2}\epsilon$. A plot of the energy and the specific heat of this model can be found in Ref. 7 and will not be reproduced here.[20] The transition temperature is given by $kT_c/\epsilon=\ln(\sqrt{2}+1)\equiv 1.13459\ldots$. We shall now consider the ferroelectric version of the model by imposing an external electric field. The transition temperature T_c is now given by

$$HV+(HV)^{-1}u^2+uHV^{-1}+uH^{-1}V$$
$$=2\max\{HV,(HV)^{-1}u^2,uHV^{-1},uH^{-1}V\}, \tag{48}$$

where $u\equiv e^{-\beta\epsilon}$.

The entire hv plane is then divided into four regions (see Fig. 3) depending on which vertex 1, 2, 3, or 4 has the lowest energy (or the largest weight). We have plotted in Fig. 3 the constant T_c contours in the hv plane. The four regions denoted by I, II, III, and IV are, respectively, the regions where the vertex 1, 2, 3, or 4 has lowest energy. Relating the vertex energies – including the external field (h, v) – to the interactions J_1 and J_2 of a rectangular Ising lattice (see Table I), it can be seen that our model is equivalent to rectangular Ising lattice with $J_1=\frac{1}{2}\epsilon+v$, $J_2=\frac{1}{2}\epsilon+h$. Hence (48) is equivalent to the well-known relation

$$\left|\sinh\frac{\epsilon+2v}{2kT_c}\,\sinh\frac{\epsilon+2h}{2kT_c}\right|=1. \tag{49}$$

For $v=0$, T_c as a function of h is plotted in Fig. 4. Finally, we plot in Fig. 5 the polarization P_x as a function of T.

VI. CONCLUSIONS

We have defined a general lattice-statistical problem and studied in detail the soluble situations. It is found that the specific heat behaves as $|T-T_c|^{-1/2}$ near T_c for the ice models and possesses the $\ln|T-T_c|$ singularity for all other soluble cases. From these results and the study of related models,[7,8] it appears that the logarithmic singularity is perhaps commonplace except in the ice models, which happen to exhibit a $|T-T_c|^{-1/2}$ singularity in the specific heat.

ACKNOWLEDGMENTS

A portion of the work reported here was carried out while both of us were at the Institute for Theoretical Physics, State University of New York at Stony Brook. We wish to thank Professor C. N. Yang for encouragement. One of us (C. F.) is also grateful to Dr. B. Sutherland and Professor J. Groeneveld for enlightening discussions. It is our pleasure to thank Dr. J. Stephenson for several useful comments on a first draft of the manuscript and for sending us preprints of his work prior to publication.

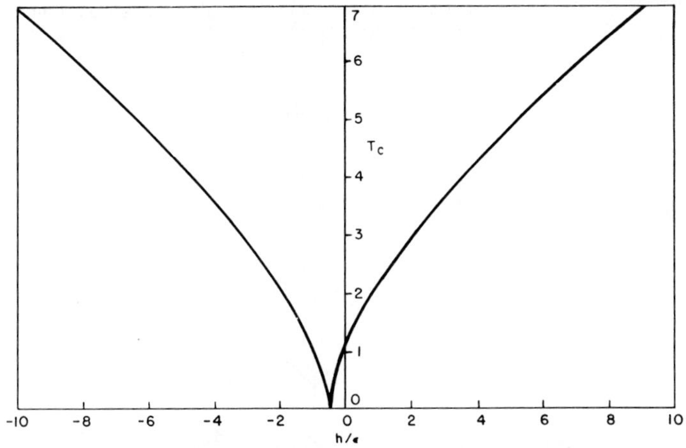

FIG. 4. The transition temperature T_c as a function of h with $v = 0$. T_c is in units ϵ/k.

APPENDIX

In this Appendix, we derive the partition function (16) for the free-fermion models using the method of dimers.[21] First, we construct a dimer lattice by expanding each vertex of the square lattice into a city of internally connected points. Provided that the structure of the city and the edge weights are chosen properly, the partition function is identical to the generating function for closely packed dimer configurations on this expanded lattice. It turns out that this trick can be accomplished by choosing the *planar* dimer city of Fig. 6 with the weights shown in the graph.[22] In Fig. 7, we list all possible dimer configurations for a dimer city. It is then easy to see that the correct vertex weights are indeed generated. For example, the first row of Fig. 7 yields

$$\frac{\omega_7}{\omega_2}\times\omega_2\times\frac{\omega_8}{\omega_2}+\frac{\omega_6-\omega_4}{\omega_2}\times\omega_2\times\frac{\omega_5-\omega_3}{\omega_2}+\frac{\omega_6-\omega_4}{\omega_2}\times\omega_3\times 1$$
$$+\frac{\omega_5-\omega_3}{\omega_2}\times\omega_4\times 1 = \frac{1}{\omega_2}\,(\omega_5\omega_6+\omega_7\omega_8-\omega_3\omega_4)=\omega_1\,.$$

The last equality follows from the free-fermion condition (15). The other vertex weights can be generated accordingly.

The next step is to direct the edges of the dimer lattice so that every closed polygon drawn on the expanded dimer lattice containing an even number of edges and enclosing an even number of points has odd numbers of arrows pointing in each direc-

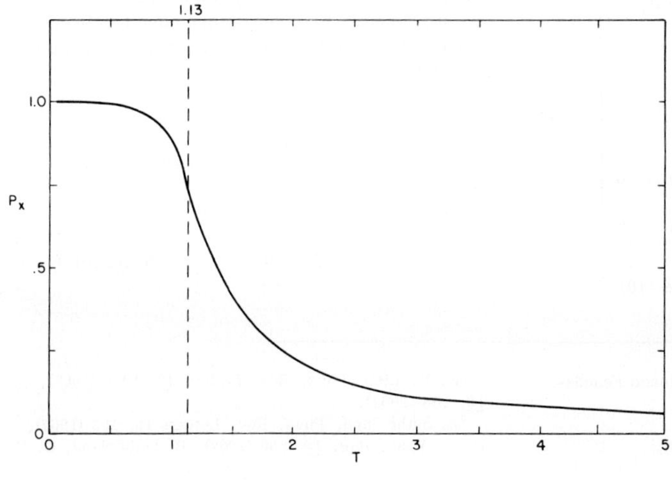

FIG. 5. Horizontal polarization as a function of temperature.

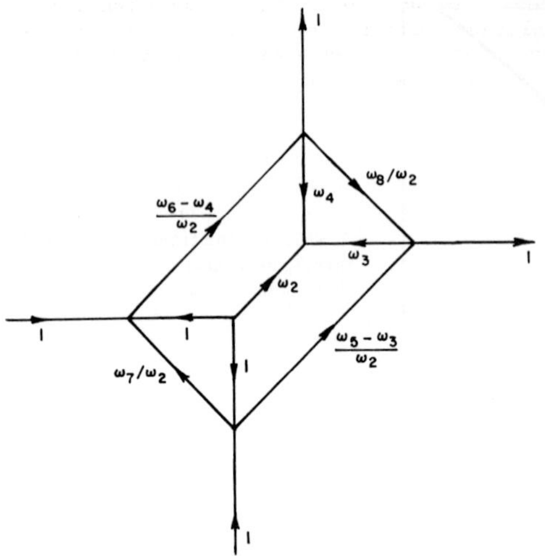

FIG. 6. The dimer city and the edge weights.

tion. This orientation can be accomplished as shown in Fig. 6. The dimer generating function, which is equal to the partition function, can now be written as a Pfaffian and evaluated accordingly.[21] For a periodic lattice, the result obtained is

$$\beta f = -\lim_{N\to\infty} \frac{1}{N} \ln Z = \frac{-1}{8\pi^2}\int_0^{2\pi} d\theta \int_0^{2\pi} d\phi \ln D, \tag{A1}$$

where D is the determinant given by

$$D = \begin{vmatrix} 0 & \frac{\omega_8}{\omega_2} & 0 & \frac{\omega_4-\omega_6}{\omega_2} & e^{i\theta} & \omega_4 \\ -\frac{\omega_8}{\omega_2} & 0 & 0 & e^{i\phi} & \frac{\omega_3-\omega_5}{\omega_2} & \omega_3 \\ 0 & 0 & 0 & 1 & 1 & \omega_2 \\ \frac{\omega_6-\omega_4}{\omega_2} & -e^{-i\phi} & -1 & 0 & -\frac{\omega_7}{\omega_2} & 0 \\ -e^{-i\theta} & \frac{\omega_5-\omega_3}{\omega_2} & -1 & \frac{\omega_7}{\omega_2} & 0 & 0 \\ -\omega_4 & -\omega_3 & -\omega_2 & 0 & 0 & 0 \end{vmatrix}. \tag{A2}$$

Substitution of (A2) into (A1) now yields (16).[23]

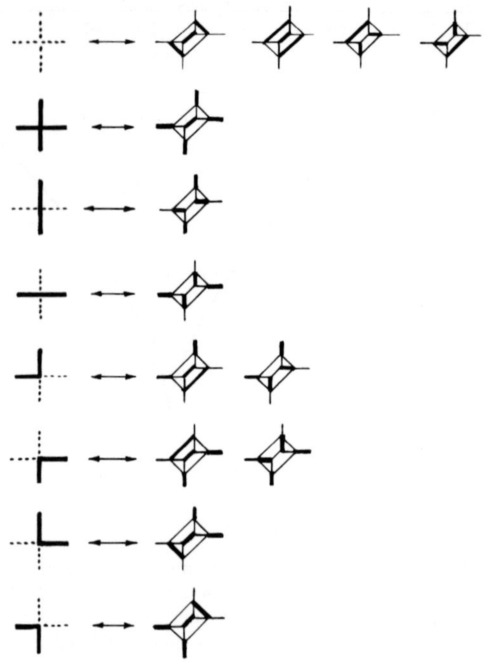

FIG. 7. The correspondences between the bond configurations and the dimer configurations.

Note added in proof. We now realize that case (2) of the free-fermion model discussed in Sec. IV is completely equivalent to the Ising model on an anistotropic triangular lattice. This makes our derivation of the logarithmic singularity in the specific heat obsolete. The formulas (31)–(40) are still useful, however, since none has been given in detail in the literature for the triangular Ising lattice.

*Work supported in part by National Science Foundation Grant No. GP-9041.

[1]E. Ising, Z. Physik 31, 253 (1925).

[2]L. Onsager, Phys. Rev. 65, 117 (1944).

[3]E. H. Lieb, Phys. Rev. Letters 18, 1046 (1967); 19, 108 (1967).

[4]B. Sutherland, Phys. Rev. Letters 19, 103 (1967); C. P. Yang, *ibid.* 19, 586 (1967); B. Sutherland, C. N.

Yang, and C. P. Yang, *ibid.* 19, 588 (1967).

[5]F. Y. Wu, Phys. Rev. Letters 18, 605 (1967).

[6]F. Y. Wu, Phys. Rev. 168, 539 (1968).

[7]F. Y. Wu, Phys. Rev. 183, 604 (1969).

[8]F. Y. Wu, Phys. Rev. Letters 22, 1174 (1969).

[9]C. Fan and F. Y. Wu, Phys. Rev. 179, 560 (1969).

[10]The symmetry relations (7a) have been obtained in Ref. 9.

[11]J. F. Nagle, J. Math. Phys. 9, 1007 (1968).

[12]Discussion in Ref. 7 was specialized to the case $u_3 = 1$, $u_4 = u_1 u_2$. The derivation of (12) and (14) presents no problem if one follows the procedures of Ref. 7 and uses the notations u_3 and u_4 in places of 1 and $u_1 u_2$.

[13]E. H. Lieb, Phys. Rev. Letters 18, 692 (1967).

[14]Except the F model for which the specific heat is continuous at T_c. But this has been identified in Ref. 8 as a limiting situation of the more familiar λ transitions.

[15]See the Appendix of Ref. 7.

[16]In the discussion of the general planar Ising models, Hurst and Green [H. S. Green and C. A. Hurst, in *Order-Disorder Phenomena*, edited by I. Preigoine (Interscience, New York, 1964), Sec. 5.3] have considered the same problem from a somewhat different point of view. They considered the simple quartic Ising lattice and introduced at each vertex point a sublattice to generate what is equivalent to our vertex weights. It can be shown that the free-fermion condition (15) is always an identity in their considerations provided that the sublattice introduced at each vertex is planar. The condition (15) therefore also reflects the solubility of planar Ising lattices.

[17]Green and Hurst (Ref. 15) have evaluated some derivatives of the partition function (16). Our discussions are simpler because of the special form of the vertex weights (2) and the use of relation (19).

[18]J. Stephenson, Can J. Phys. 47, 2621 (1969).

[19]J. Stephenson, J. Math. Phys. 11, 420 (1970).

[20]That the present model is identical to the modified F model (MF) is proved in the Appendix of Ref. 7.

[21]See, for example, E. W. Montroll, in *Applied Combinatorial Mathematics*, edited by E. F. Beckenbach (Wiley, New York, 1964), Chap. 4.

[22]The usefulness of this dimer city has also been observed by M. E. Fisher (private communication).

[23]Except with the replacement of b, c by $-b$, $-c$ in (16). But these are equivalent expressions.

PHYSICAL REVIEW B VOLUME 12, NUMBER 1 1 JULY 1975

Staggered ice-rule vertex model—The Pfaffian solution

F. Y. Wu*
Department of Physics, Northeastern University, Boston, Massachusetts 02115

K. Y. Lin†
Department of Physics, National Tsing Hua University, Hsin Chu, Taiwan, Republic of China

(Received 16 December 1974)

It is pointed out that the staggered ice-rule model contains a number of outstanding lattice statistical problems including the Ising model in a nonzero magnetic field. The most general Pfaffian solution of this staggered vertex model is studied in this paper. In special cases our solution reduces to that of two recently considered dimer models of phase transitions. It also leads to the exact solution of several vertex models as well as an exact isotherm of a general antiferroelectric model, all in the presence of both direct and staggered fields.

I. INTRODUCTION

An important recent advance in the mathematical theory of phase transitions is the solution of the ice-rule vertex models and its subsequent developments.[1] Attention in the past has been focused on models with translationally invariant vertex weights. Critical behavior of such models was found to be rather unique; there also appears to exist no apparent relationship between the ice-rule models and other problems in lattice statistics.

The situation is quite different if the vertex weights are allowed to vary from site to site. The simplest kind of variation is to allow different vertex weights for the two sublattices of a square lattice. This defines a staggered vertex model and it has been known to be related to a number of outstanding problems in lattice statistics, which include the percolation problem,[2] the Potts model,[3] and the Ashkin-Teller model.[4,5] We shall also see that the staggered ice-rule model is reducible to the Ising model in a nonzero magnetic field and some recently considered dimer models of phase transitions.[6,7] As such discussions are scattered in the literature, it is useful to collect these results for emphasis. In the absence of a general solution, it also merits to explore fully the cases soluble using the existing methods. It is the goal of the present paper to do this for the staggered ice-rule models. The study of the staggered eight-vertex model will be given subsequently.[8]

II. DEFINITION OF THE MODEL

Place arrows on the lattice edges of a square lattice L of N sites subject to the constraint that there are always two arrows in and two arrows out at each site (the ice rule). Thus only the six configurations shown in Fig. 1 are allowed at each vertex. Each vertex type is assigned a weight according to Fig. 1, or briefly

$$\begin{aligned}\{\omega\}&=\{\omega_1, \omega_2, \ldots, \omega_6\} \text{ on } A,\\ \{\omega'\}&=\{\omega_1', \omega_2', \ldots, \omega_6'\} \text{ on } B,\end{aligned} \tag{1}$$

where A and B are the two sublattices of L. The problem is to compute the generating function

$$Z=\sum\left(\prod \omega_i^{n_i}\right)\left(\prod \omega_i'^{n_i'}\right), \tag{2}$$

where the summation is extended to all allowed arrow configurations on L, and n_i (n_i') is the number of the ith-type sites on A (B). The quantities of interest are usually given in terms of the function

$$\psi=\lim_{N\to\infty}\frac{1}{N}\ln Z. \tag{3}$$

In a ferroelectric model, the vertex weights can be interpreted as the Boltzmann factors

$$\omega_i=e^{-\beta\epsilon_i}, \quad \omega_i'=e^{-\beta\epsilon_i'}, \tag{4}$$

where $\beta=1/kT$ and ϵ_i, ϵ_i' are the vertex energies. But this restriction is unnecessary and we shall consider $\{\omega\}$ and $\{\omega'\}$ general unless otherwise stated.

Very few exact results are known about this staggered ice-rule model. Besides the exact solution of two special cases, the modified KDP model[9] and an isotherm of the F model,[10] both in a staggered field, the only other available information is the spontaneous staggered polarization of the F model.[11]

Generally, there exist two methods of solution for vertex models. These are the method of the Bethe's *Ansatz*[1] and the method of Pfaffians.[12] It appears that the Bethe *Ansatz* method is not useful for models with staggered weights.[13] On the other hand, as already exampled by the two special solutions,[9,10] the validity of the Pfaffian method, while limited, can be extended to models with staggered fields. In Sec. IV, we shall obtain the most general Pfaffian solution for the staggered ice-rule model.

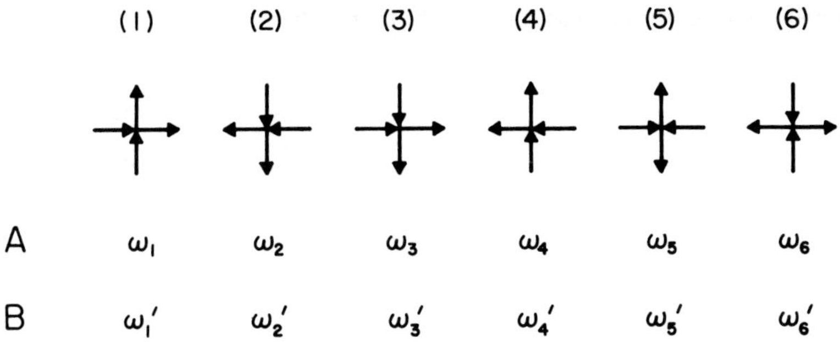

FIG. 1. Six ice-rule configurations and the associated vertex weights.

III. RELATIONSHIP WITH OTHER LATTICE STATISTICAL PROBLEMS

To visualize the relationship of the staggered ice-rule model with other lattice statistical problems, it is useful to introduce a square lattice L' so that L is the covering graph of L'.[14] The situation is shown in Fig. 2, where L' is represented by the dashed lines. Note that the edges of L' coincide with the sites of L, so L' has $\frac{1}{2}N$ sites.

(i) Ising model with a nonzero magnetic field

With

$$\{\omega\}=\{0,\ 0,\ e^{-\beta J_1},\ e^{-\beta J_1},\ e^{\beta(J_1-H/2)},\ e^{\beta(J_1+H/2)}\},$$
$$\{\omega'\}=\{e^{-\beta J_2},\ e^{-\beta J_2},\ 0,\ 0,\ e^{\beta(J_2+H/2)},\ e^{\beta(J_2-H/2)}\}, \tag{5}$$

the generating function (2) is equal to the partition function of an Ising model on L' with nearest-neighbor interactions $-J_1$, $-J_2$ and an external magnetic field H (assuming unit magnetic moment for each spin).

The simplest way to see this equivalence is to convert the ice configurations on L into bond graphs by placing a bond along each horizontal (vertical) arrow running from A to B (B to A).[15] From Eqs. (5) we see that vertices (1) and (2) cannot occur on A, and (3) and (4) not on B. The resulting bond graphs on L will then compose of unit squares which enclose sites of L'.[16] A typical bond graph is shown in Fig. 2. Let the spins of the Ising model on L' be $+1$ if and only if it is enclosed by a unit square on L. There is then a one-to-one correspondence between the spin configurations on L' and the *contributing* arrow configurations on L. Let the Ising interactions across an A (B) vertices of L be $-J_1$ ($-J_2$). It is now a simple matter to verify the equivalence of the vertex weights Eqs. (5) and the Ising Boltzman factors. This completes our proof.

(ii) The percolation problem

In an interesting paper, Temperley and Lieb[2] pointed out the connection between the bond percolation problem and a staggered ice-rule model. Let p_1 and p_2 be the respective occupation probabilities of the horizontal and vertical edges of L'. They showed that the mean number of components, c, and the mean number of circuits, s, are given by

$$c=\frac{1}{2}\left(\frac{\partial}{\partial z}\ln Z(p_1,\ p_2;\ 1,\ z)\right)_{z=1},$$
$$s=\frac{1}{2}\left(\frac{\partial}{\partial y}\ln Z(p_1,\ p_2;\ y,\ 1)\right)_{y=1}, \tag{6}$$

where $Z(p_1,\ p_2;\ y,\ z)$ is the generating function (2) with

$$\{\omega\}=\{1,\ 1,\ x,\ x,\ 1+xe^{\theta},\ 1+xe^{-\theta}\},$$
$$\{\omega'\}=\{1,\ 1,\ X,\ X,\ 1+Xe^{\theta},\ 1+Xe^{-\theta}\}. \tag{7}$$

Here

$$2\cosh\theta=yz,\quad x=p_1y/(1-p_1)z,\quad X=(1-p_2)z/p_2y. \tag{8}$$

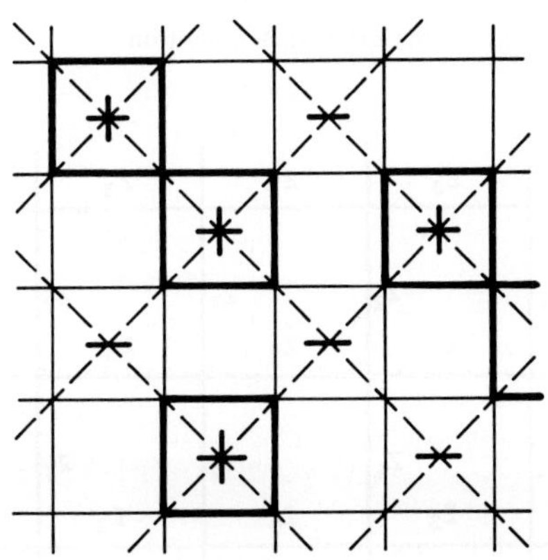

FIG. 2. Typical bond graph on L (solid lines) and the associated Ising configuration on L' (dashed lines). Ising spins are + if it is enclosed by a unit square.

(iii) Potts model

The q-component Potts model[17] has been around for many years, but no significant progress has been made toward its solution. It is therefore extremely useful and illuminating that it can be formulated as a staggered ice-rule model.[3]

Consider the Potts model defined on L'. Let the interaction energy be zero between unlike atoms and $-\epsilon_1$ $(-\epsilon_2)$ between two like atoms neighboring in the horizontal (vertical) direction. Its partition function is then related to the generating function $Z(x, X, \theta)$ defined by Eqs. (2) and (7) through

$$Z_{\mathrm{Potts}} = v_1^{N/2} Z(x, X, \theta), \tag{9}$$

with

$$2\cosh\theta = \sqrt{q}, \quad v_i = e^{\beta\epsilon_i} - 1, \quad x = v_1/\sqrt{q}, \quad X = \sqrt{q}/v_2. \tag{10}$$

Note that the number of components, q, can be treated as a *continuous* parameter in this formulation.

As a consequence of these equivalences, we can now derive the criticality conditions for the Potts model and the percolation problem on the basis of the Kramers-Wannier argument.[18]

From the obvious symmetry $Z(x, X, \theta) = Z(X, x, \theta)$ of the staggered model, we find from Eqs. (9) and (10) the following symmetry relation for the Potts model:

$$Z_{\mathrm{Potts}}(v_1, v_2) = (v_1 v_2/q)^{N/2} Z_{\mathrm{Potts}}(v_2/q, v_1/q). \tag{11}$$

This leads to the critical condition

$$v_1 v_2 = q \tag{12}$$

for the Potts model.[19] Similarly, the obvious symmetry relation $Z_{\mathrm{Potts}}(v_1, v_2) = Z_{\mathrm{Potts}}(v_2, v_1)$ of the Potts model implies

$$Z(x, X, \theta) = (xX)^{N/2} Z(1/X, 1/x, \theta). \tag{13}$$

This leads to the criticality condition

$$xX = 1 \tag{14}$$

which yields the critical probabilities[20]

$$p_1 + p_2 = 1 \tag{15}$$

for the bond percolation problem on a square lattice.

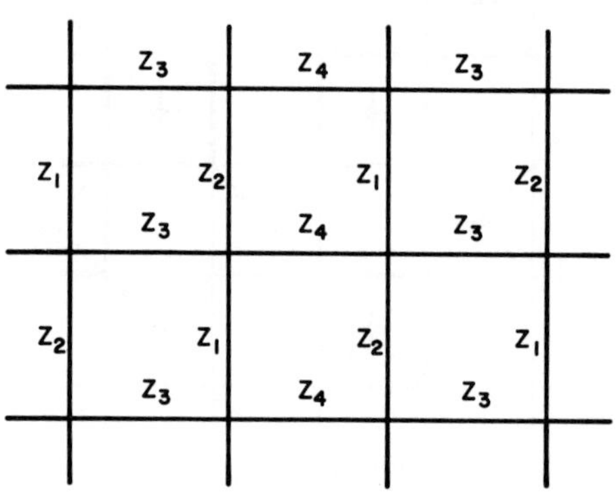

FIG. 3. Generalized dimer lattice of Ref. 6.

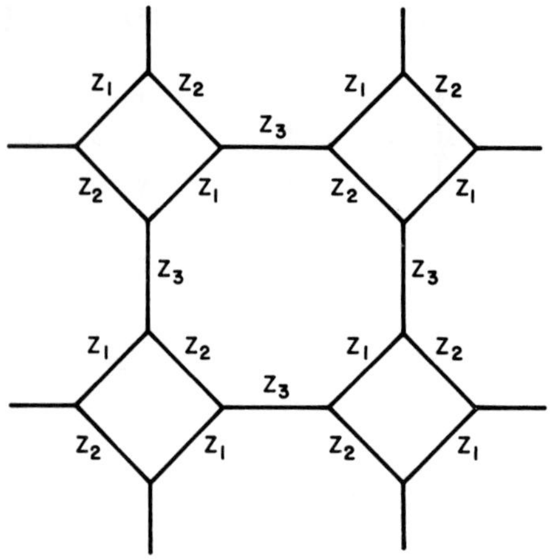

FIG. 4. Dimer model on a 4–8 lattice of Ref. 7.

(iv) General dimer lattice

In a recent study of dimer models for the phase transition in the antiferroelectric copper formate tetrahydrate, Allen[6] considered the generalized dimer lattice shown in Fig. 3. Using Baxter's method[21] of converting the dimer model into a vertex problem, it is seen that Allen's generalized dimer model is equivalent to a staggered ice-rule model with the following weights:

$$\begin{aligned} \{\omega\} &= \{z_3, z_3, z_2, z_1, 1, z_1 z_2 + z_3^2\}, \\ \{\omega'\} &= \{z_4, z_4, z_2, z_1, 1, z_1 z_2 + z_4^2\}. \end{aligned} \tag{16}$$

(v) Dimer lattice on a 4–8 lattice

Salinas and Nagle[7] have considered the dimer model on a 4–8 lattice for studying the phase transition in the layered hydrogen-bonded $SnCl_2 \cdot 2H_2O$ crystal. Their dimer model, shown in Fig. 4, is also equivalent to a staggered ice-rule model.

To see this equivalence, we observe that the squares sided by the z_1 and z_2 bonds form a square lattice L whose lattice edges are the z_3 bonds. If for each z_3 edge of L we draw an arrow from A to B (B to A) in the horizontal (vertical) direction, if the edge is covered by a z_3 dimer, otherwise an arrow from B to A (A to B), it is seen that we have established a one-to-one correspondence between the dimer configurations of the 4–8 lattice and the ice-rule configurations on L. Inspection then leads to the following staggered ice-rule vertex weights:

$$\{\omega\}=\{z_1z_3,\ z_1z_3,\ z_2z_3,\ z_2z_3,\ z_1^2+z_2^2,\ z_3^2\},$$
$$\{\omega'\}=\{z_1z_3,\ z_1z_3,\ z_2z_3,\ z_2z_3,\ z_3^2,\ z_1^2+z_2^2\}. \tag{17}$$

IV. PFAFFIAN SOLUTION

In this section, we derive and analyze the most general Pfaffian solution for the staggered ice-rule model. Generally a vertex model is soluble by the Pfaffian method if the vertex weights satisfy a local "free-fermion" condition.[22] For the staggered ice-rule model, there are two sets of weights, so the Pfaffian condition reads

$$\omega_1\omega_2+\omega_3\omega_4=\omega_5\omega_6,\quad \omega_1'\omega_2'+\omega_3'\omega_4'=\omega_5'\omega_6'. \tag{18}$$

The solution is based on the well-known fact that under the local free-fermion condition, the generating function (2) is reducible to a close-packed dimer generating function, which in turn is equal to a Pfaffian. Since the analysis is standard,[12] we leave the details in Appendix A and give here only the result:

$$\psi=\frac{1}{8\pi^2}\int_{-\pi}^{\pi}d\theta\int_{-\pi}^{\pi}d\phi\,\ln(a+be^{i\theta}+ce^{-i\theta}+fe^{i\phi}+ge^{-i\phi}), \tag{19}$$

where

$$a=\omega_5\omega_6'+\omega_5'\omega_6,\quad b=\omega_1\omega_1',\quad c=\omega_2\omega_2',$$
$$f=\omega_3\omega_3',\quad g=\omega_4\omega_4'. \tag{20}$$

It is surprising that although there are 10 independent vertex weights to begin with, the resulting expression of ψ contains only five independent parameters. We observe, in particular, that no generality is gained by taking $\omega_1\neq\omega_1'$, $\omega_2\neq\omega_2'$, $\omega_3\neq\omega_3'$, $\omega_4\neq\omega_4'$.[23] We shall also restrict a, b, c, f, g to non-negative.

In a given physical model, the parameters of physical interest such as the temperature, external fields, or activities enter through the variables a, b, c, f, g. To discuss the thermodynamics we shall therefore need the analytic behavior of $\psi(a, b, c, f, g)$. We note that the free-fermion conditions (18) imply the inequality (see Appendix B)

$$a\geq 2\sqrt{bc}+2\sqrt{fg}, \tag{21}$$

so discussion of the analytic properties of ψ is needed only in this region.

First of all, one of the two integrations in Eq. (19) can be performed. For this purpose, we need the following mathematical lemma:

Lemma. For complex A, B, C,

$$\int_{-\pi}^{\pi}d\theta\,\ln(Ae^{i\theta}+B+Ce^{-i\theta})=\begin{cases}2\pi\ln C, & \text{if } |z_1|,|z_2|\geq 1,\\ 2\pi\ln A, & \text{if } |z_1|,|z_2|\leq 1,\\ 2\pi\ln(-Az_1), & \text{if } |z_1|\geq 1\geq |z_2|,\end{cases}$$

where z_1, z_2 are the two roots of the quadratic equation

$$Az^2+Bz+C=0. \tag{22}$$

Proof of the lemma is immediate using the following result valid for complex α and β:

$$\int_{-\pi}^{\pi}d\theta\,\ln(\alpha e^{i\theta}+\beta)=\begin{cases}2\pi\ln\alpha, & \text{if } |\alpha|\geq|\beta|,\\ 2\pi\ln\beta, & \text{if } |\alpha|\leq|\beta|.\end{cases} \tag{23}$$

In order to apply the lemma to, say, the θ integration in Eq. (19), we need the sign of the function

$$s(\phi)=[1-|z_1(\phi)|\,][1-|z_2(\phi)|\,], \tag{24}$$

where $z_1(\phi)$, $z_2(\phi)$ are the roots of

$$bz^2+(a+fe^{i\phi}+ge^{-i\phi})z+c=0, \tag{25}$$

where ϕ is real. Explicitly,

$$z_1(\phi),\ z_2(\phi)=-(1/2b)[a+fe^{i\phi}+ge^{-i\phi}\ {}^{+}_{-}\Delta^{1/2}(\phi)], \tag{26}$$

with

$$\Delta(\phi)=(a+fe^{i\phi}+ge^{-i\phi})^2-4bc.$$

The branches of $\Delta^{1/2}(\phi)$ are taken so that $|z_1(\phi)|\geq|z_2(\phi)|$.

It is shown in Appendix B that $s(\phi)$ is symmetric about $\phi=0$, increasing monotonically in $\{0,\pi\}$ and decreasing monotonically in $\{-\pi,0\}$. Also

$$b[s(0)]=b+c-(a+f+g),$$
$$b[s(\pi)]=\begin{cases}b+c-|a-f-g|, & \text{if } \Delta(\pi)\geq 0\\ b+c-2\sqrt{bc}, & \text{if } \Delta(\pi)\leq 0.\end{cases} \tag{27}$$

Note that $b[s(\pi)]$ decreases from the maximum value $b+c-2\sqrt{bc}$ to 0 as $|a-f-g|$ increases from $2\sqrt{bc}$ to $b+c$. We then have the following cases to consider:

(i) $b+c>a+f+g$

In this case $s(\phi)>0$ for all ϕ. Therefore from (24), the two roots of (25) both lie outside or inside the unit circle. From the lemma we then have

$$\psi=\tfrac{1}{2}\ln\max\{b,c\}. \tag{28}$$

Now, $b=c$ cannot be realized because of (21); ψ is therefore analytic in this region.

(ii) $f+g>a+b+c$

By symmetry we have

$$\psi=\tfrac{1}{2}\ln\max\{f, g\}. \tag{29}$$

ψ is analytic in this region.

(iii) $a>b+c+f+g$

Under this condition, we have $\Delta(\pi)>0$ and $s(\pi)<0$; hence $s(\phi)<0$ for all ϕ. It follows then precisely one root of Eq. (25) lies outside the unit circle. From the lemma and Eq. (26) we then find

$$\psi=\frac{1}{4\pi}\int_{-\pi}^{\pi}d\phi\,\ln\tfrac{1}{2}[a+fe^{i\phi}+ge^{-i\phi}+\Delta^{1/2}(\phi)]. \tag{30}$$

In the complex $e^{i\phi}$ plane, the integrand in (30) has square root branch points at the four roots of

$$\Delta(\phi)=(a+fe^{i\phi}+ge^{-i\phi}+2\sqrt{bc})(a+fe^{i\phi}+ge^{-i\phi}-2\sqrt{bc}). \tag{31}$$

The inequality (21) implies that these branch points all lie on the negative real axis. Furthermore, the inequality

$$f+g<|a\pm 2\sqrt{bc}| \tag{32}$$

ensures that two of the four branch points lie outside the unit circle and two inside the unit circle. Hence generally, we can connect the two outside branch points and the two inside ones by separate branch cuts. Thus in the $e^{i\phi}$ plane, the integrand in Eq. (30) is analytic along the unit circle, the contour of integration, and hence ψ is analytic in a, b, c, f, g.

The nonanalyticity of ψ can occur only when two branch points pinch the unit circle from opposite sides so that the two branch cuts combine into a single one going through the unit circle. Inspection shows that this can happen only when the second factor in (31) has a double root or

$$a=2\sqrt{bc}+2\sqrt{fg}. \tag{33}$$

It follows then, the pinching of branch points can never happen for $a>b+c+f+g$ and can be approached on the boundary $a=b+c+f+g$ only when $b=c$, $f=g$. Therefore, ψ is analytic in $a>b+c+f+g$. The nonanalyticity of ψ at $a=b+c+f+g$ will be of a unique character when $b=c$, $f=g$. In this case, ψ can be written simply as

$$\psi=\frac{1}{8\pi^2}\int_{-\pi}^{\pi}d\phi\int_{-\pi}^{\pi}d\phi\,\ln(a+2b\cos\theta+2f\cos\phi), \tag{34}$$

nonanalytic at

$$a=2b+2f. \tag{35}$$

Expression (34) is of the form of the Onsager solution of an Ising model, and the nonanalyticity (for $b\neq 0$, $f\neq 0$) is known to be[24]

$$\text{second derivative of } \psi\sim\ln|a-2b-2f|.$$

Note that the condition $a\geq b+c+f+g$ now coincides with the restriction (21), so, in fact, Eq. (34) is the *complete* solution for $b=c$, $f=g$.

The Expression (30) for ψ can be further reduced. As a matter of fact, from (30) ψ depends on b, c only through the product bc; it would be useful to demonstrate explicitly the similar dependence on fg as required by symmetry.

Let z_1 and z_2, $|z_1|\geq|z_2|$, be the two roots of one factor in Eq. (31). Without loss of generality, let $f\geq g$. It can be easily established from $a\geq b+c+f+g$ that

$$|z_2|\leq\sqrt{(g/f)}\leq 1\leq\sqrt{(f/g)}\leq|z_1|. \tag{36}$$

Thus the roots of Eq. (31) lie outside the region

$$\sqrt{(g/f)}\leq|e^{i\phi}|\leq\sqrt{(f/g)}.$$

It follows then, in the complex ϕ plane, the integrand of Eq. (30) is analytic in

$$-U\leq\operatorname{Im}\phi\leq U\equiv\tfrac{1}{2}\ln(f/g). \tag{37}$$

Writing $fe^{i\phi}+ge^{-i\phi}=2\sqrt{fg}\,\cos(\phi-iU)$, we may then move the contour of integration in Eq. (30) from $-\pi\to\pi$ to $-\pi-iU\to\pi-iU$ and obtain

$$\psi=\frac{1}{4\pi}\int_{-\pi}^{\pi}\ln\tfrac{1}{2}\{a+2\sqrt{fg}\cos\phi+[(a+2\sqrt{fg}\cos\phi)^2-4bc]^{1/2}\}\,d\phi. \tag{38}$$

This is the desired expression. Here the branch of the square root is taken to give a larger magnitude for the quantity inside the curly brackets.

(iv) a, $b+c$, $f+g$ form a triangle

Finally, we consider the case that a, $b+c$, $f+g$ satisfy a triangular relation. Since the case $b=c$, $f=g$ has been explicitly considered in the above, we may restrict here to $b\neq c$. From Eq. (27), we see that we have always $s(0)<0$ and $s(\pi)>0$. Then by the monotonicity property of $s(\phi)$, there exists $0<\phi_1<\pi$ such that $|z_1(\phi_1)|=1$ or $s(\phi_1)=0$. Here ϕ_1 is symmetric in f and g. For definiteness assume $b>c$. We then have

$$\psi=\frac{1}{4\pi}\ln b\left(\int_{-\pi}^{-\phi_1}d\phi+\int_{\phi_1}^{\pi}d\phi\right)+\frac{1}{4\pi}\int_{-\phi_1}^{\phi_1}d\phi\,\ln[-bz_1(\phi)]$$

$$=\tfrac{1}{2}\ln b+\frac{1}{4\pi}\int_{-\phi_1}^{\phi_1}\ln[-z_1(\phi)]\,d\phi. \tag{39}$$

Note that from Eq. (27),

$$\phi_1\to 0 \quad\text{as } b+c\to a+f+g,$$
$$\phi_1\to\pi \quad\text{as } b+c\to|a-f-g|. \tag{40}$$

Also from Eq. (26),

$$z_1(\phi_1=0)=z_1(\phi_1=\pi)=-1. \tag{41}$$

To investigate the analyticity of ψ, we observe that, as discussed previously, the branch cuts of the integrand of Eq. (39) lie along the negative real axis. Hence the integrand is analytic in the complex $e^{i\phi}$ plane along the contour of integration. Therefore the integral in Eq. (39) is analytic. Any nonanalyticity of ψ can occur only at $\phi_1 = 0$ or π when the roots of Eq. (25) cross the unit circle.

It is futile to extract the singular part of Eq. (39) near $\phi_1 = 0$, π, for the integrand vanishes identical there by Eq. (41). However, since the quantities of physical interests are always given by some derivatives of ψ, we compute $\partial\psi/\partial x$, where x is some parameter contained in a, b, c, f, g. Using the fact that

$$\ln z_1(-\phi_1) + \ln z_1(\phi_1) = \ln|z_1(\phi_1)|^2 = 0\,, \tag{42}$$

the contributions that arise from taking derivatives with respect to the integration limits cancel. Thus we obtain

$$\begin{aligned}\frac{\partial\psi}{\partial x} &= \frac{1}{2}\frac{\partial}{\partial x}\ln b + \frac{1}{4\pi}\int_{-\phi_1}^{\phi_1} d\phi \frac{\partial}{\partial x}\ln[-z_1(\phi)] \\ &= \left(\frac{1}{2} - \frac{\phi_1}{2\pi}\right)\frac{\partial}{\partial x}\ln b + \frac{1}{4\pi}\int_{-\phi_1}^{\phi_1} d\phi \frac{\partial}{\partial x}\ln\tfrac{1}{2}[a + fe^{i\phi} \\ &\quad + ge^{-i\phi} + \Delta^{1/2}(\phi)]\,.\end{aligned} \tag{43}$$

The integrand in Eq. (43) does *not* vanish at $\phi_1 = 0$, π; hence, the singular part of $\partial\psi/\partial x$ behaves as ϕ_1. In particular, $\partial^2\psi/\partial x^2$ has an inverse square root singularity at $\phi_1 = 0$, π.

To summarize, we have shown that ψ given by (19) is nonanalytic in the region (21) only at

$$a + b + c + f + g = 2\max\{a,\ b + c,\ f + g\}\,. \tag{44}$$

The second derivative of ψ has an inverse-square-root singularity in Eq. (44). A special case is $b = c$, $f = g$, for which ψ has an Ising-type nonanalyticity in Eq. (44) or $a = 2b + 2f$.

V. APPLICATIONS

The results of Sec. IV are now applied to various models. First consider the implications on the related statistical problems described in Sec. II.

(i) For the Ising model with a nonzero magnetic field, the free-fermion conditions (18) lead to the rather uninteresting situation of $J_1 = J_2 = 0$. We then have $a = 2\cosh\beta H$, $b = c = f = g = 0$, and hence, directly from Eq. (19), $\psi = \frac{1}{2}\ln\cosh\beta H$.

(ii) For the percolation problem described by Eqs. (7) and (8), the free-fermion conditions (18) lead to the restriction of $yz = 0$. Similarly for the Potts models, the free-fermion conditions imply $q = 0$. Both of these conditions lie outside the regions of physical interests. Therefore the Pfaffian solution does not lead to useful results.

(iii) For the general dimer model of Fig. 3, we have

$$a = 2z_1z_2 + z_3^2 + z_4^2, \quad b = c = z_3z_4, \quad f = z_2^2, \quad g = z_1^2. \tag{45}$$

Our expression (19) for ψ now agrees with Eq. (A5) of Ref. 6. In Ref. 6, discussions were confined to the special cases $z_1 = z_2$ and $z_3 = z_4$. Here we find more generally that ψ is nonanalytic at

$$|z_1 - z_2| = |z_3 \pm z_4|\,. \tag{46}$$

Explicitly,

$$\psi = \begin{cases} \frac{1}{2}\ln\max\{z_1^2,\ z_2^2\}, & |z_1 - z_2| > z_3 + z_4, \\ \text{Eq. (39)}, & z_3 + z_4 > |z_1 - z_2| > |z_3 - z_4|, \\ \text{Eq. (38)}, & |z_3 - z_4| > |z_1 - z_2|. \end{cases} \tag{47}$$

The second derivative of ψ diverges with a square-root singularity when the nonanalytic points are approached in the middle range in Eq. (47). A special case occurs when $z_1 = z_2$ identically, the antiferroelectric model of Ref. 2. There is an Ising-type singularity at $z_3 = z_4$.

(iv) For the dimer lattice on the 4–8 lattice of Fig. 4, we have

$$a = (z_1^2 + z_2^2)^2 + z_3^4, \quad b = c = z_1^2z_3^2, \quad f = g = z_2^2z_3^2. \tag{48}$$

Our expressions (19) or (34) for ψ now agree with (2.2) of Ref. 7. The singularity is of the Ising type occurring at

$$z_1^2 + z_2^2 = z_3^2\,. \tag{49}$$

Next, we consider the exact solution of vertex models implied by the Pfaffian solution. Here we specialize the vertex weights ω_i and ω_i' to the Boltzmann factors, Eq. (4). We shall also take[25]

$$\omega_1 = \omega_1', \quad \omega_2 = \omega_2', \quad \omega_3 = \omega_3', \quad \omega_4 = \omega_4', \quad \omega_5 = \omega_6', \quad \omega_6 = \omega_5'. \tag{50}$$

For the free-fermion condition (8) to hold within the complete range of certain physical parameters, we have the following choices.

(i) Exactly soluble models

Under this category we include those models for which Eq. (18) is to hold for *all temperatures* so that the model is exactly solved. The following possibilities then arise:

(a) $\omega_1 = \omega_1' = 0, \quad \omega_3\omega_4 = \omega_5\omega_6 = \omega_3'\omega_4' = \omega_5'\omega_6'$.

This is the modified KDP model in a staggered field considered previously by one of us.[9] Our analysis now leads to the same results as in Ref. 9.

(b) $\omega_1 = \omega_2' = 0, \quad \omega_3\omega_4 = \omega_5\omega_6 = \omega_3'\omega_4' = \omega_5'\omega_6'$.

We may parametrize the vertex energies as[26]

$$\epsilon_3 = \epsilon_3' = -h + v\,, \quad \epsilon_4 = \epsilon_4' = h - v\,,$$

$$\epsilon_5=\epsilon_6'=-s, \qquad \epsilon_5'=\epsilon_6=s, \tag{51}$$

where h, v are the direct fields and s is the staggered field. The parameters in ψ are then

$$a=2\cosh S, \quad b=c=0, \quad f=e^{H-V}, \quad g=e^{-H+V}, \tag{52}$$

where $S=2\beta s$, $H=2\beta h$, $V=2\beta v$. Since $b+c=0$, regions (i) and (iv) can never be realized. Using Eqs. (29) and (38) we find

$$\psi=\tfrac{1}{2}\max\{|S|, |H-V|\}. \tag{53}$$

Thus the system is in a frozen state at all temperatures.

(c) $\omega_1=\omega_3'=0, \quad \omega_3\omega_4=\omega_5\omega_6=\omega_1'\omega_2'=\omega_5'\omega_6'$.

We may parametrize the vertex energies as

$$\begin{aligned}&\epsilon_2=\epsilon_2'=h+v, \quad \epsilon_4=\epsilon_4'=h-v,\\ &\epsilon_5=\epsilon_6'=-s, \quad \epsilon_5'=\epsilon_6=s,\end{aligned} \tag{54}$$

so that

$$a=2\cosh S, \quad b=f=0, \quad c=e^{-(H+V)}, \quad g=e^{-H+V}. \tag{55}$$

Using the results of Sec. III, we find

$$\psi=\begin{cases}\tfrac{1}{2}[|V|-H], & \cosh S\leq e^{-H}\sinh|V|,\\ \tfrac{1}{2}\ln(2\cosh S), & \cosh S\geq e^{-H}\cosh V,\\ \tfrac{1}{2}\ln c+(1/4\pi)\displaystyle\int_{-\phi_1}^{\phi_1}d\phi\,\ln[(a+ge^{-i\phi})/c], & \\ \qquad e^{-H}\sinh|V|\leq\cosh S\leq e^{-H}\cosh V, & \end{cases} \tag{56}$$

where ϕ_1 is defined by

$$|a+ge^{-i\phi_1}|=c. \tag{57}$$

Therefore a transition occurs at a temperature T_c defined by

$$\cosh S=e^{-H}\sinh|V| \quad \text{or} \quad \cosh S=e^{-H}\cosh V. \tag{58}$$

Of special interest is $v-s>h>0$, for which both equations in (58) can be satisfied and the system has two phase transitions. The system is frozen below the lower T_c with a net direct polarization, while the specific heat diverges with $\alpha'=\frac{1}{2}$ only below the higher T_c.

To compute the direct polarizations above T_c, we use Eq. (43) and note that, inside the integral sign in (43), we may replace d/dH and d/dV by

$$\frac{d}{dH}=-\frac{d}{dV}=i\frac{d}{d\phi}.$$

Thus, we find

$$P_H=2\frac{\partial\psi}{\partial H}=\left(-1+\frac{\phi_1}{\pi}\right)+\frac{i}{2\pi}\ln\left(\frac{a+ge^{-i\phi_1}}{a+ge^{i\phi_1}}\right). \tag{59}$$

Let G, A, C be the respective angles of the triangle having sides g, a, c ($\phi_1=\pi-C$); then Eq. (59) can be abbreviated as

$$P_H=-(C-G)/\pi. \tag{60a}$$

Similarly we find

$$P_V=2(\partial\psi/\partial V)=-(C+G)/\pi. \tag{60b}$$

(ii) An exact isotherm of a general staggered model

Instead of requiring the free-fermion condition (18) to hold at all temperatures, we can turn the problem around and consider Eq. (18) to define a temperature at which the Pfaffian solution is valid. An advantage of this approach is that the validity of Eq. (18) is independent of the (direct and staggered) fields, so that we have, at hand, an exact isotherm for a general staggered model. For both equations of (18) to hold, we have the following two possibilities:

(a) $\omega_1\omega_2=\omega_1'\omega_2'=u^2, \quad \omega_3\omega_4=\omega_3'\omega_4'=v^2$.

We may parametrize as follows:

$$\begin{aligned}&\omega_1=\omega_1'=ue^{(H+V)/2}, \quad \omega_2=\omega_2'=ue^{-(H+V)/2},\\ &\omega_3=\omega_3'=ve^{(H-V)/2}, \quad \omega_4=\omega_4'=ve^{-(H-V)/2},\\ &\omega_5=\omega_6'=we^{-S/2}, \quad \omega_5'=\omega_6=we^{S/2},\end{aligned} \tag{61}$$

where $w^2=u^2+u^2$. This gives

$$\begin{aligned}&a=2w^2\cosh S, \quad b=u^2e^{H+V}, \quad c=u^2e^{-(H+V)},\\ &f=v^2e^{H-V}, \quad g=v^2e^{-(H-V)}.\end{aligned} \tag{62}$$

This model is now intrinsically antiferroelectric[1] and is a generalization of the case $u^2=v^2$ considered by Baxter.[10]

Using the results of Sec. IV, we find

$$\begin{aligned}\psi&=\ln u+\tfrac{1}{2}|H+V|, && (u^2+v^2)\cosh S<u^2\cosh(H+V)-v^2\cosh(H-V),\\ &=\ln v+\tfrac{1}{2}|H-V|, && (u^2+v^2)\cosh S<v^2\cosh(H-V)-u^2\cosh(H+V),\\ &=\text{Eq. (38) independent of } H, V, && (u^2+v^2)\cosh S>u^2\cosh(H+V)+v^2\cosh(H-V),\\ &=\text{Eq. (39)}, && \text{otherwise}.\end{aligned} \tag{63}$$

There is a partial direct polarization when ψ is given by Eq. (39). To compute this polarization, we use Eq. (43) and note from Eq. (62) that inside the integral sign of Eq. (43) we may replace d/dH

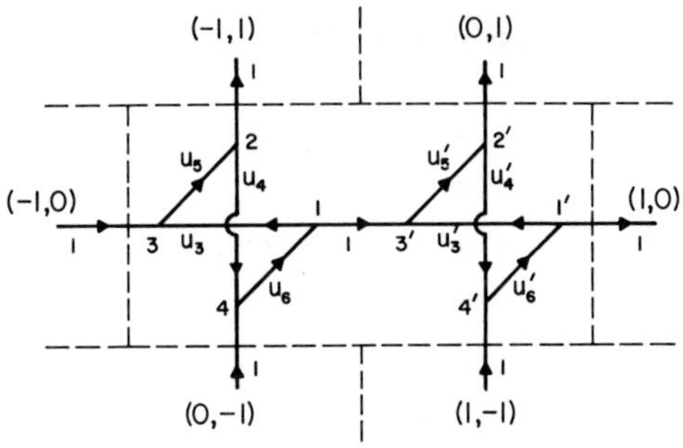

FIG. 5. Unit cell of the dimer terminal lattice L^{Δ}.

and d/dV by

$$\frac{d}{dH}=-\frac{d}{dV}=i\frac{d}{d\phi}, \tag{64}$$

so that the integration is immediately performed. Defining $z_1(\phi_1)=\exp(i\theta_1)$, where θ_1 is real, we then find

$$P_H=2\frac{\partial\psi}{\partial H}=1-(1/\pi)(\phi_1+\theta_1),$$
$$P_V=2\frac{\partial\psi}{\partial V}=1-(1/\pi)(\phi_1-\theta_1). \tag{65}$$

A special case is when there is no direct field, so that $b=c$, $f=g$. We find ψ nonanalytic at $s=0$ with the staggered susceptibility diverging as $\ln|s|$.

(b) $\omega_1\omega_2=\omega_3'\omega_4'=u^2$, $\omega_1'\omega_2'=\omega_3\omega_4=v^2$.

Clearly not all $\omega_i=\omega_i'$, $i=1, 2, 3, 4$, if $u\neq v$. We can then redefine the weights in accordance with Eq. (50) to make

$$\omega_1\omega_2=\omega_3\omega_4=\omega_1'\omega_2'=\omega_3'\omega_4'=uv.$$

The model is then the special case of (a) above, previously considered by Baxter.[10]

ACKNOWLEDGMENT

One of us (F. Y. W.) wishes to thank Dr. Shien-Siu Shu for his hospitality at the National Tsing Hua University, where this work was initiated.

APPENDIX A—PFAFFIAN SOLUTION

Divide each vertex weight of sublattice A by ω_2 and each vertex weight of sublattice B by ω_2'; we may rewrite Eq. (2) as

$$Z(\omega_i, \omega_i')=(\omega_2, \omega_2')^{N/2}Z(u_i, u_i'), \tag{A1}$$

where $u_i=\omega_i/\omega_2$, $u_i'=\omega_i'/\omega_2'$. $Z(u_i, u_i')$ is now converted into a dimer generating function Z^{Δ} as follows.

Expand each site of L into a "city" of four terminals so as to form the terminal dimer lattice L^{Δ} whose unit cell is shown in Fig. 5. Place a dimer on an edge of L^{Δ} connecting the cities if and only if the corresponding edge of L has leftward or downward arrows. Since such dimer coverings specify completely the dimer configurations of L^{Δ}, we now have a correspondence between the ice-rule configurations of L and the dimer configuration of L^{Δ}. Details of the correspondence are shown in Fig. 6. The correspondence is one-to-one, except that there exist two dimer arrangements within a city for the ice-rule configuration (1). But this degeneracy is a local property, so the two dimer weights can be grouped together at each vertex of L. Following the counterclockwise-odd rule of Kastelyn,[27] we direct the edges of L^{Δ} as shown in Fig. 5. Under this convention, the two degenerate dimer arrangements give rise to contributions of opposite signs.[27] Using the free-fermion conditions (18) which now read

$$u_5u_6-u_3u_4=u_1, \quad u_5'u_6'-u_3'u_4'=u_1', \tag{A2}$$

it is seen that the activities of the dimer configurations reproduce precisely the staggered ice-rule vertex weights. It follows then that

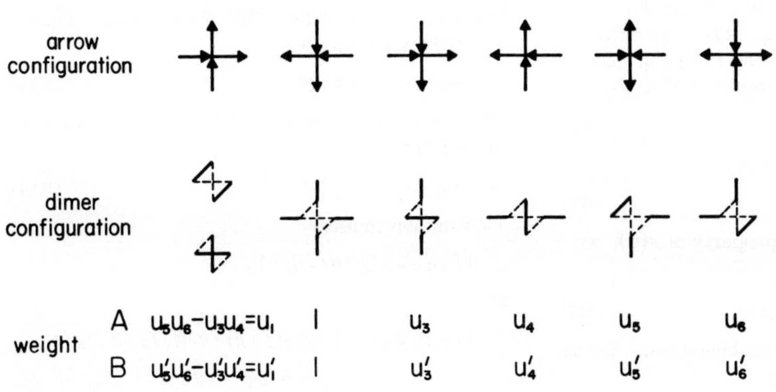

FIG. 6. Correspondences between the ice-rule configurations on L and the dimer configurations on L^{Δ}.

$$Z(\omega_i, \omega_i') = (\omega_1\omega_2')^{N/2} Z^\Delta , \tag{A3}$$

where Z^Δ is the dimer generating function for L^Δ.

Using the arrow convention of L^Δ of Fig. 5, Z^Δ can now be evaluated by standard means. The result after combining Eqs. (A3) and (3) is

$$\psi = \frac{1}{4(2\pi)^2} \int_{-\pi}^{\pi} d\alpha \int_{-\pi}^{\pi} d\beta \ln[\omega_2^2\omega_2'^2 D(\alpha, \beta)] , \tag{A4}$$

where

$$D(\alpha, \beta) = \begin{vmatrix} 0 & \underline{A} \\ -\underline{A}^* & 0 \end{vmatrix} ,$$

$$\underline{A} = \begin{vmatrix} u_3 & -u_6 & 1 & 0 \\ -u_5 & u_4 & 0 & e^{-i(\alpha-\beta)} \\ e^{i\alpha} & 0 & u_3' & -u_6' \\ 0 & e^{i\beta} & -u_5' & u_4' \end{vmatrix} . \tag{A5}$$

The factorization of the 8×8 determinant $D(\alpha, \beta)$ into $\underline{A}\,\underline{A}^*$, a unique consequence of the ice rule, is due to the fact that the sets of points $\{1, 2, 1', 2'\}$ and $\{3, 4, 3', 4'\}$ in Fig. 5 are not interconnected. Equation (A4) reduces to Eq. (19) in the text, after the change of variables $\theta = \pi - \beta$, $\phi = \pi - \alpha + \beta$ and using the fact that the right-hand side of Eq. (19) is real.

APPENDIX B—PROPERTIES OF $s(\phi)$

For a, b, c, f, g given by Eq. (20) and subject to Eq. (18), we have the inequality

$$a \geq 2\sqrt{bc} + 2\sqrt{fg} . \tag{B1}$$

Proof:

$$\begin{aligned} a^2 &= (\omega_5\omega_6' + \omega_5'\omega_6)^2 \geq 4\,\omega_5\omega_6\omega_5'\omega_6' \\ &= 4(\omega_1\omega_2 + \omega_3\omega_4)(\omega_1'\omega_2' + \omega_3'\omega_4') \\ &= 4(\omega_1\omega_1'\omega_2\omega_2' + \omega_3\omega_3'\omega_4\omega_4' + \omega_1\omega_2\omega_3'\omega_4' + \omega_1'\omega_2'\omega_3\omega_4) \\ &\geq 4[\omega_1\omega_1'\omega_2\omega_2' + \omega_3\omega_4'\omega_4\omega_4' + 2(\omega_1\omega_2\omega_3\omega_4\omega_1'\omega_2'\omega_3'\omega_4')^{1/2}] \\ &= 4(\sqrt{bc} + \sqrt{fg})^2 . \end{aligned}$$

First we compute $s(0)$ and $s(\pi)$. Using Eq. (B1), we find the results given in Eq. (27). Here the value of $s(\pi)$ is obtained by noting from Eq. (26) that both $z_1(\pi)$ and $z_2(\pi)$ are complex satisfying

$$z_1^*(\pi) = z_2(\pi), \quad \text{if } \Delta(\pi) < 0 . \tag{B2}$$

Equations (B1) and (B2) now imply

$$s(0) \leq s(\pi) . \tag{B3}$$

To prove the monotonicity property of $s(\phi)$, we consider

$$s'(\phi) = (1 - z_2/z_1^*)(d/d\phi)|z_1(\phi)| , \tag{B4}$$

where the relation $z_1z_2 = c/b$ has been used. Using Eq. (26) for $z_1(\phi)$, and letting

$$\Delta(\phi) = (x + iy)^2 ,$$

or

$$x^2 - y^2 = a^2 + 2fg - 4bc + (f^2 + g^2) \times \cos 2\phi + 2a(f+g)\cos\phi , \tag{B5}$$

$$xy = (f-g)\sin\phi[(f+g)\cos\phi + a], \tag{B6}$$

it is straightforward to find

$$\frac{d}{d\phi}|z_1(\phi)| = \frac{|z_1|}{2(x^2+y^2)}[y(f-g)\cos\phi - x(f+g)\sin\phi]. \tag{B7}$$

Thus, the extremum values of $s(\phi)$ occur at

$$(1 - z_2/z_1^*)[y(f-g)\cos\phi - x(f+g)\sin\phi] = 0 . \tag{B8}$$

Solving (B8) jointly with Eqs. (B5), (B6) for x, y, and ϕ, we find $s'(\phi) = 0$ to occur at $\phi = 0$, π and also at ϕ_0 defined by

$$\alpha\cos^2\phi_0 + \beta\cos\phi_0 + \gamma = 0 . \tag{B9}$$

Here

$$\alpha = 4afg/(f+g) > 0, \quad \gamma = a(f+g) > 0,$$

$$\beta = a^2 + 4fg - 4bc > 0 .$$

The solutions at $\phi = 0$, π lead to the acceptable results

$$\begin{aligned} &\phi = 0, \quad y = 0, \quad x^2 = \Delta(0) > 0, \\ &\phi = \pi, \quad y = 0, \quad x^2 = \Delta(\pi) \quad \text{if } \Delta(\pi) \geq 0, \\ &\quad \text{or } x = 0, \quad y^2 = -\Delta(\pi) \quad \text{if } \Delta(\pi) \leq 0. \end{aligned} \tag{B10}$$

In the latter case $[\Delta(\pi) \leq 0]$, $s'(\pi)$ vanishes due to Eq. (B2). The solution at ϕ_0 is spurious, however, as we now see.

From Eqs. (B6) and (B8), we see that for ϕ_0 to be an extremum, it is necessary to have

$$[(f+g)\cos\phi_0 + a]\cos\phi_0 > 0 . \tag{B11}$$

Now,

$$\beta^2 - 4\alpha\gamma = (a + 2\sqrt{bc} + 2\sqrt{fg})(a + 2\sqrt{bc} - 2\sqrt{fg}) \times (a - 2\sqrt{bc} + 2\sqrt{fg})(a - 2\sqrt{bc} - 2\sqrt{fg}) > 0, \tag{B12}$$

hence both roots of Eq. (B9) are negative. Equation (B11) then implies

$$\cos\phi_0 < -a/(f+g) . \tag{B13}$$

Therefore, for ϕ_0 to be real and different from π, we require

$$f + g > a . \tag{B14}$$

This in turn implies

$$(f+g)/2\sqrt{fg} > a/\sqrt{fg} > 1 ,$$

or

$$\begin{aligned} (f+g)/2\sqrt{fg} + 2\sqrt{fg}/(f+g) &> a/2\sqrt{fg} + 2\sqrt{fg}/a \\ &\geq a/2\sqrt{fg} + 2\sqrt{fg}/a - 2bc/a\sqrt{fg} , \end{aligned}$$

or

$$\alpha + \gamma > \beta . \quad \text{(B15)}$$

Using $\alpha < \gamma$ and Eq. (B14) which ensures $\beta > 2\alpha$, it is easy to verify that Eq. (B15) has the consequence that both roots of Eq. (B9) lie outside the unit circle or $|\cos\phi_0| > 1$. Thus ϕ_0 does not correspond to any physical extremum of $s(\phi)$.

Finally, since $s'(\phi)$ vanishes only at $\phi = 0$, π and $s(0) \leq s(\pi)$, we conclude that $s(\phi)$ is monotonically increasing in $\{0, \pi\}$. Changing ϕ to $-\phi$ in Eq. (25) corresponds to interchanging f and g. From Eqs. (26), (B5), and (B6), it is seen that this leaves $|z_1(\phi)|$ and $|z_2(\phi)|$ unchanged. Hence, $s(\phi)$ is symmetric about $\phi = 0$ and therefore monotonically decreasing in $\{-\pi, 0\}$.

---

*Supported in part by National Science Foundation Grant No. DMR72-03213 A01.

†Supported in part by the National Science Council, Taiwan, Republic of China.

[1]For a review, see E. H. Lieb and F. Y. Wu, in *Phase Transitions and Critical Phenomena*, edited by C. Domb and M. S. Green (Academic, London, 1972), Vol. I.

[2]H. N. V. Temperley and E. H. Lieb, Proc. R. Soc. Lond. A 322, 251 (1971).

[3]R. J. Baxter, J. Phys. C 6, L445 (1973).

[4]F. Wegner, J. Phys. C 5, L131 (1973).

[5]F. Y. Wu and K. Y. Lin, J. Phys. C 7, L181 (1974).

[6]G. R. Allen, J. Chem. Phys. 60, 3299 (1974).

[7]S. R. Salinas and J. F. Nagle, Phys. Rev. B 9, 4920 (1974).

[8]C. S. Hsue, K. Y. Lin, and F. Y. Wu, following paper, Phys. Rev. B 12, 429 (1975).

[9]F. Y. Wu, Phys. Rev. B 3, 3895 (1971).

[10]R. J. Baxter, Phys. Rev. B 1, 2199 (1970).

[11]R. J. Baxter, J. Stat. Phys. 9, 145 (1973); J. Phys. C 6, L94 (1973).

[12]See, e.g., E. W. Montroll, in *Applied Combinatorial Mathematics*, edited by E. F. Beckenback (Wiley, New York, 1964), Chap. 4.

[13]R. J. Baxter, Stud. Appl. Math. 1, 51 (1971).

[14]For a definition of covering graph, see, e.g., J. W. Essam and M. E. Fisher, Rev. Mod. Phys. 42, 271 (1970).

[15]See also Fig. 5 of Ref. 1.

[16]Assuming the sites on the upper-right-hand corners of the unit squares are on sublattice *A*.

[17]R. B. Potts, Proc. Camb. Philos. Soc. 48, 106 (1952).

[18]H. A. Kramers and G. H. Wannier, Phys. Rev. 60, 252 (1941).

[19]M. J. Stephen and L. Mittag, Phys. Lett. A 41, 357 (1972).

[20]M. F. Sykes and J. W. Essam, J. Math. Phys. 5, 1117 (1964).

[21]R. J. Baxter, Ann. Phys. (N.Y.) 70, 193 (1972).

[22]See, e.g., C. Fan and F. Y. Wu, Phys. Rev. B 2, 723 (1970).

[23]This result cannot be obtained from the ice rule alone, as is evidenced from (5).

[24]See, e.g., K. Huang, *Statistical Mechanics* (Wiley-Interscience, New York, 1973), p. 369.

[25]There is no loss of generality in so doing. Multiplying a factor x to all A vertices, a factor y to all (5) vertices, and y^{-1} to all (6) vertices, we can always make $\omega_5 = \omega_6'$ and $\omega_5' = \omega_6$ by choosing x, y properly. Then choose the new values for ω_i and ω_i', $i = 1, 2, 3, 4$, to be $(x\omega_i\omega_i')^{1/2}$.

[26]Again, no generality is gained here by introducing different fields h, v, h', v' for the two sublattices, as only the combinations of $h + h'$ and $v + v'$ will occur in Eq. (19).

[27]P. W. Kasteleyn, J. Math. Phys. 4, 287 (1963).

PHYSICAL REVIEW B VOLUME 12, NUMBER 1 1 JULY 1975

Staggered eight-vertex model

C. S. Hsue* and K. Y. Lin*
Department of Physics, National Tsing Hua University, Hsin Chu, Taiwan, Republic of China

F. Y. Wu†
Department of Physics, Northeastern University, Boston, Massachusetts 02115
(Received 3 March 1975)

An eight-vertex model with staggered (site-dependent) vertex weights is considered. The model is an extension of the usual one with translationally invariant weights and contains sixteen independent vertex weights. From its Ising representation it is seen that there are actually only eleven independent parameters. After discussing some general symmetry properties of this model, we consider in detail the soluble case of a free-fermion model. We find that the staggered free-fermion model may exhibit up to three phase transitions. Generally the specific heat has logarithmic singularities, expect in some special cases it has an exponent $\alpha = 1/2$ and the system is frozen below a unique transition point. Conditions for these special cases are given.

I. INTRODUCTION

In the consideration of vertex models in lattice statistics, [1] one usually deals with models whose vertex weights are translationally invariant. In a recent paper, [2] hereafter referred to as I, we have pointed out the importance of models with staggered, or site-dependent, weights. We have also studied the most general Pfaffian solution of the staggered six-vertex model. In this paper, we return to consider the staggered eight-vertex model.

We refer to I for the definition of a staggered model. For an eight-vertex model, the allowed vertex configurations can be conveniently described by the bond graphs shown in Fig. 1.[1] Let the vertex weights be

$$\{\omega\}=\{\omega_1, \omega_2, \ldots, \omega_8\} \text{ on subtallice A,}$$
$$\{\omega'\}=\{\omega_1', \omega_2', \ldots, \omega_8'\} \text{ on sublattice B .} \tag{1}$$

Then, as in I, our goal is to compute the "free energy"

$$\psi=\lim_{N\to\infty}\frac{1}{N}\ln Z\ , \tag{2}$$

where Z is the partition-generating function defined by Eq. (1) and N is the number of sites. In the ferroelectric language, the vertex weights are the Boltzmann factors

$$\omega_i=\exp(-\beta e_i),\ \ \omega_i'=\exp(-\beta e_i'), \tag{3}$$

where $\beta=1/kT$ and e_i, e_i' are the vertex energies. This completes the definition of the staggered eight-vertex model.

When $\omega_7=\omega_8=\omega_7'=\omega_8'=0$, this model reduces to the ice-rule case considered in I. It is also known[3] that for

$$e_1=e_2=e_1'=e_2',\ \ e_3=e_4=e_3'=e_4',$$
$$e_5=e_6=e_7'=e_8',\ \ e_7=e_8=e_5'=e_6', \tag{4}$$

the model is equivalent to the Ashkin-Teller model,[4] which generally possesses two phase transitions.[5] In other cases, as we shall see, the model may also exhibit three transitions. Thus the staggered eight-vertex model is very general and may find applications to systems with multiple phase transitions.

The outline of this paper is as follows: Some symmetry properties of the model are discussed in Sec. II. In Sec. III we consider the Ising representation of the staggered eight-vertex model, thereby showing the existence of only eleven independent parameters. In Sec. IV we study the staggered free-fermion model and summarize its thermodynamic properties. Details of analyses are found in Sec. V. A striking feature of our result is the existence of multiple phase transitions.

II. SYMMETRY RELATIONS

The partition function Z possesses a number of symmetry relations which can be obtained by quite general considerations. Obviously, Z is invariant if ω_i and ω_i', $i=1, 2, \ldots, 8$ are interchanged. This is written as

$$Z=Z(12\ldots 8;\ 1'2'\ldots 8')$$
$$=Z(1'2'\ldots 8';\ 12\ldots 8), \tag{5}$$

where i and i' denote, respectively, ω_i and ω_i'. Other symmetries can be obtained along the line of considerations for an eight-vertex model with uniform weights, the uniform model.[6] Interchanging the bonds and "holes" in the horizontal or the vertical directions, e.g., we obtain

$$Z=Z(43217856;\ 4'3'2'1'7'8'5'6'),$$
$$=Z(34128765;\ 3'4'1'2'8'7'6'5'). \tag{6}$$

A 90° clockwise rotation of the lattice leads to

$$Z = Z(12438756;\ 1'2'4'3'8'7'5'6'). \tag{7}$$

Interchanging the bonds and holes along the zigzag paths shown in Fig. 2 of Ref. 6 gives the symmetry

$$Z = Z(65782134;\ 5'6'8'7'1'2'4'3'). \tag{8}$$

The last relation reveals that the pairs of vertex weights

$$\{\omega_6, \omega_5'\},\ \{\omega_5, \omega_6'\},\ \{\omega_7, \omega_8'\},\ \{\omega_8, \omega_7'\} \tag{9a}$$

play the same role as

$$\{\omega_1, \omega_1'\},\ \{\omega_2, \omega_2'\},\ \{\omega_3, \omega_3'\},\ \{\omega_4, \omega_4'\}, \tag{9b}$$

a property also reflected in Eq. (13) below. Finally, there is the weak-graph symmetry[7] which is local property of a lattice and is therefore valid even if the weights are site dependent. If the conjugate vertex pairs[8] have equal weights

$$\begin{aligned} &\omega_1 = \omega_2 = a_1, \quad \omega_1' = \omega_2' = a_2, \\ &\omega_3 = \omega_4 = b_1, \quad \omega_3' = \omega_4' = b_2, \\ &\omega_5 = \omega_6 = c_1, \quad \omega_5' = \omega_6' = c_2, \\ &\omega_7 = \omega_8 = d_1, \quad \omega_7' = \omega_8' = d_2, \end{aligned} \tag{10}$$

this symmetry leads to, in obvious notations,

$$Z(a_1 b_1 c_1 d_1;\ a_2 b_2 c_2 d_2) = Z(\bar{a}_1 \bar{b}_1 \bar{c}_1 \bar{d}_1;\ \bar{a}_2 \bar{b}_2 \bar{c}_2 \bar{d}_2). \tag{11}$$

Here, for $i = 1, 2$,

$$\begin{aligned} &\bar{a}_i = \tfrac{1}{2}(a_i + b_i + c_i + d_i), \quad \bar{b}_i = \tfrac{1}{2}(a_i + b_i - c_i - d_i), \\ &\bar{c}_i = \tfrac{1}{2}(a_i - b_i + c_i - d_i), \quad \bar{d}_i = \tfrac{1}{2}(a_i - b_i - c_i + d_i). \end{aligned} \tag{12}$$

The case $a_1 = a_2$, $b_1 = b_2$, $c_1 = d_2$, $c_2 = d_2$ is equivalent to the Ashkin-Teller model.[3]

III. EQUIVALENCE WITH AN ISING MODEL

Although there are 16 vertex weights in the definition (1), it turns out that the vertex weights enter the free energy ψ through the combinations of

$$\begin{aligned} &v_1 = \omega_1\omega_1', \quad v_2 = \omega_2\omega_2', \quad v_3 = \omega_3\omega_3', \quad v_4 = \omega_4\omega_4', \\ &v_5 = \omega_5\omega_6', \quad v_6 = \omega_6\omega_5', \quad v_7 = \omega_7\omega_8', \quad v_8 = \omega_8\omega_7', \\ &u = \omega_1'\omega_2'\omega_3\omega_4; \quad v = \omega_1'\omega_2'\omega_5\omega_6; \quad w = \omega_1'\omega_2'\omega_7\omega_8. \end{aligned} \tag{13}$$

In other words, there are only eleven independent parameters in a staggered eight-vertex model.

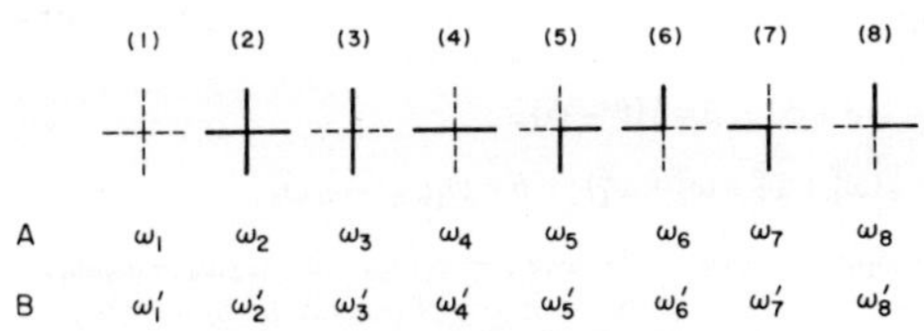

FIG. 1. The eight-vertex configurations and the associated weights.

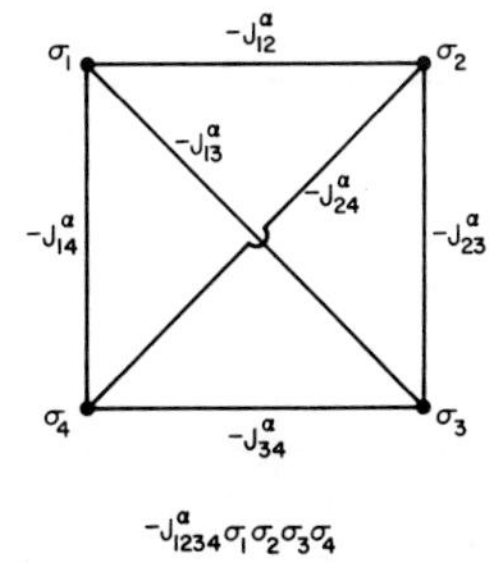

FIG. 2. Ising interactions surrounding a vertex site on the sublattice $\alpha(=A, B)$. The four-spin interaction is not shown.

To see this result, we consider the Ising representation of an eight-vertex model.[1,9] In a staggered model, the Ising interactions are also staggered. Let the four spins surrounding a site be $\sigma_1, \sigma_2, \sigma_3$, and σ_4, as shown in Fig. 2. They interact with two-spin interactions $-J^\alpha_{ij}$ and four-spin interaction $-J^\alpha_{1234}$, where the superscript $\alpha = A$ or B refers to the sublattices. Then, analogous to Eq. (29) of Ref. 1,[10] the vertex energies can be written as

$$\begin{aligned} e^\alpha(\sigma_1, \sigma_2, \sigma_3, \sigma_4) = &-J^\alpha_0 - \sum_{i<j} J^\alpha_{ij}\sigma_i\sigma_j \\ &- J^\alpha_{1234}\sigma_1\sigma_2\sigma_3\sigma_4, \quad \alpha = A, B. \end{aligned} \tag{14}$$

Here, we have adopted the convention (cf. Fig. 1) of writing $e_1 = e^A(1,1,1,1) = e^A(-1,-1,-1,-1)$, $e_1' = e^B(1,1,1,1) = e^B(-1,-1,-1,-1)$, $e_2 = e^A(1, -1, 1, -1) = e^A(-1, 1, -1, 1)$, etc. Since Eq. (14) is invariant under $\sigma_i \to -\sigma_i$, it represents sixteen independent equations. They can be solved to yield

$$\begin{aligned} J^\alpha_0 &= -\frac{1}{16}\sum_{\sigma_1\sigma_2\sigma_3\sigma_4} e^\alpha(\sigma_1, \sigma_2, \sigma_3, \sigma_4), \\ J^\alpha_{ij} &= -\frac{1}{16}\sum_{\sigma_1\sigma_2\sigma_3\sigma_4} \sigma_i\sigma_j e^\alpha(\sigma_1, \sigma_2, \sigma_3, \sigma_4), \\ J^\alpha_{1234} &= -\frac{1}{16}\sum_{\sigma_1\sigma_2\sigma_3\sigma_4} \sigma_1\sigma_2\sigma_3\sigma_4 e^\alpha(\sigma_1, \sigma_2, \sigma_3, \sigma_4). \end{aligned} \tag{15}$$

The resulting Ising lattice shown in Fig. 3 has eleven interactions:

$$\begin{aligned} &J^\alpha_{13},\ J^\alpha_{24},\ J^\alpha_{1234}, \quad (\alpha = A, B) \\ &J_0 = J^A_0 + J^B_0, \\ &J_1 = J^A_{12} + J^B_{34}, \quad J_2 = J^A_{34} + J^B_{12}, \\ &J_3 = J^A_{14} + J^B_{23}, \quad J_4 = J^A_{23} + J^B_{14}. \end{aligned} \tag{16}$$

Writing out Eqs. (16) and using the variables defined in Eqs. (13), we find explicitly

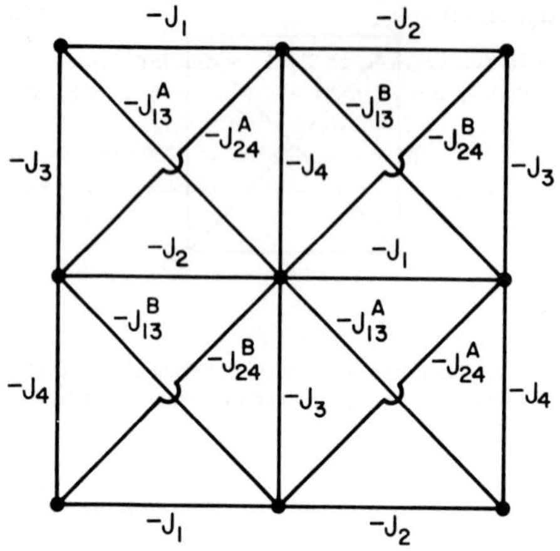

FIG. 3. Equivalent Ising lattice with staggered interactions. Four-spin interactions are not shown.

$$\exp(-8\beta J_0)=\prod_{i=1}^{8} v_i ,$$

$$\exp(8\beta J_1)=v_2v_3v_6v_8/v_1v_4v_5v_7 ,$$

$$\exp(8\beta J_2)=v_2v_3v_5v_7/v_1v_4v_6v_8 ,$$

$$\exp(8\beta J_3)=v_2v_4v_6v_7/v_1v_3v_5v_8 ,$$

$$\exp(8\beta J_4)=v_2v_4v_5v_8/v_1v_3v_6v_7 ,$$

$$\exp[8\beta(J_{24}^A+J_{24}^B)]=v_3v_4v_5v_6/v_1v_2v_7v_8 ,$$

$$\exp[8\beta(J_{13}^A+J_{13}^B)]=v_3v_4v_7v_8/v_1v_2v_5v_6 ,$$

$$\exp[8\beta(J_{1234}^A+J_{1234}^B)]=v_5v_6v_7v_8/v_1v_2v_3v_4 , \tag{17}$$

$$\exp[4\beta(J_{13}^A+J_{24}^A)]=\omega_3\omega_4/\omega_1\omega_2=u/v_1v_2 ,$$

$$\exp[4\beta(J_{24}^A+J_{1234}^A)]=\omega_5\omega_6/\omega_1\omega_2=v/v_1v_2 ,$$

$$\exp[4\beta(J_{24}^B+J_{1234}^A)]=\omega_7\omega_8/\omega_1\omega_2=w/v_1v_2 .$$

This completes the proof of our assertion (13).

IV. STAGGERED FREE-FERMION MODEL

In this section we study the most general Pfaffian solution of the staggered eight-vertex model. A vertex model is exactly soluble if a certain free-fermion condition is satisfied at each vertex. For the present model the condition reads[6]

$$\omega_1\omega_2+\omega_3\omega_4=\omega_5\omega_6+\omega_7\omega_8 ,$$
$$\omega_1'\omega_2'+\omega_3'\omega_4'=\omega_5'\omega_6'+\omega_7'\omega_8' . \tag{18}$$

Under this condition, the partition function is equal to a Pfaffian which can be evaluated in a closed form. With the details outlined in Appendix A, the result is given here,

$$\psi=\frac{1}{16\pi^2}\int_0^{2\pi}d\theta\int_0^{2\pi}d\phi\,\ln F(\theta,\phi) , \tag{19}$$

with

$$F(\theta,\phi)=F_0(\theta,\phi)-4\Delta\sin^2\phi-4\Delta'\sin^2\theta . \tag{20}$$

$$F_0(\theta,\phi)=2A-2B\cos\theta-2C\cos\phi+2D\cos(\theta-\phi)+2E\cos(\theta+\phi) ,$$

$$A=\tfrac{1}{2}(\Omega_1^2+\Omega_2^2+\Omega_3^2+\Omega_4^2),\quad B=\Omega_1\Omega_3-\Omega_2\Omega_4 ,\quad C=\Omega_1\Omega_4-\Omega_2\Omega_3 ,\quad D=\Omega_3\Omega_4-\Omega_7\Omega_8 ,\quad E=\Omega_3\Omega_4-\Omega_5\Omega_6 , \tag{21}$$

$$\Delta=(\omega_1\omega_2-\omega_5\omega_6)(\omega_1'\omega_2'-\omega_5'\omega_6'),\quad \Delta'=(\omega_3\omega_4-\omega_5\omega_6)(\omega_3'\omega_4'-\omega_5'\omega_6') ,$$

with

$$\Omega_1=v_1+v_2 ,\quad \Omega_2=v_3+v_4 ,\quad \Omega_3=v_5+v_6 ,\quad \Omega_4=v_7+v_8 ,$$

$$\Omega_5\Omega_6=v_1v_3+v_2v_4+v_5v_7+v_6v_8 ,\quad \Omega_7\Omega_8=v_1v_4+v_2v_3+v_5v_8+v_6v_7 .$$

We remark that, without the two terms involving $\sin^2\theta$ and $\sin^2\phi$ in Eq. (20), the solution is exactly of the form[6] of that of a uniform eight-vertex model with weights $\{\Omega_1,\ldots,\Omega_8\}$, which satisfy the free-fermion condition

$$\Omega_1\Omega_2+\Omega_3\Omega_4=\Omega_5\Omega_6+\Omega_7\Omega_8 . \tag{22}$$

Note that there are now only seven independent parameters in the solution (19).

It is instructive to demonstrate that the solution (19) indeed reduces to the previously known expressions. For $\omega_i=\omega_i'$, the uniform model of Ref. 6, it is readily verified that $F(\theta,\phi)$ factorizes into

$$F(\theta,\phi)=[2a+2b\cos\alpha+2c\cos\beta+2d\cos(\alpha-\beta)$$
$$+2e\cos(\alpha+\beta)][2a-2b\cos\alpha-2c\cos\beta$$
$$+2d\cos(\alpha-\beta)+2e\cos(\alpha+\beta)] , \tag{23}$$

with

$$\alpha=\tfrac{1}{2}(\theta+\phi) ,\quad \beta=\tfrac{1}{2}(\theta-\phi) ,$$

$$a=\tfrac{1}{2}(\omega_1^2+\omega_2^2+\omega_3^2+\omega_4^2) ,\quad b=\omega_1\omega_3-\omega_2\omega_4 ,$$

$$c=\omega_1\omega_4-\omega_2\omega_3 ,\quad d=\omega_3\omega_4-\omega_7\omega_8 ,\quad e=\omega_3\omega_4-\omega_5\omega_6 .$$

The solution (19) now reduces to that of Ref. 6. For $v_7=v_8=0$, the staggered ice-rule model, $F(\theta,\phi)$

can be written as

$$F(\theta,\phi)=\left|v_1 e^{i\theta}+v_2 e^{-i\theta}+v_3 e^{i(\pi-\phi)}+v_4 e^{-(\pi-\phi)}+v_5+v_6\right|^2. \tag{24}$$

This is the result of I.

To analyze the analytic properties of Eq. (19), one could, as was done in I, carry out one of the two integrations in Eq. (19). Unfortunately the resulting expression is given in terms of the roots of a quartic equation, which does not appear to be very illuminating. We shall, therefore, be less ambitious and confine our considerations to the Boltzmann weights, Eq. (3), the physical model of a ferroelectric, and require the free-fermion condition (18) to hold at *all* temperatures.

To satisfy Eq. (18) at all temperatures, we may take, without loss of generality.

$$\omega_1\omega_2=\omega_7\omega_8,\quad \omega_3\omega_4=\omega_5\omega_6. \tag{25}$$

There are now two possibilities for the second condition in Eq. (18) to hold,

$$\text{(i)}\ \ \omega_1'\omega_2'=\omega_5'\omega_6',\quad \omega_3'\omega_4'=\omega_7'\omega_8'; \tag{26a}$$

$$\text{(ii)}\ \ \omega_1'\omega_2'=\omega_7'\omega_8',\quad \omega_3'\omega_4'=\omega_5'\omega_6'. \tag{26b}$$

In both cases, we find the free energy ψ nonanalytic at $T=T_c$, defined by

$$\Omega_1+\Omega_2+\Omega_3+\Omega_4=2\max\{\Omega_1,\Omega_2,\Omega_3,\Omega_4\}. \tag{27}$$

We also find that the singular part of ψ behaves as[11]

$$\psi_{\rm sing}\sim t^2\ln|t|,\quad t=(T-T_c)/T_c\to 0, \tag{28}$$

except for (a) $\Omega_5\Omega_6\Omega_7\Omega_8=0$ in case (i), and (b) $\Omega_2\Omega_3=\Delta_1=0$ or $\Omega_1\Omega_4=\Delta_2=0$ in case (ii), where

$$\begin{aligned}\Delta_1&=\Omega_5\Omega_6\Omega_7\Omega_8-\Delta(\Omega_2+\Omega_3)^2,\\ \Delta_2&=\Omega_5\Omega_6\Omega_7\Omega_8-\Delta(\Omega_1+\Omega_4)^2.\end{aligned} \tag{29}$$

In these special cases, we find

$$\begin{aligned}\psi&=\text{frozen},\quad T<T_c,\\ \psi_{\rm sing}&\sim t^{3/2},\quad T\to T_c+.\end{aligned} \tag{30}$$

Detailed analysis leading to these results will be given in Sec. V. Note that the specific heat now has an Ising-type (logarithmic) singularity; except in cases (a) and (b), it has an exponent $\alpha=\frac{1}{2}$.[11]

One consequence of the critical condition (27) is that there may exist up to three phase transitions. To see this, let us take, without loss of generality.

$$v_1\ge v_i\ge v_j,\quad i\ne 1,\ \text{all } j\ne 1, i. \tag{31}$$

If $i=2$, then Eq. (27) reads

$$\Omega_1=\Omega_2+\Omega_3+\Omega_4, \tag{32}$$

which always has one solution. If $i\ne 2$, say $i=3$, and Ω_2 is *not* a single Boltzmann factor, then in addition to Eq. (31), (27) can also be satisfied by

$$\Omega_2=\Omega_1+\Omega_3+\Omega_4. \tag{33}$$

Since $\Omega_2<\Omega_1+\Omega_3+\Omega_4$ at $T=0,\ \infty$ and both sides of Eq. (33) are monotonic and concave in T, Eq. (33) has either 0 or 2 solutions.[11] Therefore the staggered free-fermion model will, in general, possess up to three phase transitions. Note that Eq. (33) is not valid if Ω_2 is a single Boltzmann factor. Thus, there can be only one transition in the special cases (a) and (b). A well-known example of the staggered eight–vertex model which has three phase transitions is the Ising model on the Union Jack lattice.[12] Let the first- and second-neighbor interactions be, respectively, $-J$ and $-J'$. Using Eq. (14), or more simply Eq. (29) of Ref. 1, we find the staggered vertex weights,

$$\begin{aligned}&\omega_1=\omega_1'=e^{K'+2K}, &&\omega_2=\omega_2'=e^{K'-2K}\\ &\omega_3=\omega_3'=\omega_4=\omega_4'=e^{-K'}, &&\omega_5=\omega_6=\omega_7'=\omega_8'=e^{-K'},\\ &\omega_7=\omega_8=\omega_5'=\omega_6'=e^{K'}.\end{aligned} \tag{34}$$

where $K=J/kT$, $K'=J'/kT$. It is readily verified that these weights satisfy the free-fermion condition (18). Thus Eqs. (32) and (33) read

$$\begin{aligned}e^{2K'}\cosh 4K&=e^{-2K'}+2,\\ e^{-2K'}&=e^{2K'}\cosh 4K+2.\end{aligned} \tag{35}$$

The first equation has always one solution, while the second equation has two solutions for J' in the antiferromagnetic range $-|J|<J'<-0.9068|J|$.

Before closing this section, we remark that there exists another soluble staggered eight-vertex model which exhibits two phase transitions. This is when the vertex weights are given by Eq. (10) and related by

$$a_ib_i=c_id_i,\quad i=1,2. \tag{36}$$

Although the vertex weights do not satisfy the free-fermion condition directly, the model is soluble because the Ising representation (15) decouples into a superposition of two nonequivalent nearest-neighbor Ising lattices.[1,13] Consequently, the model has two distinct transitions.

V. ANALYTIC PROPERTIES OF ψ

Details that lead to the results quoted in Sec. IV are now given. It is convenient to consider the cases (i) and (ii) or Eqs. (26a) and (26b) separately.

(i) Models satisfying the free-fermion conditions (25) and (26a)

The free energy is now given by

$$\psi=\frac{1}{16\pi^2}\int_0^{2\pi}d\theta\int_0^{2\pi}d\phi\,\ln F_0(\theta,\phi). \tag{37}$$

As we have remarked earlier, Eq. (37) is now exactly of the form of the free energy of a uniform free-fermion model,[6] the only difference being that we do not have the simplification $D=0$ or $E=0$ used in Ref. 6. Nevertheless, the properties of ψ can be similarly studied. Carry out the θ integration using, e.g., Eq. (21) of Ref. 6, we obtain

$$\psi = \frac{1}{8\pi}\int_0^{2\pi} d\phi \ln\{A - C\cos\phi + [Q_0(\phi)]^{1/2}\}, \tag{38}$$

where

$$Q_0(\phi) = [(\Omega_1\Omega_4 + \Omega_2\Omega_3)\cos\phi - \tfrac{1}{2}(\Omega_1^2 - \Omega_2^2 - \Omega_3^2 + \Omega_4^2)]^2 + 4\Omega_5\Omega_6\Omega_7\Omega_8 \sin^2\phi \geq 0. \tag{39}$$

The analytic properties of ψ will now depend on whether $Q_0(\phi)$ is a complete square.

(a) Q_0 *is a complete square.* The only possibility is

$$\Omega_5\Omega_6\Omega_7\Omega_8 = 0, \tag{40}$$

which can be realized by taking, e.g., $\omega_1 = \omega_3 = \omega_5 = \omega_7 = 0$ (Ω_1, Ω_2, Ω_3 and Ω_4 are now single Boltzmann factors). Then

$$\psi = \frac{1}{8\pi}\int_0^{2\pi} d\phi \ln \max\{\Omega_1^2 + \Omega_4^2 - 2\Omega_1\Omega_4\cos\phi,\ \Omega_2^2 + \Omega_3^2 + 2\Omega_2\Omega_3\cos\phi\}. \tag{41}$$

Now Ω_1, Ω_2, Ω_3 and Ω_4 do not form a closed polygon for $T < T_c$, where T_c is given by the critical condition (27). Then one of the two factors inside the curly brackets in Eq. (41) is always bigger and one finds

$$\psi = \tfrac{1}{2}\ln\max\{\Omega_1, \Omega_2, \Omega_3, \Omega_4\}, \quad T \leq T_c. \tag{42}$$

In other words, the system is in a frozen state below T_c. Note that in this case, there exists only one transition.

For $T > T_c$, $\Omega_1, \Omega_2, \Omega_3$ and Ω_4 do form a closed polygon. Then there exists $0 < \phi_1 < \pi$,

$$\cos\phi_1 = (\Omega_1^2 + \Omega_4^2 - \Omega_2^2 - \Omega_3^2)/2(\Omega_1\Omega_4 + \Omega_2\Omega_3), \tag{43}$$

such that the two factors in Eq. (41) are equal at ϕ_1. One finds

$$\psi = \tfrac{1}{2}\ln\max\{\Omega_1, \Omega_4\} + \frac{1}{8\pi}\int_{-\phi_1}^{\phi_1} \ln\left(\frac{\Omega_2^2 + \Omega_3^2 + 2\Omega_2\Omega_3\cos\phi}{\Omega_1^2 + \Omega_4^2 - 2\Omega_1\Omega_4\cos\phi}\right) d\phi, \quad T \geq T_c. \tag{44}$$

ψ given by Eq. (44) is analytic in $T > T_c$.

The critical case is when Ω_1, Ω_2, Ω_3 and Ω_4 just form a polygon ($T = T_c$) or $\phi_1 = 0, \pi$. To examine the singular behavior of Eq. (44) near T_c, we note that the integrand vanishes at T_c, so it is necessary to compute the derivatives of ψ.[14] Using the same argument as in I, we find the first derivatives of ψ behave as ϕ_1 near T_c. This leads to the critical behavior

$$\psi_{\rm sing} \sim t^{3/2}, \quad T \to T_c + \tag{45}$$

quoted in Eq. (30).

(b) Q_0 *is not a complete square.* $Q_0(\phi)$ is not a complete square if and only if

$$\Omega_5\Omega_6\Omega_7\Omega_8 > 0. \tag{46}$$

In this case ψ cannot be evaluated in a closed form. It can be shown,[15] however, that the first derivatives of ψ can always be expressed in terms of the complete elliptical integrals of the first and third kinds; consequently, the second derivatives of ψ have a logarithmic divergence. It is instructive that, as we shall now see, the singular behavior of ψ as well as the location of the critical point can be obtained without the recourse of actually carrying out the integration. Such an analysis will be useful in situations, as in (ii) below, when the integrals cannot be evaluated.

Generally a function $\psi(T)$ given by a double integral of the type (37) is analytic in T unless we can make

$$F_0(\theta, \phi) = 0 \tag{47}$$

at some T, θ.[16] We show in Theorem I of Appendix B that, if $Q_0(\phi)$ is not a complete square, (47) holds only at one of the four points given by (B1). For definiteness consider one of the zeros,

$$\theta = \phi = 0, \quad \Omega_1 = \Omega_2 + \Omega_3 + \Omega_4, \tag{48}$$

and focus attention on the critical region. To extract the singular part of ψ, we expand $F_0(\theta, \phi)$ about $\theta = \phi = 0$,

$$\psi_{\rm sing} \sim \int_0 d\theta \int_0 d\theta \ln[(\Omega_1 - \Omega_2 - \Omega_3 - \Omega_4)^2 + \alpha\theta^2 + \beta\theta\phi + \gamma\phi^2], \tag{49}$$

where

$$\alpha = B - D - E, \quad \beta = 2(D - E), \quad \gamma = C - D - E. \tag{50}$$

In Eq. (49), only the lower integration limits are needed, and for a given T, the integration is restricted in a region around $\theta = \phi = 0$ such that the quantity inside the square brackets is positive. To evaluate this integral, we rotate the $\{\theta, \phi\}$ coordinate by an angle $\frac{1}{2}\tan^{-1}[\beta/(\alpha - \gamma)]$ to eliminate the $\theta\phi$ term. This leads to

$$\psi_{\rm sing} \sim \int_0 d\theta \int_0 d\phi \ln[(\Omega_1 - \Omega_2 - \Omega_3 - \Omega_4)^2 + \tfrac{1}{2}(\alpha + \gamma + \delta)\theta^2 + \tfrac{1}{2}(\alpha + \gamma - \delta)\phi^2], \tag{51}$$

where

$$\delta = [\beta^2 + (\alpha - \gamma)^2]^{1/2}. \tag{52}$$

Provided that the coefficients of θ^2 and ϕ^2 in Eq. (51) are not zero, the integral (51) can be meaningfully evaluated. In order to obtain an expansion in t, we rewrite Eq. (51) in terms of

$$\tau = \Omega_1 - \Omega_2 - \Omega_3 - \Omega_4, \tag{53}$$

so that

$$\alpha+\gamma=(\Omega_2+\Omega_3)^2+(\Omega_2+\Omega_4)^2+(2\Omega_2+\Omega_3+\Omega_4)\tau\,,$$

$$\delta^2-(\alpha+\gamma)^2=-16\Omega_5\Omega_6\Omega_7\Omega_8-4(\Omega_2+\Omega_3)(\Omega_2+\Omega_4)\times(\Omega_3+\Omega_4)\tau-4(\Omega_2\Omega_3+\Omega_2\Omega_4+\Omega_3\Omega_4)\tau^2\,. \quad (54)$$

Now as $T\to T_c$, $\tau\sim t\to 0$, we see that, to the leading order in t, Eq. (51) is of the form

$$\psi_{\text{sing}}\sim\int_0 d\theta\int_0 d\phi\ln(t^2+p\theta^2+q\phi^2)\,, \quad (55)$$

where $q=O(t)$ if $\Omega_5\Omega_6\Omega_7\Omega_8=0$.

If $\Omega_5\Omega_6\Omega_7\Omega_8\neq 0$ [$Q_0(\phi)$ is not a complete square], both p and q are nonzero and Eq. (55) can be evaluated to give[17]

$$\psi_{\text{sing}}\sim t^2\ln|t|\,. \quad (56)$$

This is the result quoted in Eq. (28).[11]

If $\Omega_5\Omega_6\Omega_7\Omega_8=0$ [$Q_0(\phi)$ is a complete square], the above argument breaks down because $q=0$ at T_c [or more precisely $q=O(t)$]. Fortunately this case has been considered in (a) above. A direct way to identify this special case is to observe that $\Omega_5\Omega_6\Omega_7\Omega_8=0$ is precisely

$$\beta^2=4\alpha\gamma \quad \text{at } T_c\,.$$

Thus the condition under which (56) breaks down is when the discriminant of the quadratic form $\alpha\theta^2+\beta\theta\phi+\gamma\phi^2$ in Eq. (49) vanishes at T_c.

(ii) Models satisfying the free-fermion conditions (25) and (26b)

The free energy is now given by

$$\psi=\frac{1}{16\pi^2}\int_0^{2\pi}d\theta\int_0^{2\pi}d\phi\ln F_1(\theta,\phi)\,, \quad (57)$$

$$F_1(\theta,\phi)=F_0(\theta,\phi)-4\Delta\sin^2\phi\,. \quad (58)$$

Carrying out the θ integration, we obtain

$$\psi=\frac{1}{8\pi}\int_0^{2\pi}d\phi\ln\{A-C\cos\phi-2\Delta\sin^2\phi+[Q(\phi)]^{1/2}\}\,, \quad (59)$$

where

$$Q(\phi)=[2\Delta\sin^2\phi+(\Omega_1\Omega_4+\Omega_2\Omega_3)\cos\phi-\tfrac{1}{2}(\Omega_1^2-\Omega_2^2-\Omega_3^2+\Omega_4^2)]^2+8\Delta\Omega_2\Omega_3(1-\cos\phi)\sin^2\phi+4\Delta_1\sin^2\phi$$
$$=[2\Delta\sin^2\phi-(\Omega_1\Omega_4+\Omega_2\Omega_3)\cos\phi+\tfrac{1}{2}(\Omega_1^2-\Omega_2^2-\Omega_3^2+\Omega_4^2)]^2+8\Delta\Omega_1\Omega_4(1+\cos\phi)\sin^2\phi+4\Delta_2\sin^2\phi\geq 0\,, \quad (60)$$

with Δ_1, Δ_2 defined in Eq. (29). [The last step in Eq. (60) will be proved in Appendix B.] Again, we consider the following two cases separately.

(a) *$Q(\phi)$ is a complete square.* Excluding the case $\Omega_5\Omega_6\Omega_7\Omega_8=\Delta=0$ considered in (ia) above, $Q(\phi)$ given by Eq. (60) can be a complete square only when either

$$\Omega_2\Omega_3=\Delta_1=0 \text{ or } \Omega_1\Omega_4=\Delta_2=0\,. \quad (61)$$

Only the first case will be considered, as the two are obviously related by symmetry. The first relation in Eq. (61) can be realized by taking, e.g.,

$$\omega_3=\omega_3'=\omega_5=\omega_5'=0 \quad (62)$$

or

$$\Omega_1=v_1+v_2\,,\quad \Omega_2=v_4\,,$$
$$\Omega_3=0\,,\quad \Omega_4=v_7+v_8\,,\quad \Delta=v_1v_2=v_7v_8\,. \quad (63)$$

Here the last equality follows from the free-fermion conditions (25) and (26b). Now Eq. (59) becomes

$$\psi=\frac{1}{8\pi}\int_0^{2\pi}d\phi\ln\max\{\Omega_2^2,\ f(\phi)\}\,, \quad (64)$$

where

$$f(\phi)=\Omega_1^2+\Omega_4^2-2\Omega_1\Omega_4\cos\phi-4\Delta\sin^2\phi\,.$$

For $T<T_c$, where T_c is given by Eq. (27), Ω_1, Ω_2 and Ω_4 do not form a triangle. Then one of the two factors inside the curly brackets of Eq. (64) prevails, and the integral can be performed. After some algebra we find

$$\psi=\begin{cases}\tfrac{1}{2}\ln\Omega_2\,, & \Omega_2>\Omega_1+\Omega_4\\ \tfrac{1}{4}G(v_1,v_2,v_7,v_8)\,, & \Omega_1>\Omega_2+\Omega_4 \text{ or } \Omega_4>\Omega_2+\Omega_1\,,\end{cases} \quad (65)$$

where

$$G(v_1,v_2,v_7,v_8)=\ln(v_1v_2)+\ln\max(v_7/v_2,\ v_2/v_7)+\ln\max(v_7/v_1,\ v_1/v_7)\,.$$

Therefore, the system is in a frozen state for $T<T_c$. Note that with the weights given by Eq. (62) or (63), there can exist only one transition.

For $T>T_c$, there is a triangular relationship between Ω_1, Ω_2, and Ω_4 ($\Omega_3=0$). Since $f(\phi)$ is monotonic in ϕ in $\{0,\pi\}$, there exists ϕ_1 such that

$$\Omega_2^2=f(\phi_1)\,. \quad (66)$$

Then

$$\psi=\tfrac{1}{4}G(v_1,v_2,v_7,v_8)+\frac{1}{8\pi}\int_{-\phi_1}^{\phi_1}d\phi\ln[\Omega_2^2/f(\phi)]\,,\qquad T\geq T_c\,. \quad (67)$$

This expression is of the same form as Eq. (44); hence, following the same argument, we obtain

$$\psi_{\text{sing}}\sim t^{3/2}\,,\quad T\to T_c+\,. \quad (68)$$

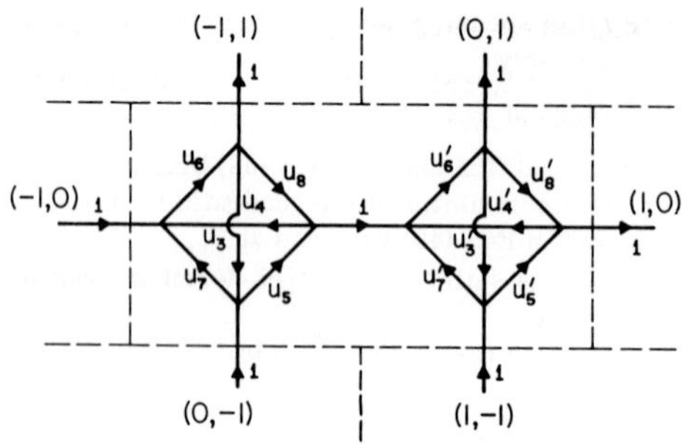

FIG. 4. Unit cell of the dimer terminal lattice L^{Δ}.

Also belonging to this category is the staggered free-fermion ice-rule model considered in I specified by

$$\begin{aligned} &\omega_1=\omega_1'=\omega_7=\omega_7'=\omega_8=\omega_8'=0\,, \\ &\omega_3\omega_4=\omega_5\omega_6\,,\quad \omega_3'\omega_4'=\omega_5'\omega_6'\,. \end{aligned} \tag{69}$$

Using the present method, we find $\Delta_2=\Omega_1\Omega_4=0$.

(b) $Q(\phi)$ *is not a complete square*. Because of the presence of the $\sin^4\phi$ term in $Q(\phi)$, ψ and its derivatives cannot be evaluated in closed forms. The method of analysis used in (ib) is now useful.

It is shown in Theorem II of Appendix B that, if $Q(\phi)$ is not a complete square, the zeros of $F_1(\theta,\phi)$ are given by Eq. (B10). Consider, e.g., the expansion of $F_1(\theta,\phi)$ about $\theta=\phi=0$ in Eq. (57). This will give us the singular behavior of ψ if $\Omega_1=\Omega_2+\Omega_3+\Omega_4$ is a critical point. Following the argument in (ib) step by step, with $\gamma=C-D-E-4\Delta>0$ and

$$\begin{aligned} \delta^2-(\alpha+\gamma)^2=-16\Delta_1-4(\Omega_2+\Omega_3)[(\Omega_2+\Omega_4)(\Omega_3+\Omega_4) \\ -4\Delta]\tau-4(\Omega_2\Omega_3+\Omega_2\Omega_4+\Omega_3\Omega_4)\tau^2\,, \end{aligned} \tag{70}$$

we see that if $\Delta_1\neq 0$ at $\tau=0$, we have $q\sim\alpha+\gamma-\delta\neq 0$ as $\tau\sim t\to 0$. The same argument used in (ib) now leads to, near the critical point $\Omega_1=\Omega_2+\Omega_3+\Omega_4$,

$$\psi_{\rm sing}\sim t^2\ln|t|\,,\quad t\to 0\,. \tag{71}$$

The argument breaks down if $\Delta_1=0$. If we also have $\Omega_2\Omega_3=0$, as given by Eqs. (62) and (63), then $Q(\phi)$ is a complete square, and the case has been considered in (iia). If $\Omega_2\Omega_3\neq 0$, it may be verified that we have either $\Delta_2=\Omega_4=0$ so $Q(\phi)$ is again a complete square, or $\Omega_1=\Omega_4$ so $\Omega_1=\Omega_2+\Omega_3+\Omega_4$ is not a critical point, and the expansion about $\theta=\phi=0$ is irrelevant.

Similarly, expansions of $F_1(\theta,\phi)$ about $\{\theta,\phi\}=\{0,\pi\}$ or $\{\pi,0\}$ lead to the singular behavior $t^2\ln|t|$ except when $\Delta_2=0$. In the latter case, we must have $\Omega_1\Omega_4=0$ to relate the expansions to the critical points. These cases have been considered in (iia).

ACKNOWLEDGMENT

One of us (F. Y. W.) wishes to thank Dr. Shien-Shu Shu for the hospitality extended to him at the National Tsing Hua University, where this work was initiated.

APPENDIX A: PFAFFIAN SOLUTION

Procedures of obtaining the Pfaffian solution for the staggered free-fermion eight-vertex model (18) follow closely that of I. First we write

$$Z(\omega_i,\omega_i')=(\omega_2\omega_2')^{N/2}Z(u_i,u_i')\,, \tag{A1}$$

where $u_i=\omega_i/\omega_2$, $u_i'=\omega_i'/\omega_2'$. $Z(u_i,u_i')$ is then converted into a dimer generating function Z^{Δ}.

To evaluate Z^{Δ}, we proceed exactly as in I; the only difference here is that a unit cell of the dimer terminal lattice is now given as shown in Fig. 4.[18] It is easily checked, as in Fig. 6 of I, that this unit cell generates all the required vertex weights. Following the same procedure, we then obtain

$$\psi=\frac{1}{4(2\pi)^2}\int_{-\pi}^{\pi}d\alpha\int_{-\pi}^{\pi}d\beta\ln[\omega_2^2\omega_2'^2D(\alpha,\beta)]\,, \tag{A2}$$

where

$$D(\alpha,\beta)=\begin{vmatrix} 0 & u_3 & -u_8 & -u_5 & 0 & 1 & 0 & 0 \\ -u_3 & 0 & u_6 & -u_7 & -e^{-i\alpha} & 0 & 0 & 0 \\ u_8 & -u_6 & 0 & u_4 & 0 & 0 & 0 & e^{-i(\alpha-\beta)} \\ u_5 & u_7 & -u_4 & 0 & 0 & 0 & -e^{-i\beta} & 0 \\ 0 & e^{i\alpha} & 0 & 0 & 0 & u_3' & -u_8' & -u_5' \\ -1 & 0 & 0 & 0 & -u_3' & 0 & u_6' & -u_7' \\ 0 & 0 & 0 & e^{i\beta} & u_8' & -u_6' & 0 & u_4' \\ 0 & 0 & -e^{i(\alpha-\beta)} & 0 & u_5' & u_7' & -u_4' & 0 \end{vmatrix}\,. \tag{A3}$$

After changing the variables $\alpha=\theta+\phi$, $\beta=\theta$, Eq. (A2) reduces to Eq. (19) in the text. Note that we can see directly from (A3) that $D(\alpha,\beta)$ factorizes if $u_7=u_8=u_7'=u_8'=0$, a result quoted in I.

APPENDIX B: ZEROS OF F_0 AND F_1

In this appendix, we determine the zeros of $F_0(\theta, \phi)$ and $F_1(\theta, \phi)$, defined, respectively, by Eqs. (21) and (58).

Theorem I

$$F_0(\theta, \phi) \geq 0 \text{ for all } \theta \text{ and } \phi,$$

and

$$F_0(\theta, \phi) = 0 \text{ if and only if},$$

(a) for $\Omega_5\Omega_6\Omega_7\Omega_8 \neq 0$, at the following points:

$$\begin{aligned} &\theta = \phi = 0, && \Omega_1 = \Omega_2 + \Omega_3 + \Omega_4, \\ &\theta = \phi = \pi, && \Omega_2 = \Omega_1 + \Omega_3 + \Omega_4, \\ &\theta = 0, \quad \phi = \pi, && \Omega_3 = \Omega_1 + \Omega_2 + \Omega_4, \\ &\theta = \pi, \quad \phi = 0, && \Omega_4 = \Omega_1 + \Omega_2 + \Omega_3; \end{aligned} \tag{B1}$$

(b) for $\Omega_5\Omega_6\Omega_7\Omega_8 = 0$, at

$$\cos\theta = (\Omega_1^2 - \Omega_2^2 - \Omega_4^2 + \Omega_3^2)/2(\Omega_1\Omega_3 + \Omega_2\Omega_4), \tag{B2}$$

$$\cos\phi = (\Omega_1^2 - \Omega_2^2 - \Omega_3^2 + \Omega_4^2)/2(\Omega_1\Omega_4 + \Omega_2\Omega_3). \tag{B3}$$

Note that Eqs. (B2) and (B3) include Eq. (B1) and have real solutions only when $\Omega_1 + \Omega_2 + \Omega_3 + \Omega_4 \geq 2\max\{\Omega_1, \Omega_2, \Omega_3, \Omega_4\}$.

Proof: We write

$$\begin{aligned} F_0(\theta, \phi) &= 2a + 2b\cos\theta + 2c\sin\theta \\ &\geq 2[a - (b^2 + c^2)^{1/2}] \end{aligned} \tag{B4}$$

where

$$\begin{aligned} a &= \tfrac{1}{2}(\Omega_1^2 + \Omega_2^2 + \Omega_3^2 + \Omega_4^2) - (\Omega_1\Omega_4 - \Omega_2\Omega_3)\cos\phi, \\ b &= \Omega_2\Omega_4 - \Omega_1\Omega_3 + (\Omega_3\Omega_4 - \Omega_1\Omega_2)\cos\phi, \\ c &= (\Omega_5\Omega_6 - \Omega_7\Omega_8)\sin\phi, \end{aligned} \tag{B5}$$

and the equal sign in Eq. (B4) holds when

$$\begin{aligned} \cos\theta &= -b/(b^2 + c^2)^{1/2}, \\ \sin\theta &= -c/(b^2 + c^2)^{1/2}. \end{aligned} \tag{B6}$$

It is readily verified that

$$a^2 - b^2 - c^2 = Q_0(\phi) \geq 0, \tag{B7}$$

where $Q_0(\phi)$ has been given in Eq. (39). Since $a \geq 0$, we conclude from Eqs. (B4) and (B7) that

$$F_0(\theta, \phi) \geq 0. \tag{B8}$$

We also see that $F_0(\theta, \phi)$ vanishes if and only if Eq. (B6) holds, and

$$Q_0(\phi) = 0. \tag{B9}$$

Using Eq. (39) for $Q_0(\phi)$, Eq. (B9) now leads to Eqs. (B1)–(B3). Q.E.D.

Theorem II

$$F_1(\theta, \phi) \geq 0 \text{ for all } \theta \text{ and } \phi,$$

and

$$F_1(\theta, \phi) = 0 \text{ if and only if},$$

(a) for $\Delta \leq 0$, and for $\Omega_2\Omega_3$ or $\Delta_1 \neq 0$ ($\Omega_1\Omega_4$ or $\Delta_2 \neq 0$ by symmetry), and $\Delta > 0$, at one of the following points:

$$\begin{aligned} &\theta = \phi = 0, && \Omega_1 = \Omega_2 + \Omega_3 + \Omega_4, \\ &\theta = \phi = \pi, && \Omega_2 = \Omega_1 + \Omega_3 + \Omega_4, \\ &\theta = 0, \quad \phi = \pi, && \Omega_3 = \Omega_1 + \Omega_2 + \Omega_4, \\ &\theta = \pi, \quad \phi = 0, && \Omega_4 = \Omega_1 + \Omega_2 + \Omega_3; \end{aligned} \tag{B10}$$

(b) for $\Omega_2\Omega_3 = 0$, $\Delta_1 = 0$, $\Delta > 0$, at

$$2\Delta\sin^2\phi + \Omega_1\Omega_4\cos\phi - \tfrac{1}{2}(\Omega_1^2 - \Omega_2^2 - \Omega_3^2 + \Omega_4^2) = 0, \tag{B11}$$

and θ given by Eq. (B6);

(c) for $\Omega_1\Omega_4 = 0$, $\Delta_2 = 0$, $\Delta > 0$, at

$$2\Delta\sin^2\phi - \Omega_2\Omega_3\cos\phi + \tfrac{1}{2}(\Omega_1^2 - \Omega_2^2 - \Omega_3^2 + \Omega_4^2) = 0, \tag{B12}$$

and θ given by Eq. (B6). Here Δ_1, Δ_2 are defined in Eq. (29). Also note that Eqs. (B11) and (B12) include (B10).

Proof: Since

$$F_1(\theta, \phi) = F_0(\theta, \phi) - 4\Delta\sin^2\phi, \tag{B13}$$

for $\Delta \leq 0$ the theorem follows directly from Theorem I. Therefore, in the following we assume $\Delta > 0$, unless otherwise noted.

We write

$$\begin{aligned} F_1(\theta, \phi) &= 2a' + 2b\cos\phi + 2c\sin\phi \\ &\geq 2[a' - (b^2 + c^2)^{1/2}], \end{aligned} \tag{B14}$$

where

$$a' = a - 2\Delta\sin^2\phi,$$

and the equal sign holds when θ is given by Eq. (B6). It is verified that, for all Δ,

$$a'^2 - b^2 - c^2 = Q(\phi) \geq 0, \tag{B15}$$

where $Q(\phi)$ is given by Eq. (60) and the non-negativeness of $Q(\phi)$ will be proved later. Now

$$a' \geq 0, \text{ for all } \phi. \tag{B16}$$

This can be seen by noting $a' \geq 0$ at $\phi = 0, \pi$; so the minimum of a', if any, occurs at $\phi = \phi_0$, $\cos\phi_0 = (\Omega_1\Omega_4 - \Omega_2\Omega_3)/4\Delta$. Hence,

$$(a')_{\min} = A - 2\Delta(1 + \cos^2\phi_0) \geq A - 4\Delta \geq 0, \tag{B17}$$

where the last step in Eq. (B17) follows from

$$A \geq 2(v_1v_2 + v_3v_4 + v_5v_6 + v_7v_8), \tag{B18}$$

and, for $\Delta > 0$,

$$v_1v_2 + v_5v_6 > \Delta, \quad v_3v_4 + v_7v_8 > \Delta. \tag{B19}$$

From Eqs. (B14)–(B16), we conclude that, for $\Delta > 0$,

$$F_1(\theta, \phi) \geq 0, \tag{B20}$$

and $F_1(\theta, \phi)$ vanishes only at

$$Q(\phi)=0, \tag{B21}$$

and θ given by Eq. (B6). Using Eq. (60) for $Q(\phi)$, Eq. (B20) now leads to the remainder ($\Delta>0$) of Theorem II.

Thus it remains only to show $Q(\phi)\geq 0$. For $\Delta\leq 0$ this follows from $Q_0(\phi)\geq 0$. For $\Delta>0$ we use (60) for $Q(\phi)$; so the proof is completed if we can show that we always have *either* $\Delta_1\geq 0$ or $\Delta_2\geq 0$. In the following we shall use only

$$\omega_1\omega_2=\omega_7\omega_8,\quad \omega_3\omega_4=\omega_5\omega_6,$$

and (B22)

$$\omega_1'\omega_2'+\omega_3'\omega_4'=\omega_5'\omega_6'+\omega_7'\omega_8';$$

so the result applies to both cases (i) and (ii) in Sec. IV. We have from Eq. (29)

$$\Delta_1=\Omega_5\Omega_6\Omega_7\Omega_8-\Delta(\Omega_2+\Omega_3)^2,$$
$$=u+v-\Delta(\Omega_2+\Omega_3)^2, \tag{B23}$$

where

$$u=(v_1v_3+v_2v_4)(v_1v_4+v_2v_3)+(v_5v_7+v_6v_8)(v_5v_8+v_6v_7)=v_3v_4(v_1-v_2)^2+v_1v_2\Omega_2^2+v_5v_6(v_7-v_8)^2+v_7v_8\Omega_3^2$$
$$\geq v_1v_2\Omega_2^2+v_7v_8\Omega_3^2, \tag{B24}$$
$$v=(v_5v_7+v_6v_8)(v_1v_4+v_2v_3)+(v_1v_3+v_2v_4)(v_5v_8+v_6v_7)=(v_1v_8+v_2v_7)(v_3v_5+v_4v_6)+(v_1v_7+v_2v_8)(v_3v_6+v_4v_5),$$

since

$$v_1v_8+v_2v_7-v_1v_2-v_7v_8+\omega_1\omega_2(\omega_3'\omega_4'+\omega_5'\omega_6')=(v_1-v_7)(v_8-v_2)+\omega_1\omega_2(\omega_3'\omega_4'+\omega_5'\omega_6')$$
$$=v_8(v_1-v_7)+v_7(v_2-v_8)+2\omega_1\omega_2\omega_3'\omega_4'=v_1(v_8-v_2)+v_2(v_7-v_1)+2\omega_1\omega_2\omega_5'\omega_6'\geq 0,\quad \text{for all } v_1,\ v_2,\ v_7,\ v_8.$$

We have

$$v\geq[v_1v_2+v_7v_8-\omega_1\omega_2(\omega_3'\omega_4'+\omega_5'\omega_6')](v_3v_5+v_4v_6+v_3v_6+v_4v_5)=2\omega_1\omega_2(\omega_1'\omega_2'-\omega_5'\omega_6')\Omega_2\Omega_3. \tag{B25}$$

Combining Eqs. (B23)–(B25), we find

$$\Delta_1\geq v_1v_2\Omega_2^2+v_7v_8\Omega_3^2+2\omega_1\omega_2(\omega_1'\omega_2'-\omega_5'\omega_6')\Omega_2\Omega_3-\Delta(\Omega_2+\Omega_3)^2=\omega_5\omega_6(\omega_1'\omega_2'-\omega_5'\omega_6')(\Omega_2+\Omega_3)^2+\omega_1\omega_2(\omega_5'\omega_6'\Omega_2^2+\omega_3'\omega_4'\Omega_3^2). \tag{B26}$$

Similarly we can show (or simply by symmetry)

$$\Delta_2\geq\omega_7\omega_8(\omega_5'\omega_6'-\omega_1'\omega_2')(\Omega_1+\Omega_4)^2$$
$$+\omega_3\omega_4(\omega_7'\omega_8'\Omega_1^2+\omega_1'\omega_2'\Omega_4^2). \tag{B27}$$

From Eqs. (B26) and (B27) we see that we have always either $\Delta_1\geq 0$ or $\Delta_2\geq 0$. Q.E.D.

---

*Supported in part by the National Science Council, Taiwan, Republic of China.

†Supported in part by the National Science Foundation.

[1] For a review, see E. H. Lieb and F. Y. Wu, in *Phase Transitions and Critical Phenomena*, edited by C. Domb and M. S. Green (Academic, London, 1972), Vol. I.

[2] F. Y. Wu and K. Y. Lin, preceding paper, Phys. Rev. B 12, 419 (1975).

[3] F. Wegner, J. Phys. C 5, L131 (1972).

[4] J. Ashkin and E. Teller, Phys. Rev. 64, 178 (1943)

[5] F. Y. Wu and K. Y. Lin, J. Phys. C 7, L181 (1974).

[6] C. Fan and F. Y. Wu, Phys. Rev. B 2, 723 (1970).

[7] J. F. Nagle and H. N. V. Temperley, J. Math. Phys. 9, 1020 (1968).

[8] Two vertex configurations are conjugate to each other if they are related by a bond-hole interchange.

[9] F. Y. Wu, Phys. Rev. B 4, 2312 (1971).

[10] There is a misprint in Fig. 11 of Ref. 1. The symbol J and J' there should be interchanged.

[11] It should be pointed out that if the two transition temperatures determined by (33) coalesce into a single one, we have near this T_c, $(\Omega_2-\Omega_1-\Omega_3-\Omega_4)^2\sim t^4$. The singular part of ψ then behaves as $\psi_{\text{sing}}\sim t^4\ln|t|$ and the specific heat is finite at T_c. This anomalous behavior, which is reminiscent of that found in a decorated Ising system [H. T. Yeh, Physica 64, 427 (1973)] and occurs only for some special vertex energies related by a pair of transcendental parametric equations, may be disregarded in physical considerations.

[12] C. G. Vaks, A. I. Larkin, and Y. N. Uvchinnikov, Sov. Phys–JETP 22, 820 (1966); see also, J. E. Sacco and F. Y. Wu, J. Phys. A (to be published).

[13] F. Y. Wu, Phys. Rev. 183, 604 (1969).

[14] Discussion here is similar to that following Eq. (39) in I.

[15] H. S. Green and C. A. Hurst, *Order-Disorder Phenomena*, edited by I. Prigogine (Interscience, New York, 1964), Sec. 5.3.

[16] The validity of Eq. (47) does not necessarily imply nonanalyticity in ψ, however. Consider e.g., $\phi=\phi_1$ defined by Eq. (43). If $Q_0(\phi)$ is a complete square, our analysis shows that Eq. (47) can hold at ϕ_1 (and appropriate θ) for all $T>T_c$, while ψ is analytic.

[17] For $pq>0$ change variables by $\theta=r\cos\alpha$, $\varphi=r\sin\alpha$. For $pq<0$, divide the integration into regions of $|p|>|q|$ and $|p|<|q|$, and change variable by $r\cosh\xi=\max\{|p|,|q|\}$, $r\sinh\xi=\min\{|p|,|q|\}$.

[18] One can also use the planar dimer city introduced in Ref. 6 and arrive at a 12×12 determinant in Eq. (A2).

Journal of Statistical Physics, Vol. 116, Nos. 1/4, August 2004 (© 2004)

The Odd Eight-Vertex Model

F. Y. Wu[1] and H. Kunz[2]

Received April 1, 2003; accepted August 14, 2003

We consider a vertex model on the simple-quartic lattice defined by line graphs on the lattice for which there is always an odd number of lines incident at a vertex. This is the odd 8-vertex model which has eight possible vertex configurations. We establish that the odd 8-vertex model is equivalent to a staggered 8-vertex model. Using this equivalence we deduce the solution of the odd 8-vertex model when the weights satisfy a free-fermion condition. It is found that the free-fermion model exhibits no phase transitions in the regime of positive vertex weights. We also establish the complete equivalence of the free-fermion odd 8-vertex model with the free-fermion 8-vertex model solved by Fan and Wu. Our analysis leads to several Ising model representations of the free-fermion model with pure 2-spin interactions.

KEY WORDS: Odd eight-vertex model; free-fermion model; exact solution.

1. INTRODUCTION

In a seminal work which opened the door to a new era of exactly solvable models in statistical mechanics, Lieb[1,2] in 1967 solved the problem of the residual entropy of the square ice. His work led soon thereafter to the solution of a host of more general lattice models of phase transitions. These include the five-vertex model,[3,4] the F model,[5] the KDP model,[6] the general six-vertex model,[7] the free-fermion model solved by Fan and Wu,[8] and the symmetric 8-vertex model solved by Baxter.[9] All these previously considered models are described by line graphs drawn on a simple-quartic lattice where the number of lines incident at each vertex is even, and therefore can be regarded as the "even" vertex models.

[1] Department of Physics, Northeastern University, Boston, Massachusetts 02115; e-mail: fywu@neu.edu

[2] Institut de Physique Théorique, École Polytechnique Fédérale, Lausanne, Switzerland.

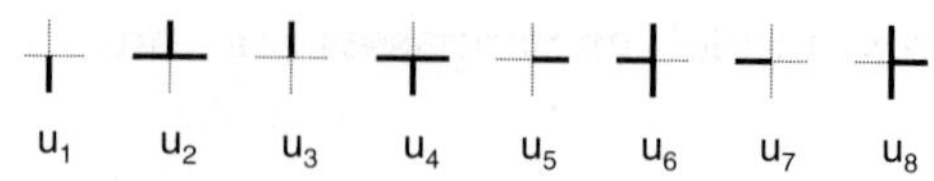

Fig. 1. Vertex configurations of the odd 8-vertex model and the associated weights.

In this paper we consider the *odd* vertex models, a problem that does not seem to have attracted much past attention. Again, one draws line graphs on the simple-quartic lattice but with the restriction that the number of lines incident at a vertex is always odd. There are again eight possible ways of drawing lines at a vertex, and this leads to the *odd 8-vertex model.* Besides being a challenging mathematical problem by itself, as we shall see the odd 8-vertex model includes some well-known unsolved lattice-statistical problems. It also finds applications in enumerating dimer configurations.[(10)]

Consider a simple-quartic lattice of N vertices and draw lines on the lattice such that the number of lines incident at a vertex is always odd, namely, 1 or 3. There are eight possible vertex configurations which are shown in Fig. 1. To vertices of type $i\ (=1, 2,..., 8)$ we associate weights $u_i > 0$. Our goal is to compute the partition function

$$Z_{12\cdots 8} \equiv Z(u_1, u_2,..., u_8) = \sum_{\text{o.l.g.}} u_1^{n_1} u_2^{n_2} \cdots u_8^{n_8} \tag{1}$$

where the summation is taken over all aforementioned odd line graphs, and n_i is the number of vertices of the type (i). The per-site "free energy" is then computed as

$$\psi = \lim_{N\to\infty} \frac{1}{N} \ln Z_{12\cdots 8}. \tag{2}$$

The partition function (1) possesses obvious symmetries. An edge can either have a line or be vacant. By reversing the line-vacancy role one obtains the symmetry

$$Z_{12345678} = Z_{21436587}. \tag{3}$$

Similarly, the left-right and up-down symmetries dictate the equivalences

$$Z_{12345678} = Z_{12347856} = Z_{34125678}, \tag{4}$$

and successive 90° counter-clockwise rotations of the lattice lead to

$$Z_{12345678} = Z_{78561243} = Z_{34127856} = Z_{56783421}. \tag{5}$$

These are intrinsic symmetries of the odd 8-vertex model.

The odd 8-vertex model encompasses an unsolved Ashkin–Teller model[11] as a special case (see below). It also generates other known solutions. For example, it is clear from Fig. 1 that by taking

$$\begin{aligned} u_1 &= y, & u_3 &= 1 \\ u_5 &= x, & u_7 &= 1 \\ u_2 &= u_4 = u_6 = u_8 = 0 \end{aligned} \tag{6}$$

(and assuming periodic boundary conditions) the line graphs generate close-packed dimer configurations on the simple-quartic lattice with activities x and y. The solution of (1) in this case is well-known.[12, 13]

2. EQUIVALENCE WITH A STAGGERED VERTEX MODEL

Our approach to the odd 8-vertex model is to explore its equivalence with a staggered 8-vertex model. We first recall the definition of a staggered 8-vertex model.[14]

A staggered 8-vertex model is an (even) 8-vertex model with sublattice-dependent vertex weights. It is defined by 16 vertex weights $\{\omega_i\}$ and $\{\omega_i'\}$, $i = 1, 2, \dots, 8$, one for each sublattice, associated with the 8 (even) line graph configurations shown in Fig. 2.

The partition function of the staggered 8-vertex model is

$$Z_{\text{stag}}(\omega_1, \omega_2, \dots, \omega_8; \omega_1', \omega_2', \dots, \omega_8') = \sum_{\text{e.l.g.}} \prod_{i=1}^{8} [\omega_i^{n_i} (\omega_i')^{n_i'}] \tag{7}$$

where the summation is taken over all even line graphs, and n_i and n_i' are, respectively, the numbers of vertices with weights ω_i and ω_i'. It is convenient to abbreviate the partition function by writing

$$Z_{\text{stag}}(\omega_1, \omega_2, \dots, \omega_8; \omega_1', \omega_2', \dots, \omega_8') \equiv Z_{\text{stag}}(12345678; 1'2'3'4'5'6'7'8'). \tag{8}$$

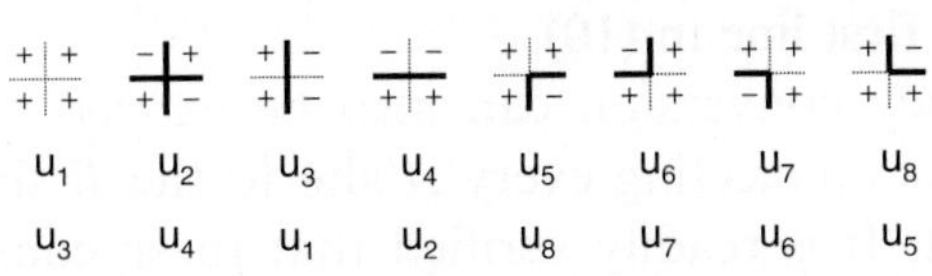

Fig. 2. An equivalent staggered 8-vertex model and the associated spin configurations on the dual.

When $\omega_i = \omega'_i$ for all i, the staggered 8-vertex model reduces to the usual 8-vertex model with uniform weights, which remains unsolved for general ω_i. When $\omega_i \neq \omega'_i$ the problem is obviously even harder. The consideration of the sublattice symmetry implies that we have

$$Z_{\text{stag}}(12345678;\ 1'2'3'4'5'6'7'8') = Z_{\text{stag}}(1'2'3'4'5'6'7'8';\ 12345678). \quad (9)$$

Returning to the odd 8-vertex model we have the following result:

Theorem. The odd 8-vertex model (1) is equivalent to a staggered 8-vertex model (8) with the equivalence

$$\begin{aligned} Z_{12\cdots 8} &= Z_{\text{stag}}(u_1, u_2, u_3, u_4, u_5, u_6, u_7, u_8;\ u_3, u_4, u_1, u_2, u_8, u_7, u_6, u_5) \\ &= Z_{\text{stag}}(u_5, u_6, u_8, u_7, u_1, u_2, u_3, u_4;\ u_7, u_8, u_6, u_5, u_4, u_3, u_1, u_2), \end{aligned}$$

or, in abbreviations,

$$\begin{aligned} Z_{12\cdots 8} &= Z_{\text{stag}}(12345678;\ 34128765) \\ &= Z_{\text{stag}}(56871243;\ 78654312). \end{aligned} \quad (10)$$

Proof. Let A and B be the two sublattices each having $N/2$ sites. Consider the set S of $N/2$ edges each of which connecting an A site to a B site immediately below it. By reversing the roles of occupation and vacancy on these edges, the vertex configurations of Fig. 1 are converted into configurations with an even number of incident lines. Because of the particular choice of S, however, the vertex weights are sublattice-dependent and we have a *staggered* 8-vertex model.

For sites on sublattice A, the conversion maps a vertex type (i) in Fig. 1 into a type (i) in Fig. 2 so that $\omega_i = u_i$ for *all i* on A. At B sites the conversion maps type (3) in Fig. 1 to type (1) in Fig. 2, (4) to (2) with $\omega'_1 = u_3$, $\omega'_2 = u_4$, etc. Writing compactly and rearranging the B weights according to configurations in Fig. 2, the mappings are

$$\begin{aligned} &\omega\{12345678\} \to u\{12345678\}, \quad &\text{at } A \text{ sites} \\ &\omega'\{12345678\} \to u\{34128765\}, \quad &\text{at } B \text{ sites.} \end{aligned} \quad (11)$$

This establishes the first line in (10).

The line-vacancy conversion can also be carried out for any of the three other edge sets connecting every A site to the B site above it, on the right, or on the left. It is readily verified that these considerations lead to the equivalence given by the second line in (10), and two others obtained from (10) by applying the sublattice symmetry (9). ∎

Remark. Further equivalences can be obtained by combining (3)–(5) with the sublattice symmetry (9).

The special case of

$$\begin{aligned} &u_1 = u_2 = u_3 = u_4 \\ &u_5 = u_6, \quad u_7 = u_8 \end{aligned} \tag{12}$$

is an Ashkin–Teller model as formulated in ref. 15 which remains unsolved. Another special case is when the weights satisfy

$$u_1 u_2 + u_3 u_4 = u_5 u_6 + u_7 u_8. \tag{13}$$

Then from (10) the staggered 8-vertex model weights satisfy the free-fermion condition

$$\begin{aligned} &\omega_1 \omega_2 + \omega_3 \omega_4 = \omega_5 \omega_6 + \omega_7 \omega_8 \\ &\omega_1' \omega_2' + \omega_3' \omega_4' = \omega_5' \omega_6' + \omega_7' \omega_8' \end{aligned} \tag{14}$$

for which the solution has been obtained in ref. 14. This case is discussed in the next section.

3. THE FREE-FERMION SOLUTION

In this section we consider the odd 8-vertex model (1) satisfying the free-fermion condition (13). In the language of the first line of the equivalence (10) we have the staggered vertex weights

$$\begin{aligned} &\omega_1 = \omega_3' = u_1, \qquad \omega_2 = \omega_4' = u_2 \\ &\omega_3 = \omega_1' = u_3, \qquad \omega_4 = \omega_2' = u_4 \\ &\omega_5 = \omega_7' = u_5, \qquad \omega_6 = \omega_8' = u_6 \\ &\omega_7 = \omega_5' = u_1, \qquad \omega_8 = \omega_6' = u_8, \end{aligned} \tag{15}$$

and hence the condition (14) is satisfied. This leads to the free-fermion staggered 8-vertex model studied in ref. 14. Using results of ref. 14 and the weights (15), we obtain after a little reduction the solution

$$\psi = \frac{1}{16\pi^2} \int_0^{2\pi} d\theta \int_0^{2\pi} d\phi \ln F(\theta, \phi) \tag{16}$$

where

$$F(\theta, \phi) = 2A + 2D\cos(\theta - \phi) + 2E\cos(\theta + \phi) + 4\Delta_1 \sin^2\phi + 4\Delta_2 \sin^2\theta \tag{17}$$

with

$$\begin{aligned} A &= (u_1u_3 + u_2u_4)^2 + (u_5u_7 + u_6u_8)^2 \\ D &= (u_5u_7)^2 + (u_6u_8)^2 - 2u_1u_2u_3u_4 \\ E &= -(u_1u_3)^2 - (u_2u_4)^2 + 2u_5u_6u_7u_8 \\ \Delta_1 &= (u_1u_2 - u_5u_6)^2 > 0 \\ \Delta_2 &= (u_3u_4 - u_5u_6)^2 > 0. \end{aligned} \tag{18}$$

As an example, specializing (16) to the weights (6) for the dimer problem, we have $A = x^2 + y^2$, $D = x^2$, $E = -y^2$, $\Delta_1 = \Delta_2 = 0$, and (16) leads to the known dimer solution[12, 13]

$$\psi_{\text{dimer}} = \frac{1}{\pi^2} \int_0^{\pi/2} d\omega \int_0^{\pi/2} d\omega' \ln(4x^2 \sin^2\omega + 4y^2 \sin^2\omega'), \tag{19}$$

which has no phase transitions. More generally for $u_i > 0$ we have $A > |D| + |E|$ and hence

$$F(\theta, \phi) > 0.$$

As a result, the free energy ψ given by (16) is analytic and there is no singularity in ψ implying that the odd free-fermion 8-vertex model has no phase transition.

4. EQUIVALENCE WITH THE FREE-FERMION MODEL OF FAN AND WU

The free energy (16) is of the form of that of the free-fermion model solved by Fan and Wu.[8] To see this we change integration variables in (16) to

$$\alpha = \theta + \phi, \qquad \beta = \theta - \phi, \tag{20}$$

the expression (16) then assumes the form

$$\begin{aligned} \psi = \frac{1}{16\pi^2} \int_0^{2\pi} d\alpha \int_0^{2\pi} d\beta \ln[&2A_1 + 2E\cos\alpha + 2D\cos\beta \\ &- 2\Delta_1 \cos(\alpha - \beta) - 2\Delta_2 \cos(\alpha + \beta)] \end{aligned} \tag{21}$$

where, after making use of (13),

$$\begin{aligned} A_1 &= A + \varDelta_1 + \varDelta_2 \\ &= (u_1u_2 + u_3u_4)^2 + (u_1u_3)^2 + (u_2u_4)^2 + (u_5u_7)^2 + (u_6u_8)^2. \end{aligned}$$

Comparing (21) with Eq. (16) of ref. 8, we find

$$\psi = \psi_{\rm FF}/2 \tag{22}$$

where $\psi_{\rm FF}$ is the per-site free energy of an 8-vertex model with *uniform* weights $w_1, w_2, \ldots, w_8$ satisfying the free-fermion condition

$$w_1w_2 + w_3w_4 = w_5w_6 + w_7w_8 \tag{23}$$

and

$$\begin{aligned} A_1 &= (w_1^2 + w_2^2 + w_3^2 + w_4^2)/2 \\ D &= w_1w_4 - w_2w_3 \\ E &= w_1w_3 - w_2w_4 \\ \varDelta_1 &= w_1w_2 - w_5w_6 \\ \varDelta_2 &= w_5w_6 - w_3w_4. \end{aligned} \tag{24}$$

We can solve for w_1, w_2, w_3, w_4, and w_5w_6 from the five equations in (24), and then determine w_7w_8 from (23).

By equating (24) with (18), it can be verified that one has

$$\begin{aligned} (-w_1 + w_2 + w_3 + w_4)^2 &= 2(A_1 - D - E - \varDelta_1 - \varDelta_2) = v_1^2 \\ (w_1 - w_2 + w_3 + w_4)^2 &= 2(A_1 + D + E - \varDelta_1 - \varDelta_2) = v_2^2 \\ (w_1 + w_2 - w_3 + w_4)^2 &= 2(A_1 + D - E + \varDelta_1 + \varDelta_2) = v_3^2 \\ (w_1 + w_2 + w_3 - w_4)^2 &= 2(A_1 - D + E + \varDelta_1 + \varDelta_2) = v_4^2, \end{aligned} \tag{25}$$

where[3]

$$\begin{aligned} v_1 &= 2(u_1u_3 + u_2u_4) \\ v_2 &= 2(u_5u_7 + u_6u_8) \\ v_3 &= 2\sqrt{(u_1u_2 + u_3u_4)^2 + (u_1u_3 - u_2u_4)^2 + (u_5u_7 - u_6u_8)^2} \\ v_4 &= 2(u_1u_2 + u_3u_4). \end{aligned} \tag{26}$$

[3] The apparent asymmetry in the expression of v_3 can be traced to the choice of the edge set S used in Section 2 in deducing the equivalent staggered 8-vertex model.

Then, taking the square root of (25), one obtains the explicit solution

$$w_i = (v_1 + v_2 + v_3 + v_4 - 2v_i)/4, \qquad i = 1, 2, 3, 4. \tag{27}$$

The 4th line of (24) now yields

$$w_5 w_6 = w_1 w_2 - (u_1 u_2 - u_5 u_6)^2, \tag{28}$$

and $w_7 w_8$ is obtained from (23).

The free-fermion model is known[8] to be critical at

$$2w_i = w_1 + w_2 + w_3 + w_4, \qquad i = 1, 2, 3, 4 \tag{29}$$

which is equivalent to $v_i = 0$. It is then clear from (26) that the critical point (29) lies outside the region $u_i > 0$ and this confirms our earlier conclusion that the free-fermion odd 8-vertex model does not exhibit a transition in the regime of positive weights. Our results also show that the model with some $u_i = 0$, e.g., $u_7 = u_8 = 0$, is critical. This is reminiscent to the known fact of the even vertex models that the 8-vertex model is critical in the 6-vertex model subspace.

Finally, we point out that the equivalence with a free-fermion model described in this section is based on the comparison of the free energies of the two models in the thermodynamic limit. It remains to be seen whether a mapping can be established which leads to (27) directly, and thus the word "equivalence" is used in a weaker sense.

5. ISING REPRESENTATIONS OF THE FREE-FERMION MODEL

The free-fermion odd 8-vertex model can be formulated as Ising models with pure 2-spin interactions in several different ways. In the preceding section we have established its equivalence with the Fan–Wu free-fermion model. Baxter[16] has shown that the Fan–Wu free-fermion model is equivalent to a checkerboard Ising model and that asymptotically it can be decomposed into four overlapping Ising models. It follows that the odd 8-vertex model possesses the same properties, namely, it is equivalent to a checkerboard Ising model and can be similarly decomposed asymptotically. We refer to ref. 16 for details of analysis.

An *alternate* Ising representation can be constructed as follows: Consider the equivalent staggered 8-vertex model given in the first line of (10). We place Ising spins on dual lattice sites as shown in Fig. 2 and write the partition function as

$$Z_{\text{Ising}} = \sum_{\text{spin config.}} \prod_{\text{A}} W(a, b, c, d) \prod_{\text{B}} W'(a, b, c, d) \tag{30}$$

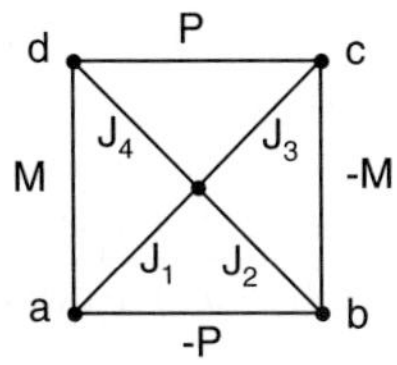

Fig. 3. Ising interactions in $W(a, b, c, d)$.

where the summation is taken over all spin configurations, and W and W' are, respectively, the Ising Boltzmann factors associated with four spins $a, b, c, d = \pm 1$ surrounding each A and B sites. Since the vertex to spin configuration mapping is 1 : 2, we have the equivalence

$$Z_{12\cdots 8} = Z_{\text{Ising}}/2. \tag{31}$$

We next require the Ising Boltzmann factors W and W' to reproduce the vertex weights ω and ω' in (10). Now to each vertex in the free-fermion model there are six independent parameters after taking into account the free-fermion condition (13) and an overall constant. We therefore need six Ising parameters which we introduce as interactions shown in Fig. 3 for $W(a, b, c, d)$ on sublattice A. Namely, we write

$$W(a, b, c, d) = 2\rho e^{M(ad-bc)/2+P(cd-ab)/2} \cosh(J_1 a + J_2 b + J_3 c + J_4 d) \tag{32}$$

where ρ is an overall constant. Explicitly, a perusal of Fig. 2 leads to the expressions

$$\begin{aligned}
u_1 &= 2\rho \cosh(J_1 + J_2 + J_3 + J_4), & u_2 &= 2\rho \cosh(J_1 - J_2 + J_3 - J_4) \\
u_3 &= 2\rho \cosh(J_1 - J_2 - J_3 + J_4), & u_4 &= 2\rho \cosh(J_1 + J_2 - J_3 - J_4) \\
u_5 &= 2\rho e^{M+P} \cosh(J_1 - J_2 + J_3 + J_4), & u_6 &= 2\rho e^{-M-P} \cosh(J_1 + J_2 + J_3 - J_4) \\
u_7 &= 2\rho e^{P-M} \cosh(-J_1 + J_2 + J_3 + J_4), & u_8 &= 2\rho e^{M-P} \cosh(J_1 + J_2 - J_3 + J_4).
\end{aligned} \tag{33}$$

These weights satisfy the free-fermion condition (13) automatically.[4]

Equation (33) can be used to solve for J_1, J_2, J_3, J_4, M, P and the overall constant ρ in terms of the weights u_i. First, using the first four

[4] Expressions in Eq. (33) are the same as Eq. (2.5) in ref. 16 except the interchange of expressions u_7 and u_8 due to the different ordering of configurations (7) and (8).

equations one solves for J_1, J_2, J_3, J_4 in terms of $\cosh^{-1}(u_i/2\rho)$, $i = 1, 2, 3, 4$. Then the overall constant ρ is solved from the equation

$$\frac{u_5 u_6}{u_7 u_8} = \frac{\cosh 2(J_1 + J_3) + \cosh 2(J_2 - J_4)}{\cosh 2(J_1 - J_3) + \cosh 2(J_2 + J_4)} \tag{34}$$

and M, P are given by

$$\begin{aligned} e^{4M} &= \left(\frac{u_5 u_8}{u_6 u_7}\right)\left[\frac{\cosh 2(J_1 - J_4) + \cosh 2(J_2 + J_3)}{\cosh 2(J_1 + J_4) + \cosh 2(J_2 - J_3)}\right], \\ e^{4P} &= \left(\frac{u_5 u_7}{u_6 u_8}\right)\left[\frac{\cosh 2(J_1 - J_2) + \cosh 2(J_3 + J_4)}{\cosh 2(J_1 + J_2) + \cosh 2(J_3 - J_4)}\right]. \end{aligned} \tag{35}$$

For B sites, we note that the weights are precisely those of A sites with the interchanges $u_1 \leftrightarrow u_3$, $u_2 \leftrightarrow u_4$, $u_5 \leftrightarrow u_8$, $u_6 \leftrightarrow u_7$. In terms of the spin configurations, these interchanges correspond to the negation of the spins b and c. Thus we have

$$\begin{aligned} W'(a, b, c, d) &= W(a, -b, -c, d) \\ &= 2\rho e^{M(ad - bc)/2 - P(cd - ab)/2} \cosh(J_1 a - J_2 b - J_3 c + J_4 d). \end{aligned} \tag{36}$$

This Boltzmann factor is the same as (30) with the same J_1, J_4, M, ρ and the negation of J_2, J_3, and P. Namely, we have

$$\begin{aligned} J_1' = J_1, \qquad J_2' = -J_2, \qquad M' = M, \qquad \rho' = \rho \\ J_3' = -J_3, \qquad J_4' = J_4, \qquad P' = -P \end{aligned} \tag{37}$$

Putting the Ising interactions together, interactions M and M' cancel and we obtain the Ising representation shown in Fig. 4. The Ising model now has five independent variables J_1, J_2, J_3, J_4, and $2P$.

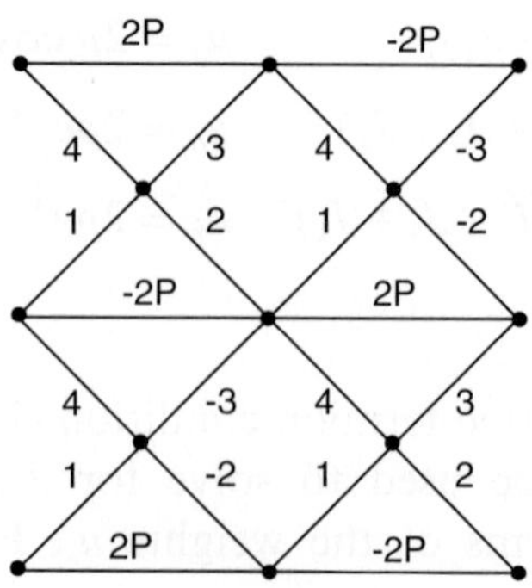

Fig. 4. An Ising model representation of the odd 8-vertex model. The number -2 stands for $-J_2$, etc.

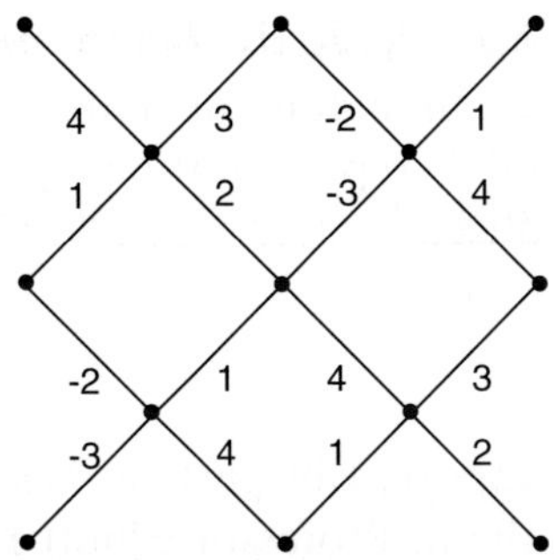

Fig. 5. An Ising model representation of the odd 8-vertex model when $u_5 = u_6$, $u_7 = u_8$. The number -2 stands for $-J_2$, etc.

If we have further

$$u_5 = u_7, \qquad u_6 = u_8, \tag{38}$$

then from the configurations in Fig. 2, we see that the weights now possess an additional up-down symmetry, namely,

$$W(a, b, c, d) = W(d, c, a, b). \tag{39}$$

Consequently we have $P = -P$ implying $P = 0$. The Ising model representation is then of the form of a simple-quartic lattice with staggered interactions as shown in Fig. 4 with $P = 0$.

If we have

$$u_5 = u_6, \qquad u_7 = u_8, \tag{40}$$

it can be seen from Fig. 2 that the A weights have the symmetry

$$W(a, b, c, d) = W(c, d, b, a) \tag{41}$$

and for B sites we have

$$W'(a, b, c, d) = W(-c, d-b, a). \tag{42}$$

In the resulting Ising model both M and P now cancel and the lattice is shown in Fig. 5.

6. SUMMARY

We have introduced an odd 8-vertex model for the simple-quartic lattice and established its equivalence with a staggered 8-vertex model. We showed that in the free-fermion case the odd 8-vertex model is completely equivalent to the free-fermion model of Fan and Wu in a noncritical regime. Several Ising model representations of the free-fermion odd 8-vertex model are also deduced.

ACKNOWLEDGMENTS

The work has been supported in part by NSF Grant DMR-9980440. The authors would like to thank Professor Elliott H. Lieb for his aspiration leading to this work. The assistance of W. T. Lu in preparing the figures is gratefully acknowledged.

REFERENCES

1. E. H. Lieb, *Phys. Rev. Lett.* **18**:692 (1967).
2. E. H. Lieb, *Phys. Rev.* **162**:162 (1967).
3. F. Y. Wu, *Phys. Rev. Lett.* **18**:605 (1967).
4. H. Y. Huang, F. Y. Wu, H. Kunz, and D. Kim, *Physica A* **228**:1 (1996).
5. E. H. Lieb, *Phys. Rev. Lett.* **18**:1046 (1967).
6. E. H. Lieb, *Phys. Rev. Lett.* **19**:108 (1967).
7. B. Sutherland, C. N. Yang, and C. P. Yang, *Phys. Rev. Lett.* **19**:588 (1967).
8. C. Fan and F. Y. Wu, *Phys. Rev. B* **2**:723 (1970).
9. R. J. Baxter, *Phys. Rev. Lett.* **26**:832 (1971).
10. F. Y. Wu, unpublished.
11. J. Ashkin and E. Teller, *Phys. Rev.* **64**:198 (1943).
12. H. N. V. Temperley and M. E. Fisher, *Phil. Mag.* **6**:1061 (1961).
13. P. W. Kasteleyn, *Physica* **27**:1209 (1961).
14. C. S. Hsue, K. Y. Lin, and F. Y. Wu, *Phys. Rev. B* **12**:429 (1975).
15. F. Y. Wu and K. Y. Lin, *J. Phys. C* **7**:L181 (1974).
16. R. J. Baxter, *Proc. R. Soc. London A* **404**:1 (1986).

Eight-vertex model on the honeycomb lattice*

F. Y. Wu†

Research School of Physical Sciences, The Australian National University, Canberra, ACT. 2601, Australia

(Received 20 December 1973)

The most general vertex model defined on a honeycomb lattice is the eight-vertex model. In this paper it is shown that the symmetric eight-vertex model reduces to an Ising model with a nonzero real or pure imaginary magnetic field H. The equivalent Ising model is either ferromagnetic with $e^{2H/kT}$ real or antiferromagnetic with $e^{2H/kT}$ unimodular. The exact transition temperature and the order of phase transition in the former case are determined. As an application of the result we verify the absence of a phase transition in the monomer-dimer system on the honeycomb lattice.

1. INTRODUCTION

The vertex model in statistical mechanics plays an important role in the study of phase transitions in lattice systems. A case of current interest is the eight-vertex model on a square lattice.[1,2] This is a rather special model in which only a limited number of the possible vertex types are allowed. The most general one on a square lattice would be the sixteen-vertex model.[3] Unfortunately, except in some special cases,[4,5] the behavior of this general model is not known.

In this paper we consider the counterpart of the sixteen-vertex model of a square lattice for the honeycomb lattice. That is, we consider an eight-vertex model defined on the hexagonal lattice. It turns out that we can say a lot more in this case. While the exact solution of this model still proves to be elusive in most cases, we can make definite statements about its phase transition. In particular, the exact transition temperature can be quite generally determined. An application of our result is the verification of the absence of a phase transition in the monomer-dimer system on the honeycomb lattice.

2. DEFINITION OF THE MODEL

In the study of a vertex model one is interested in the evaluation of a graph generating function. Consider a honeycomb lattice and draw bonds (graphs) along the lattice edges such that each edge can be independently "traced" or left "open." Denote the traced (resp. open) edges by solid (resp. broken) lines; then, as shown in Fig. 1, there are eight possible vertex configurations. With each type of vertex configuration we associate a vertex weight a, b, c, or d (see Fig. 1). Our object is to evaluate the generating partition function

$$Z=Z(a,b,c,d)=\sum_G a^{n_0}b^{n_1}c^{n_2}d^{n_3}, \tag{1}$$

where the summation is over all possible graphs on the lattice and, for a given graph G, n_i is the number of vertices having i solid lines (or bonds). This defines an "eight-vertex" model for the honeycomb lattice.

Since all possible vertex types are allowed, this eight-vertex model is the counterpart of the sixteen-vertex model of a square lattice. Note that we do not distinguish the bonds in different directions. Whereas it is possible to consider the further generalization of eight different weights, we shall not go into this complication in this paper. As a motivation we point out some special cases of interest. When $c=d=0$, the partition function (1) becomes the monomer-dimer generating function for the honeycomb lattice. When $b=d=0$, Z reduces to the partition function of a zero-field Ising model, which can be evaluated by pfaffians.

In a statistical model of phase transitions, the vertex weights are the Boltzmann factors

$$\begin{aligned} a&=\exp(-\epsilon_0/kT), \quad b=\exp(-\epsilon_1/kT),\\ c&=\exp(-\epsilon_2/kT), \quad d=\exp(-\epsilon_3/kT) \end{aligned} \tag{2}$$

where ϵ_i is the energy of a vertex having i bonds. While the weights (2) are always positive, the symmetry relations to be derived below are valid more generally for any real or complex weights.

3. SYMMETRY RELATIONS

The partition function (1) possesses a number of symmetry properties. Interchanging the solid and broken lines in Fig. 1, we obtain the symmetry relation

$$Z(a,b,c,d)=Z(d,c,b,a). \tag{3}$$

Also since both the total number of vertices, N, and the number of vertices with odd number of bonds are even, we have the negation symmetry

$$\begin{aligned} Z(a,b,c,d)&=Z(-a,-b,-c,-d)\\ &=Z(-a,b,-c,d)\\ &=Z(a,-b,c,-d). \end{aligned} \tag{4}$$

The weak graph expansion[6] yields an additional symmetry relation. For its derivation it is most convenient to use Wegner's formulation[7] of the weak-graph expansion. Denote the vertex weights by $\omega(i,j,k)$, where $i,j,k=\pm 1$ are the edge indices such that $+1$ corresponds to no bond and -1 corresponds to a bond on the edge. I.e., $\omega(+,+,+)=a$, $\omega(+,+,-)=\omega(+,-,+)=\omega(-,+,+)=b$, $\omega(+,-,-)=\omega(-,+,-)=\omega(-,-,+)=c$, and $\omega(-,-,-)=d$. Define a set of new vertex weights $\omega^*(+,+,+)=a^*$, etc. by

$$\omega^*(\alpha,\beta,\gamma)=\sum_{ijk}V_{\alpha i}V_{\beta j}V_{\gamma k}\,\omega(i,j,k), \tag{5}$$

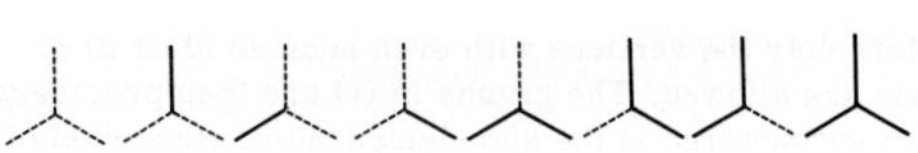

FIG. 1. The eight vertex configurations and the associated weights for a honeycomb lattice.

where the 2×2 matrix V having elements $V_{\alpha i}$ satisfies

$$V\tilde{V}=I, \tag{6}$$

I being the identity matrix. We then have the weak-graph symmetry

$$Z(a,b,c,d)=Z(a^*,b^*,c^*,d^*). \tag{7}$$

There are two possible choices for V:

$$V(y)=(1+y^2)^{-1/2}\begin{pmatrix}1 & y\\ y & -1\end{pmatrix} \tag{8}$$

or

$$U(y)=(1+y^2)^{-1/2}\begin{pmatrix}1 & y\\ -y & 1\end{pmatrix} \tag{9}$$

for aribitrary (real or complex) y. The explicit transformation generated by (8) is

$$\begin{aligned} a^* &= (1+y^2)^{-3/2}[a+3yb+3y^2c+y^3d],\\ b^* &= (1+y^2)^{-3/2}[ya-(1-2y^2)b+(y^3-2y)c-y^2d],\\ c^* &= (1+y^2)^{-3/2}[y^2a+(y^3-2y)b+(1-2y^2)c+yd],\\ d^* &= (1+y^2)^{-3/2}[y^3a-3y^2b+3yc-d]. \end{aligned} \tag{10}$$

The transformation generated by (9) leads to identical vertex weights subject to the negation symmetry $b^*\to -b^*$; $d^*\to -d^*$ hence is not independent. We shall write (10) in the short-hand notation

$$\omega^*(y)=V(y)\omega. \tag{11}$$

It is also seen that two consecutive transformations are equivalent to a single one:

$$V(x)V(y)=U\left(\frac{y-x}{1+xy}\right). \tag{12}$$

In particular we have

$$V(y)V(y)=I. \tag{13}$$

4. SPECIAL SOLUTIONS

Before we consider the model with general weights, it is useful to first consider some special cases whose solutions are known

A. $b=ua$, $c=u^2a$, $d=u^3a$

The vertex weights in this case can be converted into the bond weight u^2. Since *all* graphs are included in (1), we then obtain

$$\begin{aligned} Z&=a^NZ(1,u,u^2,u^3)\\ &=a^N(1+u^2)^{3N/2}. \end{aligned} \tag{14}$$

Here we see a simple example for which the partition function (1) does not exhibit a phase transition.

B. $b=d=0$

Here only the vertices with even number (0 or 2) of bonds are allowed. The graphs in (1) are then precisely those encountered in the high-temperature expansion of a zero-field Ising model. Writing

$$c/a=\tanh K, \tag{15}$$

we then obtain

$$\begin{aligned} Z&=Z(a,0,c,0)\\ &=a^NZ(1,0,\tanh K,0)\\ &=a^N2^{-N}(\cosh K)^{-3N/2}Z_{\text{Ising}}(0,K), \end{aligned} \tag{16}$$

where more generally $Z_{\text{Ising}}(L,K)$ is the partition function of an Ising model on the honeycomb lattice with interactions $-kTK$ and a magnetic field $-kTL$. From the known expression[8] of $Z_{\text{Ising}}(0,K)$ given by (A1) we obtain, in the large N limit,

$$\frac{1}{N}\ln Z=(16\pi^2)^{-1}\int_0^{2\pi}d\theta\int_0^{2\pi}d\phi\ln\{a^4+3c^4+2(c^4-a^2c^2)\times[\cos\theta+\cos\phi+\cos(\theta+\phi)]\}. \tag{17}$$

We remark that (17) is valid for arbitrary (real or complex) a and c, although the physical range of an Ising model is restricted to real values satisfying $|c/a|\leqslant 1$. The expression (17) is nonanalytic at

$$a/c=\pm\sqrt{3}. \tag{18}$$

Other established properties of $Z_{\text{Ising}}(L,K)$ for $L\neq 0$ are summarized in the Appendix.

C. $a=d$, $b=c$

The vertex weights are now symmetric under the interchange of the solid and the broken lines in Fig. 1. In this case we can again reduce the partition function to the form of (16). Indeed, taking $y=1$ in (10), we obtain

$$Z=Z((a+3b)/\sqrt{2},0,(a-b)/\sqrt{2},0). \tag{19}$$

The phase transition now occurs at

$$a/b=3\pm 2\sqrt{3}. \tag{20}$$

D. $ad=bc$

In this case we define the Ising parameters L and K by

$$z=\tanh K=c/a,\quad \tau=\tanh L=b/\sqrt{ac}. \tag{21}$$

Then

$$\begin{aligned} Z&=a^NZ(1,\sqrt{z}\tau,z,z^{3/2}\tau)\\ &=a^N2^{-N}(\cosh L)^{-N}(\cosh K)^{-3N/2}Z_{\text{Ising}}(L,K)\\ &=(2a^3c)^{-N}(ac-b^2)^N(a^2-c^2)^{3N/2}Z_{\text{Ising}}(L,K). \end{aligned} \tag{22}$$

Here the second step follows from the generalization of (16) to the high-temperature expansion of $Z_{\text{Ising}}(L,K)$.

E. $b^2=ac$

In this case we have

$$Z=a^NZ(1,u^{-1},u^{-2},d/a), \tag{23}$$

where $u=a/b$. The partition function on the rhs of (23) is in a form similar to that considered in Ref. 5. We then obtain in a similar fashion[9]

$$Z=(b/a)^{2N}(1+a^2/b^2)^{3N/8}(ad/bc-1)^{N/2}Z_{\text{Ising}}(L,K), \tag{24}$$

where

$$\begin{aligned} \exp(4K)&=1+a^2/b^2,\\ \exp(2L)&=(1+a^2/b^2)^{3/2}(ad/bc-1)^{-1}. \end{aligned} \tag{25}$$

We see that the Ising model is ferromagnetic for real

a/b. For the Boltzmann weights (2) (subject to $2\epsilon_1=\epsilon_0+\epsilon_2$), we find the model in general exhibits no phase transition, except for $\epsilon_0<\epsilon_1(a>b)$ and $\gamma_0<(\epsilon_3-\epsilon_0)(\epsilon_1-\epsilon_0)^{-1}<0$ the model has a first-order phase transition at $\exp(2L)=1$ or

$$(a^2+b^2)^{3/2}=a^2d-b^3. \tag{26}$$

Here $\gamma_0=3-2\ln(27+15\sqrt{3})/\ln(6+4\sqrt{3})=-0.1022204\cdots$.

5. GENERAL CASE

We are now in a position to discuss the general solution for arbitrary (positive) vertex weights (2). The idea is to introduce the weak-graph transformation (10) and choose y to make the new vertex weights satisfying either $a^*d^*=b^*c^*$ or $b^{*2}=a^*c^*$. We can then use the results of the Appendix to determine the critical behavior of the vertex model. For clarity we use subscripts 1 and 2 to distinguish the two cases. That is, in analogy to (11), we write

$$\omega_i^*\equiv\omega^*(y_i)=V(y_i)\omega,\quad i=1,2, \tag{27}$$

and consider the two cases separately.

(i) $a_1^*d_1^*=b_1^*c_1^*$: From (27) and (10) we find y_1 given by

$$y_1^2-2Ay_1-1=0, \tag{28}$$

where $A=(b^2-ac+bd-c^2)/(ad-bc)$. The new vertex weights $\omega_i^*=\{a_1^*,b_1^*,c_1^*,d_1^*\}$ are real if we take the positive solution

$$y_1=A+(A^2+1)^{1/2}>0. \tag{29}$$

Then, from (10), $a_1^*>0$. Also c_1^* is real since

$$a_1^*+c_1^*=(1+y_1^2)^{-1/2}(a+by_1+c+dy_1)>0. \tag{30}$$

The partition function is now

$$Z=(2a_1^{*3}c_1^*)^{-N}(a_1^*c_1^*-b_1^{*2})^N(a_1^{*2}-c_1^{*2})^{3N/2}\times Z_{\text{Ising}}(L_1^*,K_1^*), \tag{31}$$

where

$$\exp(2K_1^*)=(a_1^*+c_1^*)/(a_1^*-c_1^*),$$
$$\exp(2L_1^*)=[(a_1^*c_1^*)^{1/2}+b_1^*]/[(a_1^*c_1^*)^{1/2}-b_1^*]. \tag{32}$$

We observe that $\exp(2K_1^*)<1$, L_1^* = pure imaginary if $c_1^*<0$. Since not much is known about $Z_{\text{Ising}}(L_1^*,K_1^*)$ for K_1^* and L_1^* in these ranges, we shall be interested only in $c_1^*>0$. We observe in particular that, for a_1^* and c_1^* positive, $\exp(2L_1^*)\neq-1$.

(ii) $b_2^{*2}=a_2^*c_2^*$: From (27) and (10) we find y_2 given by

$$(bd-c^2)y_2^2+(ad-bc)y_2+(ac-b^2)=0. \tag{33}$$

The partition function is then

$$Z=(b_2^*/a_2^*)^{2N}(1+a_2^{*2}/b_2^{*2})^{3N/8}\times(a_2^*d_2^*/b_2^*c_2^*-1)^{N/2}Z_{\text{Ising}}(L_2^*,K_2^*). \tag{34}$$

Here the weights a_2^*,b_2^*,c_2^*,d_2^* are real if the discriminant

$$\Delta\equiv(ad-bc)^2-4(bd-c^2)(ac-b^2) \tag{35}$$

is positive. The parameters K_2^* and L_2^* are given by (25) with $a\to a_2^*$, etc. After some steps we find the simple result:

$$\exp(4K_2^*)=1+\Delta/(bd-c^2+ac-b^2)^2>0. \tag{36}$$

We shall consider $\Delta>0$ which corresponds to K_2^* being ferromagnetic. The similar expression of L_2^*, which is not needed for our discussions, is rather complicated and will not be given.

The two transformations (i) and (ii) are obviously related. To see the relationship, we observe from (27), (12), and (13) that

$$\omega_2^*=V(y_2)V(y_1)\omega_1^*=U\left(\frac{y_1-y_2}{1+y_1y_2}\right)\omega_1^*. \tag{37}$$

Since (34) is invariant under the negation of b_2^* and d_2^*, there exists a single transformation which relates ω_1^* to ω_2^*. To effect this transformation, we set $ad=bc$ in (33) and obtain $y_2=(a/c)^{1/2}$. The new weights are then

$$a_2^*=4(1+a/c)^{-3/2}(a/c)^{1/2}(b+\sqrt{ac}),$$
$$b_2^*=2(1+a/c)^{-3/2}(a/c-1)(b+\sqrt{ac}),$$
$$c_2^*=b_2^{*2}/a_2^*,$$
$$d_2^*=(1+a/c)^{-3/2}(a^{5/2}/c^{3/2}-3ab/c+3\sqrt{ac}-bc/a). \tag{38}$$

Now (36) becomes, for $ad=bc$,

$$\exp(4K_2^*)=[(a+c)/(a-c)]^2. \tag{39a}$$

Also using (38), we find

$$\exp(2L_2^*)=(\sqrt{ac}+b)/(\sqrt{ac}-b),\quad \text{if } a/c>1,$$
$$=(b+\sqrt{ac})/(b-\sqrt{ac}),\quad \text{if } a/c<1. \tag{39b}$$

Letting $a=a_1^*$, $b=b_1^*$, $c=c_1^*$, $d=d_1^*$ in (39) and comparing with (32), we then obtain the relation

$$\exp(4K_2^*)=\exp(4K_1^*),$$
$$\exp(2L_2^*)=\pm\exp(2L_1^*),\quad \text{for } a_1^*/c_1^*\gtrless1. \tag{40}$$

Note that while $\exp(2K_2^*)$ can be taken to be positive, $\exp(2K_1^*)$ can be either positive or negative. We observe from (40), (32), and (36) that $\Delta>0$ and $c_1^*>0$ are equivalent. Hence, for $\Delta>0$, K_2^* is ferromagnetic and $\exp(2L_2^*)$ is real.

Using the results of the Appendix, we conclude that, for $\Delta>0$, the nonanalyticity of Z can occur only at $\exp(2L_2^*)=+1$ or -1. To distinguish these two cases, we turn to L_1^*. Since $\exp(2K_1^*)$ may be negative, it is then convenient to consider the following situations separately:

(i) $a_1^*>c_1^*>0$: From (40) and $\exp(2L_1^*)\neq-1$, the nonanalyticity can occur only at $\exp(2L_1^*)=\exp(2L_2^*)=1$. By using (32) this is equivalent to

$$b_1^*=d_1^*=0. \tag{41}$$

A little algebra using (28) reduces (41) to

$$2(ab-cd)[(b^2-ac+bd-c^2)^2-(ad-bc)^2]+(ad-bc)(b^2-ac+bd-c^2)\times(a^2+d^2-3b^2-3c^2-2ac-2bd)=0 \tag{42}$$

which defines $T=T_c$. To see whether indeed a phase transition occurs at T_c, we observe that K_1^* and K_2^* are equal and positive. Then from the result of the Appendix we need to compute $z_c = (c_1^*/a_1^*)_{T=T_c}$. The vertex model will exhibit a first-order transition if $z_c > 1/\sqrt{3}$, a second-order transition with an infinite specific heat if $z_c = 1/\sqrt{3}$, and no transition at all if $z_c < 1/\sqrt{3}$, even if (42) has a solution. The following useful expression of z_c is obtained by combining (29), (10) and (41):

$$z_c = \left[\frac{4(ac+bd)A^2 + 4(ab-cd)A + (3b+d)(d-b) + (a+3c)(a-c)}{4(a^2+d^2)A^2 + 12(ab-cd)A + (3b+d)^2 + (a+3c)^2}\right]_{T_c}. \tag{43}$$

(ii) $c_1^* > a_1^* > 0$: In this case the nonanalyticity occurs only at $\exp(2L_2^*) = -\exp(2L_1^*) = -1$. Then T_c is again given by (41) or (42). Now $K_2^* > 0$ and $\exp(2L_2^*) = -1$; hence the vertex model always has a first-order transition. Note that we can reach the same conclusion by considering K_1^*. In this case $\exp(2K_1^*) < -1$ and $\exp(2L_1^*) = 1$. We need only to reverse the signs of $\exp(2K_1^*)$ and $\exp(2L_1^*)$ which leaves $Z_{\text{Ising}}(L_1^*, K_1^*)$ unchanged, as can be seen from the low-temperature expansion.

Combining the results in (i) and (ii), we conclude that a phase transition occurs for $\Delta > 0$ only if $z_c \geq 1/\sqrt{3}$.

A special case is that (41) or (42) is an identity. Then, for *all* Δ, $L_1^* = 0$ and Z reduces to that of a zero-field Ising model. The vertex model now exhibits the Ising-type transition (logarithmic specific heat singularity) at T_c defined by

$$\Delta/(bd - c^2 + ac - b^2)^2 = (2+\sqrt{3})^{\pm 2} - 1. \tag{44}$$

Unfortunately we are unable to make any general statement for $\Delta < 0$. For $\Delta < 0$, K_2^* is antiferromagnetic and $\exp(2L_2^*)$ is unimodular and lies on the unit circle. Presumably the zeros of an Ising antiferromagnet also distribute along the unit circle in the thermodynamic limit.[10] The vertex model then in general shows a unique transition.

6. SUMMARY

We have established the following results for the vertex model (2):

(i) If (42) is an identity, then an Ising-type transition occurs at T_c defined by (44), where Δ is given in (35).

(ii) For $\Delta \geq 0$ and (42) not an identity, a phase transition occurs at T_c defined by (42) if $z_c \geq 1/\sqrt{3}$, where z_c is given in (43). Otherwise ($z_c < 1/\sqrt{3}$) there is no phase transition. The transition is of first-order except that the specific heat diverges for $z_c = 1/\sqrt{3}$.

(iii) For $\Delta < 0$ and (42) not an identity, the vertex model is related to an Ising antiferromagnet with a pure imaginary magnetic field. Nature of the transition is not known.

It is instructive to illustrate with some examples.

(i) $a = d$, $b = c$: Since (42) is an identity, we find from (44) the critical condition

$$(a^2 + 2ab - 3b^2)/4b^2 = (2 \pm \sqrt{3})^2 - 1,$$

which agrees with (20).

(ii) $b^2 = ac$: We find $\Delta = (a^2d - b^3)^2/a^2 > 0$ and $\exp(4K_2^*) = 1 + a^2/b^2$. This is in agreement with (25). It can be verified that the condition (42) is the same as that obtained from $\exp(2L) = 1$ in (25).

(iii) $b = c = d$: We find $\Delta = b^2(a-b)^2 > 0$ and $\exp(4K_2^*) = 2$. Since K_2^* is a constant with $z_c^{-1} = 3 + 2\sqrt{2} > \sqrt{3}$, there is no phase transition.

(iv) *Monomer-dimer system*: For $c = d = 0$ the partition function (1) becomes the monomer-dimer generating function $Z_{\text{MD}}(a, b^2)$ where a and b^2 are, respectively, the monomer and dimer activities. It is known that this system does not have a phase transition.[11] We verify this by observing that $\Delta = 0$, $K_2^* = 0$. Also (42) has no solution for $c = d = 0$, $ab \neq 0$.

To obtain a closed expression for Z_{MD}, we find that, for $c = d = 0$, either $\exp(2K_2^*) = 1$, $\exp(2L_2^*) = -1$ or $\exp(2K_1^*) = -1$, $\exp(2L_1^*) = 1$. In either case the Ising partition function is identically zero. Therefore we must take the limit $c = d \to 0$ appropriately. This leads to the expression

$$Z_{\text{MD}}(a, b^2) = \lim_{c \to 0} (b/4c)^N Z_{\text{Ising}}(L_2^*, K_2^*) \tag{45}$$

where (for small c)

$$\exp(2K_2^*) = 1 + 4c/b,$$

$$\exp(2L_2^*) = -1 \pm 2a\sqrt{c}/b^{3/2}.$$

ACKNOWLEDGMENTS

I wish to thank Professor K.J. Le Couteur for his hospitality at The Australian National University, and Dr. R.J. Baxter for a discussion on the weak-graph expansion. The support of the Australian—American Educational Foundation is also gratefully acknowledged.

APPENDIX: ISING PARTITION FUNCTION

We summarize in this Appendix the relevant properties of the Ising partition function $Z_{\text{Ising}}(L, K)$.

A closed expression is known for $L = 0$. In the large N limit, one has[8]

$$\frac{1}{N}\ln Z_{\text{Ising}}(0, K) = \tfrac{3}{4}\ln 2 + \frac{1}{16\pi^2}\int_0^{2\pi} d\theta \int_0^{2\pi} d\phi \times \ln[c^3 + 1 - s^2(\cos\theta + \cos\phi + \cos(\theta+\phi))], \tag{A1}$$

where

$$c = \cosh 2K, \quad s = \sinh 2K.$$

The second derivative of (A1) diverges logarithmically at $\tanh K = \pm 1/\sqrt{3}$.

A unique property of the honeycomb lattice (coordination number = odd) is that the partition functions at $L = i\frac{1}{2}\pi$ and $L = 0$ are related. To see this connection, consider the high-temperature expansion of $Z_{\text{Ising}}(L, K)$. Using the identities for $L = i\frac{1}{2}\pi$,

$$\sum_{\sigma=\pm 1} \sigma \exp(L\sigma) = 2\sinh L = 2i, \tag{A2}$$

$$\sum_{\sigma=\pm 1} \exp(L\sigma) = 2\cosh L = 0,$$

we see that only the vertices with odd number of bonds contribute in the expansion. Thus we obtain

$$Z_{\text{Ising}}(i\tfrac{1}{2}\pi, K) = (2i)^N (\cosh K)^{3N/2} Z(0, \sqrt{z}, 0, z^{3/2}) = Z_{\text{Ising}}(0, \widetilde{K}), \tag{A3}$$

where

$$\tanh\widetilde{K}\tanh K = 1.$$

The last step follows from the symmetry relation (3) and (16). Note that $Z_{\text{Ising}}(i\frac{1}{2}\pi, K)$ is analytic for real K.

Most of the established properties for $L \neq 0$ are for ferromagnetic interactions ($K>0$). For $K>0$, $Z_{\text{Ising}}(L,K)$ can be nonanalytic in L or K only at $|\exp(2L)| = 1$.[12,13] This means $\exp(2L) = \pm 1$ for real $\exp(2L)$. At $\exp(2L) = 1$ the analyticity extends to all $0 < z < 1/\sqrt{3}$ while the first derivative w.r.t. L is discontinous for all $1/\sqrt{3} < z < 1$. At $\exp(2L) = -1$ this first derivative is presumably discontinuous for all $0 < z < 1$. This is similar to the result of a square lattice[14] and can be easily seen to hold in both the high and low temperature limits. We hope to return in the future for an exact calculation of this discontinuity.

*Work supported in part by the NSF Grant No. GH-35822 at Northeastern University.

†Senior Fulbright Scholar on leave of absence from Department of Physics, Northeastern University, Boston, Massachusetts 02115 (permanent address).

[1]C. Fan and F.Y. Wu, Phys. Rev. B 2, 723 (1970).
[2]R.J. Baxter, Ann. Phys. (N.Y.) **70**, 193 (1972).
[3]For a general introduction of the sixteen-vertex model see E.H. Lieb and F.Y. Wu, in *Phase Transitions and Critical Phenomena*, edited by C. Domb and M.S. Green (Academic, London, 1972), Vol. I.
[4]F.Y. Wu, Solid State Commun. **10**, 115 (1972).
[5]F.Y. Wu, Phys. Rev. B **6**, 1810 (1972).
[6]J.F. Nagle, J. Math. Phys. **8**, 1007 (1968).
[7]F. Wegner, Physica **68**, 570 (1973).
[8]See, e.g., Eq. (131) in C. Domb, Adv. Phys. **9**, 149 (1960).
[9]F.Y. Wu, Phys. Rev. Lett. **32**, 460 (1974).
[10]For a discussion on a plausible distribution of zeros for an Ising antiferromagnet (due to M.E. Fisher), see M. Suzuki, C. Kawabata, S. Ono, Y. Karaki, and M. Ikeda, J. Phys. Soc. Japan **29**, 837 (1970).
[11]O.J. Heilmann and E.H. Lieb, Commun. Math. Phys. **25**, 190 (1972); Phys. Rev. Lett. **24**, 1412 (1970).
[12]T.D. Lee and C.N. Yang, Phys. Rev. **87**, 410 (1952).
[13]J.L. Lebowitz and O. Penrose, Commun. Math. Phys. **11**, 99 (1968).
[14]Closed expressions for the free energy and the magnetization of a square Ising lattice at $L = i\frac{1}{2}\pi$ were first given in Ref. 12. See also B.M. McCoy and T.T. Wu, Phys. Rev. **155**, 438 (1967).

J. Phys. A: Math. Gen. **23** (1990) 375–378. Printed in the UK

COMMENT

Eight-vertex model and Ising model in a non-zero magnetic field: honeycomb lattice

F Y Wu

Department of Physics, Northeastern University, Boston, MA 02115, USA

Received 30 August 1989

Abstract. The known equivalence of the honeycomb eight-vertex model with an Ising model in a non-zero magnetic field is derived via a direct mapping. Compared with a previous derivation which uses the generalised weak-graph transformation, the new method is simpler and more direct, and can be extended to other considerations.

The eight-vertex model on the honeycomb lattice is a general lattice model playing the role of the 16-vertex model for the square lattice. The honeycomb problem was first considered by Wu [1], who used a generalised weak-graph transformation [2–4] to study its soluble cases. The honeycomb eight-vertex model has since proven to be a useful tool in deducing exact results for a number of physical problems. They include the obtaining of a closed-form expression for the critical frontier of the antiferromagnetic Ising model [5], the establishment of the effect of three-body interactions on the critical behaviour of the coexistence curve diameter of a lattice gas [6], the determination of the exact phase diagram of a spin system with two- and three-site interactions [7] and an exact analysis of the spin-1 Blume–Emery–Griffiths model [8]. A key step in all these studies is the use of the aforementioned equivalence of the eight-vertex model with an Ising model in a non-zero magnetic field. While it is fairly easy to deduce this equivalence for a special subspace of the eight-vertex model, the general equivalence of the two problems is by no means obvious. In fact, it was after considerable algebraic manipulation using a generalised weak-graph transformation that the equivalence was previously established [1, 8]. In this comment we present an alternative analysis of the eight-vertex model to arrive at the same result. The new method is very simple and direct, and can be extended to other considerations.

Consider a honeycomb lattice and draw bonds along its edges such that each edge is independently 'traced' or left 'open'. Then, there are eight different vertex configurations occurring at a vertex, which we show in figure 1. With each configuration we associate a vertex weight a, b, c or d and, as in [1], we assume all weights to be positive. The partition function of the eight-vertex model is the generating function

$$Z = Z(a, b, c, d) = \sum a^{n_0} b^{n_1} c^{n_2} d^{n_3} \tag{1}$$

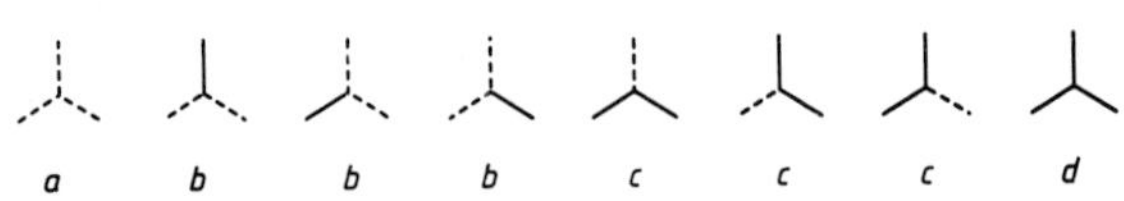

Figure 1. Vertex configurations and weights for the symmetric eight-vertex model.

where the summation is over all bond configurations of the lattice, and n_i is the number of vertices having i bonds.

Our proof that the partition function (1) is, in fact, that of an Ising model, consists of two steps. We first formulate the eight-vertex model as a decorated Ising model, and then decimate the decorating sites. The situation is illustrated in figure 2.

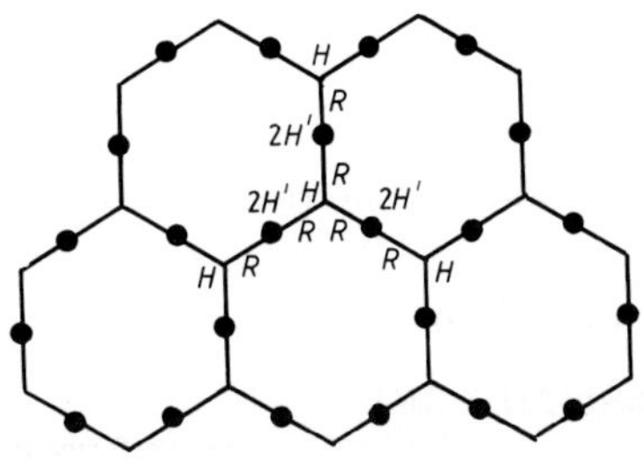

Figure 2. A decorated honeycomb lattice with the decorating sites denoted by full circles.

To formulate the eight-vertex model as a decorated Ising system, we place on each edge (of the honeycomb lattice) a decorating Ising spin σ, and let $\sigma=1$ correspond to the edge being empty and $\sigma=-1$ correspond to the edge being occupied. Then we can describe the configuration of a vertex by specifying the configurations of the three surrounding spins. It is then possible to realise the vertex weights by introducing Ising interactions R, and magnetic fields H and $2H'$ to the decorated honeycomb lattice as shown in figure 2. The tracing of a spin at a honeycomb lattice site then leads to the following realisation:

$$\begin{aligned} a &= F\,\mathrm{e}^{3H'}\cosh(H+3R) \qquad & b &= F\,\mathrm{e}^{H'}\cosh(H+R) \\ c &= F\,\mathrm{e}^{-H'}\cosh(H-R) \qquad & d &= F\,\mathrm{e}^{-3H'}\cosh(H-3R). \end{aligned} \tag{2}$$

Here F is an overall factor which does not concern us. Solving (2) for F, R, H, H', we find

$$\begin{aligned} &\cosh 2R = B/2(AC)^{1/2} \\ &\mathrm{e}^{4H'} = C/A \\ &\cosh 2H = \frac{2bc}{\sqrt{AC}}\left(\frac{B^2}{4AC}-\frac{B}{4bc}-1\right) \end{aligned} \tag{3}$$

where†

$$\begin{aligned} &A \equiv bd - c^2 = F^2\,\mathrm{e}^{-2H'}\sinh^2 2R \\ &B \equiv ad - bc = 2F^2\cosh 2R\,\sinh^2 2R \\ &C \equiv ac - b^2 = F^2\,\mathrm{e}^{2H'}\sinh^2 2R. \end{aligned} \tag{4}$$

Our next step is to decimate the decorating Ising spins, i.e. to replace the sequence of two R interactions with a magnetic field $2H'$ at the centre site, by a single interaction K with a magnetic field h at the two end sites. This decimation completes the mapping,

† The definition of A given here differs in sign from that used in [8].

and gives rise to a honeycomb Ising model with nearest-neighbour interactions K and a magnetic field

$$L = H + 3h \tag{5}$$

where H has been given in (3), and K and h are obtained from

$$\begin{aligned} f\,\mathrm{e}^{K+2h} &= \cosh(2H' + 2R) \\ f\,\mathrm{e}^{K-2h} &= \cosh(2H' - 2R) \\ f\,\mathrm{e}^{-K} &= \cosh 2H'. \end{aligned} \tag{6}$$

Here, f is another overall factor which does not concern us. Solving (6) for f, K and h, we obtain

$$\begin{aligned} \mathrm{e}^{4K} &= 1 + (B^2 - 4AC)/(A+C)^2 > 0 \\ \mathrm{e}^{4h} &= \cosh(2H' + 2R)/\cosh(2H' - 2R). \end{aligned} \tag{7}$$

Expressions (3), (5) and (7) now complete the description of the Ising parameters K and L.

The expression for e^{4K} in (7) is the same as that in [1]. However, as shown in [8], the sign of e^{2K} can be either positive or negative. The negation of e^{2K}, however, corresponds to the change $K \to K + i\pi/2$ or $\tanh K \to 1/\tanh K$, reflecting an intrinsic symmetry of the eight-vertex model. We shall therefore disregard such sign differences in our considerations. Particularly, we consider K being real, $B > 0$, $AC > 0$. We now determine the nature of the magnetic field $L = H + 3h$.

Ferromagnetic Ising model ($K > 0$). This is the case $B^2 > 4AC$. From (3) we see that both H' and R are real so that, using (7), h is also real. Consider next $\cosh 2H$ given by (3). Since this expression essentially contains two independent variables, it is convenient to parametrise by introducing $x = a/b$, $y = d/c$, $z = b/c$ which rewrite (3) as

$$\cosh 2H = \frac{1}{2\sqrt{(x-z)(y-z^{-1})}} \left(\frac{(xy-1)^2}{(x-z)(y-z^{-1})} - xy - 3 \right) \tag{8}$$

and determine the range of $\cosh 2H$ by varying z. The extremum is found to occur at $z = \sqrt{x/y}$, or $ac^3 = b^3 d$, which indeed lies in the regime $B^2 > 4AC$. This leads to the inequality $\cosh 2H > 1$. It follows that H, and hence the resulting magnetic field $L = H + 3h$, is real.

Antiferromagnetic Ising model ($K < 0$). This is the case $B^2 < 4AC$. From (3) we see that H' is real and R pure imaginary. Therefore, using (7), h is also pure imaginary. consider next the range of $\cosh 2H$. Since the extremum $z = \sqrt{xy}$ of $\cosh 2H$ determined in the above lies outside the regime $B^2 < 4AC$, a bound on $\cosh 2H$ is actually obtained by setting $B^2 = 4AC$ in (3). This consideration then leads to $|\cosh 2H| < 1$, implying H, and hence the resulting magnetic field $L = H + 3h$, is pure imaginary.

In conclusion, we have shown that the honeycomb eight-vertex model with positive vertex weights is completely equivalent to an Ising model in a non-zero magnetic field. The Ising model is either ferromagnetic with a real magnetic field, or antiferromagnetic with a magnetic field which is pure imaginary. These conclusions agree with the findings of [1, 8], but the derivation presented here is much simpler. The present approach also suggests possible extensions of our consideration. First, the method now permits straightforward extension to the asymmetric eight-vertex model, an analysis which has proven to be extremely cumbersome using the generalised weak-graph transformation [9]. Furthermore, we can also extend the analysis to other types of lattices. For a lattice of coordination number $q=4$ such as the square lattice, the corresponding vertex model is the symmetric 16-vertex model characterised by five independent vertex weights. The analogue of (2) is therefore a set of five equations containing the four variables F, H, H', R. It then follows that the vertex model is reducible to an Ising model in a four-dimensional subspace, deduced by eliminating the four variables from the five equations. This leads to results in agreement with those previously found using the generalised weak-graph transformation [9]. Finally, we point out that all these considerations, which rely only on the fact that there exists a uniform coordination number q, hold quite generally for any lattice with the same q, regardless of the spatial dimensionality.

This research was supported in part by the National Science Foundation Grant no DMR-8702596.

References

[1] Wu F Y 1974 *J. Math. Phys.* **15** 687
[2] Nagle J F 1968 *J. Math. Phys.* **8** 1008
[3] Wegner F 1973 *Physica* **68** 570
[4] Wu X N and Wu F Y 1989 *J. Phys. A: Math. Gen.* **22** L55
[5] Wu F Y, Wu X N and Blöte H W J 1989 *Phys. Rev. Lett.* **62** 2773
[6] Wu F Y and Wu X N 1989 *Phys. Rev. Lett.* **63** 465
[7] Wu X N and Wu F Y 1989 *J. Phys. A: Math. Gen.* **22** L1031
[8] Wu X N and Wu F Y 1988 *J. Stat. Phys.* **90** 41
[9] Wu X N and Wu F Y 1989 unpublished

VOLUME 32, NUMBER 9 PHYSICAL REVIEW LETTERS 4 MARCH 1974

Phase Transition in a Vertex Model in Three Dimensions

F. Y. Wu*†

Research School of Physical Sciences, The Australian National University, Canberra, Australian Capitol Territory 2600, Australia

(Received 14 December 1973)

The exact transition temperature and the order of phase transition are determined for a vertex model in three dimensions. The transition is in general of first order with a latent heat and occurs in a limited region in the parameter space. Details of the region depend on the underlying lattice and differ significantly between lattices with high (fcc and bcc) and low (simple cubic and diamond) coordination numbers.

Investigation of phase transitions in lattice systems has centered around the study of vertex models. The most general result is that of the two-dimensional eight-vertex model[1] which includes the Ising and the ferroelectric models. Little is known, however, about the critical behavior of vertex models in three dimensions.[2]

In this Letter we report on some exact results for a three-dimensional vertex model. The analysis is an extension of our earlier discussion of a sixteen-vertex model.[3] In this earlier investigation, the fact that the model is two-dimensional is explicitly used. It turns out that, with slight modifications, the argument is also applicable to three-dimensional models. We can then determine, using only elementary considerations, the exact transition temperature and the nature of phase transition for a rather general vertex model in three dimensions. It is also noteworthy that a significant difference in the critical behavior is found between lattices with high and low coordination numbers.

Consider a lattice $\mathfrak{L}$ (in any dimensionality) of N vertices (or sites) with coordination number q. Assume cyclic boundary conditions and let the $\frac{1}{2}qN$ edges of $\mathfrak{L}$ be independently covered by bonds. There are then $2^{qN/2}$ distinct bond coverages on $\mathfrak{L}$. Also at each vertex there are 2^q different bond configurations. Associate a fixed energy to each of the 2^q vertex configurations and let E be the sum of the N vertex energies for a given bond coverage, B, of $\mathfrak{L}$. The partition for this "2^q-vertex model" is then

$$Z=\sum_B e^{-E/kT} \tag{1}$$

The model we propose to consider has the following assignments for the vertex energy, E_v:

$$\begin{aligned} E_v &= n\epsilon \text{ for vertices having } n=0,1,\ldots,q-1 \text{ bonds;} \\ &= a\epsilon \text{ for vertices having } q \text{ bonds.} \end{aligned} \tag{2}$$

VOLUME 32, NUMBER 9 PHYSICAL REVIEW LETTERS 4 MARCH 1974

With these energy values, the partition function (1) can be written as

$$Z(a,u)=\sum_B u^{-2b+(q-a)s}, \tag{3}$$

where $u=e^{\epsilon/kT}$ and, for a given bond coverage B, b is the total number of bonds and s is the number of vertices having q bonds. Note that the vertex energies (2) are asymmetric and hence, for $q=4$, are different from those of the three-dimensional eight-vertex model considered in Ref. 2.

When $a=q$, the partition function (3) is trivially evaluated as

$$Z(q,u)=(1+u^{-2})^{qN/2}, \tag{4}$$

which shows no phase transition. We shall therefore consider only the cases of $a\neq q$ in the ensuing discussions.

As in Ref. 3, the partition function (3) can be related to that of an Ising ferromagnet in a nonzero magnetic field. This can be seen by considering the following bond-spin system.

Let spins of values $+1$ or -1 be located at the sites of $\mathcal{L}$ so we have a binary bond-spin system. The bond coverages B and the spin configurations σ are independent except that they are subject to the constraint that *the spins -1 can occur only at vertices having q bonds.*

For two-dimensional lattices we may introduce "bars" to the edges of the dual of $\mathcal{L}$ in place of the bonds. The bond-spin system then becomes the "bar-molecule" model of Mermin[4]; this leads to the argument in Ref. 3. But the introduction of the dual lattice is not essential. In fact, Mermin's argument for the bar-molecule model can be directly taken over to our bond-spin system. In the following we present his argument in our notations.

To each bond associate an activity x and to each spin -1 an activity z. Then the grand partition function of the binary system is

$$\Xi(x,z)=\sum_{B,\sigma}x^b z^{N_-}, \tag{5}$$

where N_- is the number of spins -1. The summation is over all B and σ subject to the constraint stated above.

First carry out the σ summation in (5) for fixed B. The constraint implies that spins $+1$ must be at all sites except at a vertex of q bonds where the spins can be either $+1$ or -1. Hence

$$\Xi(x,z)=\sum_B x^b(1+z)^s=Z(a,u), \tag{6}$$

where the last step follows from (3) with

$$x=u^{-2},\quad 1+z=u^{q-a}. \tag{7}$$

Next carry out the B summation in (5) for fixed σ. The constraint implies that all $(+,-)$ and $(-,-)$ edges must be covered with bonds, while the $(+,+)$ edges can be either covered or empty. Hence

$$\Xi(x,z)=\sum_\sigma z^{N_-}x^{N_{+-}+N_{--}}(1+x)^{N_{++}}$$

$$=z^N x^{qN/2}\sum_\sigma[(1+x)/x]^{N_{++}}z^{-N_+}, \tag{8}$$

where N_{++} is the number of $(+,+)$ neighbors, etc. We have also used the relations $N_+ + N_- = N$ and $N_{++}+N_{--}+N_+=\frac{1}{2}qN$.

Now consider an Ising model on $\mathcal{L}$ with nearest-neighbor interactions $-J$ and an external magnetic field H. Let $K=J/kT$, $L=H/kT$. It is well known that the Ising partition function can be written as[5]

$$Z_{\text{Ising}}(L,K)=\exp(\tfrac{1}{2}qNK-NL)\sum_\sigma \exp[4KN_{++}+(2L-2qK)N_+]. \tag{9}$$

Combining (6)–(9), we then obtain

$$Z(a,u)=(e^K/2\sinh 2K)^{qNJ2}e^{-NL}Z_{\text{Ising}}(L,K), \tag{10}$$

with

$$e^{4K}=1+u^2, \tag{11a}$$

$$e^{2L}=(1+u^2)^{q/2}(u^{q-a}-1)^{-1}. \tag{11b}$$

The critical behavior of the vertex model (2) now follows from that of an Ising ferromagnet ($K>0$). In particular, the temperature derivatives of the free energy

$$f(a,u)=-kT\lim_{N\to\infty}N^{-1}\ln Z(a,u)$$

are given in terms of the partial derivatives of the corresponding Ising free energy $f_{\text{Ising}}(L,K)$. Now e^{2L} is real. It follows from the Lee-Yang theorem[5] that the nonanalyticity of $f_{\text{Ising}}(L,K)$ in L can occur only at $e^{2L}=\pm1$. These are also the points where $f_{\text{Ising}}(L,K)$ can be nonanalytic in K at constant L.[6] But $e^{2L}=-1$ cannot be realized in the physical region as can be seen from (11b). Thus the singularity of $f(a,u)$ in temperature can occur only at $e^{2L}=+1$. The following results at $e^{2L}=1$ will be used: $f_{\text{Ising}}(L,K)$ is analytic in both L and K at $L=0$ for $K<K_0$,[7] and has a discontinuous first derivative $(\partial/\partial L)f_{\text{Ising}}(L,K)$ at $L=0$ for $K>K_0$. The second derivatives of $f_{\text{Ising}}(L,K)$ diverge at $L=0$ and $K=K_0$. Here K_0 is a lattice-

VOLUME 32, NUMBER 9 PHYSICAL REVIEW LETTERS 4 MARCH 1974

dependent constant given by[8]

$$
\begin{aligned}
e^{-2K_0} &= 0.8153 \text{ for fcc } (q=12),\\
&= 0.72985 \text{ for bcc } (q=8),\\
&= 0.64183 \text{ for simple cubic } (q=6),\\
&= 0.47729 \text{ for diamond } (q=4),\\
&= 3^{-1/2} \text{ for triangular } (q=6),\\
&= \sqrt{2}-1 \text{ for square } (q=4),\\
&= 2-\sqrt{3} \text{ for honeycomb } (q=3).
\end{aligned} \tag{12}
$$

Our procedure is therefore to eliminate u between $L=0$ or

$$(1+u^2)^{q/2} = u^{q-a} - 1 \quad (T=T_c) \tag{13}$$

and (11a). This gives $K=K_c(a)$. The system will exhibit a first-order transition (with a latent heat) if $K_c>K_0$, and a second-order (λ) transition with an infinite specific heat if $K_c=K_0$. In both cases the transition temperature T_c is given by (13). The system does not have a phase transition for $K_c<K_0$, even if (13) has a solution. In the first-order transition both the energy and the specific heat are finite and discontinuous at T_c. The latent heat tends to zero as $a\to a_0$, where a_0 satisfies $K_c(a_0)=K$. In this limit the specific heat at T_c diverges as $|a-a_0|^{-\theta}$, where θ equals the susceptibility exponents of the Ising model.

To determine the region of transition in the parameter space, we observe from (11a) that, for all K,

$$
\begin{aligned}
e^{2K} &\geq \sqrt{2}, \quad \epsilon \geq 0,\\
&\leq \sqrt{2}, \quad \epsilon \leq 0,
\end{aligned} \tag{14}
$$

Also from (12) we have $e^{2K_0}<\sqrt{2}$ for the fcc and bcc lattices and $e^{2K_0}>\sqrt{2}$ for all other lattices. Thus $K\geq K_0$ can be realized only for $q=8$ and 12 if $\epsilon<0$, and for all lattices if $\epsilon>0$. There is then a marked distinction in the lattices with high coordination numbers (8 or 12). The distinction can also be seen from the sign of a_0. Using (11a), (12), and (13), we find that

$$
\begin{aligned}
a_0 &= q - 2\ln(e^{2qK_0}+1)/\ln(e^{4K_0}-1)\\
&= 19.40 \text{ for fcc},\\
&= 47.672 \text{ for bcc},\\
&= -9.3299 \text{ for sc},\\
&= -0.92997 \text{ for diamond},\\
&= -3.61470\ldots \text{ for triangular},\\
&= -0.515033\ldots \text{ for square},\\
&= -0.102220\ldots \text{ for honeycomb}.
\end{aligned} \tag{15}
$$

It can be seen that, for all q, $K_c(a)$ is increasing in a. Thus there is in general no phase transition for $a<a_0$ and a first-order transition for $a>a_0$. At $a=a_0$ the system exhibits a λ transition.

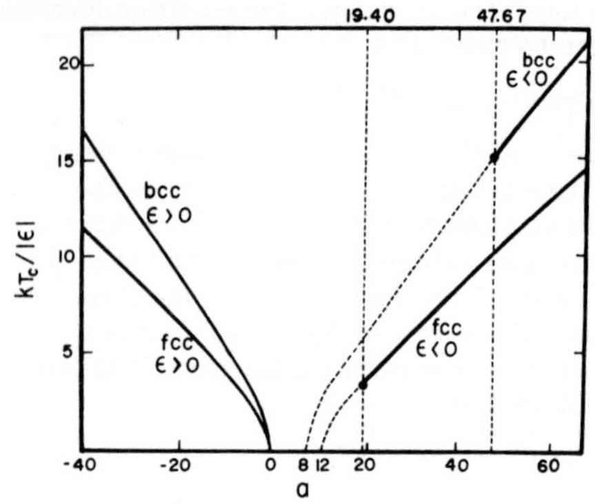

FIG. 1. The transition temperature T_c for the fcc and bcc lattices. The solid lines denote the region of first-order transition $a>a_0$ and $a<0$; the dashed lines denote the solution of (13) not corresponding to any physical transition. A λ transition occurs at $a=a_0$, denoted by the dots.

We consider the cases $\epsilon<0$ and $\epsilon>0$ separately:

(i) $\epsilon<0$. In this case (13) yields a solution for $a>q$. Since $K_c\geq K_0$ is possible only for $q=8$ or 12, only the bcc and the fcc lattices will exhibit a phase transition for $\epsilon<0$. The transition temperature for $q=8$ and 12 is shown in Fig. 1. The solid lines denote the regions of first-order transition; the broken lines denote the solution of (13) not corresponding to any physical transition. At $a=a_0$, the points denoted by the dots, a λ transition occurs.

(ii) $\epsilon>0$. In this case (13) is solvable for $a<0$ ($T_c=0$ for $a=0$). We see from (12) and (14) that, for $q=8$ and 12, $K_c>\frac{1}{4}\ln 2>K_0$. Hence for *all a* <0, the fcc and the bcc lattices exhibit a first-order transition. This situation is also shown in Fig. 1. For other lattices $a_0<0$, and $K_c>K_0$ is realized in $a_0<a<0$. This is the region of a first-order transition. The transition temperature is shown in Fig. 2 for the sc ($q=6$) and the diamond ($q=4$) lattices. Note that the transition temperature depends only on q. Thus the sc and the triangular lattices, e.g., are described by the same curve terminating at different values of a_c. Similarly, the curve for the diamond lattice in Fig. 2 is the same as that in Ref. 3 for the square lattice.[9] In all cases a λ transition occurs at $a=a_0$.

To summarize, we have determined the exact transition temperature and the order of transition in the vertex model (2) for various lattices. The transition is in general a first-order one

VOLUME 32, NUMBER 9 PHYSICAL REVIEW LETTERS 4 MARCH 1974

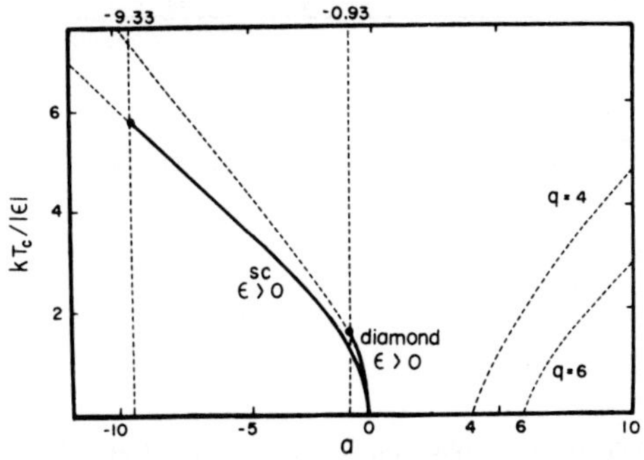

FIG. 2. The transition temperature T_c for the sc and diamond lattices. The solid lines denote the region of first-order transition $a_0 < a < 0$; the dashed lines denote the solution of (13) not corresponding to any physical transition. A λ transition occurs at $a = a_0$, denoted by the dots.

and occurs only in certain limited regions in the parameter space. Note that the transition temperature depends only on the coordination number q. The details of the lattice such as its dimensionality determine only the region of transition. This is reminiscent of the behavior of a decorated Ising system.[10] In this latter model the critical behavior is found to be different in lattices with $e^{2K_0} < 2$ and $e^{2K_0} > 2$. In the present model the distinction lies between $e^{2K_0} < \sqrt{2}$ (bcc, fcc) and $e^{2K_0} > \sqrt{2}$ (sc, diamond), leading to significantly different regions of transition.

I wish to thank Professor K. J. Le Couteur for his kind hospitality at the Australian National University, and Dr. R. J. Baxter for a useful suggestion. The support of the Australian-American Educational Foundation is also gratefully acknowledged.

*Senior Fulbright Scholar, on leave of absence from Northeastern University, Boston, Mass. 02115.

†Work supported in part by the National Science Foundation under Grant No. GH-35822 at Northeastern University.

[1]R. J. Baxter, Ann. Phys. (New York) 70, 193 (1972).

[2]A conjecture on the location of the transition temperature of the three-dimensional eight-vertex model has been given by C. Thibaudier and J. Villain, J. Phys. C: Proc. Phys. Soc., London 5, 3429 (1972). This conjecture has been made more recently by B. Sutherland, Phys. Rev. Lett. 31, 1504 (1973).

[3]F. Y. Wu, Phys. Rev. B 6, 1810 (1972).

[4]N. D. Mermin, Phys. Rev. Lett. 26, 169 (1971).

[5]T. D. Lee and C. N. Yang, Phys. Rev. 87, 410 (1952).

[6]J. L. Lebowitz and O. Penrose, Commun. Math. Phys. 11, 99 (1968).

[7]This analyticity is believed to hold, but has been proved only for sufficiently small K.

[8]See, e.g., M. E. Fisher, Rep. Progr. Phys. 30, 615 (1967).

[9]There is a misprint in the caption of Fig. 2 in Ref. 3. The condition $\epsilon < 0$ should read $\epsilon > 0$.

[10]F. Y. Wu, Phys. Rev. B 8, 4219 (1973).

Letters in Mathematical Physics **29:** 205–213, 1993. 205

Exact Solution of a Vertex Model in *d* Dimensions

F. Y. WU and H. Y. HUANG
Department of Physics, Northeastern University, Boston, MA 02115, U.S.A.

(Received: 21 July 1993)

Abstract. We consider a class of vertex models describing directed lines on a lattice in arbitrary d dimensions, and solve the model exactly for the Cartesian lattice and in the case that each loop of lines carries a fugacity -1. Our analysis, which can be carried out for arbitrary lattices, is based on an equivalence of the vertex model with a dimer problem. The dimer problem is, in turn, solved using the method of Pfaffians. It is found that the system is frozen below a critical temperature T_c with the critical exponent $\alpha = (3 - d)/2$.

Mathematics Subject Classifications (1991). 82B23, 82B27.

1. Introduction

The vertex model, which generalizes the notion of spin models in lattice statistics, plays a central role in the modern theory of critical phenomena in statistical mechanics. The first exact solution of a vertex model is that of the ice-rule models obtained by Lieb [1]. This development was soon followed by the success of solving other models [2, 3]. However, soluble models in the past have been confined mostly to two dimensions.

In this Letter, we report the exact solution of a vertex model for any lattice in arbitrary dimension. This is a class of vertex model which finds a wide range of physical applications [4–8]. Although our approach is valid quite generally, for concreteness and illustration purposes, we present explicitly the solution for the case of the Cartesian lattice $\mathscr{L}$ in d dimensions.

The vertex model is defined by placing bonds along edges of $\mathscr{L}$ with the restriction such that there are either no bond or two bonds incident at each vertex. In the latter casé, when there are two incident bonds, one of the bonds is along a positive axes direction and the other along a negative direction. Thus, we have a $(d^2 + 1)$-vertex model. To each vertex we assign a weight

$$\begin{aligned} \omega &= 1, && \text{if there is no bond,} \\ &= \sqrt{z_i z_j}, && \text{if the two bonds are in directions } i \text{ and } j, \\ &= 0, && \text{otherwise.} \end{aligned} \tag{1}$$

Alternately, the weight (1) can be regarded as bond weights z_i. From this point of view, the bonds form lines which do not intersect nor terminate, and are directed from a generally negative direction of the lattice to a positive direction. Assuming periodic boundary conditions, the lines will then loop around the lattice, with each loop cutting one or more times a plane perpendicular to the (1, 1, ..., 1) direction. To each of these loops we associate a fugacity y. The partition function is then defined to be

$$Z(y; z_1, z_2, \dots, z_d) = \sum_{\text{vertex config.}} y^l \prod_{\text{vertex}} \omega = \sum_{\text{line config.}} \prod_{\text{loops}} \left(y \prod z_i \right), \tag{2}$$

where the product inside the parentheses is taken over all bonds contained in a loop, and l is the number of loops in a given vertex or line configuration.

The partition function (2), which generalizes the usual definition for vertex models with the inclusion of the fugacity y of a global nature, can be extended in an obvious fashion to any lattice. For $y = 1$, the vertex model (2) is relevant to, among other physical applications, directed random walks [4], flux lines in superconductors [5], commensurate-imcommensurate transitions [6], and bosons in $(d - 1) + 1$ dimensions [7]. The $y = 1$ model is exactly soluble in $d = 2$ [2], and has been studied in $d = 3$ by means of mean-field [9], path-integral [10], path-integral renormalization group [11], and transfer-matrix [12] methods. There has been no exact solution, however. For $y = -1$, the model is relevant to fermions in $(d - 1) + 1$ dimensions, a physical realization to be discussed elsewhere [8]. Here, we report the result of an exact evaluation of the partition function (2) at $y = -1$.

2. An Equivalent Dimer Problem

Our approach is to first convert the vertex model (2) into a dimer problem, and then evaluate the dimer generating function at $y = -1$ using a Pfaffian. We first introduce the dimer lattice $\mathscr{L}'$.

Starting from $\mathscr{L}$, we split each lattice point of $\mathscr{L}$ into two, with one of the two points remaining attached to the d positive axes and the other to the d negative axes. The dimer lattice $\mathscr{L}'$ is then constructed by reconnecting the split pairs with new inserted edges. Further, we associate dimer weights z_i to edges originally in $\mathscr{L}$, and 1 to the new inserted edges. For the $d = 3$ simple cubic lattice, this construction is shown in Figure 1 and leads to a dimer lattice $\mathscr{L}'$ consisting of layers of honeycomb lattices interconnected by the inserted edges. More generally for arbitrary $\mathscr{L}$, this procedure leads to a dimer lattice $\mathscr{L}'$ consisting of layers of $(d - 1)$-dimensional sublattices which are bipartite, with the inserted edges connecting sites of one sublattice of a layer to sites of the other sublattice of the next layer. Particularly, the connection of layers of simple quartic sublattices according to this prescription gives rise to the dimer lattice considered by Bhattacharjee *et al.* [12].

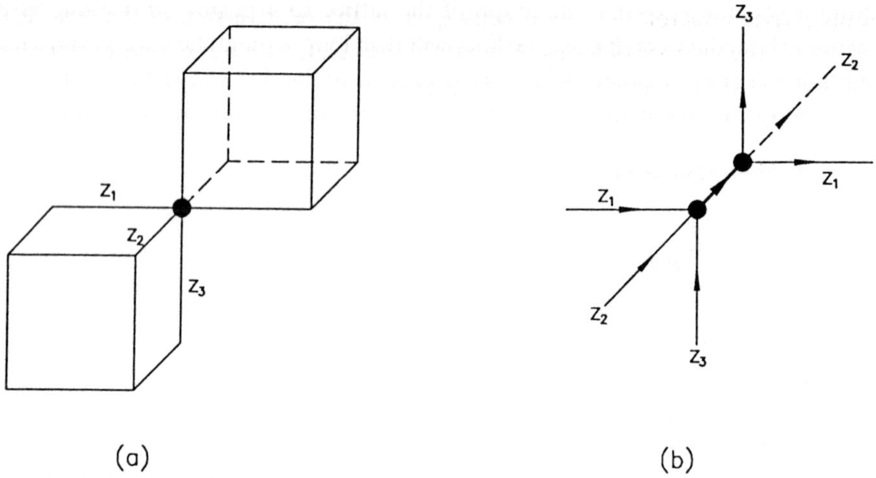

Fig. 1. (a) A lattice point of the simple cubic lattice. (b) The lattice point denoted by the solid circle in (a) is split into two and reconnected by a new edge. All edges are directed in positive directions.

We now establish a one-one correspondence between vertex configurations on $\mathscr{L}$ and dimer configurations on $\mathscr{L}'$. To see this correspondence, we superimpose any given dimer configuration C_i with a standard one, C_0, in which all inserted edges are covered by dimers. The superposition of two dimer configurations produces a graph of transition cycles, or polygons [13]. In the present case, the transition cycles belong to one of two kinds: (i) double dimers placed on new edges, and (ii) polygons looping around the lattice. In the latter case the polygons form a line configuration as in (2). Conversely, to each line configuration in (2), there exists a unique dimer configuration C_i which, when superimposed with C_0, produces the line configuration in question. This establishes the correspondence. If we further attach an overall factor y^l to each dimer configuration, where l is the number of loops, the partition function (2) can now be interpreted as a dimer generating function.

Next, we show that the dimer generating function (2), with $y = -1$, is given by a Pfaffian. Let there be $N = N_1 \times N_2 \cdots x N_d$ sites in $\mathscr{L}$, and $2N$ sites in $\mathscr{L}'$. The Pfaffian is then the square root of a $2N \times 2N$ antisymmetric determinant A whose elements are the dimer weights with appropriate $\pm$ signs. The convention of fixing the signs is to direct edges of $\mathscr{L}'$ such that the matrix element $A_{\alpha\beta}$ is positive (negative) if an edge is directed from site α to β (β to α). Now, we choose to direct all edges of $\mathscr{L}'$ along the positive direction, as shown in Figure 1 for the simple cubic lattice. Consider a typical term in the Pfaffian corresponding to a dimer configuration C_i. The sign of this term relative to the term corresponding to C_0 is the product of the signs of the transition cycles produced by the superimposition of C_i and C_0. The rule is that each transition cycle carries a sign $(-1)^{n+1}$, where n is the number of arrows pointing in a given direction when the transition cycle is traversed [13]. For transition cycles consisting of double dimers we have $n = 1$ so that the sign is always positive. For other transition cycles, or loops, we have always $n =$ even, so that each loop carries a factor -1. As a consequence, the relative sign of the term is $(-1)^l$ where l is the number of loops. This establishes that, with our choice of signs for $A_{\alpha\beta}$, the Pfaffian A is precisely the dimer generating function (2) with $y = -1$.

3. The Exact Solution

We now evaluate the Pfaffian, namely, the square root of the $2N \times 2N$ anti-symmetric determinant A with elements $\pm z_i$ and ± 1. The dimer lattice $\mathscr{L}'$ can be regarded as consisting of unit cells containing pairs of split sites, with the cells forming an $N_1 \times N_2 \times \cdots \times N_d$ simple cubic lattice. The determinant A can then be cast in terms of 2×2 determinants indexed by

$$\begin{aligned} A(n_1, \ldots, n_d | n_1, \ldots, n_d) &= \begin{pmatrix} 0 & 1 \\ -1 & 0 \end{pmatrix}, \\ A(\ldots, n_i, \ldots | \ldots, n_i + 1, \ldots) &= \begin{pmatrix} 0 & z_i \\ -z_i & 0 \end{pmatrix}, \\ A(m_1, \ldots, m_d | n_1, \ldots, n_d) &= \begin{pmatrix} 0 & 0 \\ 0 & 0 \end{pmatrix}, \text{ otherwise,} \end{aligned} \tag{4}$$

where $1 \leqslant \{m_j, n_j\} \leqslant N_j$ for $j = 1, \ldots, d$. With periodic boundary conditions, matrices in (4) are invariant when indices m_i and n_i are changed by N_i. Then the matrix A is 2×2 block-diagonal in the Fourier space. This leads to

$$\begin{aligned} &Z(-1; z_1, \ldots, z_d) \\ &= \sqrt{\det A} \\ &= \prod_{n_1=1}^{N_1} \cdots \prod_{n_d=1}^{N_d} \left[\det \begin{pmatrix} 0 & 1 + \Sigma_i z_i \, e^{2\pi i n_i/N_i} \\ -1 - \Sigma_i z_i \, e^{-2\pi i n_i/N_i} & 0 \end{pmatrix} \right]^{1/2} \\ &= \prod_{n_1=1}^{N_1} \cdots \prod_{n_d=1}^{N_d} \left| 1 + \sum_{i=1}^{d} z_i \, e^{2\pi i n_i/N_i} \right|, \end{aligned} \tag{5}$$

and the following expression for the per-site 'free energy'

$$\begin{aligned} &f(z_1, \ldots, z_d) \\ &= \lim_{N \to \infty} \frac{1}{N} \ln Z(-1; z_1, \ldots, z_d) \\ &= \frac{1}{(2\pi)^d} \int_0^{2\pi} d\theta_1 \ldots \int_0^{2\pi} d\theta_d \ln \left| 1 + \sum_{i=1}^{d} z_i \, e^{i\theta_i} \right|. \end{aligned} \tag{6}$$

For $d = 2$, the expression (6) is identical to the $y = 1$ solution [2], reflecting the equivalence of bosons and fermions in $1 + 1$ dimensions [7]. For $d > 2$, we do not have the benefit of an exact $y = 1$ solution for comparison. But by comparing with results obtained under mean-field, path-integral, and transfer-matrix methods, our exact analysis indicates that the $y = \pm 1$ solutions, while not exactly identical, belong to the same universality class in the sense that they possess the same specific heat critical exponent.

4. The Critical Behavior

The analysis of the critical behavior of the free energy is facilitated by the use of the formula

$$\frac{1}{2\pi}\int_0^{2\pi} \mathrm{d}\theta \ln|A + B\mathrm{e}^{i\theta}| = \ln \max\{|A|, |B|\} \tag{7}$$

which holds for any complex A and B. We realize bond weights by introducing bond energies ε_i, and write

$$z_i = \mathrm{e}^{-\beta\varepsilon_i}, \tag{8}$$

where $\beta = 1/kT$. There is no loss of generality to restrict to $\varepsilon_i > 0$. Then, multiplying the quantity inside the absolute signs in (6) by $\mathrm{e}^{-i\theta_1}$ and carrying out the integration over θ_1 (after renaming $\theta_i - \theta_1$ as θ_i, $i \neq 1$), we obtain

$$\begin{aligned} f(z_1, \dots, z_d) &= 0, \quad T < T_c \\ &= \frac{1}{(2\pi)^{d-1}}\int \mathrm{d}\theta_2 \dots \int_{\mathscr{R}} \mathrm{d}\theta_d \ln\left|z_1 + \sum_{j=2}^{d} z_j \mathrm{e}^{i\theta_j}\right|, \quad T > T_c, \end{aligned} \tag{9}$$

where T_c is defined by

$$z_1 + z_2 + \cdots + z_d = 1 \tag{10}$$

and $\mathscr{R}$ is the regime $|z_1 + \Sigma_{j=2}^d z_j \mathrm{e}^{i\theta_j}| > 1$. Clearly, the free energy is nonanalytic in T at T_c with the system frozen below T_c. Particularly, for $d = 3$, we can rewrite the second line in (9) as

$$\frac{4}{(2\pi)^2}\int_0^{\alpha} \mathrm{d}\theta_2 \int_0^{\theta_0(\theta_2)} \mathrm{d}\theta \ln|u(\theta_2) + z_3 \mathrm{e}^{i\theta}|, \tag{11}$$

where θ_2 and θ_0 are as shown in Figure 2, and

$$\begin{aligned} \cos\alpha &= [(1 - z_3)^2 - z_1^2 - z_2^2]/2z_1z_2, \\ \cos\theta_0(\theta_2) &= [(1 - u_2(\theta_2) - z_3^2)]/2z_3u(\theta_2), \\ u(\theta_2) &= |z_1 + z_2\,\mathrm{e}z^{i\theta_2}|. \end{aligned} \tag{12}$$

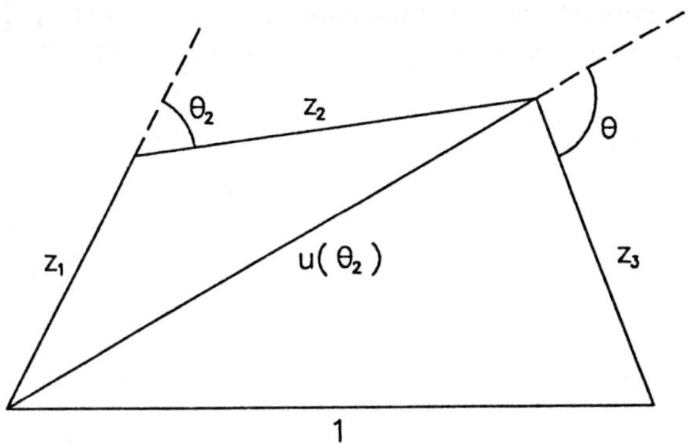

Fig. 2. Labelings of angles in (11).

To find out the nature of the transition, we evaluate the energy

$$\begin{aligned} U &= -\frac{\partial}{\partial\beta} f(z_1, \ldots, z_d) \\ &= \sum_{m=1}^{d} \varepsilon_m z_m \frac{\partial}{\partial z_m} f(z_1, \ldots, z_d) \\ &= \frac{1}{(2\pi)^{d-1}} \sum_{m=1}^{d} \varepsilon_m z_m \frac{\partial}{\partial z_m} \int_0^{2\pi} \cdots \int_0^{2\pi} \prod_{i\neq m} \mathrm{d}\theta_i \ln \max\{z_m, |x_m|\} \\ &= \frac{1}{(2\pi)^{d-1}} \sum_{m=1}^{d} \varepsilon_m \int \mathrm{d}\theta_1 \cdots \int \mathrm{d}\theta_{m-1} \int \mathrm{d}\theta_{m+1} \cdots \int_{\mathscr{R}_m} \mathrm{d}\theta_d, \end{aligned} \tag{13}$$

where

$$x_m = 1 + \sum_{j\neq m} z_j \,\mathrm{e}^{i\theta_j}, \tag{14}$$

and $\mathscr{R}_m$ is the regime $z_m > |x_m|$. In writing out the third line in (13), we have used (6) for f and carried out the integration over θ_m. The last line in (13) then follows from the fact that x_m is independent of z_m so that the regime $|x_m| > z_m$ gives rise to no contribution after taking the derivative.

Now, the volume of $\mathscr{R}_m$ is small near T_c. To find the leading behavior consider, for example, the term $m = 1$ in (13) for which we need to compare z_1 with $|x_1|$. The integral in (13) is most conveniently written in terms of angles $\phi_i = \pi - \theta_i$ shown in Figure 3. Defining angles $\phi_{n\pm}$ as the two roots of ϕ_n in

$$\begin{aligned} &|1 - z_2\,\mathrm{e}^{-i\phi_2} - \cdots - z_{n-1}\,\mathrm{e}^{-i\phi_{n-1}} - z_h \mathrm{e}^{-i\phi_n}| = \\ &= z_1 + z_{n+1} + \cdots + z_d, \quad n = 2, \ldots, d, \end{aligned} \tag{15}$$

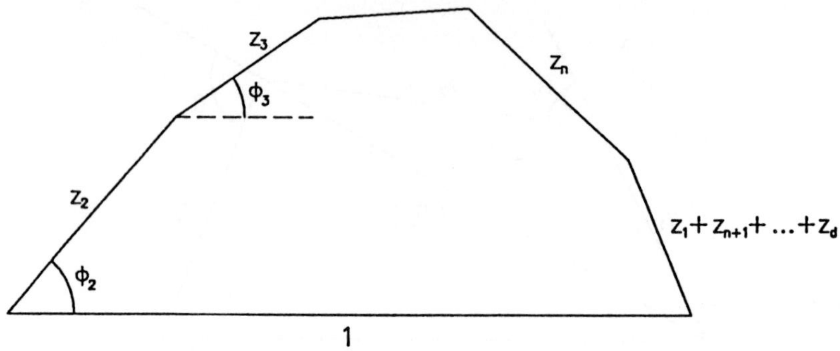

Fig. 3. Labelings of angles in (15).

we find

$$\int d\theta_2 \ldots \int_{\mathscr{R}_1} d\theta_d = \int_{\phi_{2-}}^{\phi_{2+}} d\phi_2 \ldots \int_{\phi_{d-}}^{\phi_{d+}} d\phi_d. \tag{16}$$

Now all angles including $\phi_{n\pm}$ are small near T_c. In fact, solving ϕ_n from (15) for small

$$\begin{aligned} t &= z_1 + z_2 + \cdots + z_d - 1 \\ &\sim (T - T_c)/T_c, \end{aligned} \tag{17}$$

we find

$$z_n\phi_{n\pm} = -\sum_{j=2}^{n-1} z_j\phi_j \pm \sqrt{t(z_1 + z_{n+1} + \cdots + z_d) - t^2}. \tag{18}$$

Substituting (18) into (16), we obtain the critical behavior

$$U \sim t^{(d-1)/2} \tag{19}$$

and, hence, the critical exponent

$$\alpha = (3 - d)/2. \tag{20}$$

Particularly, for $d = 3$ and isotropic $z_i = z = e^{-\beta\varepsilon}$, we find from (9) and (11)

$$\begin{aligned} U &= 0 \quad T < T_c, \\ &= \frac{2\varepsilon}{\pi^2} \int_{z-1}^{2z} \frac{du}{\sqrt{4z^2 - u^2}} \cos^{-1}\left(\frac{1 - u^2 - z^2}{2uz}\right), \quad T > T_c \end{aligned} \tag{21}$$

from which we obtain

$$c_v = \frac{dU}{dT} = \frac{3\sqrt{3}}{2\pi}(\ln 3)^2 k, \quad T \to T_c + . \tag{22}$$

Thus, the specific heat has a cusp singularity at $T_c +$.

Our analysis can be extended in a straightforward fashion to other lattices. For the ($d = 3$) diamond lattice with vertex weights $\sqrt{z_i z_j}$; $i = 1, 2; j = 3, 4$, e.g., our solution yields, analogous to (6),

$$f(z_1, z_2, z_3, z_4) = \tag{23}$$

$$= \frac{1}{(2\pi)^3} \int_0^{2\pi} d\theta_1 \int_0^{2\pi} d\theta_2 \int_0^{2\pi} d\theta_3 \ln |e^{i\theta_1} + (z_1 + z_2 e^{i\theta_2})(z_3 + z_4 e^{i\theta_3})|.$$

This leads to the same critical behavior as that of the cubic lattice and, in the isotropic case $z_i = z$, the same critical temperature.

$$T_c = \varepsilon/k \ln 2 \tag{24}$$

as that of the $y = 1$ model [11].

5. Discussions

We have evaluated exactly the partition function (2) for the d-dimensional Cartesian lattice at $y = -1$, and obtained the critical exponent $\alpha = (3 - d)/2$. This critical exponent is the same as that of the $y = 1$ model obtained previously via the mean-field [9], path-integral renormalization group [11], and transfer-matrix [12] approaches. Due to the strong constraint imposed by the line configurations occurring in the vertex model, it is perhaps not surprising that the $y = \pm 1$ models should have the same critical temperature and belong to the same universality class. However, there exist some minor differences in details of the critical behavior. In the $d = 3$ models, for example, while our exact solution for $y = -1$ leads to a jump cusp discontinuity in the specific heat, an analysis of the transfer matrix for the $y = 1$ model, which is presumably exact, suggests a logarithmic divergence [12].

Acknowledgement

This research is supported by National Science Foundation Grant DMR-9015489.

References

1. Lieb, E., *Phys. Rev. Lett.* **18**, 692, 1046 (1967); **19**, 108 (1967).
2. Wu, F. Y., *Phys. Rev. Lett.* **19**, 103 (1967); *Phys. Rev.* **168**, 539 (1968).
3. Baxter, R. J., *Phys. Rev. Lett.* **26**, 832 (1971).
4. Bhattacharjee, S. M., *Europhys. Lett.* **15**, 815 (1991).
5. Nelson, D. R., *J. Statist. Phys.* **57**, 511 (1989).
6. Pokrovsky, V. L. and Talapov, A. L., *Phys. Rev. Lett.* **42**, 65 (1979).
7. Hwa, T, private communication.
8. Hwa, T., Huang, H. Y. and Wu, F. Y. unpublished.

9. Izuyama, T. and Akutsu, Y., *J. Phys. Soc. Japan.* **51**, 50 (1982).
10. Nelson, D. R. and Seung, H. S., *Phys. Rev. B* **39**, 9153 (1989).
11. Bhattacharjee, S. M. and Rajasekaran, J. J., *Phys. Rev. A* **10**, 6202 (1991).
12. Bhattacharjee, S. M., Nagle, J. F., Huse, D. A. and Fisher, M. E., *J. Statist. Phys.* **32**, 381 (1983).
13. Kasteleyn, P. W., *Physica* **27**, 1209 (1961).

3. Duality and Gauge Transformations

Duality transformation in a many-component spin model

F. Y. Wu* and Y. K. Wang

Department of Physics, Northeastern University, Boston, Massachusetts 02115

(Received 6 October 1975)

It is shown that the duality transformation relates a spin model to its dual whose Boltzmann factors are the eigenvalues of the matrix formed by the Boltzmann factors of the original spin model. The duality relation valid for finite lattices is obtained, and applications are given.

The duality relation for two-dimensional spin models can be considered both from a topological and an algebraic point of view. A comprehensive discussion of these aspects for the Potts and the Ashkin-Teller (AT) models has been given by Mittag and Stephen.[1] More recently, Wegner[2] has reformulated the duality relation as an instance of a more general transformation. In this note we point out one further aspect of the duality transformations. Our result helps to clarify the reasoning in Wegner's formulation and also provides straightforward extensions of duality to other spin models.

Consider a q component spin model on a two-dimensional lattice L which has N sites. Let $\xi_i = 1, 2, \ldots, q$ denote the spin state of the ith site. The Hamiltonian can be generally written as

$$\mathcal{H} = -\sum_{\langle ij \rangle} J(\xi_i, \xi_j), \tag{1}$$

where $-J(\xi, \xi')$ is the interaction between the spin states ξ and ξ'. The summation in (1) is over all interacting pairs $\langle ij\rangle$ which we assume to be noncrossing. The partition function is

$$Z = \sum_{\xi_i=1}^{q} \prod_{\langle ij \rangle} u(\xi_i, \xi_j), \tag{2}$$

with

$$u(\xi, \xi') = \exp[J(\xi, \xi')/kT]. \tag{3}$$

We shall restrict our attention to the case that

$$u(\xi_i, \xi_j) = u(\xi_i - \xi_j), \quad (\text{mod } q). \tag{4}$$

Thus the matrix U whose elements are $u(\xi, \xi')$ is cyclic. It is not necessary for our discussion to further assume that U is symmetric, although in most applications this will be the case. In order to distinguish ξ_i from ξ_j for a given edge connecting sites i and j, we place an arrow on it pointing from i to j. Thus the lattice is directed. We shall also have occasion to consider the situation, such as for the AT model, that U is block-cyclic. These cases will be explored in later discussions.

We can rewrite the partition function in two different ways. First, instead of specifying the spin states by ξ_i, we may label the edge in (4) by the difference $\xi_{ij} \equiv \xi_i - \xi_j$. However, to ensure that each set of ξ_{ij} will correspond to some spin states, it is necessary (and sufficient) to require

$$\sum_{\text{cw}} \xi_{ij} = \sum_{\text{ccw}} \xi_{ij} \tag{5}$$

around each face of L. Here the summation cw (ccw) is over the edges carrying clockwise (counterclockwise) arrows around the face. Clearly, the $\xi_i \Leftrightarrow \xi_{ij}$ mapping is q to 1. Denoting the restriction (5) by a prime over the summation sign, we can now rewrite the partition sum (2) as

$$Z(u) = q \sum_{\xi_{ij}=1}^{q}{}' \prod_{\langle ij \rangle} u(\xi_{ij}). \tag{6}$$

To make connection with the partition function on the dual of L, or L^D, we now cast Z into another form. Direct the edges of L^D such that the arrows on L^D coincide with those on L if each edge of L^D is rotated 90° clockwise. The situation around a site on L is shown in Fig. 1. Now the eigenvalues of the $q \times q$ cyclic matrix U are

$$\lambda(\eta) = \sum_{\xi=1}^{q} \exp(2\pi i \xi\eta/q)\, u(\xi), \quad \eta = 1, \ldots, q, \tag{7}$$

or, conversely,

$$u(\xi_{ij}) = \sum_{\eta=1}^{q} T(\xi_i, \eta)\, \lambda(\eta)\, T^*(\xi_j, \eta), \tag{8}$$

where

$$T(\xi, \eta) = q^{-1/2} \exp(2\pi i \xi\eta/q). \tag{9}$$

We substitute (8) into (2) and carry out the sums over ξ_i. At each site of L, we have for each outgoing (incoming) arrow a factor $T(\xi, \eta)$ $[T^*(\xi, \eta)]$. Denote the spin states of the spin model on L^D by η_α and identify the η in (8) as $\eta_{\alpha\beta} \equiv \eta_\alpha - \eta_\beta$, where the arrow runs from site α to site β on the corresponding edge of L^D. Then the summation over ξ_i (cf. Fig. 1) leads to a factor

$$\sum_{\xi_i=1}^{q} T(\xi_i, \eta_{21})\, T^*(\xi_i, \eta_{23}) \cdots T(\xi_i, \eta_{1n_i})$$

$$= q^{1-n_i/2}\, \delta_{\text{Kr}}\left(\sum_{\text{cw}} \eta_{\alpha\beta} - \sum_{\text{ccw}} \eta_{\alpha\beta}\right), \tag{10}$$

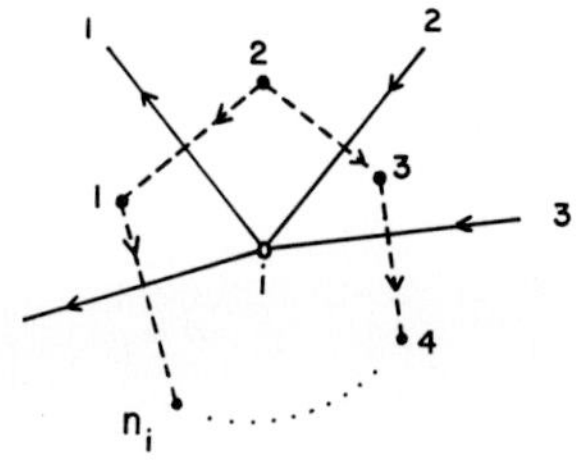

FIG. 1. The directed edges around the ith site on L. The solid (broken) lines are the edges of L (L^D).

where n_i is the number of neighbors of the ith site. The restriction imposed by the Kronecker delta on the rhs of (10) is exactly the same as in (5) for a face of L^D. Thus, after combining (8) with (2) and using (10), the partition function takes the form

$$Z = q^{N-E} \sum_{\eta_{\alpha\beta}=1}^{q}{}' \prod_{\langle\alpha\beta\rangle} \lambda(\eta_{\alpha\beta}), \tag{11}$$

where E is the number of edges of L (or L^D). Finally, by comparing (11) with (6) and using the Euler's relation for a connected planar graph,[3]

$$N + N_D = E + 2, \tag{12}$$

we obtain the identity

$$Z(u) = q^{1-N_D} Z^{(D)}(\lambda). \tag{13}$$

This is our main result and it is valid for any *finite* lattice. Here $Z^{(D)}(\lambda)$ is the partition function of the spin model on L^D whose Boltzmann factors are given by (7). While this result is implicit in Ref. 2, our discussion does bring out in a natural way the role played by the U matrix, thus clarifying the reasoning behind Wegner's formulation.

An example is the Potts model[4] with

$$U = \begin{pmatrix} e^K & 1 & \cdots & 1 \\ 1 & e^K & \cdots & 1 \\ & \cdots & \cdots & \\ 1 & 1 & \cdots & e^K \end{pmatrix}. \tag{14}$$

The eigenvalues of U are

$$\lambda_1 = e^K + q - 1, \quad \lambda_2 = \cdots = \lambda_q = e^K - 1, \tag{15}$$

so that the equivalence (13) reads

$$Z(e^K) = q^{1-N_D}(e^K - 1)^E Z^{(D)}(e^{K*}), \tag{16}$$

where

$$e^{K*} = \lambda_1/\lambda_2 = (e^K + q - 1)/(e^K - 1). \tag{17}$$

The above result is readily extended to the case where U is block-cyclic. An example is the AT model for which

$$U = \begin{pmatrix} U_1 & U_2 \\ U_2 & U_1 \end{pmatrix}, \tag{18}$$

where U_1 and U_2 are themselves 2×2 cyclic matrices. Generally we consider a matrix U which is m-fold cyclic. That is to say, U is composed of q_1 cyclic matrices, each of which in turn contains q_2 cyclic matrices, etc., the dimension of U being $q = q_1 q_2 \cdots q_m$. Thus, an element of U, which specifies the spin states of the model, is described by an m component vector $\boldsymbol{\xi} \equiv (\xi_1, \ldots, \xi_m)$ whose components can take on, respectively, $q_1, q_2, \ldots, q_m$ different values. Treating the previous ξ and η as vectors, we can carry through all the steps and again arrive at the equivalence (13), provided that in place of (7) we have

$$\lambda(\boldsymbol{\eta}) = \sum_{\boldsymbol{\xi}} \exp[2\pi i(\xi_1\eta_1/q_1 + \cdots + \xi_m\eta_m/q_m)]u(\boldsymbol{\xi}). \tag{19}$$

For the AT model we have $q_1 = q_2 = 2$, $\xi_i, \eta_i = 1, 2$. Equation (19) then leads to the duality relations derived by Ashkin and Teller.[5] As a further illustration consider the six-component spin model whose U matrix is

$$U = \begin{pmatrix} U_1 & U_2 & U_2 \\ U_2 & U_1 & U_2 \\ U_2 & U_2 & U_1 \end{pmatrix}, \tag{20}$$

where $U_1 = \begin{pmatrix} a & b \\ b & a \end{pmatrix}$ and $U_2 = \begin{pmatrix} \alpha & \beta \\ \beta & \alpha \end{pmatrix}$ are 2×2 matrices. It is easily seen that the eigenvalues of U form a similar cyclic matrix whose elements are

$$\begin{aligned} a^* &= \lambda_1 = a + b + 2(\alpha + \beta), \\ b^* &= \lambda_2 = a - b + 2(\alpha - \beta), \\ \alpha^* &= \lambda_3 = \lambda_4 = a + b - (\alpha + \beta), \\ \beta^* &= \lambda_5 = \lambda_6 = a - b - (\alpha - \beta). \end{aligned} \tag{21}$$

This is the duality transformation.

Note added in proof: Finally we remark that our result (13) is valid even if the Boltzmann factor (3) is edge-dependent. In this case the eigenvalues (7) or (19) are introduced for each edge ij and in (13) we have $u = \{u_{ij}\}$, $\lambda = \{\lambda_{ij}\}$.

*Supported in part by National Science Foundation Grant No. DMR72-03213A01.

[1]L. Mittag and J. Stephen, J. Math. Phys. 12, 441 (1971).
[2]F.J. Wegner, Physica 68, 570 (1973).
[3]We have used here $N_D = S + 1$, where S is the number of independent circuits in the graph.
[4]R. B. Rotts, Proc. Cambridge Philos. Soc. 48, 106 (1952).
[5]J. Ashkin and E. Teller, Phys. Rev. 64, 178 (1943).

J. Phys. A: Math. Gen. 22 (1989) L55–L60. Printed in the UK

LETTER TO THE EDITOR

Duality properties of a general vertex model†

X N Wu and F Y Wu

Department of Physics, Northeastern University, Boston, MA 02115, USA

Received 14 October 1988

Abstract. We consider the duality properties of a general vertex model on a lattice in any spatial dimension. The analysis is based on a generalised weak-graph transformation under which the partition function of the vertex model remains invariant. It is shown that the generalised weak-graph transformation is self-dual for lattice coordination number $q = 2$, 3, 4, 5, 6, and we conjecture that the self-dual property holds for general q. We also obtain the self-dual manifold for $q = 3$, 4, and it is found that, in an Ising subspace, the manifold coincides with the known Ising critical locus.

Consider a vertex model on a lattice $\mathscr{L}$, which can be in any spatial dimension, of E edges and with coordination number (valency) q. A line graph on $\mathscr{L}$ is a collection of a subset of the edges, which, if regarded as being covered by bonds, generates bond configurations at all vertices. With each vertex we associate a weight according to the configuration of the incident bonds. This gives rise to a 2^q-vertex model whose partition function is

$$Z = \sum_G \prod_{i=1}^{N} \omega_i \tag{1}$$

where ω_i is the weight of the ith vertex. The summation is taken over all 2^E line graphs G on $\mathscr{L}$.

The expression (1) defines a very general vertex model which encompasses many outstanding lattice statistical problems. For example, the Ising model in a non-zero magnetic field formulated in the usual high-temperature (tanh) expansion is a 2^q-vertex problem (see, e.g., Lieb and Wu 1972). It can also be shown that the eight-vertex model for $q = 3$ (Wu 1974b, Wu and Wu 1988a) as well as another special case of the general q problem (Wu 1972, 1974a) are completely equivalent to an Ising model in a non-zero magnetic field, a property that has been used to deduce the critical locus for the vertex models in question (Wu 1974a, b). However, very little is known about other properties of these vertex models.

In this letter we report some new results on duality properties for this 2^q-vertex model. We show that a generalised weak-graph transformation, which leaves the partition function unchanged, is always self-dual, and obtain the self-dual manifold (locus) for $q = 3$, 4. We further show that this self-dual locus coincides with the critical locus in the ferromagnetic Ising subspace.

† Work supported in part by the National Science Foundation Grant DMR-8702596.

0305-4470/89/020055+06$02.50

For simplicity, we consider a *symmetric* version of the model for which the vertex weight depends only on the number of bonds incident to the vertex. It should be noted this is not a severe restriction, since the analysis can be extended in a straightforward fashion to the general (asymmetric) case at the expense of a generalised weak-graph transformation of the vertex weights under which the partition function remains invariant. The weak-graph expansion was first used by Nagle (1968) in an analysis of the series expansion of six-vertex models. A general formulation of the weak-graph expansions given by Wegner (1973) permits the introduction of a free parameter into the formulation, a fact first recognised and explicitly used in the analysis of the eight-vertex model (Wu 1974b). To emphasise the extra degree of freedom introduced by the free parameter, we shall refer to the transformation containing free parameter(s) as the *generalised* weak-graph transformation†.

Consider first the case of $q=3$, namely an eight-vertex model whose vertex configurations and weights are shown in figure 1. The symmetric eight-vertex model has been considered previously (Wu 1974b, Wu and Wu 1988a), and it was established that, for a, b, c, d real, the vertex problem is completely equivalent to a ferromagnetic Ising model in a real magnetic field or an antiferromagnetic Ising model in a pure imaginary field. Using this Ising equivalence, the critical manifold of the eight-vertex model in the ferromagnetic Ising subspace is found to be‡

$$f(a, b, c, d)=0 \tag{2}$$

where

$$f(a, b, c, d) \equiv a(b^3+d^3)-d(a^3+c^3)+3(ab+bc+cd)(c^2-bd+ac-b^2). \tag{3}$$

We now show that the critical manifold (2) can also be obtained directly from an analysis of the self-dual property of the eight-vertex model.

The generalised weak-graph transformation for $q=3$ is (Wu 1974b)

$$\tilde{\boldsymbol{A}} = \boldsymbol{WA} \tag{4}$$

$$\tilde{\boldsymbol{A}} = \begin{pmatrix} \tilde{a} \\ \tilde{b} \\ \tilde{c} \\ \tilde{d} \end{pmatrix} \qquad \boldsymbol{A} = \begin{pmatrix} a \\ b \\ c \\ d \end{pmatrix}$$
$$\boldsymbol{W} = \frac{1}{(1+y^2)^{3/2}} \begin{pmatrix} 1 & 3y & 3y^2 & y^3 \\ y & 2y^2-1 & y^3-2y & -y^2 \\ y^2 & y^3-2y & 1-2y^2 & y \\ y^3 & -3y^2 & 3y & -1 \end{pmatrix} \tag{5}$$

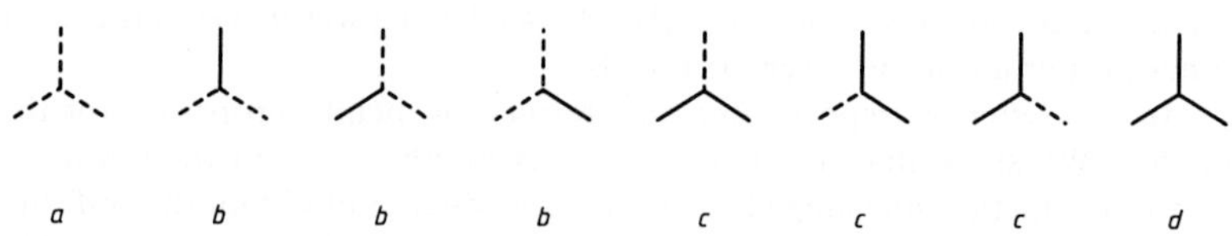

a b b b c c c d

Figure 1. Vertex configurations and weights for the symmetric eight-vertex model.

† In general, more than one parameter is needed in the analysis of the asymmetric model.

‡ See, in particular, footnote 5 of Wu and Wu (1988a).

where y is arbitrary. The partition function (1) is invariant under the transformation (4); namely we have

$$Z(\tilde{a}, \tilde{b}, \tilde{c}, \tilde{d}) = Z(a, b, c, d). \tag{6}$$

Generally, a transformation is self-dual if it possesses a fixed point, i.e. if it maps a point in the parameter space $\{a, b, c, d\}$ onto itself. For a transformation whose coefficients contain a parameter such as y in (4), we generally expect the transformation to be self-dual only for some special values of y. However, we now show that the generalised weak-graph transformation (4) is *always* self-dual, i.e. there exist fixed points for *all* y! We further determine the manifold in the parameter space containing all such fixed (self-dual) points.

Consider first the more general eigenvalue equation

$$\boldsymbol{W}\boldsymbol{A} = \lambda \boldsymbol{A} \tag{7}$$

where λ is the eigenvalue of $\boldsymbol{W}$. Combining (4) with (7), we see that the transformation $\boldsymbol{W}$ is self-dual if $\lambda = 1$. However, the transformation for $\lambda = -1$ can also be regarded as 'self-dual', since in this case the net effect of (7) is to negate all vertex weights. This introduces a factor $(-1)^N$ into the overall Boltzmann factor, and does not change anything as we generally have $N = \text{even}$.

The characteristic equation of (7) is

$$\det|W_{ij} - \lambda\delta_{ij}| = 0 \tag{8}$$

where $i, j = 1, 2, 3, 4$, and W_{ij} are elements of $\boldsymbol{W}$. After some manipulation, (8) reduces to the simple form

$$(1+y^2)^{9/2}(\lambda - 1)^2(\lambda + 1)^2 = 0. \tag{9}$$

This result is somewhat surprising. Generally, in solving an eigenvalue equation of the type of (8), we expect the eigenvalue λ to be a function of y. However, this is not the case here, and we find that solutions of $\lambda = \pm 1$ exist for *all* y. Thus, the generalised weak-graph transformation (4) is always self-dual. The location of the self-dual point will, of course, be y dependent.

The expression (9) is further revealing. It indicates that the determinant in (8) can be diagonalised by a similarity transformation into a form having diagonal elements $\lambda - 1$, $\lambda - 1$, $\lambda + 1$, $\lambda + 1$. This means that, for both $\lambda = 1$ and $\lambda = -1$, only 2 of the 4 linear equations in (7) are independent. Therefore, we can eliminate y using any two equations in (7) to obtain the self-dual manifold contaning all fixed (self-dual) points.

It is most convenient to use the first and the last equations in (7). Solving b and d from these two equations, we obtain after some algebra

$$\frac{b}{d} = \frac{a - c - \lambda c\sqrt{1+y^2}}{a + 3c - \lambda a\sqrt{1+y^2}} \qquad b + d = \frac{(a+c)y}{1 + \lambda\sqrt{1+y^2}} \qquad \lambda = \pm 1 \tag{10}$$

leading to the relations

$$\begin{aligned} y &= \frac{(b+d)(2ab + 3bc - ad)}{(a+c)(ab - cd)} \\ \lambda\sqrt{1+y^2} &= \frac{ab + 3bc - ad + cd}{ab - cd} \qquad \lambda = \pm 1. \end{aligned} \tag{11}$$

Substituting the first expression in (11) into the second and squaring both sides, we obtain the self-dual manifold

$$(2ab+3bc-ad)f(a, b, c, d)=0 \qquad \lambda=\pm 1 \tag{12}$$

where $f(a, b, c, d)$ has been given in (3). The vanishing of the first factor in (12) is equivalent to setting $y=0$, for which (4) is an identity transformation†. Therefore, the non-trivial self-dual manifold is precisely (2), obtained previously from a consideration of the Ising equivalence.

Consider next the case of $q=4$, a 16-vertex model whose vertex configurations and weights are shown in figure 2. This 16-vertex model has been considered previously in an Ising subspace (Wu 1972, 1974b). Now, the generalised weak-graph transformation is given by (4) with

$$\tilde{\boldsymbol{A}}=\begin{pmatrix}\tilde{a}\\ \tilde{b}\\ \tilde{c}\\ \tilde{d}\\ \tilde{e}\end{pmatrix} \qquad \boldsymbol{A}=\begin{pmatrix}a\\ b\\ c\\ d\\ e\end{pmatrix}$$

$$\boldsymbol{W}=\frac{1}{(1+y^2)^2}\begin{pmatrix}1 & 4y & 6y^2 & 4y^3 & y^4\\ y & 3y^2-1 & 3y^2-3y & y^4-3y^2 & -y^3\\ y^2 & 2y^3-2y & y^4-4y^2+1 & 2y-2y^3 & y^2\\ y^3 & y^4-3y^2 & 3y-3y^3 & 3y^2-1 & -y\\ y^4 & -4y^3 & 6y^2 & -4y & 1\end{pmatrix}. \tag{13}$$

Using (13), the characteristic equation (8) reduces to

$$-(1+y^2)^8(\lambda-1)^3(\lambda+1)^2=0 \tag{14}$$

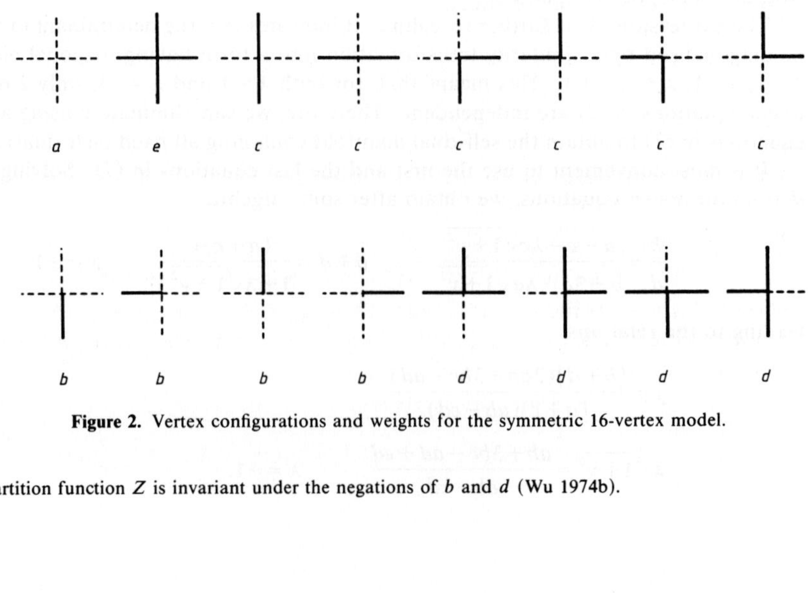

Figure 2. Vertex configurations and weights for the symmetric 16-vertex model.

† The partition function Z is invariant under the negations of b and d (Wu 1974b).

again yielding the result that the generalised weak-graph transformation (4) is self-dual for *all* y, and that (7) yields solution only for $\lambda=\pm1$.

For $\lambda=1$, (14) indicates that only two of the five equations in (7) are independent and, consequently, the self-dual manifold is obtained by eliminating y using any two of the five equations. It is again most convenient to use the first and the last equations in (7). By adding and subtracting these two equations, we obtain, respectively,

$$\begin{aligned} a+e&=2[(d-b)(y^2-1)+3cy]/y \\ a-e&=2(b+d)/y. \end{aligned} \tag{15}$$

Eliminating y from (15), we obtain the self-dual manifold

$$a^2d-be^2-3(a-e)(b+d)c+(b-d)[ae+2(b+d)^2]=0. \tag{16}$$

It can be shown (Wu and Wu 1988b) that, as in the case of $q=3$ (Wu 1974b), (16) coincides with the critical locus in the ferromagnetic Ising subspace of the vertex model. The present result establishes (16) as the self-dual locus for the whole parameter space.

For $\lambda=-1$, (14) tells us that three of the five equations in (7) are independent. Using any three equations from (7) to eliminate y, we obtain two hypersurfaces in the parameter space, and the self-dual manifold is their intersection. The difference of the first and the last equations in (7) yields

$$y=(e-a)/2(b+d) \tag{17}$$

and the hypersurfaces are then obtained by substituting (17) into any two equations in (7). In practice, however, it proves convenient to use combinations of the five equations which are factorisable after the substitution. After some algebra, we find the following factorisable expressions for the hypersurfaces:

$$\begin{aligned} &(a+2c+e)[(a-e)^2+4(b+d)^2]=0 \\ &[(a-6c+e)(a-e)^2+24c(b+d)^2+4a(3b^2-4bd-5d^2)+4e(3d^2-4bd-5b^2)] \\ &\qquad\times[(a-e)^2-4(b+d)^2]=0. \end{aligned} \tag{18}$$

Note that, unlike the case of $q=3$ for which the self-dual manifold is the same for $\lambda=\pm1$, (16) and (18) are distinct.

More generally for general q, it can be shown by following the procedure given in Wu (1974b) that the generalised weak-graph transformation (4) is

$$W_{ij}=(1+y^2)^{-q/2}\sum_{k=0}^{j}\binom{i}{k}\binom{q-i}{j-k}(-1)^k y^{i+j-2k} \qquad i,j=1,2,\ldots,q+l. \tag{19}$$

We have further evaluated the characteristic equation (8) using this W_{ij} for $q=2$, 5, 6. The results, together with those of $q=3$, 4 given in the above, can be summarised by the equality

$$\det|W_{ij}-\lambda\delta_{ij}|=(-1)^{q+1}(1+y^2)^{q^2/2}(\lambda+1)^{[(q+1)/2]}(\lambda-1)^{[(q+2)/2]} \tag{20}$$

where $[x]$ is the integral part of x, e.g., $[4]=4$, $[\frac{5}{2}]=2$. We conjecture that (20) holds for arbitrary q. It follows from (8) and (20) that the generalised weak-graph transformation (4) is always self-dual.

For $q=2n=$ even, which is the case in practice for $q>3$, there are n independent equations in (7) for $\lambda=1$ and $n+1$ independent equations for $\lambda=-1$. The self-dual manifold will then be the intersection of $n-1$ and n hypersurfaces for $\lambda=1$ and -1, respectively.

In summary, we have considered the generalised weak-graph transformation for a general vertex model in any dimension. We established that the generalised weak-graph transformation is always self-dual, and obtained the self-dual manifold for $q=3$, 4. It should be pointed out that this self-dual property is *intrinsic*, since its validity depends only on the fact that there is a uniform coordination number, q, throughout the lattice (thus applying to random lattices with uniform q as well). Consequently, one does not expect to deduce from these considerations physical properties, such as the exact critical temperature of the zero-field Ising model, which are lattice dependent.

References

Nagle J F 1968 *J. Math. Phys.* **8** 1007

Lieb E H and Wu F Y 1972 *Phase Transitions and Critical Phenomena* vol 1, ed C Domb and M S Green (New York: Academic) p 354

Wegner F 1973 *Physica* **68** 570

Wu F Y 1972 *Phys. Rev.* B **6** 1810

—— 1974a *Phys. Rev. Lett.* **32** 460

—— 1974b *J. Math. Phys.* **15** 687

Wu X N and Wu F Y 1988a *J. Stat. Phys.* **50** 41

—— 1988b unpublished

J. Phys. A: Math. Gen. 23 (1990) L131–L135. Printed in the UK

LETTER TO THE EDITOR

Algebraic invariants of the O(2) gauge transformation

J H H Perk†, F Y Wu‡ and X N Wu‡

† Department of Physics, Oklahoma State University, Stillwater, OK 74078, USA
‡ Department of Physics, Northeastern University, Boston, MA 02115, USA

Received 1 November 1989

Abstract. We consider the O(2) gauge transformation for a two-state vertex model on a lattice, and derive its fundamental algebraic invariants, the minimal set of homogeneous polynomials of the vertex weights which are invariant under O(2) transformations. Explicit expressions of the fundamental invariants are given for symmetric vertex models on lattices with coordination number $p=2, 3, 4, 5, 6$, generalising $p=3$ results obtained previously from more elaborate considerations.

In a study of the symmetry properties of discrete spin systems, Wegner [1] introduced a gauge transformation generalising the weak-graph transformation used by earlier investigators [2–4]. The gauge transformation, which describes important symmetry properties including the usual duality relation [3], is a linear transformation of the weights of a vertex model under which the partition function remains invariant. One particular symmetry property studied for over a century [5] is the construction of algebraic invariants, the homogeneous polynomials invariant under linear transformations. The problem of constructing invariants for the gauge transformation in vertex models has been studied by Hijmans *et al* [6,7] for the square lattice and, more recently, by Wu *et al* [8] and by Gwa [9] for the O(2) transformation on trivalent lattices. Specifically, Wu *et al* [8] proposed that the critical frontier of the Ising model in a non-zero magnetic field is given by the algebraic invariants of the related vertex model, and constructed the invariants by enumeration for trivalent lattices. A simpler method leading to the same invariants was later given by Gwa [9]. But the extension of both of these analyses to lattices of general coordination number p has proven to be extremely tedious, becoming almost intractable for $p>4$. Clearly, an alternative and simpler approach is needed.

In this letter we consider the O(2) gauge transformation for a two-state vertex model on a lattice of general coordination number p, and present a formulation which leads to a simple and direct determination of its algebraic invariants.

We first define the vertex model and the O(2) gauge transformation. Consider a lattice of coordination number p, with the lattice edges in one of two distinct states independently at each edge. We may regard the edges as being either 'empty' or 'covered' by a bond, so that the edge configurations generate bond graphs [10]. Introduce edge variables $s=0, 1$ so that $s=0$ ($s=1$) denotes the edge being empty (covered). With each lattice site associate a vertex weight $W(s_1, s_2, \ldots, s_p)$, where $s_1, s_2, \ldots, s_p$ indicate the states of the p incident edges. The partition function of this two-state vertex model is

$$Z(\{W\})=\sum\prod_i W(s_1, s_2, \ldots, s_p) \qquad (1)$$

where the summation is taken over all bond graphs of the lattice, and the product is taken over all vertices i.

Consider a linear transformation of the 2^p vertex weights $W(s_1, s_2, \ldots, s_p)$,

$$\tilde{W}(t_1, t_2, \ldots, t_p) = \sum_{s_1=0}^{1} \sum_{s_2=0}^{1} \cdots \sum_{s_p=0}^{1} R_{t_1 s_1} R_{t_2 s_2} \ldots R_{t_p s_p} W(s_1, s_2, \ldots, s_p). \tag{2}$$

The transformation (2) leaves the partition function invariant if R_{ts} are elements of a 2×2 matrix $\mathbf{R}$ satisfying $\tilde{\mathbf{R}}\mathbf{R} = I$, where I is the identity matrix [1]. This implies $\det|R_{ts}| = \pm 1$, and therefore the transformation (2) provides a representation of the two-dimensional orthogonal group O(2), to be referred to as the O(2) gauge transformation.

The O(2) group is generated by a rotation $\mathbf{R}^{(1)}$ or a reflection $\mathbf{R}^{(2)}$ given by

$$\mathbf{R}^{(1)} = \begin{pmatrix} \cos\theta & -\sin\theta \\ \sin\theta & \cos\theta \end{pmatrix} \qquad \mathbf{R}^{(2)} = \begin{pmatrix} \cos\theta & \sin\theta \\ \sin\theta & -\cos\theta \end{pmatrix}. \tag{3}$$

Note that $\mathbf{R}^{(2)}$ has been used exclusively in previous investigations [2,4,8].

For *symmetric* vertex models, the vertex weights depend only on the number of covered incident edges, for which we have

$$W(s_1, s_2, \ldots, s_p) = W_p(s) \qquad s = s_1 + s_2 + \ldots + s_p \tag{4}$$

where $s = 0, 1, 2 \ldots, p$ is the number of bonds incident at the vertex. We shall, however, continue to assume general vertex weights, and only below specialise the results to symmetric vertex models.

Hilbert [5, see also p 235 of Gurevich in [5]] established more than a century ago that invariants of a linear transformation are in the form of homogeneous polynomials, and that all such polynomials are expressible in terms of a minimal set of *fundamental* ones. The crux of the matter is, of course, the determination of these fundamental invariants for a given linear transformation. For the O(2) transformation, as we now show, the task can be accomplished as follows.

Introduce the change of basis

$$A_{\sigma_1 \ldots \sigma_p} = \sum_{t_1=0}^{1} \cdots \sum_{t_p=0}^{1} \left[\prod_{k=1}^{p} (\mathrm{i}\sigma_k)^{t_k} \right] W(t_1, \ldots, t_p) \tag{5}$$

where $\sigma_k = \pm 1$. For example, for $p = 2$, (5) is

$$\begin{aligned} A_{\pm\pm} &= [W(00) - W(11)] \pm \mathrm{i}[W(01)] + W(10)] \\ A_{\pm\mp} &= [W(00) + W(11)] \mp \mathrm{i}[W(01) - W(10)]. \end{aligned} \tag{6}$$

In a similar fashion we define $\tilde{A}_{\sigma_1} \ldots \sigma_p$ in terms of $\tilde{W}(s_1, \ldots, s_p)$. Then, using the identity

$$\sum_{t=0}^{1} (\mathrm{i}\sigma)^t R_{ts}^{(l)} = (-1)^{s(l-1)} \mathrm{e}^{\mathrm{i}\sigma\theta} (\mathrm{i}\sigma)^s \qquad l = 1, 2 \tag{7}$$

where $R_{ts}^{(l)}$ are elements of $\mathbf{R}^{(1)}$ or $\mathbf{R}^{(2)}$ given by (3), one obtains from (5) and (2) the following transformation property for the A:

$$\tilde{A}^{(l)}_{\sigma_1 \ldots \sigma_p} = \begin{cases} \mathrm{e}^{i(\sigma_1 + \ldots + \sigma_p)\theta} A_{\sigma_1 \ldots \sigma_p} & l = 1 \\ \mathrm{e}^{i(\sigma_1 + \ldots + \sigma_p)} \theta A^*_{\sigma_1 \ldots \sigma_p} & l = 2 \end{cases} \tag{8}$$

where $\tilde{A}^{(l)}$ is the $\tilde{A}$ obtained from (5) by using $\tilde{W}$ in (2) with $\mathbf{R}=\mathbf{R}^{(l)}$, $l=1,2$, and A^* is the complex conjugate of A. For $l=1$, the As change only by a phase factor under the O(2) transformation. It follows that any product of the As, for which the σs of all A factors in the product add to zero, is an invariant. For $l=2$, however, the products are transformed into complex conjugates, in addition to the change of a phase factor. Thus, in both cases the real parts of these products are invariant, and the imaginary parts are invariant under $\mathbf{R}^{(1)}$ while changing sign under $\mathbf{R}^{(2)}$. We shall, however, refer to both the real and imaginary parts as the invariants.

For a given p, the fundamental invariants can now be constructed by following the above prescription. For $p=2$, there are three fundamental invariants $A_{++}A_{--}$, A_{+-}, and A_{-+}, the last two being the complex conjugates of each other; any other invariant is composed of the three. For $p=3$, our consideration leads to 30 distinct products, examples of which are $A_{+++}A_{---}$, $A_{++-}A_{+--}$, $A^2_{++-}A_{+-+}A_{---}$, and $A^3_{++-}A_{---}$. This gives rise to 30 homogeneous polynomials which are invariant under $\mathbf{R}^{(1)}$, and either invariant or changing sign under $\mathbf{R}^{(2)}$.

It is shown above that fundamental invariants for the general O(2) gauge transformation (2) can be constructed in a straightforward fashion. We now specialise the consideration to the symmetric vertex model (4) for which the situation is considerably simpler.

Using (4) and (5), we have

$$\begin{aligned} A_{\sigma_1\ldots\sigma_p} &\equiv A_p(t) \\ &= \sum_{s=0}^{p} \mathrm{i}^s W_p(s) \sum_{\alpha=0}^{m} \binom{m}{\alpha} \sum_{\beta=0}^{n} \binom{n}{\beta} (-1)^\beta \delta_{\mathrm{Kr}}(\alpha+\beta, s) \end{aligned} \tag{9}$$

where $t\equiv\sigma_1+\sigma_2\ldots+\sigma_p=\pm p, \pm(p-2), \ldots, \pm 1$ or 0; $m\equiv(p+t)/2$, $n\equiv(p-t)/2$, and δ_{Kr} is the Kronecker delta function. Note that the coefficient of $W_p(s)$ in (9) is the coefficient of z^s in the expansion of $(1+\mathrm{i}z)^m(1-\mathrm{i}z)^n$. It follows that (8) becomes

$$\tilde{A}^{(l)}_{\sigma_1\ldots\sigma_p} \equiv \tilde{A}^{(l)}_p(t) = \begin{cases} \mathrm{e}^{\mathrm{i}t\theta} A_p(t) & l=1 \\ \mathrm{e}^{\mathrm{i}t\theta} A_p^*(t) & l=2. \end{cases} \tag{10}$$

It is straightforward to write down the explicit expressions of $A_p(t)$ using (9). Adopting the notation [4,8] of denoting vertex weights by $a, b, c, \ldots$ such that a is the weight of vertices having no incident bonds, b the weight of vertices having one incident bond, etc, we find for $p=2,3,\ldots,6$

$$\begin{aligned} &A_2(\pm 2) = a - c \pm 2\mathrm{i}b && A_2(0) = a + c \\ &A_3(\pm 3) = a - 3c \pm \mathrm{i}(3b - d) && A_3(\pm 1) = a + c \pm \mathrm{i}(b + d) \\ &A_4(\pm 4) = a - 6c + e \pm 4\mathrm{i}(b - d) && A_4(\pm 2) = a - e \pm 2\mathrm{i}(b + d) \\ &A_4(0) = a + 2c + e \\ &A_5(\pm 5) = a - 10c + 5e \pm \mathrm{i}(5b - 10d + f) \\ &A_5(\pm 3) = a - 2c - 3e \pm \mathrm{i}(3b + 2d - f) \\ &A_5(\pm 1) = a + 2c + e \pm \mathrm{i}(b + 2d + f) \\ &A_6(\pm 6) = a - 15c + 15e - g \pm 2\mathrm{i}(3b - 10d + 3f) \\ &A_6(\pm 4) = a - 5c - 5e + g \pm 4\mathrm{i}(b - f) \\ &A_6(\pm 2) = a + c - e - g \pm 2\mathrm{i}(b + 2d + f) \\ &A_6(0) = a + 3c + 3e + g. \end{aligned} \tag{11}$$

As dictated by (10), the fundamental invariants for each p are now constructed by forming products of $A_p(t)$, such that the sum of all t variables in the product vanishes. Adopting the notation $(t) \equiv A_p(t)$ for each p, we find the following fundamental invariants for $p = 2, 3, \ldots, 6$:

$p = 2$: $(2)(-2)$; (0)

$p = 3$: $(3)(-3)$; $(1)(-1)$; $(3)(-1)^3, (-3)(1)^3$

$p = 4$: $(4)(-4)$; $(2)(-2)$; (0); $(4)(-2)^2, (-4)(2)^2$

$p = 5$: $(5)(-5)$; $(3)(-3)$; $(1)(-1)$; $(5)(-1)^5, (-5)(1)^5$; $(3)(-1)^3, (-3)(1)^3$; $(5)(-3)(-1)^2, (-5)(3)(1)^2$; $(5)(1)(-3)^2, (-5)(-1)(3)^2$; $(5)^2(-1)(-3)^3, (-5)^2(1)(3)^3$; $(5)^3(-3)^5, (-5)^3(3)^5$

$p = 6$: $(6)(-6)$; $(4)(-4)$; $(2)(-2)$; (0); $(6)(-2)(-4), (-6)(2)(4)$; $(4)(-2)^2, (-4)(2)^2$; $(6)(-2)^3, (-6)(2)^3$; $(6)(2)(-4)^2, (-6)(-2)(4)^2$; $(6)^2(-4)^3, (-6)^2(4)^3$.

Since (t) and $(-t)$ are complex conjugates of each other, the fundamental invariants always occur in conjugate pairs (except (0) and $(t)(-t)$ which are real), and we can consider the real and imaginary parts individually. Both the real and imaginary parts are invariant under $\mathbf{R}^{(1)}$, and the real parts are invariants and the imaginary parts change sign under $\mathbf{R}^{(2)}$. For $p = 3$, e.g., there are four polynomials:

$$\begin{aligned}
I_1 &= (3)(-3) = (a-3c)^2 + (3b-d)^2 \\
I_2 &= (1)(-1) = (a+c)^2 + (b+d)^2 \\
I_3 &= \mathrm{Re}(3)(-1)^3 = \mathrm{Re}\{[a-3c+\mathrm{i}(3b-d)][a+c-\mathrm{i}(b+d)]^3\} \\
I_4 &= \mathrm{Im}(3)(-1)^3 = \mathrm{Im}\{[a-3c+\mathrm{i}(3b-d)][a+c-\mathrm{i}(b+d)]^3\}
\end{aligned} \tag{12}$$

for which I_1, I_2, and I_3 are invariant under both $\mathbf{R}^{(1)}$ and $\mathbf{R}^{(2)}$, and I_4 is invariant under $\mathbf{R}^{(1)}$ while changing sign under $\mathbf{R}^{(2)}$.

We have considered the O(2) gauge transformation for a general two-state vertex model on an arbitrary lattice, and constructed its fundamental algebraic invariants. For the symmetric vertex model on a lattice of coordination number p, our analysis shows that there are, respectively, 2, 4, 5, 15, and 14 fundamental algebraic invariants for $p = 2$, 3, 4, 5, and 6. These invariants are explicitly given in (11). In the case of $p = 3$, we have verified that the fundamental invariants P, Q, P_1, P_2 and P_3 obtained previously [8,9] can indeed be expressed in terms of those in (12). The relations are $P = (9I_2 - I_1)/8$, $Q = (I_1 - I_2)/8$, $P_1 = I_4/4$, $P_2 = (72I_3 - I_1^2 + 30I_1I_2 + 27I_2^2)/64$, and $P_3 = (-8I_3 - I_1^2 + 6I_1I_2 + 3I_2^2)/64$. Note that P_1 changes sign under $\mathbf{R}^{(2)}$, a fact previously observed [8]. For $p = 4$, we have also verified that the five fundamental invariants deduced from the ones obtained by Hijmans *et al* [6,7] can be expressed in terms of those in (11). The same set of $p = 4$ fundamental invariants have also been obtained, after considerable algebraic manipulation, by extending the analyses of Wu *et al* [8] and Gwa [9], but the method of Gwa no longer retains its simplicity for $p = 3$. Both methods, however, become almost intractable for $p > 4$.

Finally, we point out the existence of *syzygies*, polynomial relations between the linearly independent invariants. We have seen that all invariants for a given p are products of $p+1$ polynomials $A_p(t)$. It follows that there must exist relations, or syzygies, among these invariants, if the number of invariants exceeds p. Explicit expressions of syzygies are usually very difficult to construct, but they are easily identified in the present formulation. For $p=3$ and 4, e.g., the numbers of fundamental invariants are, respectively, 4 and 5, and hence there is one syzygy in each case. Explicitly, we find

$$\begin{aligned} &[(3)(-1)^3][(-3)(1)^3] = [(3)(-3)][(1)(-1)]^3 \qquad \text{for } p=3 \\ &[(4)(-2)^2][(-4)(2)^2] = [(4)(-4)][(2)(-2)]^2 \qquad \text{for } p=4. \end{aligned} \tag{13}$$

Similarly, there are ten syzygies for $p=5$ and eight for $p=6$; all can be similarly constructed.

This research was supported by National Science Foundation grants DMR-8702596 and DMR-8803678, and Dean's Incentive Grant of the Oklahoma State University.

References

[1] Wegner F J 1973 *Physica* **68** 570
[2] Nagle J F and Temperley H N V 1968 *J. Math. Phys.* **8** 1020
[3] Fan C and Wu F Y 1970 *Phys. Rev.* B **2** 723
[4] Wu F Y 1974 *J. Math. Phys.* **6** 687
[5] Hilbert D 1890 *Math. Ann.* **36** 473; 1893 *Math. Ann.* **42** 313
Gurevich G B 1964 *Foundations of the Theory of Algebraic Invariants* (Groningen: Noordhoff)
[6] Gaaff A and Hijmans J 1976 *Physica* **83A** 301, 317
[7] Schram H M and Hijmans J 1984 *Physica* **125A** 58
[8] Wu F Y, Wu X N and Blöte H W J 1989 *Phys. Rev. Lett.* **62** 2773
[9] Gwa L H 1989 *Phys. Rev. Lett.* **63** 1440
[10] Lieb E H and Wu F Y 1972 *Phase Transitions and Critical Phenomena* ed C Domb and M S Green (New York: Academic)

J. Phys. A: Math. Gen. **24** (1991) L503–L507. Printed in the UK

LETTER TO THE EDITOR

The O(3) gauge transformation and 3-state vertex models

Leh-Hun Gwa† and F Y Wu‡

† Department of Mathematics, Rutgers University, New Brunswick, NJ 08903, USA
‡ Department of Physics, Northeastern University, Boston, MA 02115, USA

Received 22 January 1991

Abstract. We consider the O(3) gauge transformation for three-state vertex models on lattices of coordination number three. Using an explicit mapping between O(3) and SL(2), we establish that there exist exactly six polynomials of the vertex weights, which are fundamentally invariant under the O(3) transformation. Explicit expressions of these fundamental invariants are obtained in the case of symmetric vertex weights.

The consideration of gauge transformations has played a central role in the study of discrete spin systems. The gauge transformation is a linear transformation of the Boltzmann weights of a spin system, such as a vertex model, which does not alter the partition function. In a classic paper Wegner [1] formulated the gauge transformation for discrete spin systems, generalizing the previously known duality and weak-graph transformations. Properties pertaining to specific spin and lattice systems remain, however, to be worked out on a case by case basis. For example, those pertaining to the O(2) transformation for the 16-vertex model on the square lattice have subsequently been studied by Hijmans *et al* [2–4].

Of particular interest in statistical mechanics is the construction of invariants of the transformation, a subject matter of great interest in mathematics at the turn of the century [5–7]. In statistical mechanics the invariants of the O(2) transformation for 2-state vertex models have been utilized to determine the criticality of the Ising models in a non-zero magnetic field [8–12]. In the case of the O(2) transformation it has been possible to explicitly construct the invariants [12, 13]. The direct construction of invariants for O(3) is more complicated, however. But the day is saved since there exists a mapping between O(3) and SL(2), and invariants for the latter are already known. In this letter we utilize this mapping to obtain invariants of the O(3) gauge transformation which is applicable to 3-state spin systems.

Consider a lattice of coordination number 3, which can be in any spatial dimension, and assume that each of the lattice edges can be independently in one of three distinct states. With each lattice site we associate a vertex weight $W(s_1, s_2, s_3)$, where $s_i = 1, 2$ and 3 specifies the states of the three incident edges. This defines a 27-vertex model and the partition function $Z = \Sigma\,\Pi\, W(s_1, s_2, s_3)$, where the summation is taken over all edge configurations of the lattice.

Wegner [1] has shown that the partition function Z remains unchanged if the vertex weights W are replaced by $\tilde{W}$ given by

$$\tilde{W}(t_1, t_2, t_3) = \sum_{s_1=1}^{3} \sum_{s_2=1}^{3} \sum_{s_3=1}^{3} R_{t_1 s_1} R_{t_2 s_2} R_{t_3 s_3} W(s_1, s_2, s_3) \tag{1}$$

0305-4470/91/100503+05$03.50

provided that R_{ts} are elements of a 3×3 matrix $\mathbf{R}$ associated with lattice edges satisfying $\tilde{\mathbf{R}}\mathbf{R}=\mathbf{I}$, where $\tilde{\mathbf{R}}$ is the transpose of $\mathbf{R}$ and $\mathbf{I}$ is the identity matrix. This implies $\det|R_{ts}|=\pm 1$, and, consequently, the transformation (1) leaves $\sum_{s_1,s_2,s_3} W^2(s_1, s_2, s_3)$ invariant and thus gives rise to a representation of the three-dimensional orthogonal group O(3). In reality the validity of the invariance of the partition function holds more generally even if $\mathbf{R}$ is edge-dependent [1]. For this reason we refer to (1) as the O(3) gauge transformation.

Explicitly, O(3) is a three-parameter group. For SO(3) or $\det|R_{ts}|=1$, e.g., we can write

$$\mathbf{R}=\begin{pmatrix} c_2c_3 & -s_1s_2c_3+c_1s_3 & c_1s_2c_3+s_1s_3 \\ -c_2s_3 & c_1c_3+s_1s_2s_3 & -c_1s_2s_3+s_1c_3 \\ -s_2 & -s_1c_2 & c_1c_2 \end{pmatrix} \tag{2}$$

where $c_i=\cos\theta_i$, $s_i=\sin\theta_i$. This can be interpreted as a rotation in the 3-space by first making a rotation θ_1 about the x axis, followed by a rotation of θ_2 about the y axis and finally a rotation θ_3 about z axis [14].

Generally, the transformation (1) forms a representation of O(3) in the space of tensors of rank 3. Let y_1, y_2, y_3 be the coordinates of the fundamental representation space of O(3). Then the general tensors of rank 3 form a 3^3-dimensional space with basis $y_m\otimes y_n\otimes y_k$, where the three y's (first, second, and third) refer to specific incident edges, and the subscripts specify the state of the incident edge.

The consideration is much simplified when the vertex weights are symmetric, i.e. $W(s_1, s_2, s_3)$ is independent of the permutation of s_1, s_2, and s_3. In this case, we can conveniently relabel the vertex weights as ω_{ijk}, where i, j, k are, respectively, the numbers of incident edges in states 1, 2, 3 subject to $i+j+k=3$. Thus, the 27 vertex weights reduce to 10 independent ones whose associated configurations are shown in figure 1, and (1) gives rise to a 10×10 matrix representation of O(3). Furthermore, the tensor product of the basis $y_m\otimes y_n\otimes y_k$ can be replaced by an ordinary product, and the vertex weights can be written as given by the polynomial representation

$$\omega_{ijk}=y_1^i y_2^j y_3^k \qquad i+j+k=3. \tag{3}$$

It is well known that the special unitary group SU(2) is two-to-one homomorphic to SO(3), a familiar example being the spinor representation of the rotation group in

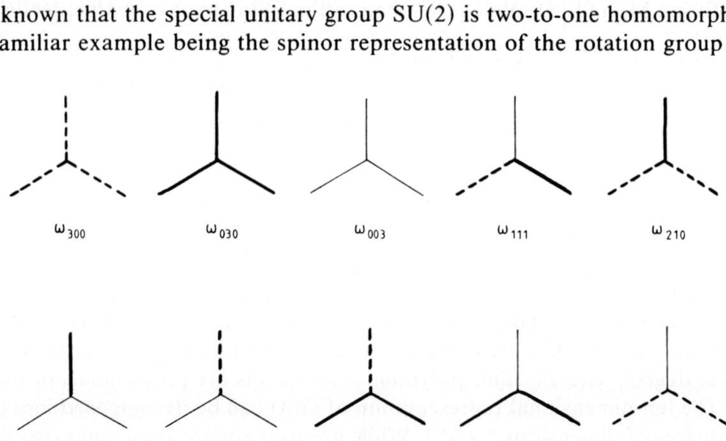

Figure 1. The ten vertex configurations and the weights of a symmetric 3-state 27-vertex model. The vertex configuration with weight ω_{ijk} is characterized by i broken, j thick, and k thin lines.

quantum mechanics. In addition, the invariants of SU(2) are identical to those of the special linear group SL(2). It follows that we can deduce the invariants of O(3) from those already known for SL(2). (Strictly speaking, this leads to invariants for SO(3), which may change sign under the odd elements of O(3).) We first describe the mapping of the representations for the two groups.

Let α_1 and α_2 be the coordinates of the fundamental representation space of SL(2). The mapping between α_1, α_2 and the coordinates y_1, y_2, y_3 of the vector representation of O(3) is

$$z_1 \equiv \alpha_1^2 = -y_1 + iy_3 \qquad z_2 \equiv \alpha_1\alpha_2 = y_2 \qquad z_3 \equiv \alpha_2^2 = y_1 + iy_3 \tag{4}$$

where z_1, z_2, z_3 form the coordinates of a rank-two symmetric tensor.

In view of (3) and (4), ω_{ijk} are raised to the sixth power of α_i and therefore invariants of O(3) must be given by tensors of rank six in $\{\alpha_1, \alpha_2\}$ with elements in the binary form

$$\begin{aligned}
&e_0 \equiv \alpha_1^6 = z_1^3 \qquad e_1 \equiv \alpha_1^5\alpha_2 = z_1^2 z_2 \\
&e_2 \equiv \alpha_1^4\alpha_2^2 = (z_1^2 z_3 + 4z_1 z_2^2)/5 \\
&e_3 \equiv \alpha_1^3\alpha_2^3 = (2z_2^3 + 3z_1 z_2 z_3)/5 \\
&e_4 \equiv \alpha_1^2\alpha_2^4 = (z_3^2 z_1 + 4z_3 z_2^2)/5 \\
&e_5 \equiv \alpha_1\alpha_2^5 = z_2 z_3^2 \qquad e_6 \equiv \alpha_2^6 = z_3^3.
\end{aligned} \tag{5}$$

Here, coefficients on the RHS are determined according to the following rules: (i) write each e_j as the average of all distinct permutations of the six α_i, (ii) for each permutation, group the six z_i into three consecutive pairs, and (iii) replace the grouped pairs by z_i using (4). For example, the first four lines of (5) are obtained from:

$$e_0 = (\alpha_1\alpha_1)(\alpha_1\alpha_1)(\alpha_1\alpha_1) = z_1^3$$

$$e_1 = \tfrac{1}{6}[(\alpha_1\alpha_1)(\alpha_1\alpha_1)(\alpha_1\alpha_2) + \text{all permutations of the six } \alpha_i] = \tfrac{1}{6}(6z_1^2 z_2) = z_1^2 z_2$$

$$e_2 = \tfrac{1}{15}[(\alpha_1\alpha_1)(\alpha_1\alpha_1)(\alpha_2\alpha_2) + \text{all permutations of the six } \alpha_i] = \tfrac{1}{15}(12z_1^2 z_2 + 3z_1^2 z_3)$$

$$e_3 = \tfrac{1}{20}[(\alpha_1\alpha_1)(\alpha_1\alpha_2)(\alpha_2\alpha_2) + \text{all permutations of the six } \alpha_i] = \tfrac{1}{20}(12z_1 z_2 z_3 + 8z_2^3).$$

The polynomial nature of symmetric tensors now makes it possible to simply substitute (4) and (3) into (5), leading to the following explicit expressions for the e_j:

$$\begin{aligned}
&e_0 = u + iv \qquad e_1 = s + it \qquad e_2 = (x + iy)/5 \\
&e_6 = -u + iv \qquad e_5 = s - it \qquad e_4 = (-x + iy)/5 \\
&e_3 = (2\omega_{030} - 3\omega_{210} - 3\omega_{012})/5
\end{aligned} \tag{6}$$

where

$$\begin{aligned}
&u = 3\omega_{102} - \omega_{300} \qquad x = \omega_{300} + \omega_{102} - 4\omega_{120} \\
&v = 3\omega_{201} - \omega_{003} \qquad y = 4\omega_{021} - \omega_{003} - \omega_{201} \\
&a = \omega_{210} - \omega_{012} \qquad t = 2\omega_{111}.
\end{aligned} \tag{7}$$

We now look for polynomials of the vertex weights (3) which are the *fundamental* invariants of the O(3) transformation, i.e. they cannot be expressed as invariants of lower degrees and all other polynomial invariants are polynomials of them.

The ten-dimensional representation of O(3) can be decomposed into two invariant subspaces of dimensions 3 and 7. While group-theoretic argument exists for its reasoning, this decomposition also arises as a consequence of the mapping between O(3) and SL(2) in the binary form (5): the presence of seven elements in (5) implies the existence of a seven-dimensional invariant subspace.

The elements of three-dimensional invariant subspace can be found easily. They are

$$\begin{aligned}\eta_1&=\omega_{300}+\omega_{102}+\omega_{120}=y_1(y_1^2+y_2^2+y_3^2)\\ \eta_2&=\omega_{030}+\omega_{012}+\omega_{210}=y_2(y_1^2+y_2^2+y_3^2)\\ \eta_3&=\omega_{003}+\omega_{021}+\omega_{201}=y_3(y_1^2+y_2^2+y_3^2).\end{aligned}\tag{8}$$

Obviously, this subspace transforms in the same way as the $\{y_1, y_2, y_3\}$ space. There is only one fundamental invariant in this subspace, namely,

$$I_0=\eta_1^2+\eta_2^2+\eta_3^2.\tag{9}$$

To find the fundamental invariants in the seven-dimensional invariant subspace mapped to SL(2), we make use of results known for SL(2). It is known [6, p 156] that the complete set of irreducible sextic invariants for SL(2) consists of five polynomials. In the mathematical literature [6, 7], these are given in concise, yet symbolic, forms as follows:

$$\begin{aligned}&I_1=(f,f)^{(6)} \qquad I_2=(i,i)^{(4)}\\ &I_3=(l,l)^{(2)} \qquad I_4=(f,l^3)^{(6)}\\ &I_5=((f,i)^{(1)},l^4)^{(8)}\end{aligned}\tag{10}$$

where

$$\begin{aligned}&f\equiv(\alpha\cdot\boldsymbol{x})^6\\ &i\equiv(f,f)^{(4)}=(\alpha\beta)^4(\alpha\cdot\boldsymbol{x})^2(\beta\cdot\boldsymbol{x})^2\\ &l\equiv(f,i)^{(4)}=(\alpha\beta)^2(\alpha\gamma)^2(\beta\gamma)^4(\alpha\cdot\boldsymbol{x})^2\\ &\alpha\cdot\boldsymbol{x}\equiv\alpha_1x_1+\alpha_2x_2 \qquad (\alpha\beta)\equiv\alpha_1\beta_2-\beta_1\alpha_2.\end{aligned}\tag{11}$$

Here, for any

$$\begin{aligned}&f=(\alpha\cdot\boldsymbol{x}_1)(\alpha\cdot\boldsymbol{x}_2)\ldots(\alpha\cdot\boldsymbol{x}_m)\\ &g=(\beta\cdot\boldsymbol{x}_1)(\beta\cdot\boldsymbol{x}_2)\ldots(\beta\cdot\boldsymbol{x}_n)\end{aligned}\tag{12}$$

we have

$$(f,g)^{(r)}\equiv C\sum_{P,Q}\frac{(\alpha_{P1}\beta_{Q1})(\alpha_{P2}\beta_{Q2})\ldots(\alpha_{Pr}\beta_{Qr})}{(\alpha\cdot\boldsymbol{x}_{P1})(\alpha\cdot\boldsymbol{x}_{P2})\ldots(\alpha\cdot\boldsymbol{x}_{Pr})(\beta\cdot\boldsymbol{x}_{Q1})(\beta\cdot\boldsymbol{x}_{Q2})\ldots(\beta\cdot\boldsymbol{x}_{Qr})}fg\tag{13}$$

where $C=[r!\binom{m}{r}\binom{n}{r}]^{-1}$ and the summation extends to all *distinct* permutations P and Q of the r integers $1, 2, \ldots, r$. In (10), the degree of the invariants as polynomials in the e_j is the same as the degree in the fs. Thus, we find I_1, I_2, I_3, I_4 and I_5 of degrees 2, 4, 6, 10 and 15, respectively, in the e_j.

We caution that the above notations are highly symbolic and should be deciphered with care. Particularly, since the αs have only symbolical meaning, they can be replaced by other symbols, i.e. $\alpha\cdot\boldsymbol{x}=\beta\cdot\boldsymbol{x}=\gamma\cdot\boldsymbol{x}$. After some reductions, we find the following explicit expressions of fundamental invariants:

$$\begin{aligned}J_1&\equiv I_2/2=e_0e_6-6e_1e_5+15e_2e_4-10e_3^2\\ J_2&\equiv I_2/24-J_1^2/36=-e_3^4+e_3^2(e_0e_6+2e_1e_5+3e_2e_4)\\ &\quad+e_0e_4^3+e_0e_2e_5^2-2e_3e_4(e_1e_4+e_0e_5)+e_2e_4(2e_1e_5-e_0e_6)\\ &\quad+e_6e_2^3+e_6e_4e_1^2-2e_3e_2(e_5e_2+e_6e_1)-2e_1^2e_5^2.\end{aligned}\tag{14}$$

Explicit expressions of I_3, I_4 and I_5, which can be worked out in a straightforward fashion, are un-illuminatively complicated, and will not be presented. It may be explicitly verified by substituting (6) and (7) into (9) and (14) that the Is and Js are invariant under the permutation of the subscripts $\{i, j, k\}$ of the vertex weights ω_{ijk}, as required by the symmetry of the three spin states of the lattice edges.

Of special interest in statistical mechanical applications is the subspace $e_1 = e_3 = e_5 = 0$ pertaining to the spin-1 Blume–Emery–Griffiths model [15]. The intersections of the six fundamental invariants in this subspace possess a much simpler form. We find, in addition to I_0 and $I_5 \equiv 0$, the following expressions:

$$\begin{aligned} J_1 &= A + 15B \\ J_2 &= C - B^2 - AB \\ J_3 &= AC + 3BC - 2B^3 - 6AB^2 \\ J_4 &= 4(5A - 9B)C^2 + (A^3 + 21A^2B - 93AB^2 + 135B^3)C \\ &\qquad + 2B^2(9A^3 - 59A^2B + 99AB^2 - 81B^3) \end{aligned} \tag{15}$$

where $A = e_0e_6$, $B = e_2e_4$, $C = e_2^3e_6 + e_4^3e_0$, $J_3 = I_3/24 + 4J_1J_2/3$, $J_4 = I_4/64$.

This work has been supported in part by the National Science Foundation Grants DMR-8918903 and DMR-9015489. We would like to thank Nolan R Wallach for bringing to our attention the useful literatures on invariants.

References

[1] Wegner F J 1973 *Physica* **68** 570
[2] Gaaff A and Hijmans J 1975 *Physica* **80A** 149; 1976 *Physica* **83A** 301, 317; 1979 *Physica* **97A** 244
[3] Hijmans J and Schram H M 1983 *Physica* **121A** 479; 1984 *Physica* **125A** 25
[4] Hijmans J 1985 *Physica* **130A** 57
[5] Hilbert D 1890 *Math. Ann.* **36** 473; 1893 *Math. Ann.* **42** 313
[6] Grace J H and Young A 1903 *The Algebra of Invariants* (Cambridge: Cambridge University Press)
[7] Glenn O E 1915 *A Treatise of the Theory of Invariants* (Boston MA: Ginn)
[8] Wu F Y, Wu X N and Blöte H W J 1989 *Phys. Rev. Lett.* **62** 2773
[9] Gwa L H 1989 *Phys. Rev. Lett.* **63** 1440
[10] Wu X N and Wu F Y 1990 *Phys. Lett.* **144A** 123
[11] Blöte H W J and Wu X N 1990 *J. Phys. A: Math. Gen.* **23** L627
[12] Gwa L H 1990 *Phys. Rev.* B **41** 7315
[13] Perk J H H, Wu F Y and Wu X N 1990 *J. Phys. A: Math. Gen.* **23** L131
[14] Wybourne B G 1974 *Classical Groups for Physicists* (New York: Wiley)
[15] Gwa L H and Wu F Y unpublished

PHYSICAL REVIEW E **67**, 026111 (2003)

Duality relation for frustrated spin models

D.-H. Lee
Department of Physics, University of California, Berkeley, California 94720

F. Y. Wu
Department of Physics, Northeastern University, Boston, Massachusetts 02115

(Received 29 October 2002; published 13 February 2003)

We consider discrete spin models on arbitrary planar graphs and lattices with frustrated interactions. We first analyze the Ising model with frustrated plaquettes. We use an algebraic approach to derive the result that an Ising model with some of its plaquettes frustrated has a dual model which is an Ising model with an external field $i\pi/2$ applied to the dual sites centered at frustrated plaquettes. In the case where all plaquettes are frustrated, this leads to the known result that the dual model has a uniform field $i\pi/2$, whose partition function can be evaluated in the thermodynamic limit for regular lattices. The analysis is extended to a Potts spin glass with analogous results obtained.

DOI: 10.1103/PhysRevE.67.026111 PACS number(s): 05.50.+q, 75.10.Hk, 75.10.−b

I. THE FRUSTRATED ISING MODEL

A central problem in the study of lattice-statistical problems is the consideration of frustrated spin systems (see, for example, Refs. [1–4]). A particularly useful tool in the study of spin systems is the consideration of duality relations (see, for example, Refs. [5,6]). Here, we apply the duality consideration to frustrated discrete spin systems.

We consider first the Ising model on an *arbitrary* planar graph G. A planar graph is a collection of vertices and (noncrossing) edges. Place Ising spins at vertices of G, which interact with competing interactions along the edges. Denote the interaction between sites i and j by $-J_{ij}=-S_{ij}J$, where $S_{ij}=\pm 1$ and $J>0$. Then the Hamiltonian is

$$\mathcal{H}(\sigma;S)=-\sum_{\langle i,j\rangle} S_{ij}J\sigma_i\sigma_j, \tag{1}$$

where $\sigma_i=\pm 1$ is the spin at the site i and the summation is taken over all interacting pairs.

The Hamiltonian (1) plays an important role in condensed matter physics and related topics. Regarding S_{ij} as a quenched random variable governed by a probability distribution, the Hamiltonian (1) leads to the Edwards-Anderson model of spin glasses [7]. By taking a different S_{ij}, the Hamiltonian becomes the Hopfield model of neural networks [8]. Here, we consider the Hamiltonian (1) with fixed plaquette frustrations.

Let G have N sites and E edges. Then it has

$$N^*=E+2-N \quad \text{(Euler relation)} \tag{2}$$

faces, including one infinite face containing the infinite region and N^*-1 internal faces which we refer to as plaquettes. The parity of a face is the product of the edge S_{ij} factors around the face which can be either $+1$ or -1. A face is *frustrated* if its parity is -1. An Ising model is frustrated if any of its plaquettes is frustrated, and is fully (totally) frustrated if every plaquette is frustrated. The fully frustrated model is also known as the odd model of the spin glass [1]. As the parity of the infinite face is the product of the parities of all plaquettes, the parity of the infinite face in a totally frustrated Ising model is -1 for $N^*=$ even and $+1$ for $N^*=$ odd. An example of a full frustration is the triangular model with $S_{ij}=-1$ for all nearest neighbor sites i,j.

The values of parity associated with all plaquettes define a "parity configuration" which we denote by Γ. The set of interactions $\{S_{ij}\}$ corresponding to a given Γ is not unique. For the triangular model, for example, any $\{S_{ij}\}$ which has either one or three $S_{ij}=-1$ edges around every plaquette is totally frustrated. For a given $\{S_{ij}\}$ and Γ, the partition function is the summation

$$Z(\{S_{ij}\})=\sum_{\sigma_1=\pm 1}\cdots\sum_{\sigma_N=\pm 1}\prod_E e^{S_{ij}J\sigma_i\sigma_j}, \tag{3}$$

where the product is taken over the E edges of G.

A. Gauge transformation

A *gauge* transformation is site-dependent redefinition of the up (down) spin directions. Mathematically, a gauge transformation transforms the spin variables according to [2]

$$\sigma_i\to\sigma_i'=w_i\sigma_i, \quad i=1,\ldots,N. \tag{4}$$

In the above, if $w_i=+1$, the original definition of up (down) spin directions is maintained, and if $w_i=-1$, the definitions of up (down) are exchanged. Under the gauge transformation, the S_{ij} in Eq. (1) transforms as follows:

$$S_{ij}\to S_{ij}'=w_iS_{ij}w_j \quad \forall i,j. \tag{5}$$

Since $w_i^2=1$, we have

$$\mathcal{H}(\sigma;S)=\mathcal{H}(\sigma';S'). \tag{6}$$

Clearly, the gauge transformation (5) leaves the parity configuration Γ unchanged, i.e.,

1063-651X/2003/67(2)/026111(5)/$20.00

$$\prod_{\text{face}} S_{ij} = \prod_{\text{face}} S'_{ij} \quad \forall \text{ face.} \tag{7}$$

For each parity configuration Γ, there are 2^{N-1} different $\{S_{ij}\}$ patterns consistent with it. To see this we note in Eq. (5), each of the 2^N choices of $\{w_i\}$ leads to a new $\{S'_{ij}\}$ except the negation of all w_i which leaves $\{S_{ij}\}$ unchanged. Conversely, any two sets of interactions $\{S_{ij}\}$ and $\{S'_{ij}\}$ for the same Γ are related by a gauge transformation which can be constructed as follows. Starting from any spin, say spin 1, assign the value $w_1 = +1$. One next builds up the graph by adding one site (and one edge) at a time. To the site 2 connected to 1 by the edge $\{12\}$, one assigns the factor $w_2 = w_1 S_{12} S'_{12}$, which yields $w_1 S_{12} w_2 = S'_{12}$ consistent to Eq. (5). Proceeding in this way around a plaquette until an edge, say $\{n1\}$, completes a plaquette. At this point, one has

$$w_n S_{n1} w_1 = \left(\prod_{\text{plaquette}} S_{ij}\right)\left(\prod_{\text{plaquette}} S'_{ij}\right) S'_{n1} = S'_{n1}, \tag{8}$$

which is again consistent to Eq. (5). Continuing in this way, one constructs all w_i which transform $\{S_{ij}\}$ into $\{S'_{ij}\}$. Note that if we had started with $w_1 = -1$, we would have resulted in the negation of all w_i. Thus, the bijection between the 2^{N-1} sets $\{S_{ij}\}$ and 2^{N-1} gauge transformations is one to one.

In addition to Eq. (7), the gauge transformation also leaves the partition function invariant [9,10], i.e.,

$$Z(\{S_{ij}\}) = \sum_{\{\sigma\}} \prod_E e^{S_{ij} J \sigma_i \sigma_j} = \sum_{\{\sigma'\}} \prod_E e^{S'_{ij} J \sigma'_i \sigma'_j} = Z(\{S'_{ij}\}). \tag{9}$$

As a result, the partition function only depends on Γ and we can rewrite Eq. (3) as

$$Z(\Gamma) = 2^{-(N-1)} \sum_{\{S_{ij}\}} Z(\{S_{ij}\}), \tag{10}$$

where the summation is over all 2^{N-1} distinct $\{S_{ij}\}$ consistent with the parity configuration Γ for the same partition function. This expression of the partition function is used to derive the duality relation in ensuing sections.

B. The fully frustrated Ising model

For our purposes, it is instructive to consider first the case of full frustration. Duality properties of fully frustrated model have previously been considered by a number of authors [2,4] for regular lattices. We present here an alternate formulation applicable to arbitrary graphs and arbitrary frustration.

The graph D dual to G has N^* sites each residing in a face of G, and E edges each intersecting an edge of G. We restrict to $N^* =$ even so that all faces of G including the infinite face are frustrated. This restriction has no effect on the taking of the thermodynamic limit in the case of regular lattices. Since the signs S_{ij} around each face are subject to the constraint $\Pi S_{ij} = -1$, we introduce in the summand of Eq. (10) a face factor $(1 - \Pi S_{ij})/2$ and sum over $S_{ij} = \pm 1$ independently. Similarly, writing $\sigma_{ij} = \sigma_i \sigma_j$, we can replace summations over $\sigma_i = \pm 1$ in Eq. (10) by $\sigma_{ij} = \pm 1$ by introducing a factor $(1 + \Pi \sigma_{ij})/2$ to each face. Thus, Eq. (10) becomes

$$Z_{\text{FF}} = \left[\frac{2 \cdot 2^{-(N-1)}}{2^{2N^*}}\right] \sum_{\{\sigma_{ij}\}} \sum_{\{S_{ij}\}} \left[\prod_E e^{S_{ij}\sigma_{ij} J}\right] \times \prod_{\text{face}} \left[\left(1 + \prod \sigma_{ij}\right)\left(1 - \prod S_{ij}\right)\right], \tag{11}$$

where the subscript FF denotes full frustration, and the extra factor 2 in Eq. (11) is due to the $2 \rightarrow 1$ mapping from σ_i to σ_{ij}.

For a face having n sides, we rewrite the face factors as

$$1 + \prod \sigma_{ij} = \sum_{\mu = \pm} \prod F(\sigma_{ij}; \mu), \tag{12}$$

$$1 - \prod S_{ij} = \sum_{\nu = \pm} \prod G(S_{ij}; \nu), \tag{13}$$

where each product has n factors

$$F(\sigma; \mu) = \delta_{\mu +} + \sigma \delta_{\mu -}, \qquad G(S; \nu) = \delta_{\nu +} + S \omega_n \delta_{\nu -}, \tag{14}$$

δ is the Kronecker delta function, and $\omega_n = (-1)^{-1/n} = e^{-i\pi/n}$.

We now regard μ and ν as indices of two Ising spins residing at each dual lattice site. After carrying out summations over σ_{ij} and S_{ij}, the partition function (11) becomes

$$Z_{FF} = 2^{-E-N^*} \sum_{\{\mu\}} \sum_{\{\nu\}} \prod_E B(\mu, \nu; \mu', \nu'), \tag{15}$$

where we have made use of the Euler relation (2) and B is a Boltzmann factor given by

$$B(\mu, \nu; \mu', \nu') = \sum_{\sigma = \pm 1} \sum_{S = \pm 1} e^{S\sigma J} F(\sigma; \mu) \times F(\sigma; \mu') G(S; \nu) G(S; \nu'). \tag{16}$$

Here, $G(S; \nu')$ is given by Eq. (14) with $\omega_n \rightarrow \omega_{n'} = e^{-i\pi/n'}$ and the two faces containing spins $\{\mu, \nu\}$ and $\{\mu', \nu'\}$ have, respectively, n and n' sides.

Substituting Eq. (14) into Eq. (16) and making use of the identities

$$\delta_{\mu +}\delta_{\mu' +} + \delta_{\mu -}\delta_{\mu' -} = (1 + \mu\mu')/2, \qquad \delta_{\mu +}\delta_{\mu' -} + \delta_{\mu -}\delta_{\mu' +} = (1 - \mu\mu')/2, \tag{17}$$

one obtains

$$B(\mu, \nu; \mu', \nu') = 2A(1 + \mu\mu')\cosh J + 2B(1 - \mu\mu')\sinh J, \tag{18}$$

where

$$A=\delta_{\nu+}\delta_{\nu'+}+\omega_n\omega_{n'}\delta_{\nu-}\delta_{\nu'-},$$
$$B=\omega_n\delta_{\nu-}\delta_{\nu'+}+\omega_{n'}\delta_{\nu+}\delta_{\nu'-}. \tag{19}$$

We number the four states $\{\mu,\nu\}=\{+,+\},\{-,-\}$, $\{-,+\},\{+,-\}$ by 1,2,3,4, respectively. The Boltzmann factor (18) can be conveniently written as a 4×4 matrix

$$B(\mu,\nu;\mu',\nu')=\begin{pmatrix} B_{11} & B_{12} & 0 & 0\\ B_{21} & B_{22} & 0 & 0\\ 0 & 0 & B_{11} & B_{12}\\ 0 & 0 & B_{21} & B_{22}\end{pmatrix}, \tag{20}$$

where

$$B_{11}=4\cosh J,\quad B_{12}=4\omega_{n'}\sinh J,$$
$$B_{21}=4\omega_n\sinh J,\quad B_{22}=4\omega_n\omega_{n'}\cosh J. \tag{21}$$

Thus, the partition function of the $\{\mu,\nu\}$ spin model is twice that of an Ising model on the dual lattice. The exchange coupling constant K and the magnetic field h in the dual model are determined by

$$B_{11}=De^{K+(h/n)+(h'/n')},\quad B_{12}=De^{-K+(h/n)-(h'/n')},$$
$$B_{21}=De^{-K-(h/n)+(h'/n')},\quad B_{22}=De^{K-(h/n)-(h'/n')}. \tag{22}$$

Here, n and n' are the number of edges incident at the two dual sites, respectively.

The solution of the above equations gives

$$e^{-2K}=\tanh J>0,\quad D=4(\omega_n\omega_{n'})^{1/2}\sqrt{\sinh J\cosh J},$$
$$e^{2(h/n)}=1/\omega_n=e^{i\pi/n},\quad e^{2(h'/n')}=1/\omega_{n'}=e^{i\pi/n'}, \tag{23}$$

or equivalently

$$K=-\frac{1}{2}\ln(\tanh J)\quad\text{and}\quad h=h'=\frac{i\pi}{2}. \tag{24}$$

Thus, we have established the equivalence

$$Z_{\rm FF}(J)=2^{N-1}i^{N^*}(\sinh J\cosh J)^{E/2}Z^{(D)}_{\rm Ising}\left(i\frac{\pi}{2},K\right), \tag{25}$$

where $Z^{(D)}_{\rm Ising}(i\pi/2,K)$ is the partition function of a ferromagnetic Ising model on D with interactions $K>0$ and an external field $i\pi/2$. In writing down Eq. (25), we have made use of the identity $2\times2^{-(E+N^*)}4^E=2^{N-1}$ and the fact that $(\omega_n\omega_{n'})^{E/2}=(-i)^{N^*}=i^{N^*}$ for $N^*=$ even. We make the following remarks:

(1) The duality relation (25) has previously been obtained by Fradkin *et al.* [2], and for the square lattice by Suzuki [4] and Sütő [11], and by Au-Yang and Perk [12] in another context.

(2) The duality relation (25) is different from the Kadanoff-Ceva-Merlini scheme [13,14] of replacing K by $K+i\pi/2$ [corresponding to $J<0$ in Eq. (23)] in the ferromagnetic model. Suzuki [4] has made the explicit use of the Kadanoff-Ceva-Merlini scheme in deriving Eq. (25) for the square lattice. For fully frustrated systems, the Suzuki method can be extended to any graph whose dual admits dimer coverings.

(3) The duality relation (25) holds for a fixed $\{S_{ij}\}$ without probability considerations and, therefore, differs intrinsically from that of a spin glass obtained recently by Nishimori and Nemoto [15] using a replica formulation.

(4) The duality relation (25) which holds for any lattice appears to support the suggestion [3] that all fully frustrated Ising models belong to the same universality class.

C. The thermodynamic limit

The partition function (25) for an Ising model in a uniform field $i\pi/2$ can be exactly evaluated for regular lattices. Defining the per-site "free energy"

$$f=\lim_{N^*\to\infty}\frac{1}{N^*}\ln Z^{(D)}_{\rm Ising}\left(i\frac{\pi}{2},K\right), \tag{26}$$

Lee and Yang [16] have obtained a closed form expression of $f(K)$ for the square lattice. Their result, which was later derived rigorously by McCoy and Wu [17] and others [14,18], is

$$f=i\frac{\pi}{2}+C+\frac{1}{16\pi^2}\int_{-\pi}^{\pi}d\theta\int_{-\pi}^{\pi}d\phi\,\ln[z+z^{-1}+2-4\cos\theta\cos\phi], \tag{27}$$

where $C=[\ln(\sinh 2K)]/2$, $z=e^{-4K}$. The free energy (27), which is the same as that obtained by Villain [1], is analytic at all nonzero temperatures.

The solution for the triangular Ising model in a field $i\pi/2$ has also been deduced previously [18,19]. However, it can also be obtained most simply by observing that the honeycomb lattice, which is the dual of the triangular lattice, has a coordination number 3. It follows that we can recast the field Boltzmann weights as $e^{i\pi\sigma_j/2}=i\sigma_j=i\sigma_j^3$ and redistribute the σ_j^3 factor at site j to its three incident edges. Then, as pointed out by Suzuki [4], each edge can be associated with a factor $i\sigma_i\sigma_j e^{K\sigma_i\sigma_j}=e^{(K+i\pi/2)\sigma_i\sigma_j}$ and the desired solution can be obtained from that of the *zero-field* honeycomb lattice with the simple replacement $K\to K+i\pi/2$. This gives

$$f=i\frac{\pi}{2}+C+\frac{1}{16\pi^2}\int_{-\pi}^{\pi}d\theta\int_{-\pi}^{\pi}d\phi\,\ln[(1+e^{4K})^2+4\cos\phi(\cos\theta+\cos\phi)], \tag{28}$$

where $C=[\ln(2\sinh 2K)]/2$ and K is the Ising interaction on the honeycomb lattice. Again, there is no finite temperature phase transition.

D. Arbitrary plaquette parities

In a similar fashion, one can extend the above analysis to Ising models with arbitrary face parities. All the steps of previous sections can be carried out, except that for faces that are not frustrated, we must replace ω_n by 1 at the corresponding dual sites. This results in a zero field (instead of a field $i\pi/2$) at these sites. Thus, for an Ising model with arbitrary parity configuration Γ, its dual model has fields 0 and $i\pi/2$, respectively, at sites in faces of parity $+1$ and -1. Explicitly, we find

$$Z(\Gamma)=2^{N-1}(-i)^{N_F}(\sinh J\cosh J)^{E/2}Z_{\text{Ising}}^{(D)}(\{h_j\},K), \tag{29}$$

where the dual partition function is

$$Z_{\text{Ising}}^{(D)}(\{h_j\},K)=\sum_{\{\mu_i\}}\prod_E e^{K\mu_i\mu_j}\prod_{\text{face}} e^{h_i\mu_i}. \tag{30}$$

Here, N_F is the number of frustrated faces and the external field at site j is $h_j=i\pi/2$ or 0 depending on whether the face associated with the site is frustrated or not. We give the following remarks:

(1) The duality relation (30) for Ising models with arbitrary frustrated plaquettes can be found as contained implicitly in Ref. [2].

(2) By writing $e^{i\pi\sigma/2}=i\sigma$ in the dual partition function, we see that the partition function of an Ising model with p frustrated faces is dual to a p-spin Ising correlation function in zero field. In particular, the $p=2$ correlation problem has been studied in detail [17], which now leads to a wealth of information on the correlation of two frustrated plaquettes.

II. POTTS SPIN GLASS

Our analysis can be extended to a q-state spin model, the Potts spin glass. First, we recall the definition of a chiral Potts model. The chiral Potts model, which was considered in Ref. [5], is a discrete q-state spin model with a cyclic Boltzmann factor $\Lambda(\xi,\xi')=\Lambda(\xi-\xi')$ between two spins at sites i and j and in states ξ_i and $\xi_j=0,1,\ldots,q-1$. The interactions are q periodic, namely, the Boltzmann factor satisfies

$$U(\xi+q)=U(\xi). \tag{31}$$

The interaction can be symmetric, namely, $U(\xi)=U(-\xi)$, as in the case of the standard Potts model [20], but in our considerations, this need not be the case.

A Potts spin glass is a chiral Potts model with random interactions. To describe the randomness, one introduces edge variables $\lambda_{ij}=\lambda_{ji}=0,1,\ldots,q-1$ and considers the partition function [15,21,22]

$$Z_{\text{Potts}}(\{\lambda_{ij}\})=\sum_{\xi_i=0}^{q-1}\prod_E U(\xi_i-\xi_j+\lambda_{ij}). \tag{32}$$

Note that if U is symmetric and $q=2$, the partition function (32) reduces to Eq. (3). A plaquette has "flux" r $(=0,1,2,\ldots,q-1)$ if [23]

$$\sum_{\text{plaq}}\lambda_{ij}=r \pmod q. \tag{33}$$

A set of $\{\lambda_{ij}\}$ leads to a flux configuration Γ, which is specified by the values of the flux for all faces.

A. Gauge transformation

A gauge transformation for the Potts spin glass is the mapping

$$\begin{gathered}\xi_i\to\xi_i'=\xi_i+\theta_i,\\ \lambda_{ij}\to\lambda_{ij}'=\lambda_{ij}+\theta_i-\theta_j,\end{gathered} \tag{34}$$

where $\theta_i=0,1,\ldots,q-1$. It is clear that this mapping leaves the flux configuration Γ unchanged, i.e.,

$$\sum_{\text{plaq}}\lambda_{ij}=\sum_{\text{plaq}}\lambda_{ij}'. \tag{35}$$

Since a global change of all ξ_i by the same amount preserves $\{\lambda_{ij}\}$, the total number of distinct $\{\lambda_{ij}\}$ consistent with a particular flux configuration Γ is q^{N-1}. Conversely, any two sets of $\{\lambda_{ij}\}$ and $\{\lambda_{ij}'\}$ giving rise to the same flux configuration are related through a gauge transformation. To see this, we start from an arbitrarily chosen site, say site 1, and set $\theta_1=0$. Next, we assign $\theta_2=\theta_1+\lambda_{12}-\lambda_{12}'$ to site 2 connected to site 1 by an edge. Continuing in this way as in the Ising case, one eventually determines a set of θ_i that transforms $\{\lambda_{ij}\}$ into $\{\lambda_{ij}'\}$, and vice versa. The bijection between the q^{N-1} configurations of λ_{ij} and gauge transformations for a given Γ is one to one.

In addition to leaving the flux configuration unchanged, gauge transformation also leaves the partition function invariant, namely,

$$Z_{\text{Potts}}(\{\lambda_{ij}\})=Z_{\text{Potts}}(\{\lambda_{ij}'\}). \tag{36}$$

Thus, analogous to Eq. (10), we have

$$Z_{\text{Potts}}(\Gamma)=q^{-(N-1)}\sum_{\{\lambda_{ij}\}}Z(\{\lambda_{ij}\}), \tag{37}$$

which is used to derive a duality relation. Again, the sum in Eq. (37) runs through all $\{\lambda_{ij}\}$ consistent with a given flux configuration Γ.

B. Duality relation

In the Potts partition function (32), we write $\xi_{ij}=\xi_i-\xi_j$, and to each face having a flux r, we introduce two factors,

$$\frac{1}{q}\sum_{\mu=0}^{q-1}e^{-i2\pi\mu(\xi_{12}+\xi_{23}+\cdots+\xi_{n1})/q}$$

$$\times\begin{cases}1 & \text{if } \xi_{12}+\cdots+\xi_{n1}=0 \pmod q\\ 0 & \text{otherwise,}\end{cases}$$

$$\frac{1}{q}\sum_{\nu=0}^{q-1} e^{-i2\pi\nu(\lambda_{12}+\lambda_{23}+\cdots+\lambda_{n1}-r)/q} \times\begin{cases}1 & \text{if } \lambda_{12}+\cdots+\lambda_{n1}=r \pmod q\\ 0 & \text{otherwise.}\end{cases} \tag{38}$$

This permits us to sum over ξ_{ij} and λ_{ij} independently. Thus, analogous to Eq. (15), we obtain

$$Z_{\text{Potts}}(\Gamma)=q^{-E-N^*}\sum_{\{\mu\}}\sum_{\{\nu\}} B(\mu,\nu;\mu',\nu'), \tag{39}$$

where

$$B(\mu,\nu;\mu',\nu')=\sum_{\lambda}\sum_{\xi} U(\xi+\lambda)\exp\left[\frac{i2\pi}{q}\left(-(\mu-\mu')\xi -(\nu-\nu')\lambda+\frac{r\nu}{n}+\frac{r'\nu'}{n'}\right)\right], \tag{40}$$

and n,n' are the numbers of sides of the two plaquettes containing $\{\mu,\nu\}$ and $\{\mu',\nu'\}$, and fluxes r and r', respectively.

We carry out the summations in Eq. (40) after introducing the Fourier transform

$$U(\xi+\lambda)=\frac{1}{q}\sum_{\eta=0}^{q-1}\Lambda(\eta)e^{i2\pi\eta(\xi+\lambda)/q}, \tag{41}$$

where $\Lambda(\eta)$ are the eigenvalues of the matrix U [5]. One obtains

$$B(\mu,\nu;\mu',\nu')=q\,\delta_{\mu-\mu',\nu-\nu'}\Lambda(\nu-\nu') \times e^{i2\pi r\nu/qn}e^{i2\pi r'\nu'/qn'}. \tag{42}$$

In the above equation, $\delta_{\mu-\mu',\nu-\nu'}$ sets $\mu-\mu'$ to $\nu-\nu'$ (mod q).

The substitution of Eq. (42) into Eq. (39) followed by summing over μ now yields the result

$$Z_{\text{Potts}}(\Gamma)=q^{1-N^*}Z_{\text{Potts}}{}^{(D)}(\{h_j\},\Lambda), \tag{43}$$

where

$$Z_{\text{Potts}}{}^{(D)}(\{h_j\},\Lambda)=\sum_{\{\nu_i\}}\left[\prod_{E}\Lambda(\nu_i,\nu_j)\right]\left[\prod_{\text{face}} e^{h_j\nu_j}\right] \tag{44}$$

is the partition function of a chiral Potts model on the dual graph D, which generalizes Eq. (29) to Potts spin glasses. The dual chiral Potts model has Boltzmann weights $\Lambda(\mu_i,\mu_j)=\Lambda(\mu_i-\mu_j)$ and external fields

$$h_j=i\frac{2\pi r_j}{q}, \quad j=1,2,\ldots,N^*, \quad r_j=0,1,\ldots,q-1 \tag{45}$$

on the spin in plaquette j which has a flux r_j. When $r_j=0$ for all j, Eq. (43) reduces to the duality relation for the zero-field chiral Potts model given by Eq. (13) in Ref. [5].

III. SUMMARY

We have obtained duality relations for planar Ising and chiral Potts models on arbitrary graphs and with fixed plaquette parity or flux configurations. Our main results are the equivalences (25) for the fully frustrated Ising model, Eq. (29) for the Ising model with arbitrary plaquette parity, and Eq. (43) for the chiral Potts model with arbitrary flux configurations. In all cases, the dual models have pure imaginary fields applied to spins in plaquettes that are frustrated and/or having a nonzero flux.

ACKNOWLEDGMENTS

We thank H. Kunz for a conversation which motivated this work. We are also indebted to C. Henley, H. Nishimori, J.H.H. Perk, L. Pryadko, and A. Sütő for calling our attention to relevant references. This work has been supported in part by NSF Grant Nos. DMR-9971503 and DMR-9980440.

[1] J. Villain, J. Phys. C **10**, 1717 (1977).
[2] E. Fradkin, B.A. Huberman, and S.H. Shenker, Phys. Rev. B **18**, 4789 (1978).
[3] G. Forgacs, Phys. Rev. B **22**, 4473 (1980).
[4] M. Suzuki, J. Phys. Soc. Jpn. **60**, 441 (1990).
[5] F.Y. Wu and Y.K. Wang, J. Math. Phys. **17**, 439 (1976).
[6] R. Savit, Rev. Mod. Phys. **52**, 453 (1980).
[7] S.F. Edwards and P.W. Anderson, J. Phys. F: Met. Phys. **5**, 965 (1975).
[8] J.J. Hopfield, Proc. Natl. Acad. Sci. U.S.A. **79**, 2554 (1982).
[9] H. Nishimori, J. Phys. C **13**, 4071 (1980).
[10] H. Nishimori, Prog. Theor. Phys. **66**, 1169 (1981).
[11] A. Sütő, Hungarian Central Research Institute for Physics Report No. KFKI-1978-95, 1978 (unpublished).
[12] H. Au-Yang and J.H.H. Perk, Phys. Lett. A **104**, 131 (1984).
[13] L.P. Kadanoff and H. Ceva, Phys. Rev. B **3**, 3918 (1971).
[14] D. Merlini, Lett. Nuovo Cimento **9**, 100 (1974).
[15] H. Nishimori and K. Nemoto, J. Phys. Soc. Jpn. **71**, 1198 (2002); Physica A (to be published).
[16] T.D. Lee and C.N. Yang, Phys. Rev. **87**, 410 (1952).
[17] B.M. McCoy and T.T. Wu, Phys. Rev. **155**, 438 (1967).
[18] K.Y. Lin and F.Y. Wu, Int. J. Mod. Phys. B **4**, 471 (1988).
[19] W.T. Lu and F.Y. Wu, J. Stat. Phys. **102**, 953 (2001).
[20] R.B. Potts, Proc. Cambridge Philos. Soc. **48**, 106 (1952).
[21] H. Nishimori and M.J. Stephen, Phys. Rev. B **27**, 5644 (1983).
[22] J.L. Jacobsen and M. Picco, Phys. Rev. E **65**, 026113 (2002).
[23] The term flux is used here in accord with the nomenclature in lattice gauge theories. See, e.g., J.B. Kogut, Rev. Mod. Phys. **51**, 659 (1979).

ELSEVIER
31 March 1997
PHYSICS LETTERS A
Physics Letters A 228 (1997) 43–47

Duality relations for Potts correlation functions

F.Y. Wu
Department of Physics, Northeastern University, Boston, MA 02115, USA
Centre Emile Borel, UMS 839 IHP (CNRS/UPMC), Paris, France

Received 18 December 1996; accepted for publication 20 January 1997
Communicated by C.R. Doering

Abstract

Duality relations are obtained for correlation functions of the q-state Potts model on any planar lattice or graph using a simple graphical analysis. For the two-point correlation we show that the correlation length is precisely the surface tension of the dual model, generalizing a result known to hold for the Ising model. For the three-point correlation an explicit expression is obtained relating the correlation function to ratios of dual partition functions under fixed boundary conditions. © 1997 Elsevier Science B.V.

1. Introduction

It is now well-known [1,2] that the correlation length of the Ising model in two dimensions is precisely the surface tension of the dual lattice. It is also known by folklore (see, for example, Ref. [3]) that a similar duality exists for the Potts model [4]. However, detailed discussion of the correlation duality for the Potts model has yet to appear in the literature. In view of the important role played by such duality relations in discussions of the equilibrium crystal shapes [5,6], a definitive understanding of this subject matter is clearly needed.

Here we take up this question and consider more generally the duality for the n-point correlation function. On the basis of a simple graphical analysis, we derive duality relations for the two- and three-point correlation functions. Particularly, we establish that the correlation length in the large lattice limit is precisely the surface tension of the dual model, thus generalizing the known Ising result, and that the three-point correlation function is given by similarly defined expressions in the dual space. Our analysis, which does not concern with the internal structure of the underlying lattice, is quite general and applies to the Potts model with arbitrary edge-dependent interactions and/or on any planar lattice or graph. It can also be extended to higher correlations.

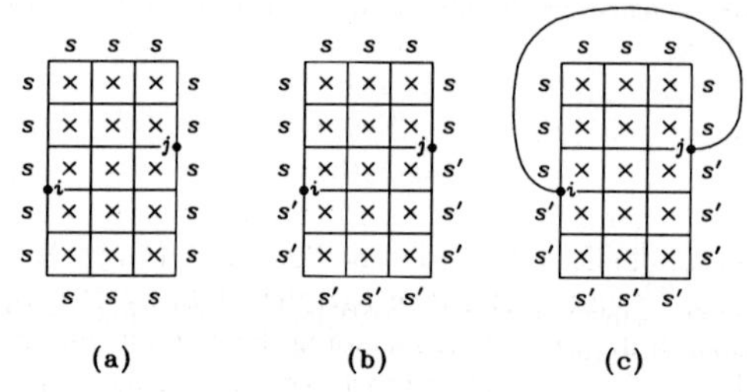

Fig. 1. (a) A 4×6 lattice L. The dual spins are denoted by crosses and the letter s. (b) The boundary condition ($s \neq s'$) for Z^*_{12}. (c) The addition of one edge to L connecting sites i and j.

Consider a q-state Potts model [4] on a two-dimensional lattice L with a free boundary. An ex-

PII S0375-9601(97)00094-7

ample of L is the rectangular 4×6 lattice shown in Fig. 1a, but more generally L can be any planar lattice or graph which does not need to be regular. The site i of L is occupied by a spin σ_i which can take on values $1, 2, \ldots, q$. Two spins at sites α and β and in states σ_α and σ_β interact with a Boltzmann factor $\exp[K_{\alpha\beta}\delta(\sigma_\alpha, \sigma_\beta)]$. Let $i, j, \ldots, n$ denote *any* n sites on the boundary. We introduce the n-point correlation function as the probability that the spins at sites $\{i, j, \ldots, n\}$ are in the respective states $\{\sigma, \sigma', \ldots, \sigma^{(n-1)}\}$,

$$P_n(\sigma, \sigma', \ldots, \sigma^{(n-1)}) = \langle \delta(\sigma_i, \sigma)\delta(\sigma_j, \sigma') \ldots \delta(\sigma_n, \sigma^{(n-1)}) \rangle, \quad (1)$$

where $\langle\ \rangle$ denotes statistical averages. It is sometimes convenient to consider the correlation function (see, for example, [7,3] for $n = 2$)

$$\begin{aligned} \Gamma_n(\sigma_i, \sigma_j, \ldots, \sigma_n) &= \langle q^{n-1}\delta(\sigma_i, \sigma_j)\delta(\sigma_j, \sigma_k) \ldots \delta(\sigma_{n-1}, \sigma_n) - 1 \rangle \\ &= q^n P_n(\sigma, \sigma, \ldots, \sigma) - 1, \end{aligned} \quad (2)$$

a quantity which vanishes identically if the n spins $\{\sigma_i, \sigma_j, \sigma_k \ldots, \sigma_{n-1}, \sigma_n\}$ are completely uncorrelated. In the case of $q = n = 2$, for example, one writes $2\delta(\sigma_i, \sigma_j) = 1 + \sigma_i\sigma_j$ where $\sigma_{i,j} = \pm 1$, then (2) becomes the usual expression $\Gamma_2(\sigma_i, \sigma_j) = \langle \sigma_i \sigma_j \rangle$ for the Ising model. It is clear that the correlation function P_n is more general than Γ_n.

Our analysis is based on the repeated use of a fundamental duality relation for the Potts model [7]. Construct L^*, the dual lattice (graph) of L, by placing spins in the *faces* of L including the exterior one. In Fig. 1a, for example, the faces of the dual lattice are denoted by crosses and the letter s, so L^* has $N^* = 15 + 1 = 16$ sites. Edges of L^* (not shown in Fig. 1a) bisect, and are in one-one correspondence with, edges of L. Let $K^*_{\alpha\beta}$ be the interaction of the edge bisecting the interaction $K_{\alpha\beta}$. Let $Z(\{K_{\alpha\beta}\})$ and $Z^*(\{K^*_{\alpha\beta}\})$ be the respective partition functions on L and L^*. Then one has the following fundamental duality relation [7],

$$Z(\{K_{\alpha\beta}\}) = qCZ^*(\{K^*_{\alpha\beta}\}), \quad (3)$$

where $K_{\alpha\beta}$ and $K^*_{\alpha\beta}$ are related by

$$(e^{K_{\alpha\beta}} - 1)(e^{K^*_{\alpha\beta}} - 1) = q, \quad (4)$$

and $C = q^{-N^*} \prod_{\text{edges}}(e^{K_{\alpha\beta}} - 1)$.

2. The two-point correlation function

Let i and j denote any two sites on the boundary of L as shown in Fig. 1a. Further let

$$\begin{aligned} Z_{11} &= Z(\sigma_i = \sigma_j = 1), \\ Z_{12} &= Z(\sigma_i = 1, \sigma_j = 2) \end{aligned} \quad (5)$$

be the partition functions with σ_i and σ_j fixed in definite states 1 and/or 2, Then, by symmetry we have

$$Z(\{K_{\alpha\beta}\}) = qZ_{11} + q(q-1)Z_{12}. \quad (6)$$

Clearly, we also have $P_2(1,1) = Z_{11}/Z$ and $P_2(1,2) = Z_{12}/Z$.

Let Z^*_{11} denote the dual partition function with the spin in the exterior face in a definite state, say, s. Alternately, from Fig. 1a, we see that Z^*_{11} can also be regarded as the dual partition function with all boundary spins interacting with the exterior spin fixed at state s. Then we have

$$Z^*(\{K_{\alpha\beta}\}) = qZ^*_{11}. \quad (7)$$

Substituting (6) and (7) into the duality relation (3), one obtains the relation

$$Z_{11} + (q-1)Z_{12} = qCZ^*_{11}. \quad (8)$$

To obtain a second relation relating Z_{11} and Z_{12}, we apply the duality relation (3) to a lattice, or graph, constructed from L by adding one additional edge connecting sites i and j with an interaction K, a situation shown in Fig. 1c. Clearly, the respective partition functions for the new lattice and its dual are now

$$\tilde{Z} = qe^K Z_{11} + q(q-1)Z_{12} \quad (9)$$

and

$$\tilde{Z}^* = qe^{K^*} Z^*_{11} + q(q-1)Z^*_{12}, \quad (10)$$

where Z^*_{12} is the partition function of the dual under the boundary condition such that the spins in the two exterior faces are in definite states $s \neq s'$ as shown in Fig. 1b. Here, K and K^* are related by (4). Substituting (9) and (10) into (3) and noting that N^* is increased by 1 in the new lattice, one obtains a second relation

$$e^K Z_{11} + (q-1)Z_{12} \\ = C(e^K - 1)[e^{K^*} Z_{11}^* + (q-1)Z_{12}^*]. \quad (11)$$

We now solve (8) and (11) for Z_{11} and Z_{12}. This leads to, after using (4) to eliminate e^{K^*},

$$Z_{11} = C[Z_{11}^* + (q-1)Z_{12}^*], \\ Z_{12} = C(Z_{11}^* - Z_{12}^*). \quad (12)$$

Note that the interaction K introduced to facilitate calculations does not enter (12). Expression (12) now leads to the desired expression for the correlation function. Particularly, using the identity $Z = q^2 C Z_{11}^*$, we obtain

$$P_2(\sigma,\sigma') = \frac{1}{q^2}\left(1 + (q\delta_{\sigma,\sigma'} - 1)\frac{Z_{12}^*}{Z_{11}^*}\right) \quad (13)$$

and

$$\Gamma_2(\sigma_i,\sigma_j) = (q-1)\frac{Z_{12}^*}{Z_{11}^*}. \quad (14)$$

For the Potts model the surface tension τ is defined by [9]

$$\tau = -\lim_{D\to\infty}\frac{1}{D}\ln\left(\frac{Z_{12}^*}{Z_{11}^*}\right), \quad (15)$$

where D is the distance between sites i and j, the two points where the boundary spin state changes from s to s' in Fig. 1b. Our result (14) now relates the correlation function on L to the surface tension on L^*. Particularly, the known exponential decay $e^{-D/\xi}$ of the two-point correlation Γ_2 above the transition temperature T_c [10], where ξ is the correlation length, leads to the identity $\xi^{-1} = \tau$, where τ is the surface tension of the dual model below T_c. This generalizes the corresponding $q = 2$ result of the Ising model [1,2] to all q.

3. The three-point correlation

Consider any three sites i, j, k on the boundary of L as shown in Fig. 2. Here, to emphasize the generality of our consideration and the fact that the internal structure of L does not enter the picture, the lattice is shown symbolically as a shaded region. Let

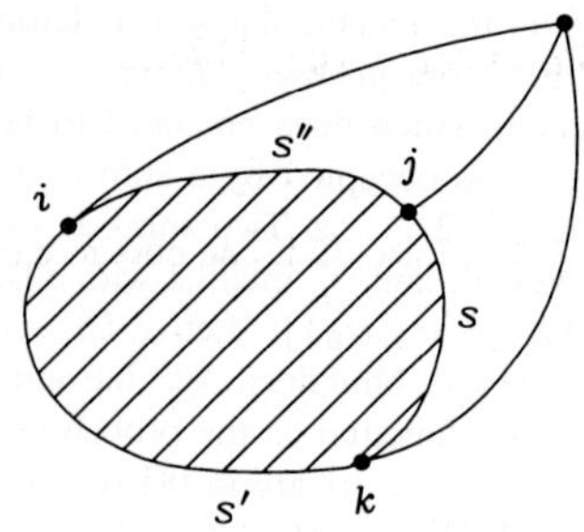

Fig. 2. The addition of three edges to L (the shaded region) connecting sites i, j, and k to a common point.

$$Z_{\sigma\sigma'\sigma''} = Z(\sigma_i = \sigma, \sigma_j = \sigma', \sigma_j = \sigma''), \\ \sigma,\sigma',\sigma'' = 1,2,\ldots,q, \quad (16)$$

denote the partition function with $\sigma_i, \sigma_j, \sigma_k$ fixed in definite states. Then, analogous to (6) the partition function can be written as

$$Z = qZ_{111} + q(q-1)(Z_{211} + Z_{121} + Z_{112}) \\ + q(q-1)(q-2)Z_{123}. \quad (17)$$

In a similar manner, let $Z_{ss's''}^*$ be the partition function of the dual model under a boundary condition such that all boundary spins between sites i and j of L interact with a fixed spin state s'', all boundary spins between j and k interact with a spin state s, and all boundary spins between k and i interact with a spin state s', a situation shown in Fig. 2. Then, the dual partition function can be written as

$$Z^* = qZ_{111}^*. \quad (18)$$

Substituting (17) and (18) into (3), one obtains the relation

$$Z_{111} + (q-1)(Z_{211} + Z_{121} + Z_{112}) \\ + (q-1)(q-2)Z_{123} = qCZ_{111}^*. \quad (19)$$

Next we modify L by connecting sites j and k with a new edge and apply the result (12). In the first line of (12), one has, by definition,

$$Z_{11} = \sum_{i=1}^{q} Z_{i11} = Z_{111} + (q-1)Z_{211},$$

$$Z^*_{11} = Z^*_{111}, \quad Z^*_{12} = Z^*_{122} = Z^*_{211}. \tag{20}$$

Substituting (20) into (12), one obtains a second relation

$$Z_{111} + (q-1)Z_{211} = C[Z^*_{111} + (q-1)Z^*_{211}]. \tag{21}$$

In a similar manner by connecting sites $\{k,i\}$ or $\{i,j\}$ with a new edge, we obtain, respectively,

$$Z_{111} + (q-1)Z_{121} = C[Z^*_{111} + (q-1)Z^*_{121}], \tag{22}$$

$$Z_{111} + (q-1)Z_{112} = C[Z^*_{111} + (q-1)Z^*_{112}]. \tag{23}$$

Finally, we apply the duality relation (3) to a lattice constructed by introducing three new edges connecting sites i, j or k to a common new point with interactions K as shown in Fig. 2. This process increases N^* by 2 and yields the duality relation

$$\begin{aligned}
&(e^{3K} + q - 1)Z_{111} \\
&\quad + (q-1)(e^{2K} + e^{K} + q - 2)(Z_{211} + Z_{121} + Z_{112}) \\
&\quad + (q-1)(q-2)(3e^{K} + q - 3)Z_{123} \\
&\quad = C(e^{K} - 1)^3 q^{-1}[e^{3K^*} Z^*_{111} \\
&\quad + (q-1)e^{K^*}(Z^*_{211} + Z^*_{121} + Z^*_{112}) \\
&\quad + (q-1)(q-2)Z^*_{123}].
\end{aligned} \tag{24}$$

From (19), (21)–(24) and using (4) to eliminate K^*, we solve Z_{111}, Z_{211}, Z_{121}, Z_{112}, and Z_{123} in terms of the dual partition functions $Z^*_{ss's''}$. This leads to

$$\begin{aligned}
Z_{111} &= (C/q)[Z^*_{111} + (q-1)(q-2)Z^*_{123} \\
&\quad + (q-1)(Z^*_{211} + Z^*_{121} + Z^*_{112})], \\
Z_{123} &= (C/q)[Z^*_{111} + 2Z^*_{123} - Z^*_{211} - Z^*_{121} - Z^*_{112}], \\
Z_{211} &= (C/q)[Z^*_{111} - (q-2)Z^*_{123} + (q-1)Z^*_{211} \\
&\quad - Z^*_{121} - Z^*_{112}], \\
Z_{121} &= (C/q)[Z^*_{111} - (q-2)Z^*_{123} - Z^*_{211} \\
&\quad + (q-1)Z^*_{121} - Z^*_{112}], \\
Z_{112} &= (C/q)[Z^*_{111} - (q-2)Z^*_{123} - Z^*_{211} - Z^*_{121} \\
&\quad + (q-1)Z^*_{112}].
\end{aligned} \tag{25}$$

This result is again independent of the parameter K used in the evaluation.

Using (25) and the identity $Z = CqZ^* = Cq^2 Z^*_{111}$, we finally obtain

$$\begin{aligned}
P_3(\sigma,\sigma',\sigma'') &= \frac{1}{q^3}[1 + 2p_{123} - (p_{211} + p_{121} + p_{112}) \\
&\quad + q(p_{211} - p_{123})\delta_{\sigma',\sigma''} + q(p_{121} - p_{123})\delta_{\sigma'',\sigma} \\
&\quad + q(p_{112} - p_{123})\delta_{\sigma,\sigma'} + q^2 p_{123}\delta_{\sigma,\sigma'}\delta_{\sigma',\sigma''}]
\end{aligned} \tag{26}$$

and

$$\begin{aligned}
&\Gamma_3(\sigma_i,\sigma_j,\sigma_k) \\
&\quad = (q-1)[p_{112} + p_{121} + p_{211} + (q-2)p_{123}] \\
&\quad = \Gamma_2(\sigma_i,\sigma_j) + \Gamma_2(\sigma_j,\sigma_k) + \Gamma_2(\sigma_k,\sigma_i) \\
&\quad + (q-1)(q-2)p_{123}.
\end{aligned} \tag{27}$$

Here,

$$p_{ss's''} = Z^*_{ss's''}/Z^*_{111} \tag{28}$$

is the ratio of dual partition functions which can be interpreted as appropriately defined surface tensions, and we have used $(q-1)p_{112} = \Gamma_2(\sigma_i,\sigma_j)$, etc. Since $Z^* = qZ^*_{111}$, the ratio $p_{ss's''}$ gives also the probability that the particular boundary condition $\{s, s', s''\}$ of Fig. 2 occurs in the dual partition function Z^*. It is readily verified that the identity

$$\sum_{\sigma''} P_3(\sigma,\sigma',\sigma'') = P_2(\sigma,\sigma') \tag{29}$$

is satisfied.

4. Summary

In summary, we have presented a graphical analysis of duality relations leading to explicit expressions for correlation functions of the q-state Potts model on any planar lattice or graph. For the two-point correlation our result (14) generalizes a relation previously known for the Ising model. For three-point correlations, our result (26) is new and can be used to compute any three-point correlation. In all cases the correlation functions are found to be given by ratios of dual partition functions under fixed boundary conditions, which can in turn be interpreted as appropriately defined surface tensions. Our consideration can be extended in a straightforward fashion to higher correlations P_n and Γ_n for $n > 3$, and to the more general

(N_α, N_β) model [11] including the $N_\alpha = N_\beta = 2$ Ashkin–Teller model [12].

Acknowledgement

I would like to thank R.K.P. Zia for suggesting this problem and enlightening discussions. I would also like to thank M.-J. Maillard for the hospitality at the Institut Henri Poincaré and H. Kunz for the hospitality at Institut de Physique Théorique, Lausanne, where this research was initiated. This work is supported in part by the National Science Foundation Grant DMR-9614170.

References

[1] P.G. Watson, J. Phys. C 1 (1968) 575.
[2] R.K.P. Zia, Phys. Lett. A 64 (1978) 345.
[3] F.Y. Wu, Rev. Mod. Phys. 54 (1982) 235.
[4] R.B. Potts, Proc. Cambr. Philos. Soc. 48 (1952) 106.
[5] C. Rottman and M. Wortis, Phys. Rev. B 24 (1981) 6274.
[6] J.E. Avron, H. van Beijeren and R.K.P. Zia, J. Phys. A 15 (1982) L81.
[7] F.Y. Wu and Y.K. Wang, J. Math. Phys. 17 (1976) 439.
[8] Y.K. Wang and F.Y. Wu, J. Phys. A 9 (1976) 593.
[9] J.R. Fontaine and Ch. Gruber, Commun. Math. Phys. 70 (1979) 243.
[10] A.H. Hintermann, H. Kunz and F.Y. Wu, J. Stat. Phys. 19 (1978) 623.
[11] E. Domany and E.K. Riedel, Phys. Rev. Lett. 40 (1978) 561.
[12] J. Ashkin and E. Teller, Phys. Rev. 64 (1943) 178.

VOLUME 79, NUMBER 25 PHYSICAL REVIEW LETTERS 22 DECEMBER 1997

Sum Rule Identities and the Duality Relation for the Potts n-Point Boundary Correlation Function

F. Y. Wu and H. Y. Huang

Department of Physics, Northeastern University, Boston, Massachusetts 02115

(Received 20 June 1997)

It is shown that certain sum rule identities exist which relate correlation functions for n Potts spins on the boundary of a planar lattice for $n \geq 4$. Explicit expressions of the identities are obtained for $n = 4$. It is also shown that the identities provide the missing link needed for a complete determination of the duality relation for the n-point boundary correlation function. The $n = 4$ duality relation is obtained explicitly. More generally we deduce the number of sum rule identities as well as a cyclic inversion relation for any n, and conjecture on the general form of the duality relation. [S0031-9007(97)04886-2]

PACS numbers: 05.50.+q, 75.10.Hk

The Potts model [1], which is a generalization of the two-component Ising model to q components for arbitrary q, has been the subject matter of intense interest in many fields ranging from condensed matter to high-energy physics. For reviews on the Potts model and its relevance, see, for example, [2,3]. However, exact results on the Potts model have proven to be extremely elusive. Rigorous results known to this date are limited, and include essentially only a closed-form evaluation of its free energy for $q = 2$, the Ising model [4], and critical properties for the square, triangular, and honeycomb lattices [5,6]. Much less is known about its correlation functions.

In this Letter we report on new sum rule identities for the Potts n-point boundary correlation function. Specifically, we show that, as a consequence of being a many-component system, the correlation functions of Potts spins on the boundary of a planar lattice must necessarily satisfy certain identities when $n \geq 4$. We further show that these identities lead to the complete determination of a correlation duality relation which, in its simplest form, relates the correlation length and the domain wall free energy and has proven to be useful in determining the equilibrium crystal shape of the Ising model [7]. Our results are very general and hold for any planar lattice or graph with arbitrary (nonuniform) edge interactions.

Consider the q-state Potts model on a planar lattice $\mathcal{L}$ with open boundary conditions, or more generally any planar graph, of N sites and E edges. Let $i, j, \ldots, m, \ell$ be n sites on the boundary ordered as shown in Fig. 1, and let σ_i denote the state of the spin at site i. Two spins of $\mathcal{L}$ at sites i' and j' interact with an interaction $K_{ij}\delta(\sigma_{i'}, \sigma_{j'})$, where $\sigma_i', \sigma_j' = 1, 2, \ldots, q$. Define the n-point correlation function [8]

$$P_n(\sigma, \sigma', \ldots, \sigma^{(n-1)}) = \langle \delta(\sigma_i, \sigma)\delta(\sigma_j, \sigma') \cdots \delta(\sigma_\ell, \sigma^{(n-1)})\rangle \quad (1)$$

as the probability that the n spins are in respective *definite* spin states $\sigma, \sigma', \ldots, \sigma^{(n)}$. In particular, the correlation function

$$\Gamma_n = q^n P_n(\sigma, \sigma, \ldots, \sigma) - 1 \quad (2)$$

vanishes identically if the n spins are completely uncorrelated.

It is convenient to write $P_{ij\cdots\ell} = P_n(i, j, \ldots, \ell) = Z_{ij\cdots\ell}/Z$, where $i, j, \ldots, \ell = 1, 2, \ldots, q$, Z is the partition function, and $Z_{ij\cdots\ell}$ the partial partition function, namely, the sum of Boltzmann factors with the boundary spin states fixed at $i, j, \ldots, \ell$. Then we have the following theorem.

Theorem: (i) The boundary correlation functions P_n, $n \geq 4$, are related by certain sum rule identities. Particularly, for $n = 4$, the identity is

$$P_{1212} = P_{1213} + P_{2131} - P_{1234}. \quad (3)$$

(ii) The number of correlation identities for a given n is $a_n = b_n - c_n$, where b_n and c_n are generated, respectively, from

$$\exp(e^t - 1) = \sum_{n=0}^{\infty} b_n t^n / n!, \quad (4)$$

$$(1 - \sqrt{1 - 4t})/2t = \sum_{n=0}^{\infty} c_n t^n. \quad (5)$$

Proof: The identity (3) is equivalent to $Z_{1212} = Z_{1213} + Z_{2131} - Z_{1234}$, which we represent graphically in Fig. 2. Consider the high-temperature expansion of $Z_{ijk\ell}$ in the form [9] of

$$Z_{ijk\ell} = \sum_G q^{n(G)} \prod_{i',j' \in G} (e^{K_{i'j'}} - 1). \quad (6)$$

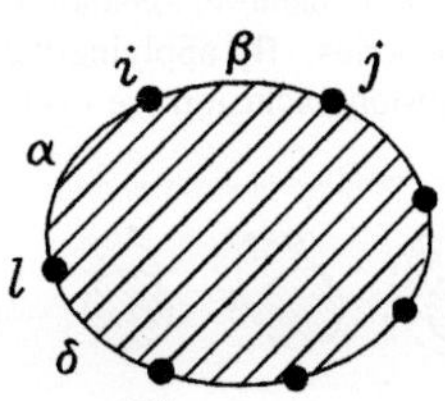

FIG. 1. A planar lattice $\mathcal{L}$ and n sites $i, j, \ldots, m, \ell$ on the boundary.

 0031-9007/97/79(25)/4954(4)$10.00

VOLUME 79, NUMBER 25 PHYSICAL REVIEW LETTERS 22 DECEMBER 1997

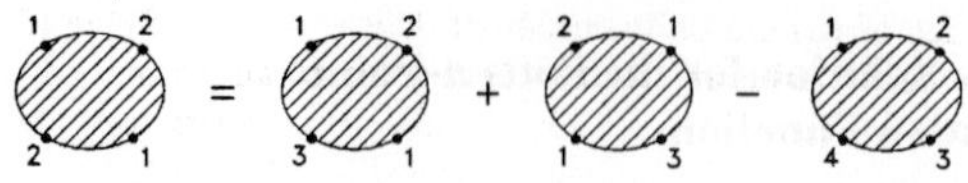

FIG. 2. Graphical representation of the sum rule identity (3).

Here, as a consequence of the fact that the four boundary sites are fixed in definite spin states, the summation is taken over all graphs $G \subseteq \mathcal{L}$ in which there are $n(G)$ clusters excluding those connected to the four boundary sites.

Apply the expansion (6) to the four Z's. It is clear that, as a consequence of sites being on the boundary and $\mathcal{L}$ being planar, we have $Z_{1212} = T_1 + T_2 + T_3$, where T_1 is the sum of graphs where sites i and k belong to the same cluster, T_2 those graphs where sites j and ℓ belong to the same cluster, and T_3 graphs i, j, k, ℓ all belong to different clusters. It is also clear that we have $Z_{1213} = T_1 + T_3$, $Z_{2131} = T_2 + T_3$, and $Z_{1234} = T_3$. The identity (3) now follows as a sum rule condition. The identity (3) can also be deduced from an application of the principle of inclusion-exclusion [10]: Z_{1212} is equal to the sum of Z_{1213} and Z_{2131} minus the overcounted terms, Z_{1234}. Clearly, the existence of (3) is a consequence of the planar connectivity topology and the fact that the sites are on the boundary. One can proceed in a similar fashion to derive sum rules for $n \geq 5$, and thus we have established (i). We remark that the sum rules manifest themselves only for $q \geq 4$.

To enumerate a_n, the number of correlation identities for a given n, it is instructive to consider the case $n = 4$. First, by enumeration we find that there are 15 distinct $Z_{ijk\ell}$. For each $Z_{ijk\ell}$ we connect sites in the same state by drawing connecting lines exterior to $\mathcal{L}$, resulting in a "connectivity" of the four points. A well-nested connectivity, or planar Z for brevity, is one in which the connecting lines do not intersect [11]. For $n = 4$, the 14 Z's shown in Fig. 3 are planar. Only Z_{1212}, which is not shown, is nonplanar.

More generally for a given n-point correlation function $Z_{ij\cdots m\ell}$, or Z for brevity, one connects sites in the same state to arrive at an n-point connectivity. Let there be altogether b_n distinct connectivities of which c_n are planar. To each Z which is nonplanar, we follow the procedure described above, namely, expanding graphically in a high-temperature series. By applying the aforementioned principle of inclusion-exclusion we eventually arrive at a

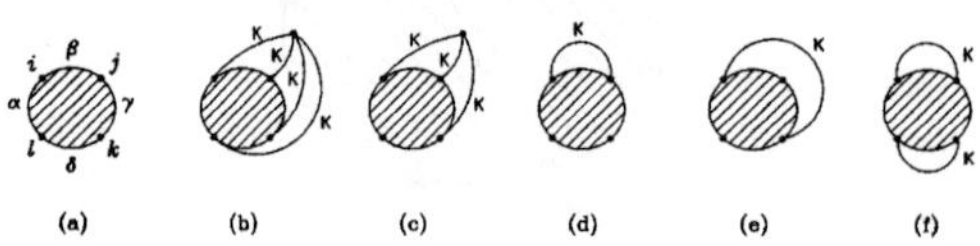

FIG. 3. The 14 planar connectivities for $n = 4$ corresponding to (a) Z_{1234}; (b) Z_{1111}; (c) Z_{1112}, occurring four times; (d) Z_{1123}, four times; (e) Z_{1213}, two times; and (f) Z_{1122}, two times.

sum rule expressing the particular correlation function in question in terms of planar ones. This gives rise to an identity for this particular Z. Furthermore, since each Z has a unique graphical expansion, all identities are distinct. It follows that the number of sum rule identities, a_n, is equal to the number of Z's which are nonplanar, namely, $b_n - c_n$.

The number c_n has been evaluated in a consideration of the transfer matrix formulation of the Potts model [12], and is found to be generated by (5). To enumerate b_n we note that it is precisely the number of ways that n objects can be partitioned into indistinguishable parts. Let there be m_ν parts of ν objects each subject to $\sum_{m_\nu=1}^{\infty} \nu m_\nu = n$. Then we have $b_n = \sum_{m_\nu=0}^{\infty} \prod_{\nu=1}^{\infty} n!/(\nu!)^{m_\nu} m_\nu!$. This leads to the generating function (4). Particularly, we find $a_4 = 15 - 14 = 1$, $a_5 = 52 - 42 = 10$, $a_6 = 203 - 132 = 71$. Q.E.D.

Duality relation for P_n.—It has been known for some time that the two-point boundary correlation function of an Ising model is related to its counterpart in the dual space. The usual derivation of this relation involves embedding expansions of the correlation functions on the lattice followed by an explicit term-by-term identification [13,14]. In a recent paper one of us [8] introduced a new approach to this problem which invokes only a repeated use of an elementary duality consideration [15]. The new approach, which is very general, also permits the extension of the duality analysis to the Potts model for $n = 2, 3$ [8]. However, an extension of the analysis of [8] to $n \geq 4$ ran into an apparent snag of inadequacy of conditions [16]. Here we show that the correlation identities derived above provide the missing link, and with the help of these identities we determine the duality relation for any n.

The consideration of [8] is based on the fundamental duality relation [15]

$$Z = qCZ^* \tag{7}$$

relating the partition function Z of any planar lattice, or graph, to the partition function Z^* on the dual. Here, $C = q^{-N^*} \prod_{\text{edges}} (e^{K_{ij}} - 1)$, with N^* being the number of sites of the dual and the product taken over all edges. The interaction K_{ij}^* dual to K_{ij} is given by $(e^{K_{ij}} - 1)(e^{K_{ij}^*} - 1) = q$.

Starting from $\mathcal{L}$ we consider a lattice $\mathcal{L}^*$ formed by introducing n spins $\alpha, \beta, \gamma, \ldots, \delta$ to the boundary of the dual of $\mathcal{L}$ (cf. Fig. 1), each interacting with neighboring dual spins within $\mathcal{L}$. (Note that $\mathcal{L}^*$ has $N^* + n - 1$ sites and is not the dual of $\mathcal{L}$.) Let $Z_{\alpha\beta\gamma\cdots\delta}^*$ be the partial dual partition function of $\mathcal{L}^*$ with the n boundary spins fixed in the respective definite states. Our goal is to obtain a duality relation in the form of a linear transformation relating the $Z_{ij\cdots m\ell}$ to $Z_{\alpha\beta\gamma\cdots\delta}^*$.

Regard the b_n planar connectivities as auxiliary lattices, and apply the fundamental duality relation to each one of them [16]. Applying the duality on $\mathcal{L}$ itself, for example, we obtain (7) which can be written as an equation relating linear combinations of the Z and Z^* [8]. Applying the duality to the planar connectivity $\mathcal{L}_n$ in which all n points are connected to a common point with interactions K as in $\mathcal{L}_4$ shown in Fig. 3(b), we obtain

$$Z_{\mathrm{aux}(n)} = Cq^{-(n-2)}(u-1)^n Z^*_{\mathrm{aux}(n)}, \quad (8)$$

where $u = e^K$, and $Z_{\mathrm{aux}(n)}$ and $Z^*_{\mathrm{aux}(n)}$ are, respectively, the partition functions of $\mathcal{L}_n$ and its dual. Now, both sides of (8) are polynomials of degree n in u. Since (8) holds for arbitrary u, the coefficients of all powers of u must be equal. However, it suffices to equate only the coefficients of the highest power of u. On the left-hand side we have $Z_{\mathrm{aux}(n)} = q(u+q-1)^n Z_{11\cdots1} +$ terms of the order of $O(u^{n-1})$. On the right-hand side we have $(u-1)^n Z^*_{\mathrm{aux}(n)} = q(u+q-1)^n Z^*_{11\cdots1} +$ other terms. This leads immediately to an expression for $Z_{11\cdots1}$. For $n=4$, for example, we obtain

$$\begin{aligned} Z_{1111} &= Cq^{-2}[Z^*_{1111} + q_1(Z^*_{2111} + Z^*_{1211} + Z^*_{1121} + Z^*_{1112}) + q_1(Z^*_{1122} + Z^*_{1221}) \\ &\quad + q_1q_2(Z^*_{1123} + Z^*_{2113} + Z^*_{2311} + Z^*_{1231}) + q_1(Z^*_{1212} + q_2Z^*_{1213} + q_2Z^*_{2131} + q_2q_3Z^*_{1234})] \\ &= \{1 + q_1(1,1,1,1) + q_1(1,1) + q_1q_2(1,1,1,1) + q_1(1,q_2,q_2,q_2q_3)\}, \end{aligned} \quad (9)$$

where $q_m = q - m$, $m = 1,2,\ldots$, and in the last line we have introduced a short-handed notation. An immediate consequence of (9) is the result

$$\begin{aligned} \Gamma_4 &= q_1(p_{2111} + p_{1211} + p_{1121} + p_{1112} + p_{1212} \\ &\quad + p_{1122} + p_{1221}) + q_1q_2(p_{1123} + p_{2113} \\ &\quad + p_{2311} + p_{1231} + p_{1213} + p_{2131}) \\ &\quad + q_1q_2q_3p_{1234}, \end{aligned} \quad (10)$$

where we have introduced (7), $Z^* = qZ^*_{1111}$, as well as $p_{\alpha\beta\gamma\delta} = Z^*_{\alpha\beta\gamma\delta}/Z^*_{1111}$. For general n the consideration of $\mathcal{L}_n$ leads to

$$\begin{aligned} \Gamma_n &= q_1 \sum p_{211\cdots1} + q_1q_2 \sum p_{231\cdots1} \\ &\quad + \cdots + q_1q_2\cdots q_{n-1}p_{123\cdots n}. \end{aligned} \quad (11)$$

Applying (7) to all c_n auxiliary lattices of planar connectivities in this fashion and equating the coefficients of the highest power of u in each case, we obtain c_n equations for the b_n unknown Z's. Combining the c_n equations with the $b_n - c_n$ sum rule identities, we have precisely b_n equations, and the duality relation can now be determined.

In the case of $n = 4$, the solution of the 15 equations leads to, in addition to (9),

$$\begin{aligned} Z_{2111} &= \{1 + (-1,-1,q_1,q_1) + (q_1,-1) + q_2(q_1,-1,-1,-1) - (1,q_2,q_2,q_2q_3)\}, \\ Z_{1122} &= \{1 + (-1,q_1,-1,q_1) - (1,1) - q_2(1,1,1,1) + (q_1,q_1q_2,-q_2,-q_2q_3)\}, \\ Z_{1123} &= \{1 - (1,-q_1,1,1) - (1,1) + (2,2,-q_2,-q_2) + (-1,-q_2,2,2q_3)\}, \\ Z_{1212} &= \{1 - (1,1,1,1) + q_1(1,1) - q_2(1,1,1,1) + Q(-1,q_1q_2,q_1q_2,q_2q_3r)\}, \\ Z_{1213} &= \{1 - (1,1,1,1) + (-1,q_1) + (2,-q_2,2,-q_2) + Q(q_1,s,s,q_3t)\}, \\ Z_{2131} &= \{1 - (1,1,1,1) + (q_1,-1) + (-q_2,2,-q_2,2) + Q(q_1,s,s,q_3t)\}, \\ Z_{1234} &= \{1 - (1,1,1,1) - (1,1) + 2(1,1,1,1) + Q[r,t,t,q_3(2-5q)]\}, \end{aligned} \quad (12)$$

where $Q = 1/(q^2 - 3q + 1)$, $r = 2q - 1$, $s = q^2 - 4q + 2$, $t = q^2 - 5q + 2$. Expressions for other $Z_{ijk\ell}$ are given by cyclic permutations.

The solutions (9) and (12) can be written in the form of a partition expansion

$$\begin{aligned} P_4(\sigma_1,\sigma_2,\sigma_3,\sigma_4) &= A_{1234} + A_{1123}\delta_{12} + A_{2113}\delta_{23} + A_{2311}\delta_{34} + A_{1231}\delta_{14} + A_{1213}\delta_{13} + A_{2131}\delta_{24} + A_{1122}\delta_{12}\delta_{34} \\ &\quad + A_{1221}\delta_{14}\delta_{23} + A_{1212}\delta_{13}\delta_{24} + A_{2111}\delta_{234} + A_{1211}\delta_{134} + A_{1121}\delta_{124} + A_{1112}\delta_{123} + A_{1111}\delta_{1234}, \end{aligned} \quad (13)$$

where $\delta_{12} = \delta(\sigma_1,\sigma_2)$, $\delta_{123} = \delta_{12}\delta_{23}$, etc. We find

$$\begin{aligned} A_{1234} &= q^{-4}[1 - (p_{2111} + p_{1211} + p_{1121} + p_{1112} + p_{1122} + p_{1221} + p_{1212}) \\ &\quad + 2(p_{1123} + p_{1231} + p_{2113} + p_{2311} + p_{1213} + p_{2131}) - 6p_{1234}], \\ A_{1123} &= q^{-3}(p_{1211} - p_{1231} - p_{2311} - p_{1213} + 2p_{1234}), \\ A_{2113} &= q^{-3}(p_{1121} - p_{1123} - p_{1231} - p_{2131} + 2p_{1234}), \\ A_{2311} &= q^{-3}(p_{1112} - p_{2113} - p_{1123} - p_{1213} + 2p_{1234}), \\ A_{1231} &= q^{-3}(p_{2111} - p_{2311} - p_{2113} - p_{2131} + 2p_{1234}), \end{aligned}$$

$$\begin{aligned}
A_{1213} &= q^{-3}(p_{1221} - p_{1231} - p_{2113} + p_{1234}), \\
A_{2131} &= q^{-3}(p_{1122} - p_{1123} - p_{2311} + p_{1234}), \\
A_{1122} &= q^{-2}(p_{1213} - p_{1234}), \qquad A_{1221} = q^{-2}(p_{2131} - p_{1234}), \\
A_{1212} &= 0, \qquad A_{2111} = q^{-2}(p_{1123} - p_{1234}), \\
A_{1211} &= q^{-2}(p_{2113} - p_{1234}), \qquad A_{1121} = q^{-2}(p_{2311} - p_{1234}), \\
A_{1112} &= q^{-2}(p_{1231} - p_{1234}), \qquad A_{1111} = q^{-1}p_{1234},
\end{aligned} \tag{14}$$

where we have used the fact that Z^* satisfies the same sum rules as the Z, including $Z^*_{1212} = Z^*_{1213} + Z^*_{2131} - Z^*_{1234}$.

For general n we write in analogous to (13) the partition expansions

$$\begin{aligned}
P_n(\sigma_1, \sigma_2, \ldots, \sigma_n) &= A_{12\cdots n} + A_{1123\cdots(n-1)}\delta_{12} \\
&\quad + \cdots + A_{11\cdots 1}\delta_{12\cdots n},
\end{aligned} \tag{15}$$

$$\begin{aligned}
p_{\alpha\beta\cdots\delta} &\equiv Z^*_{\alpha\beta\cdots\delta}/Z^*_{11\cdots 1} \\
&= B_{12\cdots n} + B_{1123\cdots(n-1)}\delta_{\alpha\beta} \\
&\quad + \cdots + B_{11\cdots 1}\delta_{\alpha\beta\cdots\delta}.
\end{aligned} \tag{16}$$

Regard the diagram in Fig. 1 as representing $A_{ij\cdots\ell}$. Construct for each A the associated connectivity as in Fig. 3, and label indices $\alpha, \beta, \ldots, \delta$ such that neighboring indices are the same if there is no line in between. Then we are led to the following *conjecture:*

$$\begin{aligned}
A_{ij\cdots\ell} &= q^{-d(ij\cdots\ell)}B_{\alpha\beta\cdots\delta} \quad \text{if the connectivity is planar,} \\
&= 0, \quad \text{otherwise,}
\end{aligned} \tag{17}$$

where $d(ij\cdots\ell)$ is the number of distinct indices in $\{i, j, \ldots, \ell\}$. The conjecture is readily verified for $n = 2, 3, 4$. In practice, for any given n, one can solve from (16) for $B_{\alpha\beta\cdots\delta}$ by applying the principle of inclusion-exclusion.

A cyclic inversion relation.—Since $\alpha\beta\gamma\cdots\delta$ are boundary sites of $\mathcal{L}^*$, the transformation relating Z^* to Z, an inversion process, is given precisely by the same transformation relating Z to Z^*. Now $\mathcal{L}^*$ has $N^* + n - 1$ sites and its dual has $N - n + 1$ sites. Also $\mathcal{L}$ and $\mathcal{L}^*$ have the same number of edges. Therefore we have

$$\begin{aligned}
Z_{ijk\cdots m\ell} &= \left(\frac{\Pi(e^{K_{ij}} - 1)}{q^{N^*+(n-2)}}\right) \\
&\quad \times \sum_{\{\alpha\beta\gamma\cdots\delta\}} \mathbf{T}_n(ij\cdots m\ell \,|\, \alpha\beta\gamma\cdots\delta)Z^*_{\alpha\beta\gamma\cdots\delta},
\end{aligned} \tag{18}$$

$$\begin{aligned}
Z^*_{\alpha\beta\gamma\cdots\delta} &= \left(\frac{\Pi(e^{K^*_{ij}} - 1)}{q^{N-n+1+(n-2)}}\right) \\
&\quad \times \sum_{\{ij\cdots m\ell\}} \mathbf{T}_n(\alpha\beta\gamma\cdots\delta \,|\, \ell ij\cdots m)Z_{\ell ij\cdots m},
\end{aligned} \tag{19}$$

where $\mathbf{T}_n$ is a $b_n \times b_n$ matrix, and the summations are over the set of b_n distinct partitions of the n indices. Substituting (19) into (18) and making use of the Euler relation $E = N + N^* - 2$, we are led to the identity, which we refer to as a cyclic inversion relation,

$$\mathbf{T}^2_n(ijk\cdots\ell \,|\, i'j'k'\cdots\ell') = q^{n-1}\delta_{ij'}\delta_{jk'}\cdots\delta_{\ell i'}. \tag{20}$$

Further discussions including properties of $\mathbf{T}_n$ and the extension to the Ashkin-Teller model will be given elsewhere [17].

We are grateful to J. L. Jacobsen for discussions and a comment [16] which has led to this investigation. This work is supported in part by National Science Foundation Grant No. DMR-9614170.

Note added.—The conjecture (17) has since been established rigorously [18].

[1] R. B. Potts, Proc. Cambridge Philos. Soc. **48**, 106 (1952).
[2] F. Y. Wu, Rev. Mod. Phys. **54**, 235 (1982).
[3] F. Y. Wu, J. Appl. Phys. **55**, 2421 (1984).
[4] L. Onsager, Phys. Rev. **65**, 117 (1944).
[5] R. J. Baxter, J. Phys. C **6**, L445 (1973).
[6] R. J. Baxter, H. N. V. Temperley, and S. E. Ashley, Proc. R. Soc. London A **358**, 535 (1978).
[7] J. E. Avron, H. van Beijeren, and R. K. P. Zia, J. Phys. A **15**, L81 (1982).
[8] F. Y. Wu, Phys. Lett. A **228**, 43 (1997).
[9] R. J. Baxter, S. B. Kelland, and F. Y. Wu, J. Phys. A **9**, 397 (1976).
[10] G. D. Birkhoff, Ann. Math. **14**, 42 (1912).
[11] H. W. J. Blöte and M. P. Nightingale, Physica (Amsterdam) **112A**, 405 (1982).
[12] H. N. V. Temperley and E. H. Lieb, Proc. R. Soc. London A **322**, 251 (1971).
[13] P. G. Watson, J. Phys. C **1**, 575 (1968).
[14] R. K. P. Zia, Phys. Lett. **64A**, 345 (1978).
[15] F. Y. Wu and Y. K. Wang, J. Math. Phys. (N.Y.) **17**, 439 (1976).
[16] J. L. Jacobsen, Phys. Lett. A **233**, 489 (1997).
[17] F. Y. Wu and W. T. Lu, Chin. J. Phys. (to be published).
[18] W. T. Lu and F. Y. Wu, e-print cond-mat/9712045.

4. The Ising Model

PHYSICAL REVIEW B VOLUME 4, NUMBER 7 1 OCTOBER 1971

Ising Model with Four-Spin Interactions*

F. Y. Wu
Department of Physics, Northeastern University, Boston, Massachusetts 02115

(Received 3 May 1971)

It is shown that Baxter's recent results on a lattice-statistical model lead to the solution of an Ising model with two- and four-spin interactions. Critical properties of this Ising model in various regions of the parameter space are given. It is argued that four-spin or crossing interactions in a two-dimensional Ising model would in general lead to a critical exponent $\alpha' \neq 0$.

The recent exact solution by Baxter[1] of a lattice-statistical model[2] constitutes a breakthrough in the study of phase transitions. The most striking feature of Baxter's solution is that the nature of the phase transition is dependent on the energy parameters of the model. While it has been known for some time that the behavior of this lattice model is quite different in the isolated soluble cases of Ising, F, and potassium dihydrogen phosphate (KDP) models, it is for the first time that a phase transition is shown to exhibit a continuously variable exponent. Baxter's solution is given in the language of a ferroelectric model. To those who are accustomed to the "magnetic" language of phase transitions, the implications of his results are perhaps not very transparent. Therefore, we wish to point out in this note the conclusions on the more familiar Ising model that can be deduced from Baxter's solution.

It can be shown[3] that the ferroelectric problem considered in Refs. 1 and 2 is equivalent to an Ising model in zero magnetic field with *finite* two- and four-spin interactions.[4] The equivalent Ising lattice, shown in Fig. 1, has first-neighbor interactions $-J_1$ and $-J_2$, second-neighbor interactions $-J$ and $-J'$, and a four-spin interaction $-J_4$ between any four spins surrounding a unit square. The Hamiltonian reads, in obvious summation notations,

$$H=-J_1\sum\sigma\sigma'-J_2\sum\sigma\sigma'-J\sum\sigma\sigma'$$

$$-J'\sum\sigma\sigma'-J_4\sum\sigma\sigma'\sigma''\sigma''' \quad . \qquad (1)$$

The energy parameters of the ferroelectric problem turn out to be, using Baxter's notation,[5]

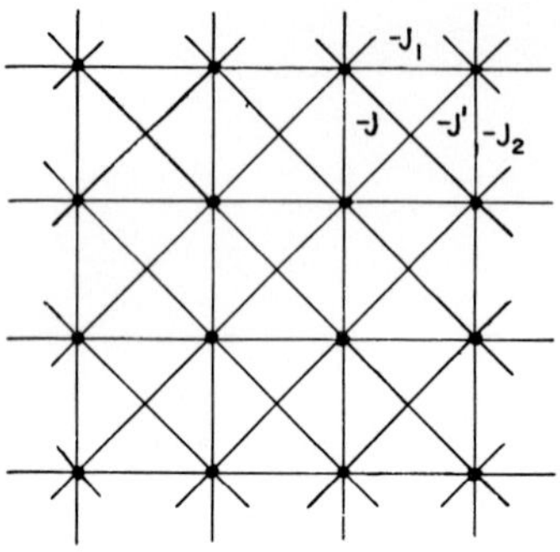

FIG. 1. Ising lattice specified by the Hamiltonian (1). Each dot denotes a spin and the four-spin interactions are not shown.

$$\epsilon_1=\epsilon_2=-J-J'-J_4\,,$$
$$\epsilon_3=\epsilon_4=J+J'-J_4\,,$$
$$\epsilon_5=\epsilon_6=J'-J+J_4\,, \qquad (2a)$$
$$\epsilon_7=\epsilon_8=-J'+J+J_4\,,$$

with an external electric field

$$(h, v)=(J_2, J_1). \qquad (2b)$$

Baxter solved the ferroelectric model with $h=v=0$. In the Ising language this corresponds to deleting the first-neighbor interactions J_1 and J_2. Therefore, the resulting Ising lattice is composed of two superimposed square lattices which are coupled together via four-spin interactions. If the four-spin interaction vanishes ($J_4=0$), the problem reduces to that of the simple square Ising lattice and can be solved by standard means.[6] The crux of the matter is that Baxter's solution can be adapted to this Ising model ($J_1=J_2=0$) for arbitrary J_4! This will be the first time that a solution is found for an Ising model with many-spin interactions.

After the hard work has been done by Baxter, it is now quite straightforward to transcribe Baxter's results to the present Ising problem (with $J_1=J_2=0$). The main results are summarized in the following.

(i) Since the partition function is invariant under the reversal of the signs of J_4 and J (or J_4 and J'), we may, without loss of generality, consider only $J_4>0$. In Fig. 2 we show the various regions in the J-J' plane defined by (a given vertex energy is favored within a region)

$$\epsilon_1<(\epsilon_3, \epsilon_5, \epsilon_7), \quad \text{region I}$$
$$\epsilon_3<(\epsilon_1, \epsilon_5, \epsilon_7), \quad \text{region II}$$
$$\epsilon_5<(\epsilon_1, \epsilon_3, \epsilon_7), \quad \text{region III}$$
$$\epsilon_7<(\epsilon_1, \epsilon_3, \epsilon_5), \quad \text{region IV}. \qquad (3)$$

(ii) The transition temperature T_c is given by

$$e^{2K_4}=|\cosh(K-K')/\sinh(K+K')|, \quad \text{regions I and II}$$
$$e^{2K_4}=|\sinh(K-K')/\cosh(K+K')|, \quad \text{regions III and IV} \qquad (4)$$

where $K=J/kT_c$, $K'=J'/kT_c$, and $K_4=J_4/kT_c$. We note that $T_c=0$ on all region boundaries.

(iii) The energy is continuous at T_c.

(iv) In regions I and II the specific heat diverges at T_c with critical components

$$0<\alpha=\alpha'=2-\pi/\mu_+<1\ . \qquad (5)$$

In regions III and IV the specific heat is continuous while the nth ($n\geq 3$) derivative of the free energy diverges as $|T-T_c|^{\pi/\mu_- -n}$ (logarithmic divergence if $\pi/\mu_-=n$), where n is the integer defined by

$$n-1<\pi/\mu_-\leq n. \qquad (6)$$

In (5) and (6),

$$\mu_\pm=\tfrac{1}{2}\pi\pm\sin^{-1}[\tanh(2K_4)]. \qquad (7)$$

(v) The case of $J_4=0$, the nearest-neighbor square Ising lattice, is a singular exception for which the specific heat has a logarithmic singularity.

Several remarks are now in order. First, we note that the critical behavior of the Ising model depends on the interactions J, J', and J_4. It is also tempting to infer from the above results that, in appropriate regions in the parameter space, the four-spin interaction will in general lead to higher than second-order transitions. We wish to point out, however, that it is also possible that this peculiar behavior is an artifact of setting

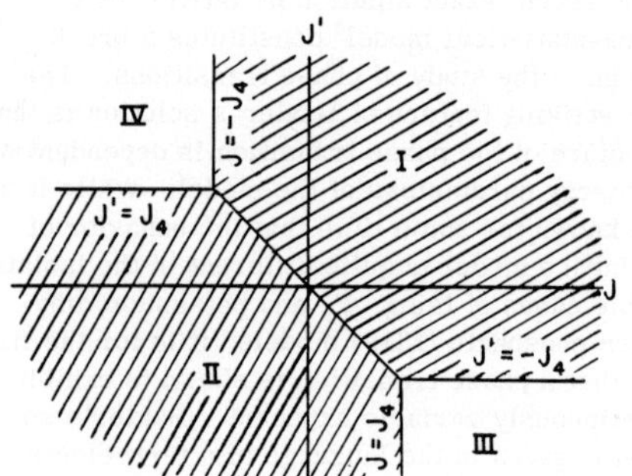

FIG. 2. Various regions in the J-J' plane for a fixed $J_4>0$. The phase transition is associated with an infinite specific heat in the shaded regions I and II, and is of higher than second order in regions III and IV.

$J_1 = J_2 = h = v = 0$ in the Hamiltonian (1). In the case of the F model, for example, it is known that the inclusion of a nonzero field (h, v) changes the infinite-order transition to a second-order one.[7] The inclusion of some nonzero values for J_1 and J_2 could have the same consequence in the present problem. It does appear safe, however, to infer that the inclusion of the four-spin interactions will in general *not* result in $\alpha = \alpha' = 0$.

The result that the nearest-neighbor square Ising lattice is a singular case with $\alpha = \alpha' = 0$ also appears somewhat disturbing, for it is generally believed that the critical exponents should depend only on the dimensionality of the model, and not on the range of interactions. We wish to present some counter arguments. First, some information is available at one particular point of the parameter space, namely, $J_1 = J_2 = J = J'$ and $J_4 = 0$. This is the square Ising lattice with equivalent first- and second-neighbor (crossing) interactions. For this model Domb and Dalton[8] and Dalton and Wood[9] have carried out numerical analyses on the high- and low-temperature series expansions. The study on the high-temperature series led to the critical exponent[8]

$$\gamma \cong 1.75 \; , \tag{8}$$

which does not differ from that of the nearest-neighbor planar Ising lattices. On the other hand, the study on the low-temperature series did not lead to such agreement. The authors of Ref. 9 attributed their results on the low-temperature exponents β and γ' to the erratic behavior of the Padé approximants. On reexamining their data on the first- and second-neighbor square lattice, we feel that unless something drastic happens in the high Padé approximants, it should be safe to infer the following bounds on the critical exponents β and γ':

$$0.80 < \gamma' < 1.30, \quad 0.13 < \beta < 0.16 \; . \tag{9}$$

Accepting (9), the Rushbrook inequality $\alpha' + 2\beta + \gamma' \geq 2$ then leads to the bound

$$\alpha' \gtrsim 0.38 \tag{10}$$

on α'. This indicates a λ transition of the type given by (5) and is definitely different from the commonly accepted value of $\alpha' = 0$ for two-dimensional lattices.[10] This result suggests that the logarithmic singularity of the nearest-neighbor Ising model is indeed a singular case. It must be noted that this is not the first time that the two-dimensional nearest-neighbor model is found to possess a unique behavior. In a recent study on the behavior of two-point correlation functions on a phase boundary, Fisher and Camp[11] showed that the planar nearest-neighbor model is unique in having a decay exponent different from the Ornstein-Zernike form. We feel that these are strong evidences which indicate that the four-spin or the crossing interactions in a planar Ising model will in general lead to a critical exponent $\alpha' \neq 0$.

---

*Work supported in part by National Science Foundation Grant No. GP-25306.

[1]R. J. Baxter, Phys. Rev. Letters 26, 832 (1971).

[2]C. Fan and F. Y. Wu, Phys. Rev. B 2, 723 (1970).

[3]The proof follows closely that given in the Appendix of F. Y. Wu, Phys. Rev. 183, 604 (1969), which will not be reproduced here. The readers are also referred to the following review article for a more comprehensive discussion: E. H. Lieb and F. Y. Wu, in *Phase Transitions and Critical Phenomena*, edited by C. Domb and M. S. Green (Academic, London, 1971).

[4]There exist other mappings between the ferroelectric and the Ising problems. [See, e.g., M. Suzuki and M. E. Fisher, J. Math. Phys. 12, 235 (1971); E. H. Lieb and F. Y. Wu, in Ref. 3.] These mappings would, however, lead to infinite Ising interactions in the present problem.

[5]The zero-energy level has been chosen to make $\epsilon_1 + \epsilon_3 + \epsilon_5 + \epsilon_7 = 0$.

[6]This is the case considered by F. Y. Wu, in Ref. 3.

[7]See E. H. Lieb and F. Y. Wu in Ref. 3.

[8]C. Domb and N. W. Dalton, Proc. Phys. Soc. (London) 89, 859 (1966).

[9]N. W. Dalton and D. W. Wood, J. Math. Phys. 10, 1271 (1969).

[10]We feel that the estimates on γ' in Ref. 9 are sufficient to indicate $\alpha' > 0$.

[11]M. E. Fisher and W. J. Camp, Phys. Rev. Letters 26, 565 (1971).

VOLUME 31, NUMBER 21 PHYSICAL REVIEW LETTERS 19 NOVEMBER 1973

Exact Solution of an Ising Model with Three-Spin Interactions on a Triangular Lattice

R. J. Baxter and F. Y. Wu*†
Research School of Physical Sciences, The Australian National University, Canberra, Australian Capitol Territory 2600, Australia
(Received 18 September 1973)

The Ising model on a triangular lattice with three-spin interactions is solved exactly. The solution, which is obtained by solving an equivalent coloring problem using the Bethe *Ansatz* method, is given in terms of a simple algebraic relation. The specific heat is found to diverge with indices $\alpha = \alpha' = \frac{2}{3}$.

An outstanding open problem in lattice statistics has been the investigation of phase transitions in Ising systems which do not possess the up-down spin-reversal symmetry.[1,2] A well-known example which remains unsolved to this date is the Ising antiferromagnet in an external field. Another problem of similar nature that has been considered recently[3-5] is the Ising model on a triangular lattice with three-body interactions. This latter model is self-dual so that its transition temperature can be conjectured[3,6] using the Kramers-Wannier argument.[7] However, the nature of the phase transition has hitherto not been known.

We have succeeded in solving this model exactly. In this paper we report on our findings. It will be seen that the results are fundamentally different from those of the nearest-neighbor Ising models. While the final expression of our solution is quite simple, the analysis is rather lengthy and involved. For continuity in reading, therefore, we shall first state the result. An outline of the steps leading to the solution will also be given.

Consider a system of N spins $\sigma_i = \pm 1$ located at the vertices of a triangular lattice L. The three spins surrounding every face interact with a three-body interaction of strength $-J$, so that the Hamiltonian reads

$$\mathcal{H} = -J \sum \sigma_i \sigma_j \sigma_k, \tag{1}$$

with the summation extending over all faces of L. Let Z be the partition function defined by (1). We find the following expression for $Z^{1/N}$ in the

VOLUME 31, NUMBER 21 PHYSICAL REVIEW LETTERS 19 NOVEMBER 1973

thermodynamic limit:

$$W = \lim_{N\to\infty} Z^{1/N} = (6yt)^{1/2}, \tag{2}$$

where $t = \sinh 2K$, with $K = |J|/kT$, and $1 \le y < \infty$ is the solution ot the algebraic equation

$$(y-1)^3(1+3y)/y^3 = 2(1-t)^4/t(1+t^2). \tag{3}$$

The function $W(t)$ is unique since y is single valued in t for $0 \le t < \infty$. It is also seen that $\ln W$ has a singular part which behaves as $|t-1|^{4/3}$ near $t=1$. It follows that a phase transition occurs at $t=1$ or $kT_c/|J| = 2/\ln(\sqrt{2}+1) = 2.269\,185\ldots$. This confirms the duality prediction[3,6] which merely reflects the invariance of the right-hand side of (3) under the transformation $t \to t^{-1}$. The energy and specific heat per spin, E and c, respectively, are plotted in Fig. 1. The energy has the critical value $E_c = -\sqrt{2}|J|$, while the specific heat diverges near T_c as

$$c \sim \tfrac{2}{9}k[\ln(\sqrt{2}+1)]^{4/3}|1-T/T_c|^{-2/3}. \tag{4}$$

This is in contradistinction to the logarithmic divergence of the nearest-neighbor models, and to the result $\alpha = \alpha' = \frac{1}{2}$ of the isotropic three-spin model on the "Union Jack" lattice.[8]

Two key steps are involved in our approach. The Ising model (1) is first converted into a coloring problem; the latter is subsequently solved using a generalization of the Bethe *Ansatz* method.[9] We now outline these two steps separately.

Conversion into a coloring problem.—The triangular lattice L is composed of three sublattices L_1, L_2, and L_3 with the property that the sites of L_i and L_j form a hexagonal lattice L_{ij}. Two such lattices, L_{13} (solid lines) and L_{23} (broken lines), are shown in Fig. 2.

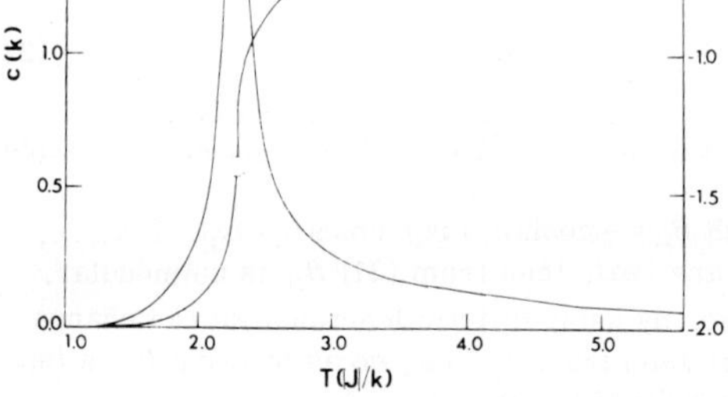

FIG. 1. Energy (E) and specific heat (c) per spin.

We first carry out a dual transformation for the spins on L_{23} while leaving the spins on L_1 unchanged. The effect of the dual transformation is to introduce an additional variable $\mu = \pm 1$ to each face of L_{23}.[10] Thus we may specify the states of the sites of L_1 by the four-valued variable (σ, μ). Furthermore, to each neighboring pair (σ_i, μ_i) and (σ_j, μ_j) of L_1, the dual transformation yields a weight factor

$$\omega_{ij} = 2^{-1/3}\{\exp[K(\sigma_i+\sigma_j)] + \mu_i\mu_j \exp[-K(\sigma_i+\sigma_j)]\}. \tag{5}$$

The partition function is then given by

$$Z = \sum \prod \omega_{ij}, \tag{6}$$

where the summation is extended over all states (σ_i, μ_i) and the product over all nearest neighbors of L_1.

Next, for each nonvanishing term in (6) we associate colors 1, 3, 5, 7 to the sites of L_1 according to the rule

$$(+,+) = \text{color } 1, \quad (-,+) = \text{color } 3, \qquad (-,-) = \text{color } 5, \quad (+,-) = \text{color } 7. \tag{7}$$

It is clear from (5) that two neighboring sites of L_1 cannot be colored 1, 5 or 3, 7. Thus we may further color the sites of L_3 with colors 2, 4, 6, 8 under the restriction that the colors of neighboring sites on L_{13} differ by exactly 1 (to modulus 8, i.e., 8 and 1 can be adjacent). Note that we have just completed a site coloring for L_{13}. Now we introduce activities to the colors:

$$z_1 = z_3 = z_5 = z_7 = 1, \qquad z_2 = 1/z_4 = z_6 = 1/z_8 = \sinh 2K. \tag{8}$$

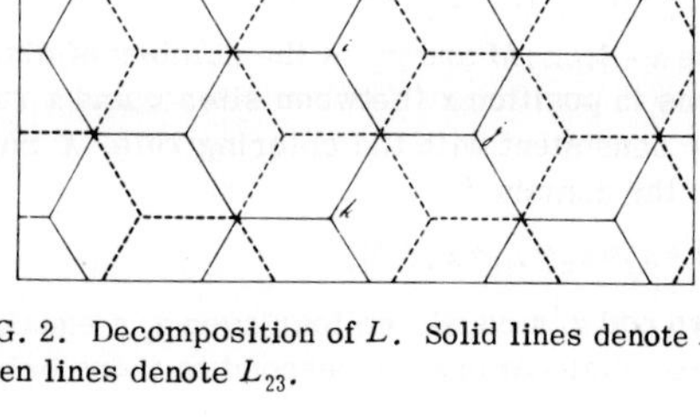

FIG. 2. Decomposition of L. Solid lines denote L_{13}, broken lines denote L_{23}.

It can be verified that the activities of the possible colorings of a site on L_3 surrounded by sites (σ_i, μ_i), (σ_j, μ_j), and (σ_k, μ_k) of L_1 (cf. Fig. 2) generate precisely the product $(2\sinh 4K)^{-1}\omega_{ij}\omega_{jk}\omega_{ki}$. Using (6), we are thus led to the identity

$$Z=(2\sinh 4K)^{N/3}Z_c, \tag{9}$$

where

$$Z_c=\sum z_1^{n_1}z_2^{n_2}\cdots z_8^{n_8} \tag{10}$$

is the generating function for the site coloring of L_{13}.

Solution of the coloring problem.—Draw arrows on the edges of the dual triangular lattice L_D of L_{13}, pointing to an observer's left (right) if the colors on L_{13} increase (decrease) as he crosses the arrow. Then at each vertex of L_D there are three arrows in and three out. Hence our coloring problem bears the same relation to the "triangular ice" model[11] as the three colorings of the square lattice[12] do to the "square ice"[13] model. In particular, since $z_1=\ldots=z_8=1$ when $T=T_c$ and $t=1$, it follows that at the critical temperature the coloring and triangular ice models are equivalent, so from (9) and Eq. (41) of Ref. 11 one obtains $W_c=\sqrt{6}$. This agrees with (2).

To obtain W for the more general activities (8), number the sites in each row of L_{13} as in Fig. 2, using cyclic boundary conditions as indicated. Let $C=\{c_1,\ldots,c_M\}$ be the coloring of the upper row of sites $1,\ldots,M$, and $C'=\{c_1',\ldots,c_M'\}$ the coloring of the lower row. Introduce the transfer matrix

$$A(C,C')=\prod_{i=1}^{M}(z_{c_i}z_{c_i'})^{1/2} \tag{11}$$

if the colors of adjacent sites differ by 1; otherwise $A(C,C')=0$.

Consider a basic sequence $\{m, m+1, m+4, m+5, m+8, m+9,\ldots\}$ of colors in a row, which corresponds to a row of arrows pointing up on L_D, and then introduce n dislocations at positions $X=\{x_1,\ldots,x_n\}$ in the sequence so as to define C by the rules

$$c_1=m, \quad c_{x+1}=c_x+2+(-1)^x-2n_x, \tag{12}$$

where $x=1,\ldots,M$ and n_x is the number of dislocations in position x (between sites x and $x+1$). To be consistent with the coloring rule, X must lie in the domain D:

$$1\le x_1\le x_2\le\ldots\le x_n\le M, \tag{13}$$

no two odd x_j's equal, no four even x_j's equal.

These dislocations correspond to there being arrows pointing down on the efges of L_D. There are the same number n in every row, so we can consider some fixed value of n between 0 and $2M$.

Let Λ and f be an eigenvalue and eigenvector of A, respectively, and $f(m,X)$ the element of f corresponding to the row coloring (12). Then the eigenvalue equation is

$$\Lambda f(m,X)=\sum_Y\left(\prod_{j=1}^{n}w(m+x_j+y_j-2j)\right)\times f(m+2,Y), \tag{14}$$

where

$$w(m)=(z_m z_{m+1})^{1/2}, \tag{15}$$

$$f(m+2,0,x_2,\ldots,x_n)=f(m,x_2,\ldots,x_n,M), \tag{16}$$

and the summation is over all $Y\in D$ such that

$$|y_j-x_j+1|\le\tfrac{1}{2}[1+(-1)^{x_j}], \tag{17}$$

and if $x_j=x_{j+1}$, then $y_j\ne y_{j+1}$. For a large lattice of $2N/3$ sites,

$$Z_c^{3/2N}=\Lambda_{\max}^{1/M}, \tag{18}$$

where $\Lambda_{\max}$ is the maximum eigenvalue of (14). We try the generalized Bethe *Ansatz*

$$f(m,X)=\sum_P a(P)\prod_{j=1}^{n}\varphi_{P_j}(m-2j,x_j), \tag{19}$$

where the summation is over all $n!$ permutations $P=\{P_1,\ldots,P_n\}$ of the integers $1,\ldots,n$. The coefficients $a(P)$ and the n functions $\varphi_j(m,x)$ are at our disposal. We require that there exist n wave numbers $k_1,\ldots,k_n$ such that

$$\varphi_j(m,x)=\varphi_j(m+4,x)=\exp(2ik_j)\varphi_j(m,x-2) \tag{20}$$

for all integers m,x. We find that the *Ansatz* works: Let E_j be defined by

$$\cosh 2E_j=\cos 2k_j+t+t^{-1}; \tag{21}$$

then Λ and $k_1,\ldots,k_n$ are given by

$$\ln\Lambda=\sum_{j=1}^{n}(E_j-ik_j), \tag{22}$$

$$\exp(iMk_j)=-\prod_{l=1}^{n}B_{jl}, \quad j=1,2,\ldots,n, \tag{23}$$

with $B_{jl}=-\cosh(E_j+ik_l)/\cosh(E_l+ik_j)$. If $k_1,\ldots,k_n$ are real, then from (21) B_{jl} is unimodular. Also, by using elliptic functions we can change variables from k_j to u_j so as to make B_{jl} a function only of u_j-u_l.

For $\Lambda_{\max}$ the equations (23) can therefore be

solved in the limit of M, n large by methods similar to those used for the normal Bethe *Ansatz*.[13,14] Using (22), (18), and (9) we then obtain the result (2).

We intend to give details elsewhere. We remark that $k_1, \ldots, k_n$ are real and distributed symmetrically over the interval $(-\frac{1}{2}\pi, \frac{1}{2}\pi)$, that $n=M$, and that $E_1, \ldots, E_n$ are real and non-negative.

One of us (F.Y.W.) wishes to thank Professor K. J. Le Couteur for his kind hospitality at the Australian National University.

Added note.—Numerical estimates of the critical indices α', β, and γ' of the present model have recently been given by H. P. Griffiths and D. W. Wood, J. Phys. C: Proc. Phys. Soc., London 6, 2533 (1973).

*Senior Fulbright Scholar on leave of absence from Northeastern University, Boston, Mass.

†Work supported in part by the National Science Foundation under Grant No. GH-35822 at Northeastern University.

[1]R. B. Griffiths, in *Statistical Mechanics and Quantum field Theory,* edited by C. DeWitt and R. Stora (Gordon and Breach, New York, 1971).

[2]G. Gallavotti, Riv. Nuovo Cimento 2, 133 (1972).

[3]D. W. Wood and H. P. Griffiths, J. Phys. C: Proc. Phys. Soc., London 5, L253 (1972).

[4]D. Merlini, A. Hintermann, and C. Gruber, to be published.

[5]D. Merlini, to be published.

[6]D. Merlini and C. Gruber, J. Math. Phys. (N.Y.) 13, 1814 (1972).

[7]H. A. Kramers and G. H. Wannier, Phys. Rev. 60, 252 (1941).

[8]A. Hintermann and D. Merlini, Phys. Lett. 41A, 208 (1972).

[9]H. A. Bethe, Z. Phys. 71, 205 (1931).

[10]F. Wegner, to be published.

[11]R. J. Baxter, J. Math. Phys. (N.Y.) 10, 1211 (1969).

[12]R. J. Baxter, J. Math. Phys. (N.Y.) 11, 3116 (1970).

[13]E. H. Lieb, Phys. Rev. 162, 162 (1967).

[14]C. N. Yang and C. P. Yang, Phys. Rev. 150, 321 (1966).

Journal of Statistical Physics, Vol. 102, Nos. 3/4, 2001

Density of the Fisher Zeroes for the Ising Model

Wentao T. Lu[1] and F. Y. Wu[1]

Received April 10, 2000

The density of the Fisher zeroes, or zeroes of the partition function in the complex temperature plane, is determined for the Ising model in zero field as well as in a pure imaginary field $i\pi/2$. Results are given for the simple-quartic, triangular, honeycomb, and the kagomé lattices. It is found that the density diverges logarithmically at points along its loci in appropriate variables.

KEY WORDS: Ising model; partition function; Fisher zeroes; density.

1. INTRODUCTION

In the analyses of lattice models in statistical mechanics such as the Ising model, the partition function is often expressed in the form of a polynomial in variables such as the external magnetic field and/or the temperature. Since properties of a polynomial are completely determined by its roots, a knowledge of the zeroes of the partition function yields all thermodynamic properties of the system. Particularly, if the zeroes lie on a certain locus, a knowledge of its density distribution along the locus is equivalent to the obtaining of the exact solution of the problem.

For the Ising model with ferromagnetic interactions, we have the remarkable Yang–Lee circle theorem[(1,2)] which states that all partition function zeroes lie on the unit circle $|z|=1$ in the complex $z=e^{2L}$ plane, where L is the reduced external magnetic field (we set $kT=1$). However, the density of the Yang–Lee zeroes on the unit circle, a knowledge of which is equivalent to solving the Ising model in a nonzero magnetic field, is known only for the Ising model in one dimension.

Fisher[(3)] has proposed that it is also meaningful to consider partition function zeroes in the complex temperature plane. Indeed, he showed that

[1] Department of Physics, Northeastern University, Boston, Massachusetts 02115.

for the zero-field Ising model on the simple-quartic lattice with nearest-neighbor reduced interactions K, the partition function zeroes lie on two circles

$$|\tanh K \pm 1| = \sqrt{2} \tag{1}$$

in the thermodynamic limit. He further showed that the known logarithmic singularity of the specific heat follows from the fact that the density vanishes linearly near the real axis. Subsequently, the Fisher loci has been determined for the infinite triangular lattice,[4] and for *finite* simple-quartic lattices which are self-dual.[5] Stephenson[6, 7] has also evaluated the density distribution on the circles in terms of a Jacobian. However, the explicit expressions of the density function of the Fisher zeroes do not appear to have been heretofore evaluated.

In this paper we complete the picture by evaluating the density function. We deduce the explicit expressions for the density of Fisher zeroes for the simple-quartic, triangular, honeycomb, and kagomé lattices. Density of the Fisher zeroes for the Ising model in a pure imaginary field $L = i\pi/2$ are also obtained.

2. THE SIMPLE-QUARTIC LATTICE

It is well-known that the bulk solution of spin models with short-range interactions is independent of the boundary conditions. For the Ising model on the simple-quartic lattice, we shall take a particular boundary condition introduced by Brascamp and Kunz[8] for which the location of the Fisher zeroes is known for any *finite* lattice. This permits us to take a a well-defined and unique bulk limit, thus avoiding a difficulty encountered in the consideration of the Ising model on a torus.[6]

Consider an $M \times 2N$ simple-quartic lattice with cylindrical boundary conditions in the N direction and fixed boundary conditions along the two edges of the cylinder. The $2N$ boundary spins on each of the two edges of the cylinder have fixed fields $\cdots + + + + + + + \cdots$ and $\cdots + - + - + - \cdots$, respectively. This is the Brascamp–Kunz boundary condition.[8] Brascamp and Kunz showed that the partition function of this Ising model is precisely

$$Z_{M,2N}(K) = 2^{2MN} \prod_{1 \leqslant i \leqslant N} \prod_{1 \leqslant j \leqslant M} [1 + z^2 - z(\cos\theta_i + \cos\phi_j)] \tag{2}$$

where

$$z = \sinh 2K, \qquad \theta_i = (2i-1)\,\pi/2N, \qquad \phi_j = j\pi/(M+1) \tag{3}$$

The per-site "free energy" in the bulk limit is then evaluated as

$$f = \lim_{M, N \to \infty} \frac{1}{2MN} \ln Z_{M, 2N}(K)$$

$$= \frac{1}{2} \ln(4z) + \frac{1}{8\pi^2} \int_{-\pi}^{\pi} d\theta \int_{-\pi}^{\pi} d\phi \ln[z + z^{-1} - (\cos\theta + \cos\phi)]$$

$$= \frac{1}{2} \ln(4z) + \frac{1}{2\pi^2} \int_{0}^{\pi} d\theta \int_{0}^{\pi} d\phi \ln[z + z^{-1} - 2\cos u \cos v] \tag{4}$$

where $u = (\theta + \phi)/2$, $v = (\theta - \phi)/2$ and we have made use of the fact that the integrands are 2π-periodic.

The partition function (2) has zeroes at the $2MN$ solutions of

$$z + z^{-1} = \cos\theta_i + \cos\phi_j, \qquad 1 \leqslant i \leqslant N, \quad 1 \leqslant j \leqslant M \tag{5}$$

The following lemma and corollaries are now used to determine the loci of the zeroes:

Lemma. The regime $-2 \leqslant z + z^{-1} \leqslant 2$ of the complex z plane, where $z + z^{-1} = \text{real}$, is the unit circle $|z| = 1$.

Proof. The lemma follows from the fact that, by writing $z = re^{i\theta}$, we have

$$z + z^{-1} = \left(r + \frac{1}{r}\right)\cos\theta + i\left(r - \frac{1}{r}\right)\sin\theta \tag{6}$$

so that $z + z^{-1} = \text{real}$ implies either $r = 1$ or $\theta = \text{integer} \times \pi$. In the latter case we have $|z + z^{-1}| = |r + r^{-1}| > 2$, which contradicts the assumption, unless $r = 1$. It follows that we have always $r = 1$, or $|z| = 1$. ∎

Corollary 1. The regime $-a \leqslant z + z^{-1} \leqslant b$, where $a, b > 2$, $z + z^{-1} = \text{real}$, of the complex z plane is the union of the unit circle $|z| = 1$ and segments $z_-(-a) \leqslant x \leqslant z_+(-a)$ and $z_-(b) \leqslant x \leqslant z_+(b)$ of the real axis, where $z_\pm(b) = (b \pm \sqrt{b^2 - 4})/2$.

Corollary 2. The regime $-a \leqslant z + z^{-1} \leqslant b$, where $a, b > 2$, $z + z^{-1} = \text{real}$, of the complex z plane, is the regime $|w| = 1$ in the complex w plane, where w is the solution of the equation

$$w + w^{-1} = \frac{4}{a + b}\left(z + z^{-1} + \frac{a - b}{2}\right) \tag{7}$$

Corollary 1 is established along the same line as in the proof of the lemma, and Corollary 2 is a consequence of the lemma since, by construction, we have $-2 \leqslant w + w^{-1} \leqslant 2$.

Returning to the partition function (2), since the right-hand side of (5) is real and lies in $[-2, 2]$, it follows from the Lemma that the $2MN$ zeroes of (2) all lie on the unit circle $|\sinh 2K| = 1$, a result which can also be obtained by simply setting the argument of the logarithm in the bulk free energy (4) equal to zero. The usefulness of this simplified procedure has been pointed out by Stephenson and Couzens[4] for the Ising model on a torus. But since the zeroes are not easily determined in that case when the lattice is finite, they termed the argument as "hand-waving." Here, the argument is made rigorous by the use of the Brascamp–Kunz boundary condition. From here on, therefore, We shall adopt the simpler approach in all subsequent considerations.

We now proceed to determine the density of the zero distribution. Let the number of zeroes in the interval $[\alpha, \alpha + d\alpha]$ be $2MNg(\alpha)\, d\alpha$ such that

$$\int_0^{2\pi} g(\alpha)\, d\alpha = 1 \tag{8}$$

and

$$f = \tfrac{1}{2} \ln(4z) + \int_0^{2\pi} d\alpha\, g(\alpha) \ln(z - e^{i\alpha}) \tag{9}$$

It is more convenient to consider the function $R(\alpha) = \int_0^\alpha g(x)\, dx$ where $2MNR(\alpha)$ gives the total number of zeroes in the interval $[0, \alpha]$ such that

$$g(\alpha) = \frac{d}{d\alpha} R(\alpha) \tag{10}$$

On the circle $|z| = 1$ writing $z = e^{i\alpha}$ and setting the argument of the logarithm in the third line of (4) equal to zero, we find α determined by

$$\cos \alpha = \cos u \cos v, \qquad 0 \leqslant \{u, v\} \leqslant \pi \tag{11}$$

Now if α_i is a solution, so are $-\alpha_i$ and $\pi - \alpha_i$, hence we have the symmetry

$$g(\alpha) = g(-\alpha) = g(\pi - \alpha) \tag{12}$$

It is therefore sufficient to consider only $0 \leqslant \{\alpha, u, v\} \leqslant \pi/2$.

The constant-α contours of (11) are constructed in Fig. 1a and are seen to be symmetric about the lines $u, v = \pm\pi/2$ in each of the 4 quadrants. Now from (3) we see that zeroes are distributed uniformly in the $\{\theta, \phi\}$-, and hence the $\{u, v\}$-plane. It follows that $R(\alpha)$ is precisely the area of the region

$$\cos\alpha > \cos u \cos v, \qquad 0 \leqslant \{\alpha, u, v\} \leqslant \pi/2 \tag{13}$$

normalized to $R(\pi/2) = 1/4$. This leads to the expression

$$R(\alpha) = \frac{1}{\pi^2} \int_0^{\alpha} \cos^{-1}\left(\frac{\cos\alpha}{\cos x}\right) dx \tag{14}$$

Using (10) and after some reduction, we obtain the following explicit expression for the density of zeroes,

$$g(\alpha) = R'(\alpha) = \frac{|\sin\alpha|}{\pi^2} K(\sin\alpha) \tag{15}$$

where $K(k) = \int_0^{\pi/2} dt(1 - k^2 \sin^2 t)^{-1/2}$ is the complete elliptic integral of the first kind. The density (15), which possesses an unexpected logarithmic divergence at $\alpha = \pm\pi/2$, is plotted in Fig. 2a. For small α, we have $g(\alpha) \approx |\alpha|/2\pi$. As pointed out by Fisher,[3] it is this linear behavior at small α which leads to the logarithmic divergence of the specific heat.

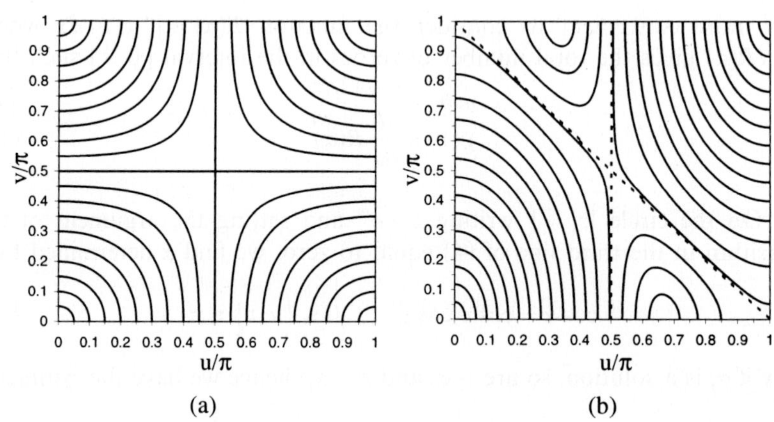

Fig. 1. Constant-α contours in the u-v plane. (a) The contour (11) for the simple-quartic lattice. Straight lines correspond to $\alpha = \pi/2$. (b) The contour (23) for the triangular lattice. Broken lines correspond to $\alpha = 2\cos^{-1}(1/3)$.

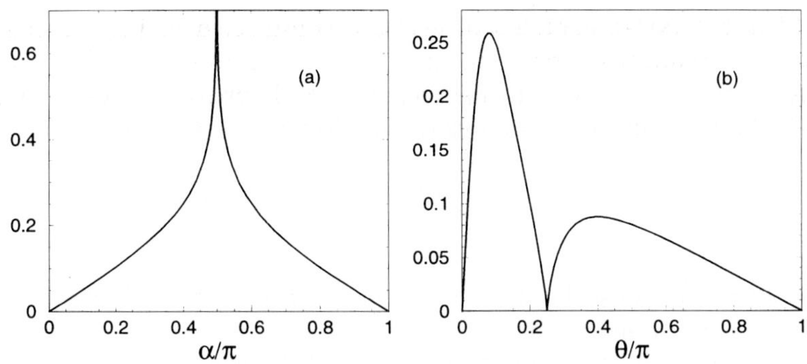

Fig. 2. Density of partition function zeroes for the simple-quartic lattice. (a) $g(\alpha)$ given by (15). (b) $g_+(\theta)$ given by (19).

We can also deduce the density of zeroes on the two Fisher circles (1) which we write as

$$\tanh K \pm 1 = \sqrt{2}\, e^{i\theta} \tag{16}$$

The angles α and θ are related by,

$$e^{i\alpha} = \pm\left(\frac{\sqrt{2} \mp e^{-i\theta}}{\sqrt{2} \mp e^{i\theta}}\right) \tag{17}$$

so that the mapping from α to θ is 1 to 2. This leads to the result

$$g(\theta) = \frac{g(\alpha)}{2}\left|\frac{d\alpha}{d\theta}\right| \tag{18}$$

Let the density of zeroes be $g_\pm(\theta)$ for the two circles (16). Then, using (17) we find

$$g_+(\theta) = g_-(\pi-\theta) = \left(\frac{k}{\pi^2}\right)\left|\frac{1-\sqrt{2}\cos\theta}{3-2\sqrt{2}\cos\theta}\right| K(k) \tag{19}$$

where

$$k = \frac{2\,|\sin\theta|\,(\sqrt{2}-\cos\theta)}{3-2\sqrt{2}\cos\theta} \tag{20}$$

The density (19) is plotted as Fig. 2b. Note that the divergence in the density distribution in (15) on the unit circle is removed in (19) for the two Fisher circles. This is due to the fact that $d\alpha/d\theta$ vanishes linearly at $\alpha = \pm\pi/2$. We have also $g_+(\pi/4) = g_-(3\pi/4) = 0$, and for small θ we find

$$g_\pm(\theta) = \left(\frac{3 \pm 2\sqrt{2}}{\pi}\right) |\theta| \tag{21}$$

Here, again, the linear behavior of $g_+(\theta)$ at $\theta = 0$ leads to the logarithmic singularity of the specific heat.

It is also of interest to consider zeroes of the Ising model in the Potts variable $x = (e^{2K} - 1)/\sqrt{2}$. In the complex x plane it is known[9] that the partition function zeroes are on two unit circles centered at $x = 1$ and $x = -\sqrt{2}$. We find the density along the two circles to be, respectively, $g_-(\theta)$ and $g_+(\theta)$.

3. THE TRIANGULAR LATTICE

For the triangular Ising model with nearest-neighbor interactions K, the free energy assumes the form[10, 11]

$$\begin{aligned} f &= C + \frac{1}{8\pi^2} \int_{-\pi}^{\pi} d\theta \int_{-\pi}^{\pi} d\phi \ln[z + z^{-1} + 1 - [\cos\theta + \cos\phi + \cos(\theta + \phi)]] \\ &= C + \frac{1}{2\pi^2} \int_0^{\pi} du \int_0^{\pi} dv \ln[z + z^{-1} + 2 - 2\cos u(\cos u + \cos v)] \end{aligned} \tag{22}$$

where $C = [\ln(4z)]/2$, $z = (e^{4K} - 1)/2$, and we have introduced variables $u = (\theta + \phi)/2$, $v = (\theta - \phi)/2$. Now the value of the sum of the three cosines in (22) lies between $-3/2$ and 3. It then follows from Corollary 1 that in the complex z plane the zeroes lie on the union of the unit circle $|z| = 1$ and the line segment $[-2, -1/2]$ of the real axis, a result first obtained by Stephenson and Couzens.[4]

The density of the zero distribution can now be computed in the same manner as described in the preceding section. For z on the unit circle we write $z = e^{i\alpha}$. Then α is determined by

$$\cos\alpha = -1 + \cos u(\cos u + \cos v), \qquad 0 \leqslant \{u, v\} \leqslant \pi \tag{23}$$

and $R(\alpha)$ is the area of the region

$$\cos\alpha > -1 + \cos u(\cos u + \cos v) \tag{24}$$

Clearly, we have the symmetry $g_{\rm cir}(\alpha) = g_{\rm cir}(\pi - \alpha)$ and we need only to consider $0 \leqslant \alpha \leqslant \pi$. From a consideration of the constant-α contours of (23) shown in Fig. 1b, we obtain after some algebra the result

$$g_{\rm cir}(\alpha) = \frac{|\sin \alpha|}{\pi^2 \sqrt{A(\alpha)}} K(k) \tag{25}$$

where $A(\alpha) = (5 + 4 \cos \alpha)^{1/2}$ and

$$\begin{aligned} k^2 &= F[A(\alpha)] \\ F(x) &\equiv \frac{1}{16}\left(\frac{3}{x} - 1\right)(1+x)^3 \end{aligned} \tag{26}$$

Particularly, for small α, we find $g_{\rm cir}(\alpha) \approx |\alpha|/2\sqrt{3}\,\pi$.

In a similar fashion we find, on the line segment $z \in [-2, -1/2]$, we write $z = -e^{\lambda}$ and obtain

$$g_{\rm line}(\lambda) = \frac{|\sinh \lambda|}{\pi^2 k \sqrt{B(\lambda)}} K(k^{-1}), \qquad -\ln 2 \leqslant \lambda \leqslant \ln 2 \tag{27}$$

where $B(\lambda) = [5 - 4 \cosh \lambda]^{1/2}$ and

$$k^2 = F[B(\lambda)] \tag{28}$$

While the density of zeroes is everywhere finite, the logarithmic divergence is recovered if the zeroes are all mapped onto a unit circle (see (38) below). Specifically, we have $g_{\rm cir}(\pi) = g_{\rm line}(0) = 0$, and $g_{\rm line}(\pm \ln 2) = \sqrt{3}/2\pi$. The densities (25) and (27) are plotted in Fig. 3.

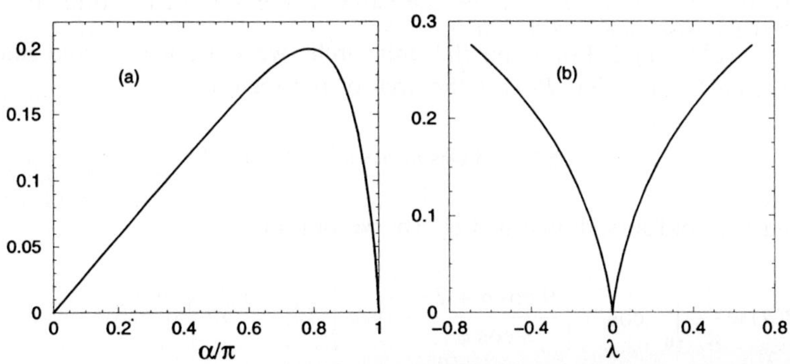

Fig. 3. Density of partition function zeroes for the triangular lattice. (a) $g_{\rm cir}(\alpha)$ given by (25). (b) $g_{\rm line}(z)$ given by (27).

Matveev and Shrock[18] have discussed zeroes of the triangular Ising model in the complex $u=e^{-4K}$ plane, for which the zeroes are distributed on the union of the circle

$$u=\tfrac{1}{3}(2e^{i\alpha}-1), \qquad -\pi<\alpha\leqslant\pi \tag{29}$$

and the line segment

$$-\infty<u\leqslant-\tfrac{1}{3} \tag{30}$$

Using our results we find the respective densities

$$g_{\text{cir}}(\alpha)=\frac{|\sin\phi|}{9\pi^2}\,[C(\alpha)]^{7/2}\,K(k) \tag{31}$$

where $C(\alpha)=3(5-4\cos\phi)^{-1/2}$, $k^2=F[C(\alpha)]$, and

$$g_{\text{line}}(u)=\left|\frac{(1+u)(1-3u)}{4\pi^2u^2(1-u)^2\,k\,\sqrt{D(u)}}\right|\,K(k^{-1}) \tag{32}$$

where $D(u)=\sqrt{(1+3u)/u(1-u)}$ and $k^2=F[D(u)]$. At the end point we have $g_{\text{line}}(-1/3)=9\sqrt{3}/8\pi$.

The density of zeroes assumes a simpler form if we use Corollary 2 to map all zeroes onto a unit circle in the complex w plane, where w is root of the quadratic equation

$$w+w^{-1}=\tfrac{4}{7}(z+z^{-1}+\tfrac{3}{4}) \tag{33}$$

and $z=(e^{4K}-1)/2$. For w on the unit circle, we write $w=e^{i\alpha}$ and find in analogous to (13) that $R(\alpha)$ is the area of the region

$$\cos\alpha>\tfrac{1}{9}[8\cos u(\cos u+\cos v)-7] \tag{34}$$

Using the contours shown in Fig. 1b, we obtain

$$\begin{aligned}R(\alpha)&=\frac{1}{\pi^2}\int_0^{\phi_0}\cos^{-1}\left[\frac{9\cos\alpha+7}{8\cos\phi}-\cos\phi\right]d\phi, && \alpha\in[0,\alpha_0]\\ &=\frac{1}{2}-\frac{1}{\pi^2}\int_{\phi_0}^{\phi_1}\cos^{-1}\left[\frac{9\cos\alpha+7}{8\cos\phi}-\cos\phi\right]d\phi, && \alpha\in[\alpha_0,\pi]\end{aligned} \tag{35}$$

where $\alpha_0 = 2\cos^{-1}(1/3)$ and

$$\begin{aligned} \phi_0 &= \cos^{-1}\left[\frac{3}{2}\cos\frac{\alpha}{2} - \frac{1}{2}\right] \\ \phi_1 &= \pi - \cos^{-1}\left[\frac{3}{2}\cos\frac{\alpha}{2} + \frac{1}{2}\right], \qquad \text{for} \quad \cos\frac{\alpha}{2} \leqslant \frac{1}{3} \end{aligned} \tag{36}$$

Note that we have $R(\alpha_0) = 3/8$, $R(\pi) = 1/2$.

Finally, using (10), we obtain

$$\begin{aligned} g(\alpha) &= \frac{9\sin\alpha}{8\pi^2} \int_0^{\phi_0} \frac{d\phi}{\sqrt{(\cos^2\phi - \cos^2\phi_0)(\Delta^2 - \cos^2\phi)}}, \qquad \alpha \in [0, \alpha_0] \\ &= \frac{9\sin\alpha}{8\pi^2} \int_{\phi_0}^{\phi_1} \frac{d\phi}{\sqrt{(\cos^2\phi - \cos^2\phi_0)(\cos^2\phi_1 - \cos^2\phi)}}, \qquad \alpha \in [\alpha_0, \pi] \end{aligned} \tag{37}$$

where $\Delta = [1 + 3\cos(\alpha/2)]/2$. After some manipulation and making use of integral identities (A1) and (A2) derived in the Appendix, we obtain

$$\begin{aligned} g(\alpha) &= \frac{3\sqrt{3}}{8\pi^2} |\sin\alpha| \sqrt{\sec\frac{\alpha}{2}}\, K(k), \qquad \alpha \in [0, \alpha_0] \\ &= \frac{3\sqrt{3}}{8\pi^2} \left|\frac{\sin\alpha}{k}\right| \sqrt{\sec\frac{\alpha}{2}}\, K(k^{-1}), \qquad \alpha \in [\alpha_0, \pi] \end{aligned} \tag{38}$$

where

$$k^2 = \frac{1}{16}\left(\sec\frac{\alpha}{2} - 1\right)\left(1 + 3\cos\frac{\alpha}{2}\right)^3 \tag{39}$$

Note that $g(\alpha)$ diverges logarithmically at $\alpha = \pm\alpha_0$.

4. SIMPLE-QUARTIC ISING MODEL IN A FIELD $i\pi/2$

The two-dimensional Ising model can be solved when there is an external magnetic field $i\pi/2$. The solution for the simple-quartic lattice was first given by Lee and Yang[(2)] and a rigorous derivation of which was given later by McCoy and Wu.[(12)] In 1988 Lin and Wu[(13)] gave a general prescription for writing down the solution of the Ising model in a field $i\pi/2$ by transcribing the solution in a zero field.

The most general known solution of the Ising model in a field $i\pi/2$ is a model with a generalized checkerboard type interactions.[14] Matveev and Shrock[15] have also studied the zeroes for the simple-quartic Ising model in a field $i\pi/2$.

For the simple-quartic lattice Lee and Yang[2] gave the free energy in a field $i\pi/2$ as

$$f = i\frac{\pi}{2} + C + \frac{1}{16\pi^2}\int_{-\pi}^{\pi} d\theta \int_{-\pi}^{\pi} d\phi \ln[z + z^{-1} + 2 - 4\cos\theta\cos\phi] \tag{40}$$

where $C = (\ln \sinh 2K)/2$, $z = e^{-4K}$. Setting the argument of the logarithm in (40) equal to zero we have $-6 \leqslant z + z^{-1} \leqslant 2$ and hence from Corollary 2 we see that in the complex z plane zeroes of the partition function lie on the unit circle $|z| = 1$ and the line segment $-3 - 2\sqrt{2} \leqslant z \leqslant -3 + 2\sqrt{2}$ of the real axis.

On the unit circle $|z| = 1$ we write $z = e^{i\alpha}$ and find the density

$$g_{\text{cir}}(\alpha) = \frac{|\sin\alpha|}{2\pi^2} K(k) \tag{41}$$

where

$$k^2 = (3 + \cos\alpha)(1 - \cos\alpha)/4 \tag{42}$$

On the line segment, we write $z = -e^{\lambda}$ with $-2\ln(1 + \sqrt{2}) \leqslant \lambda \leqslant 2\ln(1 + \sqrt{2})$, we find the density

$$g_{\text{line}}(\lambda) = \frac{|\sinh\lambda|}{2\pi^2} K(k) \tag{43}$$

where

$$k^2 = (3 - \cosh\lambda)(1 + \cosh\lambda)/4 \tag{44}$$

At the end points we have $g_{\text{line}}(\pm 2\ln(1 + \sqrt{2})) = 1/\sqrt{2}\,\pi$. The density functions (41) and (43) are plotted in Fig. 4.

5. TRIANGULAR ISING MODEL IN A FIELD $i\pi/2$

The solution for the triangular model in a field $i\pi/2$ was first obtained in ref. 13 by applying a transformation in conjunction with the solution of a staggered 8-vertex model. Here, for completeness, we present an alternate and more direct derivation.

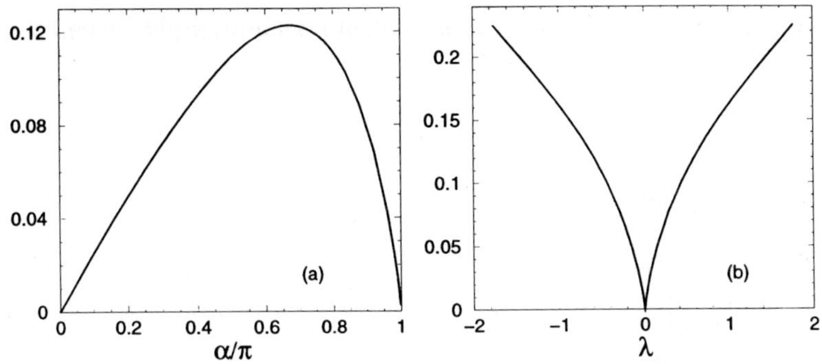

Fig. 4. Density of partition function zeroes for the simple-quartic lattice Ising model in a field $i\pi/2$. (a) $g_{\mathrm{cir}}(\alpha)$ given by (41). (b) $g_{\mathrm{line}}(z)$ given by (43).

Consider a triangular Ising lattice of N sites whose sites are arranged as shown in Fig. 5a. After making use of the identity $e^{i\pi\sigma/2} = i\sigma$, the partition function assumes the form

$$Z_N = i^N \sum_{\sigma_i = \pm 1} \prod_{\mathrm{nn}} e^{K\sigma_i\sigma_j} \prod_j \sigma_j \tag{45}$$

where the first product is over all nearest neighbors, and the second product over all sites. Now it is known that the triangular Ising model can be mapped into an 8-vertex model on the dual of the square lattice,[16] also of N sites. However, in order to properly treat the factor $\prod_j \sigma_j$ in (45), we need to divide the N "cells" of the lattice, where a cell is shown in Fig. 5b, into two sublattices, A and B, and associate two σ_j's to each cell belonging to one sublattice, say, B. This permits us to rewrite (45) as

$$Z_N = i^N \sum_{\sigma_i = \pm 1} \prod_{\mathrm{cells}} W_{\mathrm{stg}}(\sigma_1, \sigma_2, \sigma_3, \sigma_4) \tag{46}$$

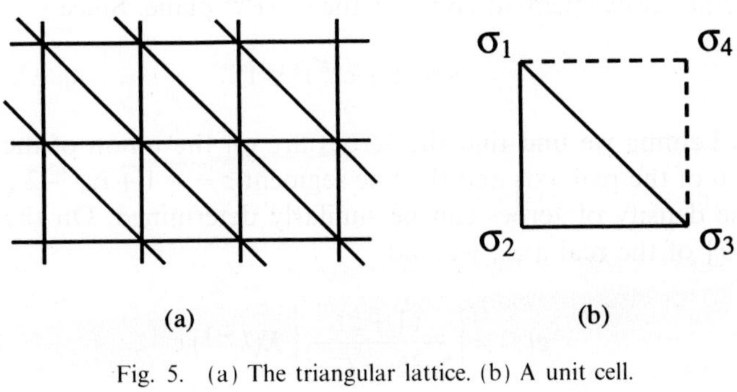

Fig. 5. (a) The triangular lattice. (b) A unit cell.

where $W_{\mathrm{stg}}(\sigma_1, \sigma_2, \sigma_3, \sigma_4)$ is a staggered Boltzmann weight given by

$$\begin{aligned} W_{\mathrm{stg}}(\sigma_1, \sigma_2, \sigma_3, \sigma_4) &= e^{K(\sigma_1\sigma_2+\sigma_2\sigma_3+\sigma_3\sigma_1)} && \text{for } A \\ &= (\sigma_1\sigma_2)\, e^{K(\sigma_1\sigma_2+\sigma_2\sigma_3+\sigma_3\sigma_1)} && \text{for } B \end{aligned} \tag{47}$$

The 8-vertex weights are

$$\begin{aligned} \{\omega_1,..., \omega_8\} &= \{e^{3K}, e^{-K}, e^{-K}, e^{-K}, e^{-K}, e^{-K}, e^{-K}, e^{3K}\} \\ \{\omega'_1,..., \omega'_8\} &= \{e^{3K}, -e^{-K}, -e^{-K}, e^{-K}, e^{-K}, -e^{-K}, e^{-K}, -e^{3K}\} \end{aligned} \tag{48}$$

Furthermore, from the mapping convention of Fig. 1 of ref. 17, we see that the mapping between the spin and 8-vertex configurations is 2 to 1. This leads to

$$Z_N = 2i^N Z_N(\{\omega\}, \{\omega'\}) \tag{49}$$

which is an exact equivalence between Z_N and the partition function $Z_N(\{\omega\}, \{\omega'\})$ of the staggered 8-vertex model.

Now the weights (48) satisfy the free-fermion condition[16] for which $Z_N(\{\omega\}, \{\omega'\})$ has already been evaluated.[17] Using Eq. (19) of ref. 17 and after some reduction, one obtains the following expression for the per-site free energy,

$$f = i\frac{\pi}{2} + C + \frac{1}{4\pi^2}\int_0^\pi d\theta \int_0^\pi d\phi\, \ln[(1+e^{4K})^2 + 4\cos\phi(\cos\theta + \cos\phi)] \tag{50}$$

where $C = [\ln(2\sinh 2K)]/2$. As a result, the partition function zeroes are located at

$$(1+e^{4K})^2 = -4\cos\phi(\cos\theta + \cos\phi), \qquad 0 \leqslant \{\theta, \phi\} \leqslant \pi \tag{51}$$

It is therefore convenient to consider the $z = e^{4K}$ plane. Since

$$-8 \leqslant (1+e^{4K})^2 \leqslant 1 \tag{52}$$

using the Lemma we find that the zeroes are on the union of the segment $-2 \leqslant z \leqslant 0$ of the real axis and the line segment $z = -1 + iy$, $-2\sqrt{2} \leqslant y \leqslant 2\sqrt{2}$. The density of zeroes can be similarly determined. On the segment $z \in [-2, 0]$ of the real axis, we find

$$g(z) = \left| \frac{(1+z)}{2\pi^2 k \sqrt{E(z)}} \right| K(k^{-1}) \tag{53}$$

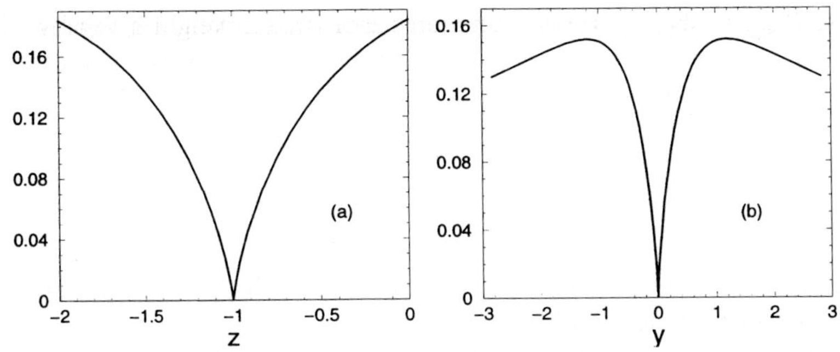

Fig. 6. Density of partition function zeroes for the triangular Ising model in a field $i\pi/2$. (a) $g(z)$ given by (53). (b) $g(y)$ given by (54).

where $E(z)=\sqrt{-z(2+z)}$ and $k^2=F[E(z)]$. Particularly, we have $g(0)=g(-2)=1/\sqrt{3}\,\pi$ and $g(-1)=0$. On the line segment $z=-1+iy$, we find

$$g(y)=\left|\frac{y}{2\pi^2\sqrt{H(y)}}\right| K(k) \tag{54}$$

where $H(y)=\sqrt{1+y^2}$ and $k^2=F[H(y)]$. Particularly, we have $g(0)=0$ and $g(\pm 2\sqrt{2})=1/\sqrt{6}\,\pi$. These results are plotted in Fig. 6.

We remark that in the complex $x=e^{-4K}$ plane considered in ref. 18, the segment $-2\leqslant z\leqslant 0$ of the real axis maps onto $-\infty\leqslant x\leqslant -1/2$ while the line segment $z=-1+iy$, $-2\sqrt{2}\leqslant y\leqslant 2\sqrt{2}$, is mapped onto the circular arc $x^{-1}=\frac{1}{2}(-1+e^{i\theta})$, $\theta_0=\tan^{-1}(4\sqrt{2/7})\leqslant|\theta|\leqslant\pi$. The density of zeroes on the arc is found to be

$$g_{\mathrm{arc}}(\theta)=\frac{(1+\cos\theta)^2}{2\pi^2\sqrt{I(\theta)}\,|\sin^3\theta|}\,K(k) \tag{55}$$

where $I(\theta)=[2(1+\cos\theta)]^{1/2}/|\sin\theta|$ and $k^2=F[B(i\theta)]$ with $B(i\theta)=[5-4\cos\theta]^{1/2}$. The densities at the end points of the arc are $g_{\mathrm{arc}}(\pm\theta_0)=3\sqrt{3}/2\sqrt{2}\,\pi$.

6. THE HONEYCOMB AND KAGOMÉ LATTICES

The partition function of an Ising model on a planar lattice with interactions K is proportional to the partition function on the dual lattice with interactions K^*,[19] where K and K^* are related by

$$e^{-2K^*}=\tanh K \tag{56}$$

Consequently, their partition function zeroes coincide when expressed in terms of appropriate variables. Now the honeycomb and triangular lattices are mutually dual, it follows that for the honeycomb lattice with interactions K, in the complex

$$z=\tfrac{1}{2}(e^{4K^*}-1)=(\cosh 2K-1)^{-1} \tag{57}$$

plane, zeroes of the partition function coincides with those of the triangular lattice partition function (22).

For the honeycomb Ising model in an external field $i\pi/2$, the free energy can be obtained from that in a zero field via a simple transformation.[13, 18] Writing the partition function in the form of (45) and replacing the product $\prod_i \sigma_i$ by $\prod_i \sigma_i^3$, it is clear that, besides the factor i^N, the partition function is the same as that in a zero field with the replacement

$$e^{K(\sigma_i\sigma_j-1)} \to (\sigma_i\sigma_j)\, e^{K(\sigma_i\sigma_j-1)} \tag{58}$$

or, equivalently, $e^{2K} \to -e^{2K}$. It follows that in the complex

$$z=(-\cosh 2K-1)^{-1} \tag{59}$$

plane, the zeroes coincide with those of the triangular lattice partition function (22).

The Ising model on the kagomé lattice with interactions K can be mapped to that on an honeycomb lattice with interactions J, by applying a star-triangle transformation followed by a spin decimation. The procedure, which is standard[20] and will not be repeated here, leads to the relation

$$e^{2J}=(e^{4K}+1)/2 \tag{60}$$

As a result, we conclude that, in the complex

$$z=(\cosh 2J-1)^{-1}=2(1-\tanh 2K)/\tanh^2 2K \tag{61}$$

plane, zeroes of the kagomé partition function coincides with those of the triangular lattice partition function (22). The evaluation of the kagomé partition function in an external field $i\pi/2$ remains unresolved, however.

APPENDIX. TWO INTEGRATION IDENTITIES

In this Appendix we derive the integration identities

$$I_1 = \int_0^{\pi/2} \frac{dt}{\sqrt{(1-a^2\sin^2 t)(b^2+a^2\sin^2 t)}} = \frac{1}{\sqrt{a^2+b^2}} K\left(a\sqrt{\frac{1+b^2}{a^2+b^2}}\right) \tag{A1}$$

$$I_2 = \int_a^b \frac{dx}{\sqrt{(1-x^2)(x^2-a^2)(b^2-x^2)}} = \frac{1}{b\sqrt{1-a^2}} K\left(\frac{1}{b}\sqrt{\frac{b^2-a^2}{1-a^2}}\right) \tag{A2}$$

which do not appear to have previously been given.

To obtain (A1), we expand the integrand using the binomial expansion

$$(1-x)^{-\alpha} = \sum_{k=0}^{\infty} \frac{(\alpha)_k}{k!} x^k \tag{A3}$$

where $(\alpha)_k = \alpha(\alpha+1)\cdots(\alpha+k-1) = \Gamma(\alpha+k)/\Gamma(\alpha)$, and carry out the integration term by term using the formula

$$\frac{2}{\pi}\int_0^{\pi/2} \sin^{2m} t\, dt = \frac{(1/2)_m}{m!} \tag{A4}$$

This yields

$$\begin{aligned} I_1 &= \frac{\pi}{2b} \sum_{j,k=0}^{\infty} \frac{(1/2)_j (1/2)_k (1/2)_{j+k}}{j!\,k!\,(j+k)!} a^{2j}\left(-\frac{a^2}{b^2}\right)^k \\ &\equiv \frac{\pi}{2b} F_1\left(\frac{1}{2};\frac{1}{2},\frac{1}{2};1;a^2,-\frac{a^2}{b^2}\right) \end{aligned} \tag{A5}$$

where the hypergeometric function of two variables is (cf. 9.180.1 of ref. 21)

$$F_1(\alpha;\beta,\beta';\gamma;x,y) = \sum_{j=0}^{\infty}\sum_{k=0}^{\infty} \frac{(\alpha)_{j+k}(\beta)_j(\beta')_k}{j!\,k!\,(\gamma)_{j+k}} x^j y^k \tag{A6}$$

This leads to the integration formula (A1) after making use of the identity (cf. 9.182.1 of ref. 21)

$$F_1(\alpha;\beta,\beta';\beta+\beta';x,y) = (1-y)^{-\alpha} F\left(\alpha,\beta;\beta+\beta';\frac{x-y}{1-y}\right) \tag{A7}$$

where F is the hypergeometric function (cf. 9.100 of ref. 21)

$$F(\alpha,\beta;\gamma;z) = \sum_{j=0}^{\infty} \frac{(\alpha)_j(\beta)_j}{j!\,(\gamma)_j} z^j \tag{A8}$$

and the identity (cf. 8.113.1 of ref. 21)

$$K(k) = \frac{\pi}{2} F\left(\frac{1}{2}, \frac{1}{2}; 1; k^2\right) \tag{A9}$$

The integral (A2) is obtained by introducing the change of variable $x^2 = (b^2 - a^2) \sin^2 t + a^2$, which yields

$$I_2 = \frac{1}{1-a^2} \int_0^{\pi/2} \frac{dt}{\sqrt{(1 - c^2 \sin^2 t)[a^2/(1-a^2) + c^2 \sin^2 t)}} \tag{A10}$$

where $c^2 = (b^2 - a^2)/(1 - a^2)$. The integral I_2 is now of the form of I_1 and (A2) is obtained after applying (A1).

ACKNOWLEDGMENTS

Work has been supported in part by NSF grants DMR-9614170 and DMR-9980440.

REFERENCES

1. C. N. Yang and T. D. Lee, Statistical theory of equations of states and phase transitions I. Theory of condensation, *Phys. Rev.* **87**:404–409 (1952).
2. T. D. Lee and C. N. Yang, Statistical theory of equations of state and phase transitions. II. Lattice gas and Ising model, *Phys. Rev.* **87**:410–419 (1952).
3. M. E. Fisher, The Nature of critical points, *in Lecture Notes in Theoretical Physics*, Vol. 7c, W. E. Brittin, ed. (University of Colorado Press, Boulder, 1965), pp. 1–159.
4. J. Stephenson and R. Couzens, Partition function zeros for the two-dimensional Ising model, *Physica A* **129**:201–210 (1984).
5. W. T. Lu and F. Y. Wu, Partition function zeroes of a self-dual Ising model, *Physica A* **258**:157–170 (1998).
6. J. Stephenson, Partition function zeros for the two-dimensional Ising model. II, *Physica A* **136**:147–159 (1986).
7. J. Stephenson, On the density of partition function zeros, *J. Phys. A* **20**:4513–4519 (1987).
8. H. J. Brascamp and H. Kunz, Zeroes of the partition function for the Ising model in the complex temperature plane, *J. Math. Phys.* **15**:65–66 (1974).
9. C. N. Chen, C. K. Hu, and F. Y. Wu, Partition function zeros of the square lattice Potts model, *Phys. Rev. Lett.* **76**:169–172 (1996).
10. R. M. F. Houtappel, Order–disorder in hexagonal lattice, *Physica* **16**:425–455 (1950).
11. G. H. Wannier, Antiferromagnetism: The triangular Ising Net, *Phys. Rev.* **79**:357–364 (1950).
12. V. Matveev and R. Shrock, Complex-temperature properties of the 2D Ising model for nonzero magnetic field, *Phys. Rev. E* **53**:254–267 (1996).
13. B. M. McCoy and T. T. Wu, Theory of Toeplitz determinants and the spin correlations of the two-dimensional Ising model. II, *Phys. Rev.* **155**:438–452 (1967).
14. K. Y. Lin and F. Y. Wu, Ising model in the magnetic field $i\pi kT/2$, *Int. J. Mod. Phys. B* **4**:471–481 (1988).

15. F. Y. Wu, Two-dimensional Ising model with crossing and four-spin interactions and a magnetic field $i\pi kT/2$, *J. Stat. Phys.* **44**:455–463 (1986).
16. V. Matveev and R. Shrock, Complex temperature properties of the 2*D* Ising model with $\beta H = i\pi/2$, *J. Phys. A* **28**:4859–4882 (1995).
17. C. Fan and F. Y. Wu, General lattice model of phase transitions, *Phys. Rev. B* **2**:723–733 (1970).
18. C. S. Hsue, K. Y. Lin, and F. Y. Wu, Staggered eight-vertex model, *Phys. Rev. B* **12**:429–437 (1975).
19. F. Y. Wu and Y. K. Wang, Duality transformation in a many component spin model, *J. Math. Phys.* **17**:439–440 (1976).
20. I. Syozi, Transformation of Ising Models, in *Phase Transition and Critical Phenomena*, Vol. 1, C. Domb and M. Green, eds. (Academic Press, London, 1970), pp. 269–329.
21. I. S. Gradshteyn and I. M. Ryzhik, *Table of Integrals, Series and Products*, 5th ed. (Academic Press, New York, 1994).

PHYSICAL REVIEW E, VOLUME 63, 026107

Ising model on nonorientable surfaces: Exact solution for the Möbius strip and the Klein bottle

Wentao T. Lu and F. Y. Wu
Department of Physics, Northeastern University, Boston, Massachusetts 02115

(Received 20 July 2000; published 22 January 2001)

Closed-form expressions are obtained for the partition function of the Ising model on an $\mathcal{M}\times\mathcal{N}$ simple-quartic lattice embedded on a Möbius strip and a Klein bottle. The solutions all lead to the same bulk free energy, but for finite $\mathcal{M}$ and $\mathcal{N}$ the expressions are different depending on whether the strip width $\mathcal{M}$ is odd or even. Finite-size corrections at criticality are analyzed and compared with those under cylindrical and toroidal boundary conditions. Our results are consistent with the conformal field prediction of a central charge $c=1/2$, provided that the twisted Möbius boundary condition is regarded as a free or fixed boundary.

DOI: 10.1103/PhysRevE.63.026107 PACS number(s): 05.50.+q, 05.20.−y, 04.20.Jb, 02.10.Ab

I. INTRODUCTION

There has been considerable recent interest [1–3] in studying lattice models on nonorientable surfaces, both as new challenging unsolved lattice-statistical problems and as a realization and testing of predictions of the conformal field theory [4]. In a recent paper [1] we presented the solution of dimers on the Möbius strip and Klein bottle, and studied its finite-size corrections. In this paper we consider the Ising model.

The Ising model in two dimensions was first solved in 1944 by Onsager [5], who obtained a closed-form expression of the partition function for a simple-quartic $\mathcal{M}\times\mathcal{N}$ lattice wrapped on a cylinder. The exact solution for an $\mathcal{M}\times\mathcal{N}$ lattice on a torus, namely, with periodic boundary conditions in both directions, was obtained by Kaufman four years later [6]. Onsager and Kaufman used spinor analysis to derive the solutions, and the solution under the cylindrical boundary condition was rederived later by McCoy and Wu [7] using the method of dimers. More recently, we obtained the solution for a finite Ising lattice with a self-dual boundary condition [8]. As far as we know, these are the only known solutions of the two-dimensional Ising model on finite lattices with conventional boundary conditions.

Here, using the method of dimers, we derive exact expressions for the partition function of the Ising model on finite Möbius strips and Klein bottles. As we shall see, as a consequence of the Möbius topology, the solution assumes a form which depends on whether the width of the Möbius strip is even or odd. However, all solutions yield the same bulk free energy. We also present results of finite-size analyses for corrections to the bulk solution, and compare with those deduced under other boundary conditions. Our explicit calculations confirm that the central charge is $c=1/2$, provided that the twisted Möbius boundary condition is regarded as a free or fixed boundary.

The organization of this paper is as follows. The partition function for a $2M\times N$ Möbius strip is evaluated in Sec. II, with details given in Sec. III. The partition function for a $(2M-1)\times N$ Möbius strip is evaluated in Sec. IV, and results for the Klein bottle are given in Sec. V. In Sec. VI we carry out a finite-size analysis for large lattices, and results at criticality are given for $2M\times N$ lattices.

II. $2M\times N$ MÖBIUS STRIP

To begin with, we consider a $2M\times N$ simple-quartic Ising lattice $\mathcal{L}$ embedded on a Möbius strip, where M and N are integers, $2M$ is the width, and N is the length of the strip, which can be either even or odd. The Ising model has anisotropic reduced interactions K_h along the (horizontal) length direction, and K_v along the (vertical) width direction. The example of a 4×5 Möbius strip $\mathcal{L}$ is shown in Fig. 1.

To facilitate considerations, it is convenient to let the row of N vertical edges located in the middle of the strip take on a different interaction K_1, as shown. The desired result is then obtained by setting $K_1=K_v$. In addition, by setting $K_1=0$ the Möbius strip reduces to an $M\times 2N$ strip with a "cylindrical" boundary condition, namely, periodic in one direction and free in the other, for which the partition function was evaluated by McCoy and Wu [7]. By setting $K_1=\infty$ the two center rows of spins coalesce into a single row with an (additive) interaction, which in this case is $2K_h$. These are two key elements of our consideration.

Following standard procedures [7] we write the partition function of the Ising model on $\mathcal{L}$ as

$$Z^{\mathrm{Mob}}_{2M,N}(K_h,K_v,K_1)=2^{2MN}(\cosh K_h)^{2MN}(\cosh K_v)^{2(M-1)N}\times(\cosh K_1)^N G(z_h,z_v,z_1), \tag{1}$$

where $z_h=\tanh K_h$, $z_v=\tanh K_v$, and $z_1=\tanh K_1$, and

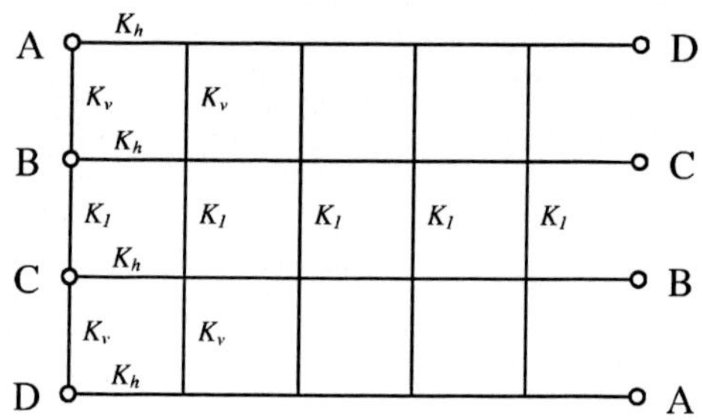

FIG. 1. A 4×5 Möbius strip $\mathcal{L}$. Vertices labeled A,B,C, and D are repeated sites.

1063-651X/2001/63(2)/026107(9)/$15.00

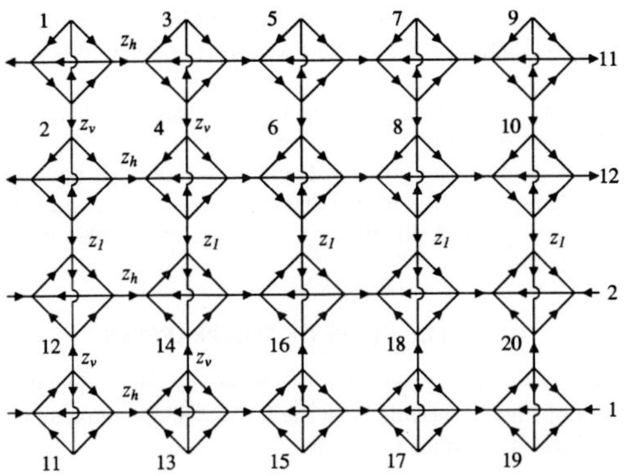

FIG. 2. The dimer lattice $\mathcal{L}_D$ corresponding to the 4×5 Möbius strip. The cities are numbered from 1 to 20, and the six edges within each city all carry the weight 1.

$$G(z_h,z_v,z_1)=\sum_{\text{closed polygons}} z_h^{n_h} z_v^{n_h} z_1^{n_1} \tag{2}$$

is the generating function of all closed polygonal graphs on $\mathcal{L}$ with edge weights z_h, z_v, and z_1. Here, n_h, n_v, and n_1 are the numbers of polygonal edges with weights z_h, z_v, and z_1, respectively.

The generating function $G(z_h,z_v,z_1)$ is a multinomial in z_h, z_v, and z_1 and, due to the Möbius topology, the integer n_1 can take on any value in $\{0,N\}$. Thus we have

$$G(z_h,z_v,z_1)=\sum_{n_1=0}^{N} T_{n_1}(z_h,z_v)z_1^{n_1}, \tag{3}$$

where $T_{n_1}(z_h,z_v)$ are polynomials in z_h and z_v with strictly positive coefficients.

To evaluate $G(z_h,z_v,z_1)$, we again follow the usual procedure of mapping polygonal configurations on $\mathcal{L}$ onto dimer configurations on a dimer lattice $\mathcal{L}_D$ of $8MN$ sites, constructed by expanding each site of $\mathcal{L}$ into a "city" of four sites [7–9]. The resulting $\mathcal{L}_D$ for the 4×5 $\mathcal{L}$ is shown in Fig. 2.

Since the deletion of all z_1 edges reduces the lattice to one with a cylindrical boundary condition solved in Ref. [7], we orient all edges with weights z_h, z_v, and 1 as in Ref. [7]. In addition, all z_1 edges are oriented in the direction shown in Fig. 2. Then we have the following theorem:

Theorem: Let A be the $8MN\times 8MN$ antisymmetric determinant defined by the lattice edge orientation shown in Fig. 2, and let

$$\mathrm{Pf}A(z_h,z_v,z_1)=\sqrt{\det A(z_h,z_v,z_1)} \tag{4}$$

denote the Pfaffian of A. Then

$$\mathrm{Pf}A(z_h,z_v,z_1)=\sum_{n_1=0}^{N} \epsilon_{n_1} T_{n_1}(z_h,z_v)z_1^{n_1}, \tag{5}$$

where $\epsilon_{4m}=\epsilon_{4m+1}=1$ and $\epsilon_{4m+2}=\epsilon_{4m+3}=-1$ for any integer $m\geqslant 0$.

Remark: Define

$$X_p=\sum_{m=0}^{[N/4]} T_{4m+p}(z_h,z_v)z_1^{4m+p}, \quad p=0,1,2, \text{ and } 3, \tag{6}$$

where $[N/4]$ is the integral part of $N/4$, so that $G(z_h,z_v,z_1)=X_0+X_1+X_2+X_3$. It then follows from Eq. (5) that we have

$$\mathrm{Pf}A(z_h,z_v,\pm iz_1)=X_0+X_2\pm i(X_1+X_3). \tag{7}$$

As a consequence, we obtain

$$G(z_h,z_v,z_1)=\tfrac{1}{2}[(1-i)\mathrm{Pf}A(z_h,z_v,iz_1) +(1+i)\mathrm{Pf}A(z_h,z_v,-iz_1)], \tag{8}$$

where, as evaluated in the next section, the Pfaffian is given by

$$\mathrm{Pf}A(z_h,z_v,z_1) =[z_v(1-z_h^2)]^{MN} \times\prod_{n=1}^{N}\left[\frac{\sinh(M+1)t(\phi_n)-c(z_1)\sinh Mt(\phi_n)}{\sinh t(\phi_n)}\right], \tag{9}$$

with

$$c(z_1)=\frac{z_v^2(1+z_h^2+2z_h\cos\phi_n)+2(-1)^n z_1 z_h \sin\phi_n}{z_v(1-z_h^2)},$$

$$\cosh t(\phi)=\frac{\cosh 2K_h\cosh 2K_v-\sinh 2K_h\cos\phi}{\sinh 2K_v}, \tag{10}$$

$$\phi_n=(2n-1)\pi/2N.$$

Here we have used the fact that $\Pi_{n=1}^{N}=\Pi_{n=N+1}^{2N}$ in the product in Eq. (9). Substituting these results into Eq. (1), and setting $K_1=K_v$, we are led to the following explicit expression for the partition function:

$$Z_{2M,N}^{\mathrm{Mob}}(K_h,K_v,K_v)=\tfrac{1}{2}(2\sinh 2K_v)^{MN}(\cosh K_v)^{-N} \times[(1-i)F_+ +(1+i)F_-], \tag{11}$$

where

$$F_\pm=\prod_{n=1}^{N}\left[e^{Mt(\phi_n)}\left(\frac{e^{t(\phi_n)}-c(\pm iz_v)}{2\sinh t(\phi_n)}\right) -e^{-Mt(\phi_n)}\left(\frac{e^{-t(\phi_n)}-c(\pm iz_v)}{2\sinh t(\phi_n)}\right)\right]. \tag{12}$$

This completes the evaluation of the Ising partition function for the $2M\times N$ Möbius strip. For example, for a 2×5 Möbius strip, this leads to

$$\begin{aligned}\mathrm{Pf}\,A(z_h,z_v,z_1)&=1+z_h^{10}+10z_1z_h^5-5z_1^2z_h^2(1+z_h^2+z_h^4+z_h^6)\\&\quad-20z_1^3z_h^5+5z_1^4z_h^4(1+z_h^2)+2z_1^5z_h^5,\\G(z_h,z_v,z_1)&=1+z_h^{10}+10z_1z_h^5+5z_1^2z_h^2(1+z_h^2+z_h^4+z_h^6)\\&\quad+20z_1^3z_h^5+5z_1^4z_h^4(1+z_h^2)+2z_1^5z_h^5,\end{aligned}\tag{13}$$

which can be verified by explicit enumerations.

Note that we have $\cosh t(\phi_n)\geq 1$, so we can always take $t(\phi_n)\geq 0$. For large M, the leading contribution in Eq. (12) is therefore

$$F_\pm \sim \prod_{n=1}^{N} e^{Mt(\phi_n)},\tag{14}$$

and, hence, from Eq. (11),

$$\frac{1}{2MN}\ln Z_{2M,N}^{\mathrm{Mob}}(K_h,K_v,K_v)\sim\frac{1}{2}\ln(2\sinh 2K_v)+\frac{1}{2N}\sum_{n=1}^{N} t(\phi_n).\tag{15}$$

We now prove the theorem.

Considered as a multinomial in z_h, z_v, and z_1, there exists a one-to-one correspondence between terms in the dimer generating function $G(z_h,z_v,z_1)$ and (linear combinations of) terms in the Pfaffian [Eq. (4)]. However, while all terms in $G(z_h,z_v,z_1)$ are positive, terms in the Pfaffian do not necessarily possess the same sign. The crux of the matter is to find an appropriate linear combination of Pfaffians to yield the desired $G(z_h,z_v,z_1)$. For this purpose it is convenient to compare an arbitrary term C_1 in the Pfaffian with a standard one C_0. We choose C_0 to be a term in which no z_h, z_v, and z_1 dimers are present.

The superposition of two dimer configurations represented by C_0 and C_1 produces superposition polygons. Kasteleyn [10] showed that the two terms will have the same sign if edges of $\mathcal{L}_D$ can be oriented such that all superposition polygons are oriented "clockwise-odd," namely, that there be an odd number of edges oriented in the clockwise direction.

Now since all z_h, z_v, and 1 edges of $\mathcal{L}_D$ are oriented as in Ref. [7], terms in the Pfaffian with no z_1 edges ($n_1=0$) will have the same sign as C_0. To determine the sign of a term when z_1 edges are present, we associate a + sign with each clockwise-odd superposition polygon, and a − sign to each clockwise-even superposition polygon. Then the sign of C_1 relative to C_0 is the product of the signs of all superposition polygons. The following elementary facts can be readily verified.

(i) Deformations of the borders of a superposition polygon always change m_1, the number of *its* z_1 edges, by multiples of 2.

(ii) The sign of a superposition polygon is reversed under border deformations which change m_1 by 2.

(iii) Superposition polygons having 0 or 1 z_1 edges have a sign +.

(iv) There can be at most one superposition polygon having an odd number of z_1 edges, a property unique to nonorientable surfaces.

Let $m_1=4m+p$, where m is an integer and $p=0,1,2$, and 3. Because of point (iv), we need only to consider the presence of at most one polygon having $p=1$ or 3. It now follows from points (i) and (iii) that $\epsilon_{4m}=\epsilon_{4m+1}=+$, and from points (i), (ii), and (iii) that $\epsilon_{4m+2}=\epsilon_{4m+3}=-$. This establishes the theorem.

III. EVALUATION OF THE PFAFFIAN

We now derive expression (9). From the edge orientation of $\mathcal{L}_D$ of Fig. 2, one finds that the $8MN\times 8MN$ antisymmetric matrix A assumes the form

$$\begin{aligned}A(z_h,z_v,z_1)&=A_0(z_v)\otimes I_{2N}+A_+(z_h)\otimes J_{2N}+A_-(z_h)\otimes J_{2N}^T\\&\quad+A_1(z_1)\otimes H_{2N},\end{aligned}\tag{16}$$

where A_0, A_+, A_-, and A_1 are $4M\times 4M$ matrices, I_{2N} is the $2N\times 2N$ identity matrix, and J_{2N} and H_{2N} are $2N\times 2N$ matrices:

$$J_{2N}=\begin{pmatrix}0&1&0&\cdots&0\\0&0&1&\cdots&0\\\vdots&\vdots&\vdots&\ddots&\vdots\\0&0&0&\cdots&1\\-1&0&0&\cdots&0\end{pmatrix},\quad H_{2N}=\begin{pmatrix}0&I_N\\-I_N&0\end{pmatrix}.\tag{17}$$

In addition, one has

$$\begin{aligned}A_0(z_v)&=a_{0,0}\otimes I_M+a_{0,1}(z_v)\otimes F_M+a_{0,-1}(z_v)\otimes F_M^T,\\A_\pm(z_h)&=a_{\pm 1,0}(z_h)\otimes I_M,\\A_1(z_1)&=a(z_1)\otimes G_M,\end{aligned}\tag{18}$$

where F_M and G_M are $M\times M$ matrices:

$$F_M=\begin{pmatrix}0&1&0&\cdots&0\\0&0&1&\cdots&0\\\vdots&\vdots&\vdots&\ddots&\vdots\\0&0&0&\cdots&1\\0&0&0&\cdots&0\end{pmatrix},$$

$$G_M=\begin{pmatrix}0&0&\cdots&0&0\\0&0&\cdots&0&0\\\vdots&\vdots&\ddots&\vdots&\vdots\\0&0&\cdots&0&0\\0&0&\cdots&0&1\end{pmatrix},\tag{19}$$

F_M^T is the transpose of F_M, and

$$a_{0,0}=\begin{pmatrix} 0 & 1 & -1 & -1 \\ -1 & 0 & 1 & -1 \\ 1 & -1 & 0 & 1 \\ 1 & 1 & -1 & 0 \end{pmatrix},\quad a(z)=\begin{pmatrix} 0 & 0 & 0 & 0 \\ 0 & 0 & 0 & 0 \\ 0 & 0 & z & 0 \\ 0 & 0 & 0 & 0 \end{pmatrix},$$

$$a_{1,0}(z)=\begin{pmatrix} 0 & z & 0 & 0 \\ 0 & 0 & 0 & 0 \\ 0 & 0 & 0 & 0 \\ 0 & 0 & 0 & 0 \end{pmatrix},\quad a_{0,1}(z)=\begin{pmatrix} 0 & 0 & 0 & 0 \\ 0 & 0 & 0 & 0 \\ 0 & 0 & 0 & z \\ 0 & 0 & 0 & 0 \end{pmatrix},\tag{20}$$

$$a_{-1,0}(z)=-a^T_{1,0}(z),\quad a_{0,-1}(z)=-a^T_{0,1}(z).$$

We use the fact that the determinant in Eq. (4) is equal to the product of the eigenvalues of matrix A. To evaluate the latter, we note that J_{2N}, J^T_{2N}, and H_{2N} mutually commute, so that they can be diagonalized simultaneously. This leads to the respective eigenvalues $e^{i\phi_n}, e^{-i\phi_n}$ and $i(-1)^{n+1}$ and the expression

$$\det A(z_h,z_v,z_1)=\prod_{n=1}^{2N}\det A_M(z_h,z_v,z_1;\phi_n),\tag{21}$$

where

$$A_M(z_h,z_v,z_1;\phi_n)=A_0(z_v)+A_+(z_h)e^{i\phi_n}+A_-(z_h)e^{-i\phi_n}+i(-1)^{n+1}A_1(z_1)\tag{22}$$

is a $4M\times 4M$ matrix. Writing this out explicitly, we have

$$A_M(z_h,z_v,z_1;\phi_n)=\begin{pmatrix} B(z_h) & a_{0,1}(z_v) & 0 & & \\ a_{0,-1}(z_v) & B(z_h) & a_{0,1}(z_v) & 0 & \\ \vdots & \vdots & \ddots & \vdots & \vdots \\ & 0 & a_{0,-1}(z_v) & B(z_h) & a_{0,1}(z_v) \\ & & 0 & a_{0,-1}(z_v) & C(z_h,z_1) \end{pmatrix},\tag{23}$$

where $C(z,z_1)=B(z)+i(-1)^{n+1}a(z_1)$, and

$$B(z)=\begin{pmatrix} 0 & 1+ze^{i\phi_n} & -1 & -1 \\ -(1+ze^{-i\phi_n}) & 0 & 1 & -1 \\ 1 & -1 & 0 & 1 \\ 1 & 1 & -1 & 0 \end{pmatrix}.\tag{24}$$

The evaluation of $\det A_M(z_h,z_v,z_1;\phi_n)$ can be carried out by using a recursive procedure introduced in Ref. [8]. Specifically, let $B_M=B_M(\phi_n)=\det A_M(z_h,z_v,z_1;\phi_n)$, and D_M be the determinant of the matrix $A_M(z_h,z_v,z_1;\phi_n)$ with the fourth row and fourth column removed. Then we have

$$\mathrm{Pf}\,A(z_h,z_v,z_1)=\sqrt{\det A(z_h,z_v,z_1)}=\prod_{n=1}^{2N}\sqrt{B_M(\phi_n)}.\tag{25}$$

Furthermore, by expanding the determinants one finds the recursion relation (which is the same as that in Ref. [8] when $z_h=z_v=z$),

$$\begin{pmatrix} B_M \\ D_M \end{pmatrix}=\begin{pmatrix} a_{11} & a_{12} \\ a_{21} & a_{22} \end{pmatrix}\begin{pmatrix} B_{M-1} \\ D_{M-1} \end{pmatrix},\quad M\geq 2,\tag{26}$$

with

$$a_{11}=1+z_h^2-2z_h\cos\phi_n,$$

$$a_{12}=-2iz_v^2z_h\sin\phi_n,\tag{27}$$

$$a_{21}=2iz_h\sin\phi_n,$$

$$a_{22}=z_v^2(1+z_h^2+2z_h\cos\phi_n),$$

and the initial conditions (which are different from Ref. [8])

$$B_1\equiv B_1(z_h,z_1)=1+z_h^2-2z_h\cos\phi_n-2(-1)^n z_1z_h\sin\phi_n,$$

$$D_1\equiv D_1(z_h,z_1)=2iz_h\sin\phi_n-i(-1)^n z_1(1+z_h^2+2z_h\cos\phi_n).\tag{28}$$

This leads to the solutions

$$B_M(\phi_n)=B_1\frac{\lambda_+^M-\lambda_-^M}{\lambda_+-\lambda_-}-(a_{22}B_1-a_{12}D_1)\frac{\lambda_+^{M-1}-\lambda_-^{M-1}}{\lambda_+-\lambda_-},$$

$$D_M(\phi_n)=D_1\frac{\lambda_+^M-\lambda_-^M}{\lambda_+-\lambda_-}-(a_{11}D_1-a_{21}B_1)\frac{\lambda_+^{M-1}-\lambda_-^{M-1}}{\lambda_+-\lambda_-},\tag{29}$$

where $\lambda_\pm=z_v(1-z_h^2)e^{\pm t(\phi_n)}$ are the eigenvalues of the 2×2 matrix in Eq. (26). After some algebraic manipulation, from Eqs. (25) and (29) we obtain the expression [Eq. (9)] quoted in Sec. II.

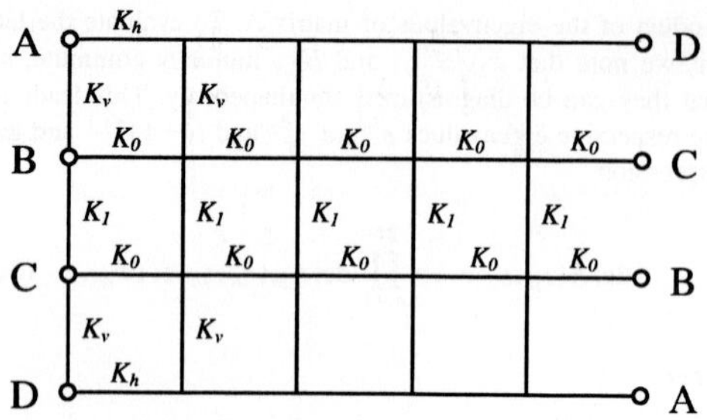

FIG. 3. Labelings of a 4×5 Möbius strip which reduces to a 3×5 Möbius strip upon taking $K_1=\infty$.

IV. $(2M-1)\times N$ MÖBIUS STRIP

The $(2M-1)\times N$ Möbius strip is considered in this section. In order to make use of results of the preceding sections, we start from the $2M\times N$ Möbius strip of Sec. II, and let spins in the two center rows of the strip [the Mth and $(M+1)$th rows] having interactions $K_0=K_h/2$. The example of a 4×5 lattice with these interactions is shown in Fig. 3. Then, by taking $K_1=\infty$ $(z_1=1)$, as described in Sec. I, this lattice reduces to the desired $(2M-1)\times N$ Möbius strip of uniform interactions K_h and K_v.

Following this procedure, we have

$$Z^{\mathrm{Mob}}_{2M-1,N}(K_h,K_v)=2^{(2M-1)N}(\cosh K_h\cosh K_v)^{2(M-1)N}\times\cosh^{2N}(K_h/2)G(z_h,z_v,z_0,z_1)|_{z_1=1},\quad(30)$$

where $z_0=\tanh(K_h/2)$, and $G(z_h,z_v,z_0,z_1)$ is the generating function of closed polygons on the $2M\times N$ Möbius net with edge weights as described above.

The generating function $G(z_h,z_v,z_0,z_1)$ can be evaluated as in the previous sections. In place of Eq. (25), we now have

$$\mathrm{Pf}A(z_h,z_v,z_0,z_1)=\prod_{n=1}^{2N}\sqrt{\det A_M(z_h,z_v,z_0,z_1;\phi_n)},\quad(31)$$

where

$$A_M(z_h,z_v,z_0,z_1;\phi_n)=\begin{pmatrix} B(z_h) & a_{0,1}(z_v) & 0 & & \\ a_{0,-1}(z_v) & B(z_h) & a_{0,1}(z_v) & 0 & \\ \vdots & \vdots & \ddots & \vdots & \vdots \\ & 0 & a_{0,-1}(z_v) & B(z_h) & a_{0,1}(z_v) \\ & & 0 & a_{0,-1}(z_v) & C(z_0,z_1) \end{pmatrix}.\quad(32)$$

Then Eq. (8) becomes

$$G(z_h,z_v,z_0,z_1)=\tfrac{1}{2}[(1-i)\mathrm{Pf}A(z_h,z_v,z_0,iz_1)+(1+i)\mathrm{Pf}A(z_h,z_v,z_0,-iz_1)].\quad(33)$$

The evaluation of $\det A_M(z_h,z_v,z_0,z_1;\phi_n)$ can again be done recursively. As before, define $B_M=\det A_M(z_h,z_v,z_0,z_1;\phi_n)$, and let D_M be the determinant of A_M with the fourth row and column removed; one again obtains recursion relations (26), and arrives at precisely the same solution [Eq. (29)], but now with different initial conditions

$$B_1=B_1(z_0,z_1),\quad D_1=D_1(z_0,z_1),\quad(34)$$

where the functions B_1 and D_1 are as defined in Eq. (28). After some algebra, this leads to the solution

$$\mathrm{Pf}A(z_h,z_v,z_0,z_1)=[z_v(1-z_h^2)]^{(M-1)N}\times\prod_{n=1}^{N}\left[\frac{c_1\sinh Mt(\phi_n)-c_2\sinh(M-1)t(\phi_n)}{\sinh t(\phi_n)}\right],\quad(35)$$

where

$$c_1=\frac{2z_0}{z_h}\{1-z_h[\cos\phi_n+(-1)^n z_1\sin\phi_n]\},$$

$$c_2=2z_0z_v\left\{\frac{1}{z_h}+[\cos\phi_n+(-1)^n z_1\sin\phi_n]\right\}.\quad(36)$$

The substitution of Eq. (35) into Eqs. (33) and (30) now completes the evaluation of the partition function for a $(2M-1)\times N$ Möbius strip.

V. KLEIN BOTTLE

The Ising model on a Klein bottle can be considered similarly. We first consider a $2M\times N$ lattice $\mathcal{L}$, constructed by connecting the upper and lower edges of the Möbius strip of Fig. 1 in a periodic fashion with N extra vertical edges. As in the case of the Möbius strip, it is convenient to let the extra edges have interactions K_2. The desired solution is obtained at the end by setting $K_1=K_2=K_v$.

The Ising partition function for the Klein bottle now assumes the form

$$Z^{\text{Kln}}_{2M,N}(K_h,K_v,K_1,K_2) =2^{2MN}(\cosh K_h)^{2MN}(\cosh K_v)^{2(M-1)N} \times(\cosh K_1\cosh K_2)^N G^{\text{Kln}}(z_h,z_v,z_1,z_2), \tag{37}$$

where

$$G^{\text{Kln}}(z_h,z_v,z_1,z_2)=\sum_{\text{closed polygons}} z_h^{n_h}z_v^{n_v}z_1^{n_1}z_2^{n_2} \tag{38}$$

generates all closed polygons on the $2M\times N$ lattice $\mathcal{L}$ with edge weights $z_i=\tanh K_i$, $i=h,v,1$, and 2. The desired partition function is then given by

$$Z^{\text{Kln}}_{2M,N}(K_h,K_v,K_v,K_v) =2^{2MN}(\cosh K_h\cosh K_v)^{2MN}G^{\text{Kln}}(z_h,z_v,z_v,z_v). \tag{39}$$

Again, it is convenient to first write $G^{\text{Kln}}(z_h,z_v,z_1,z_2)$ as a multinomial in z_h,z_v,z_1, and z_2 in the form of

$$G^{\text{Kln}}(z_h,z_v,z_1,z_2)=\sum_{m,n=0}^{N} T_{m,n}(z_h,z_v)z_1^m z_2^n, \tag{40}$$

where $T_{m,n}(z_h,z_v)$ are polynomials in z_h and z_v with strictly positive coefficients.

The evaluation of $G^{\text{Kln}}(z_h,z_v,z_1,z_2)$ parallels that of $G(z_h,z_v,z_1)$ for the Möbius strip. One first maps the lattice $\mathcal{L}$ into a dimer lattice $\mathcal{L}_D$ by expanding each site into a city of four sites, as shown in Fig. 2. Orient all k_h, k_v, and k_1 edges of $\mathcal{L}_D$ as shown, and orient all k_2 edges in the same (downward) direction as the k_1 edges. Then this defines an $8MN\times 8MN$ antisymmetric matrix obtained by adding an extra term to $A(z_h,z_v,z_1)$ given by Eq. (16), namely,

$$A^{\text{Kln}}(z_h,z_v,z_1,z_2)=A(z_h,z_v,z_1)+b(z_2)\otimes G'_M\otimes H_{2N}. \tag{41}$$

Here

$$b(z_2)=\begin{pmatrix} 0&0&0&0\\ 0&0&0&0\\ 0&0&0&0\\ 0&0&0&-z_2\end{pmatrix},$$

$$G'_M=\begin{pmatrix} 1&0&\cdots&0&0\\ 0&0&\cdots&0&0\\ \vdots&\vdots&\ddots&\vdots&\vdots\\ 0&0&\cdots&0&0\\ 0&0&\cdots&0&0\end{pmatrix}. \tag{42}$$

Then, in place of theorem (5), we now have

$$\text{Pf}\,A^{\text{Kln}}(z_h,z_v,z_1,z_2)=\sum_{m,n=0}^{N}\epsilon_m\epsilon_n T_{m,n}(z_h,z_v)z_1^m z_2^n, \tag{43}$$

from which, in a similar manner, one obtains the result

$$\begin{aligned}G^{\text{Kln}}(z_h,z_v,z_1,z_2)=\tfrac{1}{2}[&\text{Pf}\,A^{\text{Kln}}(z_h,z_v,iz_1,-iz_2)\\ &+\text{Pf}\,A^{\text{Kln}}(z_h,z_v,-iz_1,iz_2)\\ &-i\,\text{Pf}\,A^{\text{Kln}}(z_h,z_v,iz_1,iz_2)\\ &+i\,\text{Pf}\,A^{\text{Kln}}(z_h,z_v,-iz_1,-iz_2)].\end{aligned} \tag{44}$$

To evaluate the Pfaffian (43), we note that matrix (41) can again be diagonalized in the $\{2N\}$ subspace, yielding

$$\text{Pf}\,A^{\text{Kln}}(z_h,z_v,z_1,z_2)=\prod_{n=1}^{2N}\sqrt{\det A^{\text{Kln}}_M(z_h,z_v,z_1,z_2;\phi_n)}, \tag{45}$$

where

$$A^{\text{Kln}}_M(z_h,z_v,z_1,z_2;\phi_n)=A_M(z_h,z_v,z_1;\phi_n) +i(-1)^{n+1}b(z_2)\otimes G'_M. \tag{46}$$

Now we expand $\det A^{\text{Kln}}_M$ in z_2. Since, upon setting $z_2=0$, the determinant is precisely B_M and the term linear in z_2, the $\{4,4\}$ element of the determinant, is by definition D_M, one obtains

$$\det A^{\text{Kln}}_M(z_h,z_v,z_1,z_2;\phi_n)=B_M+i(-1)^n z_2 D_M,\quad M\geq 2, \tag{47}$$

where B_M and D_M were already computed in Eq. (29). This leads to

$$\text{Pf}\,A^{\text{Kln}}(z_h,z_v,z_1,z_2) =\left(1+\frac{z_1z_2}{z_v^2}\right)^N[z_v(1-z_h^2)]^{MN}\times \prod_{n=1}^{N}\left[\frac{\sinh(M+1)t-c(z_1,z_2)\sinh Mt}{\sinh t}\right], \tag{48}$$

where

$$c(z_1,z_2)=\frac{1}{z_v(1-z_h^2)(z_v^2+z_1z_2)}[(1+z_h^2)(z_v^4+z_1z_2)+2z_h(z_v^4-z_1z_2)\cos\phi_n+2(-1)^n\times(z_1+z_2)z_hz_v^2\sin\phi_n]. \tag{49}$$

Setting $z_1=z_2=z_v$ in Eq. (44) and using Eq. (48), after some algebra one obtains

$$G^{\mathrm{Kln}}(z_h,z_v,z_v,z_v)=[z_v(1-z_h^2)]^{MN}\left[\prod_{n=1}^{N}2\cosh Mt(\phi_n)+\mathrm{Im}\prod_{n=1}^{N}\left(\frac{\sinh Mt(\phi_n)}{\sinh t(\phi_n)}D(\phi_n)\right)\right], \tag{50}$$

where

$$D(\phi_n)=\frac{1}{z_v(1-z_h^2)}[(1+z_h^2)(1-z_v^2)-2z_h(1+z_v^2)\cos\phi_n-4i(-1)^nz_hz_v\sin\phi_n], \tag{51}$$

and Im denotes the imaginary part. The substitution of Eq. (50) into Eq. (39) now completes the evaluation of the partition function for a $2M\times N$ Klein bottle.

For a 2×2 Klein bottle, for example, one finds

$$\mathrm{Pf}A^{\mathrm{Kln}}(z_h,z_v,z_1,z_2)=1+z_h^4+4(z_1+z_2)z_h^2-2(z_1^2+z_2^2)z_h^2+2z_1z_2(1+z_h^2)^2-4z_1z_2(z_1+z_2)z_h^2+z_1^2z_2^2(1+z_h^4),$$

$$G^{\mathrm{Kln}}(z_h,z_v,z_1,z_2)=1+z_h^4+4(z_1+z_2)z_h^2+2(z_1^2+z_2^2)z_h^2+2z_1z_2(1+z_h^2)^2+4z_1z_2(z_1+z_2)z_h^2+z_1^2z_2^2(1+z_h^4), \tag{52}$$

which can be verified by explicit enumerations.

For a $(2M-1)\times N$ Klein bottle we can proceed as above by first considering a $2M\times N$ Klein bottle with interactions K_h,K_v,K_1, and K_2 and, within the center two rows, interactions $K_0=K_h/2$, as shown in Fig. 3. This is followed by taking $K_1\to\infty$ and $K_2=K_v$. Thus, in place of Eq. (39), we have

$$Z^{\mathrm{Kln}}_{2M-1,N}(K_h,K_v)=2^{(2M-1)N}(\cosh K_h)^{2MN}\times(\cosh K_v)^{(2M-3)N}\cosh^{2N}(K_h/2)\times G^{\mathrm{Kln}}(z_h,z_v,z_0,1,z_v), \tag{53}$$

where $z_0=\tanh(K_h/2)$, and $G^{\mathrm{Kln}}(z_h,z_v,z_0,z_1,z_2)$ generates polygonal configurations on the $2M\times N$ lattice with weights as shown. Then, as in the above, we find

$$G^{\mathrm{Kln}}(z_h,z_v,z_0,z_1,z_2)=\tfrac{1}{2}[\mathrm{Pf}A^{\mathrm{Kln}}(z_h,z_v,z_0,iz_1,-iz_2)+\mathrm{Pf}A^{\mathrm{Kln}}(z_h,z_v,z_0,-iz_1,iz_2)-i\,\mathrm{Pf}A^{\mathrm{Kln}}(z_h,z_v,z_0,iz_1,iz_2)+i\,\mathrm{Pf}A^{\mathrm{Kln}}(z_h,z_v,z_0,-iz_1,-iz_2)], \tag{54}$$

where $\mathrm{Pf}A^{\mathrm{Kln}}(z_h,z_v,z_0,z_1,z_2)$ is found to be given by the right-hand side of Eq. (35), but now with

$$c_1=(1+z_0^2)(1-z_1z_2)-2z_0(1+z_1z_2)\times\cos\phi_n-2(-1)^n(z_1+z_2)z_0\sin\phi_n,$$

$$c_2=\frac{1}{z_v(1-z_h^2)}\{(z_v^2+z_1z_2)[(1-z_hz_0)^2+(z_h-z_0)^2]+2(z_h-z_0)(1-z_hz_0)[(z_v^2-z_1z_2)\cos\phi_n+(-1)^n(z_v^2z_1+z_2)\sin\phi_n]\}, \tag{55}$$

expressions which are valid for arbitrary z_h,z_v,z_0,z_1, and z_2. For $z_0=\tanh(K_h/2)$, the case we are considering, Eq. (55) reduces to

$$c_1=\frac{2z_0}{z_h}[1-z_1z_2-z_h(1+z_1z_2)\cos\phi_n-(-1)^nz_h(z_1+z_2)\sin\phi_n],$$

$$c_2=\frac{2z_0}{z_hz_v}[z_v^2+z_1z_2+z_h(z_v^2-z_1z_2)\cos\phi_n+(-1)^nz_h(z_v^2z_1+z_2)\sin\phi_n] \tag{56}$$

[which reduces further to Eq. (36) after setting $z_2=0$]. The explicit expression for the partition function is now obtained by substituting Eq. (54) into Eq. (53).

VI. BULK LIMIT AND FINITE-SIZE CORRECTIONS

In the thermodynamic limit, our solutions of the Ising partition function give rise to a bulk "free energy"

$$f_{\mathrm{bulk}}(K_h,K_v)=\lim_{M,N\to\infty}\frac{1}{2MN}\ln Z(K_h,K_v). \tag{57}$$

Here, $Z(K_h,K_v)$ is any one of the four partition functions. For example, using the solution $Z^{\mathrm{Mob}}_{2M,N}$ given by Eq. (15) for the $2M\times N$ Möbius strip, one obtains

$$f_{\mathrm{bulk}}(K_h,K_v)=\frac{1}{2}\ln(2\sinh 2K_v)+\frac{1}{2\pi}\int_0^{\pi}d\phi\,t(\phi), \tag{58}$$

where $t(\phi)$ is given by Eq. (10). This leads to the Onsager solution

$$f_{\text{bulk}}(K_h,K_v)=\ln 2+\frac{1}{2\pi^2}\int_0^\pi d\phi\int_0^\pi d\theta \times\ln[\cosh 2K_h\cosh 2K_v-\sinh 2K_h\cos\theta -\sinh 2K_v\cos\phi)]. \tag{59}$$

Steps leading from Eq. (58) to Eq. (59) can be found, for example, in Ref. [11]. The bulk free energy $f_{\text{bulk}}(K_h,K_v)$ is nonanalytic at the critical point $\sinh 2K_h \sinh 2K_v=1$.

For large M and N, one can use the Euler-MacLaurin summation formula to evaluate corrections to the bulk free energy. For the purpose of comparing with the conformal field predictions [4], it is of particular interest to analyze corrections at the critical point. We have carried out such an analysis for $2M\times N$ lattices with isotropic interactions $K_h=K_v=K$. In this case the critical point is $\sinh 2K_c=1$ or, equivalently, $2K_c=\ln(\sqrt{2}+1)$ at which we expect to have the expansion

$$\ln Z_{2M,N}(K_c)=2MNf_{\text{bulk}}(K_c)+Nc_1(\xi,K_c)+2Mc_2(\xi,K_c)+c_3(\xi,K_c)+\cdots, \tag{60}$$

where $\xi=N/2M$ is the aspect ratio of the lattice.

The evaluation of terms in Eq. (60) was first carried out by Ferdinand and Fisher [12] for toroidal boundary conditions. Following Ref. [12], as well as similar analyses for dimer systems [1,13], we have evaluated Eq. (60) for other boundary conditions. For the $2M\times N$ Möbius strip, for example, one starts with an explicit expression [Eq. (11)] for the partition function, and uses the Euler-MacLaurin formula to evaluate the summations. The analysis is lengthy, even at the critical point. We shall give details elsewhere [14], and quote only the results, here

$$c_1(\xi,K_c)\equiv c_1^{\text{Mob}}=I-K_c=-0.087\,618\ldots,$$

$$c_2(\xi,K_c)=0, \tag{61}$$

$$c_3(\xi,K_c)=-\frac{1}{2}\ln 2+\frac{1}{12}\ln\left[\frac{2\vartheta_3^2(0|i\xi)}{\vartheta_2(0|i\xi)\vartheta_4(0|i\xi)}\right]+\frac{1}{2}\ln\left[1+\frac{\vartheta_3(0|i\xi/2)-\vartheta_4(0|i\xi/2)}{2\vartheta_3(0|i\xi)}\right]$$

where

$$I=\frac{1}{2\pi}\int_0^\pi \ln(\sqrt{2}\sin\phi+\sqrt{1+\sin^2\phi})d\phi=0.353\,068\ldots,$$

and $\vartheta_i(u|\tau), i=2,3,$ and 4, are the Jacobi theta functions [15]

$$\vartheta_2(u|\tau)=2\sum_{n=1}^{\infty} q^{[n-(1/2)]^2}\cos(2n-1)u,$$

$$\vartheta_3(u|\tau)=1+2\sum_{n=1}^{\infty} q^{n^2}\cos 2nu, \tag{62}$$

$$\vartheta_4(u|\tau)=1+2\sum_{n=1}^{\infty} (-1)^n q^{n^2}\cos 2nu,$$

with $q=e^{i\pi\tau}$. For the $2M\times N$ Klein bottle, we find, similarly,

$$c_1(\xi,K_c)=0,$$

$$c_2(\xi,K_c)=0, \tag{63}$$

$$c_3(\xi,K_c)=\frac{1}{6}\ln\left[\frac{2\vartheta_3^2(0|2i\xi)}{\vartheta_2(0|2i\xi)\vartheta_4(0|2i\xi)}\right]+\ln\left[1+\sqrt{\frac{\vartheta_2(0|2i\xi)}{2\vartheta_3(0|2i\xi)}}\right].$$

If one takes the limit of $N\to\infty$ $(M\to\infty)$ first in Eq. (60), while keeping M (N) finite, one obtains

$$\lim_{N\to\infty}\frac{1}{N}\ln Z_{2M,N}(K_c)=2Mf_{\text{bulk}}(K_c)+c_1+\Delta_1/2M+O(1/M^2),$$

$$\lim_{M\to\infty}\frac{1}{2M}\ln Z_{2M,N}(K_c)=Nf_{\text{bulk}}(K_c)+c_2+\Delta_2/N+O(1/N^2), \tag{64}$$

where c_1, c_2, Δ_1, and Δ_2 are constants. Results of our findings are listed in Table I above. Also listed are values for toroidal boundary conditions taken from Ref. [12], and values for the cylindrical boundary conditions computed by us using the solution given in Ref. [7].

TABLE I. Results of our findings for different configurations.

	Cylindrical	Toroidal	Möbius	Klein
c_1	c_1^{Mob}	0	c_1^{Mob}	0
c_2	0	0	0	0
Δ_1	$\pi/48$	$\pi/12$	$\pi/48$	$\pi/12$
Δ_2	$\pi/12$	$\pi/12$	$\pi/48$	$\pi/48$

For a lattice strip of infinite length and finite width, whose free energy is of the form of Eq. (64), the conformal field theory [4] predicts $\{\Delta_1,\Delta_2\}=\pi c/24$ (or $\pi c/6$), where c is the central charge, when the boundary condition in the finite width direction is free or fixed (or periodic). Thus numbers in the first two columns of Table I give rise to the value of

$$c=1/2. \tag{65}$$

Likewise, numbers in the last two columns also yield $c=1/2$, provided that the (twisted) Möbius boundary condition is regarded as a free boundary.

VII. SUMMARY

We have solved and obtained closed-form expressions for the partition function of an Ising model on finite Möbius strips and Klein bottles. The solution assumes different forms depending on whether the width of the lattice is even or odd. For a $2M\times N$ Möbius strip, where $2M$ is its width, the partition function $Z_{2M,N}^{\rm Mob}$ is given by Eq. (11), with $F_\pm$ given by Eq. (12). For a $(2M-1)\times N$ Möbius strip, we employ a trick by first considering a $2M\times N$ lattice and then "fusing" it into the desired lattice by coalescing two rows of spins. The resulting partition function $Z_{2M-1,N}^{\rm Mob}$ is given by Eq. (30), in which the generating function $G(z_h,z_v,z_0,z_1)$ is Eq. (33) with the Pfaffians given by Eq. (35).

For a $2M\times N$ Klein bottle the partition function $Z_{2M,N}^{\rm K\,ln}$ is given by Eq. (39) in which the generating function $G^{\rm K\,ln}(z_h,z_v,z_v,z_v)$ is given by Eq. (50). For a $(2M-1)\times N$ Klein bottle the partition function $Z_{2M-1,N}^{\rm K\,ln}$ is given by Eq. (53) in which the generating function $G^{\rm K\,ln}(z_h,z_v,z_0,1,z_v)$, $z_0=\tanh(K_h/2)$, is computed using Eq. (54). All solutions yield the same Onsager bulk free energy [Eq. (59)].

We have also carried out finite-size analyses of all solutions including that of the Ising model under cylindrical boundary conditions at criticality. The analyses yield a central charge $c=1/2$, in agreement with the conformal field prediction [4], provided that the (twisted) Möbius boundary condition is regarded as a free or fixed boundary.

ACKNOWLEDGMENT

This work was supported in part by National Science Foundation Grant No. DMR-9980440.

[1] W.T. Lu and F.Y. Wu, Phys. Lett. A **259**, 108 (1999).

[2] W.-J. Tzeng and F.Y. Wu, Appl. Math. Lett. **13**, 19 (2000).

[3] G. Tesler, J. Comb. Theory, Ser. B **78**, 198 (2000).

[4] H.W.J. Blöte, J.L. Cardy, and M.P. Nightingale, Phys. Rev. Lett. **56**, 742 (1986).

[5] L. Onsager, Phys. Rev. **65**, 117 (1944).

[6] B. Kaufman, Phys. Rev. **76**, 1232 (1949).

[7] B.M. McCoy and T.T. Wu, *The Two-dimensional Ising Model* (Harvard University Press, Cambridge, MA, 1973).

[8] W.T. Lu and F.Y. Wu, Physica A **258**, 157 (1998).

[9] P.W. Kasteleyn, J. Math. Phys. **4**, 287 (1963).

[10] P.W. Kasteleyn, Physica (Amsterdam) **27**, 1209 (1961).

[11] See, for example, K. Huang, *Statistical Mechanics* (Wiley, New York, 1987), p. 387.

[12] A.S. Ferdinand and M.E. Fisher, Phys. Rev. **185**, 832 (1969).

[13] A.E. Ferdinand, J. Math. Phys. **8**, 2332 (1967).

[14] W.T. Lu, Ph.D thesis, Northeastern University, 2001.

[15] I. S. Gradshteyn and I. M. Ryzhik, *Table of Integrals, Series, and Products* (Academic Press, New York, 1994), 8.180.

5. The Potts Model

J. Phys. A: Math. Gen., Vol. 9, No. 3, 1976. Printed in Great Britain. © 1976

Equivalence of the Potts model or Whitney polynomial with an ice-type model

R J Baxter†, S B Kelland† and F Y Wu‡§

†Research School of Physical Sciences, The Australian National University, Canberra, ACT 2600, Australia

‡Department of Physics, Northeastern University, Boston, Massachusetts 02115, USA

Received 3 November 1975

Abstract. The partition function of the Potts model on any lattice can readily be written as a Whitney polynomial. Temperley and Lieb have used operator methods to show that, for a square lattice, this problem is in turn equivalent to a staggered ice-type model. Here we rederive this equivalence by a graphical method, which we believe to be simpler, and which applies to any planar lattice. For instance, we also show that the Potts model on the triangular or honeycomb lattice is equivalent to an ice-type model on a Kagomé lattice.

1. Introduction

There is at present considerable interest amongst statistical mechanics and combinatorialists in the evaluation of the Whitney polynomial of a graph. This is because this problem is the same as obtaining the partition function of the Potts (1952) model on the graph (Kasteleyn and Fortuin 1969, Fortuin and Kasteleyn 1972, Baxter 1973), while in addition it contains the percolation and colouring problem as special cases.

Some exact results are available for the square lattice graph. In particular, when the associated Potts model has two states per spin, it becomes the Ising model and the problem is soluble. Also, Temperley and Lieb (1971) have established a remarkable equivalence between the Whitney polynomial for a square lattice $\mathscr{L}$ and the partition function of a staggered ice-type model on a related square lattice $\mathscr{L}'$.

Although the polynomial has not yet been evaluated exactly for the square lattice, it is tempting to think that it may be. The problem has therefore attracted attention amongst theoreticians, to the extent that we feel it worthwhile presenting a re-derivation of the equivalence established by Temperley and Lieb. We use graphical methods which we believe to be simpler than the operator method of Temperley and Lieb. Further, they apply to any planar lattice, regular or not.

2. Potts model and Whitney polynomial

First we define the q-state Potts model. There is more than one model by this name (Potts 1952, Domb 1974); the one we use here is the 'scalar' rather than the 'vector' model.

§ Supported in part by National Science Foundation Grant No. DMR 72-03213 A01.

Let $\mathscr{L}$ be any lattice and with each site i associate a spin σ_i with values $1, 2, \ldots, q$. Let nearest-neighbour spins have interaction energy $-\epsilon$ if they are alike, zero if they are different. Then the partition function is

$$Z=\sum \exp\left(\beta\epsilon \sum_{(i,j)} \delta(\sigma_i, \sigma_j)\right) \tag{1}$$

where the summation inside the exponential is over all nearest-neighbour pairs (i, j) on the lattice. The summation outside is over all states of the spins. If there are N sites, then there are q^N spin states of the system.

Set

$$v=e^{\beta\epsilon}-1, \tag{2}$$

then

$$Z=\sum \prod_{(i,j)} (1+v\delta(\sigma_i, \sigma_j)). \tag{3}$$

Let E be the number of edges of the lattice. Then the summand in (3) is a product of E factors, and expanding the product gives 2^E terms.

To each term we associate a bond-graph in $\mathscr{L}$ by placing bonds on edges where we have taken the corresponding $v\delta(\sigma_i, \sigma_j)$ term in the expansion. If we take the unit term, we leave the corresponding edge empty.

This gives a one-to-one correspondence between terms in the expansion of the summand of (3), and graphs on $\mathscr{L}$.

Consider a typical graph G, containing l bonds and C connected components (regarding an isolated site as a component). Then the corresponding term in (3) contains a factor v^l, and the effect of the delta functions is that all sites within a component must have the same spin σ. Summing over all independent spins therefore gives

$$Z=\sum_G q^C v^l, \tag{4}$$

where the summation is over all the 2^E graphs G that can be drawn on $\mathscr{L}$. The expression (4) is a Whitney (1932) polynomial.

It is easy to see that (4) contains the percolation and colouring problems as special cases. In particular,

$$\left(\frac{\partial}{\partial q}\ln Z\right)_{q=1}$$

is the mean number of components of the percolation problem. Also, if $\epsilon=-\infty$ and $v=-1$, then the spins (or colours) of adjacent sites must be different, and Z becomes the q-colouring polynomial of the lattice.

The edges of regular lattices can be grouped naturally into certain classes. For instance the square lattice has edges which are either horizontal or vertical. It is then natural and convenient to generalize (1)–(4) so as to allow different values of the interaction energy $-\epsilon$, according to which class the corresponding edge belongs. If ϵ_r is the value of ϵ for edges of class r, and

$$v_r=\exp(\beta\epsilon_r)-1, \tag{5}$$

then the required generalization of (4) is easily seen to be:

$$Z = \sum_G q^C v_1^{l_1} v_2^{l_2} v_3^{l_3} \ldots \tag{6}$$

where the summation is over all graphs G, C is the number of connected components in G, and l_r is the number of bonds on edges of class r ($r = 1, 2, 3, \ldots$).

3. Planar lattices: the surrounding lattice $\mathscr{L}'$

The remarks of § 2 apply to any lattice $\mathscr{L}$, whatever its structure or dimensionality. From now on we specialize to $\mathscr{L}$ being a planar lattice. It does not have to be regular, but can be any finite set of points (sites) and straight edges linking pairs of points. Points which are linked by an edge are said to be 'neighbours' or 'adjacent'. Planar means that no two edges cross.

We associate with $\mathscr{L}$ another planar lattice $\mathscr{L}'$, as follows.

Draw simple polygons surrounding each site of $\mathscr{L}$ such that:

(i) no polygons overlap, and no polygon surrounds another;
(ii) polygons of non-adjacent sites have no common corner;
(iii) polygons of adjacent sites i and j have one and only one common corner. This corner lies on the edge (i, j).

We take the corners of these polygons to be the sites of $\mathscr{L}'$, and the edges to be the edges of $\mathscr{L}'$. Hereinafter we call these polygons the 'basic polygons' of $\mathscr{L}'$.

We see that there are two types of sites of $\mathscr{L}'$. Firstly, those common to two basic polygons. These lie on edges of $\mathscr{L}$ and have four neighbours in $\mathscr{L}'$. We term these 'internal' sites. Secondly, there can be sites lying on only one basic polygon. These have two neighbours and we term them 'external' sites. (The reason for this terminology will become apparent when we explicitly consider the regular lattices.)

The above rules do not determine $\mathscr{L}'$ uniquely, in that its shape can be altered, and external sites can be added on any edge. However, the topology of the linkages between internal sites is invariant, and the general argument of the following sections applies to any allowed choice of $\mathscr{L}'$. (For the regular lattices there is an obvious natural choice.) In figure 1 we show an irregular lattice $\mathscr{L}$ and its surrounding graph $\mathscr{L}'$.

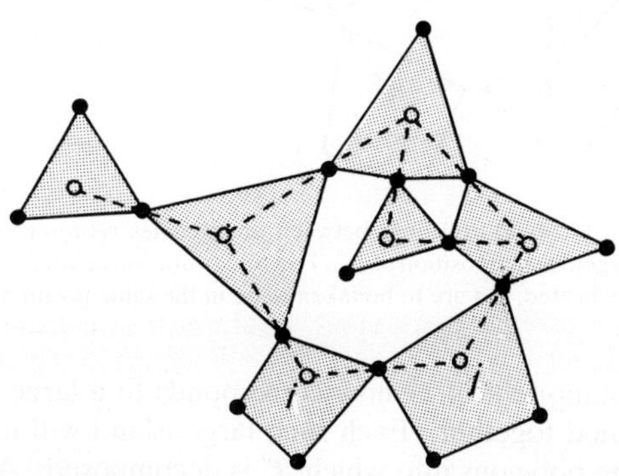

Figure 1. An irregular lattice $\mathscr{L}$ (open circles and broken lines) and its surrounding lattice $\mathscr{L}'$ (full circles and lines). The interior of each basic polygon is shaded, denoting 'land'.

It is helpful to shade the interior of each basic polygon, as in figure 1, and to regard such shaded areas as 'land', unshaded areas as 'water'. Then $\mathscr{L}'$ consists of a number of 'islands'. Each island contains a site of $\mathscr{L}$. Islands touch on edges of $\mathscr{L}$, at internal sites of $\mathscr{L}'$.

4. Polygon decompositions of $\mathscr{L}'$

We now make a one-to-one correspondence between graphs G on $\mathscr{L}$ and decompositions of $\mathscr{L}'$ as follows.

If G does not contain a bond on an edge (i, j), then at the corresponding internal site of $\mathscr{L}'$ separate two edges from the other two so as to separate the islands i and j, as in figure 2(a). If G contains a bond, separate the edges so as to join the islands, as in figure 2(b). Do this for all edges of $\mathscr{L}$.

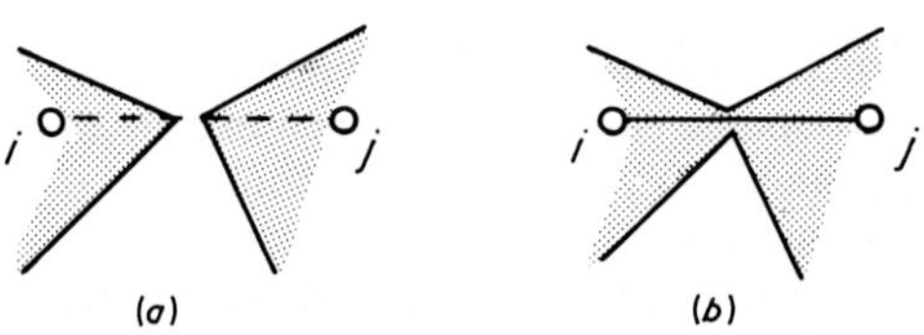

Figure 2. The two possible separations of the edges at an internal site of $\mathscr{L}'$ (lying on the edge (i, j) of $\mathscr{L}$). The first represents no bond between i and j, the second a bond.

The effect of this is to decompose $\mathscr{L}'$ into a set of disjoint polygons, an example being given in figure 3. (We now use 'polygon' to mean any simple closed polygonal path on $\mathscr{L}'$.)

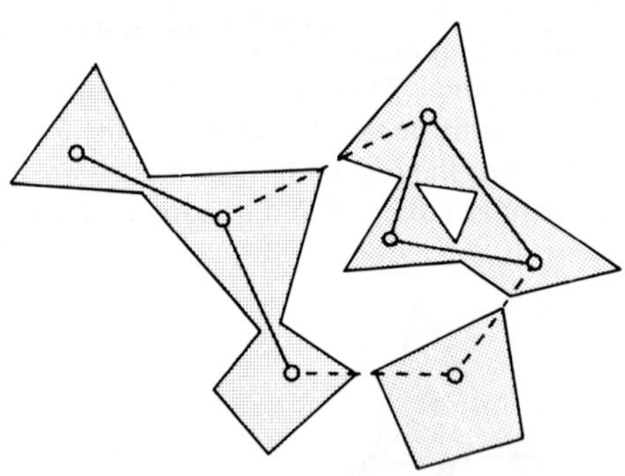

Figure 3. A graph G on $\mathscr{L}$ (full lines between open circles represent bonds), and the corresponding polygon decomposition of $\mathscr{L}'$. To avoid confusion at internal sites, sites of $\mathscr{L}'$ are not explicitly indicated, but are to be taken to be in the same positions as in figure 1.

Clearly any connected component of G now corresponds to a large island in $\mathscr{L}'$, made up of basic islands joined together. Each such large island will have an outer perimeter, which is one of the polygons into which $\mathscr{L}'$ is decomposed. A large island may also contain lakes within; these correspond to circuits of G and also have a polygon as outer perimeter.

Each polygon is of one of these two types. Thus $\mathscr{L}'$ is broken into p polygons, where

$$p = C + S, \tag{7}$$

and C and S are, respectively, the number of connected components and circuits in G. If $\mathscr{L}$ has N sites, then Euler's relation gives

$$S = C - N + l_1 + l_2 + l_3 + \ldots. \tag{8}$$

Eliminating S and C from the above equations (6). (7), (8), it follows that

$$Z = q^{N/2} \sum q^{p/2} x_1^{l_1} x_2^{l_2} x_3^{l_3} \ldots, \tag{9}$$

where

$$x_r = q^{-1/2} v_r \tag{10}$$

and we now take the summation to be over all polygon decompositions of $\mathscr{L}'$. Here p is the number of polygons in the decomposition, and l_r is the number of internal sites of class r where the edges have been separated as in figure 2(b).

5. Equivalent ice-type model on $\mathscr{L}'$

In this section we first define an ice-type model (Lieb 1967) on the lattice $\mathscr{L}'$, and state that its partition function is $q^{-N/2}Z$. We then prove this equivalence.

Let θ and z be two parameters given by

$$q^{1/2} = 2\cosh\theta, \qquad z = \exp(\theta/2\pi). \tag{11}$$

Then the ice-type model is defined as follows.

(a) Place arrows on the edges of $\mathscr{L}'$ so that at each site an equal number of arrows point in and out.

(b) With each external site associate a weight z^{α} if an observer moving in the direction of the arrows turns through an angle α to his left, or an angle $-\alpha$ to his right, as he goes through the site. This angle α is shown in figure 4.

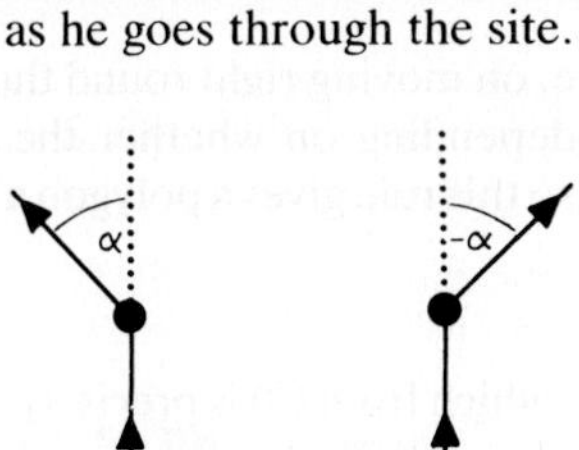

Figure 4. External sites of $\mathscr{L}'$ at which an observer moving in the direction of the arrows turns through an angle α to his left, or equivalently an angle $-\alpha$ to his right. Note that $-\pi < \alpha < \pi$, and the angle between the edges is $\pi - |\alpha|$.

(c) There are six possible arrangements of arrows at an internal site, as shown in figure 5. With arrangement k on a site of class r associate a weight ω_k, where

$$\omega_1, \ldots, \omega_6 = z^{\alpha-\gamma},\ z^{\gamma-\alpha},\ x_r z^{\beta-\delta},\ x_r z^{\delta-\beta},\ z^{-\beta-\delta} + x_r z^{\alpha+\gamma},\ z^{\beta+\delta} + x_r z^{-\alpha-\gamma} \tag{12}$$

and the angles α, β, γ, δ are those shown in figure 5.

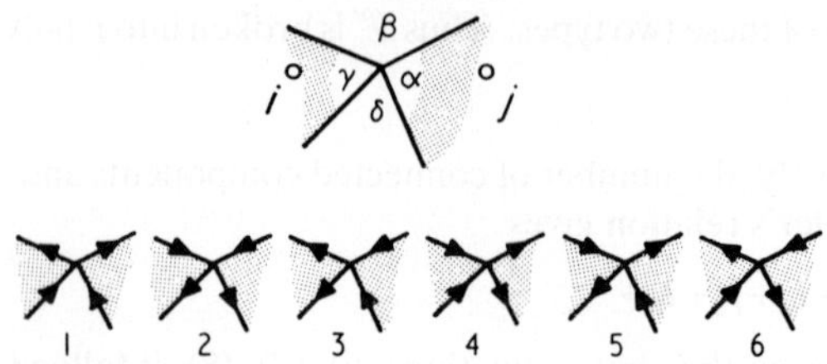

Figure 5. A typical internal site of $\mathscr{L}'$, showing the angles between edges and the six possible arrangements of arrows. When using (12) and this figure it is important to note that the angles α and γ lie inside basic polygons ('islands') of $\mathscr{L}'$, while β and δ lie outside.

The partition function of this ice-type model is

$$Z' = \sum \prod (\text{weights}), \tag{13}$$

where the product is over all sites of $\mathscr{L}'$ and the summation over all allowed arrow configurations on $\mathscr{L}'$. We shall prove that

$$Z = q^{N/2} Z', \tag{14}$$

where Z is the Potts model partition function (or Whitney polynomial) defined and discussed in the preceding sections.

5.1. *Proof of equivalence*

Take a polygon decomposition of $\mathscr{L}'$ and place arrows on the edges so that at each corner there is one pointing in and one pointing out. Give a polygon corner a weight z^{α}, where α is the angle to the left through which an observer moving in the direction of the arrows turns when passing through the corner. Since edges cannot overlap, α must lie in the interval $-\pi < \alpha < \pi$.

Since each polygon is a simple closed curve, on moving right round the polygon this observer turns through a total angle $\pm 2\pi$, depending on whether the arrows point anticlockwise or clockwise. Both cases occur, so this rule gives a polygon a total weight

$$z^{2\pi} + z^{-2\pi}. \tag{15}$$

The conditions (11) ensure that this is $q^{1/2}$, which from (9) is precisely the weight we wish to associate with each polygon. It follows from (9) that $q^{-N/2}Z$ can be obtained by the following procedure.

(*A*) At each internal site of $\mathscr{L}'$ separate the edges, either as in figure 2(*a*) or (*b*). If the latter, associate a weight x_r, where r is the class of the site.

(*B*) Place arrows on the edges round each site (internal and external), so as to follow one another round the polygon corners. Associate the appropriate weight z^{α} with each corner.

(*C*) Do (*A*) and (*B*) *independently for each site*. Then require that on an edge (i, j) there cannot be an arrow pointing into (or out of) both sites i and j. Reject all configurations that fail this requirement on any edge.

(*D*) Sum over all remaining configurations thus obtained, weighted by the product of the individual weights.

Since (A) and (B) can be performed independently for each site, they can be combined into:

(AB) Place arrows on the edges surrounding each site in any configuration that is generated by rules (A) and (B). Give each such configuration a weight equal to the product of the x_r and corner weights given by rules (A) and (B), *summed over all ways* (A) *and* (B) *give this arrow configuration.*

Since (A) and (B) can only give configurations with an equal number of arrows into and out of each site, the new rules (AB), (C), (D) define an ice-type model on $\mathscr{L}'$, with site weights given by (AB).

At external sites rule (B) immediately gives the rule (b) that we wish to establish. At internal sites (A) and (B) give eight possibilities, as shown in figure 6, but we note that the last two resulting arrow configurations are the same in the top and bottom rows. Calculating the weights from rules (A) and (B), the resulting total weights for the six distinct configurations are those given by equation (12) of rule (c). Thus the ice-type model defined by (AB), (C), (D) is the same as that given by (a), (b), (c). This proves the equation (14) and establishes the desired equivalence.

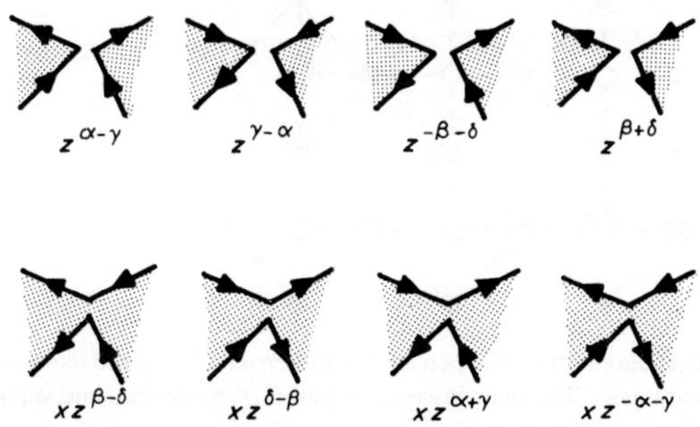

Figure 6. The two possible separations of edges at an internal site of $\mathscr{L}'$, and the eight allowed arrangements of arrows thereon. The product of the weights given by rules (A) and (B) is shown underneath, using the notation of figure 5 and omitting the suffix of x_r.

5.2. Four-colouring problem

It is fascinating to wonder whether this equivalence brings us any closer to a proof of the famous four-colour conjecture. From (11), $q=4$ is a 'critical' case, since for $q<4$ the parameter θ is pure imaginary, while for $q>4$ it is real.

At $q=4$, $z=1$ and for the colouring polynomial $x=-\frac{1}{2}$. Thus the weights (12) are real and the last two, as well as the first, are positive. However, the third and fourth are negative. For a bipartite lattice $\mathscr{L}$ it is not difficult to show that configurations 3 and 4 occur in pairs, so can both be replaced by positive values, which proves that in this case there is a positive number of four-colourings of $\mathscr{L}$. However, since a bipartite lattice can by definition be coloured with two colours, this is not a significant result! One needs a proof that Z is positive for any lattice when θ is real.

Another intriguing point which suggests that our transformation may be relevant is the following. It is conjectured from numerical and other studies that the real zeros of the colouring polynomial of an arbitrary planar lattice tend to limits as the lattice becomes large (Kasteleyn 1975). These limits are supposed to occur at the 'Beraha numbers' $q=[2\cos(\pi/n)]^2$, $n=2,3,4,\ldots$ (Tutte 1974).

From (11) we see that this corresponds to our parameter θ having the simple set of values $\theta = \mathrm{i}\pi/n$, $n = 2, 3, 4, \ldots$.

6. The regular lattices

For the interior of regular lattices there is an obvious natural choice of $\mathscr{L}'$, namely to take the sites of $\mathscr{L}'$ to be the midpoints of the edges of $\mathscr{L}$, and to take two sites of $\mathscr{L}'$ to be adjacent if and only if the corresponding edges of $\mathscr{L}$ meet at a common site and bound a common face. All sites of $\mathscr{L}'$ are then 'internal' except at the boundaries, which is the reason for our terminology. In figures 7 and 8 we show the square and triangular lattices and their surrounding lattices (square and Kagomé, respectively).

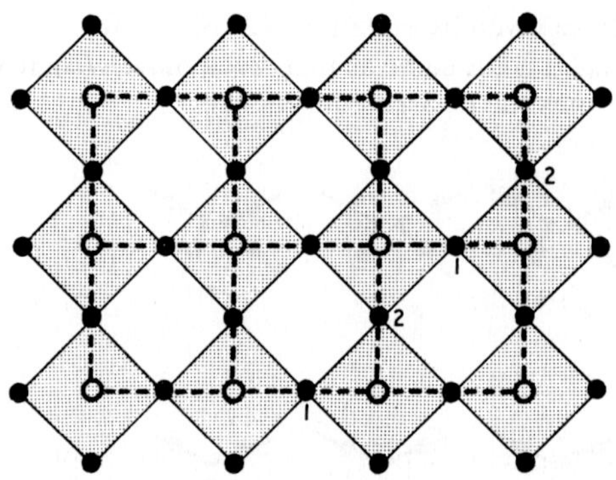

Figure 7. The square lattice $\mathscr{L}$ (open circles and broken lines) and its surrounding lattice $\mathscr{L}'$ (full circles and lines). The two classes of edges of $\mathscr{L}$, horizontal and vertical, are indicated by the numbers 1 and 2, respectively.

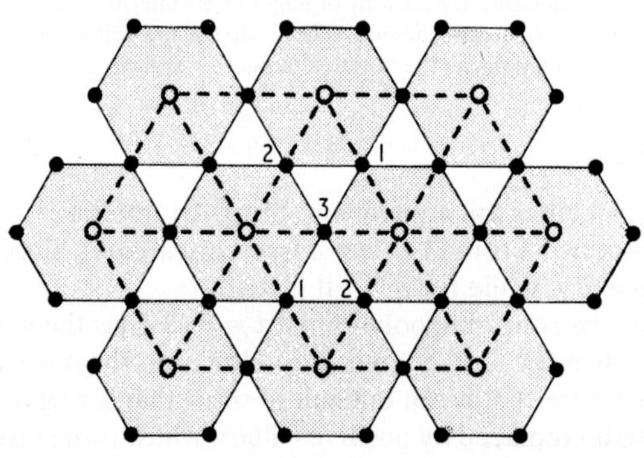

Figure 8. The triangular lattice $\mathscr{L}$ (open circles and broken lines) and its surrounding Kagomé lattice $\mathscr{L}'$ (full circles and lines).

The square lattice has two classes of edges, horizontal and vertical, which we call classes 1 and 2, respectively. The triangular lattice has three classes (1, 2 and 3), as shown in figure 8. Setting

$$s = \mathrm{e}^{\theta/2}, \qquad t = \mathrm{e}^{\theta/3}, \tag{16}$$

it follows from equation (12) and figure 5 that for the square lattice the vertex weights of an internal site of $\mathscr{L}'$ of class r are

$$\omega_1, \ldots, \omega_6 = 1, 1, x_r, x_r, s^{-1}+x_r s, s+x_r s^{-1}, \tag{17}$$

while for the triangular lattice they are

$$\omega_1, \ldots, \omega_6 = 1, 1, x_r, x_r, t^{-1}+x_r t^2, t+x_r t^{-2}. \tag{18}$$

The arrow configurations are still labelled $1, \ldots, 6$ as in figure 5, where the adjacent sites of $\mathscr{L}$ are drawn to the left and the right. Thus to obtain the weight of an arrow configuration at a site of $\mathscr{L}'$ on a vertical edge of the square lattice $\mathscr{L}$, it is necessary to turn figure 7 through 90° before using figure 5 and equation (17). Similar rotations are necessary for edges 1 and 2 of the triangular lattice.

For the square (triangular) lattice, external sites of $\mathscr{L}'$ have weight $s^{1/2}(t^{1/2})$ if the arrows turn to the left through the site, and weight $s^{-1/2}(t^{-1/2})$ if they turn to the right.

If we wish, we can eliminate fractional powers of e^{θ} by associating additional mutually inverse weights with the tips and tails of some arrows. Consider for instance the surrounding lattice $\mathscr{L}'$ of the square lattice, shown in figure 7. With every arrow pointing up and to the right associate a further weight $s^{-1}(s)$ with the site into (out of) which it points. This leaves the partition function unchanged, but on sites of type 1 multiplies ω_5, ω_6 by s, s^{-1}, respectively. On sites of type 2 it multiplies them by s^{-1}, s. We can then verify that our ice-type model for the square lattice is the same as that of Temperley and Lieb (their figure 1 and table 2, reversing vertical arrows and rotating through 135° clockwise). The only difference is that we have included the boundary conditions.

6.1. Duality

The Potts model is known to have a duality property (Potts 1952, Kihara *et al* 1954, Wu and Wang 1976). Our methods provide another way of re-deriving this.

Let $\mathscr{L}_D$ be the lattice dual to a lattice $\mathscr{L}$. Then from figures 7 and 8 it is apparent that the surrounding lattice $\mathscr{L}'$ of $\mathscr{L}$ is also the surrounding lattice of $\mathscr{L}_D$. (Here we do ignore boundary conditions.)

From (9) we can regard the Potts model partition function Z as a function of q and $x_1, x_2, x_3, \ldots$. Let us also define a Potts model on the dual lattice $\mathscr{L}_D$, with partition function Z_D and parameters $q, y_1, y_2, y_3, \ldots$ (using the obvious one-to-one correspondence between edges of $\mathscr{L}$ and $\mathscr{L}_D$). We can repeat the above reasoning to obtain the ice-type model on $\mathscr{L}'$ that is equivalent to the dual model. We find that it is again given by rules (a), (b), (c), except that in (12) x_r disappears and the terms that originally did not contain a factor x_r now contain y_r.

This is equivalent to first dividing the weights (12) by x_r, then replacing x_r by y_r^{-1}. Thus we obtain the duality relation

$$Z(q; x_1, x_2, \ldots) = x_1^{e_1} x_2^{e_2} \ldots Z_D(q; x_1^{-1}, x_2^{-1}, \ldots), \tag{19}$$

where e_r is the number of edges in $\mathscr{L}$ (or $\mathscr{L}_D$) of class r.

The Potts model on the honeycomb lattice is therefore also equivalent to an ice-type model on the Kagomé lattice.

6.2. Cyclic boundary conditions

The equivalence can be extended to lattices wound on a cylinder. In this case the rule leading to (15) gives a polygon that winds around the cylinder an incorrect weight $1+1=2$. To correct this, we draw a 'seam' through the lattice $\mathscr{L}'$, avoiding all sites, from the bottom of the cylinder to the top. We then associate an extra weight $e^{\theta}(e^{-\theta})$ with arrows pointing to the left (right) on edges that cross the seam. This ensures that *all* polygons give the correct contribution (15).

It is not clear that the equivalence can be further generalized to toroidal boundary conditions. In one sense this does not matter, since in lattice statistics we are usually interested in the 'thermodynamic limit' of a large lattice, when boundary conditions are expected to be irrelevant. However, we feel it does clarify the equivalence to establish it exactly for finite lattices.

Note added in proof. Our 'surrounding' lattice is the same as the 'medial' graph in graph theory (Ore 1967, pp 47 and 124). The 'Whitney' polynomial is also known as the 'dichromatic' polynomial (Tutte 1967). Nagle (1968) used the method of § 2 to show the equivalence of the colouring problem with a Whitney polynomial, and in 1969 defined a staggered ice-type model similar to those that occur here.

References

Baxter R J 1973 *J. Phys. C: Solid St. Phys.* **6** 1445–8
Domb C 1974 *J. Phys. A: Math., Nucl. Gen.* **7** 1335–48
Fortuin C M and Kasteleyn P W 1972 *Physica* **57** 536–64
Kasteleyn P W 1975 *Talk at the Science Research Council Rencontre on Combinatorics, Aberdeen, 6–12 July, 1975*
Kasteleyn P W and Fortuin C M 1969 *J. Phys. Soc. Japan Suppl.* **26** 11–14
Kihara T, Midzuno Y and Shizume T 1954 *J. Phys. Soc. Japan* **9** 681–7
Lieb E H 1967 *Phys. Rev.* **162** 162–72
Nagle J F 1968 *J. Math. Phys.* **9** 1007–19
—— 1969 *J. Chem. Phys.* **50** 2813–8
Ore O 1967 *The Four-Color Problem* (New York and London: Academic)
Potts R B 1952 *Proc. Camb. Phil. Soc.* **48** 106–9
Temperley H N V and Lieb E H 1971 *Proc. R. Soc.* A **322** 251–80
Tutte W T 1967 *J. Comb. Theory* **2** 301–20
—— 1974 *Can. J. Math.* **26** 893–907
Whitney H 1932 *Ann. Math., NY* **33** 688–718
Wu F Y and Wang Y K 1976 *J. Math. Phys.* submitted for publication

The Potts model

F. Y. Wu*

Institut für Festkörperforschung der Kernforschungsanlage Jülich, D-5170 Jülich, West Germany

This is a tutorial review on the Potts model aimed at bringing out in an organized fashion the essential and important properties of the standard Potts model. Emphasis is placed on exact and rigorous results, but other aspects of the problem are also described to achieve a unified perspective. Topics reviewed include the mean-field theory, duality relations, series expansions, critical properties, experimental realizations, and the relationship of the Potts model with other lattice-statistical problems.

CONTENTS

*Permanent and present address: Department of Physics, Northeastern University, Boston, MA 02115.

I. INTRODUCTION

This is a tutorial review on the static properties of the Potts model (Potts, 1952). The Potts model is a generalization of the Ising model (Ising, 1925) to more-than-two components, and has been a subject of increasingly intense research interest in recent years. Historically, a four-component version of the model was first studied by Ashkin and Teller (1943). But the model of general q components bears its current name after it was proposed by Domb [see Domb (1974a)] three decades ago to his then research student Potts as a thesis topic (Potts, 1951). The problem was also considered two years later in an independent study by Kihara *et al.* (1954).

The problem attracted little attention in its early years. But in the last ten years or so there has been a strong surge of interest, largely because the model has proven to be very rich in its contents. It is now known that the Potts model is related to a number of outstanding problems in lattice statistics; the critical behavior has also been shown to be richer and more general than that of the Ising model. In the ensuing efforts to explore its properties, the Potts model has become an important testing ground for the different methods and approaches in the study of the critical point theory.

There is now a vast number of research papers published on the subject; it has come to the point that it is difficult for new students to grasp the problem. It is with this intended readership in mind that the present article is written.

The main aim of this review is to bring out the important properties and results of the Potts model in a logical order. Derivations will be given when they are simple and illustrate a point, but these will be kept at a minimum to preserve continuity in reading. As the frontier of this active research field is fast advancing, it is not possible to give an up-to-date account of every facet of it. Instead, I shall concentrate on those aspects that are least likely to change in time. Here the choice of materials is to some extent arbitrary, and my emphasis will be on exact and rigorous results. I shall, in particular, not recount every development of the renormalization group and numerical studies, although the relevant results will be described in appropriate places as they arise in discussions. Another topic not covered in this review is the planar Potts model, a subject at the very frontier of current research interest; several reviews already exist on this and the related planar rotator problem (Kosterlitz and Thouless, 1978; Nelson, 1979; Barber, 1980; Frölich and Spencer, 1981). Also, the topic of series expansions will be discussed only to the extent of developing the expansions for the partition function and for finite graphs (and lattices). Specific techniques for the more efficient series generation for infinite lattices will not be discussed.

With these guidelines in mind, I have organized this review as follows.

In Secs. I.A and I.B the various versions of the Potts model are defined. Section I.C gives a mean-field solution of the (standard) Potts model including a discussion on the critical dimensionality. Section I.D reviews the status on experimental realizations of the Potts model, and Sec. I.E gives the solution for the Bethe lattice.

Section II discusses the exact duality relation for the various Potts model, with derivations given in some cases. In Sec. III some graphical aspects of the series expansions are reviewed. Section IV discusses the relationship of the Potts model with other lattice-statistical problems. These include the ice-rule model, the percolation problem, and the resistor network.

The critical properties of the Potts model are reviewed in Sec. V, where the exact as well as conjectured results are presented. Results on numerical analyses and experiments are also discussed to give a unified view on the critical properties.

Section VI discusses the random-bond Potts model, a problem that has been of very recent interest. A list of unsolved problems suggested in the course of this review is provided in Sec. VII.

A. The Potts model

The problem originally proposed by Domb was to regard the Ising model as a system of interacting spins that can be either parallel or antiparallel. Then an appropriate generalization would be to consider a system of spins confined in a plane, with each spin pointing to one of the q equally spaced directions specified by the angles

$$\Theta_n = 2\pi n/q, \quad n = 0, 1, \ldots, q-1 . \tag{1.1}$$

In the most general form the nearest-neighbor interaction depends only on the relative angle between the two vectors. This is quite generally known as a system of $Z(q)$ symmetry whose Hamiltonian reads

$$\mathscr{H} = -\sum_{\langle ij \rangle} J(\Theta_{ij}) , \tag{1.2}$$

where the function $J(\Theta)$ is 2π periodic and $\Theta_{ij} = \Theta_{n_i} - \Theta_{n_j}$ is the angle between the two spins at neighboring sites i and j. The $Z(q)$ model plays an important role in the lattice gauge theories and has attracted a growing interest [see, for example, a review by Kogut (1979)].

The model suggested by Domb (Potts, 1952) is to choose

$$J(\Theta) = -\varepsilon_1 \cos\Theta . \tag{1.3}$$

Using a Kramer-Wannier (1941) type analysis, Potts was able to determine the critical point of this model on the square lattice for $q = 2,3,4$. While unable to extend this finding to $q > 4$, Potts reported as a remark at the end of his paper (Potts, 1952) the critical point for all q of the following model:

$$J(\Theta_{ij}) = \varepsilon_2 \delta_{Kr}(n_i, n_j) . \tag{1.4}$$

It is this version of a q-component model that has attracted the most attention.

Following the suggestion of Domb (1974a), we shall name the model (1.3) the *planar* Potts model and the model (1.4) the *standard* Potts model, or simply the *Potts model.* Other names of these models have also appeared in the literature. The planar Potts model has been referred to as the vector Potts model and also as the clock model in recent literature; the standard Potts model has often been called the Ashkin-Teller-Potts model for historical reasons. It appears that Domb's suggestion is the simplest, which should be adopted in conjunction with using the name of the Ashkin-Teller model for the four-component model with (and without) symmetry breakings.

The (standard) Potts model is ferromagnetic if $\varepsilon_2 > 0$ and antiferromagnetic if $\varepsilon_2 < 0$. The interaction (1.4) can be alternately formulated to reflect its full symmetry in a $q-1$ dimensional space. This is achieved by writing in (1.4)

$$\delta_{Kr}(\alpha,\beta) = \frac{1}{q}[1 + (q-1)\mathbf{e}^{\alpha} \cdot \mathbf{e}^{\beta}] , \tag{1.5}$$

where $\mathbf{e}^{\alpha}$, $\alpha = 0, 1, \ldots, q-1$ are q unit vectors pointing in the q symmetric directions of a hypertetrahedron in $q-1$ dimensions. Examples of these vectors for $q = 2,3,4$ are shown in Fig. 1. The Hamiltonian in the form of (1.4) and (1.5) has proven convenient to use in the continuous-spin formulation of the Potts model [see, for example, Zia and Wallace (1975)].

The planar and the standard models are identical for $q = 2$ (Ising) and $q = 3$ with $\varepsilon_2 = 2\varepsilon_1$ and $\varepsilon_2 = 3\varepsilon_1/2$, respectively. Also, the four-state planar model is reducible to the $q = 2$ models (Betts, 1964) and this equivalence is valid for arbitrary lattices (Kasteleyn, 1964). There

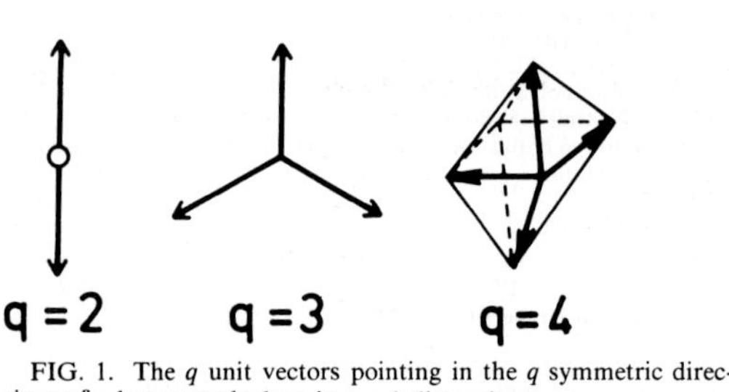

FIG. 1. The q unit vectors pointing in the q symmetric directions of a hypertetrahedron in $q-1$ dimensions.

exist no apparent relations between the planar and standard models for $q>4$.

In addition to the two-site interactions, there can also be multisite interactions as well as external fields. For a Potts model on a lattice G of N sites, the Hamiltonian $\mathscr{H}$ generally takes the form

$$-\beta\mathscr{H}=L\sum_i \delta_{\mathrm{Kr}}(\sigma_i,0)+K\sum_{(i,j)}\delta_{\mathrm{Kr}}(\sigma_i,\sigma_j)+K_3\sum_{(i,j,k)}\delta_{\mathrm{Kr}}(\sigma_i,\sigma_j,\sigma_k)+\cdots , \quad (1.6)$$

where $\beta=1/kT$, and $\sigma_i=0,1,\ldots,q-1$ specifies the spin states at the ith site and

$$\delta_{\mathrm{Kr}}(\sigma_i,\ldots,\sigma_k)=1, \text{ if } \sigma_i=\ldots=\sigma_k$$
$$=0, \text{ otherwise}.$$

Here $K=\beta\varepsilon_2$, K_n, $n\geq 3$, is the strength of the n-site interactions, and L is an external field applied to the spin state 0. [See Eqs. (1.18) and (3.1) for further forms of the Hamiltonian.] The partition function is

$$Z_G(q;L,K,K_n)=\sum_{\sigma_i=0}^{q-1}e^{-\beta\mathscr{H}} . \quad (1.7)$$

The physical properties of the system are derived in the usual way by taking the thermodynamic limit. Relevant thermodynamic quantities include the per site "free energy"

$$f(q,;L,K,K_n)=\lim_{N\to\infty}\frac{1}{N}\ln Z_G(q;L,K,K_n) , \quad (1.8)$$

the per site energy

$$E(q;L,K,K_n)=-\frac{\partial}{\partial\beta}f(q;L,K,K_n) , \quad (1.9)$$

and the per site "magnetization,"

$$M(q;L,K,K_n)=-\frac{\partial}{\partial L}f(q;L,K,K_n) . \quad (1.10)$$

The order parameter m, which takes the values 0 and 1 for completely disordered and ordered systems, respectively, is defined to be (Straley and Fisher, 1973; Binder, 1981)

$$m(q,L,K,K_n)=(qM-1)/(q-1) . \quad (1.11)$$

A ferromagnetic transition is then accompanied with the onset of a spontaneous ordering

$$m_0=m(q;0+,K,K_n) . \quad (1.12)$$

The critical exponents α, α', β, γ, γ', δ, ... can be defined in the usual fashion [see, for example, Fisher (1967)] from the singular behavior of these thermodynamic quantities near the critical temperature T_c.

The two-point correlation function $\Gamma_{\alpha\alpha}(\mathbf{r}_1,\mathbf{r}_2)$ of the zero-field Potts model is (Wang and Wu, 1976)

$$\Gamma_{\alpha\alpha}(\mathbf{r}_1,\mathbf{r}_2)=P_{\alpha\alpha}(\mathbf{r}_1,\mathbf{r}_2)-q^{-2} , \quad (1.13)$$

where $P_{\alpha\alpha}(\mathbf{r}_1,\mathbf{r}_2)$ is the probability that the sites at $\mathbf{r}_1$ and $\mathbf{r}_2$ are both in the same spin state α. Clearly, $\Gamma_{\alpha\alpha}$ takes the respective values 0 and $(q-1)/q^2$ for completely disordered and completely ordered systems. This then suggests the following relation between the large distance correlation and the spontaneous ordering:

$$\lim_{|\mathbf{r}_1-\mathbf{r}_2|\to\infty}\Gamma_{\alpha\alpha}(\mathbf{r}_1,\mathbf{r}_2)=(q-1)(m_0/q)^2 . \quad (1.14)$$

Indeed, the relation (1.14), which first appeared as a footnote in Potts and Ward (1955) for $q=2$, can be established by a decomposition of the correlation function into those of the extremum states (Kunz, 1981). It has also been established rigorously that $\Gamma_{\alpha\alpha}$ decays exponentially above the critical temperature T_c (Hintermann *et al.*, 1978). The decay of $\Gamma_{\alpha\alpha}$ for $T\leq T_c$ is not known except for $q=2$ (McCoy and Wu, 1973). Furthermore, the surface tension for the generalized Potts model has been discussed by Fontaine and Gruber (1979). It can be shown that, in two dimensions, the surface tension is related to the two-point correlation function of the dual model.

As we shall see, the analysis of the Potts model is closely related to the problem of graph colorings, so it is useful to introduce here the needed definitions.

Let $P_G(q)$ be the number of ways that the vertices of a graph G can be colored in q different colors such that no two vertices connected by an edge bear the same color. Then $P_G(q)$ is a function of q and is known as the chromatic function for the graph G.

Consider next an antiferromagnetic Potts model on G with pure two-site interactions $K<0$. Consider further the zero-temperature limit of $K\to-\infty$. It is clear that in this limit the partition function (1.7) reduces to

$$Z_G(q;K=-\infty)=P_G(q) . \quad (1.15a)$$

This simple connection between the Potts partition function and the chromatic function is valid for G in any dimension. In addition, a graphical interpretation of $P_G(q)$ for $q=-1$ has been given by Stanley (1973).

For a lattice G of N sites, the free energy (1.8) in the zero-temperature limit of $K\to-\infty$ becomes the ground-state entropy

$$W_G(q)\equiv\lim_{N\to\infty}\frac{1}{N}\ln P_G(q) . \quad (1.15b)$$

The existence of this limit has been discussed by Biggs (1975).

There are three exact results on $W_G(q)$ for $q\geq 3$. These are the values for the $q=3$ square lattice (Lieb, 1967a, 1967b), $q=4$ triangular lattice (Baxter, 1970), and the $q=3$ Kagomé lattice (Baxter, 1970):

$$W_{\mathrm{sq}}(3)=(4/3)^{3/2}$$
$$=1.539\,60\ldots,$$
$$W_{\mathrm{tri}}(4)=\prod_{n=1}^{\infty}\left[\frac{(3n-1)^2}{(3n)(3n-2)}\right]$$
$$=1.460\,99\ldots,$$
$$W_{\mathrm{Kagom\acute{e}}}(3)=[W_{\mathrm{tri}}(4)]^{1/3} \quad (1.15c)$$
$$=1.134\,70\ldots .$$

B. The dilute model

If vacancies can occur on the lattice, then we have a site-diluted Potts model, or a Potts lattice gas (Berker *et al.*, 1978), for which the lattice sites are randomly occupied with Potts spins. Consideration of this dilute Potts model has proven fruitful in the renormalization group studies of the Potts model (Nienhuis *et al.*, 1979); it also generates other statistical mechanical models including those of polymer gelation (Coniglio *et al.*, 1979) and the problem of site percolation in a lattice gas (Murata, 1979).

As in the usual consideration of random systems, the dilution in the Potts model can be either quenched, in which the vacancies are fixed in positions, or annealed, in which the vacancies can move around and are in thermoequilibrium with the surroundings. Very little is known about the quenched site-diluted system; it is the annealed system that has received the most attention.

The Hamiltonian $\mathscr{H}$ for an annealed site-diluted model reads

$$-\beta\mathscr{H}=\sum_{i,j}t_it_j[K'+K\delta_{Kr}(\sigma_i,\sigma_j)]+\sum_i(1-t_i)\ln z_i\ , \qquad (1.16)$$

where z_i is the fugacity of the vacancy at the ith site, and $t_i=0(1)$ indicates that the ith site is vacant (occupied). The partition function of the dilute model is

$$Z^{(D)}(q;,K',K,z_i)=\sum_{t_i=0}^{1}\sum_{\sigma_i=0}^{q-1}e^{-\beta\mathscr{H}}\ , \qquad (1.17)$$

where the summation over σ_i is for $t_i=1$ only.

If the vacancies are considered as being a spin state, then the dilute model can also be regarded as an (undiluted) Potts model of $(q+1)$ components. The Hamiltonian of this $(q+1)$-state model is

$$-\beta\mathscr{H}_{q+1}=K\sum_{(i,j)}\delta_{\mathrm{Kr}}(\sigma_i,\sigma_j)+\sum_i L_i\delta_{Kr}(\sigma_i,0)+M\sum_{(i,j)}\delta_{\mathrm{Kr}}(\sigma_i,0)\delta_{\mathrm{Kr}}(\sigma_j,0)\ , \qquad (1.18)$$

where, in addition to the field L_i at site i, an additional field M is introduced which applies to neighboring sites that are both in the spin state 0. Writing

$$Z(q+1;K,M,L_i)=\sum_{\sigma_i=0}^{q}e^{-\beta\mathscr{H}_{q+1}}\ , \qquad (1.19)$$

we then have the identity

$$Z^{(D)}(q;K',K,z_i)=e^{EK'}Z(q+1;K,M,L_i)\ , \qquad (1.20)$$

with

$$M=K'-K\ ,$$

$$e^{L_i}=z_ie^{-\gamma_iK'}\ .$$

Here E is the total number of edges of the lattice and γ_i is the valence of the ith site.

For bipartite lattices it is possible to consider dilute models in which the vacancies are restricted to occurring at only one of the two sublattices. A special class of such lattices is those with bond decorations with vacancies restricted to the decorating sites. The critical properties of this diluted model can be derived from those of the underlying undiluted model, and have led to some unique features, including the existence of a two-phase region for $q>4$ (Wu, 1980). Similar results have also been obtained for the regular (undecorated) honeycomb lattice (Wu and Zia, 1981; Kondo and Temesvári, 1981).

C. The mean-field solution

It is well known that the mean-field description of the Ising model gives a qualitatively correct picture of the phase transition. In the absence of an exact solution, it is therefore natural first to examine the q-component Potts model in the mean-field approximation.

Such a study was first carried out by Kihara *et al.* (1954) under the Bragg-Williams approximation (Bragg and Williams, 1934). They found the transition to be of first order for all $q>2$, and, apparently without realizing the importance attached to this implication, dismissed the result as "being far from reality."

The mean-field theory was considered again by Mittag and Stephen (1974) [see also Straley and Fisher (1973)]. With the guide of the known exact critical properties of the two-dimensional model (Baxter, 1973a), they showed that the mean-field result is exact to the leading order in the large q expansion in $d=2$ dimensions. In fact, the exact result in $d=2$ shows a first-order transition for $q>4$ (Sec. V.B). We then expect, more generally, the existence of a critical value $q_c(d)$ such that, in d dimensions, the mean-field theory is valid for $q>q_c(d)$. We shall look at this point briefly before going on to the mean-field solution.

Regarding q and d as being both continuous, the critical value of $q_c(d)$ implies the existence of a critical dimensionality $d_c(q)$ such that the mean-field behavior prevails in $d>d_c(q)$. The known points are $d_c(2)=4$ and $q_c(2)=4$. It has also been suggested (Toulouse, 1974), and subsequently verified by Monte Carlo simulation (Kirkpatrick, 1976) and by series analyses (Gaunt *et al.*, 1976; Gaunt and Ruskin, 1978), that the critical dimensionality $d_c(1)$ for the percolation process (see Sec. IV.B) is 6. A schematic plot of $q_c(d)$ is thus made in Fig. 2, where we have also incorporated the renormalization-group results of $q_c(1+\varepsilon)\sim\exp(2/\varepsilon)$ for small ε (Andelman and Berker, 1981; Nienhuis *et al.*, 1981), $q_c(d)=2$ for $d>4$ (Aharony and Pytte, 1981), and assumed first-order transition at the point $q=3$, $d=3$ (see Sec. V.B). A plot of the first-order region similar to Fig. 2 can be found in Riedel (1981) and Nienhuis, *et al.* (1981).

We now describe a mean-field theory of the Potts model equivalent to that of Kihara *et al.* (1954).

We start from the mean-field Hamiltonian (Husimi, 1953; Temperley, 1954; Kac, 1968)

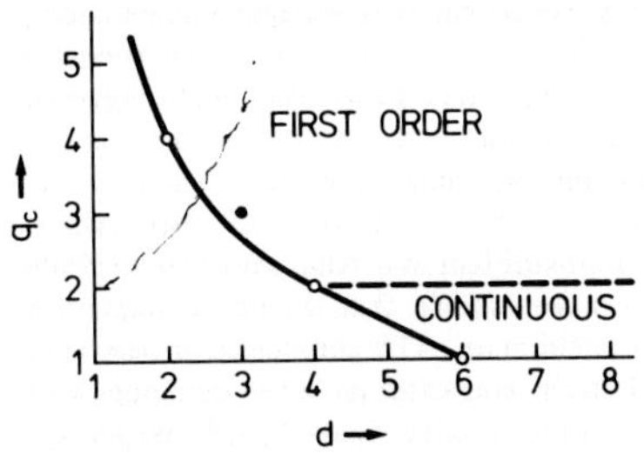

FIG. 2. Schematic plot of $q_c(d)$, the critical value of q beyond which the transition is mean-field—like (first order for $q>2$ and continuous for $q \le 2$). The known points $q_c(2)=4$, $q_c(4)=2$, and $q_c(6)=1$ are denoted by open circles. The black circle indicates the assumed first-order transition for $d=3$, $q=3$.

$$\mathscr{H} = -\frac{1}{N}\gamma\varepsilon_2 \sum_{i<j} \delta_{\mathrm{Kr}}(\sigma_i,\sigma_j) \tag{1.21}$$

for a system of N spins, each of which interacts with the other $N-1$ spins via an equal strength of $\gamma\varepsilon_2/N$, γ being the coordination number of the lattice.

Let x_i be the fraction of spins that are in the spin state $i=0,1,\ldots,q-1$, subject to

$$\sum_i x_i = 1 . \tag{1.22}$$

Then, to the leading order in N, the energy and entropy per spin are

$$\frac{E}{N} = -\tfrac{1}{2}\gamma\varepsilon_2 \sum_i x_i^2 ,$$

$$\frac{S}{N} = -k\sum_i x_i \ln x_i , \tag{1.23}$$

and the free energy per spin, A, is given by the expression

$$\beta A = \sum_i (x_i \ln x_i - \tfrac{1}{2}\gamma K x_i^2) , \tag{1.24}$$

where $K=\beta\varepsilon_2$.

For ferromagnetic interactions ($\varepsilon_2>0$) we look for a solution in the form of

$$x_0 = \frac{1}{q}[1+(q-1)s] ,$$

$$x_i = \frac{1}{q}(1-s), \quad i=1,2,\ldots,q-1 , \tag{1.25}$$

where the order parameter $0 \le s \le 1$ is to take the value s_0 which minimizes the free energy. It follows that a long-range order exists ($x_0 > x_i$) in the system if $s_0>0$.

What actually happens can be readily seen from the expansion of $A(s)$ for small s. Using (1.24) and (1.25), we find

$$\begin{aligned}\beta[A(s)-A(0)] &= \frac{1+(q-1)s}{q}\ln[1+(q-1)s] + \frac{q-1}{q}(1-s)\ln(1-s) - \frac{q-1}{2q}\gamma K s^2 \\ &= \frac{q-1}{2q}(q-\gamma K)s^2 - \tfrac{1}{6}(q-1)(q-2)s^3 + \cdots .\end{aligned} \tag{1.26}$$

It is the existence of a negative coefficient in the cubic term for $q>2$ which signifies the occurrence of a first-order transition (Harris *et al.*, 1975; de Gennes, 1971).

The order parameter s_0 is to be determined as a function of temperature T from $A'(s_0)=0$. It is seen that $s_0=0$ is always a solution, but at sufficiently low temperatures other solutions of $s_0>0$ emerge which may actually yield a lower free energy. The critical point is then defined to be the temperature T_c at which this shift of minimum free energy occurs.

For $q=2$ this leads to the usual mean-field consideration of the Ising model, namely,

$$\ln[(1+s_0)/(1-s_0)] = \gamma K s_0 . \tag{1.27}$$

From (1.27) we see that the critical point is

$$\gamma K_c = 2 . \tag{1.28}$$

The transition is continuous since $s_0=0$ at T_c.

The situation is different for $q>2$ because the order parameter jumps from 0 to a value $s_c>0$ discontinuously at T_c. In this case the critical parameters s_c and T_c are solved jointly from $A'(s_c)=0$ and $A(s_c)=A(0)$. One finds

$$\gamma K_c = \frac{2(q-1)}{q-2}\ln(q-1) , \tag{1.29}$$

$$s_c = (q-2)/(q-1) . \tag{1.30}$$

Using (1.23) we can also compute the latent heat per spin L, yielding the result

$$L = \gamma\varepsilon_2 (q-2)^2/2q(q-1) . \tag{1.31}$$

Other critical parameters can be similarly obtained. As we have already remarked, these expressions agree with the exact results in $d=2$ dimensions (Sec. V.B) to the leading order in the large q expansion.

D. Experimental realizations

For many years the Potts model was considered a system exhibiting an order-disorder transition primarily of theoretical interest. However, it has been recognized in recent years that it is also possible to realize the Potts model in experiments. Substances and experimental systems which can be regarded as realizations of the various Potts models have been suggested and identified; relevant experiments have been performed. It is through the combined effort in both theory and experiments that a converging picture in understanding the Potts transition has begun to emerge.

The underlying principle in the experimental realization of a spin system is the principle of universality, from which one is led to seek for real systems belonging to the same universality class, i.e., having the same set of critical exponents, as the spin model in question. For the Potts model one is guided by its Landau-Ginzburg-Wilson (LGW) Hamiltonian [Zia and Wallace (1975) and Amit (1976) for general q; Golner (1973) Amit and Shcherbakov (1974) and Rudnick (1973) for $q=3$]. An example is the transition occurring in monolayers and submonolayers adsorbed on crystal surfaces. The transitions in these systems have long been known (Somorja, 1973). But Domany *et al.* (1977) showed that the adsorbed systems can be classified and catalogorized using the Landau theory and the LGW Hamiltonian of the adatoms regarded as a lattice gas. It has since been shown (Domany *et al.*, 1978; Domany and Riedel, 1978; Domany and Schick, 1979) that transitions belonging to the various universality classes of the two-dimensional spin models can be realized by appropriately choosing the substrate array and the adatom coverage; some of these suggestions have indeed been verified in experiments.

1. $q=2$ (Ising) systems

Magnetic substances that are well approximated by simple Ising systems are numerous and well known (see, for example, a review by de Jongh and Miedema (1974)]. We mention here only the most notable examples, $CoCs_2Br_5$ in $d=2$ (Wielinga *et al.*, 1967; Mess *et al.*, 1967), $CoCs_3Cl_5$ (Wielinga *et al.*, 1967) and $DyPO_4$ (Wright *et al.*, 1971) in $d=3$.

The possibility of realizing the $d=2$ Ising model in adsorbed systems was suggested by Domany and Schick (1979), who showed that, at 1/2 coverage, an adsorbed system on a substrate of honeycomb array should exhibit an Ising-type behavior. This prediction has since been confirmed by the careful specific heat measurement (Tejwani *et al.*, 1980) of the adsorbed ^{4}He atoms on krypton preplated graphite.

2. $q=3$ systems

The critical behavior of the three-state Potts model, especially in $d=3$, provides a clear-cut test of the mean-field prediction and has been a subject of considerable interest. On the experimental side Mukamel *et al.* (1976) have suggested that in a diagonal magnetic field a cubic ferromagnet with three easy axes can be regarded as the $q=3$ Potts model, thus providing an experimentally accessible realization in $d=3$. Experimental study on one of such cubic ferromagnets, $DyAl_2$, has since been carried out (Barbara *et al.*, 1978), and the finding of a first-order transition is consistent with current understanding (see Sec. V.B). Other variants of the three-state model in cubic rare-earth compounds have also been suggested (Kim *et al.*, 1975). In addition, the first-order structural transition occurring in some substances such as the stressed $SrTiO_3$ is in the $q=3$ universality class (Aharony *et al.*, 1977; Blankschtein and Aharony, 1980a, 1980b, 1981). It has also been shown that the phase diagram of the structural transition in A15 compound in the presence of internal strain and external stress coincides with that of the $q=3$ Potts model (Szabo, 1975; Weger and Goldberg, 1973). A fluid mixture of five (suitably chosen) components can also be regarded as a realization of the $q=3$ system, and experiment on one such mixture, ethylene glycol + water + lauryl alcohol + nitromethane + nitroethane, also indicated a first-order transition (Das and Griffiths, 1979).

The relevance of the adsorbed monolayers in the $q=3$, $d=2$ Potts model was first pointed out by Alexander (1975). Specifically, it was suggested that the adsorption of ^{4}He atoms on graphite at $\frac{1}{3}$ coverage provides a realization of the three-state model. Such adsorbed systems have since been the subject of careful experimental studies (Bretz, 1977; Tejwani *et al*, 1980); the experimental results are in agreement with the theoretical predictions (see Sec. V.C). Other possible realizations of the $q=3$ systems in adsorptions have been discussed by Domany and Riedel (1978), Domany *et al.* (1978), and Domany and Schick (1979). The adsorption of krypton on graphite as a three-component Potts model has also been considered by Berker *et al.* (1978). It has also been suggested that the structural ordering observed in silver β alumina is a realization of the $q=3$, $d=2$ Potts model (Gouyet *et al.*, 1980; Gouyet, 1980).

3. $q=4$ systems

The general discussion on the classification scheme of the adsorbed systems (Domany *et al.*, 1978; Domany and Schick, 1979; Domany and Riedel, 1978) has led to a variety of possible realizations of the $q=4$ model in $d=2$. It was suggested, in particular, that N_2 adsorbed on krypton-plated graphite should exhibit a critical behavior as the $q=4$ Potts model (Domany *et al.*, 1977). In addition, Park *et al.* (1980) have studied O_2 adsorbed on the surface of nickel as a realization of the $q=4$ model. In three dimensions the realization of the $q=4$ (and $q=3$) model in type I fcc antiferromagnets (such as CeAs) has been suggested recently by Domany *et al.* (1981).

4. $0 \leqslant q \leqslant 1$ systems

It has been shown (Lubensky and Isaacon, 1978) that transitions in the gelation and vulcanization processes in branched polymers are in the same universality class of the $0 \le q \le 1$ Potts model. This suggests that by properly choosing the polyfunctional units which are allowed to interact in a polymeric solution, Potts models of different values of q between zero and one may be realized in the polymer systems.

E. The Bethe lattice

The Potts model is exactly soluble on the Bethe lattice. As in the case of the Ising model (Eggarter, 1974; von Heimburg and Thomas, 1974; Matsuda, 1974), one finds a phase transition characterized by a diverging susceptibility without a long-range order (Wang and Wu, 1976).

A Bethe lattice is a Cayley tree [for definitions of graphical terms see, for example, Essam and Fisher (1970)] having the same valence γ at all interior sites. Then for the Potts model (1.4) the free energy (1.8) is trivially evaluated to yield

$$f(q,K)=\ln(e^K+q-1)\ , \tag{1.32}$$

which is analytic in the temperature T. The correlation function (1.13) can also be evaluated, yielding

$$\Gamma_{\alpha\alpha}(\mathbf{r}_1,\mathbf{r}_2)=\frac{q-1}{q^2}\left[\frac{e^K-1}{e^K+q-1}\right]^{|\mathbf{r}_1-\mathbf{r}_2|}\ , \tag{1.33}$$

where $|\mathbf{r}_1-\mathbf{r}_2|$ is the distance between $\mathbf{r}_1$ and $\mathbf{r}_2$. Consequently, there exists no large distance correlation.

To compute the zero-field susceptibility χ one explicitly carries out the summation in the fluctuation relation

$$\chi(q;K)=\lim_{N\to\infty} N^{-1}\sum_{\mathbf{r}_1,\mathbf{r}_2}\Gamma_{\alpha\alpha}(\mathbf{r}_1,\mathbf{r}_2) \tag{1.34}$$

and finds that χ diverges for $T\le T_c[\sqrt{(\gamma-1)}]$, where $T_c(x)$ is defined by

$$\begin{aligned}|K_c(x)| &=\ln[(q+x-1)/(x-1)],\ \ \varepsilon_2>0 \\ &=\ln[(x-1)/(x+1-q)],\ \ \varepsilon_2<0,\ \ q<x+1\ .\end{aligned} \tag{1.35}$$

It is noteworthy that the critical behavior on the Bethe lattice is different from that of the mean-field solution of Sec. I.C, and that of the Bethe-Peierls approximation [see, for example, Huang (1963)]. The latter solutions yield a nonzero long-range order and corresponds to the critical behavior occurring in the interior of a Bethe lattice.

II. DUALITY RELATIONS

Duality relations are useful in obtaining exact information on spin systems. For a review of duality in field theory and in statistical systems in general see Savit (1980). See also Gruber *et al.* (1977) for a discussion of duality transformation for general q-component models. Here we shall be concerned with the explicit formulation of duality for Potts models in two dimensions. These are relations connecting the partition functions in the high- and low-temperature regions.

A. Models with two-site interactions

A duality relation for the Potts model was first derived for the square lattice with pure two-site interactions on the basis of the transfer matrix approach [Potts (1952); see also Kihara *et al.*, (1954)]. The duality relation has since been rederived from other considerations and generalized to all planar lattices [see, Mittag and Stephen (1971); Wu and Wang (1976)]. The following derivation is based on a simple theorem in graph theory known to mathematicians for many years (Whitney, 1932).

Write the partition function (1.7) with pure two-site interactions in the form of

$$Z_G(q;K)=\sum_{\sigma_i=0}^{q-1}\prod_{\langle ij\rangle}[1+v\delta_{\mathrm{Kr}}(\sigma_i,\sigma_j)]\ , \tag{2.1}$$

where

$$v=e^K-1\ . \tag{2.2}$$

Next multiply out the product and represent the terms in the product by the subgraphs $G'\subseteq G$ whose edge sets correspond to the v factors in the terms. Let $b(G')$ be the number of edges (bonds) and $n(G')$ the number of connected components (clusters), including isolated points, in G'. The partition function then takes the following simple form after carrying out the spin summations (Baxter, 1973):

$$Z_G(q;K)=\sum_{G'\subseteq G} v^{b(G')}q^{n(G')}\ . \tag{2.3}$$

The expression (2.3) is the starting point of various formulations of the Potts model (see Sec. V); it also serves as a natural extension of the Potts model to nonintegral values of q. The latter formulation [in the form of (2.3)] leads to the random cluster model of Kasteleyn and Fortuin (1969) and forms the basis of studying the Potts model for general values of q (see, for example, Blöte *et al.*, 1981). An immediate application of (2.3) is to combine with (1.15a), yielding the chromatic function $P_G(q)$ in the form of a polynomial in q:

$$P_G(q)=\sum_{G'\subseteq G}(-1)^{b(G')}q^{n(G')}\ . \tag{2.4}$$

This is the well-known Birkhoff (1912) formula for the chromatic function.

In a consideration of the problem of map colorings, Whitney (1932) introduced the following Whitney rank function:

$$W_G(x,y)=\sum_{G'\subseteq G}x^{b(G')}(y/x)^{c(G')}\ , \tag{2.5}$$

where $c(G')$ is the number of independent circuits in G'. Now $c(G')$ is related to $b(G')$ and $n(G')$ through the Euler relation (valid for any G' not necessarily planar)

$$c(G')=b(G')+n(G')-N\ , \tag{2.6}$$

where N is the number of vertices in G. Substituting (2.6) into (2.5) and comparing with (2.3), we obtain the identity

$$Z_G(q;K)=q^N W_G\left[\frac{v}{q},v\right]. \tag{2.7}$$

The duality for Z_G then follows from a similar relation for W_G (Whitney, 1932). We now cast Whitney's derivation of this duality in the language of the Potts variables.

Let D be the dual of G, and to each $G'\subseteq G$ introduce a $D'\subseteq D$ whose edge set complements that of G'. For example, if G is 4×4 lattice of 16 sites, then D is a graph having 16 faces and 10 sites, including one site residing exterior to G. An example of the correspondence between a typical G' and D' for this G and D is shown in Fig. 3. Since by construction each circuit of G' encircles a cluster of D', and vice versa, we have

$$n(D')=c(G')+1\ ,$$
$$n(G')=c(D')+1\ . \tag{2.8}$$

We also have

$$b(G')+b(D')=E\ , \tag{2.9}$$

where E is the total number of edges of G (or D). Combining (2.3) with (2.6) and (2.8), we obtain

$$Z_G(q;K)=v^E q^{1-N_D}\sum_{D'\subseteq D}\left[\frac{q}{v}\right]^{v(D')}q^{n(D')}$$
$$=v^E q^{1-N_D}Z_D(q;K^*)\ , \tag{2.10}$$

where

$$(e^K-1)(e^{K^*}-1)=q\ , \tag{2.11}$$

and $N_D=E+2-N$ is the number of sites of D. Thus (2.10) relates the partition functions of the Potts models on G and D and maps the high- and low-temperature partition functions of the dual models onto each other.

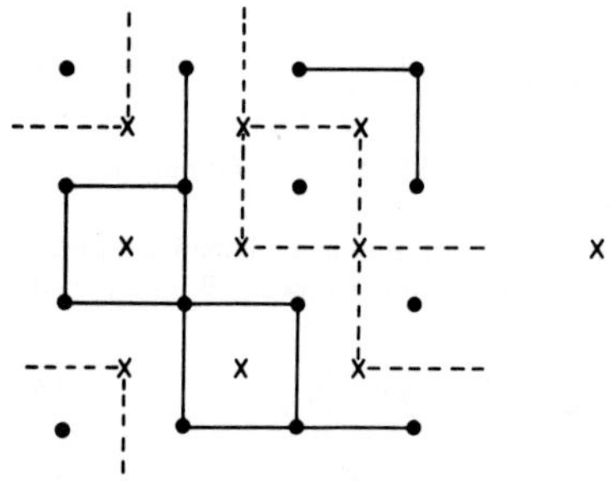

FIG. 3. Example of a mapping between a subgraph G' on a 4×4 lattice G and subgraph D' on D, the dual of G. The $N=16$ sites of G are denoted by the dots and the $N_D=10$ sites of D are denoted by the crosses. The single site of D residing exterior to G is connected to eight other sites of D. The edges in $G'(D')$ are denoted by the solid (broken) lines. In the configuration shown we have $b(G')=12$, $n(G')=6$, $c(G')=2$, $b(D')=12$, $n(D')=3$, $c(D')=5$.

The duality relation (2.11) is a local property in the sense that it also holds for edge-dependent interactions. A discussion of this generalization can be found in Wu (1978).

B. Models with multisite interactions

Essam (1979) has extended the duality relation to Potts models with multisite interactions [see also Kasai *et al.* (1980)]. For the purpose of discussing this generalization, it is convenient first to extend the definition of a dual to a special class of lattices or graphs.

Consider a lattice $G=(V,E)$ whose vertex set V is bipartite. That is, we can write $V=(S,I)$ such that vertices in S neighbor only vertices in I, and vice versa. The edge set E consists of the lines connecting these neighboring (S and I) vertices. Examples of such lattices are shown in Fig. 4.

Let S^* denote the set of vertices residing in the faces of G. Then we define the dual of G, $G^*=(V^*,E^*)$, in a similar way, i.e., the vertex set is $V^*=(S^*,I)$ and the edge set E^* consists of the lines connecting the neighboring (S^* and I) sites. Thus the lattice of Fig. 4(a) is self-dual, while those of Fig. 4(b) and Fig. 4(c) are dual to each other.

Now consider a Potts model on G with the spins located at the vertices of S and interactions specified by the vertices of I. Specifically, the γ spins surrounding a vertex of I interact with a γ-site interaction (of strength K_γ, say) of the form of (1.6). The dual model is similarly defined with spins located at S^* and interacting with γ-site interactions (of strength K_γ^*) surrounding the sites of I. (Note that the I vertices serve no purpose in the two models other than specifying the interactions.) The following duality relation exists between the two Potts models (Essam, 1979):

$$Z_G(q;K_\gamma)=(e^{K_\gamma}-1)^{N_I}q^{1+N_{S^*}}Z_{G^*}(q;K^*)\ , \tag{2.12}$$

where

$$(e^{K_\gamma}-1)(e^{K_\gamma^*}-1)=q^{\gamma-1} \tag{2.13}$$

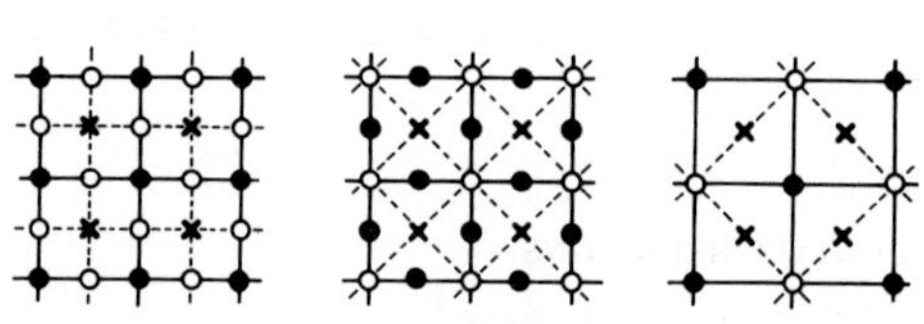

FIG. 4. Examples of lattices G consisting of bipartite vertex set $V=(S,I)$ and edge set (connecting S and I) denoted by the solid lines. The S vertices are denoted by the black circles and the I vertices by the open circles. The dual lattice G^* consists of the vertex set $V^*=(S^*,I)$ and the edge set (connecting S^* and I) denoted by the broken lines. The S^* vertices are denoted by the crosses. The lattice (a) is self-dual, and the lattices (b) and (c) are dual to each other.

and N_I, N_{S^*} are, respectively, the numbers of vertices in the sets I and S^*.

For pure two-site interactions, i.e., $\gamma=2$, (2.12) reduces readily to (2.10). Here again the duality relation (2.13) is valid more generally for position-dependent interactions and nonuniform values of γ. We now outline the proof of (2.12) for the case of uniform values of K_γ and γ. Generalization to nonuniform values is immediate.

To facilitate the proof we define subgraphs $G'\subseteq G$ as the set consisting of the S vertices and randomly chosen $I(G')$ of the I vertices, together with the $\gamma I(G')$ edges incident to the $I(G')$ vertices. A typical subgraph G' obtained in this fashion for the lattice G of Fig. 4(b) is shown in Fig. 5. These subgraphs are useful in representing terms in the expansion of the partition function Z_G. As in (2.1) and (2.3), we obtain the expression

$$Z_G(q;K)=\sum_{G'\subseteq G}(e^{K_\gamma}-1)^{I(G')}q^{n(G')}, \tag{2.14}$$

where $n(G')$ is the number of clusters, including isolated S vertices, in G'.

Again, to each $G'\subseteq G$ a dual subgraph $G^{*\prime}\subseteq G^*$ is constructed (and vice versa) by taking the complement of G' [i.e., $G^{*\prime}$ consists of the S^* vertices and the remaining $N_I-I(G')=I(G^{*\prime})$ of the I vertices together with their $\gamma I(G^{*\prime})$ incident edges]. An example of this construction is shown in Fig. 5.

It is now straightforward to proceed as in the case of $\gamma=2$ to obtain a duality relation. In place of (2.6) and (2.9) we now have the Euler relation

$$n(G')=N_S-(\gamma-1)I(G')+c(G') \tag{2.15}$$

and

$$I(G')+I(G^{*\prime})=N_I\ , \tag{2.16}$$

while relations (2.8) remain unchanged. Using these identities and further the equality

$$N_S+N_{S^*}=(\gamma-1)N_I+2\ , \tag{2.17}$$

one obtains (2.12) from (2.14).

As we have already remarked, the duality relation (2.12) is more general, being valid for nonuniform interactions and/or γ's. As an example, consider a Potts model on the square lattice which, in addition to the usual two-site interactions K_2, also possesses four-site interactions K_4 in *every* face of the square lattice. This Potts model and the associated G lattice are shown in Fig. 6. In Fig. 7 we see the lattice G^* dual to G and the associated dual Potts model. It is seen that the dual Potts model has two-site interactions K_2^* and four-site interactions K_4^* with a different topology. The special case of $K_2=0$ ($K_2^*=\infty$) of this relation has been considered by Burkhardt (1979).

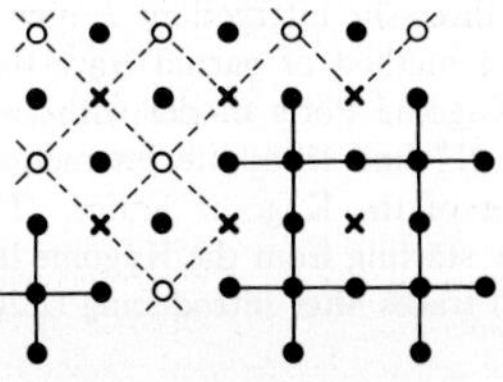

FIG. 5. A subgraph G' for the lattice G of Fig. 4(b) and the associated subgraph $G^{*\prime}\subseteq G^*$. G' is denoted by the solid circles and the solid lines, and $G^{*\prime}$ consists of the vertices denoted by the open circles and the crosses, and the broken lines. In the configuration shown $I(G')=5$, $I(G^{*\prime})=7$.

C. Other duality relations

In addition to the duality relations described above, there exist other relations, notable for lattices with triangular symmetry, which do not fall into the same scheme. While these latter duality relations can be derived from a number of different considerations (Baxter *et al.*, 1978; Wu and Lin, 1980), we describe here a derivation using a method due to Burkhardt (1979), which has the advantage of being easily adapted to other lattices.

Consider a Potts model on a lattice G whose dual is bipartite. The interactions, which may consist two-site and multisite components, are restricted to spins surrounding every *other* face of G. Then, by introducing dual spins residing in the faces where there are no interactions, a partial trace can be taken in the partition function resulting in a Potts model represented by these dual spins. This is the essential idea of Burkhardt (1979).

Consider, for instance, a triangular Potts model which has anisotropic two-site interactions K_1, K_2, K_3 and a three-site interaction L in every up-pointing triangle. The situation is shown in Fig. 8, where the shaded triangles denote the Potts interactions. The partition function takes the form

$$Z_\Delta(q;K_1,K_2,K_3,L)=\sum_{\sigma_i=0}^{q-1}\prod_\Delta e^{H(\sigma_1,\sigma_2,\sigma_3)}\ , \tag{2.18}$$

where

$$H(\sigma_1,\sigma_2,\sigma_3)=K_1\delta_{23}+K_2\delta_{31}+K_3\delta_{12}+L\delta_{12}\delta_{23}\ , \tag{2.19}$$

$$\delta_{ij}=\delta_{Kr}(\sigma_i,\sigma_j)\ ,$$

and the product is taken over all up-pointing triangles.

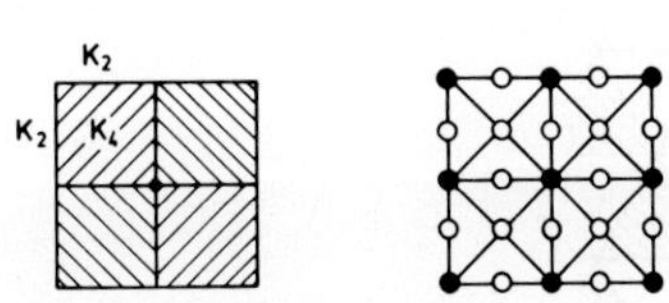

FIG. 6. A square Potts lattice with two-site interactions K_2 and four-site interactions K_4; the associate $G=(S,I)$ lattice is shown on the right. The solid circles denote the S vertices locating the spins; the open circles denote the I vertices defining the interactions.

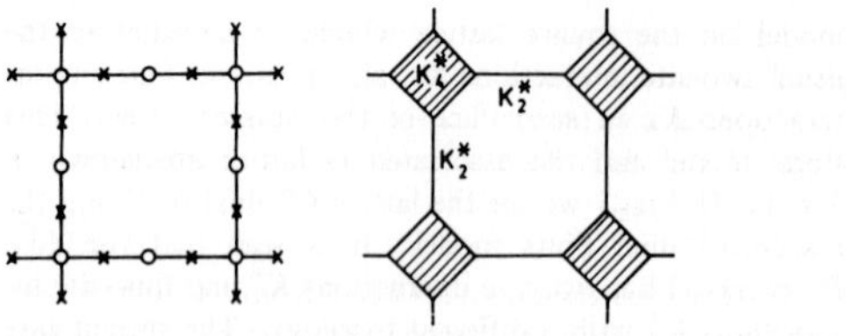

FIG. 7. The dual lattice G^* and its associated Potts lattice. The Potts lattice has two-site interactions K_2^* and four-site interactions K_4^* as shown.

Since $H(\sigma_1,\sigma_2,\sigma_3)$ depends only on the differences $\sigma_{12}=\sigma_1-\sigma_2$, $\sigma_{23}=\sigma_2-\sigma_3$, $\sigma_{31}=\sigma_3-\sigma_1$ (mod$q$), it is convenient to regard the variables σ_{ij} as being independent. This is permitted if we introduce to each (up- and down-pointing) triangular face a factor (hence a variable τ)

$$\sum_{l=-\infty}^{\infty} \delta(\sigma_{12}+\sigma_{23}+\sigma_{31},lq) = \frac{1}{q}\sum_{\tau=0}^{q-1} \exp\left[\frac{2\pi i}{q}\tau(\sigma_{12}+\sigma_{23}+\sigma_{31})\right] . \tag{2.20}$$

With these factors in place, the σ_{ij} summations in the partition function can now be carried out. If, in addition, we also take the partial trace over the τ-variables over the up-pointing triangles, we are then left with the expression

$$Z_\Delta(q;K_1,K_2,K_3,L)=\sum_{\tau_i=0}^{q-1}\prod_{\nabla} e^{H^*(\tau_1,\tau_2,\tau_3)} , \tag{2.21}$$

where

$$e^{H^*(\tau_1,\tau_2,\tau_3)}=\frac{1}{q^2}\sum_{\sigma_1\sigma_2\sigma_3} e^{H(\sigma_1,\sigma_2,\sigma_3)} e^{(2\pi i/q)(\tau_1\sigma_{23}+\tau_2\sigma_{31}+\tau_3\sigma_{12})} . \tag{2.22}$$

The product in the right-hand side of (2.21) is taken over every down-pointing triangle of a triangular lattice of the same size. The evaluation of (2.22) is facilitated by writing

$$e^{H(\sigma_1,\sigma_2,\sigma_3)}=1+v_1\delta_{23}+v_2\delta_{31}+v_3\delta_{12}+y\delta_{12}\delta_{23} , \tag{2.23}$$

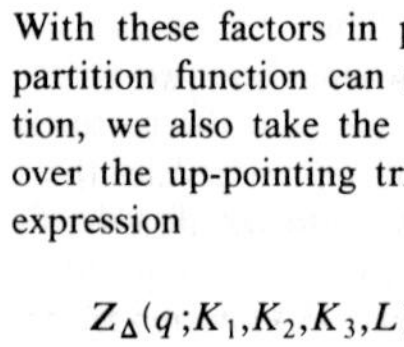

FIG. 8. Triangular Potts model with two-site interactions K_1, K_2, K_3, and three-site interactions L in alternate triangular faces.

where

$$\begin{aligned} v_i&=e^{K_i}-1 \\ y&=v_1v_2+v_2v_3+v_3v_1+v_1v_2v_3 \\ &\quad +(e^L-1)(1+v_1)(1+v_2)(1+v_3) \\ &=e^{L+K_1+K_2+K_3}-e^{K_1}-e^{K_2}-e^{K_3}+2 . \end{aligned} \tag{2.24}$$

Thus we obtain from (2.22)

$$e^{H^*(\sigma_1,\sigma_2,\sigma_3)}=\frac{y}{q}\left[1+\frac{qv_1}{y}\delta_{23}+\frac{qv_2}{y}\delta_{31} + \frac{qv_3}{y}\delta_{12}+\frac{q^2}{y}\delta_{12}\delta_{23}\right] , \tag{2.25}$$

and hence from (2.21) the duality relation

$$Z_\Delta(q;K_1,K_2,K_3,L)=\left[\frac{y}{q}\right]^N Z_\nabla(q;K_1^*,K_2^*,K_3^*,L^*) , \tag{2.26}$$

with

$$e^{K_i^*}-1=qv_i/y ,$$

$$y^*=q^2/y ,$$

where y^* is defined in terms of K_i^* and L^* as in (2.24). In (2.26), N is the number of sites of the triangular lattice and Z_∇ is the partition function of the same model with interactions in every down-pointing triangle.

Since by symmetry $Z_\Delta(q;K_i,L)=Z_\nabla(q;K_i,L)$, the partition function (2.18) is self-dual about the point

$$y=q . \tag{2.27}$$

The duality relation (2.26) was first observed by Kim and Joseph (1974) in the special case of $L=0$. The full duality (2.26) was first derived by Baxter *et al.* (1978) using an algebraic method, and later rederived by Enting (1978c) and by Wu and Lin (1980) from graphical considerations.

This method of taking partial traces can be readily adapted to other lattices. Applications to the square lattice, including a rederivation of the Essam duality (2.12) for pure four-site interactions mentioned before, have already been given by Burkhardt (1979). Here we state the result of another application (Enting and Wu, 1982).

Consider the triangular Potts model with two-site interactions K_1,K_2,K_3 and three-site interactions L now in *every* triangular face. The method of partial trace then relates this model to a Kagomé Potts model with two-site interactions K_1^*, K_2^*, K_3^* and three-site interactions L^* in the triangular faces of the Kagomé lattice. The equivalence is best seen by starting from the Kagomé lattice and taking the partial traces after introducing (2.20). The result leads to

$$Z_{\text{Triangle}}(q;K_1,K_2,K_3,L) = \left[\frac{y}{q}\right]^N Z_{\text{Kagomé}}(q;K_1^*,K_2^*,K_3^*,L^*) \tag{2.28}$$

with

$$e^{K_i^*}-1=qv_i/y\ ,$$

$$y^*=q^2/y\ , \tag{2.29}$$

where y^* is defined in terms of K_i^*,L^* as in (2.24), while y is similarly defined in terms of $K_i/2$ and L (Enting and Wu, 1982).

Finally, it should be noted that Enting (1975c) has considered a "quasi" q-state Potts model on the triangular lattice with three-site interactions. He showed that this model, which is an extension of the $q=2$ three-spin Ising model of Baxter and Wu (1973), possesses an exact duality relation.

D. The $Z(q)$ model

The $Z(q)$ model (1.2) plays an important role in the lattice gauge theories, and has already been eloquently reviewed in this perspective (Kogut, 1979; Einhorn *et al.*, 1980). Here we describe an exact duality relation valid for the $Z(q)$ model in two dimensions (Wu and Wang, 1976).

For the interaction (1.2) the nearest-neighbor Boltzmann factor reads

$$u(n_i-n_j)=\exp\{\beta J[2\pi(n_i-n_j)/q]\}\ , \tag{2.30}$$

where the interaction $J(\Theta)$ is 2π periodic. Denote the partition function with the nearest-neighbor Boltzmann factor (2.30) by $Z(u)$. It then follows from a simple geometric consideration (Wu and Wang, 1976) that $Z(u)$ is related to a partition function $Z^{(D)}(\lambda)$ similarly defined on the dual lattice. This exact duality reads

$$Z(u)=q^{1-N_D}Z^{(D)}(\lambda)\ , \tag{2.31}$$

where N_D is the number of sites of the dual lattice, and the λ's are the nearest-neighbor Boltzmann factors of the dual model given by

$$\lambda(m)=\sum_{n=0}^{q-1}\exp(2\pi imn/q)u(n)\ ,$$

$$m=0,1,\ \cdots\ ,q-1\ . \tag{2.32}$$

[In fact, the q λ's are the eigenvalues of the $q\times q$ cyclic matrix whose elements are (2.30).] The duality relation (2.31) has proven to be useful in constructing the phase diagram of the $Z(q)$ model [see, for example, Wu, (1979a), Cardy, (1980), Alcaraz and Köberle, (1980)]. Note that the duality (2.31) includes the duality (2.10) of the (standard) Potts model as a special case. Here again the duality (2.31) is valid more generally for models with edge-dependent interactions.

E. The dilute model

Extending the idea of duality in terms of graphical representations as presented in Sec. II.A, it is straightforward to derive a dual model for the dilute Potts model in the form of a graph-generating function. We refer to Wu (1981) for details of this extension.

Of special interest is a constrained version of the dilute model (1.16) whose parameters satisfy the relation

$$e^{K'}=q/(e^{K}+q-1)\ . \tag{2.33}$$

Under this constraint the dual of (1.16) is a Potts model with two-site and multisite interactions. The exact equivalence reads (Wu, 1981)

$$Z(q;K',K,z_i)=q^{N-1}e^{-EK^*}Z^{(D)}(q;K^*,L_i)\ , \tag{2.34}$$

with

$$e^{K}=(e^{K^*}+q-1)/(e^{K^*}-1)\ ,$$

$$e^{K'}=1-e^{-K^*}\ , \tag{2.35}$$

$$z_i=q(e^{L_i}-1)\ .$$

Here Z and $Z^{(D)}$ are, respectively, the partition functions of the dilute and the dual models. The dual model has nearest-neighbor interactions K^* and multisite interactions L_i among the spins surrounding the ith site of the original lattice.

III. SERIES EXPANSIONS

In the absence of an exact solution, series expansions and analyses remain as one of the most useful tools in the investigation of the critical properties of a model system. We describe in this section the various series expansions that can be developed for the Potts partition function. Specifically, we consider the Potts model defined on a finite graph G, and study the various subgraph expansions of the partition function. It should be pointed out that while one can always extract from these expansions the series for infinite lattices by taking G as a lattice, as is done in Kihara *et al.* (1954) and Straley and Fisher (1973), the use of sophisticated techniques is more efficient in generating long series. We shall not discuss the details of these developments.

The techniques and methods for generating long series are very much q-dependent. The $q=1$ and $q=2$ systems are special, and have been the subject of intense research interests for many years. For reviews of these developments see Essam (1980) for the $q=1$ (percolation) model, Domb (1974b) for the $q=2$ (Ising) model, and Gaunt and Guttman (1974) for series analyses.

Development of expansions for the general q problem was initiated by Kihara *et al.* (1954) from a "primitive" consideration (as described in Domb, 1960) of the partition function series. Modern techniques applicable to the general q problem have since been developed, largely due to the effort of Enting. The low-temperature, high-field series have been generated by the use of the methods of partial generating functions of Sykes *et al.* (1965), the linkage rule of Sykes and Gaunt (1973) (Enting, 1974a, 1974b, 1975b, 1978a), and more recently by the finite-lattice methods (de Neef, 1975; de Neef and Enting, 1977; Enting, 1978a, 1978b, 1980b). The high-temperature series for the square lattice have been generated in a similar fashion (de Neef, 1975; de Neef and Enting, 1977; Enting and Baxter, 1977; Enting, 1978a, 1978b). For specific values of q, the finite lattice methods used in conjunction with a high-speed digital computer have proven to be capable of producing series of lengths otherwise difficult to achieve.

In the following we consider a Potts model defined on a finite graph G, which can also be a lattice, and study the various subgraph expansions of the partition function. According to the expansion parameter to be used, these expansions can be classified as the low- and high-temperature series.

A. Low-temperature expansion

The low-temperature expansion for the Potts model with ferromagnetic nearest-neighbor interactions ($K>0$) can be generated by explicitly enumerating the spin configurations, and this can be done for any finite graph G. Starting from a configuration in which all spins are in the same state, one can generate other spin configurations one at a time by considering states with one spin different, two spins different, etc. This procedure also has the advantage of including fugacities, or external fields, to the different individual states. Thus one obtains quite generally an expansion of the form

$$Z_G(q;K,z_1,\ldots,z_n)=e^{EK}\sum_{\substack{n_1\cdots n_q\\ n_1+\cdots+n_q=N}} a(n_1,\ldots,n_q,s)\times z_n^{n_1}\cdots z_q^{n_q}e^{-sK}, \tag{3.1}$$

where $a(n_1,\ldots,n_q,s)$ is the number of spin configurations in which there are n_i spins in the state i and s edges connecting neighboring spins in different states; z_i is the fugacity for the ith spin state. Terms in (3.1) can be further grouped according to the relative importance of the expansion parameters of interest, and this has led to the various low-temperature series expansions.

In zero fields ($z_1=\cdots=z_q=1$) the expansion (3.1) simplifies to

$$Z_G(q;K)=qe^{EK}\left[1+\sum_{s=\gamma}^{E}a_se^{-sK}\right], \tag{3.2}$$

where $a_s=\sum_{n_i}a(n_1,\ldots,n_q,s)$ and γ is the coordination number of G.

Despite its simple form, the usefulness of (3.2) is limited by the extent to which the numbers a_s can be computed. However, an alternate expression of (3.2) can be generated as follows: For planar G, introduce the dual lattice D and draw bonds along the edges of D separating spins in different states. It is clear that the bonds form subgraphs $D'\subseteq D$ that are *closed,* i.e., without vertices of degree 1. [I shall denote the summation over such subgraphs by the superscript (c).] Furthermore, to each D' there exist precisely $P_{D'}^{(F)}(q)$ spin configurations, where $P_{D'}^{(F)}(q)$ is the number of q-colorings of the *faces* of D'. Thus (3.2) can be rewritten as

$$Z_G(q;K)=e^{EK}\sum_{D'\subseteq D}{}^{(c)}P_{D'}^{(F)}(q)e^{-b(D')K}, \tag{3.3}$$

where $b(D')$ is the number of bonds in D'. In this form the low-temperature expansion can be more conveniently enumerated.

The generalization of (3.3) to higher dimensions is straightforward but more tedious. One needs to keep track of the "partitions" separating regions of different spin states as well as the number of q colorings of these regions. In this way low-temperature expansion can be in principle generated for any dimension d. [See Sykes (1979) for $q=2$, $d=4$, and Ditzian and Kadanoff (1979) for $q=4$, $d=4$ expansions].

B. High-temperature expansions

The expansion (2.3) for the Potts partition function is already in the form of a high-temperature expansion. [The corresponding expansion for models with multisite interactions is (2.14).] Since in this form the partition function is expanded over *all* subgraphs $G'\subseteq G$ where G is the lattice, the expansion is rather inefficient in generating high-order terms.

To remedy this situation, one can rewrite the partition function (2.1) in the form of (Domb, 1974a)

$$Z_G(q;K)=\sum_{\sigma_i=0}^{q-1}\prod_{\langle ij\rangle}[t(1+f_{ij})], \tag{3.4}$$

where

$$t=(q+v)/q, \tag{3.5}$$

$$f_{ij}=\frac{v}{q+v}[-1+q\delta_{\mathrm{Kr}}(\sigma_i,\sigma_j)],$$

and proceed to expand (3.4) graphically as in (2.1). It can be readily verified that

$$\sum_{\sigma_j=0}^{q-1}f_{ij}=0, \tag{3.6}$$

and, consequently, all subgraphs with one or more vertices of degree 1 give rise to zero contributions. The number of subgraphs that occur in the expansion is therefore greatly reduced. Thus one obtains

$$Z_G(q;K)=t^E\sum_{G'\subseteq G}{}^{(c)}w(G'), \tag{3.7}$$

where the superscript (c) has the same meaning as in (3.3), i.e., summation over subgraphs without vertices of degree 1. Also $w(G')=\sum\prod_{G'}f_{ij}$ is a weight factor associated with the subgraph G'.

Domb (1974a) noticed that the weight factor $w(G')$ depends essentially on the topology of G' and, consequently, it is necessary to consider only those subgraphs of star topology. He then proceeded to determine $w(G')$

for the leading star graphs.

An expression of $w(G')$ for general G' can be obtained by further expanding in $w(G')$ the product $\prod f_{ij}$ (Wu, 1978). This procedure leads to, as in Domb (1974a), the consideration of G_s', the star graph which is topologically isomorphic to G'. (G_s' is obtained from G' by disregarding *all* vertices of degree 2.) This analysis (Wu, 1978) leads to the following general expression for $w(G')$:

$$w(G')=q^{N-N(G_s')}\left[\frac{v}{v+q}\right]^{b(G')}\times(-1)^{b(G_s')}Z_{G_s'}(q;e^K=1-q)\ . \tag{3.8}$$

Here $b(G_s')$ and $N(G_s')$ are, respectively, the numbers of bonds and sites in G_s', and $Z_{G_s'}$ is the partition function of a Potts model on G_s'. For example, the weight factor for the G' of Θ topology shown in Fig. 9 is [for a definition of graph topology see, for example, Domb (1974b)]

$$\begin{aligned} w(G')&=q^{N-1}\left[\frac{v}{v+q}\right]^{b(G')}(-1)^3q[(1-q)^3+q-1]\\ &=q^N(q-1)(q-2)\left[\frac{v}{v+q}\right]^{b(G')}\ . \end{aligned} \tag{3.9}$$

The expression (3.7) with $w(G')$ given by (3.8) is again a high-temperature expansion and is valid for arbitrary G. Note that the terms in the expansion are of the form of $[v/(v+q)]^{b(G')}$ with coefficients determined purely by the topology of G'. This expansion also reveals a curious "recursion" relation for the Potts partition function. The expansion (3.7) was first used by Nagle (1971) in a computation of the chromatic polynomial, the special case of $v=-1$. However, his procedure was rather elaborate and the explicit expression (3.8) for the graph weights was not made apparent. The extension of Nagle's procedure to general v was later pointed out by Temperley (1976).

The expression (3.8) for $w(G')$ can be further reduced if G', hence G_s', is planar (G is not necessarily planar). This is accomplished by introducing the duality relation (2.10) to rewrite $Z_{G_s'}(q;e^K=1-q)$. This leads to, upon using (1.15a),

$$w(G')=q^{N-1}P_{G_s'}^{(F)}(q)\left[\frac{v}{v+q}\right]^{b(G')}\quad(\text{planar } G')\ . \tag{3.10}$$

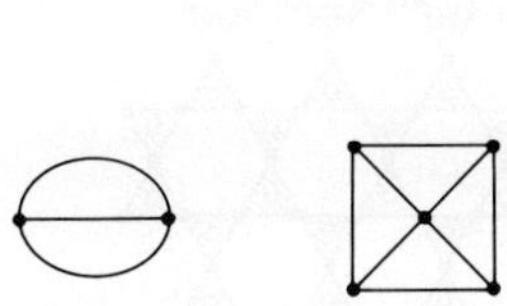

FIG. 9. Examples of star graph G_s'. The numbers of sites and bonds of the two graphs shown are $N(\Theta)=2$, $b(\Theta)=3$, $N(F)=5$, $b(F)=8$.

Furthermore, since G' and G_s' are topologically isomorphic, we have $P_{G_s'}^{(F)}(q)=P_{G'}^{(F)}(q)$ so that the coefficient in (3.10) can be quite easily generated in practice. For example, the numbers of face colorings of the subgraphs G' represented by the Θ and F topologies shown in Fig. 9 are

$$P_\Theta^{(F)}(q)=q(q-1)(q-2)$$

and

$$\begin{aligned} P_F^{(F)}(q)&=q(q-1)[(q-3)^2+q-2]\\ &=q(q-1)(q^2-5q+7)\ . \end{aligned} \tag{3.11}$$

Substitution of these numbers into (3.10) then leads to the graph weights which have previously been obtained by Domb (1974a) from a more elaborate procedure.

The high-temperature expansions (2.3) and (3.7) are useful in that the subgraphs are on G and are valid for G in any dimension. For planar G, subgraphs $G'\subseteq G$ are planar. Then we can always combine (3.10) with (3.7), and this leads to

$$Z_G(q,K)=v^Eq^{1-N_D}e^{EK^*}\sum_{G'\subseteq G}{}^{(c)}P_{G'}^{(F)}(q)e^{-b(G')K^*}\quad(\text{planar } G)\ , \tag{3.12}$$

where $e^{-K^*}=v/(v+q)$. We recognize that this is precisely the "low-temperature" expansion (3.3) on the dual graph D. In fact, we could have derived (3.12) more directly by combining the duality relation (2.10) with (3.3), and it was using this procedure that Kihara *et al.* (1954) first generated the high-temperature series for the square lattice.

The high-temperature expansion (2.3) can also be extended to include external fields as in (3.1). We refer to Kim and Joseph (1975) for a discussion on this formulation.

C. Series developments

I summarize in this section the present status on series developments for the $q>2$ Potts models on infinite lattices. Description on the results of series analyses will be found in Sec. V.

1. Square lattice

Series expansion for the q-dependent zero-field partition function was first developed by Kihara *et al.* (1954) up to terms of u^{16}, where u can be either the low-temperature variable e^{-K} or the high-temperature variable e^{-K^*} related by the duality relation (2.11). Enting (1977) has pointed out, however, that their coefficient of u^{16} is in error [see also de Neef (1975)]. The series has been extended to terms of u^{31} for $q=3$ by Enting (1980a) using the finite lattice method. In addition, Enting (1980a) has also generated the $q=2$ series for the order parameter to u^{31}.

The q-dependent low-temperature expansion (3.1) which includes external fields has been developed by Straley and Fisher (1973) to the order of u^{13}. For specific values of q, the high-field low-temperature series have been developed for $q=3$ (Enting, 1974a) and for $q=4,5,6$ (Enting, 1974b). The zero-field low-temperature series have also been obtained in various lengths by Zwanzig and Ramshaw (1977) for $q=2,3,4$, and by de Neef and Enting (1977) for $q=3$.

The q-dependent high-temperature series (2.3) including an external field has been formulated by Kim and Joseph (1975). From this formulation they obtained the susceptibility series for $q=3,4,5,6$.

2. Triangular lattice

Series expansions for the triangular lattice have been derived mostly for $q=3$. The high-field expansion was first studied by Enting (1974a). Series expansions for the zero-field model with two-site and/or three-site interactions in half of the triangles have been considered by Enting (1978c, 1980c); Enting and Wu (1982) have generated series for models with pure three-site interactions in every triangle and for the antiferromagnetic model. The high-field low-temperature expansion for $q=4$ has been given by Enting (1975). In addition, the high-temperature susceptibility series has been given by Kim and Joseph (1975) for $q=3,4,\ldots,8$.

3. Honeycomb lattice

It is to be noted that some results of the honeycomb lattice are related to those of the triangular lattice. The only independent series for the honeycomb lattice appears to be the low-temperature, high-field series for the $q=3$ model (Enting, 1974b).

4. Lattices in $d>2$ dimensions

Series developments for three-dimensional lattices have been generated mostly for the $q=3$ models. The high-temperature, low-field and the low-temperature, high-field expansions for the simple cubic lattice have been considered by Straley (1974). The high-field series have been further extended by Enting (1974a) for the sc, fcc, and bcc lattices; Ditzian and Oitmaa (1974) also considered the $q=3$ series for the fcc lattice. In addition, the $q=3$ high-temperature susceptibility series for the bcc lattice has been given by Kim and Joseph (1975). The most recent high-field expansions for the $q=3$ sc and bcc lattices have been given by Miyashita *et al.* (1979).

For the $q=4$ model Ditzian and Kadanoff (1979) have generated the high-temperature series for the hypercubic lattices for $d\geq 2$ up to $d=10$ dimensions. In addition, they also obtained the low-temperature series for the $q=4$, $d=4$ hypercubic lattice.

IV. RELATION WITH OTHER PROBLEMS

The Potts model is related to a number of other outstanding problems in lattice statistics. While most of these other problems are also unsolved, the connection with the Potts model has made it possible to explore their properties from the known information on the Potts model or vice versa. It is from this consideration that most of the known properties of the critical behavior of the two-dimensional Potts model have been established.

A. Vertex model

The Potts model in two dimensions is equivalent to an ice-rule vertex model. This representation of the Potts model, first pointed out by Temperley and Lieb (1971) for the square lattice, has been extended to arbitrary planar lattices (Baxter *et al.*, 1976). Here I shall state only the result.

Consider a Potts model on a planar lattice (or graph) $\mathscr{L}$ of N sites. Then this Potts model is related to an ice-rule vertex model defined on a related lattice (or graph) $\mathscr{L}'$ through the simple relation

$$Z_{\text{Potts}}=q^{N/2}Z_{\text{vertex}}\,, \tag{4.1}$$

where Z_{Potts} and Z_{vertex} are the respective partition functions.

For a given $\mathscr{L}$, the related lattice $\mathscr{L}'$ is not necessarily unique. The basic properties of $\mathscr{L}'$ are that (i) the faces of $\mathscr{L}'$ are bipartite, and (ii) the lattice $\mathscr{L}$ can be embedded in the faces of $\mathscr{L}'$ such that the sites of $\mathscr{L}$ occupy one set of the bipartite faces. For Potts models with pure two-site interactions, one such construction of $\mathscr{L}'$ is the surrounding lattice (or medial graph) of $\mathscr{L}$, obtained by connecting the neighboring midpoints of the edges of $\mathscr{L}$. For example, the surrounding lattice of a square lattice is a square lattice, and that of a honeycomb (and triangular) lattice is a Kagomé lattice. These situations are shown in Fig. 10. Note that the coordination number of the surrounding lattice $\mathscr{L}'$ is always 4. Moreover, it proves convenient to shade those faces of $\mathscr{L}'$ containing sites of $\mathscr{L}$ for the purpose of distinction (there are always two shaded and two unshaded faces intersecting at a site of $\mathscr{L}'$).

The ice-rule vertex problem on $\mathscr{L}'$ is defined as follows: Attach arrows to the edges of $\mathscr{L}'$ such that there

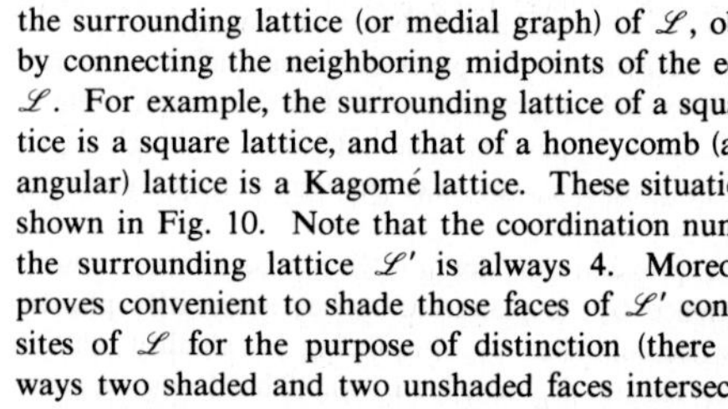

FIG. 10. Examples of a planar lattice $\mathscr{L}$ (open circles) and the associated surrounding lattice $\mathscr{L}'$ (solid circles).

are always two arrows in and two arrows out at a site of $\mathscr{L}'$ (the ice rule). The six ice-rule vertices are shown in Fig. 11. Vertex weights are then assigned according to the vertex arrow configurations. In the most general case the weights depend on the angles between the four incident edges relative to the face shading (Baxter *et al.*, 1976). For the square, triangular, and honeycomb lattices the weights are given by

$$(\omega_1,\ldots,\omega_6)=(1,1,x_r,x_r,A_r,B_r)\,, \tag{4.2}$$

where

$$\begin{aligned}(A_r,B_r)&=(s^{-1}+x_r s,s+x_r s^{-1}) \quad \text{square}\\ &=(t^{-1}+x_r t^2,t+x_r t^{-2}) \quad \text{triangular}\\ &=(t^{-2}+x_r t,t^2+x_r t^{-1}) \quad \text{honeycomb}\,,\end{aligned} \tag{4.3}$$

with

$$\begin{aligned}&s=e^{\Theta/2},\quad t=e^{\Theta/3},\quad 2\cosh\Theta=\sqrt{q}\\ &x_r=(e^{K_r}-1)/\sqrt{q}\,.\end{aligned} \tag{4.4}$$

Here we have allowed different Potts interactions along the different lattice axes. It should be pointed out that the equivalence (4.1) holds only for lattices $\mathscr{L}$ and $\mathscr{L}'$ that are both planar with special boundary conditions. It is *not* generally valid for lattices with toroidal periodic boundary conditions (Baxter, 1982a, 1982b).

The vertex weights (4.2) can be transformed into a more symmetric form (Hintermann *et al.*, 1978):

$$(A_r,B_r)=(c_r z,c_r z^{-1})\,, \tag{4.5}$$

with

$$\begin{aligned}&c_r^2=A_r B_r=1+x_r^2+\sqrt{q}\,x_r\\ &z=(A_1A_2/B_1B_2)^{1/2} \quad \text{square}\\ &\quad=(A_1A_2A_3/B_1B_2B_3)^{1/3} \quad \text{triangular and honeycomb}\,.\end{aligned} \tag{4.6}$$

In this form the variable $\ln z$ can be regarded as a staggered field applied to the system.

For the Potts model on the triangular $\mathscr{L}$, another choice of $\mathscr{L}'$ is shown in Fig. 12, for which $\mathscr{L}'$ is again a triangular lattice. One is thus led to the consideration of a 20-vertex model on the triangular lattice. The equivalence of the triangular Potts model with such an (ice-rule) 20-vertex model was first established by Baxter *et al.* (1978), and a graphical analysis was later given by Wu and Lin (1980). One novel point of this choice of $\mathscr{L}'$ is the possibility of including three-site interactions in alternate triangles in the Potts model. Details of this equivalence can be found in Baxter *et al.* (1978).

As in (2.3) the vertex-model representation also serves as a natural continuation of the Potts model to continuous values of q. It is to be noted, however, that the vertex weights (4.2) are real for $q\geq 4$ and complex for $q<4$.

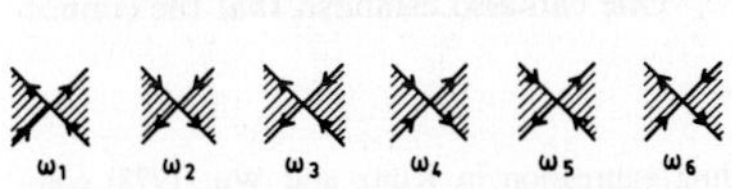

FIG. 11. The six ice-rule configurations at a vertex of the surrounding lattice and the associated vertex weights.

B. Percolation (q = 1 limit)

The percolation process provides a simple picture of a critical point transition that has been of theoretical interest for some years [see, for example, reviews by Essam (1972, 1980)]. It was first pointed out by Kasteleyn and Fortuin (1969) that the problem of the bond percolation can be formulated in terms of the Potts model. This formulation has been used in, for example, the renormalization group studies of the percolation problem (Harris, *et al.*, 1975; Dasgupta, 1976). The method of Kasteleyn and Fortuin has since been elucidated by Stephen (1977) and by Wu (1978), and extended further to the problem of site percolation (Giri *et al.*, 1977; Kunz and Wu, 1978). Murata (1979) has similarly formulated the site percolation in a lattice gas as a dilute Potts model.

1. Bond percolation

In a bond percolation process there is a probability p for each edge of an (infinite) lattice G to be "occupied" and a probability $1-p$ for it to be "vacant." Two sites that are connected through a chain of occupied edges are said to be in the same cluster. Then various questions can be asked concerning the clusters distribution (Essam, 1972). Among others, one is interested in the percolation probability $P(p)$ that a given point, say, the origin, of the lattice belongs to an infinite cluster, and the mean size $S(p)$ of the finite cluster that contains the origin. In the latter instance the cluster size can be measured by either the site or the edge content.

Consider a nearest-neighbor q-component Potts model whose Hamiltonian $-\beta\mathscr{H}_q$ is given by (1.18). A straightforward high-temperature expansion of its partition function as in (2.3) leads to the expression[1] (Wu, 1978)

$$Z(q;K,M,L)=\sum_{G'\subseteq G} v^{b(G')}\prod_c\left(e^{Ls_c(G')+L_1b_c(G')}+q-1\right)\,, \tag{4.7}$$

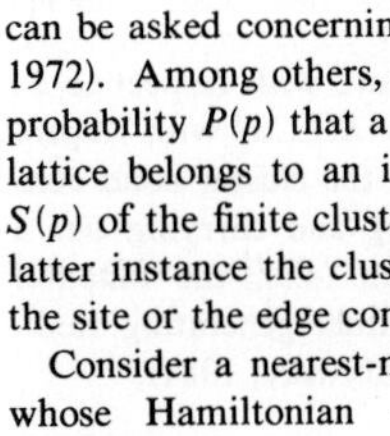

FIG. 12. Triangular $\mathscr{L}'$ (solid circles) for the triangular lattice $\mathscr{L}$ (open circles).

---

[1]The corresponding expression in Wu (1978) contains a misprint. The phase after Eq. (35) should read "where $e^{L_1}=(e^{K+H_1/kT}-1)/(e^K-1)$."

where

$$e^{L_1}=(e^{K+M}-1)/(e^K-1) \tag{4.8}$$

and we have taken $L_i=L$ in the Hamiltonian (1.18). The product in (4.7) is over all connected clusters of G', including isolated sites, and $s_c(G')$, $b_c(G')$ are respectively the numbers of sites and occupied edges of a cluster.

Defining the per site free energy $f(q;K,L,M)$ as in (1.8), one then has

$$\begin{aligned} h(K,L,M) &\equiv \left[\frac{\partial}{\partial q} f(q;K,L,M)\right]_{q=1} \\ &= \langle \sum_c e^{-Ls_c-L_1b_c} \rangle_0 , \end{aligned} \tag{4.9}$$

where

$$\begin{aligned} \langle A \rangle_0 &= \lim_{N\to\infty} N^{-1}\langle A \rangle \\ \langle A \rangle &= \sum_{G'\subseteq G} p^{b(G')}(1-p)^{E-b(G')}A(G') \\ p &= 1-e^{-(K+M)} . \end{aligned} \tag{4.10}$$

Now the right-hand side of (4.9) is precisely the generating function for quantities of interests in the percolation problem. For example,

$$\begin{aligned} P(p) &= 1+h'(K,0+,0) , \\ S(p) &= h''(K,0+,0) , \end{aligned} \tag{4.11}$$

where the derivatives of $h(K,L,0)$ are with respect to L. It is also clear that derivatives of $h(K,L,M)$ with respect to L_1 generate quantities involving the cluster bond contents. Furthermore, by rearranging and carrying out a partial summation of the terms in (4.9), the function $h(K,L,M)$ reduces to the *bond*-animal generating function for G as follows (Harris and Lubensky, 1981):

$$h(K,L,M)=\sum_A p^b q^t z^s , \tag{4.9'}$$

where

$$p=e^{-M}(1-e^{-K}) ,$$

$$q=e^{-(K+M)} ,$$

$$z=e^{-L} .$$

The summation in (4.9′) is taken over all bond animals that pass through a given point—say, the origin—of G, and the symbols b,s,t denote, respectively, the numbers of bonds, sites, and perimeter bonds. The bond percolation is generated, as in (4.11), by taking $p+q=1$ or $M=0$, and the pure animal problem is generated by $q=1$ or $K+M=0$.

One can show in a similar fashion (Wu, 1978) that the connectivity $c(\mathbf{r},p)$, defined to be the probability that the sites at the origin and at $\mathbf{r}$ belong to the same cluster, is given by

$$c(\mathbf{r},p)=\left[\frac{\partial}{\partial q}\Gamma_{\alpha\alpha}(\mathbf{O},\mathbf{r})\right]_{q=1} , \tag{4.12}$$

where $\Gamma_{\alpha\alpha}(\mathbf{r}_1,\mathbf{r}_2)$ is the two-point correlation function (1.13) of the Potts model. Thus a knowledge of the Potts model for general q will yield the solution of the bond percolation problem. This is the result of Kasteleyn and Fortuin (1969).

2. Site percolation

In a site percolation process each site of an infinite lattice is occupied independently with a probability s. A cluster is then a set of occupied sites connected by the lattice edges. One can ask the same kind of questions regarding the cluster distributions as in the case of the bond percolation (Essam, 1972).

The site percolation problem can be formulated as the $q=1$ limit of a Potts model with multisite interactions (Giri *et al.*, 1977; Kunz and Wu, 1978). In addition to the multisite interactions as given in (1.6), one also introduces a multisite external field as in (1.6). Quite generally, to describe site percolation on a lattice G of N sites and coordination number γ, one considers a Potts model on the covering lattice G_c defined with its $\frac{1}{2}\gamma N$ sites located on the edges of G. The Potts model has the Hamiltonian

$$-\beta\mathscr{H}=\sum_i\left[K\sum_\alpha\delta_\alpha(i)+M\delta_0(i)\right] , \tag{4.13}$$

where $\delta_\alpha(i)=1$ if all γ sites of G_c surrounding the ith site of G are in the same state α, $\alpha=0,1,\ldots,q-1$ and $\delta_\alpha(i)=0$ otherwise. Let $f(q;K,M)$ be the free energy (1.8) for the Hamiltonian (4.13) and define

$$h(K,M)=\left[\frac{\partial}{\partial q}f(q;K,M)\right]_{q=1} . \tag{4.14}$$

Then it is straightforward to show[2] (Kunz and Wu, 1978)

$$h(K,M)=\langle b\rangle_0-(\gamma-\tfrac{1}{2})s+\left\langle\sum_c e^{-Ls_c}\right\rangle_0 , \tag{4.15}$$

where b is the number of pairs of neighboring occupied sites and $\langle\ \rangle_0$ is an average defined as in (4.10) over site occupations. Additionally,

$$\begin{aligned} e^L &= (e^{K+M}-1)/(e^K-1) , \\ s &= 1-e^{-(K+L)} . \end{aligned} \tag{4.16}$$

The function $h(K,M)$ now generates the site percolation on G. In particular, the percolation probability is

$$P(s)=1+\left[\frac{\partial}{\partial L}h(K,M)\right]_{L=O+} \tag{4.17}$$

with $s=1-e^{-K}$. One can also establish that the connec-

[2]The corresponding expression in Kunz and Wu (1978) contains a misprint. The last expression in the phrase after Eq. (3) should read "$e^L=(e^{K+H/kT}-1)/(e^K-1)$."

tivity in the percolation problem is connected to the Potts correlation function as in (4.12).

For site percolation on planar lattices, the generating Potts model is not necessarily unique (Temperley and Ashley, 1981). The situation parallels the formulation of the Potts model as a vertex model described in Sec. IV.A. Generally, the site percolation on any planar lattice $\mathcal{L}$ can be formulated as a Potts model on the related lattice $\mathcal{L}'$ as defined in Sec. IV.A. One observes that the covering lattice G_c is precisely the surrounding lattice of Sec. IV.A. But other choices of $\mathcal{L}'$ may also be possible. Examples of pairs of $\mathcal{L}$ and $\mathcal{L}'$ have been given in Figs. 10 and 12.

3. Site-bond percolation

In a mixed site-bond percolation each site of a lattice G is occupied with a probability s and each bond of G is occupied with a probability p. Two sites are in the same cluster if they are connected through a sequence of occupied sites and edges. The pure bond and site percolations are then recovered by taking the respective special cases of $s=1$ and $p=1$.

For the mixed problem the percolation threshold is defined to be the phase boundary

$$g(s,p)=0\ , \tag{4.18}$$

beyond which there is a nonzero probability that a given site belongs to an infinite cluster.

It is straightforward to generalize the previous considerations to this site-bond percolation problem. Instead of giving this formulation in its most general form, we focus on the connection of the percolation boundary (4.18) with the corresponding Potts critical point.

Consider, for example, the site-bond percolation on the square lattice. The appropriate choice of the Potts lattice is that of Fig. 7. Let

$$f(q;e^{-K_2^*},e^{-K_4^*})=0 \tag{4.19}$$

denote the critical point of this Potts model, and then the previous considerations will lead to the following expression for the percolation threshold (4.18):

$$f(1;1-p,1-s)=0\ . \tag{4.20}$$

The extension of (4.20) to lattices in general is obvious. Generally for the threshold of site-bond percolation on any lattice G (or $\mathcal{L}$), we consider in a similar way the critical condition of the Potts model on the covering lattice G_c (or $\mathcal{L}'$) which has *two* sites on each edge of G. The two spins on an edge interact with an interaction of strength K_2^*, while the γ spins immediately surrounding a vertex of G interact with an interaction of strength K_γ^*. The threshold *percolation* is then obtained from the Potts critical condition with the substitution of $e^{-K_2^*}=1-p$, $e^{-K_\gamma^*}=1-s$ at $q=1$. This result is valid for G in any dimension.

For planar G the duality relation (2.12) and (2.13) can be used at this point to yield valuable information. Particularly, the threshold for the site-bond percolation with respective probabilities (s,p) can also be derived from the critical condition of the Potts model on the dual lattice G^* (as defined in Sec. II.B). [This result was first derived for the honeycomb G by Kondor (1980), and the generalization was later given by Wu (1981).] For pure bond percolation $(s=1)$ the resulting Potts model can also be interpreted as generating a bond percolation on the dual lattice D (as defined in Sec. II.A); (2.13) then relates the critical probabilities p_c and p_c^* for a pair of dual lattices by

$$p_c+p_c^*=1\ . \tag{4.21}$$

For pure site percolation $(p=1)$ a similar consideration using (2.13) relates the critical probabilities s_c and s_c^* for a pair of *matching* lattices by

$$s_c+s_c^*=1\ . \tag{4.22}$$

Both (4.21) and (4.22) are well-known results in percolation processes [see Essam (1972, 1979)].

C. Resistor network ($q=0$ limit)

1. Result of Fortuin and Kasteleyn

The problem of finding the effective resistance between two node points of a network of linear resistors was solved by Kirchhoff (1847) a century and a half ago. But Fortuin and Kasteleyn (1972) showed that Kirchhoff's solution can be expressed as a $q=0$ limit of the Potts partition function. Here we shall examine this connection.

The Potts model is defined on a lattice G whose sites coincide the node points of the network and whose edges coincide the resistors. Thus, the Potts spins at sites i and j interact with an interaction of strength $-kTK_{ij}$ if the two nodes are connected via a resistor of resistance r_{ij}. A consistent picture is achieved if we take $K_{ij}\sim r_{ij}^{-1}$.

To obtain the effective resistance R_{kl} between two node points k and l, consider also the lattice $\bar{G}$ derived from G by adding an edge connecting the sites k and l. [$\bar{G}\equiv G$ if the edge (kl) is already present in G.] Denote the partition function of the Potts model on G by $Z_G(q;v_{ij})$, where $v_{ij}=\exp(K_{ij})-1$, and similarly define $Z_{\bar{G}}(q;,v_{ij})$. Then the result of Fortuin and Kasteleyn can be stated simply as

$$R_{kl}=\lim_{q\to 0}\left[\frac{\partial}{2x_{kl}}Z_{\bar{G}}(q;q^\alpha x_{ij})/Z_G(q;q^\alpha x_{ij})\right], \tag{4.23}$$

where $0<\alpha<1$ and $x_{ij}=r_{ij}^{-1}$.

Note that the value of R_{kl} in (4.23) is independent of the value of $0<\alpha<1$ ($\alpha=\frac{1}{2}$ in Fortuin and Kasteleyn 1972). If the edge (kl) is already in G, then (4.23) becomes

$$R_{kl}=\lim_{q\to 0}\frac{\partial}{\partial x_{kl}}\ln Z_G(q;q^\alpha x_{ij}),\quad 0<\alpha<1\ , \tag{4.24}$$

and is the nearest-neighbor correlation of the $q \to 0$ Potts model.

To prove (4.23) we first show that $Z_G(q;q^\alpha x_{ij})$ generates spanning trees on G in the $q \to 0$ limit [Fortuin and Kasteleyn (1972); see also Stephen (1976); Wu (1977)].

In analogy to (2.3), we write

$$Z_G(q;q^\alpha x_{ij}) = \sum_{G' \subseteq G} q^{n(G')+ab(G')} \prod_{E(G')} x_{ij} = q^{aN} \sum_{G' \subseteq G} q^{ac(G')+(1-\alpha)n(G')} \prod_{E(G')} x_{ij} , \tag{4.25}$$

where the product is over the edge set of G', N is the total number of sites of G, and $c(G')$ is the number of independent circuits in G' as defined by (2.6). For $0<\alpha<1$ and in the limit of $q \to 0$, the leading contributions in (4.25) are those terms represented by the spanning trees of G. A spanning tree T' is a connected subgraph which covers all sites $[n(T')=1]$ and has no circuit $[c(T')=0]$. Equation (4.25) then leads to the following expression for the spanning tree polynomial:

$$T_G(x_{ij}) \equiv \sum_{T' \subseteq G} \prod_{E(T')} x_{ij} = \lim_{q \to 0} q^{\alpha(1-N)-1} Z_G(q;q^\alpha x_{ij}), \quad 0<\alpha<1 . \tag{4.26}$$

The expression (4.23) now follows immediately from (4.26) and the result established by Kirchhoff (1847) which states, in present notation,

$$R_{kl} = \left[\frac{\partial}{\partial x_{kl}} T_{\bar{G}}(x_{ij}) \right] \Big/ T_G(x_{ij}) . \tag{4.27}$$

2. Result of Kirchhoff

For completeness, I outline a proof of the Kirchhoff result (4.27) [see also den Nijs (1979c)].

Let V_i be the potential at the ith node. The well-known Kirchoff's law states that there is a net current of magnitude

$$I_i = \sum_{j \neq i} x_{ij}(V_i - V_j) \tag{4.28}$$

flowing into the network at the ith node.

Equation (4.28) can be written in the more revealing form

$$I_i = \sum_{j=1}^{N} A_{ij} V_j, \quad i=1,2,\ldots,N \tag{4.29}$$

where

$$A_{ij} = \sum_{k \neq i} x_{ik}, \quad i=j$$
$$= -x_{ij}, \quad i \neq j \tag{4.30}$$

are the elements of the tree matrix for G [see, for example Hararay (1969) p. 158]. Then

$$R_{kl} = \frac{V_k - V_l}{I} , \tag{4.31}$$

where V_k and V_l are obtained from the solution of (4.29), with

$$I_i = I(\delta_{ik} - \delta_{il}) . \tag{4.32}$$

The tree matrix A has the property that the sum of each row or each column is identically zero. This has the consequence that all cofactors of A are identical and are equal to the spanning tree polynomial $T_G(x_{ij})$ (Hararay, 1969).[3] It also implies that we need only to consider the $N-1$ independent equations

$$I\delta_{ik} = \sum_{j=1}^{N} A_{ij}(V_j - V_l), \quad i,j \neq l . \tag{4.33}$$

It follows that

$$R_{kl} = |A^{(kl)}| / |A^{(l)}| , \tag{4.34}$$

where $|A^{(l)}|$ is the determinant of A with the lth row and column removed, and $|A^{(kl)}|$ is the same determinant but with the lth and the kth rows and columns removed.

Now $|A^{(l)}| = T_G(x_{ij})$, since it is a cofactor of A. It is also apparent upon a moment's reflection that $|A^{(kl)}|$ is the coefficient of x_{kl} in the cofactor $|A^{(k)}|$ (or $|A^{(l)}|$) if the nodes k and l are connected by a resistor. If there is no resistor between k and l, we simply add such a resistor, and $|A^{(kl)}|$ is again the coefficient of x_{kl} in $|A^{(k)}|$ or $|A^{(l)}|$. Thus

$$|A^{(kl)}| = \frac{\partial}{\partial x_{kl}} T_{\bar{G}}(x_{ij}) .$$

This reduces (4.34) to (4.27), thus completing the proof.

3. Remarks

The number of spanning trees on G can be obtained from (4.26) at $x_{ij}=1$. For $\alpha=\frac{1}{2}$ the right-hand side of (4.26) can be exactly evaluated for the square, triangular, and honeycomb lattices. Using this formulation, Wu (1977) has computed the number of spanning trees for these three lattices. The number of oriented spanning trees (complete self-avoiding walks) for the square lattice with a given orientation has been computed by Kasteleyn (1963).

It has been pointed out to this author that the spanning trees can be counted more directly by evaluating the infinite limit of the determinant $|A(x_{ij})|$ at $x_{ij}=1$, and that this procedure is valid for lattices in any dimension.[4]

---

[3]This fact has been used by Temperley (1958) to obtain a numerical estimate of $T_G(1)$, the number of spanning trees, for an infinite square lattice.

[4]I am indebted to P. W. Kasteleyn and H. Kunz for this comment.

Indeed, the correct values for $T_G(1)$ are generated from this procedure for the three two-dimensional lattices considered by Wu (1977). However, it is to be noted that the determinant $|A(x_{ij})|\equiv 0$ identically for G finite. If we denote the eigenvalues of $A(x_{ij})$ by $0,\lambda_2,\lambda_3,\ldots,\lambda_N$, then for finite G we have

$$T_G(x_{ij})=\prod_{n=2}^{N}\lambda_n \ .$$

It is only in the limit of infinite G that this result is identical to that obtained by directly evaluating the determinant $|A(x_{ij})|$.

With the choice of $\alpha=1$, Eq. (4.25) generates forests on G (Stephen, 1976; Wu, 1977). A forest is a subgraph F' without loops $[c(F')=0]$. This is described by

$$\begin{aligned} F_G(x_{ij}) &= \sum_{F'\subseteq GE(F')}\prod x_{ij} \\ &= \lim_{q\to 0} q^{-N} Z_G(q,qx_{ij}) \ . \end{aligned} \tag{4.35}$$

More generally, the Potts partition function (2.3) is precisely the dichromatic polynomial (Tutte, 1954) of G, which generates forest weighted according to a specific description. One such possible weighting has been described by Temperley and Lieb (1971) [see also description in Fortuin and Kasteleyn (1972)].

Finally we remark that the connection (4.23) of the resistance R_{kl} with the Potts partition can be formally extended to networks containing nonresistive impediments. If capacitances and inductances are present, the only complication in the formulation is that the corresponding x_{ij} will generally be complex. This does not change the form of (4.23) and the result is valid in all cases.

D. Dilute spin glass ($q=\frac{1}{2}$ limit)

Consider a spin glass (Edwards and Anderson, 1975) described by the Ising Hamiltonian

$$\mathcal{H}=-\sum_{(i,j)}J_{ij}S_iS_j \ , \tag{4.36}$$

where $S_i=\pm 1$, and each of the exchange interactions J_{ij} has an independent probability distribution

$$P(J_{ij})=p[\delta(J_{ij}-J)+\delta(J_{ij}+J)]+r\delta(J_{ij}) \ , \tag{4.37}$$

with

$$2p+r=1 \ . \tag{4.38}$$

This describes a dilute spin glass (Aharony, 1978) for which a transition from a paramagnetic magnetic phase to a spin glass phase is expected in the ground state. Some aspects of this problem are related to the Potts model in the $q=\frac{1}{2}$ limit (Aharony, 1978; Aharony and Pfeuty, 1979). In particular, an exact critical concentration can be deduced.

To begin with, we start from the Potts partition function (4.7) with $M=0$, which now reads

$$Z(q;K,L)=\sum_{G'\subseteq G} v^{b(G')}\prod_{c}\left[e^{Ls_c(G')}+q-1\right] \ , \tag{4.39}$$

where $v=e^K-1$. The zero-field susceptibility of the Potts model, χ, can be obtained straightforwardly by further differentiating the magnetization (1.10). This yields, after using (4.39) and the identity $\sum_{G'}s_c(G')=N$,

$$\chi(q;K)=\frac{1-q}{q^2}\left\langle \sum_c s_c^2\right\rangle_q \ , \tag{4.40}$$

with

$$\langle A\rangle_q=\sum_{G'\subseteq G} w_q(G')A(G')\Big/\sum_{G'\subseteq G} w_q(G') \ , \tag{4.41}$$

$$w_q(G')=\left[\frac{v}{q}\right]^{b(G')} q^{c(G')} \ . \tag{4.42}$$

Here, as in Sec. II.A, $b(G')$ and $c(G')$ are, respectively, the numbers of edges (bonds) and independent circuits (plaquettes) in G'. We also obtain from (2.3) and (2.6)

$$Z(q;K,0)=q^N\sum_{G'\subseteq G} w_q(G') \ . \tag{4.43}$$

Now specialize these results to $q=\frac{1}{2}$. Write

$$2v=(1-r)/r \tag{4.44}$$

and identify $1-r$ as the probability that a given edge is occupied (by either $+J$ or $-J$) in the dilute spin glass. Since the interactions $\pm J$ occur with equal probabilities, the probability that the sign of the product $\prod_P J_{ij}$ over the interactions around any plaquette is positive is exactly $\frac{1}{2}$. It follows that $r^E w_{1/2}(G')$ is the probability that the configuration G' occurs with $\mathrm{sgn}(\prod_P J_{ij})=+$ around *all* plaquettes.

Now we return to the spin glass problem and discuss its ground-state properties. Toulouse (1977) has introduced the idea of frustration which describes a plaquette as being "frustrated" if $\mathrm{sgn}(\prod_P J_{ij})=-$. It is then clear that if we retain only those configurations in the dilute spin glass in which no plaquette is frustrated [the Mattis (1976) spin glass], then the relevant configurations occur with probabilities $r^E w_{1/2}(G')$, and $q^N r^E Z(\frac{1}{2};K,0)$ gives the overall probability that the system has only nonfrustrated graphs. Similarly, the susceptibility (4.40) describes the "Mattis" spin glass ordering

$$\left\langle \sum_{(i,j)\in c} S_iS_j\right\rangle_{1/2}$$

in an equivalent ferromagnetic ground state (Aharony and Pfeuty, 1979).

From the above we see that the $q=\frac{1}{2}$ Potts model describes a dilute spin glass in which all frustrations are excluded. The critical concentration at which the system changes from a paramagnetic to a spin-glass phase is now obtained from (4.44):

$$r_c=1/(e^{2K_c}-1) \ , \tag{4.45}$$

where K_c is the critical point of the corresponding $q=\frac{1}{2}$ Potts model. For two-dimensional lattices this value can be obtained from the (presumed) exact $q=\frac{1}{2}$ Potts critical condition (see Sec. V.A.1).

E. Classical spin systems

The Potts model can be formulated as a problem of classical interacting spins. An example is the $q=3$ model described by the Hamiltonian [cf. (1.18)]:

$$-\beta\mathscr{H}_3=\sum_{(i,j)}[K\delta_{Kr}(\sigma_i,\sigma_j)+M\delta_{Kr}(\sigma_i,0)\delta_{Kr}(\sigma_i,0)\delta_{Kr}(\sigma_j,0)]+L\sum_i\delta_{Kr}(\sigma_i,0)\ , \qquad (4.46)$$

where $\sigma_i=0,1,2$. This Hamiltonian can be regarded as that of a spin-1 system whose spin variables are $S_i=-1,0,1$. In terms of the new variables we write

$$\delta_{Kr}(S_i,S_j)=1+\tfrac{1}{2}S_iS_j+\tfrac{3}{2}S_i^2S_j^2-S_i^2-S_j^2 \qquad (4.47)$$

and

$$\delta_{Kr}(S_i,0)=1-S_i^2 \qquad (4.48a)$$

or

$$\delta_{Kr}(S_i,\pm 1)=\tfrac{1}{2}S_i(1\pm S_i)\ , \qquad (4.48b)$$

depending on which S_i value is to be identified as the Potts spin state $\sigma_i=0$. If the state $S_i=0$ is identified as the Potts state $\sigma_i=0$, we use (4.48a) and obtain

$$-\beta\mathscr{H}_3=K_0+\sum_{(i,j)}(JS_iS_j+K^*S_i^2S_j^2)-\Delta\sum_i S_i^2\ , \qquad (4.49)$$

with

$$K_0=K+2L+M\ ,$$
$$J=K/2\ ,$$
$$K^*=M+3K/2\ ,$$
$$\Delta=\gamma K+L+\gamma M\ ,$$

where γ is the coordination number of the lattice. The expression (4.49) is of the form of the Hamiltonian of the Blume-Capel model (Blume, 1966; Capel, 1966) and has been studied extensively [see, for example, Blume *et al.*, 1971; Berker and Wortis, 1976). In particular, the zero-field ($M=L=0$) Potts model corresponds to a Blume-Capel model with parameters satisfying $J{:}K^*{:}\Delta=1{:}3{:}2\gamma$.

If the state $S_i=1$ or $S_i=-1$ is identified as the $\sigma_i=0$ Potts spin, then we use (4.48b) and the resulting Hamiltonian will take a different form containing terms proportional to S_i+S_j and $S_iS_j^2+S_i^2S_j$. The equivalence of these different forms of the Hamiltonian reflects a general symmetry under the relabelling of the states (Berker and Wortis, 1976).

More generally, any system of classical q-state spins, the Potts models included, can be formulated as a spin $(q-1)/2$ system. The spin Hamiltonian will generally take the form

$$-\beta\mathscr{H}_q=\sum_{(i,j)}\sum_{\alpha\geq\beta}K_{\alpha\beta}(S_i^\alpha S_j^\beta+S_i^\beta S_j^\alpha)+\sum_i\sum_\alpha L_\alpha S_i^\alpha\ , \qquad (4.50)$$

where the powers α and β run from 0 to $q-1$, and $S_i=-(q-1)/2,\ -(q-3)/2,\ldots,(q-1)/2$. Higher powers of α,β do not appear in (4.50) due to the fact that they can be eliminated using the identity

$$[S_i+(q-1)/2][S_i+(q-3)/2]\cdots\times[S_i-(q-1)/2]=0\ . \qquad (4.51)$$

There are $q(q+1)/2$ independent interactions in (4.50) which, for a given spin model, can always be determined arbitrary to the identification (permutation) of the spin states (as in the example of $q=3$). This arbitrariness again reflects a general symmetry of the spin Hamiltonian (4.50).

V. CRITICAL PROPERTIES

The only exact solution of the Potts model known to this date is the Onsager (1944) solution of the $q=2$ (Ising) model in $d=2$ dimensions (McCoy and Wu, 1973). However, a large body of information, in both exact as well as numerical forms, has also been accumulated on the critical properties of the various Potts models. These results are surveyed in this section.

A. Location of the critical point

1. Two-dimensional lattices

The critical point of the ferromagnetic Potts model is now rigorously known for the square, triangular, and honeycomb lattices for all $q\geq 4$ (Hintermann *et al.*, 1978) and for $q=2$ (Onsager, 1944). The critical condition can be stated simply as

$$z=1\ , \qquad (5.1a)$$

where z is given by (4.6), or, more explicitly,

$$\begin{aligned}&x_1x_2=1 && \text{square}\\ &\sqrt{q}\,x_1x_2x_3+x_1x_2+x_2x_3+x_3x_1=1 && \text{triangular}\\ &\sqrt{q}+x_1+x_2+x_3=x_1x_2x_3 && \text{honeycomb}\ .\end{aligned} \qquad (5.1b)$$

Here the variable x_i is defined in (4.4).

The derivation of (5.1) (for $q\geq 4$) follows essentially from a circle theorem (Suzuki and Fisher, 1971) for the vertex model equivalence (4.2) of the Potts model. The theorem states that, for $q\geq 4$ and regarding $\ln z$ as an external field, the zeros of the Potts partition can occur only at zero external field. This then leads to (5.1).

Since the critical point (5.1) agrees with the exact Ising $q=2$ result (Onsager, 1944), it is expected that (5.1) is also exact for $q=3$, although a rigorous proof of this assumption is still lacking. The expression (5.1b) for the critical point of the Potts model was first conjectured by Potts (1952) for the square lattice [see also Kihara *et al.* (1954)]. The conjecture makes use of the duality relation (2.10) and is based on a Kramers-Wannier (1941) type argument, which determines the transition point at the self-dual point $x_1 x_2 = 1$. The extension of the conjecture to isotropic triangular and honeycomb lattices was first suggested by Kim and Joseph (1974), and later extended to anisotropic lattices by Baxter *et al.* (1978) [see also Burkhardt and Southern (1978)]. Baxter (1973a) and Baxter *et al.* (1978) have shown that a first-order transition indeed occurs at the conjectured points for all $q>4$; and the uniqueness of this transition has subsequently been established by Hintermann *et al.* (1978). There has been no convincing proof of the validity of the critical point (5.1b) for $q<4$, except in the isolated case of $q=2$.

The critical point of the generalized (checkerboard) square lattice of Fig. 13 has been conjectured to be (Wu, 1979)

$$\sqrt{q}+x_1+x_2+x_3+x_4=x_1x_2x_3+x_2x_3x_4+x_3x_4x_1+x_4x_1x_2+\sqrt{q}\,x_1x_2x_3x_4 \ . \tag{5.2}$$

The conjecture (Wu, 1979) on the critical point for the Kagomé and diced lattices has since been shown to be incorrect (Enting and Wu, 1982).

For the Potts model on the triangular lattice which has two- and three-site interactions in half of the triangular faces (see Fig. 8), the partition function satisfies the self-dual relation (2.26). On the basis of this duality Baxter *et al.* (1978) proposed that the critical point is the self-dual point

$$e^{L+K_1+K_2+K_3}-e^{K_1}-e^{K_2}-e^{K_3}+2=q \ . \tag{5.3}$$

Wu and Zia (1981) have subsequently shown from a rigorous continuity and uniqueness argument that the transition point is indeed (5.3) for $q\geq 4$ in the ferromagnetic region $K_i\geq 0$, $L+K_1+K_2+K_3\geq 0$. They also showed that (5.3) is valid for $q=2$, regardless of the nature of the interactions. It is expected that (5.3) is also valid for the $q=3$ ferromagnetic transition.

For isotropic lattice ($K_1=K_2=K_3$) and zero three-site interactions ($L=0$), (5.3) can be solved giving explicitly

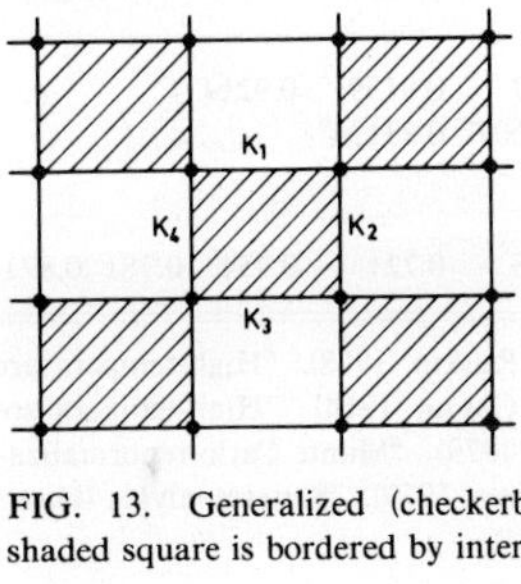

FIG. 13. Generalized (checkerboard) square lattice. Each shaded square is bordered by interactions K_1, K_2, K_3, and K_4.

$$\begin{aligned} e^{K_c}&=2\cos\left[\tfrac{2}{3}\cos^{-1}\frac{\sqrt{q}}{2}\right], \quad q\leq 4 \\ &=2\cosh\left\{\tfrac{2}{3}\ln\left[\frac{\sqrt{q}}{2}+\left[\frac{q}{4}-1\right]^{1/2}\right]\right\}, \quad q\geq 4 \ . \end{aligned} \tag{5.4}$$

The exact critical point for the triangular model where there is a three-site interaction L in every triangular face remains unknown except for $q=2$, for which the problem reduces to the nearest-neighbor Ising model, and for $L=0$ and $K_i\geq 0$, for which the critical condition is (5.3). However, Enting and Wu (1982) have shown that a special limit of the isotropic model ($K_1=K_2=K_3=K$) reduces to the hard hexagon lattice gas solved by Baxter (1980). This leads to the critical point

$$z_c=\tfrac{1}{2}(11+5\sqrt{5}) \tag{5.5}$$

after first taking the $K\to\infty$, $L\to-\infty$ limit with $e^{K+L}=[(q-1)/z]^{1/6}$ held constant, followed with the limit of $q\to\infty$.

Of special interest is the $q=3$ triangular model which, with appropriate interactions, admits ferromagnetic and/or antiferromagnetic ground-state orderings. Enting and Wu (1982) have obtained a rigorous lower bound on the critical point for this model from a Peierls-type argument. Numerical estimates of the critical point has been obtained by position-space renormalization group (Schick and Griffiths, 1977), series analysis (Enting and Wu, 1982), and Monte Carlo simulation (Saito, 1982). These results are summarized in Table I.

Finally, by summing over the spin states of the decorating sites of a decorated lattice, the critical properties of a dilute Potts model on the decorated lattice can be determined from those of the underlying lattice. This is a generalization of the Syozi model (Syozi, 1965; Syozi and Miyazima, 1966), and in this way the critical point of the dilute decorated two-dimensional models can be exactly determined (Wu, 1980).

TABLE I. Numerical estimates of the critical point for the three-state triangular Potts lattice with two- and three-site interaction $\{K,L\}$. I. Three-site interactions ($K=0, L>0$). II. Coexistence line ($K=-2L/3<0$). III. Antiferromagnetic two-site interactions ($L=0, K<0$). IV. Ferromagnetic two-site interactions ($L=0, K>0$).

	I	II	III	IV
Exact value	e^{-L_c}	e^{K_c}	e^{K_c}	$e^{-K_c}=0.53208\ldots$ [a]
Renormalization group[b]	0.55	0.066	0.210	0.59
Series analysis[c]	0.5038	0.137	0.204	
Monte Carlo[d]	0.5058	0.1360	0.2050	

[a] $e^{K_c}=2\cos(\pi/9)$ for model IV from (5.4).
[b] Schick and Griffiths (1977).
[c] Enting and Wu (1982).
[d] Saito (1982).

2. Three-dimensional lattices

There is no exact result in three dimensions and the critical point can be located only by numerical means. The estimate of the transition temperature for the three-dimensional Ising ($q=2$) model has long been known [see, for example, Fisher (1967)]; recent investigation of the three-dimensional Potts model has been mostly for the $q=3$ and $q=4$ models. Table II lists the various estimates on the critical points for three-dimensional lattices (series estimates are based on the assumption of a continuous transition).

3. Lattices in $d>3$ dimensions

For lattices in higher-than-three dimensions we are again guided by numerical studies only. Besides the Monte Carlo renormalization group study for $d=4$, $q=3$ (Blöte and Swendsen, 1979), series expansion in general d dimensions was first considered for the $q=2$ (Ising) model by Fisher and Gaunt (1964). This line of work has been extended further for $d=4$ (Sykes, 1979; Gaunt *et al.*, 1979) and for $q=4$ (Ditzian and Kadanoff, 1979). Table III lists the estimates on the critical point obtained in these studies [the Ising results are also included for comparisons].

B. Nature of transition

The mean-field solution of the Potts model has been discussed in Sec. I.C, where we have also examined the question on the existence of the critical value $q_c(d)$ above which the transition is mean-field—like in d dimension. We now turn to the question on the nature of transition and critical properties at the transition point.

1. Two dimensions

It is very remarkable that one now knows quite precisely the behavior of the two-dimensional Potts model at the critical point, even though there is no exact solution. Specifically, exact information can be obtained for the Potts model at the critical point for the square, triangular, and honeycomb lattices. One finds that the transition is of first order for $q>4$, and is continuous for $q\le 4$ (Baxter, 1973a; Baxter *et al.*, 1978).

The analysis makes use of the ice-rule vertex model formulation of the Potts model formulated in Sec. IV.A. For the Potts model on the square, triangular, and honeycomb lattices, the weights (4.2) of the ice-rule vertex model take the simple form

$$(\omega_1,\ldots,\omega_6)=(1,1,x_r,x_r,c_r,c_r) \tag{5.6}$$

at the critical point (5.1). Now this last vertex model is exactly soluble (Lieb, 1967; Lieb and Wu, 1972; Kelland, 1974); therefore the Potts partition function can be evaluated exactly at the critical point.

The exact solution of the vertex model (5.6) shows a transition occurring at the point

$$c_r=1+x_r \tag{5.7}$$

or, upon using (4.4) and (4.6),

$$q=4\,, \tag{5.8}$$

with $q>4$ corresponding to $T<T_c(c_r>1+x_r)$ in the vertex model. Now, regarding the vertex model (5.6) as the Potts model at the critical point (for which q is free to vary), the transition suggests that the critical properties of the Potts model will exhibit a change at $q=4$.

To see what kind of changes occurs in the critical properties, one evaluates further the internal energy (1.9) of the Potts model. From (4.2) and (4.5) it is clear that

TABLE II. Numerical estimates of the critical point for three-dimensional lattices.

	sc	bcc
$q=1$	$e^{-K_c}=0.753$[a]	
$q=2$	0.64816[b]	0.72993[b]
$q=3$	0.5784[c]	0.6747[c]
	0.5769[d]	0.669[f]
	0.577[e]	
	0.571[f]	
	0.575[h]	
$q=4$	0.523[f]	0.625[f]
	0.524[g]	
	0.532[h]	
$q=6$	0.472[h]	

[a]Series analyses (Gaunt and Ruskin, 1978). [b]Series analyses (Sykes *et al.*, 1972). [c]Low-temperature series analysis (Miyashita *et al.*, 1979). [d]Monte Carlo (Hermann, 1979). [e]Monte Carlo renormalization group (Blöte and Swendsen, 1979). [f]High-temperature series analysis (Kim and Joseph, 1975). [g]High-temperature series analysis (Ditzian and Kadanoff, 1979). [h]Monte Carlo (Ono and Ito, 1982).

TABLE III. Numerical estimates of the critical point for the hypercubic lattice in d dimensions.

$d=$	4	5	6	7	8	10
$q=1$[a]	$e^{-K_c}=0.839$	0.882	0.9159	0.9214		
$q=2$	0.74100[b]	0.79607[b]	0.83134[b]			
	0.74132[c]					
$q=3$	0.6788[d]					
$q=4$[e]	0.620	0.678	0.721	0.754	0.781	0.821

[a]Series analyses (Gaunt and Ruskin, 1978). [b]High-temperature series analysis (Fisher and Gaunt, 1964). [c]High-temperature series analysis (Gaunt *et al.*, 1979). [d]Monte Carlo renormalization group (Blöte and Swendsen, 1979). [e]Series analyses (Ditzian and Kadanoff, 1979).

the *critical* internal energy is related to the *zero-field* (staggered) polarization induced by the external field lnz. That is, expression of the Potts critical internal energy will include a term proportional to the zero-field (staggered) polarization.

For the vertex model (4.2) and (4.5) on the square lattice, Baxter (1973b, 1973c) has shown that a spontaneous (staggered) polarization exists for $T<T_c$. Baxter further argues that other terms occurring in the internal energy are continuous at the critical point. It follows that the $q>4$ Potts critical internal energy is discontinuous by an amount proportional to the zero-field (staggered) polarization. This then implies the existence of a nonzero latent heat for $q>4$, and that the transition at (5.1b), if any, is continuous for $q\leq 4$. This line of analysis has been extended to the triangular and honeycomb lattices (Baxter *et al.*, 1978), reaching the same conclusion regarding the nature of transition.

For completeness and convenience for references, I give the relevant results on the Potts model at the critical point.

For the isotropic square lattice, the free energy (1.8) at the critical temperature T_c is given by the expression

$$f(q;T_c)=\tfrac{1}{2}\ln q+\Theta+2\sum_{n=1}^{\infty}n^{-1}e^{-n\Theta}\tanh(n\Theta)\,,\quad q\geq 4 \tag{5.9a}$$

$$=\ln 2+4\ln[\Gamma(\tfrac{1}{4})/2\Gamma(\tfrac{3}{4})]\,,\quad q=4 \tag{5.9b}$$

$$=\tfrac{1}{2}\ln q+\int_{-\infty}^{\infty}\frac{dx}{x}\tanh(\mu x)\frac{\sinh(\pi-\mu)x}{\sinh(\pi x)}\,,\quad q\leq 4 \tag{5.9c}$$

where

$$\cosh\Theta=\sqrt{q}\,/2,\quad \Theta\geq 0,\quad q\geq 4$$

$$\cos\mu=\sqrt{q}\,/2,\quad 0\leq\mu<\frac{1}{2}\pi,\quad q\leq 4\,. \tag{5.10}$$

The internal energy (1.9) at the critical point is

$$E(q,T_c\pm)=\varepsilon_2(1+q^{-1/2})\times\left[-1\pm\Delta(q)\tanh\left[\tfrac{1}{2}\Theta\right]\right], \tag{5.11}$$

where

$$\Delta(q)=0,\quad q\leq 4$$

$$=\prod_{n=1}^{\infty}(\tanh n\Theta)^2,\quad q\geq 4\,. \tag{5.12}$$

For the isotropic triangular lattice the results are

$$f(q;T_c)=\tfrac{1}{2}\ln q+\Theta+3\sum_{n=1}^{\infty}n^{-1}e^{-n\Theta}\sinh\left[\frac{2n}{3}\Theta\right]\Big/\cosh(n\Theta),\quad q\geq 4 \tag{5.13a}$$

$$=\ln 2+3\ln\left[\frac{\Gamma(\frac{7}{6})\Gamma(\frac{1}{3})}{\Gamma(\frac{2}{3})\Gamma(\frac{5}{6})}\right],\quad q=4 \tag{5.13b}$$

$$=\tfrac{1}{2}\ln q+\tfrac{3}{2}\int_{-\infty}^{\infty}\frac{\sinh(\pi-\mu)x\,\sinh(2\mu x/3)}{x\sinh(\pi x)\cosh(\mu x)}dx,\quad q\leq 4 \tag{5.13c}$$

$$E(q;T_c\pm)=3\varepsilon_2[\sinh(\tfrac{4}{3}\Theta)\sinh(\tfrac{2}{3}\Theta)\big/\sinh(2\Theta)]\left[-1-2\sum_{n=1}^{\infty}e^{-n\Theta}\cosh(\tfrac{2}{3}n\Theta)/\cosh(n\Theta)\pm\Delta(q)\right],\quad q\geq 4 \tag{5.14a}$$

$$=-4\varepsilon_2[1-\tfrac{2}{3}\ln 2],\quad q=4 \tag{5.14b}$$

$$=-3\varepsilon_2\sin(\tfrac{4}{3}\mu)\sin(\tfrac{2}{3}\mu)\csc(2\mu)\int_{-\infty}^{\infty}\frac{\sinh[(\pi-\mu)x]\cosh(\frac{2}{3}\mu x)}{\sinh(\pi x)\cosh(\mu x)}dx,\quad q\leq 4 \tag{5.14c}$$

Corresponding expressions for the honeycomb lattice can be deduced from (5.13), (5.14), and the duality relation (2.10). The latent heat in all cases is given by, for $q\geq 4$,

$$L(q)=E(q,T_c+)-E(q,T_c-)$$

$$\sim(q-4)^{1/2}\exp[-g(q-4)^{-1/2}],\quad q=4+\,, \tag{5.15}$$

displaying an essential singularity at $q=4$ (g is a constant).

These results can be extended to the triangular lattice with anisotropic interactions (Baxter *et al.*, 1978). It is noteworthy that the general expressions of the relevant quantities are of the form $\psi(q,x_1)+\psi(q,x_2)+\psi(q,x_3)$, where the x's are defined in (4.4) and related by the critical condition (5.1b). The results (5.10)–(5.15) can then be obtained from these general expressions by taking the special cases of $x_1=x_2, x_3=0$ (square) and $x_1=x_2=x_3$ (triangular). For completeness I include in Table IV results of numerical evaluations (Sarbach and Wu, 1981b) of (5.11), (5.14) and (5.15) for $q=1,2,\ldots,10$.

Owing to the very fact that the critical behavior is precisely known, the $d=2$ Potts model has become an important testing ground in the modern theory of the criti-

TABLE IV. Numerical evaluations of the critical parameters.

	q	1	2	3	4	5	6	7	8	9	10
Square	e^{K_c}					$1+\sqrt{q}$					
	$E(q;T_c)$					$\frac{1}{2}\left[1+\frac{1}{\sqrt{q}}\right]$					
	$L(q)$	0	0	0	0	0.0265	0.1007	0.1766	0.2432	0.2998	0.3480
Triangular	e^{K_c}	1.5321	1.7321	1.8794	2.0000	2.1038	2.1958	2.2790	2.3553	2.4260	2.4920
	$E(q;T_c)$	1.0000	0.8333	0.7603	0.7172	0.6881	0.6669	0.6506	0.6377	0.6271	0.6183
	$L(q)$	0	0	0	0	0.0310	0.1172	0.2042	0.2795	0.3429	0.3962

cal point. For example, the success in predicting the known first-order transition has been crucial to the testing of the various approaches. The following developments are noted in this connection.

Renormalization group studies of the "ε-expansion" type, where $\varepsilon=4-d$ [see, for example, Golner (1973); Rudnick (1975)] led to a first-order transition for small ε. But early attempts in the position space renormalization group have invariably failed to yield the known first-order transition [see, for example, Burkhardt *et al.* (1976); Dasgupta (1977); den Nijs and Knops (1978); den Nijs (1979)]. However, Nienhuis *et al.* (1979, 1980a) have shown that, by including a dilution in the Potts model as described in Sec. I.B., the first-order transition can be seen in this enlarged parameter space as a crossover of the critical behavior into tricritical (for $q \le q_c$) at q_c. In this way, a variational renormalization group study (Nienhuis *et al.*, 1980a) has yielded the excellent value of $q_c=4.08$ versus the exact value $q_c=4$. This renormalization group description of the Potts (and the cubic) model has been reviewed by Riedel (1981). The exact critical free energy for the $q=4$ model has also been reproduced quite accurately by a variational renormalization group calculation (Ashley, 1978; Temperley and Ashley, 1979).

For the triangular Potts model with both two- and three-site interactions, the position space renormalization group calculation yielded a continuous transition in both the ferromagnetic and antiferromagnetic models (Schick and Griffiths, 1977), while the inclusion of a dilution into this problem does not appear to lead to a consistent prediction (Kinzel, 1981). However, both series analysis (Enting and Wu, 1982) and Monte Carlo simulation (Saito, 1982) indicate that the transitions along the ferro- and antiferromagnetic coexistence line (Model II in Table I) and the antiferromagnetic model (Model III in Table I) are actually first order. This finding is in line with the fact that the ground states of these two models have a higher symmetry and are, respectively, ninefold and sixfold degenerate.

The $d=2$ Potts model has also been studied in a Monte Carlo simulation of its dynamic as well as static properties (Binder, 1981). Excellent agreement with the known exact results for $q=3,4,5,6$ has been observed. Some of the theoretical predictions have also been verified by the experimental investigations of systems realizing the $d=2$ Potts models (Sec. I.D).

2. Three dimensions

No exact results are known for the Potts model in three dimensions. Here, one is especially interested in elucidating the nature of transition in the $q=3$ model which resides close to the border of the validity of the mean-field scheme (see Sec. I.C).

Renormalization group studies in $d=3$ are inconclusive. While calculations of the "ε-expansion" type predicted a first-order transition for $q=3$ (see, for example, Rudnick, 1975), the real space renormalization group yielded a continuous transition [see, for example, Burkhardt *et al.* (1976)]. Series analyses did not fare much better either: Miyashita *et al.* (1979) found the $q=3$ low-temperature series inadequate to identify the nature of transition, although earlier work on the high-temperature series has indicated that the transition is first order for all $q \ge 3$ (Kim and Joseph, 1975). But a recent (numerical) study using the variational renormalization group has indicated that the transition in the $q=3$, $d=3$ model is definitely first order (Nienhuis *et al.*, 1981).

A more positive identification of the nature of transition in $d=3$ is provided by Monte Carlo investigations. Herrmann (1979) studied the $q=3,4$ models and Knak Jensen and Mouritsen (1979) studied the $q=3$ model by Monte Carlo simulations; Blöte and Swendsen (1979) investigated the $q=3$ model by the Monte Carlo renormalization group. In all cases, clear indications were obtained that the transition is first order. The cluster variation method (Levy and Sudano, 1978) also predicted a first-order transition. In addition, experiments on systems belonging to the same universality class as the $q=3$ model indicated the transition being of first order (Sec. I.D). The current belief based on these considerations is that the $q=3$ Potts model in three dimensions posseses a first-order transition, an assumption we have already taken into account in constructing Fig. 2.

3. General d dimensions

Only a few results are available for the Potts model in higher than three dimensions. The Monte Carlo renormalization group indicated that the transition in the $d=4$, $q=3$ model is first order (Blöte and Swendsen, 1979). Ditzian and Kadanoff reached the same conclusion for the $d=4$, $q=4$ model from analyzing the high- and low-temperature series. Extending the consideration of the dilute model of Nienhuis *et al.* (1979) to general d dimensions, Andelman and Berker (1981) obtained from a simple variational renormalization group analysis estimate on the value of $q_c(d)$ for continuous values of d. Their finding is consistent with the picture that the transition in higher dimensions is first order for all $q \geq 2$. The picture of the merging of the critical and tricritical lines at $q_c(d)$ for general d has also been confirmed in an analysis of the differential renormalization equation for the dilute Potts model by Nauenberg and Scalapino (1980). Their analysis also led to an essential singularity in $(q-q_c)^{1/2}$ in the latent heat, thus extending (5.15) to all d, and a logarithmic correction to the power-law behavior in the free energy near T_c. The dilute Potts model has also been studied by Berker *et al.* (1980) in the infinite-state limit in one dimension. Using a Migdal-Kadanoff renormalization scheme argued to be exact in the limit of $d \to 1+$, $q \to \infty$, with $l=(d-1)\ln q$ finite, they uncovered a variety of phase transitions and a "singularity" in the critical properties at $l=\ln 4$.

C. Phase diagram

We are now in a position to discuss the structure of the phase diagram of the Potts model in light of the foregoing discussions. In this regard the $q=3$ and $q=4$ models are special due to the fact that the phase diagram is dimension-dependent. The situation for the $q=3$ model, which has been alluded to by Straley and Fisher (1973), is as follows.

Consider the $q=3$ model described by the general partition function (3.1) in which external fields $H_i = kT \ln z_i$ are applied to spin state i $(=0,1,2)$. The structure of the phase diagram in the full (T,H_0,H_1,H_2) space is best seen in the subspace

$$H_0+H_1+H_2=0\ , \tag{5.16}$$

which retains the full symmetry of the model. This leads to the "triangle diagram" shown in Fig. 14 and 15.

Straley and Fisher (1973) argue that a planar coexistence surface, $H_1=H_2$, exists in the region where one of the external fields, say, H_0, is large and negative. This coexistence surface is bound by a line of critical point since the transition is essentially Ising-like. By symmetry there exist two other similar coexistence planes, and the three planes must meet at the line of symmetry $H_1=H_2=H_3=0$, $T<T_c$ (a triple point line), since the three ordered phases can coexist below the zero-field transition temperature T_c. The construction of the remaining portion of the phase diagram is now dictated by the nature of transition.

If the zero-field transition is critical (in the sense of divergent fluctuations) as found in $d=2$, then the three critical lines come in to meet at the zero-field transition point, turning it into an "anomalous" tricritical point. This situation is shown in Fig. 14. If the zero-field transition is first order, as believed to be the case in $d=3$, then the zero-field transition point is not "critical." Instead, it is a quadruple point where the three ordered phases and the disordered phase can coexist. Then the full phase diagram is expected to be as shown in Fig. 15. Note that there now exist three additional weblike first-order surfaces, also terminating at lines of critical points. The six critical lines now join at three tricritical points of the "normal" type (in the sense that the three joining critical lines meet tangentially).

The phase diagram of the $q=4$ model can be discussed in a similar way by considereing a "tetrahedron diagram" in a four-dimensional space, with a comparable difference expected between the $d=2$ and $d=3$ models.

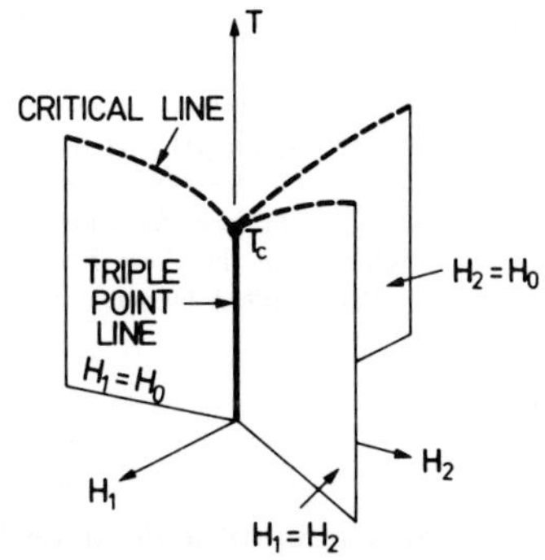

FIG. 14. Schematic phase diagram for the three-state Potts model in two dimensions. The three coexistence planes meet at the triple point line (solid line) and terminate at three critical lines (broken curves). The three critical lines meet at the zero-field transition point at T_c forming an "anomalous" tricritical point.

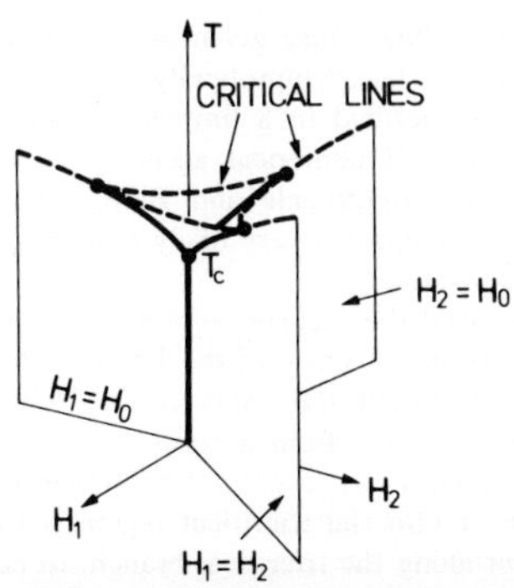

FIG. 15. Schematic phase diagram for the three-state Potts model in three dimensions. The three planar and the three weblike coexistence planes meet at the triple point lines (solid curves) and terminate at the critical lines (broken curves). The zero-field transition point at T_c is a quadrupole point, and the critical lines meet at three "ordinary" tricritical points.

D. Critical exponents

The critical exponents of the Potts model are well defined for the $d=2$, $q\leq 4$ system which exhibits a continuous transition. As in the usual description of the thermodynamics near a critical point [see, for example, Fisher (1967)], the critical behavior of the Potts free energy $f(q;K,L)$ in $d=2$ is characterized by the "dominant" singularities

$$f(q;K,0)\sim|K-K_c|^{2/y_t},\quad K\sim K_c \tag{5.17}$$

$$f(q;K_c,L)\sim|L|^{2/y_h},\quad L\sim 0\,. \tag{5.18}$$

These two expressions also serve to define the thermal and magnetic exponents y_t and y_h. The critical exponents are then obtained from the relations

$$\begin{aligned} 2-\alpha&=2/y_t\,,\\ 1+1/\delta&=2/y_h\,, \end{aligned} \tag{5.19}$$

and the usual scalings and hyperscaling.

In order to obtain the explicit q dependences of y_t and y_h for the two-dimensional model, it is necessary to solve the vertex model (4.2), or any other equivalent formulation of the Potts model, at temperatures slightly off the critical point (5.1) or with a small field. This has not been accomplished to this date. However, on the basis of a consideration of the vertex model formulation and its connection with the Baxter (1971) eight-vertex model and the Ashkin-Teller (1943) model, den Nijs (1979b) has made the following conjecture on the thermal exponent:

$$y_t=3(1-u)/(2-u),\quad q\leq 4 \tag{5.20}$$

with

$$0\leq u\equiv(2/\pi)\cos^{-1}(\sqrt{q}/2)\leq 1\,. \tag{5.21}$$

Black and Emery (1981) have since given an argument showing the conjecture to be asymptotically exact; the conjecture has also been verified in a finite-size scaling analysis to a high degree of numerical accuracy for a wide range values of q (Nightingale and Blöte, 1980; Blöte *et al.*, 1981). It now appears very likely that (5.20) is, in fact, the exact expression.

In Sec. V.B.1 I described the occurrence of a tricritical line in the enlarged parameter space of the Potts model when a dilution is introduced (Nienhuis *et al.*, 1979). Nienhuis *et al.* (1979) suggested from a consideration of the renormalization topology that a natural continuation of the thermal exponent into the tricritical region is to take $y_t^{(T)}$, the exponent along the tricritical branch, to be given by (5.20), as well, provided that one takes $-1\leq u\leq 0$ in (5.21). This picture has been further substantiated by Kadanoff variational renormalization calculations (Nienhuis *et al.*, 1980a; Burkhardt, 1980).

A conjecture similar to (5.20) has been made on the critical and tricritical magnetic exponents y_h and $y_h^{(T)}$. The conjecture

$$y_h,y_h^{(T)}=(3-u)(5-u)/4(2-u)\,, \tag{5.22}$$

with $0\leq u\leq 1$ for y_h and $-1\leq u\leq 0$ for $y_h^{(T)}$, is obtained independently by Nienhuis *et al.* (1980b) from a consideration of renormalization group results and by Pearson (1980) from a pure numeral fitting. But the validity of (5.22) has again been verified numerically to a high degree of accuracy (Nightingale and Blöte, 1980; Blöte *et al.*, 1981).

Using the conjectured expression for the temperature and magnetic exponents, it is then a simple matter to write down all critical and tricritical exponents of the Potts model. One obtains

$$\begin{aligned} \alpha&=\alpha'=2(1-2u)/3(1-u)\,,\\ \beta&=(1+u)/12\,,\\ \gamma&=\gamma'=(7-4u+u^2)/6(1-u)\,,\\ \delta&=(3-u)(5-u)/(1-u^2)\,,\\ \nu&=\nu'=(2-u)/3(1-u)\,,\\ \eta&=(1-u^2)/2(2-u)\,, \end{aligned} \tag{5.23}$$

where $u>0$ $(u<0)$ for the critical (tricritical) exponents. For convenience we list in Table V the predicted critical exponents for $q=0,1,2,3,4$.

First we compare the conjectured values in Table V with the known exact results, which are unfortunately limited in numbers.

The value of $y_h=2$ for $q=0$ agrees with the exact value obtained by Kunz (1981).

The $q=2$ values in Table V are in agreement with the known Ising results. In addition, the $q=3$ Potts model is believed to be in the same universality class of the hard hexagon lattice gas (Alexander, 1975), and the predicted values of $\alpha=\frac{1}{3}$, $\beta=\frac{1}{9}$ are confirmed by the exact solution of the hard hexagon problem (Baxter, 1980). The $q=4$ Potts model is considered in the same universality class of the Baxter-Wu model (Enting, 1975; Domany and Riedel, 1978); the predicted values of $\alpha=\frac{2}{3}$, $\beta=\frac{1}{12}$ again agree with the exact exponents (Baxter and Wu, 1973; Baxter *et al.*, 1975). These exact results lend firm support to the correctness of the conjectures.

On the other hand, it is fruitful and illuminating to compare the conjectured values with those obtained by various numerical means. This comparison is done in Table VI for $q=1,3,4$. [A more complete summary of the numerical results for $q=1$ can be found in Essam (1980).] It is seen that the agreement is generally good, except that a consistent difference is found in the case of $q=4$, the region where the finite-size scaling estimates (Blöte *et al.*, 1981) and the Monte Carlo renormalization group analysis (Rebbi and Swendsen, 1980) exhibit slow convergence. Presumably, this difficulty is due to the presence of a strong (logarithmic) confluent singularity associated with a marginal exponent at $q=4$ (Nauenberg and Scalapino, 1980; Cardy *et al.*, 1981). It is noteworthy that a finite-size analysis of an associated

TABLE V. Critical exponents (5.23) for the Potts model in two dimensions.

q	y_t	y_h	u	$\alpha=\alpha'$	β	$\gamma=\gamma'$	δ	ν	η
0	0	2	1	$-\infty$	$\frac{1}{6}$	∞	∞	∞	0
1	$\frac{3}{4}$	$\frac{51}{48}$	$\frac{2}{3}$	$-\frac{2}{3}$	$\frac{5}{36}$	$2\frac{7}{18}$	$18\frac{1}{5}$	$\frac{4}{3}$	$\frac{5}{24}$
2	1	$\frac{15}{8}$	$\frac{1}{2}$	0	$\frac{1}{8}$	$\frac{7}{4}$	15	1	$\frac{1}{4}$
3	$\frac{6}{5}$	$\frac{28}{15}$	$\frac{1}{3}$	$\frac{1}{3}$	$\frac{1}{9}$	$\frac{13}{9}$	14	$\frac{5}{6}$	$\frac{4}{15}$
4	$\frac{3}{2}$	$\frac{15}{8}$	0	$\frac{2}{3}$	$\frac{1}{12}$	$\frac{7}{6}$	15	$\frac{2}{3}$	$\frac{1}{2}$

one-dimensional quantum system (Hermann, 1981) leads to evidence supporting this correction.

The three-state models on the triangular lattice with *pure* three-site interactions have also been analyzed by series studies. For the model where the three-site interactions are present in half of the triangles (see Fig. 8), series analysis based on the (presumed) exact critical point (5.3) yielded the exponents $\alpha=\frac{1}{3}$, $\beta=\frac{1}{9}$ (Enting, 1980c). The same set of exponents is also indicated for the model with three-site interactions in every triangle (Enting and Wu, 1982). These findings are consistent with the predictions of the universality argument.

Few results are available for the critical exponents of Potts models in higher dimensions. However, both the thermal and the magnetic exponents have been computed numerically as functions of q by Nienhuis *et al.* (1981) at

TABLE VI. Numerical estimates on the critical exponents of the q-state Potts model in two dimensions. Error bars in estimations are not included in this table. LT is low temperature, HT is high temperature, RG is renormalization group.

q	Method	$\alpha=2(1-y_t^{-1})$	β	γ	$\delta=y_h/(2-y_h)$	ν
1	Conjectured value	$-\frac{2}{3}$	$\frac{5}{36}$	$2\frac{7}{18}$	$18\frac{1}{5}$	$\frac{4}{3}$
	Monte Carlo (Kirkpatrick, 1976)		$0.136\leq\beta\leq0.15$	2.3		
	Series expansion (Dunn *et al.*, 1975)		0.15	2.38		1.34
	Series expansion (Sykes *et al.*, 1976b, 1976a; Gaunt and Sykes, 1976)		0.138	2.43	18.0	
	Series expansion (Domb and Pearse, 1976)	−0.668				
	Real space RG (Reynolds *et al.*, 1977, 1978)	−0.712	0.138	2.435	18.6	1.356
	Real space RG (Lobb and Karasek, 1980)	−0.685				
	Kadanoff variational RG (Dasgupta, 1976)	−0.686	0.140	2.406	18.25	1.343
	Monte Carlo RG (Eschbach *et al.*, 1981)	−0.666				
3	Conjectured value	$\frac{1}{3}$	$\frac{1}{9}$	$\frac{13}{9}$	14	$\frac{5}{6}$
	HT series expansion (Kim and Joseph, 1975)			1.42		
	Series expansion (Zwanzig and Ramshaw, 1977)	0.296				
	Series expansion (de Neef and Enting, 1977)	0.42				
	HT series expansion (Miyashita *et al.*, 1979)		0.1064	1.50	15.5	
	Lt series expansion (Enting, 1980a)	$\sim\frac{1}{3}$	0.109			
	Kadanoff variational RG (Burkhardt *et al.*, 1976)	0.3365	0.1061	1.451	14.68	
	Kadanoff variational RG (Dasgupta, 1977)	0.326	0.107	1.460	14.64	0.837
	Cumulant and variational RG (Shenker *et al.*, 1979)	0.210				0.895
	Monte Carlo RG (Swendsen, 1979; Rebbi and Swendsen, 1980)	0.352	0.101	1.445	15.26	0.824
4	Conjectured value	$\frac{2}{3}$	$\frac{1}{12}$	$\frac{7}{6}$	15	$\frac{2}{3}$
	HT series expansion (Kim and Joseph, 1975)			1.20		
	Series expansion (Zwanzig and Ramshaw, 1977)	0.45				
	LT series expansion (Enting, 1975a)	0.64	0.089			
	HT series expansion (Ditzian and Kadanoff, 1979)	0.5		1.17		
	Kadanoff variational RG (Dasgupta, 1977)	0.488	0.091	1.330	15.53	0.756
	Cumulant and variational RG (Schenker *et al.*, 1979)	0.358				0.821
	Duality invariant RG (Hu, 1980)	0.4870				0.7565
	Monte Carlo RG (Eschbach *et al.*, 1981)	0.507				
	Analysis of one-dimensional quantum system (Herrmann, 1981)	0.649				
	Monte Carlo RG (Swendsen *et al.*, 1982)	0.660				

$d=1.58,2,2.32$ (using the variational renormalization group) and for continuous values of d in $1 \le d \le 5$ (using the Migdal bond-moving approximation). The more interesting case is the $q=1$ (percolation) model for which the transition is continuous for all $d \le d_c(1)=6$. There have been a number of numerical estimates on the exponents for the $q=1$ model in $d=3,4,5$. For a comprehensive summary of these results see Essam (1980).

E. The antiferromagnetic model

In an antiferromagnetic Potts model ($K<0$) it is energetically favorable for two neighboring spins to be in distinct spin states. As a consequence, the ground state of the $q \ge 3$ model on bipartite lattices (and the $q=2$ model if the lattice is not bipartite) has a nonzero entropy. Then the argument can be made as in Wannier (1950) that a transition of the usual type accompanying the onset of a long-range order will not arise.

However, Berker and Kadanoff (1980) have argued from a rescaling argument that in such systems a distinctive low-temperature phase in which correlations decay algebraically can exist. For the q-state antiferromagnetic Potts model this behavior is permitted when the spatial dimensionality d is sufficiently high, or, for a fixed d, when q is less than a cutoff value $q_0(d)$. While it remains to be seen whether such a phase indeed occurs in such systems, it is noteworthy that an approximate Migdal-Kadanoff renormalization carried out by Berker and Kadanoff (1980) yields the cutoff values $q_0(2)=2.3$ and $q_0(3)=3.3$, predicting the existence of such a phase in the $q=3$ model in three dimensions. Monte Carlo simulations, however, indicate the existence of an ordered low-temperature phase in three dimensions for both $q=3$ and $q=4$ (Banavar *et al.*, 1980). Monte Carlo simulations have also been carried out for the square lattice with antiferromagnetic nearest-neighbor coupling and ferromagnetic next-nearest-coupling for $q \ge 3$ (Grest and Banavar, 1981); the result shows a variety of unusual transitions.

For the square lattice it is known that the $q=2$ antiferromagnetic (Ising) system exhibits a transition at $e^{k_c}=\sqrt{2}-1$. While this transition may be an isolated singularity, more likely it is one point lying on a singular trajectory (Kim and Enting, 1979). A good indication of how this trajectory might behave can be inferred from the exact result of the antiferromagnetic model on the decorated lattice [Fig. 16(a)]. For antiferromagnetic interactions ($K<0$) this decorated model should exhibit the general features of a system with a nonzero entropy.

Taking the partial traces over the bond-decorating sites leads to an effective square lattice, as shown in Fig. 16(b). This Potts lattice has ferromagnetic interactions K^* given by

$$e^{K^*}=(e^{2K}+q-1)/(2e^{K}+q-2)\,. \tag{5.24}$$

Using the exact critical point (5.1b), or

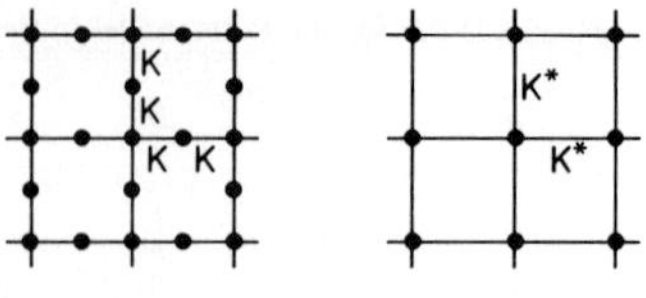

FIG. 16. (a) Decorated square lattice with interactions K. (b) Equivalent lattice with interactions K^*.

$$e^{K_c^*}=1+\sqrt{q} \tag{5.25}$$

for the square lattice, one obtains the following exact critical point for the antiferromagnetic ($K<0$) decorated model [see also Wu (1980)]:

$$e^{K_c}=1+\sqrt{q}-\sqrt{q}\,(1+\sqrt{q}\,)^{1/2}\,. \tag{5.26}$$

The expression (5.26) is highly instructive, for it shows e^{K_c} decreasing monotonically from 1 to 0 in the range between $q=0$ and $q=q_0=\frac{1}{2}(3+\sqrt{5})$. This cutoff value of $q_0(2)=2.618\ldots$ is close to the value 2.3 of the rescaling prediction. [It is noteworthy that the same $q_0=2.618\ldots$ is found in a site-diluted antiferromagnetic Potts model on the honeycomb lattice (Kondor and Temesvári, 1981).] A similar behavior in the square-lattice model is therefore also expected. Indeed, Kim and Enting (1979) have analyzed the series expansion of the chromatic function (1.15a) for the square lattice. Their finding of a singularity at $q=q_0\approx 2.22$ on the line $e^K=0$ conforms with the above reasonings.

Putting these pieces of information together, we then expect the line of singularity to behave in a fashion shown schematically in Fig. 17. Whether K_c jumps from 0 to a nonzero value at q_0, as implied by the rescaling argument, remains to be seen. But the general behavior of the singularity trajectory should be as indicated. This contrasts with the conjecture

$$e^{K_c}=[q(q-1)]^{1/2}+1-q \tag{5.27}$$

made by Ramshaw (1979) shown by the thin broken line in Fig. 17. Ramshaw's conjecture permits a transition

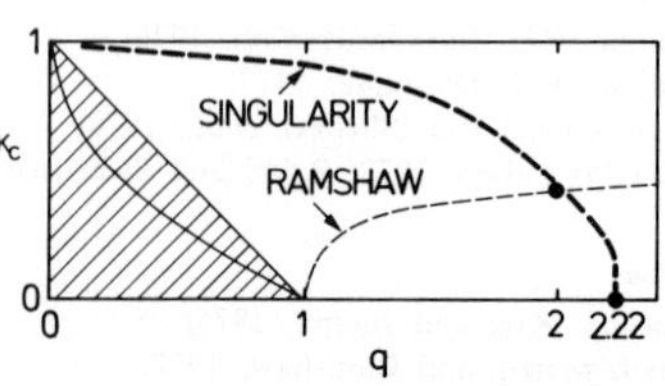

FIG. 17. Schematic plot of the singularity trajectory (heavy broken line) of the antiferromagnetic Potts model on the square lattice. The trajectory passes through the points (0,1), $(2,\sqrt{2}-1)$, (2.22,0), and may have a jump discontinuity at $q\approx 2.22$ as shown. The shaded region is self-dual with the solid line denoting the self-dual point (5.28). The Ramshaw conjecture is given by (5.27). [See note added in proof below: The singularity trajectory should pass through the point (3,0), instead of the point (2.22,0).]

for all $q>1$. It should be noted that the antiferromagnetic model on the square lattice is self-dual in the region $0\le q<1$, $0<e^{K}<1-q$, indicated by the shaded area in Fig. 17. If a unique transition exists in this region, then it must occur at the self-dual point deduced from (2.11), or

$$e^{K_c}=1-\sqrt{q}\,,\quad 0\le q<1\,. \tag{5.28}$$

Clearly, our discussion precludes the existence of this transition.

Also of interest is the square-lattice Potts model with mixed ferromagnetic ($K_x>0$) and antiferromagnetic ($K_y<0$) interactions considered by Kinzel *et al.* (1981). The Monte Carlo simulation suggests that the transition in this model, if any, is of an unconventional type, and a Migdal-Kadanoff transformation determines this transition point at

$$(1+e^{K_x^{(c)}})(1-e^{K_y^{(c)}})=q\,, \tag{5.29}$$

a result know to be exact at $q=2$.

Note added in proof: Baxter (1982b) has shown that the $q\le 4$ antiferromagnetic model ($K_x<0$, $K_y<0$) on the square lattice is soluble at

$$(e^{K_x}+1)(e^{K_y}+1)=4-q\,. \tag{5.30}$$

He also concluded that the antiferromagnetic model exhibits a continuous transition at this point. This implies that the singularity trajectory in Fig. 17 should cross the q axis at $q=3$, instead of $q=2.22$ as shown. This crossing point is also predicted by a phenomenological renormalization group calculation (Nightingale and Schick, 1981).

VI. RANDOM-BOND MODEL

A. Model definition

A random Potts model that has been of interest recently is the random-bond problem in which each interaction takes on values subject to an uncorrelated probability distribution. Thus the Hamiltonian takes the form

$$\mathscr{H}=-\sum_{(i,j)}J_{ij}\delta_{\mathrm{Kr}}(\sigma_i,\sigma_j)\,, \tag{6.1}$$

where J_{ij} is a random variable governed by a distribution $P(J_{ij})$. As a realistic spin model the randomness is quenched, or frozen, in positions. One would like to investigate the properties of this system as a function of the parameters contained in $P(J)$.

A simple choice of $P(J)$ is the two-valued discrete distribution

$$P(J)=p\delta(J-J_1)+(1-p)\delta(J-J_2)\,, \tag{6.2}$$

where $0\le p\le 1$. For $q=2$ and $J_1+J_2=0$, this becomes the spin glass problem (Edwards and Anderson, 1975); for $J_2=0$ and general q, this defines the bond-diluted Potts model (as versus the site-diluted model of Sec. I.B) where each bond has a probability p of possessing an interaction $-J_1$ and a probability $1-p$ of being vacant. We shall consider this random-bond Potts model in this section.

In a quenched system the thermodynamic quantities of interest are computed for each random configuration; only after this computation is the average over the random bond distribution taken. As an example, the per site free energy for a lattice G of N sites and E edges (bonds) is taken to be

$$f_G(q;p,K_1,K_2)=\frac{1}{N}\sum_{M=0}^{E}p^M(1-p)^{E-M}F_G^{(M)}(q;K_1,K_2)\,, \tag{6.3}$$

where $K_i=\beta J_i$, and

$$F_G^{(M)}(q;K_1,K_2)=\sum_{\{M\}}\ln Z_G^{\{M\}}(q;K_1,K_2) \tag{6.4}$$

is a sum over all $\binom{E}{M}$ configurations $\{M\}$ for which there are M bonds of interaction $-J_1$ and $E-M$ bonds $-J_2$; $Z_G^{\{M\}}(q;K_1,K_2)$ is the partition function for a fixed configuration $\{M\}$.

Evaluation of averages of the type given by (6.3) is often effected (and also compounded) by the use of the n-replica trick (Emery, 1975). But as we shall see, it is not always necessary to use this trick to extract the needed information.

B. Duality relation

Following the route of our discussion of the regular Potts model, we now derive a duality relation for the random-bond model (6.1) on planar lattices.

As we have already pointed out in Sec. II.A, the duality relation (2.10) is valid quite generally for edge-dependent interactions. This means that we can write (2.10) for each of the partition functions $Z_G^{\{M\}}$ in (6.4). This leads to

$$Z_G^{\{M\}}(q;K_1,K_2)=q^{1-N_D}(e^{K_1}-1)^M(e^{K_2}-1)^{E-M}\times Z_D^{\{M\}}(q;K_1^*,K_2^*)\,, \tag{6.5}$$

where

$$(e^{K_i}-1)(e^{K_i^*}-1)=q \tag{6.6}$$

and $Z_D^{\{M\}}$ is the corresponding partition function on the dual lattice D specified by the same bond configuration $\{M\}$.

It is now a simple matter to substitute (6.5) into (6.4) and (6.3) to obtain the following duality relation:

$$f_G(q;p,K_1,K_2)=\frac{1-N_D}{N}\ln q+p\frac{E}{N}\ln(e^{K_1}-1)+(1-p)\frac{E}{N}\ln(e^{K_2}-1)+\frac{N_D}{N}f_D(q;p,K_1^*,K_2^*)\,. \tag{6.7}$$

The generalization of (6.7) which is valid for any finite G, to arbitrary distributions $P(J)$ has been given by Sarbach and Wu (1981a) and by Jauslin and Swendsen (1981).

The duality relation (6.7) for the free energy f was first given by Schwartz (1979) [see also Fisch (1978)] for the $q=2$ random-bond Ising problem. The general q formulation has since been discussed using the n-replica technique by Southern and Thorpe (1979), generalizing an earlier $q=2$ result by Domany (1978), and by Aharony and Stephen (1980).

C. Location of the critical point

For an infinite lattice G the free energy (6.7) will become singular along a certain trajectory, $T=T_c(p)$, in the (p,T) space. This trajectory then defines the critical point in the random-bond model.

It is therefore pertinent to inquire whether the duality relation (6.7) is useful in determining this critical point in the case of planar lattices, especially for the square lattice since it is self-dual.

The answer to this inquiry is negative, since, even in the case of the square lattice, the duality (6.7) simply describes a symmetry of the free energy about a point in the (K_1,K_2) space for fixed p. But the square-lattice free energy possesses an additional symmetry

$$f_{\rm sq}(q;p,K_1,K_2)=f_{\rm sq}(q;1-p,K_2,K_1)\ . \tag{6.8}$$

Therefore, at $p=\frac{1}{2}$, the singularity in the free energy is preserved under the transformation $(K_1,K_2)\rightarrow(K_2,K_1)\rightarrow(K_2^*,K_1^*)$. Then, if a unique transition exists in this system, it must occur at $K_1=K_2^*=K_1^c$, $K_2=K_1^*=K_2^c$, or

$$(e^{K_1^c}-1)(e^{K_2^c}-1)=q \quad (p=\tfrac{1}{2})\ . \tag{6.9}$$

This exact critical point was first obtained by Fisch (1978) for $q=2$ and extended to general q by Kinzel and Domany (1981).

There has been no exact result on the location of the critical point for general p. The conjectured expressions on the $q=2$ square lattice critical point for the bond-diluted model ($K_1=K$, $K_2=0$) (Nishimori, 1979a) and for the square-lattice model with arbitrary $P(J)$ (Nishimori, 1979b) have shown to be incorrect (Aharony and Stephen, 1980). A similar determination of the general q critical point for the bond-diluted model (Southern, 1980),

$$e^{-K_c}=[q^{1-p_c/p}-1]/(q-1),\quad p\geq p_c\ , \tag{6.10}$$

where p_c is the bond percolation threshold, is presumably also inexact, although it does give the correct limit for $q=1$ (Yeomans and Stinchcombe, 1980):

$$e^{-K_c}=1-p_c/p,\quad p\geq p_c\ . \tag{6.11}$$

In the bond-diluted model ($K_1=K$, $K_2=0$), we generally expect $T_c(p)$ to vanish for $p\leq p_c$, since below p_c only finite clusters are present and there can be no ferromagnetic ordering. For $q=1$ the model describes a percolation process on the already diluted lattice with the exact critical point $p(1-e^{-K_c})=p_c$ or (6.11). Therefore we expect $T_c(p)$ to behave as shown schematically in Fig. 18. Note that increasing the value of q corresponds to decreasing the "effective" ferromagnetic interactions; $T_c(p)$ goes down as a consequence. In addition, the behavior of $T_c(p)$ in the small dilution limit has been investigated in a cumulant expansion analysis for $q=2$ (Harris, 1974). The result is

$$T_c(p)=T_c(1)[1-a(1-p)],\quad q=2,\quad p\approx 1 \tag{6.12}$$

with $a=1.329$ and 1.060, respectively, for the square and the simple cubic lattices. [See also Sarbach and Wu (1981a)].

D. Critical behavior

Consider the bond-diluted ($K_2=0$) system whose phase diagram is shown in Fig. 18. The behavior of such (bond- or site-) diluted systems near the point Q ($p=p_c$, $T=0$) has been of considerable theoretical interests. Stauffer (1975) has argued in the case of $q=2$ that the point Q should be viewed as a type of higher-order critical point. The transition is percolationlike if approached along the $T=0$ path, and thermally driven if approached along $p=p_c$ (Stanley *et al.*, 1976). With the application of scaling, a crossover from the percolation problem to thermal ordering is then expected in the critical region (the vicinity of the point Q). In particular, one is led to consider the crossover exponents $\phi=\nu_p/\nu_t$, where ν_p and ν_t are the respective percolation and thermal correlation exponents. This scaling argument has been extended to spin systems of general q components (Lubensky, 1977). Wallace and Young (1977) have shown rigorously that $\phi=1$ for the continuous Potts model in the limit of $q\rightarrow 1$. Using a renormalization procedure which is exact near $T=0$, Coniglio (1981) has been able to establish that $\phi=1$ for any q and spatial dimensionality d.

The $d=2$ bond-diluted Potts model has been studied by the position-space renormalization group (Yeomans and Stinchcombe, 1980; Kinzel and Domany, 1981).

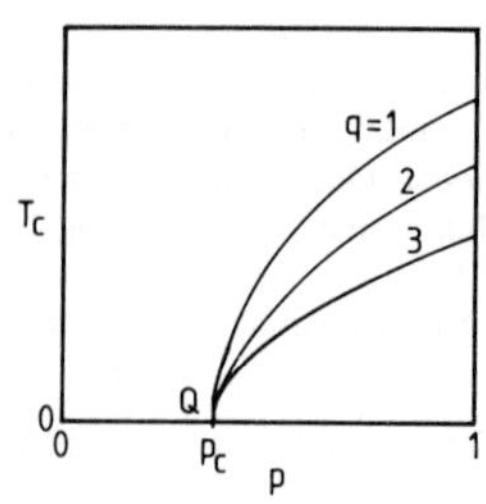

FIG. 18. Schematic plot of $T_c(p)$, the critical temperature as a function of the bond concentration p, of the bond-diluted model for different values of q. p_c is the bond percolation threshold.

This has led to numerical results on phase diagrams and thermodynamic functions. In particular, the prediction of Harris (1974) that the critical behavior of the dilute system deviates from that of the pure system only for $q>2$ (when the specific heat of the pure system diverges) is verified. The bond-diluted system has also been studied under an "effective interaction approximation" (Turban, 1980).

VII. UNSOLVED PROBLEMS

It is customary to include in an introductory review a list of unsolved problems to exemplify topics for further research. The following is a partial list of such problems as suggested in the presentation of this review. Here, again, emphasis has been placed on problems which require rigorous or exact treatments. But I have excluded the obviously over-ambitious problems such as the exact evaluation of the free energy (1.8).

1. Rigorous establishment of the validity of the critical condition (5.1b) for the $d=2$, $q<4$, $q\neq 2$ models. Among other implications, this would provide a rigorous proof on the ($q=1$) bond percolation thresholds, which has been lacking to this date.
2. Determination of the critical point for two-dimensional lattices other than those described by (5.1b). This includes the checkerboard lattice [conjecture (5.2)] and the Kagomé lattice.
3. Decay of the correlation function at $T=T_c$ and for $T<T_c$.
4. Exact or rigorous results on the dilute model with quenched site dilution (Sec. I.B). In particular, one is interested in the crossover behavior between the thermal and percolation exponents near the percolation threshold where the transition temperature vanishes.
5. Determination of the critical point for the triangular lattice model with two- and three-site interactions. A solution to this problem will lead to, among others, the solution of two long outstanding unsolved problems: the exact percolation threshold for site-percolation on the honeycomb lattice and the Potts critical point for the Kagomé lattice. It also leads to a direct determination of the critical fugacity (5.5) for the hard hexagon lattice gas.
6. The antiferromagnetic model requires further understanding including the nature of its transition for $q<q_0(d)$.
7. Rigorous proof on the validity of the conjectures (5.20) and (5.22) for the thermal and magnetic exponents for the two-dimensional models.
8. A rigorous argument proving (or disproving) that the $d=3$, $q=3$ transition is first order.
9. Exact determination of the critical point for the quenched bond-diluted model in two dimensions.

ACKNOWLEDGMENTS

This work is based in part on lectures I gave at Laboratorium voor Technische Natuurkunde, Delft, thus forming part of the research programme of the "Stichting voor Fundamenteel Onderzoek der Materie (FOM)," which is financially supported by the "Nederlandse Organisatie voor Zuiver-Wetenschappelijk Onderzoek (ZWO)." I wish to thank my colleagues at Institut für Festkörperforschung, Kernforschungsanlage Jülich, Instituut Lorentz, Leiden, and Laboratorium voor Technische Natuurkunde, Delft, for encouragement, and A. Aharony, A. N. Berker, E. Domany, H. J. Herrmann, P. W. Kasteleyn, I. Ono, E. K. Riedel, H. E. Stanley, and R. K. P. Zia for bringing relevant references to my attention. I am especially grateful to I. G. Enting and H. N. V. Temperley for critical readings of the manuscript and for their most helpful comments and criticisms. This work was supported in part by the National Science Foundation under Grant No. DMR-78-18808.

REFERENCES

Aharony, A., 1978, J. Phys. C **11**, L457.
Aharony, A., and P. Pfeuty, 1979, J. Phys. C **12**, L125.
Aharony, A., and E. Pytte, 1981, Phys. Rev. B **23**, 362.
Aharony, A., and M. J. Stephen, 1980, J. Phys. C **13**, L407.
Aharony, A., K. A. Müller and W. Berlinger, 1977, Phys. Rev. Lett. **38**, 33.
Alcaraz, F. C., and R. Köberle, 1980, J. Phys. A **13**, L153.
Alexander, S., 1975, Phys. Lett. A **54**, 353.
Amit, D. J., 1976, J. Phys. A **9**, 1441.
Amit, D. J., and A. Shcherbakov, 1974, J. Phys. C **7**, L96.
Andelman, D., and A. N. Berker, 1981, J. Phys. A **14**, L91.
Ashkin, J., and E. Teller, 1943, Phys. Rev. **64**, 178.
Ashley, S. E., 1978, J. Phys. A **11**, 2015.
Banavar, J. R., G. S. Grest, and D. Jasnow, 1980, Phys. Rev. Lett. **45**, 1424.
Barbara, B., M. F. Rossignol, and P. Bak, 1978, J. Phys. C **11**, L183.
Barber, M. N., 1980, Phys. Rep. **59**, 376.
Baxter, R. J., 1970, J. Math. Phys. **11**, 784.
Baxter, R. J., 1971, Phys. Rev. Lett. **26**, 832.
Baxter, R. J., 1973a, J. Phys. C **6**, L445.
Baxter, R. J., 1973b, J. Stat. Phys. **9**, 145.
Baxter, R. J., 1973c, J. Phys. C **6**, L94.
Baxter, R. J., 1980, J. Phys. A **13**, L61.
Baxter, R. J., 1982a, "The Inversion Relation Method for Some Two-Dimensional Exactly Solved Models in Lattice Statistics," to appear in J. Stat. Phys.
Baxter, R. J., 1982b, "Critical Antiferromagnetic Square Lattice Potts Model," to appear in Proc. R. Soc. London.
Baxter, R. J., and F. Y. Wu, 1973, Phys. Rev. Lett. **31**, 1294.
Baxter, R. J., M. F. Sykes, and M. G. Watt, 1975, J. Phys. A **8**, 245.
Baxter, R. J., S. B. Kelland, and F. Y. Wu, 1976, J. Phys. A **9**, 397.
Baxter, R. J., H. N. V. Temperley, and S. E. Ashley, 1978, Proc. Roy. Soc. London, Ser. A **358**, 535.
Berker, A. N., and L. P. Kadanoff, 1980, J. Phys. A **13**, L259.
Berker, A. N., and M. Wortis, 1976, Phys. Rev. B **14**, 4946.
Berker, A. N., S. Oslund, and F. A. Putnam, 1978, Phys. Rev. B **17**, 3650.
Berker, A. N., D. Andelman, and A. Aharony, 1980, J. Phys. A **13**, L413.
Betts, D. D., 1964, Can. J. Phys. **42**, 1564.
Biggs, N. L., 1975, J. Phys. A **8**, L110.

Binder, K., 1981, J. Stat. Phys. **24**, 69.
Birkhoff, G. D., 1912, Ann. Math. **14**, 42.
Black, J. L., and V. J. Emery, 1981, Phys. Rev. B **23**, 429.
Blankschtein, D., and A. Aharony, 1980a, J. Phys. C **13**, 4635.
Blankschtein, D., and A. Aharony, 1980b, Phys. Rev. B **22**, 5549.
Blankschtein, D., and A. Aharony, 1981, J. Phys. C **14**, 1919.
Blöte, H. W. J., M. P. Nightingale, and B. Derrida, 1981, J. Phys. A **14**, L45.
Blöte, H. W. J., and R. H. Swendsen, 1979, Phys. Rev. Lett. **43**, 799.
Blume, M., 1966, Phys. Rev. **141**, 517.
Blume, M., V. J. Emery, and R. B. Griffiths, 1971, Phys. Rev. A **4**, 1071.
Bragg, W. L., and E. J. Williams, 1934, Proc. Roy. Soc. London, Ser. A **145**, 699.
Bretz, M., 1977, Phys. Rev. Lett. **38**, 501.
Burkhardt, T. W., 1979, Phys. Rev. B **20**, 2905.
Burkhardt, T. W., 1980, Z. Phys. B **39**, 159.
Burkhardt, T. W., and B. W. Southern, 1978, J. Phys. A **11**, L247.
Burkhardt, T. W., H. J. F. Knops, and M. den Nijs, 1976, J. Phys. A **9**, L179.
Capel, H. W., 1966, Physica **32**, 966.
Cardy, J. L. 1980, J. Phys. A **13**, 1507.
Cardy, J. L., M. Nauenberg, and D. J. Scalapino, 1981, Phys. Rev. B **22**, 2560.
Coniglio A., 1981, Phys. Rev. Lett. **46**, 250.
Coniglio, A., H. E. Stanley, and W. Klein, 1979, Phys. Rev. Lett. **42**, 518.
Das, B. K., and R. B. Griffiths, 1979, J. Chem. Phys. **70**, 5555.
Dasgupta, C., 1976, Phys. Rev. B **14**, 1221.
Dasgupta, C., 1977, Phys. Rev. B **15**, 3460.
Ditzian, R. V., and L. P. Kadanoff, 1979, J. Phys. A **12**, L229.
Ditzian, R. V., and J. Oitman, 1974, J. Phys. A **7**, L61.
Domany, E., 1978, J. Phys. C **11**, L337.
Domany, E., M. Schick, and J. S. Walker, 1977, Phys. Rev. Lett. **38**, 1148.
Domany, E., and E. K. Riedel, 1978, J. Appl. Phys. **49**, 1315.
Domany, E., and M. Schick, 1979, Phys. Rev. B **20**, 3828.
Domany, E., and M. Schick, J. S. Walker, and R. B. Griffiths, 1978, Phys. Rev. B **18**, 2209.
Domany, E., Y. Schnidman, and D. Mukamel, 1981, "Type I fcc Antiferromagnets in a Field: A realization of the $q=3$- and $q=4$-state Potts Models," preprint.
Domb, C., 1960, Adv. Phys. **9**, 149.
Domb, C., 1974a, J. Phys. A **7**, 1335.
Domb, C., 1974b, in *Phase Transitions and Critical Phenomena*, edited by C. Domb and M. S. Green (Academic, London), Vol. 3, p. 1.
Domb, C., and C. J. Pearse, 1976, J. Phys. A **9**, L137.
Dunn, A. G., J. W. Essam, and D. S. Ritchie, 1975, J. Phys. C **8**, 4219.
Edwards, S. F., and P. W. Anderson, 1975, J. Phys. F **5**, 965.
Eggarter, T. P., 1974, Phys. Rev. B **9**, 2989.
Einhorn, M. B., R. Sauit, and E. Rabinovici, 1980, Nucl. Phys. B **170**, 16.
Emery, V. J., 1975, Phys. Rev. B **11**, 239.
Enting, I. G., 1974a, J. Phys. A **7**, 1617.
Enting, I. G., 1974b, J. Phys. A **7**, 2181.
Enting, I. G., 1975a, J. Phys. A **8**, L35.
Enting, I. G., 1975b, J. Phys. A **8**, 727.
Enting, I. G., 1975c, J. Phys. A **8**, 1690.
Enting, I. G., 1977, J. Phys. A **10**, 325.
Enting, I. G., 1978a, Austr. J. Phys. **31**, 515.
Enting, I. G., 1978b, J. Phys. A **11**, 563.
Enting, I. G., 1978c. Austr. J. Phys. **31**, 383.
Enting, I. G., 1980a, J. Phys. A **13**, L133.
Enting, I. G., 1980b, J. Phys. A **13**, L279.
Enting, I. G., 1980c, J. Phys. A **13**, L409.
Enting, I. G., and R. J. Baxter, 1977, J. Phys. A **10**, L117.
Enting, I. G., and F. Y. Wu, 1982, Triangular Lattice Potts Models, submitted to J. Stat. Phys.
Eschbach, P. D., D. Stauffer, and H. J. Herrmann, 1981, Phys. Rev. B **23**, 422.
Essam, J. W., 1972, in *Phase Transitions and Critical Phenomena* edited by C. Domb and M. S. Green (Academic, London), Vol. 2, p. 197.
Essam, J. W., 1979, J. Math. Phys. **20**, 1769.
Essam, J. W., 1980, Rep. Prog. Phys. **43**, 53.
Essam, J. W., and M. E. Fisher, 1970, Rev. Mod. Phys. **42**, 272.
Fisch, R., 1978, J. Stat. Phys. **18**, 111.
Fisher, M. E. 1967, Rep. Prog. Phys. **30**, 615.
Fisher, M. E., and D. S. Gaunt, 1964, Phys. Rev. A **133**, 224.
Fontaine, J. R., and Ch. Gruber, 1979, Commun. Math. Phys. **70**, 243.
Fortuin, C. M., and P. W. Kasteleyn, 1972, Physica **57**, 536.
Fröhlich, J., and T. Spencer, 1981, "The Kosterlitz-Thouless transition in two-dimensional Abelian spin systems and the Coulomb gas," preprint, I.H.E.S.
Gaunt, D. S., and A. J. Guttmann, 1974, in *Phase Transitions and Critical Phenomena,* edited by C. Domb and M. S. Green (Academic, London), Vol. 3, p. 181.
Gaunt, D. S., and H. Ruskin, 1978, J. Phys. A **11**, 1369.
Gaunt, D. S., and M. F. Sykes, 1976, J. Phys. A **9**, 1104.
Gaunt, D. S., M. F. Sykes, and H. Ruskin, 1976, J. Phys. A **9**, 1899.
Gaunt, D. S., M. F. Sykes, and S. McKenzie, 1979, J. Phys. A **12**, 871.
de Gennes, P. G., 1971, Mol. Cryst. Liq. Cryst. **12**, 193.
Giri, M. R., M. J. Stephen, and G. S. Grest, 1977, Phys. Rev. B **16**, 4971.
Golner, G., 1973, Phys. Rev. B **8**, 3419.
Gouyet, J. F., 1980, in *Ordering in Two Dimensions,* edited by S. K. Sinha (North-Holland, New York), p. 355.
Gouyet, J. F., B. Sapoval, and P. Pfeuty, 1980, J. Phys. Lett. **41**, L115.
Grest, G. S., and J. R. Banavar, 1981, Phys. Rev. Lett. **46**, 1458.
Griffiths, R. B., and J. C. Wheeler, 1970, Phys. Rev. A **2**, 1047.
Gruber, Ch., A. Hintermann, and D. Merlini, 1977, "The duality transformation," Ch. 4 of "Arbitrary Spin Lattice Systems, " Part III of *Group Analysis of Classical Lattice Systems,* Lecture Notes in Physics (Springer, Berlin), Vol. 60, p. 293.
Hararay, F., 1969, *Graph Theory* (Addison-Wesley, Reading).
Harris, A. B., 1974, J. Phys. C **7**, 1671.
Harris, A. B., and T. C. Lubensky, 1981, Phys. Rev. B **24**, 2656.
Harris, A. B., T. C. Lubensky, W. K. Holcomb, and C. Dasgupta, 1975, Phys. Rev. Lett. **35**, 327.
von Heimberg, J., and H. Thomas, 1974, J. Phys. C **7**, 3433.
Herrmann, H. J., 1979, Z. Phys. B. **35**, 171.
Herrmann, H. J., 1981, "The Potts Model as a One-Dimensional Quantum System," preprint.

Hintermann, A., H. Kunz, and F. Y. Wu, 1978, J. Stat. Phys. **19**, 623.
Hu, B., 1980, J. Phys. A **13**, L321.
Huang, K., 1963, *Statistical Mechanics* (Wiley, New York).
Husimi, K., 1953, "Statistical mechanics of condensation," in *Proceedings of an International Conference on Theoretical Physics,* edited by H. Yukawa (Science Council of Japan, Tokyo), p. 531.
Ising, E., 1925, Z. Phys. **21**, 613.
Jauslin, H. R., and R. H. Swendsen, 1981, Phys. Rev. B **24**, 313.
de Jongh, L. J., and A. R. Miedema, 1974, Adv. Phys. **23**, 1.
Kac, M., 1968, in *Statistical Physics, Phase Transitions and Superfluidity,* edited by M. Chrétien, E. P. Gross, and S. Deser (Gordon and Breach, New York), Vol. 1, p. 241.
Kasai, Y., T. Takano, and I. Syozi, 1980, Prog. Theor. Phys. **63**, 1917.
Kasteleyn, P. W., 1963, Physica **29**, 1329.
Kasteleyn, P. W., 1964, Report in Aachen Conference on Statistical Mechanics.
Kasteleyn, P. W., and C. M. Fortuin, 1969, J. Phys. Soc. Jpn. **26**, (Suppl.), 11.
Kelland, S. B., 1974, Aust. J. Phys. **27**, 813.
Kihiara, T., Y. Midzuno, and J. Shizume, 1954, J. Phys. Soc. Jpn. **9**, 681.
Kim, D., and I. G. Enting, 1979, J. Comb. Theory B **26**, 327.
Kim, D., and R. J. Joseph, 1974, J. Phys. C **7**, L167.
Kim, D., and R. J. Joseph, 1975, J. Phys. A **8**, 891.
Kim, D., P. M. Levy, and L. F. Uffer, 1975, Phys. Rev. B **12**, 989.
Kinzel, W., 1981, private communication.
Kinzel, W., and E. Domany, 1981, Phys. Rev. B **23**, 3421.
Kinzel, W., W. Selke, and F. Y. Wu, 1981, J. Phys. A **14**, L399.
Kirchhoff, G., 1847, Ann. Phys. Chem. **72**, 497.
Kirkpatrick, S., 1976, Phys. Rev. Lett. **36**, 69.
Knak Jensen, S. J., and O. G. Mouritsen, 1979, Phys. Rev. Lett. **43**, 1736.
Kogut, J. B., 1979, Rev. Mod. Phys. **51**, 659.
Kondor, I., 1980, J. Phys. C **13**, L531.
Kondor, I., and T. Temesvári, 1981, J. Phys. C **14**, L97.
Kosterlitz, J. M., and K. J. Thouless, 1978, Prog. Low Temp. Phys. **7B**, 371.
Kramers, H. A., and G. H. Wannier, 1941, Phys. Rev. **60**, 252.
Kunz, H., 1981, unpublished.
Kunz, H., and F. Y. Wu, 1978, J. Phys. C **11**, L1.
Levy, P. M., and J. J. Sudano, 1978, Phys. Rev. B **18**, 5087.
Lieb, E. H., 1967a, Phys. Rev. Lett. **18**, 692.
Lieb, E. H., 1967b, Phys. Rev. **162**, 162.
Lieb, E. H., 1967c, Phys. Rev. Lett. **18**, 1046.
Lieb, E. H. and F. Y. Wu, 1972, in *Phase Transitions and Critical Phenomena,* edited by C. Domb and M. S. Green (Academic, London), Vol. 1, p. 332.
Lobb, C. J., and K. A. Karasek, 1980, J. Phys. C **13**, L245.
Lubensky, T. C., 1977, Phys. Rev. B **15**, 311.
Lubensky, T. C., and J. Isaacson, 1978, Phys. Rev. Lett. **41**, 829.
Matsuda, M., 1974, Prog. Theor. Phys. **51**, 1053.
Mattis, D. C., 1976, Phys. Lett. **56A**, 421.
McCoy, B. M., and T. T. Wu, 1973, *The Two-Dimensional Ising Model* (Harvard University, Cambridge Press).
Mess, K. W., E. Lagendijk, D. A. Curtis, and W. J. Huiskamp, 1967, Physica **34**, 126.
Mittag, L., and J. Stephen, 1971, J. Math. Phys. **12**, 441.
Mittag, L., and J. Stephen, 1974, J. Phys. A **7**, L109.
Miyashita, S., D. D. Betts, and C. J. Elliott, 1979, J. Phys. A **12**, 1605.
Mukamel, D., M. E. Fisher, and E. Domany, 1976, Phys. Rev. Lett. **37**, 565.
Murata, K. K. 1979, J. Phys. A **12**, 81.
Nagle, J. F., 1971, J. Comb. Theory B **10**, 42.
Nauenberg, M., and D. J. Scalapino, 1980, Phys. Rev. Lett. **44**, 837.
de Neef, T., 1975, Ph. D. Dissertation, Einhoven, University of Technology.
de Neef, T., and I. G. Enting, 1977, J. Phys. A **10**, 801.
Nelson, D. R., 1979, Phys. Rep. **49**, 255.
Nienhuis, B., A. N. Berker, E. K. Riedel, and M. Schick, 1979, Phys. Rev. Lett. **43**, 737.
Nienhuis, B., E. K. Riedel, and M. Schick, 1980a, J. Phys. A **13**, L31.
Nienhuis, B., E. K. Riedel, and M. Schick, 1980b, J. Phys. A **13**, L189.
Nienhuis, B., E. K. Riedel, and M. Schick, 1981, Phys. Rev. B **23**, 6055.
Nightingale, M. P., and H. W. J. Blöte, 1980, Physica A **104**, 352.
Nightingale, M. P., and M. Schick, 1981, "Three-State Square Lattice Potts Antiferromagnet," to appear in J. Phys. A.
den Nijs, M.P.M., 1979a, Physica A **95**, 449.
den Nijs, M.P.M., 1979b, J. Phys. A **12**, 1857.
den Nijs, M.P.M., 1979c, Ph.D. thesis, Catholic University, Nijmegen.
den Nijs, M.P.M., and H. J. F. Knops, 1978, Physica A **93**, 441.
Nishiromi, H., 1979a, J. Phys. C **12**, L641.
Nishiromi, H., 1979b, J. Phys. C **12**, L905.
Ono, I. and K. Ito, 1982, "Monte Carlo Simulation and Pair Approximation on the Phase Transition of Restricted Orientational Lattice Model for Liquid Crystal," preprint.
Onsager, L., 1944, Phys. Rev. **65**, 117.
Park, R. L., T. L. Einstein, A. R. Kortan, and L. D. Roelofs, 1980, in *Ordering in Two Dimensions,* edited by S. K. Sinha (North-Holland, New York), p. 17.
Pearson, R. B., 1980, Phys. Rev. B **22**, 2579.
Potts, R. B., 1951, Ph.D. thesis, University of Oxford.
Potts, R. B., 1952, Proc. Camb. Phil. Soc. **48**, 106.
Potts, R. B., and J. C. Ward, 1955, Prog. Theor. Phys. **13**, 38.
Ramshaw, J. D., 1979, J. Phys. A **12**, L271.
Rebbi, C., and R. H. Swendsen, 1980, Phys. Rev. B **21**, 4094.
Reynolds, P. J., W. Klein, and H. E. Stanley, 1977, J. Phys. C **10**, L167.
Reynolds, P. J., H. E. Stanley, and W. Klein, 1978, J. Phys. A **11**, L199.
Riedel, E. K., 1981, "The Potts and Cubic Models in Two Dimensions. A Renormalization-Group Description," to appear in Physica.
Rudnick, J., 1975, J. Phys. A **8**, 1125.
Saito, Y., 1982, "Monte Carlo Study of the Three-State Potts Model with Two- and Three-Body Interaction," submitted to J. Phys. A.
Sarbach, S., and F. Y. Wu, 1981a, "Exact Results on the Random Potts Model," to appear in Z. Phys. B.
Sarbach, S., and F. Y. Wu, 1981b, unpublished.
Savit, R., 1980, Rev. Mod. Phys. **52**, 453.
Schick, M., and R. B. Griffiths, 1977, J. Phys. A **10**, 2123.

Schwartz, M., 1979, Phys. Lett. A **75**, 102.
Shenker, S. J., L. P. Kadanoff, and A. M. M. Pruisken, 1979, J. Phys. A **12**, 91.
Somorjai, G. A., 1973, Surf. Sci. **34**, 156.
Southern, B. M., 1980, J. Phys. C **13**, L285.
Southern, B. M., and M. F. Thorpe, 1979, J. Phys. C **12**, 5351.
Stanley, H. E., R. J. Birgeneau, P. J. Reynolds, and J. F. Nicoll, 1976, J. Phys. C **9**, L553.
Stanley, R. P., 1973, *Disc. Math.* **5**, 171.
Stauffer, 1975, Z. Phys. B **22**, 161.
Stephen, M. J., 1976, Phys. Lett. A **56**, 149.
Stephen, M. J., 1977, Phys. Rev. B **15**, 5674.
Straley, J. P., 1974, J. Phys. A **7**, 2173.
Straley, J. P., and M. E. Fisher, 1973, J. Phys. A **6**, 1310.
Suzuki, M., and M. E. Fisher, 1971, J. Math. Phys. A **9**, 397.
Swendsen, R. H., 1979, Phys. Rev. Lett. **42**, 859.
Swendsen, R. H., D. Andelman, and A. N. Berker, 1982, Critical Exponents and Marginality of the Four-state Potts Model: Monte Carlo Renormalization Group, preprint.
Sykes, M. F., 1979, J. Phys. A **12**, 879.
Sykes, M. F., J. W. Essam, and D. S. Gaunt, 1965, J. Math. Phys. **6**, 283.
Sykes, M. F. and D. S. Gaunt, 1973, J. Phys. A **6**, 643.
Sykes, M. F., D. S. Gaunt, and M. Glen, 1976a, J. Phys. A **9**, 97.
Sykes, M. F., D. S. Gaunt, and M. Glen, 1976b, J. Phys. A **9**, 725.
Sykes, M. F., D. S. Gaunt, P. D. Roberts, and J. A. Wyles, 1972, J. Phys. A **5**, 640.
Syozi, I., 1965, Prog. Theor. Phys. **34**, 189.
Syozi, I., and S. Miyazima, 1966, Prog. Theor. Phys. **36**, 1083.
Szabo, N., 1975, J. Phys. C **8**, L397.
Tejwani, M. J., O. Ferreira, and O. E. Vilches, 1980, Phys. Rev. Lett. **44**, 152.
Temperley, H. N. V., 1954, Proc. R. Soc. London **67**, 233.
Temperley, H. N. V., 1958, Faraday Soc. Discuss. **25**, 92.
Temperley, H. N. V., 1976, Report at Aberdeen Recontre on Combinatorial Mathematics and Its Application.
Temperley, H. N. V., and S. E. Ashley, 1979, Proc. R. Soc. London Ser. A **365**, 371.
Temperley, H. N. V., and S. E. Ashley, 1981, "Site Percolation Problems and Multisite Potts Models," to appear in J. Phys. A.
Temperley, H. N. V., and E. H. Lieb, 1971, Proc. R. Soc. London Ser. A **322**, 251.
Toulouse, G., 1974, Nuovo Cim. B **23**, 234.
Toulouse, G., 1977, Commun. Phys. **2**, 115.
Turban, L., 1980, J. Phys. C **13**, L13.
Tutte, W. T., 1954, Can. J. Math. **6**, 80.
Wallance, D. J., and A. P. Young, 1977, Phys. Rev. B **17**, 2384.
Wang, Y. K., and F. Y. Wu, 1976, J. Phys. A **9**, 593.
Wannier, G. H., 1950, Phys. Rev. **79**, 357.
Weger, M. and I. B. Goldberg, 1973, Solid State Phys. **28**, 1.
Whitney, H., 1932, Ann. Math. **33**, 688.
Wielinga, R. F., H. W. J. Blöte, J. A. Roest, and W. J. Huiskamp, 1967, Physica **34**, 223.
Wright, J. C., H. W. Moos, J. H. Colwell, B. W. Magnum, and D. D. Thornton, 1971, Phys. Rev. B **3**, 843.
Wu, F. Y., 1977, J. Phys. A **10**, L113.
Wu, F. Y., 1978, in *Studies in Foundations and Combinatories,* edited by G.-C. Rota, Adv. in Math. Supp. Stud., Vol. 1, p. 151.
Wu, F. Y., 1979a, J. Phys. C **12**, L317.
Wu, F. Y., 1979b, J. Phys. C **12**, L645.
Wu, F. Y., 1980, J. Stat. Phys. **23**, 733.
Wu, F. Y., 1981, J. Phys. A **14**, L39.
Wu, F. Y., and K. Y. Lin, 1980, J. Phys. A **13**, 629.
Wu, F. Y., and Y. K. Wang, 1976, J. Math. Phys. **17**, 439.
Wu, F. Y., and R. K. P. Zia, 1981, J. Phys. A **14**, 721.
Yeomans, J. M., and R. B. Stinchcombe, 1980, J. Phys. C **13**, L239.
Zia, R. K. P., and D. Wallace, 1975, J. Phys. A **8**, 1495.
Zwanzig, R., and J. Ramshaw, 1977, J. Phys. A **10**, 65.

Critical Phenomena II

D. Hone, Chairperson

Potts model of magnetism (invited)

F. Y. Wu[a]
Division of Materials Research, National Science Foundation, Washington D. C. 20550

The Potts model is a generalization of the Ising model of magnetism to more-than-two components. First considered by Potts in 1952, the problem has aroused considerable interest in recent years. It has been shown that the model is very rich in its content and, in addition, the extra degree of freedom exhibited by the number of components permits the model to be realized in a wide range of physical systems. In this paper we review those aspects of the Potts model related to its contents as a model of magnetism, focusing particular attention to the results obtained since a previous review was written. Topics reviewed include the upper and lower critical dimensionalities, critical properties, and some exact and rigorous results, for both the ferromagnetic and antiferromagnetic models.

PACS numbers: 05.50. + q, 75.10.Hk

I. INTRODUCTION

The Potts model, introduced by Potts[1] more than thirty years ago as a generalization of the Ising model of magnetism, has attracted increasing recent attention. it is now known that the critical behavior of the Potts model is very rich and more general than that of the Ising model, and that the model can be realized in many different physical systems. It is also known that the Potts model is related to a number of outstanding statistical problems. Discussions of these and other related topics of the Potts model have been given in a recent review[2] which summarizes the status of our understanding of the problem as of 1981. A large quantity of new results on the Potts model have since been accumulated, and it would not be possible to discuss all these developments without engaging a major endeavor. The purpose of this paper is to present a self-contained, albeit limited, review of the Potts model in its role as a model of magnetism, paying particular attention to those developments since the previous review was written. Due to the limitation of space, however, derivations of the results will not be given; a number of other pertinent developments, such as those related to the chiral Potts models not directly related to magnetism, will also be omitted.

Consider a system of N spins located on a lattice such that each spin can have q values (states) described by, say, $\sigma = 1,2,..., q$. The nearest-neighbor spins have an interaction energy $-\epsilon$ if their spin values are alike, zero if they are different. Thus the partition function is

$$Z(q,K) = \sum \exp\Big[K\sum_{nn}\delta(\sigma,\sigma')\Big], \tag{1}$$

where $K = \epsilon/kT$. Here the first summation is over all spin configurations, and the summation in the exponential is over all nearest-neighor pairs of the lattice.

[a] Permanent Address: Department of Physics, Northeastern University, Boston, Massachusetts 02115.

The system is ferromagnetic if $K > 0$. In this case the ground state consists of configurations in which all spins are in the same state. Furthermore, the system will exhibit, at sufficiently low temperatures, a spontaneous magnetization showing an ordering in one of the q spin states. For $K < 0$ the system is antiferromagnetic for which the ground state is one in which two nearest neighbors have distinct spin values. Thus, from the viewpoint of the ground-state orderings (and disorderings), the Potts model generalizes the two-component Ising model to q components.

The thermodynamics of the Potts model are derived from the "free energy" defined by taking the thermodynamic limit of the logarithmic partition function

$$f(q,K) = \lim_{N\to\infty} N^{-1} \ln Z(q,K). \tag{2}$$

From Eq. (2) one obtains the energy and specific heat, respectively,

$$E(q, K) = -\epsilon \frac{\partial}{\partial K} f(q, K), \tag{3}$$

$$C(q, K) = kK^2 \frac{\partial^2}{\partial K^2} f(q, K), \tag{4}$$

which, in turn, determine the nature of the transition. The spontaneous magnetization is defined by[2]

$$M = [q\langle\delta(\sigma,1)\rangle - 1]/(q-1), \tag{5}$$

where σ is a spin located in the interior of the lattice, and $\langle\,\rangle$ denotes the thermal average taken with all spins at the boundary fixed in the spin state 1. The spontaneous magnetization M vanishes identically for $T > T_c$, where T_c is the critical temperature, and its exact expression for $T < T_c$ is known for the two-dimensional $q = 2$ (Ising) model only.

While the parameter q enters Eq. (1) as an integer, one can always analytically continue the expressions (2)–(5) to arbitrary, even complex, values of q. This can be accomplished by continuing q, for example, after the summations in the partition function have been taken so that q appears in

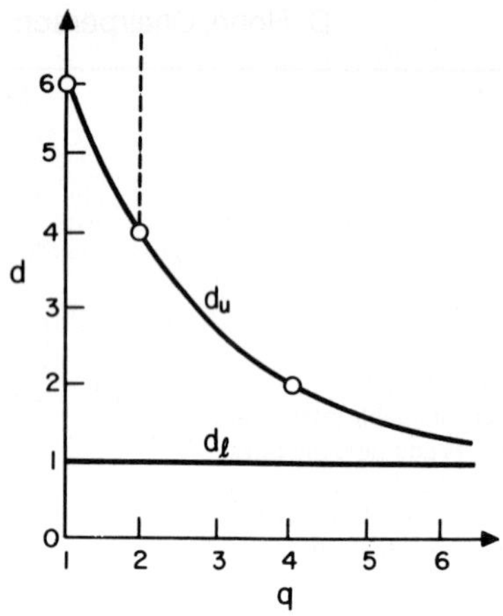

FIG. 1. Upper and lower critical dimensions of the ferromagnetic Potts model. The circles denote the exactly known points.

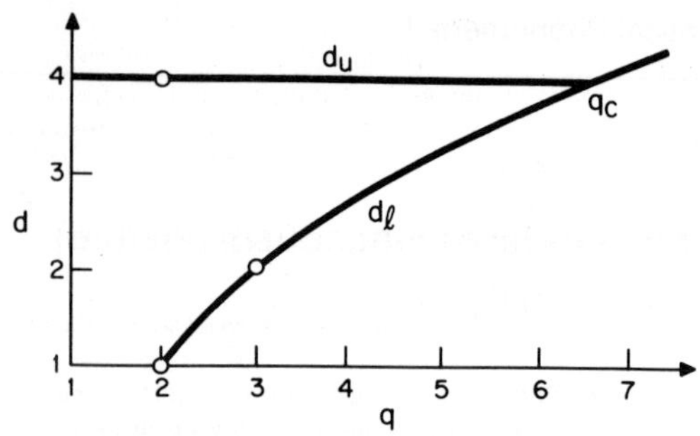

FIG. 2. Upper and lower critical dimensions of the antiferromagnetic Potts model. The circles denote the exactly known points. The two critical dimensionalities coincide for $q \geqslant q_c$.

$Z(q,K)$ as a parameter, rather than a summation index. This procedure considerably generalizes the Potts model (1).

II. CRITICAL DIMENSIONS

It is illuminating and often useful to regard the thermodynamic properties of a system as functions of the dimensionality d of the underlying lattice. One then defines the critical dimensions as the values of d at which the critical behavior experiences a change or simplification. This critical dimension can, in principle, take on integral as well as nonintegral values.

There exists in general two critical dimensions in systems which exhibit some kind of critical behavior: A lower critical dimension d_ℓ characterized by the fact that the system no longer goes critical whenever $d \leqslant d_\ell$, and an upper critical dimension d_u characterized by the fact that the critical behavior suffers a change, such as becoming mean-field-like, for $d \geqslant d_u$. For the Potts model we would generally expect, unless otherwise indicated, both d_ℓ and d_u to be q dependent.

Consider first the ferromagnetic case. Since it is exactly known that the ferromagnetic model exhibits a transition in two dimensions,[3] for $q > 1$ at least, and no transition in one dimension, it is not unreasonable to regard, as is verified by the renormalization group result for $q = 2$,[4] that the lower critical dimension to be one for all q. This constant $d_\ell = 1$ is shown by the horizontal line in Fig. 1.

The upper critical dimension for the ferromagnetic model is the dimension beyond which the system becomes mean-field-like, or the fluctuation becomes unimportant. Now the mean-field solution leads to a phase transition which is continuous for $q \leqslant 2$ and first order for $q > 2$.[2] Thus $q = 2$ is a border in the mean-field regime and is depicted by the broken line segment in Fig. 1. Also shown in Fig. 1 is a (schematic) plot of the upper critical dimension $d_u(q)$ passing through three known exactly points, $(q, d) = (1,6),(2, 4), (4, 2)$.[2] It should be noted that the transition is always continuous for $1 < q < 2$ and $d > 1$, with the upper critical dimension separating the classical (cusp singularity in the specific heat) and non-classical (divergent specific heat) regions.

Consider next the case of the antiferromagnetic model. Berker and Kadanoff[5] have argued on the basis of a rescaling consideration that a q-dependent lower critical dimension $d_\ell(q)$ should exist, and obtained its numerical estimates. Phenomenological[6] and Monte Carlo[7] renormalization group studies in two dimensions and subsequently an exact analysis for the square lattice[8] have established the exact result that $d_\ell(3) = 2$ (see Sec. IV below). In addition, it has been further established that $d_\ell(2) = 1$.[9] Thus, we now have two exact points, $(q,d) = (2, 1), (3, 2)$ for the lower critical dimension. It is also expected the lower critical dimension to behalf as,[10] for large q,

$$d_\ell(q) \sim \ln q, \quad q \to \infty. \tag{6}$$

A (schematic) plot of $d_\ell(q)$ reflecting these behaviors is shown in Fig. 2.

Consider now the upper critical dimension $d_u(q)$. We again regard it to be the dimension beyond which the system becomes mean-field-like. Now the mean-field solution of the antiferromagnetic model leads to a continuous transition for all q.[11] Further, for bipartite lattices, the $q = 2$ ferromagnetic and antiferromagnetic models are isomorphic. These considerations lead to an exact point $d_u(2) = 4$.

For other values of q we use the fact that, *if* a transition exists and *if* the transition is continuous in the antiferromagnetic Potts model, then it is in the same university class of the $O(n)$ model with $n = q - 1$.[12,13] Now the upper critical dimension of the $O(n)$ model is 4 for all $n(\geqslant 1)$.[14] It is then very plausible that, as q increases from 2 (and n increases from 1), the upper critical dimension of the q-state antiferromagnetic Potts model remains to be 4 until q reaches a critical value q_c defined by

$$d_\ell(q) = 4. \tag{7}$$

Beyond q_c, the upper and lower critical dimensions coalesce and the transition is always classical (mean-field-like). This conjectured behavior of the upper critical dimension for bipartite lattices is shown in Fig. 2.

III. CRITICAL PROPERTIES

The ferromagnetic Potts model exhibits a phase transition in two dimensions, and that the transition is continuous with nonclassical exponents for $q\leqslant 4$, and is first order, i.e., accompanied with a latent heat, for $q>4$. A summary of these and other relevant critical properties, including a list of critical exponents and expressions for critical quantities can be found in Ref. 2. More recently using the corner transfer matrix approach, Baxter[15] has shown that the spontaneous magnetization M is discontinuous at the critical temperature T_c for $q>4$, jumping from the value zero for $T>T_c$ to a nonzero value $M(T_c-)$ at T_c. The exact expression for this jump discontinuity obtained by Baxter is

$$M(T_{C}-) = 0 \qquad q\leqslant 4$$
$$= \prod_{n=1}^{\infty}[(1-e^{-(2n-1)\theta})/(1+e^{-2n\theta})] \qquad q\geqslant 4, \tag{8}$$

where

$$q=2\cosh\theta. \tag{9}$$

It is remarkable that the expression (8), which depends only on q, not on the interactions K, is valid for all two-dimensional lattices,[15] a fact first conjectured by Kim.[16] The expression (8) possesses an essential singularity at $q=4$ near and above which it behaves as

$$M(T_c-)\sim 2\exp[-\pi^2/8(q-4)^{1/2}]. \tag{10}$$

This is in agreement with the renormalization group prediction.[17]

A first-order transition exists in the Potts model in $d\geqslant 2$ dimensions when q is sufficiently large. This fact, while expected intuitively on the basis of the mean-field analysis, has recently been rigorously proved.[18] One can also use this fact to establish the existence of a first-order transition in certain Ising models which are equivalent to a Potts model of $q=2^n$, $n=1,2,\ldots$, components.[19]

The critical exponents of the Potts model are well defined when the transition is continuous. To obtain the two leading thermal exponents we expand the singular part of the free energy (2) about the critical point K_c:

$$f_{\text{sing}}(q,K)\sim|K-K_c|^{2-\alpha}[1+a_1|K-K_c|^{\Delta_1}+\ldots]. \tag{11}$$

The thermal exponents are obtained from Eq. (11) using the relations $y_1=d/(2-\alpha)$ and $\Delta_1=-y_2/y_1$. In a similar fashion for the two leading magnetic exponents we write, for r large, the critical ($K=K_c$) two-point correlation function $\Gamma_c(r)$ as[20]

$$\Gamma_c(r)\sim r^{-(d-2+\eta)}[1+a_2r^{-\Delta_2}+\ldots],\quad r\to\infty. \tag{12}$$

The exact value of the leading thermal exponent for $d=2$ is[21]

$$\alpha=2(1-2u)/3(1-u), \tag{13}$$

where

$$0\leqslant u=(2/\pi)\cos^{-1}(\sqrt{q}/2)\leqslant 1,\quad 0\leqslant q\leqslant 4. \tag{14}$$

Nienhuis[22] has further computed the correction to scaling and obtained

$$\Delta_1=4u/3(1-u). \tag{15}$$

The exact values of the two leading magnetic exponents for the two-dimensional models have been derived by den Nijs[20] from a consideration of the Coulomb-gas representation of the Potts model. The results are

$$\eta=(1-u^2)/2(2-u), \tag{16}$$

$$\Delta_2=4/(2-u). \tag{17}$$

Thus, $\Delta_1=4, 8/3, 4/3, 2/3, 0$, $\Delta_2=1, 3, 8/3, 12/5, 2$, for $q=0, 1, 2, 3, 4$ respectively. The $q=3$ value of $\Delta_1=2/3$ is to be compared with the numerical estimates of $\Delta_1\sim 0.6$ from a low-temperature series analysis.[23]

The partition function of the Potts model satisfies an inversion functional relation which is most easily derived by considering its transfer matrix.[24,25] For the square lattice with anisotropic interactions K_1 and K_2, the inversion relation reads

$$Z(q,K_1,K_2)Z[q,-K_1,\ln(2-q-e^{K_2})]$$
$$=(e^{K_1}-1)(1-q-e^{-K_2}), \tag{18}$$

where $Z(q,K_1,K_2)$ is the corresponding partition function of the anisotropic model. The validity of Eq. (18) has been verified by perturbative large-q expansions.[24,26] The inversion relation (18), which is based on the transfer matrix formalism, can be readily generalized to other, including the checkerboard and the simple cubic, lattices.[27,28] However, unlike the $q=2$ Ising case for which the inversion relation can be used in conjunction with an analytic assumption to uniquely derive the partition function,[29] the inversion relation for general q does not seem to lead to any determination of the Potts partition function. There also appears to be some profound differences between the inversion relations of the $d=2$ and $d=3$ models.[28]

The critical point of the Potts model are exactly known for the square, triangular, and honeycomb lattices.[2] But using the inversion relation one can further locate the critical point for the checkerboard lattice with interactions K_1,K_2,K_3, and K_4. This leads to the following critical manifold[27]:

$$\prod_{i=1}^{4}[(e^{K_i}+e^{2\theta})/(e^{K_i+\theta}+e^{-\theta})]=1. \tag{19}$$

The expression (19) confirms an earlier conjecture.[30] It is of some interest to note that Svrakic[31] used a heuristic argument to deduce the exact critical point for some two-dimensional models, but the argument does not reproduce Eq. (19) when applied to the checkerboard lattice.

For three-dimensional lattices Park and Kim[32] have obtained accurate estimates of the critical point from an analysis of the large-q series expansions of the susceptibility and magnetization. More generally, Hajdukovic[33] proposed an expression for the critical point of the q-state model on a d-dimensional hypercubic lattice. His conjecture,

$$e^{dK}-2^{1/(d-1)}e^{dK/2}-q+1=0, \tag{20}$$

is indeed exact for $d=2$ and $d=1$, and is fairly accurate for $q\geqslant 2$ in $d\geqslant 3$. But it gives an erroneous critical probability of $p_c=1-2^{-1/3}=0.2063$ for the bond percolation ($q=1$) on the simple cubic lattice ($d=3$). Since this prediction differs appreciably from the value $p_c=0.247$ derived from accurate numerical analyses,[34] the expression (20) cannot be correct for general q and d.

IV. ANTIFERROMAGNETIC MODEL

The antiferromagnetic Potts model has received considerable attention in the last two years. It has been studied under different approaches including the finite-size[6] and Monte Carlo[7] renormalization groups, phenomenological scaling transformation,[35] mean-field analysis,[11] Monte Carlo simulations,[11,12,36] and exact analysis.[8] Particularly, Baxter[8] observed that the two-dimensional antiferromagnetic model (1) is exactly solvable at

$$(e^K + 1)^2 = 4 - q \tag{21}$$

and argued that, as in the case of the ferromagnetic model, the solubility condition (21) should coincide with the critical point and that the transition at Eq. (21) is continuous. We see from Eq. (21) that the transition occurs at real temperatures only for $q \leqslant 3$. The vanishing of the transition temperature at $q = 3$ establishes the exact result of $q_c(3) = 2$ quoted in Sec. II.

Most of the known results of the Potts model are derived from its equivalence with a six-vertex model[37] and the established results of the latter problem. However, in order to establish the equivalence of the two models, it is necessary to assume a boundary condition for the Potts model which is periodic in no-more-than-one direction. On the other hand, the known results of the six-vertex model are always derived by assuming periodicity in two directions. Therefore, care must be taken in transcribing the results. Indeed, one can even see in one dimension that the antiferromagnetic Potts model is very sensitive to the boundary conditions. The introduction of a periodic boundary condition to a one-dimensional chain, e.g., induces a transition for $1 < q < 2$.[9] In two dimensions, Baxter[8] observed that the solution of the six-vertex model with periodic boundary conditions does *not* always yield the correct critical free energy for the antiferromagnetic Potts model with free boundaries. While this particular difficulty is circumvented by the use of the inversion relation, there is a clear need for alternate approaches to the antiferromagnetic problem.

In the absence of further exact data, the antiferromagnetic model is best studied by Monte Carlo simulations.[11,12] It has been established that, for bipartite lattices and except for $q = 3$ and $d = 2$ for which there is no transition, a low-temperature ordered phase exists with distinct spin states favored on each of the two sublattice. This is the antiferromagnetic (AF) phase, and the transition into which is found to be continuous. However, the critical properties associated with this transition remain undetermined. In this connection, the universality-class equivalence of the q-state antiferromagnetic model and the $0(q-1)$ model (cf. Sec. II) should prove useful in further studies. It should also be borne in mind that the lattice symmetry, whether bipartite or tripartite, also plays a crucial role in determining the critical properties.

The picture is more complicated when there are next-nearest-neighbor competing ferromagnetic interactions. Finite-size scaling analysis of the 3-state model has led to some understanding of the phase diagram as well as numerical estimates of the critical exponents in two dimension.[38] It also suggests that the model is in the same universality class of the 6-state clock (planar Potts) model.[38] In three dimensions, Monte Carlo study[11] of the 4-state model indicates a mean-field-like phase diagram. That is, for small ferromagnetic interactions the system exhibits a sequence of three transitions. At low temperatures the system possesses a net "ferromagnetic" ordering. As the temperature rises, the system first changes the ordering to one charaterized by a broken sublattice symmetry (BSS) for which one spin state is favored on one sublattice and the other $q - 1$ states on the other sublattice. The system next goes into the AF phase, and eventually becoming paramagnetic (P) at high temperatures. For larger ferromagnetic interactions, the AF and BSS phases disappear in succession, leaving finally a single transition between the ferromagnetic and paramagnetic phases. The mean-field theory predicts that all, except the AF-P, transitons are first order. This prediction appears to be already realized in the $q = 4$, $d = 3$ model.[11] But the critical exponents associated with the continuous AF-P transition remain undetermined.

ACKNOWLEDGMENTS

I would like to thank J. R. Banavar, M. E. Fisher, M. P. Nightingale, L. Nosanow, E. Riedel, and H. E. Stanley for useful conversations. This research was performed as part of the NSF Independent Research Program. However, any of the opinions expressed herein are those of the author and do not necessarily reflect the views of the NSF.

[1]R. B. Potts, Proc. Camb. Philos. Soc. **48**, 106 (1952).
[2]F. Y. Wu, Rev. Mod. Phys. **54**, 235 (1982).
[3]R. J. Baxter, J. Phys. C **6**, L445 (1973).
[4]D. R. Nelson and M. E. Fisher, Ann. Phys. **91**, 226 (1975).
[5]A. N. Berker and L. P. Kadanoff, J. Phys. A **13**, L259 (1980).
[6]M. P. Nightingale and M. Schick, J. Phys. A **15**, L39 (1982).
[7]C. Jayaprakash and J. Toblchnick, Phys. Rev. B **25**, 4890 (1982).
[8]R. J. Baxter, Proc. R. Soc. London Ser. A **383**, 43 (1982).
[9]F. Y. Wu, report in 15th IUPAP Stat. Phys. Conference (1983).
[10]A. N. Berker (private comunication).
[11]J. R. Banavar and F. Y. Wu (to be published).
[12]J. R. Banavar, G. S. Grest, and D. Jasnow, Phys. Rev. Lett. **45**, 1424 (1980); J. R. Banavar, G. S. Grest, and D. Jasnow, Phys. Rev. B **25**, 4639 (1982).
[13]J. R. Banavar (private communication).
[14]K. G. Wilson and M. E. Fisher, Phys. Rev. Lett. **28**, 240 (1972).
[15]R. J. Baxter, J. Phys. A **15**, 3329 (1982).
[16]D. Kim. Phys. Lett. A **87**, 127 (1981).
[17]J. L. Cardy, M. Nauenberg and D. J. Scalapino, Phys. Rev. B **22**, 2560 (1980).
[18]R. Kotecky and S. B. Sholsman, Commun. Math. Phys. **83**, 493 (1982).
[19]J. Bricmont, A. Messager, and J. L. Lebowitz, Phys. Lett. A **95**, 169 (1983).
[20]M. P. M. den Nijs, Phys. Rev. B **27**, 1674 (1983).
[21]J. L. Black and V. J. Emery, Phys. Rev. B **23**, 429 (1981); M. P. M. den Nijs, J. Phys. A **12**, 1857 (1979).
[22]B. Nienhuis, J. Phys. A **15**, 199 (1982).
[23]J. Adler, I. G. Enting, and V. Privman, J. Phys. A **16**, 1967 (1983).
[24]M. T. Jaekel and J. M. Maillard, J. Phys. A **15**, 2241 (1982).
[25]R. J. Baxter, J. Stat. Phys. **28**, 1 (1982).
[26]M. T. Jaekel and J. M. Maillard, J. Phys. A **15**, 2509 (1982).
[27]J. M. Maillard and R. Rammal, J. Phys. A **16**, 1073 (1983).
[28]J. M. Maillard and R. Rammel, J. Phys. A **16**, 353 (1983).
[29]R. J. Baxter, in *Fundamental Problems in Statistical Mechanics V: Pro-*

ceedings of Enschede Summer School, edited by E. G. D. Cohen (North Holland, Amsterdam, 1982).
[30]F. Y. Wu, J. Phys. C **12**, L645 (1979).
[31]N. M. Svrakic, Phys. Lett. A **80**, 43 (1980).
[32]H. Park and D. Kim. J. Korean Phys. Soc. **15**, 55 (1982).
[33]D. Hajdukovic, J. Phys. A **16**, L193 (1983).
[34]D. Stauffer, Phys. Reports, **54**, 1 (1979).
[35]A. H. Osbaldestin and D. W. Wood, J. Phys. A **15**, 3593 (1982).
[36]F. Fucito, J. Phys. A **16**, L541 (1983).
[37]R. J. Baxter, S. B. Kelland, and F. Y. Wu, J. Phys. A **9**, 397 (1976).
[38]M. P. M. den Nijs, M. P. Nightingale, and M. Schick, Phys. Rev. B **26**, 2490 (1982).

Journal of Statistical Physics, Vol. 19, No. 6, 1978

Exact Results for the Potts Model in Two Dimensions

A. Hintermann,[1] H. Kunz,[2,3] and F. Y. Wu[4,5]

Received August 18, 1978

By considering the zeros of the partition function, we establish the following results for the Potts model on the square, triangular, and honeycomb lattices: (i) We show that there exists only one phase transition; (ii) we give an exact determination of the critical point; (iii) we prove the exponential decay of the correlation functions, in one direction at least, for all temperatures above the critical point. The results are established for $q \geqslant 4$, where q is the number of components.

KEY WORDS: Potts model; critical point; zeros of partition function; correlation function.

1. INTRODUCTION

The critical point of the Potts model was first conjectured for the square lattice by Potts[(1)] using the Kramers–Wannier argument[(2)] in conjunction with an assumption of the uniqueness of the transition. The conjecture has since been extended to the triangular and honeycomb lattices[(3,4)] under more stringent conditions. While it has been established that a transition indeed occurs at these conjectured points,[(4,5)] the uniqueness of the transition has not been proven and, consequently, the determination of the Potts critical point remains very much an unsettled question.

We present in this paper an analysis of the Potts model which leads to an exact determination of its critical point for the square, triangular, and honeycomb lattices. In addition to confirming the previous conjectures, our analysis also establishes the uniqueness of the transition. We also prove that the correlation functions decay exponentially above the critical point.

[1] Swiss Institute for Nuclear Research (SIN), Villigen, Switzerland.

[2] Laboratoire de Physique Théorique, Ecole Polytechnique Fédérale, Lausanne, Switzerland.

[3] Work supported by the Fond. National Suisse de la Recherche Scientifique.

[4] Department of Physics, Northeastern University, Boston, Massachusetts.

[5] Work supported in part by NSF Grant No. DMR 76-20643.

0022-4715/78/1200-0623$05.00/0

Our analysis is based on the consideration of the zeros of the Potts partition function. Generally, the zeros of the partition function of a thermodynamic system trace a certain locus in the complex inverse temperature $\beta = 1/kT$ plane, and a phase transition occurs at the point where the locus crosses the positive β axis.[6] The strategy of our consideration is then to determine the region in the neighborhood of the positive β axis that is free of zeros. To carry out such an analysis we first convert the Potts model into a vertex model[5,7] for which some information on the zero distribution is known.[8] Since there is only one complex variable, namely β, arising in the Potts model, the fugacity z in the vertex model is actually temperature dependent. We next let z become independent and consider more generally the partition function $Z(\beta, z)$ of the vertex model. This permits the use of the circle theorem due to Suzuki and Fisher,[8] which states that $Z(\beta, z)$ is free of zeros for β real and $|z| \neq 1$. We further establish in the appendix that $Z(\beta, z)$ is free of zeros for small $|z|$ and β in a neighborhood of the positive axis. Combining these two results and making use of the Lebowitz–Penrose Lemma,[9] we then deduce that $Z(\beta, z)$ is in fact free of zeros for all $|z| \neq 1$ and β in a neighborhood of the positive axis. Returning now to the Potts model for which $z = z(\beta)$, this implies that the Potts free energy can be nonanalytic in β only at $|z(\beta)| = 1$; this in turn leads to a unique critical point. In Section 3 we sketch a proof which establishes the exponential decay of the correlation functions for all temperatures above the critical point.

Due to technical reasons, our results are established strictly only for $q \geqslant 4$, where q is the number of components in the Potts model. Since the $q = 2$ (Ising) model is exactly soluble, this leaves only the $q = 3$ Potts model unsettled for the time being.

2. POTTS CRITICAL POINT

It was established by Baxter *et al.*[7] that the partition function Z_N of the Potts model on a planar lattice L of N sites is related to the partition function Z_N' of an ice-rule vertex model on a related medial lattice L' by the relation

$$Z_N = q^{N/2} Z_N' \tag{1}$$

The medial lattice is essentially the covering lattice of L, constructed by connecting the midpoints of adjacent edges of L. For example, as illustrated in Figs. 7 and 8 of Ref. 7, the medial lattice of a square lattice is also square, and that of the triangular and honeycomb lattices is the Kagomé lattice. The vertex configurations of the ice-rule model are shown in Fig. 1a, where the

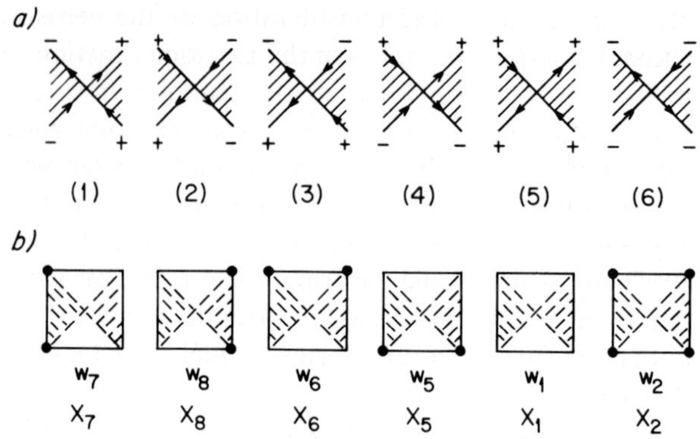

Fig. 1. (a) The ice-rule vertex configurations of the medial lattice and the associated spin configurations. The shaded regions denote the faces occupied by the sites of the Potts lattice. (b) The corresponding Ising configurations X_j and their weights w_j as defined in Ref. 17.

shaded regions denote the faces of L' occupied by the sites of L. In the notations of Ref. 7, the vertices have the following weights:

$$\{\omega_1,..., \omega_6\} = \{1, 1, x_r, x_r, A_r, B_r\} \tag{2}$$

where

$$\begin{aligned}(A_r, B_r) &= (s^{-1} + x_r s, s + x_r s^{-1}) \quad \text{square} \\ &= (t^{-1} + x_r t^2, t + x_r t^{-2}) \quad \text{triangular}\end{aligned} \tag{3}$$

with

$$s = e^{\theta/2}, \qquad t = e^{\theta/3}, \qquad \cosh\theta = \sqrt{q}/2$$

$$x_r = u_r/\sqrt{q}, \qquad u_r = \exp(\beta\epsilon_r) - 1 \tag{4}$$

Here, $\epsilon_r > 0$ is the interaction in the Potts model between neighboring sites on edges of L in a given direction r, $r = 1, 2$ $(=1, 2, 3)$ for the square (triangular) lattice. There is no need to consider the honeycomb lattice separately since the triangular and honeycomb Potts models are related by a duality relation.[10] We have also included in Fig. 1a a spin representation of the vertex configurations, obtained by assigning a spin σ to each arrow such that $\sigma = +1$ (-1) if, crossing the arrow from the shaded region to the unshaded region, the arrow points toward one's right (left). We also remark that the vertex model (2) offers a natural extension of the Potts model to nonintegral values of q, which we shall assume to be the case.

Next we generalize the vertex weights (2) into a form reflecting the ice-rule restriction. Observe that if all arrows on the edges in a given direction are reversed, vertices (5) and (6) are converted to either the source or the sink of arrows. The conservation of arrows then implies the following relation:

$$n_{r5} + n_{r'6} = n_{r6} + n_{r'5}, \qquad r \neq r' \tag{5}$$

where n_{r5} (n_{r6}) denotes the number of the (5) [(6)] vertices on the type r edge of L. As a consequence of (5), the partition function Z_N' is unchanged if we use the following vertex weights in place of (2):

$$\begin{aligned}\{\omega_1,\ldots,\omega_6\} &= \{1, 1, x_r, x_r, u_rA_r, u_r^{-1}B_r\} \\ &\equiv \{1, 1, x_r, x_r, c_rz, c_rz^{-1}\}\end{aligned} \tag{6}$$

provided that we take

$$\begin{aligned} u_1u_2 &= 1 \qquad \text{square} \\ u_1u_2u_3 &= 1 \qquad \text{triangular}\end{aligned} \tag{7}$$

A similar argument leading to (6) can be found in Ref. 4.

Now, the variables c_r and z given by

$$c_r^2 = A_rB_r \tag{8}$$

$$\begin{aligned} z^4 &= A_1A_2/B_1B_2 \qquad \text{square} \\ z^6 &= A_1A_2A_3/B_1B_2B_3 \qquad \text{triangular}\end{aligned} \tag{9}$$

are both functions of the inverse temperature β of the Potts model. In order to make use of an established theorem on the zeros of a partition function, we now generalize the partition function Z_N' by regarding z in (6) as an independent variable and consider $Z_N' = Z_N'(\beta, z)$. Any conclusion so reached for $Z_N'(\beta, z)$ can obviously be specialized to the Potts model by once again introducing (9). Note that $Z_N'(\beta, z)$ is invariant under the change $z \rightarrow z^{-1}$, since a reversal of all arrows results in only an interchange of the weights ω_5 and ω_6 in (6).

To locate the zeros of $Z_N'(\beta, z)$ in the complex z plane for real β, we make use of a generalized Lee–Yang circle theorem due to Suzuki and Fisher.[8] Identifying z in (6) as the same variable z appearing in Eq. (2.3) of Ref. 8, and using the spin representation of the vertex configurations shown in Fig. 1a, we see that the partition function $Z_N'(\beta, z)$ is in precisely the form of that occurring in the Suzuki–Fisher (SF) theorem. For real β and

$$q > 4 \tag{10}$$

the variables x_r and c_r are both real, so that the condition (A) of the SF theorem is fulfilled. It is also readily verified that the condition (B) of the SF theorem holds under the same conditions. It then follows from the SF theorem that the zeros of $Z_N'(\beta, z)$ lie on the unit circle in the complex z plane for real β and $q > 4$.

Since $Z_N'(\beta, z) \sim |z|^{-M}$ for small $|z|$, where $M = 2N$ and $3N$, respectively, for the square and triangular lattices L, it proves convenient to consider, instead of Z_N', the function

$$F_N(\beta, z) = z^M Z_N'(\beta, z) \tag{11}$$

which is a polynomial of degree $2M$ in z. Using (6), we see in particular that

$$F_N(\beta, 0) = C^N \tag{12}$$

where

$$C = \prod_r c_r \tag{13}$$

This permits us to write

$$F_N(\beta, z) = C^N \prod_{i=1}^{2M} (1 - z/z_i) \tag{14}$$

where z_i are the $2M$ zeros of $F_N(\beta, z)$ satisfying $|z_i| = 1$ for real β and $q > 4$.

Consider now the function

$$G_N(\beta, z) = [F_N(\beta, z)]^{-1} \tag{15}$$

We have established that:

(i) $F_N(\beta, z) \neq 0$ for $|z| \neq 1$, β real, and $q > 4$.

We shall also establish in the appendix that:

(ii) $F_N(\beta, z) \neq 0$ for all $|z| < \delta$ and $\operatorname{Re} \beta \geqslant 0$, $|\operatorname{Im} \beta| < \pi/2\epsilon$, $q > 4$, where $\epsilon = \sup_r \epsilon_r$ and δ is some strictly positive constant depending only on $\{\epsilon_r\}$.

Furthermore, (14) implies the following bound on $G_N(\beta, z)$:

(iii)
$$\begin{aligned} G_N(\beta, z) &\leqslant C^{-N} \prod_i (1 - |z/z_i|) \\ &= C^{-N}(1 - |z|)^{-2M} \end{aligned} \tag{16}$$

The function $G_N(\beta, z)$ now satisfies precisely the conditions of the Lebowitz–Penrose Lemma[(9)] for a function of two variables. Applying the Lemma, we conclude that:

(iv) $F_N(\beta, z) \neq 0$ for all $|z| \neq 1$, $q > 4$, and β in some neighborhood of the positive real axis, the region of the neighborhood being uniform with regard to N.

Now $z(\beta)$ is real analytic in β. It follows from (iv) that for $q > 4$, the partition function $Z_N(\beta)$ of the Potts model is free of zeros in β, when β is in a complex neighborhood D of $[0, \beta_c)$ of $(\beta_c, \infty]$, where $\beta_c = \beta_c(q)$ is given by $z(\beta_c) = 1$. In fact, this conclusion holds also for $q = 4$, provided that we take

$$\beta_c(4) = \lim_{q \to 4+} \beta_c(q) \tag{17}$$

This is permitted because $Z_N(\beta)$ is a polynomial in e^β whose coefficients are continuous functions of q. Consequently, the zeros of the polynomial also depend on q continuously.

Following the standard arguments,[(6)] we now conclude that, for $q \geqslant 4$, the free energy of the Potts model,

$$f(\beta) = \lim_{N \to \infty} (1/M) \ln Z_N(\beta) \tag{18}$$

is a real analytic function of β when β is positive, except possibly at β_c. To determine β_c, we use (9) and the condition $z(\beta_c) = 1$ to obtain

$$\begin{aligned} x_1 x_2 &= 1 && \text{square} \\ \sqrt{q}\, x_1 x_2 x_3 + x_1 x_2 + x_2 x_3 + x_3 x_1 &= 1 && \text{triangular} \\ \sqrt{q} + x_1 + x_2 + x_3 &= x_1 x_2 x_3 && \text{honeycomb} \end{aligned} \tag{19}$$

Here we have used the duality relation[(10)] $x_r x_r^* = 1$ to relate the triangular and the honeycomb lattices.

Two comments are in order at this point. First we comment on the limitation of our results to $q \geqslant 4$. For $0 < q < 4$, conditions (A) and (B) of the Suzuki–Fisher theorem no longer hold and the locus of the zeros of $Z_N'(\beta, z)$ is not known. However, numerical results[(11)] indicate that the zeros do leave the unit circle, and, in fact, $z(\beta)$ lies on the unit circle for β real. It is clear that the strategy of the proof would be very different. This seems to confirm the change of the analytic properties of the Potts model found to exist at $q = 4$.[(4,5)] We wish to point out, however, that the critical point (19) does coincide with the exact (Ising) result at $q = 2$, and agrees with the previously conjectured critical point[(3,4)] including the $q = 1$ limit of the bond percolation.[(12)]

Finally, we comment that, strictly speaking, our analysis establishes only the fact that the nonanalyticity of $f(\beta)$, if any, can occur only at β_c. Now it has been explicitly established that $f(\beta)$ is indeed nonanalytic in β at β_c.[(4,5)] It follows that the Potts model has only one critical point, and that the critical point is given by (19).

3. CORRELATION FUNCTIONS

An interesting consequence of our analysis is that it allows us to establish the exponential decay of the correlation functions, for all temperatures above the critical temperature. We outline here the main steps of the proof.

First of all, our result on the zeros of the partition function remains true if, instead of using the free boundary conditions, we take a boundary condition such that the lattice is periodic in one direction. For the lattice is then still planar and all steps of our proof including the adoption of the result of Ref. 7 remain unchanged. In particular, a transfer matrix formulation of the partition function can be formulated, and there exists a domain D in the $u = e^{\beta} - 1$ plane containing the origin and the segment $[0, \exp(\beta_c) - 1]$ such that $Z_N(\beta) \neq 0$ when $u \in D$. Also, it has been shown by Israel[13] that when $|u| < c$, where c is some constant, $Z_N(\beta) \neq 0$ and the correlation functions decay exponentially. These two facts now permit the use of a theorem due to Penrose and Lebowitz[14] to conclude that the gap between the largest and the second largest eigenvalues of the transfer matrix remains nonzero, uniformly in N, when $u \in D$. But since this gap is a lower bound to the coherence length in the direction of periodicity along which the transfer matrix is defined, it follows that the correlation functions decay exponentially in the direction of periodicity for all $u \in D$, and hence for all $0 < \beta < \beta_c$. This establishes the stated result. Details of the proof follow closely that of Ref. 14 for the lattice gas, and will not be reproduced.

APPENDIX. PROOF OF PROPERTY (ii)

We establish in this appendix the property (ii) on the zeros of the function $F_N(\beta, z)$. The strategy here is to use the spin representation of the six-vertex model shown in Fig. 1a, and consider this as a constrained Ising model. The Asano contraction technique[15] is then applied to yield the desired property.

The idea of the Asano contraction is to obtain $F_N(\beta, z)$ by "contraction" of polynomials in few variables so as to relate the properties of zeros of the small polynomials to the zeros of $F_N(\beta, z)$. In the present case of a six-vertex model, the main problem of finding the small polynomials to build up $F_N(\beta, z)$ has already been solved in a more general context by Hintermann and Gruber.[16,17] The following discussion uses results established in Ref. 17.

The first step is to conform with the notations of Ref. 17. Associate a dot to each spin $\sigma = -1$ as indicated in Fig. 1b and compare the resulting configurations with those shown on p. 189 of Ref. 17. We then find the following relationships between the vertex weights w_j of Ref. 17 and the vertex weights ω_j defined in Section 2:

$$\begin{aligned} \{w_1, w_2, \dots, w_8\} &= \{\omega_5, \omega_6, 0, 0, \omega_4, \omega_3, \omega_1, \omega_2\}, \qquad r = 1 \\ &= \{\omega_5, \omega_6, 0, 0, \omega_1, \omega_2, \omega_3, \omega_4\}, \qquad r = 2 \qquad \text{(A1)} \\ &= \{\omega_5, \omega_6, 0, 0, \omega_3, \omega_4, \omega_2, \omega_1\}, \qquad r = 3 \end{aligned}$$

where $r = 1, 2$ (1, 2, 3) for the square (triangular) lattice.

Following Ref. 17, we put $w_j = \exp(-\beta e_j)$ and adopt the convention $e_j = 0$ if $w_j = 0$, $j = 1, 2,..., 8$. Then from (6) we have

$$\begin{aligned} -\beta\{e_1,..., e_8\} &= \{\ln c_1 z, \ln c_1 z^{-1}, 0, 0, \ln x_1, \ln x_1, 0, 0\}, \qquad r = 1 \\ &= \{\ln c_2 z, \ln c_2 z^{-1}, 0, 0, 0, 0, \ln x_2, \ln x_2\}, \qquad r = 2 \\ &= \{\ln c_3 z, \ln c_3 z^{-1}, 0, 0, \ln x_3, \ln x_3, 0, 0\}, \qquad r = 3 \end{aligned} \tag{A2}$$

Notice that the energies e_j for $r = 1$ and 3 are identical except for the difference in the subscripts. Next, as in the conventions given in p. 190 of Ref. 17, we associate to each vertex a local Hamiltonian $-H_c$ such that

$$H_c(X_j) = e_j, \qquad j = 1, 2,..., 8 \tag{A3}$$

where X_j refers to the spin configurations. Write

$$-H_c = J_0 + \sum_{i=1}^{4} J_i\sigma_i + \tfrac{1}{2} \sum_{i\neq k=1}^{4} J_{ik}\sigma_{ik} \tag{A4}$$

where $\sigma_1,..., \sigma_4$ are the four spins surrounding a vertex and $\sigma_{ik} = \sigma_i\sigma_k$. We find with

$$\begin{aligned} J_B &= -\frac{1}{8} \sum_{j=1}^{8} \sigma_B(X_j)e_j \\ \sigma_B(X_j) &= (-1)^{|B\cap X_j|}, \qquad B \in \{(i), (i, k)\}_{i,k=1,...,4} \end{aligned} \tag{A5}$$

$$\begin{aligned} &(4\beta)\{J_0, J_1 = J_2 = J_3 = J_4, J_{12} = J_{34}, J_{13} = J_{24}, J_{14} = J_{23}\} \\ &\quad= \{\ln c_1x_1, \ln z, \ln c_1x_1, \ln(c_1/x_1), \ln(c_1/x_1)\}, \qquad r = 1 \\ &\quad= \{\ln c_2x_2, \ln z, \ln(c_2/x_2), \ln(c_2/x_2), \ln c_2x_2\}, \qquad r = 2 \\ &\quad= \{\ln c_3x_3, \ln z, \ln c_3x_3, \ln(c_3/x_3), \ln(c_3/x_3)\}, \qquad r = 3 \end{aligned} \tag{A6}$$

We now have the identity

$$Z_N' = \tilde{Z}_N \tag{A7}$$

where $\tilde{Z}_N$ is the partition function of the constrained Ising model described by the Hamiltonian (A4)–(A6) and in which only the configurations with $w_j \neq 0$ are allowed. As in Ref. 17, since each lattice site of the Ising model belongs to two constraints, each of which giving a field contribution $J_i = (4\beta)^{-1} \ln z$, the field activity variable is simply $z_i = \exp(-4\beta J_i) = z^{-1}$, $i = 1, 2, 3, 4$. Similarly, the two-body activities are

$$z_{ij} = \exp(-2\beta J_{ij}) \in \{(c_rx_r)^{-1/2}, (x_r/c_r)^{1/2}\}$$

and we have

$$\tilde{Z}_N = \tilde{Z}_N(z, z_{ij}) \tag{A8}$$

The next step is to study $\tilde{Z}_N$ for independent z, c_r, x_r. This can be achieved by means of the Asano contractions of small polynomials.[15] Since, in the notations of Ref. 17, Eq. (A4) implies $\mathscr{B} \equiv \mathscr{B}_\infty$ and we have trivially $\overline{\mathscr{B}} \supset \mathscr{B}_\infty$, we can use the prescription of p. 233 of Ref. 17 to find the following small polynomial:

$$M_c = 1 + z_{14}z_{13}z_{24}z_{23}(z_1z_2 + z_3z_4) + z_{12}z_{13}z_{24}z_{34}(z_1z_4 + z_2z_3) + z_1z_2z_3z_4 \tag{A9}$$

associated with the constraints. Since only the variables z_i undergo one contraction and the two-body activities undergo no contractions, we can consider the products of the two-body activities as complex parameters. It is then necessary to study only the following type of local polynomial:

$$M_c(z_1, z_2, z_3, z_4) = 1 + u_1(z_1z_2 + z_3z_4) + u_2(z_1z_4 + z_2z_3) + z_1z_2z_3z_4 \tag{A10}$$

with

$$\begin{aligned} \{u_1, u_2\} &= \{x_1c_1^{-1}, c_1^{-1}\}, \qquad r = 1 \\ &= \{c_2^{-1}, x_2c_2^{-1}\}, \qquad r = 2 \\ &= \{x_3c_3^{-1}, c_3^{-1}\}, \qquad r = 3 \end{aligned} \tag{A11}$$

Let $u_1, u_2 \in \mathbb{C}$. Since

$$\min_{|z_i|,|z_j|<\rho} \operatorname{Re}(u_kz_iz_j) = -|u_k|\rho^2$$

we have

$$\operatorname{Re} M_c(z_1, z_2, z_3, z_4) \geqslant 1 - 2(|u_1| + |u_2|)\rho^2 - \rho^4$$

for $|z_i| < \rho$, $i = 1, 2, 3, 4$. It follows that

$$M_c(z_1, z_2, z_3, z_4) \neq 0$$

for

$$|z_i|^2 < [1 + (|u_1| + |u_2|)^2]^{1/2} - |u_1| - |u_2| \tag{A12}$$

The case in which we are interested concerns two independent complex variables z and β. A straightforward calculation shows that there exists a $\delta > 0$ such that

$$\delta < [1 + (|u_1| + |u_2|)^2]^{1/2} - |u_1| - |u_2|$$

whenever $\operatorname{Re}\beta \geqslant 0$, $|\operatorname{Im}\beta| \leqslant \pi/2\epsilon$, where $\epsilon = \sup_r \epsilon_r$ and $q > 4$. This establishes property (ii) for Z' and, consequently, for $F_N(\beta, z)$.

ACKNOWLEDGMENT

One of us (FYW) wishes to thank Prof. Ph. Choquard for his kind hospitality at Laboratoire de Physique Théorique, Ecole Polytechnique Fédérale, where part of this research was done.

REFERENCES

1. R. B. Potts, *Proc. Camb. Phil. Soc.* **48**:106 (1952).
2. H. A. Kramers and G. H. Wannier, *Phys. Rev.* **60**:252 (1941).
3. D. Kim and R. I. Joseph, *J. Phys. C* **7**:L167 (1974).
4. R. J. Baxter, H. N. V. Temperley, and S. E. Ashley, *Proc. Roy. Soc. Lond. A* **358**:535 (1978).
5. R. J. Baxter, *J. Phys. C* **6**:L445 (1973).
6. C. N. Yang and T. D. Lee, *Phys. Rev.* **87**:404 (1952).
7. R. J. Baxter, S. B. Kelland, and F. Y. Wu, *J. Phys. A* **9**:397 (1976).
8. M. Suzuki and M. E. Fisher, *J. Math. Phys.* **12**:235 (1971).
9. J. L. Lebowitz and O. Penrose, *Comm. Math. Phys.* **11**:99 (1968).
10. F. Y. Wu, *J. Math. Phys.* **18**:611 (1977).
11. K. S. Chang, S. Y. Wang, and F. Y. Wu, *Phys. Rev. A* **4**:2324 (1971).
12. M. F. Sykes and J. W. Essam, *J. Math. Phys.* **5**:1117 (1964).
13. R. B. Israel, *Comm. Math. Phys.* **50**:245 (1976).
14. O. Penrose and J. L. Lebowitz, *Comm. Math. Phys.* **39**:165 (1974).
15. T. Asano, *Phys. Rev. A* **4**:1409 (1970).
16. A. Hintermann and C. Gruber, *Physica* **84A**:101 (1976).
17. C. Gruber, A. Hintermann, and D. Merlini, in *Lecture Notes in Physics, Vol. 60*, J. Ehlers and K. Hepp, eds. (Springer-Verlag, 1977).

VOLUME 76, NUMBER 2 PHYSICAL REVIEW LETTERS 8 JANUARY 1996

Partition Function Zeros of the Square Lattice Potts Model

Chi-Ning Chen and Chin-Kun Hu
Institute of Physics, Academia Sinica, Nankang, Taipei 11529, Taiwan

F. Y. Wu
Department of Physics, Northeastern University, Boston, Massachusetts 02115
(Received 13 February 1995)

We have evaluated numerically the zeros of the partition function of the q-state Potts model on the square lattice with reduced interactions K. On the basis of our numerical results, we conjecture that, both for finite planar self-dual lattices and for lattices with free or periodic boundary conditions in the thermodynamic limit, the zeros in the $\mathrm{Re}(x) > 0$ region of the complex $x = (e^K - 1)/\sqrt{q}$ plane are located on the unit circle $|x| = 1$.

PACS numbers: 05.50.+q, 75.10.–b

In 1952 Yang and Lee [1] introduced the concept of considering the zeros of the grand partition function of statistical mechanical systems, a consideration that has since opened new avenues to the study of phase transitions. While Yang and Lee considered the zeros in the complex fugacity plane, or equivalently the complex magnetic field plane in the case of spin systems, Fisher [2] in 1964 called attention to the relevance of the zeros of the canonical partition function in the complex temperature plane. Using the square lattice Ising model as an example, he showed that the partition function zeros are distributed on circles in the thermodynamic limit, and that the logarithmic singularity of the two-dimensional model arises as a consequence of the zero distribution. Since the consideration of zeros in the temperature plane is conceptually simpler, there have been numerous studies of the temperature zeros of spin systems. For example, the Ising partition function zeros have further been considered for the triangular [3], kagomé [4], and the simple cubic [5] lattices. Similarly, partition function zeros have been examined numerically for the square lattice Potts model [6,7], the three-state triangular Potts model [8,9], and the Z_n models [10,11]. Specifically, the distribution of zeros of the three-state Potts model appears to follow a simple geometric locus in the ferromagnetic region [6,12], and the loci for the four-state Potts model appear to include a unit circle [7]. The partition function zeros have also been analyzed for lattices of $m \times \infty$ strips using a transfer matrix formalism [13]. However, except in the case of the triangular Potts model with pure three-spin interactions [9], there appears to have been no definite statement on the zero distributions, which is supported by numerical or exact results.

In this paper we follow up on the consideration of the partition function zeros of the q-state Potts model on the square lattice [6,7,12,13], and make a conjecture on their distribution. We first determine numerically the zeros in the complex temperature plane for small lattices under a special self-dual boundary condition. On the basis of our numerical results, we conjecture that, for *finite* planar self-dual lattices as well as for lattices with free or periodic boundary conditions in the thermodynamic limit, the zeros in the ferromagnetic regime are located on a unit circle. Unlike the Yang-Lee zeros of the Ising model for which the zeros are on a unit circle but with a density distribution which crosses the positive real axis only for temperatures $T \le T_c$, where T_c is the critical temperature, the zero distribution of the Potts partition function crosses the positive real axis for all $q > 1$. In fact, it is the density distribution near the positive real axis that determines the critical behavior of spin systems [2].

Consider the q-state Potts model on a lattice, or graph, G, of linear dimension L and having N vertices and E edges. Let the nearest-neighbor interaction be $J\delta_{\mathrm{Kr}}(\sigma_i, \sigma_j)$, where $\sigma_i, \sigma_j = 1, ..., q$ denote the spin states at vertices i and j connected by an edge and q is an integer. The partition function can be written as [14]

$$Z \equiv Z_G(q, K) = \sum_{G' \subseteq G} (e^K - 1)^{b(G')} q^{n(G')}, \qquad (1)$$

where $K = J/kT$, the summation is taken over all subgraphs $G' \subseteq G$, and $b(G')$ and $n(G')$ are, respectively, the numbers of edges and clusters, including isolated vertices, of G'.

Introducing the variable

$$x = (e^K - 1)/\sqrt{q}, \qquad (2)$$

we rewrite (1) as a polynomial in x,

$$Z \equiv P_G(q, x) = \sum_{b=0}^{E} c_b(q) x^b, \qquad (3)$$

where

$$c_b(q) = q^{b/2} {\sum_{G' \subseteq G}}' q^{n(G')}, \qquad (4)$$

and the prime denotes that the summation is taken over all $G' \subseteq G$ for fixed $b(G') = b$. Then, for planar G, the polynomial P_G possesses the duality relation [15]

$$P_G(q, x) = q^{N-1-E/2} x^E P_D(q, x^{-1}), \qquad (5)$$

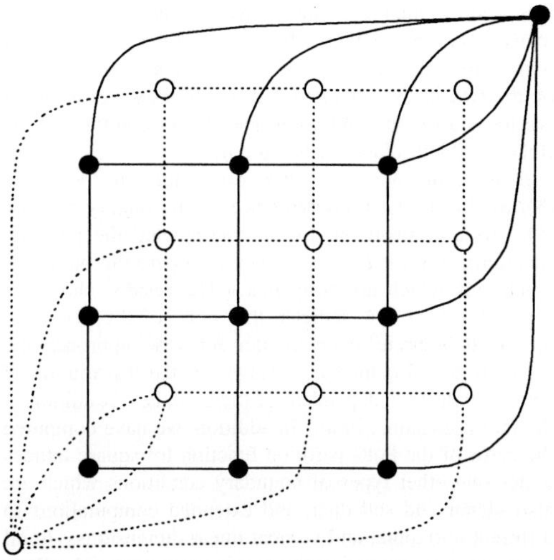

FIG. 1. An $L \times L$ self-dual lattice in solid lines and solid circles with $L = 3$. The dual lattice is denoted by dotted lines and open circles.

where D is the graph dual to G. In the case of the square lattice for which D is identical to G in the thermodynamic limit regardless of boundary conditions, (5) implies that the system is critical at

$$x_c = 1 . \tag{6}$$

To take full advantage of the duality relation (5), we consider a planar self-dual square lattice G, which is an $L \times L$ square lattice with $N = L^2 + 1, E = 2L^2$, and the special boundary condition shown in Fig. 1. While this lattice is planar and self-dual for any finite L, there is no difference between this lattice and square lattices with other boundary conditions in the thermodynamic limit.

We have used a fast algorithm proposed recently by two of us [16] to generate the partition function $P_G(q,x)$ for $L = 2$, 3, 4, 5, 6, and 7 [17]. The planar self-dual property (5) now implies the reciprocal relation

$$c_b(q) = c_{E-b}(q) , \tag{7}$$

and, as a result, the roots occur in pairs of x_i and x_i^{-1}. We then computed the zeros of $P_G(q,x)$ in the complex x plane and tracked their movement as q increases from 1. At $q = 1$, all roots are found to be located at $x = -1$. As q increases, some roots begin to spread into an arc of

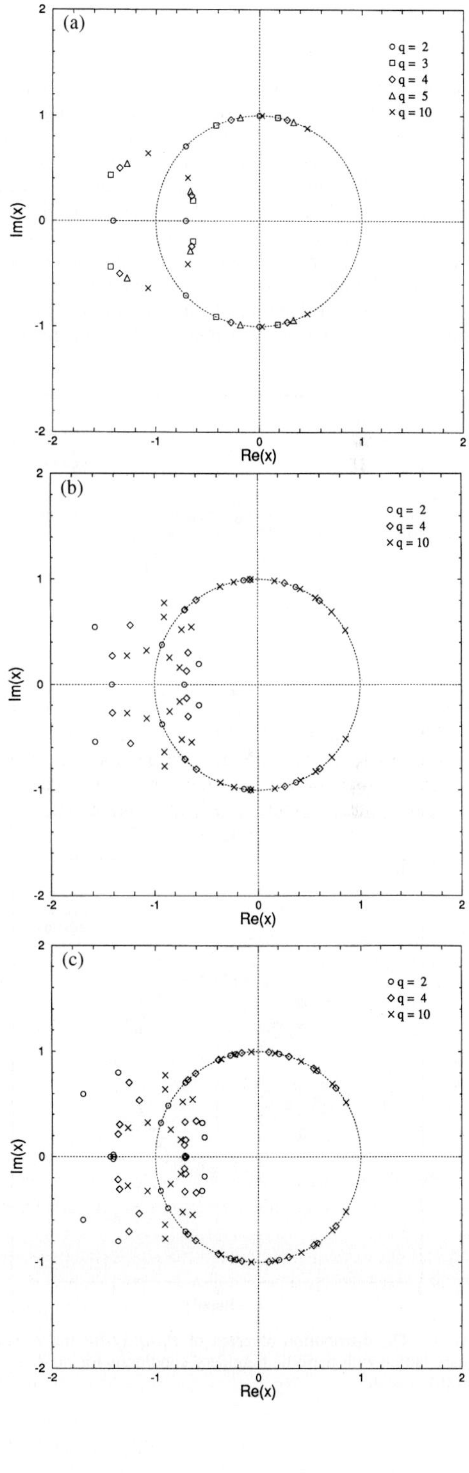

FIG. 2. The distribution of zeros of $P_G(q,x)$ for the $L \times L$ self-dual square lattice of Fig. 1 in the complex x plane for (a) $L = 2$, (b) $L = 3$, and (c) $L = 4$.

the unit circle

$$|x| = 1\,, \tag{8}$$

with the arc centered about $x = -1$. As q continues to increase, more and more roots appear on a larger arc and all zeros on the circle move on the circle toward the positive real axis, while others wander within the $\mathrm{Re}(x) < 0$ half plane. When q reaches a certain critical value $q_c(L)$ which depends on L, all zeros are located at the unit circle $|x| = 1$. This implies that all roots of the Potts partition function are located on the unit circle in the limit of infinite q and any finite L. We have established this latter result rigorously [18]. Typical results for $L = 2, 3, 4$ are shown in Fig. 3 (results for $L = 5, 6, 7$ are similar and not shown because they involve many more data points). For all practical purposes and within numerical errors, many roots are located precisely on the circle. We also find that, in all cases, zeros which are located off the unit circle are always confined in the $\mathrm{Re}(x) < 0$ half plane for integral q.

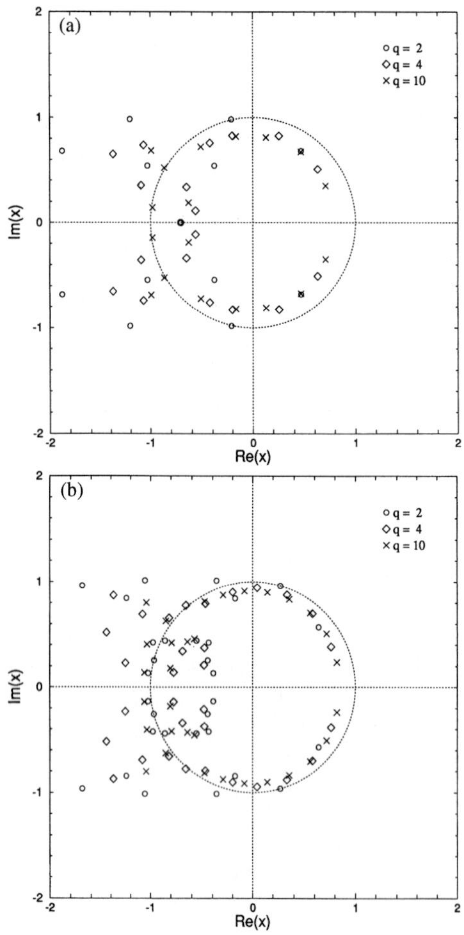

FIG. 3. The distribution of zeros of $P_G(q,x)$ for the $L \times L$ square lattice with periodic boundary conditions for (a) $L = 3$, and (b) $L = 4$.

It is significant that zeros do reside on the circle (8), as this is *not* a consequence of the duality relation (5). As a comparison, we have computed the partition function zeros for $L \times L$ lattices with periodic boundary conditions which are nonplanar. The results, shown in Fig. 4 for $L = 3, 4$, indicate that none of the zeros are on the unit circle, even though zeros do approach the circle as L and q increase. However, the distribution of zeros should be independent of the boundary condition in the thermodynamic limit. In addition, we have computed the zeros of the Potts partition function for square lattices under two other types of boundary conditions which are also planar and self-dual, and extended computations to different horizontal and vertical linear dimensions. In all cases we have arrived at the same conclusion: The Potts partition function zeros in the $\mathrm{Re}(x) > 0$ half plane all reside on the unit circle $|x| = 1$. These findings now lead us to make the following conjecture.

Conjecture: For finite planar self-dual lattices and for square lattice with free or periodic boundary conditions in the thermodynamic limit, the Potts partition function zeros in the $\mathrm{Re}(x) > 0$ half plane are located on the unit circle $|x| = 1$.

It is a curious fact that the self-dual feature of a planar lattice somehow forces many roots to locate on the circle, even for small lattices. Furthermore, our conjecture is consistent with a similar conjecture on the zero distribution of the Potts model with pure three-site interactions [9], which also possesses a duality relation similar to (5). The key appears to lie in the validity of the duality relation (5).

For $q = 2$, it is known [2] that, in the thermodynamic limit, zeros also lie on the circle

$$|x + \sqrt{2}| = 1\,. \tag{9}$$

Maillard and Rammal [6] have suggested on the basis of an inversion relation consideration that, for $q < 4$, the circle

$$\left| x + \frac{2}{\sqrt{q}} \right| = \sqrt{\frac{4}{q} - 1} \tag{10}$$

can be a good candidate as the generalization of (9). However, (10) does not appear to be in agreement with our numerical data. It should also be pointed out that our conjecture is consistent with prior numerical studies [6,7] as well as results of certain algebraic approximations for $m \times \infty$ strips [13].

The reduced per-site free energy of the Potts model is now given by the expression

$$f = N^{-1} \sum_i \ln(1 - x/x_i) + \text{const}, \tag{11}$$

where the summation is taken over all roots x_i of the polynomial (3). Fisher [2] has pointed out that the critical behavior near the critical point x_c is determined solely by the root distribution in the regime near the positive real axis. Thus, for x near $x_c = 1$, we collect those zeros along an arc of the unit circle intersecting the positive real axis or, equivalently, the zeros $x_i = e^{i\theta_i}, \theta_i$ small. Let the zeros be distributed with a density $Ng(\theta)$. We can rewrite, in the thermodynamic limit, the singular part of (11) as

$$f_{\text{sing}} = \frac{1}{2\pi} \int_{-\Delta}^{\Delta} g(\theta) \ln(t + i\theta)\, d\theta\,, \tag{12}$$

where $t = x_c - x$ and Δ is a small number. Note that we have $g(\theta) = g(-\theta)$ since $c_b(q)$ is real. Fisher [2] has shown that the density $g(\theta) = a|\theta|$ near $\theta = 0$, where a is a constant, yields the logarithmic singularity of the Ising model. Along the same line, the small θ density distribution [2,19]

$$g(\theta) = \begin{cases} a|\theta|^{1-\alpha(q)}, & q \le 4, \\ \epsilon(q), & q > 4, \end{cases} \tag{13}$$

leads to the specific singularity $|t|^{-\alpha(q)}$ for $q \le 4$ and a jump discontinuity of amount $\epsilon(q)$ in U for $q > 4$. These are the known critical behaviors of the Potts model.

We thank J.M. Maillard for calling our attention to Ref. [6] and useful comments. Two of us (C.N.C. and C.K.H.) are supported by the National Science Council of the Republic of China (Taiwan) under Grants No. NSC 84-2112-M-001-013 Y, No. 84-0501-I-001-037-B12, and No. NSC 84-2112-M-001-048. The work by F.Y.W. is supported in part by the National Science Foundation through Grants No. DMR-9313648 and No. INT-920788261; he would also like to thank T.T. Tsong for the kind hospitality extended to him at the Academia Sinica where this work was completed.

[1] C.N. Yang and T.D. Lee, Phys. Rev. **87**, 404 (1952).
[2] M.E. Fisher, in *Lecture Notes in Theoretical Physics,* edited by W.E. Brittin (University of Colorado Press, Boulder, 1965), Vol. 7c.
[3] J. Stephenson, J. Phys. A **13**, 4513 (1987).
[4] R. Abe, T. Dotera, and T. Ogawa, Prog. Theor. Phys. **85**, 509 (1991).
[5] R.B. Pearson, Phys. Rev. B **26**, 6285 (1982).
[6] J.M. Maillard and R. Rammal, J. Phys. A **16**, 353 (1983).
[7] P.P. Martin, J. Phys. A **19**, 3267 (1986).
[8] P.P. Martin and J.M. Maillard, J. Phys. A **19**, L547 (1986).
[9] J.C.A. d'Auriac, J.M. Maillard, G. Rollet, and F.Y. Wu, Physica (Amsterdam) **206A**, 441 (1994).
[10] P.P. Martin, J. Phys. A **21**, 4415 (1988).
[11] B. Bonnier and Y. Leroyer, Phys. Rev. B **44**, 9700 (1991).
[12] P.P. Martin, in *Integrable Systems in Statistical Mechanics,* edited by G.M. d'Ariano, A. Montorsi, and M.G. Rasetti (World Scientific, Singapore, 1985).
[13] D.W. Wood, R.W. Turnbull, and J.K. Ball, J. Phys. A **20**, 3495 (1987).
[14] See, for example, F.Y. Wu, Rev. Mod. Phys. **54**, 235 (1982).
[15] F.Y. Wu and Y.K. Wang, J. Math. Phys. (N.Y.) **17**, 439 (1976).
[16] C.N. Chen and C.K. Hu, Phys. Rev. B **43**, 11519 (1991).
[17] The series for the partition function $P_G(q,x)$ for $L = 2, 3, ..., 7$ can be obtained from CNC at the e-mail address "chencn@phys.sinica.edu.tw".
[18] F.Y. Wu, G. Rollet, H.Y. Huang, J.M. Maillard, C.-K. Hu and C.-N. Chen, following Letter, Phys. Rev. Lett. **76**, 173 (1996).
[19] P.P. Martin, *Potts Models and Related Problems in Statistical Mechanics* (World Scientific, Singapore, 1991).

6. Critical Frontiers

J. Phys. C: Solid State Phys., Vol. 12, 1979. Printed in Great Britain. © 1979

LETTER TO THE EDITOR

Critical point of planar Potts models†

F Y Wu

Department of Physics, Northeastern University, Boston, Massachusetts 02115, USA

Received 31 May 1979

Abstract. The critical point of the Potts model is conjectured for the following planar lattices: (i) generalised (chequerboard) square lattice; (ii) Kagomé lattice; (iii) triangular lattice with two- and three-site interactions. As a result, the critical probability for bond percolation on the Kagomé lattice is determined to be $p_c = 0{\cdot}524430$.

Properties of the Potts (1952) model are much more difficult to deduce than those of the Ising model. While the exact solution of the Ising model is now known for all planar lattices (for a list of the solutions for various Ising lattices, see Domb 1960 and Syozi 1972), the critical properties of the Potts model are only partially known, and are confined to specific lattices (Baxter 1973, Baxter *et al* 1978). In particular, the critical temperature has been established for the square, triangular and honeycomb lattices only (Hintermann *et al* 1978).

For the square lattice, the Potts critical point can be determined in a straightforward way from a duality argument (Potts 1952), while for the triangular and honeycomb lattices the argument is more complicated and involves additional steps (Kim and Joseph 1974, Baxter *et al* 1978, Hintermann *et al* 1978). It turns out that none of these analyses can be extended to other lattices.

For the purpose of providing useful reference points as well as for completing the list, it is desirable to have a knowledge of the Potts critical point for other lattices. We consider this problem in this Letter and make several conjectures. The conjectures are based on established results and plausible arguments, and are shown to be correct in various limits.

We shall consider q-component Potts models with ferromagnetic interactions. To facilitate discussions, we first summarise the existing known results.

The established critical point for the nearest-neighbour (ferromagnetic) Potts model on the square, triangular and honeycomb lattices are (Hintermann *et al* 1978)

$$x_1 x_2 = 1 \qquad \text{(square)} \tag{1}$$

$$\sqrt{q}x_1x_2x_3 + x_1x_2 + x_2x_3 + x_3x_1 = 1 \qquad \text{(triangular)} \tag{2}$$

$$\sqrt{q} + x_1 + x_2 + x_3 = x_1x_2x_3 \qquad \text{(honeycomb)} \tag{3}$$

† Work supported in part by the National Science Foundation.

where $x_r = [\exp(K_r) - 1]/\sqrt{q}$, $K_r = \epsilon_r/kT$, $\epsilon_r \geqslant 0$ being the interaction in the spatial direction $r = 1, 2, 3$. Note that the expressions (2) and (3) are related by the duality relation (Wu 1977)

$$x_r x_r^* = 1. \tag{4}$$

The critical condition for the triangular lattice has recently been extended (Baxter *et al* 1978, Wu and Lin 1979) to include three-site interactions $-\epsilon\delta_{ij}\delta_{jk}$ among the sites i, j, k surrounding every *other* triangular face. Here $\delta_{ij} = 1$ if the sites i and j are in the same state and $\delta_{ij} = 0$ otherwise. On the basis of a duality argument, the critical point of this model is located at

$$\exp(K_1 + K_2 + K_3 + K) = \exp(K_1) + \exp(K_2) + \exp(K_3) + q - 2 \tag{5}$$

where $K = \epsilon/kT$. For $K = 0$, equation (5) reduces to equation (2), as it should.

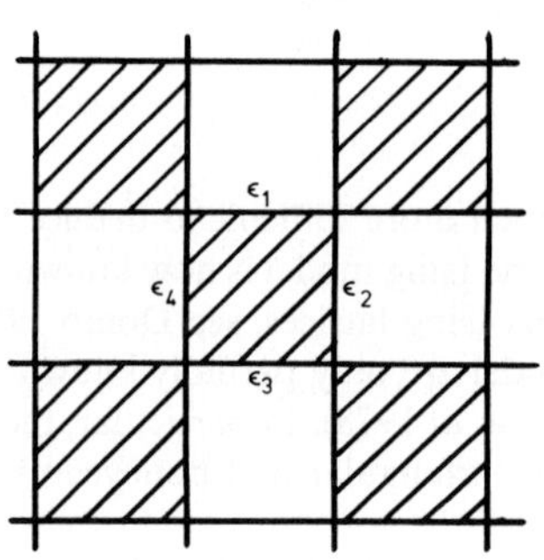

Figure 1. Generalised (chequerboard) square lattice. Each shaded square is bordered by interactions ϵ_1, ϵ_2, ϵ_3 and ϵ_4.

Generalised (chequerboard) square lattice. Consider the generalised (chequerboard) square lattice shown in figure 1. This lattice reduces to the honeycomb and triangular lattices respectively if one of the interactions is taken to be 0 or ∞. Thus its critical condition should reflect the same limits. Furthermore, since the generalised square lattice is self-dual, it follows from equation (4) that the criticality is invariant under the transformation $x_r \to x_r^{-1}$. These considerations then suggest the following expression for its critical point:

$$\sqrt{q} + x_1 + x_2 + x_3 + x_4 = x_1x_2x_3 + x_2x_3x_4 + x_3x_4x_1 + x_4x_1x_2 + \sqrt{q}x_1x_2x_3x_4. \tag{6}$$

Indeed, this is the only expression which is self-dual and which reduces to equations (3) and (2) upon taking $x_4 = 0$ and ∞ respectively. We conjecture that equation (6) is the correct critical condition.

The conjecture (6) is verified for $q = 2$ (the Ising model). In this case the exact critical point is known (Utiyama 1951, Domb 1960, Syozi 1972) to be

$$gdK_1 + gdK_2 + gdK_3 + gdK_4 = \pi \tag{7}$$

where $g\mathrm{d}K = 2\tan^{-1}\exp(K) - \pi/2$. It may be verified that equation (6) indeed reduces to equation (7) upon taking $q = 2$.

Triangular lattice with 2- and 3-site interactions. Consider next the triangular Potts model with the Hamiltonian

$$\mathscr{H} = -\sum_r \epsilon_r \sum_{ij} \delta_{ij} - \epsilon \sum_{\Delta} \delta_{ij}\delta_{jk} - \epsilon' \sum_{\nabla} \delta_{ij}\delta_{jk}. \tag{8}$$

Here, in addition to the two-site interactions ϵ_r, there are three-site interactions ϵ (ϵ') around each up-pointing (down-pointing) triangular face. This model has been studied by the renormalisation group technique (Schick and Griffiths 1977).

By symmetry we expect the critical condition of this model to be symmetric in ϵ_1, ϵ_2, ϵ_3 and also in ϵ and ϵ'. Now for $\epsilon' = 0$ the critical condition (for ferromagnetic interactions) is given by equation (5). The logical generalisation to $K' > 0$ is then

$$\exp(K_1 + K_2 + K_3 + K + K') = \exp(K_1) + \exp(K_2) + \exp(K_3) + q - 2. \tag{9}$$

We conjecture that equation (9) gives the critical point for the Potts model (8) for ferromagnetic interactions.

The conjecture (9) is again verified for $q = 2$. In this case the three-site interactions are reducible to two-site interactions (see e.g. Wu and Lin 1979). It is readily seen that, for $q = 2$, equation (9) agrees with the Ising exact result.

Kagomé lattice. Properties of the Potts model on the Kagomé lattice (figure 2) appear to be very elusive, and nothing is known at present. We shall, however, deduce its critical point from the conjecture (9).

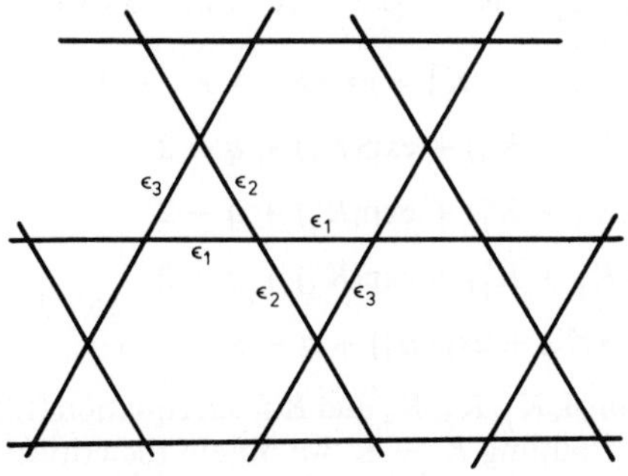

Figure 2. The Kagomé lattice with anisotropic interactions.

Consider first $\epsilon = \epsilon'$ in the Hamiltonian (8) and carry out a star–triangle transformation over every triangular face. This leads to the diced lattice as shown in figure 3. The transformation is well defined (Kim and Joseph 1974). Specifically, split each two-site interaction into two halves, each belonging to a neighbouring triangular face. As shown

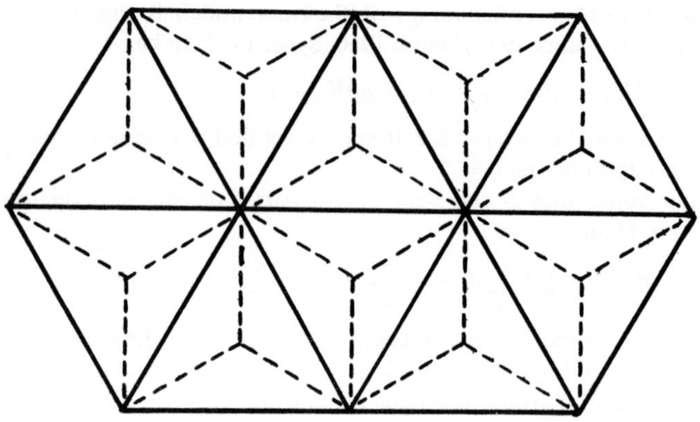

Figure 3. Star–triangle transformation relating the triangular (solid lines and the diced (broken lines) lattices.

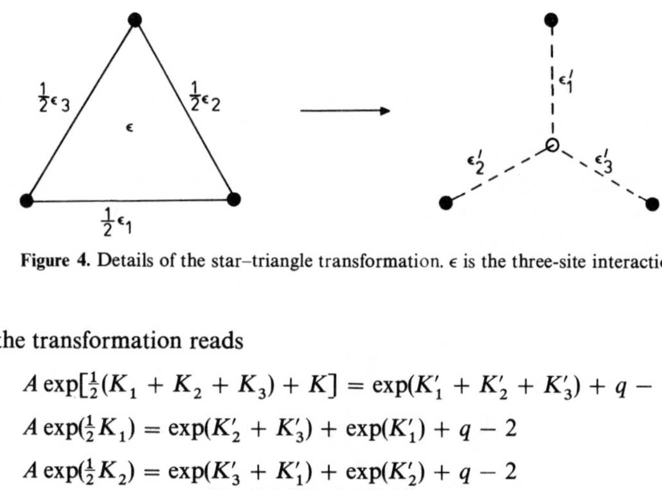

Figure 4. Details of the star–triangle transformation. ϵ is the three-site interaction.

in figure 4, the transformation reads

$$\begin{aligned}
&A\exp[\tfrac{1}{2}(K_1+K_2+K_3)+K] = \exp(K_1'+K_2'+K_3')+q-1\\
&A\exp(\tfrac{1}{2}K_1) = \exp(K_2'+K_3')+\exp(K_1')+q-2\\
&A\exp(\tfrac{1}{2}K_2) = \exp(K_3'+K_1')+\exp(K_2')+q-2\\
&A\exp(\tfrac{1}{2}K_3) = \exp(K_1'+K_2')+\exp(K_3')+q-2\\
&A = \exp(K_1')+\exp(K_2')+\exp(K_3')+q-3
\end{aligned} \tag{10}$$

where $K_r' = \epsilon_r'/kT$. We can solve for A, K_1, K_2, K_3 and K from equation (10). Substituting the solution into equation (9) and putting $K' = K$, we obtain the critical condition for the diced lattice. Finally, since the Kagomé and the diced lattices are mutually dual, the critical condition for the Kagomé lattice is deduced by applying the duality relation (4). The algebra is straightforward and we give here only the final expression:

$$\begin{aligned}
&qx_1^2x_2^2x_3^2 + 2\sqrt{q}x_1x_2x_3(x_1x_2+x_2x_3+x_3x_1) + 2x_1x_2x_3(x_1+x_2+x_3)\\
&\quad + x_1^2x_2^2 + x_2^2x_3^2 + x_3^2x_1^2\\
&\quad = 2\sqrt{q}x_1x_2x_3 + 4(x_1x_2+x_2x_3+x_3x_1)\\
&\quad + 2\sqrt{q}(x_1+x_2+x_3) + q.
\end{aligned} \tag{11}$$

We conjecture that equation (11) gives the critical point of the Kagomé Potts model.

The conjectured critical condition (11) is verified in two instances. As can be readily seen, for $q = 2$ equation (11) agrees with the exact transition point of the Ising model on the Kagomé lattice (Kano and Naya 1953, Domb 1960, Syozi 1972), which reads

$$\cosh K_1 \cosh K_2 \cosh K_3 + \sinh K_1 \sinh K_2 \sinh K_3 = \cosh K_1 + \cosh K_2 + \cosh K_3. \quad (12)$$

Also, if one of the interactions ϵ_r vanishes, one can take the partial trace over two-thirds of the spins, and the result is a square Potts lattice. It is easily seen that in this instance equation (11) again leads to the exact result (1).

We plot in figure 5 the dependence of the critical temperature on q for various isotropic planar Potts lattices. For the square lattice the critical parameter $\exp(K) = 1 + \sqrt{q}$ is linear in $\sqrt{q}$, while for others the dependence on $\sqrt{q}$ is seen to be close to linear. The plot also shows a strong dependence of $\exp(K)$ on the coordination number z.

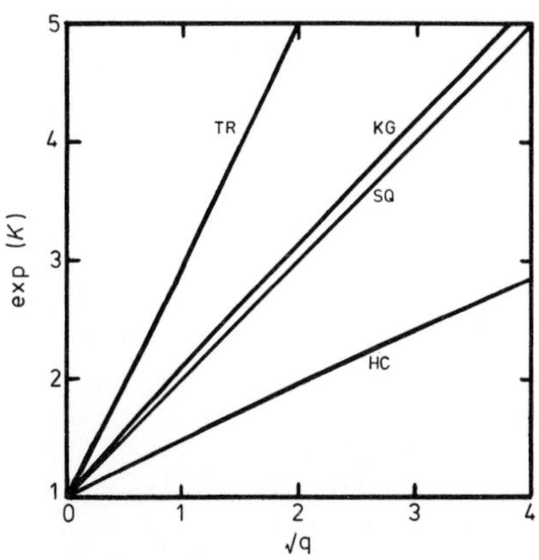

Figure 5. Critical parameter $\exp(K)$ as a function of $\sqrt{q}$ for the triangular (TR), Kagomé (KG), square (SQ) and honeycomb (HC) lattices.

For $q = 1$, the critical condition leads to the critical probability $p_c[=1 - \exp(-K)]$ of bond percolations (Kasteleyn and Fortuin 1969, Wu 1978). The result is listed in the following:

$$\begin{aligned} p_c &= 0{\cdot}652704 \qquad (z = 3, \text{honeycomb}) \\ &= 0{\cdot}524430 \qquad (z = 4, \text{Kagomé}) \\ &= 0{\cdot}5 \qquad (z = 4, \text{square}) \\ &= 0{\cdot}347296 \qquad (z = 6, \text{triangular}). \end{aligned} \quad (13)$$

The value of p_c for the Kagomé lattice is new; other values in equation (13) have previously been obtained by Sykes and Essam (1964). The values of p_c indicate that the square lattice is somehow more 'close-packed' than the Kagomé lattice, although they both have the same coordination number $z = 4$.

References

Baxter R J 1973 *J. Phys. C: Solid St. Phys.* **6** L445–8
Baxter R J, Temperley H N V and Ashley S E 1978 *Proc. R. Soc.* A**358** 535–59
Domb C 1960 *Adv. Phys.* **9** 149–361
Hintermann A, Kunz H and Wu F Y 1978 *J. Stat. Phys.* **19** 623–32
Kano K and Naya S 1953 *Prog. Theor. Phys.* **10** 158–72
Kasteleyn P W and Fortuin C M 1969 *J. Phys. Soc. Japan (suppl.)* **26** 11–4
Kim D and Joseph R I 1974 *J. Phys. C: Solid St. Phys.* **7** L167–9
Potts R B 1952 *Proc. Camb. Phil. Soc.* **48** 106–9
Schick M and Griffiths R B 1977 *J. Phys. A: Math. Gen.* **10** 2123–31
Sykes M F and Essam J W 1964 *J. Math. Phys.* **8** 1117–27
Syozi I 1972 *Phase Transitions and Critical Phenomena* ed C Domb and M S Green Vol. 1 (London: Academic Press)
Utiyama T 1951 *Prog. Theor. Phys.* **6** 907–9
Wu F Y 1977 *J. Math. Phys.* **18** 611–2
—— 1978 *J. Stat. Phys.* **18** 115–23
Wu F Y and Lin K Y 1979 *J. Phys. A: Math. Gen.* to be published

PRL 96, 090602 (2006) PHYSICAL REVIEW LETTERS week ending 10 MARCH 2006

New Critical Frontiers for the Potts and Percolation Models

F. Y. Wu

Northeastern University, Boston, Massachusetts 02115, USA

(Received 8 January 2006; published 8 March 2006)

We obtain the critical threshold for a host of Potts and percolation models on lattices having a structure which permits a duality consideration. The consideration generalizes the recently obtained thresholds of Scullard and Ziff for bond and site percolation on the martini and related lattices to the Potts model and to other lattices.

DOI: 10.1103/PhysRevLett.96.090602 PACS numbers: 05.50.+q, 02.50.−r, 64.60.Cn

The Potts model [1] has been in the forefront of active research for many years. Despite concerted efforts, however, very few exact results are known [2]. Unlike the Ising model for which the exact solution is known for all two-dimensional lattices, the relatively simple question of locating the critical frontier of the Potts model has been resolved only for the square, triangular, and honeycomb lattices [3–5]. The determination of the Potts critical frontier for other two-dimensional lattices has remained very much an open problem [6].

In two recent papers using a star-triangle relation and a dual transformation, Scullard [7] and Ziff [8] succeeded to determine the critical thresholds of site and bond percolation processes for several new two-dimensional lattices. As percolation problems are realized in the $q = 1$ limit of the q-state Potts model [9,10], the new percolation results suggest the possibility that similar thresholds can also be determined for the Potts model.

In this Letter we report this extension. We derive more generally the exact critical frontier of the Potts model for a large class of two-dimensional lattices including those considered in [7,8], and obtain the corresponding percolation thresholds.

Consider a lattice having the structure shown in Fig. 1, where each shaded triangle denotes a network connected to its exterior through 3 spins $\sigma_1, \sigma_2, \sigma_3$. It was established by Baxter, Temperley, and Ashley [3] using an algebraic approach that this Potts model possesses a duality relation and a self-dual trajectory. A graphical proof of the duality relation was later given by Wu and Lin [11], and subsequently Wu and Zia [12] established rigorously that in the ferromagnetic regime of the parameter space the critical threshold is indeed the self-dual trajectory.

Specifically, write the Boltzmann factor for the shaded triangle as

$$F(\sigma_1, \sigma_2, \sigma_3) = A + B_1\delta_{23} + B_2\delta_{31} + B_3\delta_{12} + C\delta_{123}, \tag{1}$$

where $\delta_{ij} = \delta_{\sigma_i,\sigma_j}$, $\delta_{ijk} = \delta_{ij}\delta_{jk}\delta_{ki}$. Then the model possesses a duality relation in the parameter space $\{A, B_1, B_2, B_3, C\}$. In the ferromagnetic regime

$$B_1 + B_2 + B_3 + C > 0, \qquad B_i + B_j + C > 0, \qquad i \neq j \tag{2}$$

the critical frontier of the Potts model is given by the self-dual trajectory

$$qA - C = 0. \tag{3}$$

By realizing the shaded network as a simple triangle, for example, one recovers from (3) the critical point for the Potts and bond percolation models on the square, triangular, and honeycomb lattices [3]. Another realization of the Boltzmann factor (1) is the random cluster model [9] with 2- and 3-site interactions [11]. The isotropic version of the random cluster model has been analyzed very recently by Chayes and Lei [13] who established on a rigorous ground the duality relation and the self-dual trajectory (3). Our new results concern with other realizations of (1).

Consider the network shown in Fig. 2 as an instance of the shaded triangle in Fig. 1. This gives rise to the martini lattice shown in Fig. 3 [7,8]. The Boltzmann factor for the network is

$$F(\sigma_1, \sigma_2, \sigma_3) = \sum_{\{\sigma_4,\sigma_5,\sigma_6\}=1}^{q} \exp[V_1\delta_{14} + V_2\delta_{25} + V_3\delta_{36} + W_1\delta_{56} + W_2\delta_{46} + W_3\delta_{45} + M\delta_{456}], \tag{4}$$

where V_i and W_i are 2-site Potts interactions and M a 3-site interaction.

It is straightforward to cast (4) in the form of (1) [14,15] to obtain

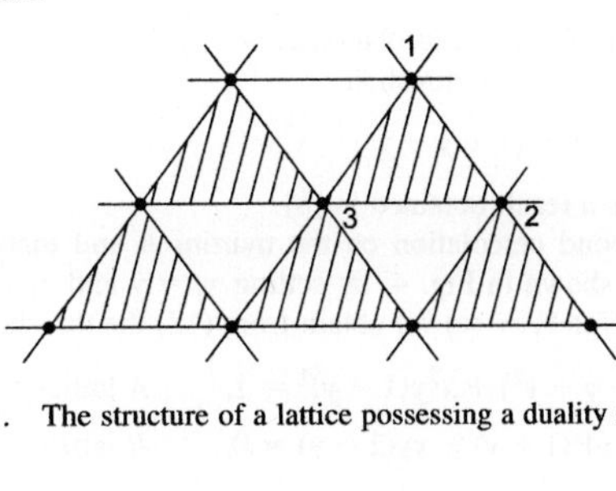

FIG. 1. The structure of a lattice possessing a duality relation.

PRL **96,** 090602 (2006) PHYSICAL REVIEW LETTERS week ending 10 MARCH 2006

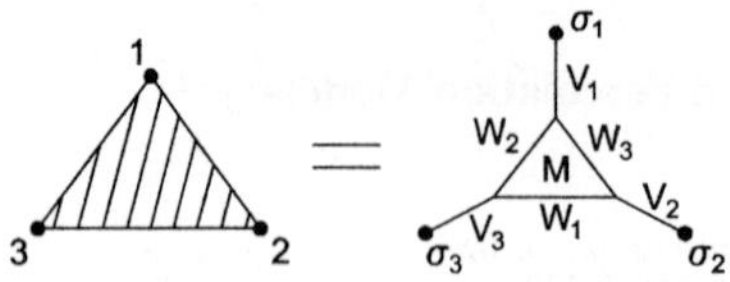

FIG. 2. The realization of Fig. 1 for the martini lattice.

$$
\begin{aligned}
A &= v_1v_2v_3 + v_1v_2(q + w_1 + w_2) + v_2v_3(q + w_2 + w_3) \\
&\quad + v_3v_1(q + w_3 + w_1) + (q + v_1 + v_2 + v_3) \\
&\quad \times [q^2 + q(w_1 + w_2 + w_3) + h] \\
B_i &= v_jv_k[h + (q + v_i)w_i], \qquad i \neq j \neq k \neq j \\
C &= v_1v_2v_3h,
\end{aligned} \tag{5}
$$

where

$$
\begin{aligned}
v_i &= e^{V_i} - 1, \qquad w_i = e^{W_i} - 1 \\
h &= e^{M+W_1+W_2+W_3} - e^{W_1} - e^{W_2} - e^{W_3} + 2.
\end{aligned} \tag{6}
$$

As alluded to in the above, in the ferromagnetic regime $W_i \geq 0$, $V_i \geq 0$, $M \geq 0$ satisfying (2), the critical frontier of this Potts model is the self-dual trajectory (3) which now reads

$$
\begin{aligned}
&q(q + v_1 + v_2 + v_3)[q^2 + q(w_1 + w_2 + w_3) + h] \\
&+ q[v_1v_2v_3 + v_1v_2(w_1 + w_2 + q) + v_2v_3(w_2 + w_3 + q) \\
&\qquad + v_3v_1(w_3 + w_1 + q)] - v_1v_2v_3h = 0.
\end{aligned} \tag{7}
$$

The critical frontier (7) is a new result for the Potts model.

For $M = \infty$ one retains only terms linear in h and (7) reduces to the critical frontier $q^2 + q(v_1 + v_2 + v_3) = v_1v_2v_3$ of the honeycomb lattice. For $M = 0$, $V_1 = V_2 = V_3 = V$, $W_1 = W_2 = W_3 = W$, which is the isotropic model with pure 2-site interactions, (7) becomes

$$
\begin{aligned}
&q(q + 3v)(q^2 + 3qw + 3w^2 + w^3) \\
&\qquad + qv^2(v + 6w + 3q) - v^3(3w^2 + w^3) = 0,
\end{aligned} \tag{8}
$$

where $v = e^V - 1$, $w = e^W - 1$. For $w = v$ it reduces further to the expression

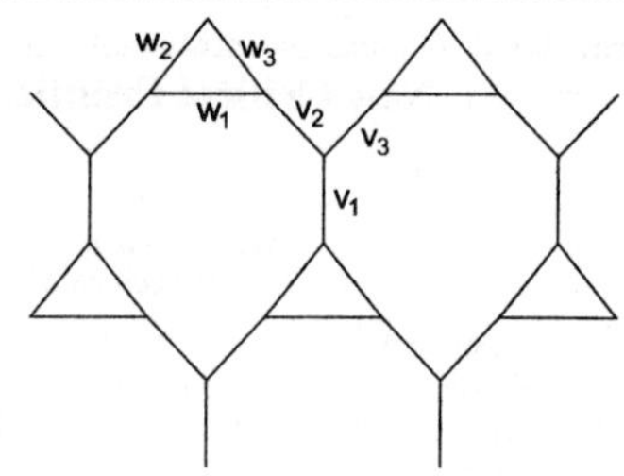

FIG. 3. The martini lattice.

$$
\begin{aligned}
&q^4 + 6q^2v + q^2v^2(15 + v) \\
&\qquad + qv^3(16 + 3v) - v^5(3 + v) = 0.
\end{aligned} \tag{9}
$$

One variation of the martini lattice is the *A* lattice [7,8] shown in Fig. 4(a) obtained from the martini lattice by setting $v_1 = \infty$, $v_2 = v_3 = v$, $w_1 = w_2 = w_3 = w$. This gives rise to the Potts critical frontier

$$
\begin{aligned}
&q^3 + q^2(2v + 3w) + q(v^2 + 4wv + 3w^2 + w^3) \\
&\qquad - v^2(3w^2 + w^3) = 0, \qquad A \text{ lattice},
\end{aligned} \tag{10}
$$

Another variation of the martini lattice is the *B* lattice [7,8] shown in Fig. 4(b) obtained from the martini lattice by setting $v_2 = v_3 = \infty$, $v_1 = v$, $w_1 = w_2 = w_3 = w$. This leads to the Potts critical frontier

$$
q^2 + q(v + 2w) - vw^2(3 + w) = 0, \qquad B \text{ lattice}. \tag{11}
$$

Both expressions (10) and (11) are new.

We now specialize the above results to percolation. It is well known that bond percolation is realized by taking the $q = 1$ limit of the q-state Potts model with 2-site interactions [9,16]. For bond percolation on the martini lattice in Fig. 3, we set $q = 1$ and introduce bond occupation probabilities $x_i = 1 - e^{-V_i}$, $y_i = 1 - e^{-W_i}$. The percolation (7) then assumes the form

$$
\begin{aligned}
&x_1x_2(y_3 + y_1y_2 - y_1y_2y_3) + x_2x_3(y_1 + y_2y_3 - y_1y_2y_3) + x_3x_1(y_2 + y_3y_1 - y_1y_2y_3) \\
&\qquad - x_1x_2x_3(y_1y_2 + y_2y_3 + y_3y_1 - 2y_1y_2y_3) = 1 + (e^M - 1)(1 - x_1x_2 - x_2x_3 - x_3x_1 + x_1x_2x_3).
\end{aligned} \tag{12}
$$

For isotropic bond percolation $x_i = x$, $y_i = y$, and $M = 0$, this reduces to the threshold

$$
3x^2y(1 + y - y^2) - x^3y^2(3 - 2y) = 1, \tag{13}
$$

which is a result obtained in [8].

For bond percolation on the martini *A* and martini *B* lattices shown in Fig. 4, by setting $y_i = y$ and $x_i = x$ or $x_i = 1$ (for $V_i = \infty$) we obtain from (12) the thresholds

$$
\begin{aligned}
&2xy(1 + y - y^2) + x^2y(1 - y)^2 = 1, \qquad A \text{ lattice}, \\
&(1 - y)^2(1 + y) - xy(2 - y) = 0, \qquad B \text{ lattice}.
\end{aligned} \tag{14}
$$

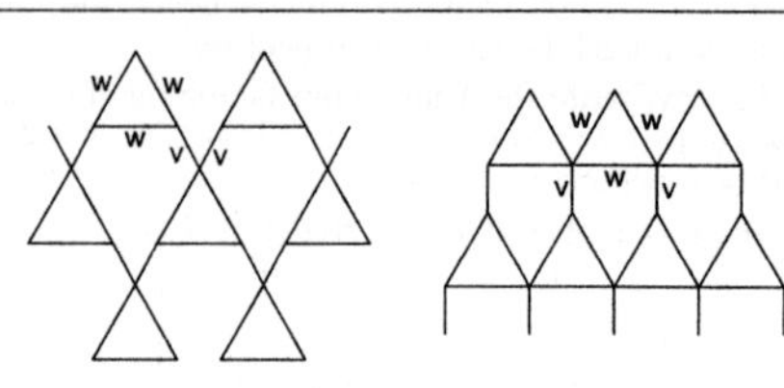

(a) (b)

FIG. 4. (a) The martini-*A* lattice. (b) The martini-*B* lattice.

For uniform bond occupation probability $x = y = p$, Eqs. (13) and (14) yield the thresholds $p_c = 1/\sqrt{2}$, $0.625\,457\cdots$ [8] and $1/2$ [7], respectively. Note that the thresholds (13) and (14) can also be deduced from (8), (10), and (11), by setting $q = 1$, $v = x/(1-x)$, $w = y/(1-y)$.

Consider next a correlated bond-site percolation process on the honeycomb lattice with edge occupation probabilities x_1, x_2, x_3 and alternate site occupation probabilities s and 1. Now the site percolation is realized in the $q = 1$ limit of the q-state Potts model with multisite interactions [10]. Therefore, by setting $y_i = 0$, $s = 1 - e^{-M}$ we obtain from (12) the critical frontier for this site-bond percolation,

$$s(x_1x_2 + x_2x_3 + x_3x_1 - x_1x_2x_3) = 1. \qquad (15)$$

The expression (15), which generalizes an early result due to Kondor [17] for $x_1 = x_2 = x_3$, is the central result of [7] derived from a star-triangle consideration. Here, it is deduced as the result of an application of our general formulation.

As pointed out by Scullard [7] and Ziff [8], the expression (15) also gives the threshold for site percolation on the martini lattice of Fig. 3, where x_1, x_2, x_3 are occupation probabilities of the three sites around a triangle and s is the occupation probability of the site at the center of the Y. For uniform occupation probability $x_1 = x_2 = x_3 = s$, (15) yields the threshold $s_c = 0.764\,826\cdots$ for site percolation on the martini lattice [7].

Setting $x_3 = 1$ in (15) we obtain the threshold for site percolation on the martini A lattice of Fig. 4(a) as

$$s(x_1 + x_2) = 1, \qquad \text{site percolation-}A\text{ lattice}, \qquad (16)$$

where x_1, x_2 are occupation probabilities of the 3-coordinated sites and s the occupation probability of the 4-coordinated sites. For uniform occupation probability $x_1 = x_2 = s$, (16) yields the threshold $s_c = 1/\sqrt{2}$ for site percolation on the A lattice. Likewise setting $x_2 = x_3 = 1$ in (15), we obtain the threshold for site percolation on the martini B lattice of Fig. 4(b),

$$s(1 + x) = 1, \qquad \text{site percolation-}B\text{ lattice}, \qquad (17)$$

where $x = x_1$ and s are, respectively, the occupation probabilities of the 5-coordinated sites and 3-coordinated sites. For uniform occupation probability $x = s$, (17) yields the threshold $s_c = (\sqrt{5} - 1)/2$ for site percolation on the B lattice. These results have been reported in [7,8].

As another example of our formulation, consider the Potts model on the lattice in Fig. 5 with pure 2-site interactions $U, V, W \geq 0$. Writing $u = e^U - 1$, $v = e^V - 1$, $w = e^W - 1$, we obtain after a little algebra the Boltzmann factor (1) with

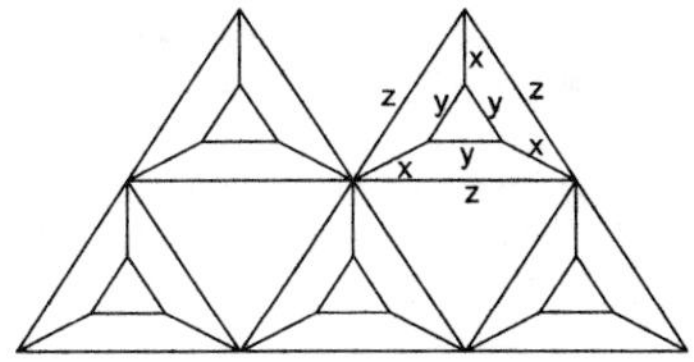

FIG. 5. A lattice with Potts interactions U, V, W. Labels shown are the corresponding bond percolation probabilities $x = 1 - e^{-V}$, $y = 1 - e^{-W}$, $z = 1 - e^{-U}$.

$$\begin{aligned} A &= v^3 + 3v^2(q + 2w) + (3v + q)(q^2 + 3qw + 3w^2 + w^3) \\ B &= uA + v^2[3w^2 + w^3 + (q + v)w] \\ C &= u^2(u + 3)A + 3uv^2(u + 1)(u + 2) \\ &\quad \times [3w^2 + w^3 + (q + v)w] + (u + 1)^3v^3(3w^2 + w^3). \end{aligned} \qquad (18)$$

The critical frontier is again the self-dual trajectory $qA - C = 0$.

The resulting self-dual trajectory assumes a simpler form for the percolation problem. For bond percolation we set $q = 1$, $u = z/(1-z)$, $v = x/(1-x)$, $w = y/(1-y)$, where x, y, x are the respective bond occupation probabilities shown in Fig. 5. This yields the bond percolation critical threshold

$$\begin{aligned} &1 - 3z + z^3 - (1 - z^2)[3x^2y(1 + y - y^2) \\ &\quad \times (1 + z) + x^3y^2(3 - 2y)(1 + 2z)] = 0. \end{aligned} \qquad (19)$$

Setting $z = 0$ in (19), it reduces to the bond percolation threshold (13) of the martini lattice. Setting $y = 1$ (19) gives the bond percolation threshold

$$1 - 3z + z^3 - (1 - z^2)[3x^2(1 + z) - x^3(1 + 2z)] = 0 \qquad (20)$$

for the dual of the martini lattice, which is the lattice in Fig. 5 with all small triangles shrunk into single points.

For uniform bond percolation probabilities $x = y = z = p$, (19) becomes

$$1 - 3p - 2p^3 + 12p^5 - 5p^6 - 15p^7 + 15p^8 - 4p^9 = 0 \qquad (21)$$

yielding the threshold $p_c = 0.321\,808\cdots$. Compared with the threshold $p_c = 0.707\,106\cdots$ for the martini lattice, it confirms the expectation that percolation threshold decreases as the lattice becomes more connected.

In summary, we have shown that the critical frontier of a host of Potts models with 2- and multisite interactions on lattices having the structure depicted in Fig. 1 can be explicitly determined. The resulting critical frontier assumes the very simple form $qA - C = 0$, where A and C are parameters defined in (1). The corresponding threshold for bond and/or site percolation are next deduced by setting

$q = 1$. Specializations of our formulation to the martini, the A, B, and other lattices are presented.

I would like to thank R. M. Ziff for sending me a copy of [8] prior to publication and for a useful conversation. I am indebted to H. Y. Huang for assistance in the preparation of the Letter.

[1] R. B. Potts, Proc. Cambridge Philos. Soc. **48**, 106 (1952).
[2] For a review of the Potts model, see F. Y. Wu, Rev. Mod. Phys. **54**, 235 (1982).
[3] R. J. Baxter, H. N. V. Temperley, and S. E. Ashley, Proc. R. Soc. A **358**, 535 (1978).
[4] A. Hinterman, H. Kunz, and F. Y. Wu, J. Stat. Phys. **19**, 623 (1978).
[5] F. Y. Wu, J. Phys. C **12**, L645 (1979).
[6] For a review in the case of the kagome lattice, see C. A. Chen, C. K. Hu, and F. Y. Wu, J. Phys. A **31**, 7855 (1998).
[7] C. Scullard, Phys. Rev. E **73**, 016107 (2006).
[8] R. M. Ziff, Phys. Rev. E **73**, 016134 (2006).
[9] P. W. Kasteleyn and C. M. Fortuin, J. Phys. Soc. Jpn. Suppl. **26**, 11 (1969); C. M. Fortuin and P. W. Kasteleyn, Physica (Amsterdam) **57**, 536 (1972).
[10] H. Kunz and F. Y. Wu, J. Phys. C **11**, L1 (1978); **11**, L357 (1978).
[11] F. Y. Wu and K. Y. Lin, J. Phys. A **13**, 629 (1980).
[12] F. Y. Wu and R. K. P. Zia, J. Phys. A **14**, 721 (1981).
[13] L. Chayes and H. K. Lei, cond-mat/0508254.
[14] J. M. Maillard, G. Rollet, and F. Y. Wu, J. Phys. A **26**, L495 (1993).
[15] C. King and F. Y. Wu, Int. J. Mod. Phys. B **11**, 51 (1997).
[16] F. Y. Wu, J. Stat. Phys. **18**, 115 (1978).
[17] I. Kondor, J. Phys. C **13**, L531 (1980).

VOLUME 62, NUMBER 24 PHYSICAL REVIEW LETTERS 12 JUNE 1989

Critical Frontier of the Antiferromagnetic Ising Model in a Magnetic Field: The Honeycomb Lattice

F. Y. Wu and X. N. Wu

Department of Physics, Northeastern University, Boston, Massachusetts 02115

H. W. J. Blöte

Laboratorium voor Technische Natuurkunde, Postbus 5046, 2600 GA Delft, The Netherlands

(Received 13 March 1989)

A closed-form expression is proposed for the critical frontier, or the critical line, of the antiferromagnetic Ising model on the honeycomb lattice in a nonzero external magnetic field. We formulate the Ising model as an 8-vertex model and identify the critical frontier as a locus invariant under a generalized weak-graph transformation. In its simplest form the locus is an expression containing two unknown constants which we determine numerically. The resulting critical frontier lies very close to results of a finite-size analysis.

PACS numbers: 05.50.+q

The Ising model in a nonzero magnetic field has been standing, for many years, as one of the most intriguing and outstanding unsolved problems in statistical physics. While a wealth of exact information is now known for the Ising model in the absence of an external field,[1] its properties in a nonzero field H have proven to be illusive and remain largely unknown. Of particular interest is the location of the critical frontier Γ, or the critical line, along which the per-site partition function becomes singular (in H and the temperature T). For Ising ferromagnets it is now well established that the critical frontier is the $T \leq T_c$ segment of the $H=0$ axis,[2-4] where T_c is the zero-field critical temperature. For antiferromagnets it is expected that a critical frontier exists in the (H,T) plane separating regimes of ordered and disordered phases. However, there have been few analytic studies on its precise location, despite efforts of closed-form approximations[5,6] and other analyses.[7-9]

In this Letter we consider the antiferromagnetic Ising model on the honeycomb lattice and construct a closed-form expression for its critical frontier. Our analysis is based on a consideration of the invariance property of the critical frontier and our results of a finite-size scaling analysis.[10] We first formulate the Ising model as an 8-vertex model and inquire more generally about the critical frontier Γ_{8v} of the 8-vertex model in its own parameter space. The partition function of the 8-vertex model is known to be invariant under a generalized weak-graph transformation.[11-13] Assuming that Γ_{8v} is also invariant under this transformation and that its analytic expression takes the simplest possible algebraic forms, we are led to explicit closed-form expressions for Γ_{8v}, including one known to correspond to the $H=0$ line of the Ising ferromagnet. The other trajectories generated in the antiferromagnetic Ising subspace can be identified as the possible locations for the critical frontier Γ. The simplest such expression, given by (14) below, contains two unknown constants, which we determine using a finite-size scaling analysis. The resulting closed-form expression of the critical frontier permits us to compute the critical fugacity of the nearest-neighbor exclusion lattice gas on the honeycomb lattice.

Consider a lattice of N sites, and coordination number $q=3$ with periodic boundary conditions. This can be, e.g., the 2D honeycomb or the 3D hydrogen-peroxide[14,15] lattice. We start from the Ising Hamiltonian

$$\mathcal{H} = -J\sum \sigma_i\sigma_j - H\sum \sigma_i \,, \tag{1}$$

where the first summation is over all interacting pairs and the second summation is over all lattice sites. The partition function is $Z(K,L) = \sum_{\sigma_i = \pm 1} e^{-\mathcal{H}/kT}$, where $K=J/kT$, $L=H/kT$. We inquire about the location of Γ, the critical frontier, of the per-site partition function $\kappa(K,L) = \lim_{N\to\infty}[Z(K,L)]^{1/N}$.

The starting point of our analysis is the high-temperature expansion of the partition function[1,11]

$$Z(K,L) = (2\cosh L)^N (\cosh K)^{3N/2} Z_{8v}(a,b,c,d) \,, \tag{2}$$

where $Z_{8v}(a,b,c,d)$ is the partition function of an 8-vertex model on the same lattice whose vertex weights are shown in Fig. 1 with

$$a=1, \quad b=v^{1/2}h \quad c=v, \quad d=v^{3/1}h \,. \tag{3}$$

Here $v=\tanh K$ and $h=\tanh L$. Hence, b and d are pure imaginary for antiferromagnetic interactions $K<0$. It is therefore convenient to consider the more general problem of locating the critical frontier Γ_{8v} for the per-site

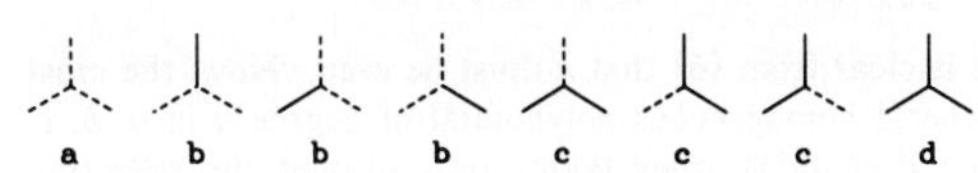

FIG. 1. The 8-vertex configurations and the associated vertex weights.

partition function $\kappa_{8r}(a,b,c,d)=\lim_{N\to\infty}[Z_{8r}(a,b,c,d)]^{1/N}$ for generally complex a, b, c, and d. Once this is done, the desired Ising results can be obtained by specializing to (3). It should be pointed out here that it is an artifact that b and d are pure imaginary. Since vertices of one and three solid lines (cf. Fig. 1) always occur in pairs, the overall vertex weight contains factors b^2, bd, and d^2 and is always real.

The partition function $\kappa_{8r}(a,b,c,d)$ satisfies both the reflection symmetry[11]

$$\kappa_{8r}(a,b,c,d)=\kappa_{8r}(d,c,b,a) \tag{4}$$

and the weak-graph symmetry[11,12]

$$\kappa_{8r}(\tilde{a},\tilde{b},\tilde{c},\tilde{d})=\kappa_{8r}(a,b,c,d)\,, \tag{5}$$

where y is a parameter whose value is arbitrary and

$$\begin{aligned}\tilde{a}&=(a+3yb+3y^2c+y^3d)(1+y^2)^{-3/2}\,,\\ \tilde{b}&=[ya-(1-2y^2)b-(2y-y^3)c-y^2d](1+y^2)^{-3/2}\,,\\ \tilde{c}&=[y^2a-(2y-y^3)b+(1-2y^2)c+yd](1+y^2)^{-3/2}\,,\\ \tilde{d}&=(y^3a-3y^2b+3yc-d)(1+y^2)^{-3/2}\,.\end{aligned} \tag{6}$$

As a consequence of the weak-graph symmetry (5), we expect Γ_{8r} to be invariant under the transformation (6). Indeed, it is known that one branch of the critical frontier Γ_{8r} of the 8-vertex model is[11,12,16]

$$P_1\equiv a(b^3+d^3)-d(a^3+c^3)+3(ab+bc+dc)(c^2-bd-b^2+ac)=0\,, \tag{7}$$

and it is readily verified that (7) is invariant under (6). In fact, all points on (7) are fixed points of the transformation (for some y).[13] We now inquire whether there exist other loci in the generally complex parameter space that are also invariant.

Let the critical frontier Γ_{8r} be

$$f(a,b,c,d)=0\,. \tag{8}$$

For (8) to remain invariant under (6), the function $f(a,b,c,d)$ must transform like $f(\tilde{a},\tilde{b},\tilde{c},\tilde{d})=\alpha f(a,b,c,d)$, where α is a constant depending at most on y. The identity $\{\tilde{\tilde{a}},\tilde{\tilde{b}},\tilde{\tilde{c}},\tilde{\tilde{d}}\}=\{a,b,c,d\}$ then implies $\alpha=\pm1$. Furthermore, since the transformation (6) is linear, we expect the function $f(a,b,c,d)$ to be of the form of a homogeneous polynomial in a, b, c, and d as in (7).[17]

Let $f_{n\pm}$ be a homogeneous polynomial of a, b, c, and d of degree n having this property, namely, it satisfies

$$f_{n\pm}(\tilde{a},\tilde{b},\tilde{c},\tilde{d})=\pm f_{n\pm}(a,b,c,d)\,. \tag{9}$$

It is clear from (6) that n must be even. Now, the most general homogeneous polynomial of degree n in a, b, c, and d contains, after taking into account the reflection symmetry (4), 6, 19, 44, ... independent coefficients for $n=2, 4, 6, \ldots$, respectively. The substitution of (6) into (9) with $f_{n\pm}(a,b,c,d)$ in the form of a polynomial yields [after multiplying a factor $(1+y^2)^{6n}$ throughout and collecting terms] a new polynomial in a, b, c, d, and y. For (9) to hold for arbitrary a, b, c, d, and y, it is necessary that coefficients of *all* terms of this new polynomial vanish.

This appears to be almost impossible at first glance, since the number of terms in the polynomial of a, b, c, d, and y is $(6n+1)4^n$, which far exceeds the number of independent coefficients in $f_{n\pm}(a,b,c,d)$ so that we have at hand a set of overdetermined equations. However, after some algebra, we find that many of the equations are redundant and all, except a few, coefficients in $f_{n\pm}(a,b,c,d)$ are uniquely determined. We find that, for $n=2,4,6$, all invariant polynomials must take the following forms:

$$\begin{aligned}f_{2+}(a,b,c,d)&=c_1P+c_2Q\,,\\ f_{4-}(a,b,c,d)&=c_1P_1\,,\\ f_{4+}(a,b,c,d)&=c_1P_2+c_2P^2+c_3PQ+c_4Q^2\,,\\ f_{6-}(a,b,c,d)&=f_{2+}P_1\,,\\ f_{6+}(a,b,c,d)&=f_{2+}P_2+c_3P^3+c_4P^2Q+c_5PQ^2+c_6Q^3\,,\end{aligned} \tag{10}$$

where c_i are arbitrary constants, P_1 has been given in (7), and

$$\begin{aligned}P&=a^2+3ac+3bd+d^2\,,\\ Q&=b^2-ac+c^2-bd\,,\\ P_2&=2(a^4+d^4)-6(a^2c^2+b^2d^2)+12(a^2b^2+c^2d^2)\\ &\quad-5a^2d^2+27b^2c^2+36(ab+cd)bc+18abcd\,.\end{aligned} \tag{11}$$

It is gratifying to see that $f_{4-}(a,b,c,d)=0$ gives rise to the known singularity (8). We now examine other loci implied by (10) and (11).

It is easy to see (from the Ising realization) that $f_{2+}(a,b,c,d)=0$ cannot be a critical frontier. The next choice is therefore $n=4$, for which we have already seen that $f_{4-}(a,b,c,d)=0$ gives rise to the known critical frontier (7). The other $n=4$ critical frontier is $f_{4+}(a,b,c,d)=0$, or, equivalently,

$$c_1P_2+c_2P^2+c_3PQ+c_4Q^2=0\,. \tag{12}$$

As one of the four constants c_i can be chosen arbitrarily, only three constants in (12) are undetermined. It should be noted that, at this point, our arguments hold very generally and apply to any (including random) lattice of coordination number $q=3$. It remains, however, to determine the constants c_i which will now be lattice dependent.

VOLUME 62, NUMBER 24 PHYSICAL REVIEW LETTERS 12 JUNE 1989

TABLE I. Computation of the critical field $L_c = H_c/kT$. Numbers in parentheses are estimated error bars implied by the finite-size extrapolation.

	L_c	
K	Finite-size analysis	Equation (14)
−0.7	0.582431408(5)	0.582429186
−0.8	1.119884213(5)	1.119888647
−0.9	1.520604370(7)	1.520610887
−1.0	1.875990455(10)	1.875996047
−1.2	2.530154228(10)	2.530156031
−1.5	3.458129780(10)	3.458127977

In the Ising subspace (3), we have from (11),

$$P_2 = 2[1-3v^2+2v(3+9v+10v^2+9v^3+3v^4)h^2 - v^4(3-v^2)h^4], \tag{13}$$
$$P = 1+3v+v^2(3+v)h^2,$$
$$Q = v(1-v)(-1+h^2).$$

The substitution of (13) into (12) leads to a quadratic equation in h^2 and, solving this equation for $x=1-h^2$, we obtain

$$\cosh^2 L = \frac{v}{(1+v)^3}\left[v^2 + \frac{(6c_2-c_3)v+12c_1+c_3}{2(2c_1+c_2)} - \frac{1}{2(2c_1+c_2)}\sqrt{X}\right], \tag{14}$$

where $v = \tanh K < 0$ and

$$X = [(12c_1+c_3)^2 - 4c_4(2c_1 \pm c_2)] \times (1-v)^2 + 144c_1(2c_1+c_2)v .$$

Here, we have chosen the minus sign in front of $\sqrt{X}$ as dictated by numerical results below. Near T_c, (14) leads to $H \sim (T_c - T)^{1/2}$. Furthermore, at low temperatures, the critical frontier (14) terminates at $H = \pm 3J$ with a finite slope. These are the main features of the critical frontier for the Ising antiferromagnet, and thus permit us to identify (14) as the critical frontier for the antiferromagnet.

We now proceed to determine the constants c_i in (14). For the honeycomb lattice the zero-field critical point is known[18] to occur at $v_c = -1/\sqrt{3}$. We therefore require the left-hand side of (12) to yield a factor $\sqrt{3}v+1$ upon setting $h=0$. This leads to the constraint $c_4 = 3\alpha c_3 - 9\alpha^2 c_2$, where $\alpha = 2-\sqrt{3}$.

To determine the remaining two constants, say c_2 and c_3 assuming $c_1 = 1$, we have carried out a finite-size scaling analysis of the magnetic correlation length, details of which will be published elsewhere. Using strips of $m\times\infty$ lattices, $m \le 20$, we have determined numerically the critical field $L_c(K)$ for various temperatures $T = -1/kK$. The results are shown in the second column of Table I together with estimations of the error bars. Using these figures and a least-squares fit, we obtain the values $c_2 = -1.5153435316$ and $c_3 = -1.7953179207$. However, it turns out that the critical field computed from (14) is somewhat insensitive to small constrained variations of the values of c_2 and c_3 used. Utilizing this fact, we then tried to fit c_2 and c_3 into expressions containing the factor $\sqrt{3}$ which yield critical fields as accurate as those produced by using the least-squares values. After an extensive search, the following choice emerges:

$$c_1 = 1, \quad c_2 = -(4+3\sqrt{3})/6, \quad c_3 = -(1-9\sqrt{3})/8, \quad c_4 = -3(3-\sqrt{3})/8 . \tag{15}$$

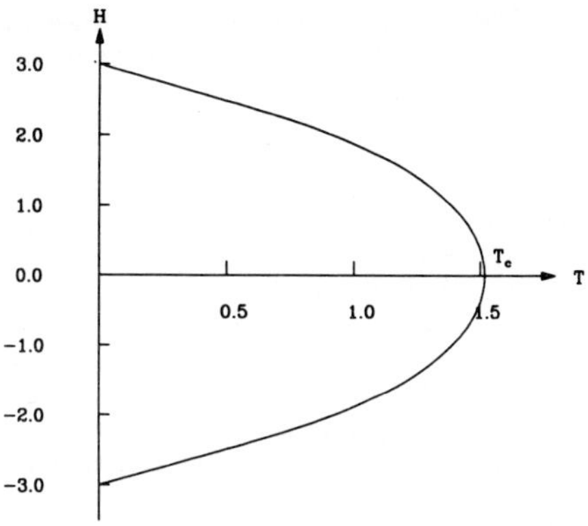

FIG. 2. The critical frontier (14) for the antiferromagnetic honeycomb Ising lattice, where H is in units of $|J|$ and T in units of $|J|/k$.

A plot of (14) with c_i given by (15) is shown in Fig. 2.

In the third column of Table I we list values of the critical field computed using (14). While the computed values are not entirely within the error bars implied by our finite-size data, differences between the two sets of data are very small. This suggests that, for all practical purposes, the expression (14) with c_i's given by (15) can be used as an accurate representation of the exact critical frontier.

The initial slope of the critical frontier is $-\frac{1}{2}\ln z_c$, where z_c is the critical fugacity of a nearest-neighbor exclusion lattice gas.[3] Expanding (14) about $T=0$ and using (15), we obtain explicitly the expression

$$z_c = \frac{2(2c_1+c_2)}{2c_2-8c_1-c_3+[(12c_1+c_3)^2-(2c_1+c_2)(36c_1+4c_4)]^{1/2}} = 7.85178004\ldots . \tag{16}$$

As an independent check, we have also carried out a direct finite-size analysis of the nearest-neighbor exclusion lattice gas, and obtained the critical fugacity numerically as $z_c = 7.851\,721\,75(13)$. The difference between the two figures is again very small.

Finally, we remark that an agreement entirely within the error bars can be achieved by adopting a locus of a higher n, say, $n=6$. In that case the critical field is given by $f_{6+}(a,b,c,d)=0$, where a, b, c, and d are given by (3), leading to a cubic equation for determining L_c and the least-squares values $c_1=1$, $c_2=-0.366\,782\,364\,27$, $c_3=-2.166\,369\,5118$, $c_4=-2.813\,389\,2132$, and $c_5=-1.096\,379\,6403$. This formula may very well be the exact one, apart from the numerical uncertainties in the c_i's.

This research is supported in part by the National Science Foundation Grant No. DMR-8702596 and the NATO Grant No. 198/84. One of us (H.W.J.B.) wishes to thank M. P. Nightingale for the hospitality extended to him at the University of Rhode Island where a portion of this work was carried out.

[1]For a general review of information available for Ising models, see C. Domb, in *Phase Transitions and Critical Phenomena,* edited by C. Domb and M. S. Green (Academic, New York, 1974), Vol. 3.

[2]C. N. Yang and T. D. Lee, Phys. Rev. **87**, 404 (1952).

[3]T. D. Lee and C. N. Yang, Phys. Rev. **87**, 410 (1952).

[4]J. L. Lebowitz and O. Penrose, Commun. Math. Phys. **11**, 99 (1968).

[5]E. Müller-Hartmann and J. Zittartz, Z. Phys. B **27**, 261 (1977).

[6]K. Y. Lin and F. Y. Wu, Z. Phys. B **33**, 181 (1979).

[7]W. Kinzel and M. Schick, Phys. Rev. B **23**, 3435 (1981).

[8]B. Nienhuis, H. J. Hilhorst, and H. W. J. Blöte, J. Phys. A **17**, 3559 (1984).

[9]H. W. J. Blöte and M. P. M. den Nijs, Phys. Rev. B **37**, 1766 (1988).

[10]For reviews of finite-size analysis, see M. N. Barber, in *Phase Transitions and Critical Phenomena,* edited by C. Domb and J. L. Lebowitz (Academic, New York, 1983), Vol. 8; M. P. Nightingale, J. Appl. Phys. **53**, 7927 (1982).

[11]F. Y. Wu, J. Math. Phys. (N.Y.) **15**, 687 (1974).

[12]X. N. Wu and F. Y. Wu, J. Stat. Phys. **90**, 41 (1988).

[13]X. N. Wu and F. Y. Wu, J. Phys. A **22**, L55 (1989).

[14]H. Heesch and F. Laves, A. Krist. **85**, 443 (1933).

[15]A. F. Wells, Acta Crystallogr. **7**, 535 (1954).

[16]It can be shown that (7) is the only critical frontier of the 8-vertex model in the real parameter space.

[17]This assumption is verified by all known results of vertex models, including the Baxter model.

[18]G. H. Wannier, Rev. Mod. Phys. **17**, 50 (1945).

RAPID COMMUNICATIONS

PHYSICAL REVIEW B VOLUME 43, NUMBER 16 1 JUNE 1991

Critical surface of the Blume-Emery-Griffiths model on the honeycomb lattice

Leh-Hun Gwa

Department of Mathematics, Rutgers University, New Brunswick, New Jersey 08903

F. Y. Wu

Department of Physics, Northeastern University, Boston, Massachusetts 02115

(Received 14 March 1991)

We consider the Blume-Emery-Griffiths (BEG) model on the honeycomb lattice and obtain a closed-form expression for the critical surface of second-order transitions. The BEG model is first formulated as a three-state vertex model. Using the fact that the BEG critical surface coincides with that of a general three-state vertex model, we construct critical surfaces by forming polynomial combinations of vertex weights that are invariant under an O(3) gauge transformation. We then carry out a finite-size analysis of the BEG model, and use data so obtained to determine coefficients appearing in the polynomial combination. This procedure leads to a closed-form expression of the critical surface which reproduces all numerical data accurately.

The Blume-Emery-Griffiths (BEG) model[1] is a spin-1 system described by the (reduced) Hamiltonian

$$-\mathcal{H}/kT=J\sum_{\langle ij\rangle}S_iS_j+K\sum_{\langle ij\rangle}S_i^2S_j^2-\Delta\sum_i S_i^2\,, \qquad (1)$$

where $S_i=0,\pm1$. The model was first proposed to explain certain magnetic transitions.[2-4] It has also proven to be useful for modeling of the λ transition in ^{3}He-^{4}He mixtures[1] and the phase changes in a microemulsion.[5] An important feature of the critical behavior of the BEG model is the occurrence of a multicritical phenomenon accompanied with the onset of first- and second-order transitions.[6] However, studies of its phase diagram carried out in the past have been mostly by approximations, including renormalization-group[7] and mean-field[1,8,9] analyses, and Monte Carlo simulations.[10] An exact determination of its phase diagram has proven to be elusive, and has been limited to the subspaces $J=0$,[11,12] and $K=-\ln\cosh J$.[13-16] In this paper, we present results on a precise determination of the second-order phase surface for the BEG model (1) on the honeycomb lattice.

Our approach parallels that of recent progress made in the determination of the phase diagram for antiferromagnetic Ising models.[17-21] By using an invariance property in conjunction with results of a finite-size analysis, it has been possible to obtain closed-form expressions for the phase boundaries of the Ising models, which agree with all numerical data to an extremely high degree of precision.[17-19] For spin-1 systems such as the BEG model, the underlying invariance is that of an O(3) gauge transformation, whose properties have recently been studied.[22] Here we make use of these invariance properties and results of a finite-size analysis, which we carry out, to obtain closed-form expressions of the second-order transition phase boundary for the honeycomb BEG model.

We first formulate the BEG model as a three-state vertex model. Starting from the partition function of the BEG model,

$$Z_{\rm BEG}=\sum_{S_i=0,\pm1}\prod_{\langle ij\rangle}\exp(JS_iS_j+KS_i^2S_j^2)\prod_i e^{-\Delta S_i^2}\,, \qquad (2)$$

we write

$$\exp(JS_iS_j+KS_i^2S_j^2)=1+zS_iS_j+tS_i^2S_j^2\,, \qquad (3)$$

where

$$z\equiv e^K\sinh J,\quad t\equiv e^K\cosh J-1\,, \qquad (4)$$

and expand the product $\prod_{\langle ij\rangle}$ in Eq. (2). For each term in the expansion for which the factor 1, zS_iS_j, or $tS_i^2S_j^2$ is taken, we draw, respectively, a dotted, heavy, or thin line over the corresponding lattice edge. Then $Z_{\rm BEG}$ generates graphs on the underlying lattice. Next we associate weights to lattice sites (vertices), obtained by carrying out the summations $\sum_{S_i=0,\pm1}$. Using the identities

$$\sum_{S_i=0,\pm1}S_i^ne^{-\Delta S_i^2}=\begin{cases}2e^{-\Delta}+1, & n=0\\ 2e^{-\Delta}, & n=\text{even}\\ 0, & n=\text{odd}\,,\end{cases} \qquad (5)$$

we see that only those vertices having an even number of incident heavy lines have nonzero weights. For the honeycomb lattice, this leads to a 14-vertex model shown in Fig. 1 with the weights

$$\begin{aligned}&a=\omega_{300}=1+\tfrac{1}{2}e^{\Delta},\ b=\omega_{201}=t^{1/2},\ c=\omega_{102}=t\,,\\ &d=\omega_{003}=t^{3/2},\ f=\omega_{120}=z,\ g=\omega_{021}=zt^{1/2}\,.\end{aligned} \qquad (6)$$

Here, for convenience, we have multiplied a factor $e^{\Delta}/2$ to all vertex weights, and introduced the abbreviated notation ω_{ijk} to represent the weight of a vertex configuration with respective i,j,k dotted, heavy, and thin lines.

The most general three-state vertex model for the honeycomb lattice is a 27-vertex model[22] which, in addition to the 14 vertices shown in Fig. 1, includes vertices with weights ω_{030}, ω_{111}, ω_{210}, and ω_{012}. The partition function of this general model is invariant under an O(3) gauge transformation. Furthermore, it has been shown that for threefold-coordinated lattices the O(3) gauge transformation possesses six fundamental invariants, and

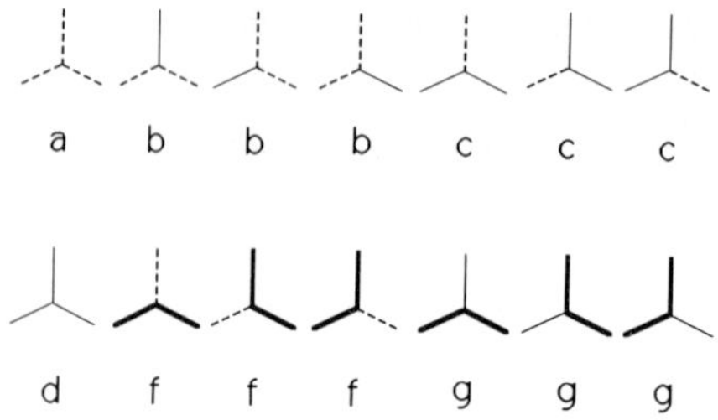

FIG. 1. Vertex configurations of the three-state 14-vertex model for the BEG model.

their explicit expressions have been given.[22] In the BEG subspace $\omega_{030}=\omega_{111}=\omega_{210}=\omega_{012}=0$, one of these fundamental invariants vanishes identically and the remaining five reduce to

$$\begin{aligned} I_0 &= (a+c+f)^2+(b+d+g)^2 , \\ J_1 &= A+15B , \\ J_2 &= C-B^2-AB , \\ J_3 &= AC+3BC-2B^3-6AB^2 , \\ J_4 &= 4(5A-9B)C^2+(A^3+21A^2B-93AB^2+135B^3) \\ &\quad +2B^2(9A^3-59A^2B+99AB^2-81B^3) , \end{aligned} \tag{7}$$

where

$$\begin{aligned} A &= -e_0e_0^*, \quad B=-e_2e_2^*, \quad C=-e_0^*e_2^3-e_0(e_2^*)^3 , \\ e_0 &= -a+3c+i(3b-d) , \\ e_2 &= \tfrac{1}{5}[(a+c-4f)-i(b+d-4g)] . \end{aligned} \tag{8}$$

It should be emphasized that Eqs. (7) are intersections of fundamental invariants of the 27-vertex model in the BEG subspace, which are not themselves invariants for the BEG model. We shall, however, simply call A, B, and C the invariants for convenience in ensuing discussions.

Generally, we expect critical surfaces of the 27-vertex model to be invariant under the O(3) gauge transformation. It follows that they must lie on subspaces whose explicit expressions are formed from the six fundamental invariants. For the BEG model, in particular, the phase boundaries are then given by expressions formed by the invariants I_0, A, B, and C. If we further assume that phase boundaries are given in terms of these expressions as homogeneous polynomials of the vertex weights, we are led to the following possible expressions for the phase boundaries:

$$\begin{aligned} P_2 &\equiv I_0+c_1A+c_2B=0 , \\ P_4 &\equiv C+c_1I_0^2+c_2A^2+c_3B^2+c_4I_0A+c_5I_0B+c_6AB=0 , \end{aligned} \tag{9}$$

and similar higher-degree polynomials, where the c_j's are constants yet to be determined. Here, P_n in Eq. (9) is a polynomial homogeneous in the nth degree of the vertex weights, and we apply the method of finite-size analysis to compute the coefficients c_j's. First, for fixed ratios of Δ/K and J/K, we use finite-size scaling to determine the critical temperature $K_c=1/kT_c$. Values of K_c so determined are then used in conjunction with (9) to determine the coefficients c_j's from a least-squares fit.

The finite-size theory[23–25] of second-order transitions predicts that the correlation length at a critical point scales linearly with the linear dimension of the system n. The correlation length can be found by computing the first- and second-largest eigenvalues of the transfer matrix, λ_1 and λ_2, for a row of n lattice sites. More precisely, we have

$$n\zeta \ln(\lambda_1/\lambda_2)=2\pi x_H , \tag{10}$$

with ζ a geometric factor and x_H the magnetic scaling dimension. We compute λ_1 and λ_2 for the transfer matrix of the BEG model on a honeycomb lattice, taken in a direction perpendicular to one of the lattice edges.[26] The transfer matrix in this direction is symmetric and the geometric factor takes the value $\zeta=1/\sqrt{3}$.

A conjugate-gradient algorithm is used to search for the leading eigenvalues and eigenvectors. The leading eigenvector is contained in the subspace with all elements positive, as guaranteed by Perron-Frobenius theorem; the next largest eigenvalue is computed by confining the algorithm to the subspace orthogonal to the leading eigenvector. The latter procedure is carried out by choosing a random vector, and subtracting from it the component along the leading eigenvector, to ensure that the search covers the maximal subspace. This is in contrast to the usual procedure for the Ising model, where a symmetry known to hold in the next leading eigenvector is used explicitly in its construction.

A total of 205 data points for $n=2$, 4, 6 are collected from the finite-size calculation, using $x_H=\frac{1}{8}$ for the Ising universality class. A least-squares fit is then performed on the $n=6$ data to obtain the best values for the constants in the polynomial equations $P_2=0$ and $P_4=0$. The results are $c_1=-0.5996$ and $c_2=47.08$ for the second-degree equation, and $c_1=0.005\,260$, $c_2=0.095\,34$, $c_3=5.601$, $c_4=-0.070\,96$, $c_5=0.3708$, and $c_6=-3.374$ for the fourth-degree equation. The finite-size data and curves obtained from the polynomial equations (9) with coefficients determined in the above are plotted in Fig. 2. Considering the facts that this is a surface fitting and that there are no more than six adjustable constants, the fit is remarkably good. As a comparison, we have carried out a similar least-squares fit using fourth- and sixth-degree homogeneous polynomials constructed from the invariants (7). In this case, we find the best fits marked by ruptures across smooth data points, indicating that the fits are unsatisfactory. This finding also indicates that we are on the right track in choosing I_0, A, B, and C as the building blocks of BEG invariants. We therefore suggest that the equation $P_4=0$ can be used as a good closed-form approximation for the second-order critical surface of the BEG model.

As an independent check of the accuracy of our results,

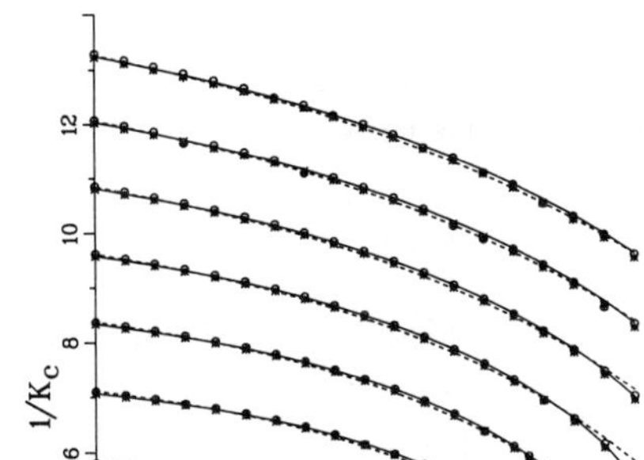

FIG. 2. Second-order phase boundary from finite-size analysis (○: $n=2$; ×: $n=4$; +: $n=6$), and from the polynomial equations $P_2=0$ (dashed line) and $P_4=0$ (solid line). The 12 branches are for $J/K=0.6$, 0.8, 1, 2, 3, . . . , 10, in the order of increasing $1/K_c$.

we have used (9) to compute the critical point at $\Delta/K=2$, $J/K=\frac{1}{3}$, for which the BEG model (1) is known[27] to be equivalent to the three-state Potts model with nearest-neighbor interactions $2K/3$ and the exact critical point $K_c^{-1}=0.449\,175$. Using $P_4=0$ we obtain $K_c^{-1}=0.458\,756$. Considering that this point lies well beyond the regime of our data points and close to the first-order boundary, the agreement is very good.

Guided by the behavior of the phase diagram obtained under the mean-field approximation,[1] we expect, as Δ/K increases, the second-order lines in Fig. 2 to terminate and change into first-order lines at multicritical points. Indeed, results of our numerical fitting show that calculated values of the coefficients are very sensitive to input boundary data at the lower-right-hand corner of Fig. 2, signifying a changeover into a first-order surface (for which the finite-size scaling needs to be reformulated). Thus, the second-order surface given by (9) should terminate at a first-order surface rising in the lower-right-hand regime of Fig. 2.

Finally, we remark on how one might locate this first-order surface. Generally, again by invariance arguments, we expect the first-order surface to be also given in the form of (9). Furthermore, the intersection of the surface with the $J=0$ plane is exactly known.[11,12] In the plane $J=0$, the BEG model (1) is completely equivalent to an Ising model with interactions $K_I=K/4$ and a magnetic field $(\frac{3}{2}K-\Delta+\ln 2)/2$. The first-order boundary is then precisely the line segment $J=0$, $e^{\Delta}=2e^{3K/2}$, $e^{2K_I}=e^{K/2}\leq 2+\sqrt{3}$. In addition, from a ground-state energy analysis, one finds that the first-order surface contains the zero-temperature phase boundary $J/K+1=2\Delta/3K$. These exact intersections together with any precision determination of a few first-order transition points will then enable one to complete the picture of the full phase diagram.

This work is supported in part by NSF Grants No. DMR-8918903 and No. DMR-9015489. We would like to thank L. Y. Zhu for his interest during the initial stage of this work. We would also like to thank X. N. Wu for helpful discussions on the numerical procedure.

[1]M. Blume, V. J. Emery, and R. B. Griffiths, Phys. Rev. B **4**, 1071 (1971).
[2]M. Blume, Phys. Rev. **141**, 517 (1966).
[3]H. W. Capel, Physica **32**, 966 (1966); **33**, 795 (1967); **37**, 423 (1967).
[4]M. Blume and R. E. Watson, J. Appl. Phys. **38**, 991 (1967).
[5]M. Schick and W. Shih, Phys. Rev. B **34**, 1797 (1986).
[6]For a review of the tricritical behavior including that in the BEG model, see L. D. Lawrie and S. Sarbach, in *Phase Transitions and Critical Phenomena,* edited by C. Domb and J. L. Lebowitz (Academic, New York, 1984), Vol. 9.
[7]A. N. Berker and M. Wortis, Phys. Rev. B **14**, 4946 (1976).
[8]D. Furman, S. Dattagupta, and R. B. Griffiths, Phys. Rev. B **15**, 441 (1977).
[9]Y. L. Wang and D. Rauchwarger, Phys. Lett. **59A**, 73 (1976).
[10]J. D. Kimel, S. Black, P. Carter, and Y. L. Wang, Phys. Rev. B **35**, 3347 (1987).
[11]R. B. Griffiths, Physica **33**, 690 (1967).
[12]F. Y. Wu, Chin. J. Phys. **16**, 153 (1976).
[13]T. Horiguchi, Phys. Lett. **113A**, 425 (1986).
[14]F. Y. Wu, Phys. Lett. A **116**, 245 (1986).
[15]R. Shankar, Phys. Lett. A **117**, 365 (1986).
[16]X. N. Wu and F. Y. Wu, J. Stat. Phys. **50**, 41 (1988).
[17]F. Y. Wu, X. N. Wu, and H. W. J. Blöte, Phys. Rev. Lett. **62**, 2773 (1989).
[18]X. N. Wu and F. Y. Wu, Phys. Lett. A **144**, 123 (1990).
[19]H. W. J. Blöte and X. N. Wu, J. Phys. A **23**, L627 (1990).
[20]L. H. Gwa, Phys. Rev. Lett. **63**, 1440 (1989).
[21]L. H. Gwa, Phys. Rev. B **41**, 7315 (1990).
[22]L. H. Gwa and F. Y. Wu, J. Phys. A (to be published).
[23]M. N. Barber, in *Phase Transitions and Critical Phenomena,* edited by C. Domb and J. L. Lebowtiz (Academic, New York, 1983), Vol. 8.
[24]V. Privman and M. E. Fisher, Phys. Rev. B **30**, 1766 (1988).
[25]M. E. Fisher, in *Critical Phenomena,* Proceedings of the International School of Physics, "Enrico Fermi," Course 51, edited by M. S. Green (Academic, New York, 1971).
[26]H. W. J. Blöte, F. Y. Wu, and X. N. Wu, Int. J. Mod. Phys. B **4**, 619 (1990).
[27]F. Y. Wu, Rev. Mod. Phys. **54**, 235 (1982).

J. Phys. C: Solid State Phys., Vol. 7, 1974. Printed in Great Britain. © 1974

LETTER TO THE EDITOR

Two phase transitions in the Ashkin–Teller model†

F Y Wu‡ and K Y Lin

Department of Physics, National Tsing Hua University, Hsinchu, Taiwan, Republic of China

Received 19 March 1974

Abstract. Assuming a continuous dependence of the criticality on the energy parameters, it is conjectured that the Ashkin–Teller model has in general two phase transitions. The two critical points coalesce into a single one when the two middle energies in the model are degenerate.

Some 30 years ago Ashkin and Teller (1943) introduced a lattice statistical model, now known as the Ashkin–Teller (AT) model, as a generalization of the Ising model to a four-component system. Using the Kramers–Wannier (1941) argument and the assumption of a unique transition, they conjectured the location of the critical point for a special case of this model in which three of the four components are degenerate. Their conjecture has recently been extended to the general AT model (Fan 1972a), but Wegner (1972) showed that the argument does not in general locate the critical point. It is therefore worthwhile to examine more closely the problem of the existence of phase transitions in the AT model. Here, on the basis of established results and a plausible continuity assumption on the criticality, we conjecture that the AT model has in general two phase transitions. Only in a special case do the two transitions coalesce into a single one.

The AT model is defined on a square lattice whose sites are occupied by any of the four kinds of atoms A, B, C and D. Two neighbouring atoms interact with an energy: ϵ_0 for AA, BB, CC, DD; ϵ_1 for AB, CD; ϵ_2 for AC, BD; and ϵ_3 for AD, BC. By relabelling the atoms, on a sublattice basis if necessary, one sees that the four ϵ's can be freely permuted. This is a basic symmetry of the AT model.

A useful spin representation (Fan 1972b) of the AT model is as follows. At each site of the square lattice one introduces two Ising spins so as to form a two-layer lattice. The four nearest-neighbouring spins interact with a four-body interaction $-J_3 = \frac{1}{4}(\epsilon_0 + \epsilon_3 - \epsilon_1 - \epsilon_2)$ and two-body interactions $-J_1 = \frac{1}{4}(\epsilon_0 + \epsilon_1 - \epsilon_2 - \epsilon_3)$ and $-J_2 = \frac{1}{4}(\epsilon_0 + \epsilon_2 - \epsilon_1 - \epsilon_3)$ within each layer. The basic symmetry of the ϵ's then implies that J_1, J_2 and J_3 can also be permuted.

By performing a dual transformation for the Ising spins in one layer and interpreting the result as a vertex model (Wu 1971, Kadanoff and Wegner 1971), Wegner (1972)

† Supported in part by the National Science Council, Republic of China.

‡ On leave from Department of Physics, Northeastern University, Boston, Massachusetts, USA, and supported in part by the National Science Foundation Grant No. GH-35822 at Northeastern University.

showed that the AT model is equivalent to a *staggered* eight-vertex model. To be specific, the equivalent vertex model has vertex weights a, b, c_+, d_+ on sublattice A and a, b, d_+, c_+ on sublattice B, with

$$a = 2^{-1/2}(\omega_0 + \omega_1), \qquad b = 2^{-1/2}(\omega_2 - \omega_3),$$
$$c_+ = 2^{-1/2}(\omega_2 + \omega_3), \qquad d_+ = 2^{-1/2}(\omega_0 - \omega_1). \tag{1}$$

Here $\omega_i = \exp(-\epsilon_i/kT)$ and we have adopted Baxter's notations (Baxter 1971) a, b, c, d for the vertex weights.

From the basic symmetry we may take, without loss of generality, $\omega_0 = 1$ and $\omega_1, \omega_2, \omega_3, \leqslant 1$. As the temperature varies from 0 to ∞, the point $(\omega_1, \omega_2, \omega_3)$ traces in the ω-space a curve, the thermodynamic path Γ, from (0, 0, 0) to (1, 1, 1). Assuming a continuous dependence of the criticality on the parameters ω_1, ω_2 and ω_3 (the continuity assumption), the critical point of the AT model will trace a surface σ in the ω-space. A phase transition occurs whenever Γ intersects the surface σ. This continuity assumption appears to have been first clearly stated by Thibaudier and Villain (1972) in connection with locating the critical point for the three-dimensional eight-vertex model.

Some exact information is available for the critical surface σ. When any one of the Ising interactions J_1, J_2 and J_3 vanishes, the Ising representation of the AT model decouples into two independent nearest-neighbour models. From the exact result on the Ising model (Onsager 1944) we conclude that the points

$$J_i = 0, \qquad \exp(-2|J_j|/kT) = r \equiv \sqrt{2} - 1, \qquad i \neq j \tag{2}$$

lie on σ. A little algebra shows that in the ω-space these are the straight lines L, shown in figure 1, connecting the points $(r, 0, 0)$, $(r, 1, r)$, $(0, 0, r)$, $(1, r, r)$, $(0, r, 0)$, $(r, r, 1)$, $(r, 0, 0)$ in succession.

From the eight-vertex representation (1) of the AT model, it is known from Baxter's work (Baxter 1971) that when $c_+ = d_+$ the critical condition is $a = |b| + c_+ + d_+$.

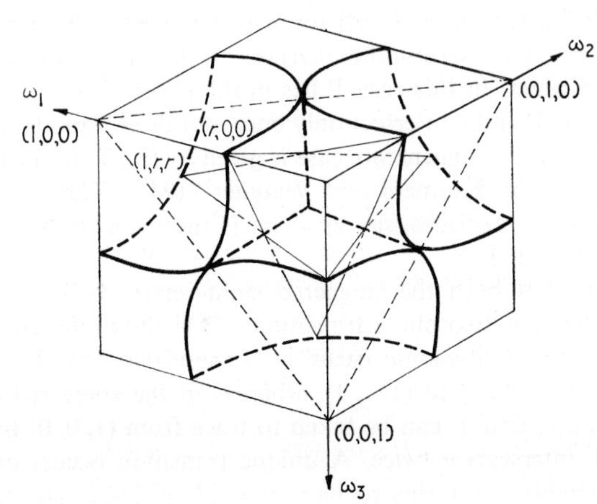

Figure 1. Schematic plot of the critical surface σ in the ω-space. The light lines L lie on σ, and the heavy broken lines L_1 are the intersections of σ with the plane $\omega_1 + \omega_2 + \omega_3 = 1$. The cube $0 \leqslant \omega_i \leqslant 1$ indicates the physical region.

Coupled with the basic symmetry of the AT model we then conclude that σ intersects the plane

$$\omega_1 + \omega_2 + \omega_3 = 1 \tag{3}$$

at the line segments L_1

$$\omega_i = \omega_j \geqslant \omega_k, \qquad i, j, k \text{ distinct.} \tag{4}$$

These segments are also shown in figure 1. Note that L_1 and two of the lines L meet at the common points (r^2, r, r), (r, r^2, r) and (r, r, r^2).

When $\epsilon_1 = \epsilon_2 = \infty$ the AT model reduces to a nearest-neighbour Ising model and is exactly soluble. [Γ now traces along the axis from (0, 0, 0) to (0, 0, 1).] Hence σ intersects the axes of the ω-space only at $(0, 0, r)$, $(0, r, 0)$ and $(r, 0, 0)$. Similar consideration for $\epsilon_1 = 0$, $\epsilon_2 = \epsilon_3$ shows that σ intersects the line $\omega_1 = 1$, $\omega_2 = \omega_3$ at $(1, r, r)$, etc.

Finally the $c_+ \leftrightarrow d_+$ symmetry of the staggered eight-vertex model (1) implies the invariance of σ under the transformation

$$\begin{aligned}
\omega'_1 &= (1 + \omega_1 - \omega_2 - \omega_3)/(1 + \omega_1 + \omega_2 + \omega_3) \\
\omega'_2 &= (1 - \omega_1 + \omega_2 - \omega_3)/(1 + \omega_1 + \omega_2 + \omega_3) \\
\omega'_3 &= (1 - \omega_1 - \omega_2 + \omega_3)/(1 + \omega_1 + \omega_2 + \omega_3).
\end{aligned} \tag{5}$$

This is a 'reflection' symmetry about the plane (3), relating the two portions of σ separated by (3). Note that the points $(\omega_1, \omega_2, \omega_3)$ and $(\omega'_1, \omega'_2, \omega'_3)$ are colinear with $(-1, -1, -1)$.

With this information, the geometry of the critical surface σ can be pictured as three bowl-shaped pieces sewn together at the line segments L_1 as shown schematically in figure 1. We conjecture that this is the case. Note that σ is known to pass through the light lines L and the heavy broken lines L_1 in figure 1. Now ω_1, ω_2 and ω_3 are monotonic along the thermodynamic path Γ; also Γ lies within one of the regions $\omega_i \geqslant \omega_j \geqslant \omega_k$. Assuming a regular-shaped σ surface as shown in figure 1, then Γ will in general intersect σ at two points, one on each side of the plane (3). Consequently, the general AT model has two phase transitions; the two transition temperatures are related by (5). A special situation arises when the two middle energies of the AT model are equal, ie $\epsilon_i = \epsilon_j \leqslant \epsilon_k$ with i, j, k distinct. In this case Γ lies in the plane $\omega_i = \omega_j$ and intersects σ at one of the lines L_1. Thus Γ and σ intersect only once and the AT model has only one transition. The critical point in this degenerate case is given by (3). This includes the case $\epsilon_1 = \epsilon_2 = \epsilon_3 > 0$ considered by Kramers and Wannier (1941). The conclusions are unchanged when one of the ω's vanishes, say $\epsilon_1 = \infty$. [Γ now lies in the $\omega_1 = 0$ plane and traces from (0, 0, 0) to (0, 1, 1).]

Our conjecture implies that both the staggered eight-vertex model and the Ising equivalent of the AT model have two phase transitions. This situation can also be seen from the trajectory of the thermodynamic paths Γ of these two models. In the Ising model, Γ again traces from (0, 0, 0) to (1, 1, 1), whereas in the staggered eight-vertex model a little reflection shows that Γ can be taken to trace from (1, 0, 0) to (0, 1, 0) in figure 1. In either case Γ intersects σ twice. A unique transition occurs only when Γ intersects σ at L_1. The condition for this to happen is $|J_i| = |J_j| \geqslant |J_k|$ for the Ising model and $c_+ = d_+$ for the staggered eight-vertex model. The former condition has previously been given by Wegner (1972) from symmetry considerations.

We point out in conclusion that the continuity assumption should always be taken with caution. While there exist other vertex models for which it is known to be invalid

(Wu 1972, 1974), the assumption seems plausible in the present model and is certainly consistent with all the known results. Indeed, from our analysis it would be most suprising if the general AT model does *not* exhibit two phase transitions.

References

Ashkin J and Teller E 1943 *Phys. Rev.* **64** 178
Baxter R J 1971 *Phys. Rev. Lett.* **26** 832
Fan C 1972a *Phys. Rev.* B **6** 902
——1972b *Phys. Lett.* **39A** 136
Kadanoff L P and Wegner F J 1971 *Phys. Rev.* **B 4** 3989
Kramers H A and Wannier G H 1941 *Phys. Rev.* **60** 252
Onsager L 1944 *Phys. Rev.* **65** 117
Thibaudier C and Villain J 1972 *J. Phys.* C: *Solid St. Phys.* **5** 3429
Wegner F 1972 *J. Phys.* C: *Solid St. Phys.* **5** L131
Wu F Y 1971 *Phys. Rev.* B **4** 2312
——1972 *Phys. Rev.* B **6** 1810
——1974 *Phys.Rev. Lett.* **32** 460

7. Percolation

Journal of Statistical Physics, Vol. 18, No. 2, 1978

Percolation and the Potts Model

F. Y. Wu[1]

Received July 13, 1977

The Kasteleyn–Fortuin formulation of bond percolation as a lattice statistical model is rederived using an alternate approach. It is shown that the quantities of interest arising in the percolation problem, including the critical exponents, can be obtained from the solution of the Potts model. We also establish the Griffith inequality for critical exponents for the bond percolation problem.

KEY WORDS: Percolation; Potts model; Griffiths inequality.

1. INTRODUCTION

The percolation process provides a simple picture of a critical point transition and has been of increasing recent theoretical interest. We refer to several review articles[(1–3)] for a general survey of the subject. An important development first established by Kasteleyn and Fortuin[(4,5)] is the connection between bond percolation and a lattice statistical model. This consideration leads to a formulation of the percolation problem which is extremely useful, for many of the techniques readily available in statistical mechanics can now be applied to percolation (see, e.g., Ref. 6). However, much of this otherwise elegant result appears to be masked under the formality of the graph-theoretic approach of Kasteleyn and Fortuin, and we feel it worthwhile to have an alternate derivation to elucidate the situation. We present an approach to the Kasteleyn–Fortuin formulation of bond percolation which we believe to be simpler and more direct; it also permits a straightforward extension of the Griffiths inequality to the percolation problem. In Section 2 we review the bond percolation problem for the purpose of establishing the notation. The Potts model is introduced in Section 3, and we show that the quantities of interest arising in the percolation problem, including the critical exponents,

Work supported in part by NSF Grant No. DMR 76-20643.

[1] Department of Physics, Northeastern University, Boston, Massachusetts.

can be obtained from the solution of the Potts model. Expressions showing these relationships are explicitly given. The Griffiths exponent inequality for the percolation process is established in Section 4.

2. BOND PERCOLATION

Consider a lattice (or graph) G composed of N sites (vertices) and M edges (lines). The graph does not have to be regular, although consideration in practice is always confined to regular lattices and for M and N large, with the ratio

$$z = \lim_{M,N \to \infty} M/N \tag{1}$$

finite. An example is $z = 2$ for the square lattice.

In a bond percolation process, there is a probability p for each edge of G to be "occupied" independently and a probability $1 - p$ for it to be "vacant." It is convenient to picture the occupied edges as covered by bonds placed along the edges. Two sites that are connected by a chain of occupied edges, or bonds, are said to belong to the same cluster. Then, one of the first questions that can be raised about this percolation process is the probability that a randomly chosen site, say, the origin, of an infinite lattice belongs to a cluster of infinite size. This is the percolation probability $P(p)$.

It is clear that $P(0) = 0$ and $P(1) = 1$. The interesting property of $P(p)$ is that it remains zero for p less than a certain critical value p_c, and rises sharply at p_c with the behavior[3]

$$P(p) \sim (p - p_c)^{\beta}, \qquad p \simeq p_c+ \tag{2}$$

for some positive β. This defines the critical exponent β for the percolation process.

To formulate the problem mathematically, observe that each bond configuration of G is conveniently represented by a subgraph G' of G whose edge set contains precisely the occupied edges. Let $A \equiv A(G')$ be a number associated with some property of the subgraph G'. Examples are

$$\begin{aligned} e &\equiv e(G') = \text{number of occupied edges in } G' \\ n &\equiv n(G') = \text{number of clusters in } G' \end{aligned} \tag{3}$$

The probability for the configuration G' to occur is

$$\pi(G') = p^e(1 - p)^{M-e} \tag{4}$$

The average of the quantity A is then defined to be

$$\langle A \rangle = \sum_{G'} \pi(G')A(G') \tag{5}$$

As in statistical mechanics, the average quantities of interest are usually extensive, and we shall need the "thermodynamic" limit

$$\langle A\rangle_0 = \lim_{N\to\infty} N^{-1}\langle A\rangle \tag{6}$$

Generally let $A_c \equiv A_c(G')$ denote some property of a cluster c in G', and $A_0 \equiv A_0(G')$ the property of the cluster containing the origin. Examples are s_c and s_0, the number of sites in a cluster, and b_c and b_0, the number of bonds in a cluster. Then, by translational symmetry, we have the identity[7]

$$\langle A_0\rangle = \left\langle \sum_c s_c A_c \right\rangle_0 \tag{7}$$

The factor s_c in (7) arises because there are precisely s_c ways for the origin to lie within a cluster of s_c sites. If we define

$$\begin{aligned} \gamma_0 &= 1, \qquad \text{if } s_0 \text{ is finite} \\ &= 0, \qquad \text{otherwise} \end{aligned}$$

It then follows from our definitions that

$$P(p) \equiv 1 - \langle \gamma_0\rangle = 1 - \left\langle \sum_c{}' s_c \right\rangle_0 \tag{8}$$

where the prime restricts the summation to finite clusters.

Other quantities of interest include $S(p)$, the mean size of the *finite* cluster that contains the origin. In our notation,

$$S(p) \equiv \langle s_0\gamma_0\rangle = \left\langle \sum_c{}' s_c{}^2 \right\rangle_0 \tag{9}$$

Here, we have adopted the convention of specifying the cluster size by its site content. Alternately, as is customary in bond percolation considerations, we may also specify the cluster size by its bond content. The mean cluster size is then taken to be

$$S^{(B)}(p) = (zp)^{-1}\left\langle \sum_c{}' b_c{}^2 \right\rangle_0 \tag{10}$$

It can be shown[8] that (9) and (10) give rise to the same exponents γ and γ':

$$\begin{aligned} S(p) &\sim (p - p_c)^{-\gamma}, \qquad p \simeq p_c- \\ &\sim (p_c - p)^{-\gamma'}, \qquad p \simeq p_c+ \end{aligned} \tag{11}$$

Here the primed exponent refers to $p \simeq p_c+$, which is analogous to $T \simeq T_c-$ of the ordinary critical point.

Another quantity of interest is the mean number of clusters per site

$$G(p) = \langle n\rangle_0 \tag{12}$$

Since infinite clusters, whenever they appear, are limited in numbers, there is presumably no need to distinguish $G(p)$ and $G^{(F)}(p)$, the mean number of *finite* clusters. This is certainly so for $p \leqslant p_c$. In evaluating (12), the isolated sites are considered as individual clusters. The behavior of the singular part of $G(p)$ now defines the critical exponents α and α':

$$G_{\text{sing}}(p) \sim |p - p_c|^{2-\alpha, 2-\alpha'}, \qquad p \simeq p_c \tag{13}$$

Finally, the pair connectivity $c(\mathbf{r}, p)$ is defined to be the probability that the sites at the origin and at $\mathbf{r}$ are connected (i.e., they belong to the same cluster). This connectivity can be written as

$$c(\mathbf{r}, p) = \langle \gamma(\mathbf{r}) \rangle \tag{14}$$

where

$$\begin{aligned} \gamma(\mathbf{r}) &= 1, \qquad \text{if the sites at the origin and } \mathbf{r} \text{ are connected} \\ &= 0, \qquad \text{otherwise} \end{aligned}$$

The decay of the connectivity at p_c defines the exponent η for the percolation process:

$$c(\mathbf{r}, p_c) \sim r^{-(d-2+\eta)}, \qquad r \to \infty \tag{15}$$

Here d is the dimensionality of the lattice.

It is also customary to consider $c^{(F)}(\mathbf{r}, p)$, the probability that the sites at the origin and at $\mathbf{r}$ belong to the same *finite* cluster.[7] We expect $c(\mathbf{r}, p)$ and $c^{(F)}(\mathbf{r}, p)$ to be identical for $p \leqslant p_c$.

The Fourier transform of the pair connectivity takes the form

$$\tilde{c}(k, p) \equiv \sum_{\mathbf{r}} c(\mathbf{r}, p) \exp(-i\mathbf{k}\cdot\mathbf{r}) = \left\langle \sum_{\mathbf{r}\in c_0} \exp(-i\mathbf{k}\cdot\mathbf{r}) \right\rangle \tag{16}$$

where the summation in (16) extends over all sites in the cluster c_0 containing the origin. By expanding the exponential and averaging over all directions of $\mathbf{k}$, one obtains the moment expansion[9]

$$\tilde{c}_{\text{av}}(k, p) = \sum_{n=0}^{\infty} \frac{(-1)^n k^{2n}}{(2n+1)!} \mu_{2n}(p) \tag{17}$$

where

$$\mu_n(p) = \left\langle \sum_{\mathbf{r}\in c_0} r^n \right\rangle \tag{18}$$

There is strong evidence that $\mu_n(p)$ diverges as[9]

$$\mu_n(p) \sim |p_c - p|^{-\gamma - n\nu}, \qquad p \approx p_c - \tag{19}$$

which defines the critical exponents ν. The exponent ν' can be defined similarly from the critical behavior of the corresponding $\mu_n^{(F)}(p)$ at p_c+. The definitions of the exponents δ, Δ, and Δ' for the percolation process will be given in the next section.

3. THE POTTS MODEL

The Potts model[10] is a generalization of the Ising model so that the spins can be in one of q states ($q = 2$ is the Ising model). Let the spin state at the ith site be specified by $\xi_i = 1, 2, \ldots, q$. The Hamiltonian for the Potts model in an external field $-H$ is

$$\mathscr{H} = -\epsilon \sum_{\langle ij \rangle} \delta_{kr}(\xi_i, \xi_i) - H \sum_i \delta_{kr}(\xi_i, \alpha) \tag{20}$$

where the first summation extends over the M edges of G, and the external field $-H$ is applied to the spin state α. The partition function of the Potts model now reads

$$Z(q; k, L) = \sum_{\xi_i = 1}^{q} \prod_{\langle ij \rangle} [1 + v\delta_{kr}(\xi_i, \xi_j)] \prod_i [1 + u\delta_{kr}(\xi_i, \alpha)] \tag{21}$$

where

$$v = e^K - 1, \qquad u = e^L - 1, \qquad K = \epsilon/kT, \qquad L = H/kT \tag{22}$$

Following Baxter[11] and Ref. 5, we expand the first product in (21) and use the subgraphs of G to represent the terms in the expansion. Each term in the expansion is conveniently represented by a subgraph G' whose edge set coincides with the $v\delta_{kr}(\xi_i, \xi_j)$ factors contained in the term. For a given G' of the expansion, we further expand the second product in (21) for *each cluster*. For the first term, viz. 1, in the expansion for a cluster, the summations in (21) yield a factor q. For the remaining $2^{s_c} - 1$ terms of the cluster, which contains s_c sites, the summations yield a factor $(1 + u)^{s_c} - 1 = e^{Ls_c} - 1$. It follows that (21) takes the form

$$Z(q; K, L) = \sum_{G'} v^e \prod_c (e^{Ls_c} + q - 1) \tag{23}$$

Comparing (23) with (5), we see that we can write

$$Z(q; K, L) = e^{MK} \left\langle \prod_c (e^{Ls_c} + q - 1) \right\rangle \tag{24}$$

provided that, of course, we take $v = p/(1 - p)$, or

$$p = 1 - e^{-K} \tag{25}$$

Equation (24) is the basic relation connecting the Potts model with the bond percolation problem. Note that while the Potts partition function (21) is defined strictly for positive, integral q, the expression (23) or (24) provides a natural continuation of the partition function to other values of q. This leads to the random cluster model of Refs. 4 and 5. For our purposes, it suffices to start from the solution of the Potts model and simply treat the parameter q occurring in the solution as a continuous variable. This permits us to carry out operations such as derivatives with respect to q. Now write the free energy per site of the Potts model as

$$f(q; K, L) = \lim_{N\to\infty} N^{-1} \ln Z(q; K, L) \tag{26}$$

and further define

$$h(K, L) = \left[\frac{\partial}{\partial q} f(q; K, L)\right]_{q=1} \tag{27}$$

It is easy to verify that, after an interchange of the order of the derivative and the thermodynamic limit,

$$h(K, L) = \left\langle \sum_c e^{-Ls_c} \right\rangle_0 = \left\langle \sum_c{}' e^{-Ls_c} \right\rangle_0, \qquad L > 0 \tag{28}$$

While for $L = 0$ the summation in (28) ranges over all clusters, the summation is, in effect, restricted to clusters of finite size for any $L > 0$. We then obtain from (12) and (28) the identities

$$G(p) = h(K, 0), \qquad G^{(F)}(p) = h(K, 0+) \tag{29}$$

Therefore, $G(p) = G^{(F)}(p)$ if and only if $h(K, L)$ is continuous at $L = 0$. Further define

$$P(p, L) \equiv 1 + (\partial/\partial L) h(K, L) \tag{30}$$

Comparison of (8) and (28) then establishes the identity

$$P(p) = P(p, 0+) \tag{31}$$

Similarly, (9) leads to the expression

$$S(p) = \left[\frac{\partial^2}{\partial L^2} h(K, L)\right]_{L=0+} \tag{32}$$

It is now seen that $h(K, L)$ plays the role of the free energy of a statistical model and we are led to the correspondences[4,5]

$$G(p) \leftrightarrow \text{free energy}$$

$$P(p) \leftrightarrow \text{magnetization}$$

$$S(p) \leftrightarrow \text{susceptibility}$$

Pursuing the analogy further, it is now possible to define the exponent δ for the percolation process from the relation[12]

$$P(p_c, L) \sim L^{1/\delta}, \qquad L \simeq 0 \tag{33}$$

Similarly, we can define the gap exponents Δ and Δ' using

$$\left[\frac{\partial^3}{\partial L^3} h(K, L)\right]_{L=0+} \sim |p - p_c|^{-\gamma-\Delta, -\gamma-\Delta'} \tag{34}$$

The above analysis can be extended if an external field of the form

$$-H_1 \sum_{\langle ij \rangle} \delta_{kr}(\xi_i, \alpha)\, \delta_{kr}(\xi_j, \alpha)$$

is included in the Potts Hamiltonian (20). This changes (24) into

$$Z(q; K, L, L_1) = e^{MK}\left\langle \prod_c (e^{Ls_c + L_1 b_c} + q - 1) \right\rangle \tag{35}$$

where $L_1 = H_1/kT$. Quantities involving the averages of b_c, such as $S^{(B)}(p)$ in (10), can be conveniently expressed as the derivatives of the free energy per site defined by (35). In fact, the analogy between the Potts model and the bond percolation may be developed by considering L_1 instead of L as the external field. Since this does not lead to any new values for the critical exponents,[8] we shall not consider it any further.

Finally, we consider the pair connectivity $c(\mathbf{r}, p)$. Let $P_{\alpha\alpha}(\mathbf{r})$ be the probability in the zero-field Potts model that the sites at the origin and $\mathbf{r}$ are both in the spin state α. Since $P_{\alpha\alpha}(\mathbf{r}) = q^{-2}$ when there is no correlation, the correlation function can be defined to be[13]

$$\Gamma_{\alpha\alpha}(\mathbf{r}) = P_{\alpha\alpha}(\mathbf{r}) - q^{-2} \tag{36}$$

Now

$$P_{\alpha\alpha}(\mathbf{r}) = \sum{}' \prod_{\langle ij \rangle} [1 + v\delta_{kr}(\xi_i, \xi_j)]/Z(q; K, 0) \tag{37}$$

where the prime over the summation denotes that the spins at the origin and at $\mathbf{r}$ are held at state α. Again, we use a graphical representation of terms in the expansion of the numerator in (37). However, it is now necessary to distinguish whether the sites at the origin and at $\mathbf{r}$ are connected. We thus obtain

$$P_{\alpha\alpha}(\mathbf{r}) = q^{-1}\frac{\langle \gamma(\mathbf{r}) q^n \rangle}{\langle q^n \rangle} + q^{-2}\frac{\langle [1 - \gamma(\mathbf{r})] q^n \rangle}{\langle q^n \rangle} \tag{38}$$

It follows then that

$$q^2(q-1)^{-1}\Gamma_{\alpha\alpha}(F) = \langle \gamma(\mathbf{r}) q^n \rangle / \langle q^n \rangle \tag{39}$$

or, using (14),

$$c(\mathbf{r}, p) = \left[\frac{\partial}{\partial q}\Gamma_{\alpha\alpha}(\mathbf{r})\right]_{q=1} \tag{40}$$

Similarly, we expect

$$c^{(F)}(\mathbf{r}, p) = \left[\frac{\partial}{\partial q}\Gamma_{\alpha\alpha}(\mathbf{r}, 0+)\right]_{q=1} \tag{41}$$

where $\Gamma_{\alpha\alpha}(\mathbf{r}, L)$ is an extension of (36) using the Hamiltonian (20). However, the proof of (41) requires an identity for the Potts model similar to that of the equivalence of the large-distance, two-spin correlation and the square of the spontaneous magnetization of the Ising model. This identity has not been rigorously established.

The relations (40) suggests the further analogy[4,5]

$$c(\mathbf{r}, p) \leftrightarrow \text{correlation function}$$

The exponents η, ν, and ν' for the percolation may then be extracted from the Potts correlation function.

4. GRIFFITHS EXPONENT INEQUALITY

Kasteleyn and Fortuin[4,5] have established the Rushbrooke inequality[14]

$$\alpha' + 2\beta + \gamma' \geqslant 2 \tag{42}$$

for the bond percolation. We now show that our formulation of the percolation permits a straightforward derivation of the Griffiths inequality[15]

$$\alpha' + \beta(1 + \delta) \geqslant 2 \tag{43}$$

for the critical exponents for the bond percolation.

We observe from (28) that $g(K, L) \equiv L + h(K, L)$ is convex in L. This implies that the Legendre transformation

$$A(p, P) \equiv g[K, L(K, P)] - L(K, P)P \tag{44}$$

of $g(K, L)$ is concave in P, where $p = 1 - e^{-K}$ and $L(K, P)$ is to be derived from (30). Further, from the existence of $P(p) = P(p, L = 0+)$ and the relation $L = \partial A/\partial P$ we know that $A(p, P)$ is a constant for $p > p_c$ and $P \leqslant P(p)$. These are the basic ingredients needed in the derivation of the Griffiths inequality for a thermodynamic system. Following the standard argument,[16] we are thus led to the Griffiths inequality (43) for the percolation problem. The appearance of α' in the equality is spurious wherever $\alpha' < 0$, which is the case for some percolation problems in two dimensions.[17] For such systems, the Griffiths equality should read

$$\beta(1 + \delta) \geqslant 2 \qquad (\alpha' \leqslant 0) \tag{45}$$

NOTE ADDED IN PROOF

It has been recently established that the site percolation can be formulated as a Potts model with many-body interactions (H. Kunz and F. Y. Wu, to be published in *J. Phys. C*).

ACKNOWLEDGMENT

I wish to thank J. W. Essam for an illuminating discussion.

REFERENCES

1. H. L. Frisch and J. M. Hammersley, *J. Soc. Indust. Appl. Math.* **11**:894 (1963).
2. V. K. S. Shante and S. Kirkpatrick, *Adv. Phys.* **20**:325 (1961).
3. J. W. Essam, in *Phase Transitions and Critical Phenomena*, C. Domb and M. S. Green, eds. (Academic, London, 1972), Vol. 2.
4. P. W. Kasteleyn and C. M. Fortuin, *J. Phys. Soc. Japan* **26** (Suppl.):11 (1969).
5. C. M. Fortuin and P. W. Kasteleyn, *Physica* **57**:536 (1972).
6. A. B. Harris, T. C. Lubensky, W. K. Holcomb, and C. Dasgupta, *Phys. Rev. Lett.* **35**:327 (1975).
7. A. G. Dunn, J. W. Essam, and J. M. Loveluck, *J. Phys. C* **8**:743 (1975).
8. J. W. Essam, K. M. Gwilym, and J. M. Loveluck, *J. Phys. C* **9**:365 (1976).
9. A. G. Dunn, J. W. Essam, and D. S. Ritchie, *J. Phys. C* **8**:4219 (1975).
10. R. B. Potts, *Proc. Camb. Phil. Soc.* **48**:106 (1952).
11. R. J. Baxter, *J. Phys. C* **6**:L445 (1973).
12. D. S. Gaunt and M. F. Sykes, *J. Phys. A* **9**:1109 (1976).
13. Y. K. Wang and F. Y. Wu, *J. Phys. A* **9**:593 (1976).
14. G. S. Rushbrooke, *J. Chem. Phys.* **39**:842 (1963).
15. R. B. Griffiths, *Phys. Rev. Lett.* **14**:623 (1965).
16. M. E. Fisher, *Rep. Prog. Phys.* **30**:615 (1967).
17. C. Domb and C. J. Pearce, *J. Phys. A* **9**:L137 (1976).

An infinite-range bond percolation[a)]

F. Y. Wu

Northeastern University, Boston, Massachusetts 02115

A bond percolation on a lattice in which all pairs of vertices are connected is considered. The percolation problem is treated by carrying out the Kasteleyn-Fourtuin formulation of taking the $q = 1$ limit of a related q-component Potts model; the latter is exactly solved. For a lattice of N sites, an average of pN occupied bonds and N large, it is found that the system is percolating for $p > 1/2$. Closed-form expression is also obtained for the moment-generating function of the cluster size. The analysis yields the mean-field exponents $\beta = \gamma = \gamma' = 1$.

PACS numbers: 05.50. + q, 05.70.Jk

It is customary to regard the Bethe-lattice solution of the percolation problem as its mean-field approximation [1]. While the Bethe-lattice consideration has led to yield the mean-field exponents, it has not been very helpful in providing an useful picture in understanding the mean-field nature of the percolation transition. In this paper we consider a model of a bond percolation, which is an extension of the usual mean-field spin models. By solving this percolation problem exactly we have a picture of a mean-field percolation which is on the same footing as those of the spin systems.

It is well-known that a meaningful mean-field model for spin systems is one in which all spins interact with equal strength of the order of 1/N, N being the total number of spins [2]. In the same spirit we therefore picture a mean-field model for bond percolation as a percolation process is which all pairs of sites are connected by bonds with an independent probability, also of the order of 1/N. Specifically, consider a system of N sites subject to a percolation process in which every pair of sites can be connected by a bond with a probability 2p/N; then on the average there are pN bonds. Thus this is a percolation with long-range interations. It is known that such a percolation describes the problem of random graphs [3], and can be treated using a purely probabilistic approach [4]. Here we utilize the method of statistical physics by first considering a related Potts model [5]. The Potts model is then solved to give results on the percolation problem.

Consider a system of N q-state Potts spins whose Hamiltonian reads

$$H = \frac{K}{N}\sum_{<i,j>} \delta(\sigma_i,\sigma_j) + L \sum_i \delta(\sigma_i,1) \tag{1}$$

where we have taken $kT = -1$, $\sigma_i = 1,2,\ldots,q$ denotes the spin states at the ith site, $i = 1,2,\ldots,N$ and δ is the Kronecker delta function. The summations $<i,j>$ are taken between all pairs of spins.

It is well-known [5,6] that the $q = 1$ limit of a Potts model generates a percolation. Particularly the Potts Hamiltonian (1) generates the long-range percolation under consideration with the following cluster-size generating function [6]:

$$G(L) = \frac{1}{N}\left\langle \sum_c \exp(-Ls_c)\right\rangle$$

$$= \left[\frac{\partial}{\partial q} A(q;K,L)\right]_{q=1} \tag{2}$$

where

$$2p/N = 1 - e^{-K/N} \tag{3}$$

Here s_c is the number of sites in a cluster c, $A(q;K,L)$ is the free energy of (1).

To compute $A(q;K,L)$, let $x_i, i=1,2,\ldots,q$, denote the fraction of spins in the ith spin state. Then, in the limit of $N\to\infty$, we have

$$A(q;K,L) = \max_{\{x_i\}}[E(x_i) - \sum_i x_i \ln x_i] \tag{4}$$

where

$$E(x_i) = \frac{K}{2}\sum x_i^2 + Lx_1 \tag{5}$$

is the "energy" of the system computed from (1) and subject to $\sum x_i = 1$.

It is now straightforward to compute $A(q;K,L)$ from (4) and (5). The result yields [7]

$$G(L) = 1 - s_o - p(1-s_o)^2 \tag{6}$$

where s_o is determined from

$$2ps_o + L + \ln(1-s_o) = 0 \tag{7}$$

Quantities of interest in percolation can now be computed by taking the respective derivatives of G(L). Thus, we find the percolation probability, P(p), and the mean cluster size, S(p), to be given by [6]

$$P(p) = 1 + G'(0+) = s_o \tag{8}$$

$$S(p) = G''(0+) = (1-s_o)/(1-2p+2ps_o) \tag{9}$$

Now, for $p<1/2$, (7) with $L=0$ has only one solution $s_o=0$, hence $P(p)=0$ identically. For $p>1/2$, however, a second solution $s_o>0$ arises and there is a nonzero percolation probability $P(p)=s_o$. The average cluster size is given by (9). Near the threshold we find

$$P(p) \simeq p - 1/2 \tag{10}$$

$$S(p) \simeq |p - 1/2|^{-1} \tag{11}$$

This leads to the mean-field sxponents $\beta=\gamma=\gamma'=1$, in agreement with the finding of [1]. We also note that the critical value of $p = 1/2$ concides with the finding of the purely probabilistic approach [4].

REFERENCES

a) Supported in part by the National Science Foundation

1. M.E. Fisher and J.W. Essam, J. Math. Phys. 2, 609 (1961).
2. M. Kac in Statistical Physics, Phase Transitions and Superfluidity, Eds. M. Crêtien, E.P. Gross and S. Deser, Gordon and Breach, New York, 1968.
3. D.J.A. Welsh, Sci. Prog. Oxf. 64, 65 (1977).
4. P. Erdos and A. Rényi, Publ. Math. Inst. Hung. Acad. Sci. 5, 17 (1960).
5. P.W. Kasteleyn and C.M. Fortuin, J. Phys. Soc. Japan 26, (Suppl) 11 (1969).
6. See, e.g., F.Y. Wu, Rev. Mod. Phys. 54, 235 (1982).
7. F.Y. Wu, J. Phys. A15, L395 (1982).

PHYSICAL REVIEW LETTERS

VOLUME 48 22 MARCH 1982 NUMBER 12

Domany-Kinzel Model of Directed Percolation: Formulation as a Random-Walk Problem and Some Exact Results

F. Y. Wu
Department of Physics, Northeastern University, Boston, Massachusetts 02115

and

H. Eugene Stanley
Center for Polymer Studies, Boston University, Boston, Massachusetts 02215
(Received 28 December 1981)

It is shown that the directed percolation on certain two-dimensional lattices, in which the occupation probability is unity along one spatial direction, is related to a random-walk problem, and is therefore exactly solvable. As an example, the case of the triangular lattice is solved. It is also shown that the square-lattice solution obtained previously by Domany and Kinzel can be derived using Minkowski's "taxicab geometry."

PACS numbers: 05.70.Jk, 05.70.+q, 64.60.Fr, 05.50.+q

Directed percolation[1] has aroused considerable recent interest among workers from many fields of physics, because of its applications ranging from Reggeon field theory[2] to Markov processes involving branching, recombination, and absorption that arise in chemistry and biology.[3] The combination of renormalization-group, Monte Carlo computer-simulation, and series-expansion procedures has led to a great deal of progress.[4-10]

Relatively little is known in the way of exact solutions for the directed percolation problem. However, in a recent Letter, Domany and Kinzel[11] have proposed a particularly elegant model of directed percolation for a square lattice which is amenable to exact solution. Consider a bond percolation process for which the horizontal and vertical bonds are intact (occupied) with respective probabilities p_H and p_V. Adopt the "sun-belt" convention of placing westward and southward arrows, respectively, on all horizontal and vertical bonds.[12] Domany and Kinzel considered general p_H, p_V and also obtained for $p_H=1$, $p_V=p$ a closed form expression for the probability, $P(\vec{R},p)$, that a site $\vec{R}$ located to the south and west of the origin could be reached by one or more connected paths. They found that for large R, there exists a $p_c(\vec{R}/|\vec{R}|)$ such that $P(\vec{R},p \geq p_c) = 1$ and that $P(\vec{R},p \to p_c^-) \sim \exp(-R/\xi)$ with $\xi = (p_c - p)^{-2}$.

Here we present the following further exact results on the Domany-Kinzel problem:

(i) We show that the Domany-Kinzel model of directed percolation is related to a random-walk problem.

(ii) We show more generally that directed percolation on certain two-dimensional nets in which the occupation probability is unity along one spatial direction can also be formulated as a random-walk problem, leading to a simple derivation and analysis of the solution. As an example, the triangular lattice is treated.

Consider first the Domany-Kinzel problem of an infinite square lattice whose sites are denoted by the coordinates (i,j), and let $\vec{O}=(0,0)$, $\vec{R}=(N-1, L)$, so that point $\vec{R}$ is $N-1$ units to the west

VOLUME 48, NUMBER 12 PHYSICAL REVIEW LETTERS 22 MARCH 1982

of the origin and L units to the south of the origin.[12] A bond configuration of the lattice is percolating if there exists at least one directed path running from $\vec{O}$ to $\vec{R}$. Then the key to the Domany-Kinzel solution lies in the fact that a *unique* path can be singled out for each percolating configuration. This can be accomplished by adopting the convention of following the downward arrow whenever possible. Thus, starting from $\vec{O}$, one traverses horizontally, unless there is a down arrow originating from $\vec{O}$, in which case one follows the down arrow immediately. Generally, one follows the first down arrow en route to the next row, and repeats the process. Clearly, a unique path connecting $\vec{O}$ to $\vec{R}$ will be singled out by this process in each percolating configuration. (The path shown in Fig. 1 of Ref. 11 follows the opposite convention, going from $\vec{R}$ to $\vec{O}$, but the effect is the same.)

Since $p_H = 1$, a given configuration must be percolating as soon as the path reaches row L at any point (n, L) with $0 \leq n \leq N-1$. Hence one can write

$$P(\vec{R}, p) = \sum_{n=0}^{N-1} p W_{n, L-1}, \tag{1}$$

where $W_{n, L-1}$ is the probability that the path shall reach the point $(n, L-1)$ on row $L-1$. In writing (1), we have already summed over all percolating configurations corresponding to the same path. Consider now the paths running from $(0, 0)$ to $(n, L-1)$. There are precisely n horizontal and $L-1$ vertical arrows in such paths, with each vertical arrow carrying a weight (probability) p and each horizontal arrow a weight (probability) $q = 1 - p$. It follows that

$$W_{n, L-1} = p^{L-1} q^n C_{n, L-1}, \tag{2}$$

where $C_{n, L-1}$ is the number of distinct paths connecting $(0, 0)$ and $(n, L-1)$. Since the vertical and horizontal arrows can occur in any order, we have

$$C_{n, L-1} = \binom{n+L-1}{L-1}. \tag{3}$$

This is the result of Domany and Kinzel who derived it using a different (and more involved) method of counting and analyzed it using a method whose generalization to other lattices is not apparent.

It is of interest to point out here that the number $C_{n, L-1}$ also arises in taxicab geometry, a metric system first proposed by Minkowski over 70 years ago,[13] as the number of "straight" lines between two fixed points.[14] Specifically, $C_{n, L-1}$ is the total number of distinct "taxicab routes" from point $(0, 0)$ to point $(n, L-1)$ on a directed lattice; that $C_{n, L-1}$ is simply given by Eq. (3) is demonstrated clearly in a recent popular account of Minkowski's taxicab geometry.[15]

The paths connecting $(0, 0)$ and $(n, L-1)$ can also be regarded as those traced by a random walker on a directed lattice. Then $W_{n, L-1}$ is the probability that the walker will eventually reach $(n, L-1)$. The formulation as a random-walk problem offers a natural and clean way to analyze the results (2) and (3); it can also be extended to other two-dimensional lattices when the occupation probability is unity along one spatial direction.

As an example, consider the directed percolation problem on a triangular lattice in which the horizontal bonds are present with probabilities $p_H = 1$, the vertical bonds with probabilities $p_V = p$, and the diagonal bonds with probabilities $p_D = p'$. All bonds are directed in the south, west, and southwest directions as shown in Fig. 1. We again compute the probability $P(\vec{R}, p, p')$ that the sites $\vec{O} = (0, 0)$ and $\vec{R} = (N-1, L)$ are connected by at least one directed path. The Domany-Kinzel case is recovered by taking $p' = 0$.

As in the Domany-Kinzel problem, a key step of the solution is to devise a convention which will generate a unique path connecting $\vec{O}$ and $\vec{R}$ in percolating configurations. For this purpose we adopt the convention of following the arrows in the order of vertical, diagonal, and horizontal at each site. Thus, starting from $\vec{O}$ and following arrows according to the order just described, we shall always reach $\vec{R}$ in configurations which are percolating.[16] This convention also assigns the weights p, qp', and qq', respectively, to

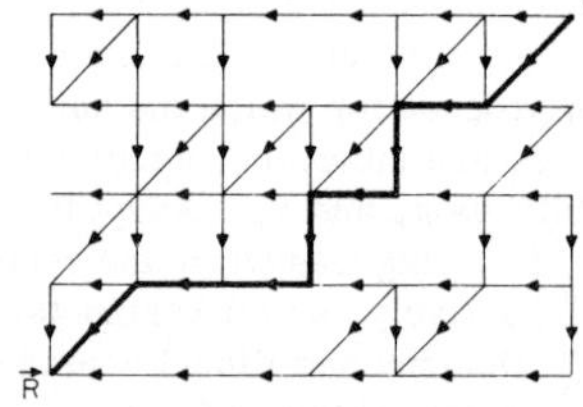

FIG. 1. A typical percolating configuration for the triangular lattice with $N-1 = 6$ and $L = 4$. The bonds are all oriented, and are intact with respective probabilities $p_H = 1$, $p_V = p$, and $p_D = p'$. The heavy lines denote the *unique* path associated with this configuration.

the vertical, diagonal, and horizontal arrows along the path, where $q=1-p$ and $q'=1-p'$.

In analogy to (1), we now have

$$P(\vec{R},p,p')=(1-qq')\sum_{n=0}^{N-2} W_{n,L-1}+pW_{N-1,L-1}=1-(1-qq')\sum_{n=N-1}^{\infty} W_{n,L-1}+pW_{N-1,L-1}, \tag{4}$$

where we have distinguished the case $n=N-1$ from the cases $0\leq n\leq N-2$. The second equality follows from the elementary fact that the point (∞, L) is connected to the origin with probability 1.

To proceed further, we now regard $W_{n,L-1}$ as the probability that a walker will reach $(n, L-1)$ from (0, 0) in a random walk on the triangular lattice with anisotropic probabilities 0, 0, 0, qq', p, qp' for the six directions. Then $W_{n,L-1}$ can be computed by standard means, leading to the expression[17]

$$W_{n,L-1}=\frac{1}{(2\pi)^2}\iint_{-\pi}^{\pi} d\varphi_1 d\varphi_2 \frac{\exp[-in\varphi_1-i(L-1)\varphi_2]}{1-qq'\exp(i\varphi_1)-p\exp(i\varphi_2)-qp'\exp[i(\varphi_1+\varphi_2)]}. \tag{5}$$

Simply stated, $W_{n,L-1}$ is the coefficient of $x^n y^{L-1}$ in the expansion of $[1-qq'x-py-qp'xy]^{-1}$. If we introduce $z_1=\exp(-i\varphi_1)$ and $z_2=\exp(-i\varphi_2)$, the integrations in (5) become contour integrals around the unit circles in the z_1 and z_2 planes. It is readily established that the simple pole in the z_2 plane is always within the unit circle $|z_2|=1$. Therefore, after carrying out the z_2 integration, we obtain

$$W_{n,L-1}=(\tfrac{1}{2}i)\oint_{|z|=1} F_{n,L}(z)\,dz, \tag{6}$$

where

$$F_{n,L}(z)=z^n(pz+qp')^{L-1}/(z-qq')^L. \tag{7}$$

For the Domany-Kinzel case, $p'=0$, $q'=1$, and (7) reduces to $F_{n,L}(z)=p^{L-1}z^{n+L-1}/(z-q)^L$, so that a straightforward integration of (6) leads to the result (2) and (3). For the general problem, a direct evaluation of (6) and (4) yields a double series which is not easily analyzed. But we can substitute (6) into (4) and deform the contour in the first term to $|z|=r$, $qq'<r<1$, to permit carrying out the summation. This leads to

$$P(\vec{R},p,p')=1-\frac{1-qq'}{2\pi i}\oint_{|z|=r}\frac{dz}{1-z}F_{N-1,L}(z)+\frac{p}{2\pi i}\oint_{|z|=1}dz\,F_{N-1,L}(z). \tag{8}$$

The integrals in (8) can be evaluated by the method of steepest descent. For $N=\alpha L$ large, α finite, we write $F_{N,L}(z)\simeq[f_\alpha(z)]^L$, $f_\alpha(z)=z^\alpha(pz+qp')/(z-qq')$, and find that the integrand is stationary at z_0 determined by $f'(z_0)=0$. It can be verified that $f_\alpha(z_0)$ attains its maximum value of 1 at $z_0=1$ and that $z_0\gtrless 1$ for $\alpha\lessgtr\alpha_c$, where $\alpha_c=q/(1-qq')$. Therefore $F_{N,L}(z_0)=0$ for $z_0\neq 1$ and $F_{N,L}(1)=(1-qq')^{-1}$. In the second integral in (8), we can always deform the contour to pass z_0 so that it gives rise to zero contribution after using the method of steepest descent. The first integral is again zero for $\alpha>\alpha_c$, since in this case $z_0<1$ and the contour can be deformed continuously to the stationary point. But for $\alpha<\alpha_c$ this deformation sweeps past the simple pole at $z=1$, generating a residue $-(1-qq')F_{N-1,L}(1)=-1$. For $\alpha=\alpha_c$ only half of the residue is generated. Thus we find[18]

$$\begin{aligned} P(\vec{R},p,p')&=1, \quad \alpha>\alpha_c,\\ &=0, \quad \alpha<\alpha_c,\\ &=\tfrac{1}{2}, \quad \alpha=\alpha_c. \end{aligned} \tag{9}$$

In fact, for $\alpha<\alpha_c$, we find after Taylor expanding about $z_0=1$, $P(\vec{R},p,p')\simeq[f_\alpha(z_0)]^L\sim\exp(-R/\xi)$, with $\xi\sim(\alpha_c-\alpha)^{-2}$. This leads to the same critical exponent $\nu=2$ as in the Domany-Kinzel solution,[19] reflecting a Gaussian distribution of the profile of the percolation cone.[20] It is also instructive to note that for $p=0$ ($q=1$) the lattice is again simple quadratic with $\vec{R}$ situated at $(N-L-1, L)$. The resulting critical value then reads $\alpha_c-1=q'/(1-q')$, in agreement with that of Domany and Kinzel.

The connection of directed percolation with a random-walk problem is more general and appears to be applicable whenever a *unique* path can be associated with a percolating configuration. In particular, it is applicable to the percolation problem for which *any* number of bonds are present between the sites (i,j) and $(i-k, j+1)$, $k=-1, 0, 1, 2, 3, \ldots$, with respective probabilities p_k, in addition to $p_H=1$. In such cases $P(\vec{R},\{p_k\})$ can always be computed by considering a corresponding random-walk problem.

The authors thank E. Domany, W. Klein, and S. Redner for comments on this manuscript. This work has been supported in part by grants from the National Science Foundation, the U. S.

Volume 48, Number 12 PHYSICAL REVIEW LETTERS 22 March 1982

Office of Naval Research, and the U. S. Army Research Office. One of us (H.E.S.) thanks the Guggenheim Foundation for a Fellowship.

[1]S. R. Broadbent and J. M. Hammersley, Proc. Cambridge Philos. Soc. 53, 629 (1957).

[2]J. L. Cardy and R. L. Sugar, J. Phys. A 13, L423 (1980).

[3]F. Schlögl, Z. Phys. 253, 147 (1972).

[4]S. P. Obukhov, Physica (Utrecht) 101A, 145 (1980).

[5]J. Kertész and T. Vicsek, J. Phys. C 13, L343 (1980).

[6]W. Kinzel and J. M. Yeomans, J. Phys. A 14, L163 (1981).

[7]S. Redner, J. Phys. A 14, L349 (1981); S. Redner and A. C. Brown, J. Phys. A 14, L285 (1981).

[8]W. Klein and W. Kinzel, J. Phys. A 14, L405 (1981).

[9]J. W. Essam and K. De'Bell, J. Phys. A 14, L459 (1981).

[10]D. Dhar, M. Barma, and M. K. Phani, Phys. Rev. Lett. 47, 1238 (1981).

[11]E. Domany and W. Kinzel, Phys. Rev. Lett. 47, 5 (1981).

[12]We adopt the conventions of Ref. 11; other conventions exist in the directed-percolation literature.

[13]H. Minkowski, *Gesammelte Abhandlungen* (Chelsea, New York, 1967).

[14]F. Sheid, Mathematics Teacher 54, 307 (1961).

[15]M. Gardner, Sci. Am. 243, No. 5, 18 (1980).

[16]Other conventions (such as the order of diagonal, vertical, and horizontal) may lead to a "trap," even though the configuration is percolating.

[17]See, e.g., E. W. Montroll, in *Applied Combinatorial Mathematics*, edited by E. F. Beckenbach (Wiley, New York, 1964), Chap. 4.

[18]Note that our analysis does not involve the cancellation of large factors and the use of the Stirling approximation; it is always tricky to perform such cancellations when the end result is a number of the order of magnitude of unity [cf. the discussion in Ref. 11 following Eq. (8)].

[19]The result $\nu = 2$ was also recently obtained with use of position-space renormalization-group arguments (W. Klein, to be published).

[20]B. C. Harms and J. P. Straley, "Directed percolation: Shape of the percolation cone, conductivity exponents, high dimensionality behavior, and the nature of the phase diagram" (to be published).

8. Graph Theory

Reprinted from *Printed in U.S.A.*
JOURNAL OF MATHEMATICAL PHYSICS VOLUME 4, NUMBER 11 NOVEMBER 1963

Cluster Development in an N-Body Problem*

F. Y. Wu†
Washington University, Saint Louis, Missouri
(Received 25 June 1963)

The procedure for generating useful cluster development in problems dealing with the Jastrow wavefunction, as proposed by Wu and Feenberg, is discussed in detail. The existence of the expansion is proved to all orders; also a simple rule is given for computing the expansion coefficients. The result can be considered as a generalization of the Ursell–Mayer formulas.

I. INTRODUCTION

In the statistical and quantum mechanical treatment of an N-body system, one often needs the technique of the Ursell–Mayer-type cluster expansion[1-3] in evaluating physical quantities of interest. In the general case the quantity considered may be a function of N distinct indices. For example, in case of fermions, when a trial wavefunction involving N single-particle orbitals is used to describe the system, all quantities calculated for this system depend on the N orbitals explicitly. This kind of problem was first taken up by Jastrow[4] who introduced a correlated wavefunction containing a factor describing the correlation among particles in addition to a Slater determinant composed of plane-wave orbitals. The mathematics of the cluster expansion in problems involving such wavefunction has been discussed exhaustively by Hartogh and Tolhoek[5] in a series of papers in which general expansion theorems are fully developed. On the other hand, a much simpler formalism which is more general, in the sense that the problem of Jastrow wavefunction appears as a special case of its application, was introduced by Iwamoto and Yamada.[6] Considering the complicated notation and the large quantity of combinatorial algebra involved in the discussion of Hartogh and Tolhoek, an alternative rigorous treatment along the more general lines indicated by Iwamoto and Yamada seems desirable. However it is rather difficult to demonstrate the general character of the cluster expansion by the method of Iwamoto and Yamada; also the existence of the expansion for arbitrary order is not easily proved. An alternative procedure which retains the essential simplicity and directness of the Iwamoto–Yamada approach has been found by Wu and Feenberg and used by them to compute numerical results in a study of the fermion liquid.[7] It is therefore the purpose of this note to present a further detailed discussion of the latter procedure. The existence of the expansion is proved to all orders; also a simple rule is given for computing the expansion coefficients. The result reduces to the Ursell–Mayer formulas if one neglects the difference between all indices and, therefore, our formula is, in a sense, a generalization of the Ursell–Mayer formalism.

II. PROCEDURE OF CLUSTER DEVELOPMENT

As is well-known, the cluster-expansion procedure is usually used in evaluating the logarithm of quantity which behaves like exponential function of N. In applications, the quantity of the order of $e^{\alpha N}$ often has an integral form

$$K_{12\cdots N} = \int W_{12\cdots N}(\tau_1, \tau_2, \cdots, \tau_N)\, d\tau_1\, d\tau_2 \cdots d\tau_N, \tag{1}$$

in which the indices $1, 2, \cdots, N$ refer to, e.g., the N different single-particle orbitals involved in the trial wavefunction. The function W is symmetric in all coordinates $\tau_1, \tau_2, \cdots, \tau_N$, of the N particles and the subscripts $1, 2, \cdots, N$. A systematic way to handle the problem, i.e., to evaluate $\ln K_{12\cdots N}$ is as follows.[5,7]

First one defines the reduced K quantities by specifying certain rules by which one generates the K quantities of one index, two indices, etc. For example if $K_{12\cdots N}$ is given by Eq. (1), one way to define the reduced K's is[5]

$$K_l = \int W_l(\tau_1)\, d\tau_1, \tag{2}$$

---

* The study was supported in part by the Air Force Office of Scientific Research Grant AFOSR–62–412.

† Present Address: Department of Physics, Virginia Polytechnic Institute, Blacksburg, Virginia.

[1] H. D. Ursell, Proc. Cambridge Phil. Soc. **23**, 685 (1927).

[2] J. E. Mayer, J. Chem. Phys. **5**, 67 (1937).

[3] B. Kahn and G. E. Uhlenbeck, Physica **5**, 399 (1938).

[4] R. Jastrow, Phys. Rev. **98**, 1479 (1955).

[5] F. Iwamoto and M. Yamada, Progr. Theoret. Phys. (Kyoto) **17**, 543 (1957); **18**, 345 (1957).

[6] C. D. Hartogh and H. A. Tolhoek, Physics **24**, 721, 875, 896 (1958).

[7] F. Y. Wu and E. Feenberg, Phys. Rev. **128**, 943 (1962).

$$K_{mn} = \int W_{mn}(\tau_1, \tau_2)\, d\tau_1\, d\tau_2, \text{ etc.}, \tag{3}$$

with $l, m, n \cdots$ referring to different indices in the set $1, 2, \cdots, N$. The range of integration is the same in all these integrals. Then the following relations generate a set of X quantities, or a set of cluster integrals:

$$K_l = X_l,$$
$$K_{lm} = X_l X_m + X_{lm},$$
$$K_{lmn} = X_l X_m X_n + X_l X_{mn} + X_m X_{nl} + X_n X_{lm} + X_{lmn}, \cdots, \tag{4}$$

until finally one reaches

$$K_{12\cdots N} = \sum_{\sum_s s m_s = N} \{\underbrace{\cdots X_i \cdots X_j \cdots}_{m_1 \text{ factors}}\} \times \{\underbrace{\cdots X_{kl} \cdots X_{mn} \cdots}_{m_2 \text{ factors}}\} \cdots, \tag{5}$$

in which the summation extends over all possible products subject to conditions (i) no repeated indices; (ii) permutation within a bracket not distinguishable. Each K, hence X, is symmetric with respect to its indices.

As stated earlier, in application we are interested only in the cases

$$\ln K_{12\cdots N} \sim O(N). \tag{6}$$

In fact if the rule that specifies the reduced K's is taken properly, one always has

$$X_{(m \text{ indices})} \sim O(N^{1-m}), \tag{7}$$

and Eq. (6) follows as a result of Eq. (7), as we shall see. Both Eqs. (6) and (7) are checked easily in the special case of Mayer's cluster expansion for a classical imperfect gas for which all X's having the same number of indices are identical.

It is convenient to write

$$K_{12\cdots N} = X_1 X_2 \cdots X_N I, \tag{8}$$

$$x_{lm\cdots n} = X_{lm\cdots n}/X_l X_m \cdots X_n, \tag{9}$$

$$I = 1 + \sum_{m<n} x_{mn} + \sum_{l<m<n} x_{lmn} + \sum_{h<l<m<n} (x_{hl}x_{mn} + x_{hm}x_{nl} + x_{hn}x_{lm} + x_{hlmn}) + \cdots; \tag{10}$$

then

$$\ln K_{12\cdots N} = \sum_{l=1}^{N} \ln X_l + \ln I. \tag{11}$$

The problem is now reduced to the evaluation of $\ln I$. At this stage, Iwamoto and Yamada[5] write down a set of differential equations and solve the problem by constructing successive approximate solutions. The extension of their method to arbitrary order is difficult to carry out. In particular, the differential equations must be written down with great care in order to not overcount many terms.[8] In the following, we shall first outline the alternative procedures suggested by Wu and Freenberg[7] and then extend the discussion to all orders.

Let $I_{lm\cdots n}$ denote the function generated by omitting all terms in Eq. (10) containing indices in the set $l, m, \cdots, n$. Then it is easy to see that in Eq. (10) the coefficient of the factor x_{qm} is exactly I_{qm}, of x_{qmn} is I_{qmn}, etc. It follows that Eq. (10) can be rewritten as

$$I = I_q + \sum_m x_{qm} I_{qm} + \sum_{m<n} x_{qmn} I_{qmn} \cdots + \sum_{m<n<\cdots<l} x_{qmn\cdots l} I_{qmn\cdots l} \cdots + x_{12\cdots N}. \tag{12}$$

Now we write

$$\ln I = G,$$
$$\ln I_q = G_q = G - H_q, \tag{13}$$
$$\ln I_{qm} = G_{qm} = (G - H_q - H_m)[1 + O(1/N)], \text{ etc.},$$

where $G_{lm\cdots n}$ denotes the function generated by omitting all terms in G containing indices in the set $l, m, \cdots, n$. Retaining only the leading terms in the exponentials, we transform Eq. (12) into

$$e^{H_q} = 1 + \sum_m x_{qm} e^{-H_m} + \frac{1}{2!} \sum_{mn} x_{qmn} e^{-(H_m + H_n)} + \cdots. \tag{14}$$

Here the convention that any x with repeated indices vanishes, i.e. $x_{\cdots i \cdots i \cdots} = 0$, is introduced. We emphasize that the substitutions Eq. (13) will be justified by actually determining G.

Both sides of Eq. (14) and also each multiple sum are now independent of N so that a formal expansion of the exponentials is permissible. Taking the logarithms of the left- and right-hand members of Eq. (14), one obtains

$$H_q = W - \tfrac{1}{2}W^2 + \tfrac{1}{3}W^3 - \cdots, \tag{15a}$$

$$W = \sum_m x_{qm}\left[1 - H_m + \frac{1}{2!}H_m^2 - \cdots\right] + \frac{1}{2!}\sum_{mn} x_{qmn} \times \left[1 - (H_m + H_n) + \frac{1}{2!}(H_m + H_n)^2 - \cdots\right] + \cdots. \tag{15b}$$

As the equation stands, H_q can be generated by an

[8] We are indebted to Dr. Iwamoto for this remark.

 $= \sum_{mn} x_{qm} x_{mn}$ (a)

 $= \sum_{mn} x_{mq} x_{qn}$ (b)

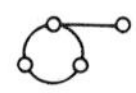 $= \sum_{hlmn} x_{hl}\, x_{lmn}$ (c)

FIG. 1. Meaning of some simple diagrams.

obvious iteration procedure. Indeed, we get for the first few terms

$$H_q = \sum_m x_{qm} + \sum_{mn} \left(\frac{1}{2!} x_{qmn} - \frac{1}{2} x_{mq} x_{qn} - x_{qm} x_{mn}\right) + \cdots, \tag{16}$$

which, in turn, produces

$$G = \frac{1}{2!} \sum_{mn} x_{mn} + \sum_{lmn} \left(\frac{1}{3!} x_{lmn} - \frac{1}{2} x_{lm} x_{mn}\right) + \cdots, \tag{17}$$

as observed from the relation

$$H_q = G - G_q. \tag{18}$$

This observation is best demonstrated if one starts from the expression of G given by Eq. (17) and uses Eq. (18) to compute H_q; then

$$\begin{aligned} H_q &= \frac{1}{2!} \sum_{mn} x_{mn} + \sum_{lmn} \left(\frac{1}{3!} x_{lmn} - \frac{1}{2} x_{lm} x_{mn}\right) + \cdots \\ &\quad - \left[\frac{1}{2!} \sum_{mn \neq q} x_{mn} + \sum_{lmn \neq q} \left(\frac{1}{3!} x_{lmn} \right.\right. \\ &\quad \left.\left. - \frac{1}{2} x_{lm} x_{mn}\right) + \cdots\right] = \frac{1}{2!} \sum_m (x_{qm} + x_{mq}) \\ &\quad + \sum_{mn} \left[\frac{1}{3!} (x_{qmn} + x_{mqn} + x_{mnq}) \right. \\ &\quad \left. - \frac{1}{2} (x_{qm} x_{mn} + x_{mq} x_{qn} + x_{mn} x_{nq})\right] + \cdots, \end{aligned} \tag{16'}$$

which is identically Eq. (16) on taking account of the symmetry of the x's in the indices. Clearly the ratio between the coefficients of $\sum_{mn} x_{mq}\, x_{qn}$ and $\sum_{mn} x_{qm}\, x_{mn}$ in Eq. (16) must be 1 : 2 in order to generate the term $\sum_{lmn} x_{lm}\, x_{mn}$ in G. The necessity of such correlation between the coefficients in the expansion of H_q persists to all orders. The explicit statement of such correlation will be made after we introduce the diagrammatical representation below. Also one easily checks, with the help of Eqs. (7) and (9), that the first three terms of Eq. (17) are of the order $O(N)$. This linear dependence on N also persists to all terms of G. This fact follows from the structure of Eq. (15b) which tells us that the inclusion of each m index $x \sim O(N^{1-m})$ brings in exactly $m - 1$ free summations, and hence a factor $N^{1-m} \cdot N^{m-1} \sim O(1)$.

A diagrammatical representation which will facilitate our discussion is now introduced. In the following, the x's are called elements. An m-index element is imagined as a rigid frame with m holes (or vertices) attached to it: open holes represent the dummy indices of a summation, and a black or solid hole (indexed by q) is not summed over. No meaning is attached to the order of holes in an element (symmetry of an element in all indices). A collection of elements (i.e., a product of x's) is called a diagram. The meaning of some simple diagrams are given by Fig. 1. A diagram is singly connected if all elements of the diagram are connected without forming any closed path; i.e., each hole common to two or more elements is an articulation which, if omitted, would dissociate the diagram into disconnected parts or branches. The degeneracy S of a hole with respect to a diagram is the number of holes equivalent to it because of the symmetry of the elements in the indices. In counting the degeneracy, we make no distinction between the black and white holes. Thus $S = 2, 1$ for the black holes of the diagrams of Figs. 1(a) and 1(b), respectively. Also $S = m$ for the holes of a single m-vertex element. It is also convenient to define the symmetry number[9] T of a diagram as the number of ways one can permute a definite set of distinct numbers attached to the open holes of the diagram without changing the topology of the diagram. For example, $T = 1, 2, 2$ for the diagrams of Figs. 1(a), 1(b), 1(c), respectively. Also $T = m!$ for a single m-vertex element with no black hole, and $T = (m - 1)!$ for the same element when one of the holes is black. It is clear that if in a diagram a black hole is replaced by an open hole, the symmetry number is changed from T to ST where S is the degeneracy of the hole under consideration.

With this diagrammatical notation, it is readily observed from the structure of Eq. (15) that each term in the expression of H_q can be represented by a singly connected diagram with one black hole. In fact one can always find terms in Eq. (16) corresponding to an arbitrary diagram. Therefore we write

$$H_q = \sum h_\alpha \begin{bmatrix}\text{all distinct singly connected} \\ \text{diagrams with one black hole}\end{bmatrix}, \tag{19}$$

with appropriate coefficient h_α for each diagram. The correlation among the coefficients h_α to ensure the existence of G is now clear: consider two diagrams composed of the same collection of elements which differ only by the different positions taken by the black holes [e.g., Figs. 1(a) and 1(b)], the criterion is simply to require the ratio of their h coefficients be [see Eqs. (16) and (16′)]

$$h_a/h_b = S_a/S_b = T_b/T_a, \tag{20}$$

with S_a, T_a; S_b, T_b referring to, respectively, the corresponding degeneracies of the black holes and the symmetry numbers. In the last step of Eq. (20) we have used the relation

$$S_aT_a = \begin{pmatrix}\text{the symmetry number of the diagram if}\\ \text{the black hole is replaced by an open hole}\end{pmatrix} = S_bT_b. \tag{21}$$

Once Eq. (20) is established, one has immediately

$$G = g \begin{pmatrix}\text{all distinct singly connected}\\ \text{diagrams with no black holes}\end{pmatrix}, \tag{22}$$

in which the coefficient g of an arbitrary diagram is given by the following procedure. First one changes (any) one of the holes in this diagram into a black one and looks for the coefficient h_α of this new diagram in Eq. (19). If the degeneracy of the black hole is denoted by S_α, then

$$g = h_\alpha/S_\alpha. \tag{23}$$

Therefore our problem is to show the validity of Eq. (20) and to obtain an explicit expression for h_α. To this end we state the following Lemmas to be proved in the Appendix.

Lemma A. In the expansion of H_a, let $h_1, h_2, \cdots$ denote the coefficients of diagrams in which the black hole is not an articulation, and h_a the coefficient of a diagram in which the black hole is an articulation having $n_1, n_2, \cdots$ identical branches with coefficients $h_1, h_2, \cdots$, respectively. Then

$$h_a = (-1)^{n-1}(n-1)!\prod_l h_l^{n_l}/n_l!, \tag{24}$$

where $n = \sum_l n_l$ is the total number of branches at the articulation.

Lemma B. The coefficient h_α of an arbitrary diagram in Eq. (19) is given by

$$h_\alpha = \frac{1}{T_\alpha}\prod_{\begin{pmatrix}\text{all holes of}\\ \text{the diagram}\end{pmatrix}}(-1)^{n_i-1}(n_i-1)!, \tag{25}$$

in which T_α is the symmetry number of this diagram and n_i the number of elements connected by the ith hole ($n_i = 1$ if the hole is not an articulation).

An immediate consequence of Lemma B is Eq. (20). Therefore we have completed the proof that the expansion of G exists. Finally upon combining

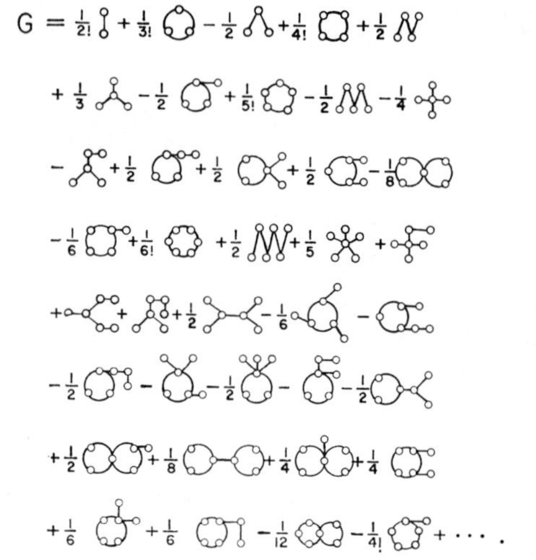

FIG. 2. Diagrammatical equation for G up to terms involving six indices.

Eqs. (23), (25), and (21), the coefficient g belonging to an arbitrary diagram in Eq. (22) has the explicit form

$$g = \left(\frac{1}{\text{symmetry number of the diagram}}\right) \times \prod_{\begin{pmatrix}\text{all holes of}\\ \text{the diagram}\end{pmatrix}}(-1)^{n_i-1}(n_i-1)!. \tag{26}$$

This concludes the derivation of the expansion formula for $G = \ln I$. We give in Fig. 2 the explicit expression for G up to terms involving six indices. As checked easily, this result reduces to the Ursell–Mayer expansion if the cluster integrals involving the same number of indices are all identical.

III. CONCLUSION

We have shown that the logarithm of the quantity I of Eq. (10) can be expanded into a sum of terms represented by singly connected diagrams in the form of Eq. (22), with Eq. (26) furnishing an easy way to determine the expansion coefficients. We must note that terms down by a factor $1/N$ are neglected in the result. Also, the expansion is useful only when it converges fast enough so that the leading terms produce a good approximation. This seems, indeed, to be the cases in application.[5,7]

ACKNOWLEDGMENTS

The author wishes to express his thanks to Professor Eugene Feenberg for suggesting this problem, as well as for a critical reading of the manuscript.

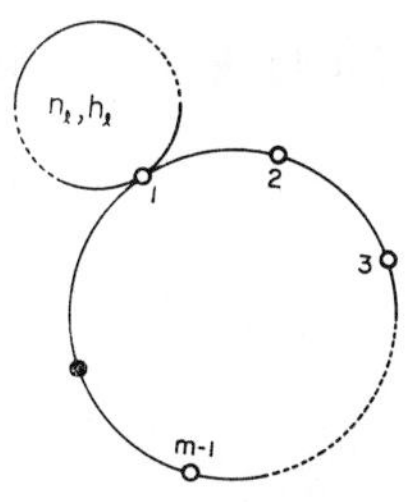

FIG. 3. Schematical diagram of an m-hole element with only one hole attaching branches.

He is also indebted to Dr. Tohru Morita for illuminating comments and for calling attention to the resemblance of this expansion to the fugacity expansion formula for the classical fluids under the presence of external field.[9] The expansion formulas obtained here, as pointed out by Dr. Morita, can also be derived on the basis of the technique developed in references 9 and 10. Helpful discussions with C. T. Chen–Tsai is also appreciated.

APPENDIX

Proof of Lemma A: The diagram with the black hole having n branches clearly comes from the term $(-W)^n/n$ of Eq. (15a). The multiplicity for occurence of such a diagram in the expansion of W^n is $n!/\prod n_l!$, with coefficient $\prod h_l^{n_l}$. Upon combining these with the factor $(-)^n/n$, Eq. (24) is derived. Q.E.D.

Proof of Lemma B: Lemma B will be proved by induction. First, it is obvious from Eq. (15) that Lemma B holds for a single m-vertex element for which

$$h_\alpha = 1/(m-1)! = 1/T_\alpha. \tag{27}$$

Therefore it remains only to show that Lemma B holds for any diagram if it applies to arbitrary diagram composed of fewer elements.

First let us consider diagrams having articulation black holes. Let the branches of the diagram be specified by the set $\{n_l, h_l\}$ as stated in Lemma A; then the symmetry number of the diagram is given by

$$T_\alpha = \prod_l T_l^{n_l} n_l!, \tag{28}$$

where T_l denote the symmetry numbers of the branches. Now, by assumption, Lemma B applies to diagrams composed of fewer elements so that the coefficient h_l of each branch is given by

$$h_l = \frac{1}{T_l} \prod_{\left[\substack{\text{all holes of}\\ \text{the branch}}\right]} (-1)^{n_i-1}(n_i-1)!. \tag{29}$$

The substitution of Eqs. (29) and (28) into Eq. (24) now yields

$$h_\alpha = \frac{1}{T_\alpha} \prod_{\left[\substack{\text{all holes of}\\ \text{the diagram}}\right]} (-1)^{n_i-1}(n_i-1)!.$$

Q.E.D.

Next consider the diagram in which the black hole is not an articulation. In the most general case, the black hole sits on an m-hole element, with p of the m holes attaching branches. In order to illustrate the essential points of the proof, we shall consider the case of $p = 1$ only. The proof for the general case can be constructed in a completely similar fashion.[11]

Consider the diagram shown schematically in Fig. 3 in which the black hole sits on an m-hole element with one open hole attaching n branches denoted again by the set $\{n_l, h_l, T_l\}$. The symmetry number of the diagram is now given by

$$T_\alpha = (m-2)! \prod_l T_l^{n_l} n_l!, \tag{30}$$

with the symmetry numbers T_l of the individual branches related to h_l through Eq. (29). The coefficient h_α of this diagram comes from the term W of Eq. (15a). More specifically it comes from the following terms:

$$\frac{1}{(m-1)!} \sum_{12\cdots m-1} x_{q12\cdots m-1} \sum_{i=1}^{m-1} \sum_{k=1}^{n} \frac{(-H_i)^k}{k!}$$
$$= \frac{1}{(m-2)!} \sum_{12\cdots m-1} x_{q12\cdots m-1} \sum_{k=1}^{n} \frac{(-H_1)^k}{k!}. \tag{31}$$

Comparison of Eq. (31) with Fig. 3 indicates that we need to collect terms represented by the set of branches $\{n_l, h_l, T_l\}$ in the expansion of H_1^k. The first term H_1 contains only one such term. The second term H_1^2 contains more than one contribution. In fact, for each distinct way that the n branches are divided into two groups, there corresponds a contribution with one factor of H_1 contributing to one group of branches, and another H_1 to the other group. The multiplicity for occurence of such terms in the expansion of H_1^2 is 1 if the two groups are identical, and 2 otherwise. In general for the term H_1^k we consider all the distinct ways that the n branches are divided into k groups of which $\beta_1, \beta_2, \cdots, \beta_r$ are identical. Let the h coefficients of these groups be denoted by $\bar{h}_1, \bar{h}_2, \cdots, \bar{h}_r$, respectively. Then for each distinct way that the

[9] T. Morita and K. Hiroike, Progr. Theoret. Phys. (Kyoto) **25**, 537 (1961).

[10] T. Morita, Progr. Theoret. Phys. (Kyoto) **21**, 501 (1959).

[11] F. Y. Wu, Dissertation, Washington University (1963).

n branches are divided, there corresponds, in the expansion of H_1^k, a contribution to the coefficient h_α with multiplicity $k!/\beta_1!\beta_2! \cdots \beta_r!$. It follows then from Eq. (31)

$$h_\alpha = \frac{1}{(m-2)!} \sum_{k=1}^{n} \frac{(-1)^k}{k!} \times \sum_{(\beta_1\beta_2\cdots\beta_r)} \frac{k!}{\beta_1!\,\beta_2!\cdots\beta_r!} \bar{h}_1^{\beta_1} \cdots \bar{h}_r^{\beta_r}, \tag{32}$$

in which $(\beta_1\beta_2 \cdots \beta_r)$ denotes the summation taken under the restriction $\beta_1 + \beta_2 + \cdots + \beta_r = k$. Now the coefficients $\bar{h}_1, \bar{h}_2, \cdots \bar{h}_r$ can be obtained by Lemma A. However it proves useful at this point to note that the factor $(n-1)!/\prod n_l!$ appearing in Eq. (24) is just the number of distinct ways to perform cyclic permutations on the group of n branches (among which $n_1, n_2, \cdots$ are identical). Using this interpretation of Lemma A for the expressions of $\bar{h}_1, \bar{h}_2, \cdots, \bar{h}_r$, we get

$$\bar{h}_1^{\beta_1}\bar{h}_2^{\beta_2} \cdots \bar{h}_r^{\beta_r} = (-1)^{n-k} \begin{Bmatrix}\text{product of the number of}\\ \text{distinct cyclic paramutations}\\ \text{of the branches in each group}\end{Bmatrix} \prod_l h_l^{n_l}. \tag{33}$$

Substituting Eq. (33) into Eq. (32), we have

$$h_\alpha = \frac{1}{(m-2)!}(-1)^n M \prod_l h_l^{n_l}, \tag{34}$$

where

$$M = \sum_{k=1}^{n} \sum_{(\beta_1\beta_2\cdots\beta_r)} \frac{1}{\beta_1!\,\beta_2!\cdots\beta_r!} \begin{Bmatrix}\text{product of the num-}\\ \text{ber of distinct cyclic}\\ \text{permutations of the}\\ \text{branches in each}\\ \text{group}\end{Bmatrix}$$

$$= \sum_{k=1}^{n} \sum_{(\beta_1\beta_2\cdots\beta_r)} \begin{Bmatrix}\text{the number of distinct ways one can}\\ \text{permute the } n \text{ branches under the}\\ \text{particular grouping by performing}\\ \text{cyclic permutations within each}\\ \text{group while the groups are un-}\\ \text{numbered}\end{Bmatrix}$$

$$= \begin{Bmatrix}\text{the number of distinct ways one}\\ \text{can permute the } n \text{ branches by first}\\ \text{dividing into unnumbered groups and}\\ \text{then performing cyclic permutations}\\ \text{within each group}\end{Bmatrix}.$$

It is well-known that each permutation of a collection of objects can be analyzed into groups of cyclic permutations in an unique way. Therefore also

$$M = \begin{Bmatrix}\text{the number of distinct permutations of}\\ \text{the } n \text{ branches}\end{Bmatrix} = n!/\prod n_l!. \tag{35}$$

The substitution of Eqs. (29) and (35) into Eq. (34) and the introduction of Eq. (30) now yields Eq. (25).

Q.E.D.

Reprinted from JOURNAL OF STATISTICAL PHYSICS

Vol. 52, Nos. 1/2, July 1988
Printed in Belgium

Potts Model and Graph Theory

F. Y. Wu[1]

Received December 15, 1987

Elementary exposition is given of some recent developments in studies of graph-theoretic aspects of the Potts model. Topics discussed include graphical expansions of the Potts partition function and correlation functions and their relationships with the chromatic, dichromatic, and flow polynomials occurring in graph theory. It is also shown that the Potts model realization of these classic graph-theoretic problems provides alternate and direct proofs of properties established heretofore only in the context of formal graph theory.

KEY WORDS: Potts model; partition function; correlation function; graph theory; chromatic function; flow polynomial.

1. INTRODUCTION

Studies of the Potts model[1] are often facilitated by the use of graphical terms and graphical analyses. The connection of the Potts model with graph theory was first formulated by Kasteleyn and Fortuin,[2,3] who treated the bond percolation, resistor network, spanning trees, and other problems of graph-theoretic nature as a Potts model. Conversely, graph-theoretic considerations have led to formulations of the Potts model leading to results in statistical mechanics otherwise difficult to see. (For reviews of the Potts model see Wu.[4,5]) More recently, Essam and Tsallis[6] uncovered the connection of the Potts model with the flow polynomial in graph theory,[7,8] and this consideration has since been extended to multisite correlation functions[9] and their applications[10] and the associated duality relations.[11] However, many of these findings are presented in formal mathematical language and often rely on theorems established in graph

[1] Department of Physics, University of Washington, Seattle, Washington 98195. Permanent address for reprint requests: Department of Physics, Northeastern University, Boston, Massachusetts 02115.

theory. As a result, the significance of these developments does not appear to have been generally appreciated.

In this paper we present a self-contained, albeit elementary, exposition of these recent developments. While most of the results presented here are not new, our derivations are less formal and in many instances more direct than those previously given, thus shedding new light to the role played by these classic graph-theoretic problems in statistical physics.

Definitions useful to our discussions are given in Section 2. In Section 3 we describe high-temperature expansions of the Potts partition function and their associated graphical representations, and show that they lead to the chromatic, dichromatic, and flow polynomials in graph theory. In Section 4 we discuss properties of the flow polynomial and show that the Potts model formulation leads to alternate and simple proofs of these properties. In Section 5 graphical considerations are extended to correlation functions.

2. DEFINITIONS

Consider a standard Potts model on a graph G of N sites and E edges (we assume that G does not contain single-edge loops). The spin at the ith site can take on q distinct values $\sigma_i = 1, 2, \ldots, q$, and the Hamiltonian is

$$-\beta\mathscr{H} = K \sum_{e \in G} \delta_{\mathrm{Kr}}(\sigma_i, \sigma_j) \tag{1}$$

where the summation extends to all edges in G. It should be noted that, while we have assumed the same interaction K along all edges, our discussions and results can be extended to include edge-dependent interactions. We chose not to consider this generalization, however, for the sake of retaining clarity of discussions. A concise summary of results in the general case can be found in ref. 10.

It is often convenient to regard the spin σ_i as being represented by a unit vector $\hat{\mathbf{s}}_i$ pointing in one of the q symmetric directions of a hypertetrahedron in $q-1$ dimensions. The connection to (1) is then made by using

$$\delta_{\mathrm{Kr}}(\sigma_i, \sigma_j) = \frac{1}{q}\left[1 + (q-1)\,\hat{\mathbf{s}}_i \cdot \hat{\mathbf{s}}_j\right] \tag{2}$$

The partition function is

$$Z_G = \mathrm{Tr}\, e^{-\beta\mathscr{H}} \tag{3}$$

where the spin sum $\sum_{\sigma_i = 1, 2, \ldots, q}$ has been denoted by taking the trace. When $K = -\infty$, all pairs of neighbors connected by edges must be in different states. Then in this limit the partition function becomes the chromatic function,

$$Z_G|_{e^K=0} = P(q, G) \tag{4}$$

which gives the number of q-colorings of G, i.e., the number of ways that the N vertices of G can be colored with q colors such that two vertices connected by an edge always bear different colors.

The m-spin correlation function for spins at sites $1, 2,..., m$ is the probability that vectors $\hat{\mathbf{s}}_1,..., \hat{\mathbf{s}}_m$ point in the same direction. This probability is given by

$$\begin{aligned} \Gamma_{12\cdots m} &\equiv Z_G^{-1}\,\mathrm{Tr}[s_{1\alpha}s_{2\alpha}\cdots s_{m\alpha}\exp(-\beta\mathscr{H})] \\ &\equiv \langle s_{1\alpha}s_{2\alpha}\cdots s_{m\alpha}\rangle^T \end{aligned} \tag{5}$$

where the superscript T denotes the thermal average dictated by taking the trace. Here $s_{i\alpha} \equiv \hat{\mathbf{s}}_i \cdot \hat{\mathbf{e}}_\alpha$, and $\hat{\mathbf{e}}_\alpha$ is a unit vector pointing in the direction α of the hypertetrahedron. In particular, the one-spin correlation function

$$\Gamma_1 = \langle s_{i\alpha}\rangle^T = \frac{1}{q-1}\langle q\delta(\sigma_i, \alpha) - 1\rangle^T \tag{6}$$

is the order parameter of the ferromagnetic Potts model, whose numerical value lies between 0 and 1.

More generally, one defines a *partitioned* m-spin correlation function[9] as the probability that spins at vertices $1, 2,..., m$ are partitioned into b ($\leqslant q$) blocks such that (i) all spins within a block are in the same spin state, and (ii) the spin states of the b blocks are all distinct. For each block partition $P(m)$ of the m integers $1, 2,..., m$, the corresponding partitioned correlation function is

$$\Gamma_{P(m)} = \left\langle \prod_{B\in P} \prod_{i\in B}{}' s_{i\alpha_B} \right\rangle^T \tag{7}$$

where B is a block index and the prime of the second product indicates the restriction (ii), namely, spin states of the b blocks are all different. Clearly, (7) becomes (5) when there is only one block, so that $b = 1$ and $P(m) = (12\cdots m)$.

Graphical representations that we shall encounter are derived on the basis of the Mayer expansion,[12] which converts a product of edge factors into a summation of products over subgraphs. The basic identity, which we shall use repeatedly, is

$$\prod_{e\in G}(1 + h_{ij}) = \sum_{G'\subseteq G}\prod_{e\in G'} h_{ij} \tag{8}$$

Here, the edge factor connecting vertices i and j is written as $1+h_{ij}$, and the summation of the rhs of (8) is taken over all subgraphs $G' \subseteq G$ covering the same vertex set.

It is convenient to introduce a "percolation" average as follows: Consider subgraphs $G' \subseteq G$, which we regard as representing percolation configurations with bond occupation probability p. Then, the percolation average of any quantity $X(g)$ considered as a function of the subgraph g is defined to be

$$\langle X \rangle_p \equiv \sum_{g \subseteq G} X(g)\, p^{b(g)}(1-p)^{E-b(g)} \tag{9}$$

where $b(g)$ is the number of edges in g.

We may rewrite (9) as a power series in p. Expanding $(1-p)^{E-b(g)}$ in (9), we obtain

$$\langle X \rangle_p = \sum_{g \subseteq G} X(g)\, p^{b(g)} \sum_{g'} (-p)^{b(g')} \tag{10}$$

where g' is a set of edges not in g. The union of the two edge sets g and g' constitutes a subgraph of G. Call this subgraph $G' \subseteq G$, so that

$$b(G') = b(g) + b(g') \tag{11}$$

and regard g as a subgraph of G'. It folows that (10) can be rewritten as

$$\begin{aligned} \langle X \rangle_p &= \sum_{G' \subseteq G} p^{b(G')} \sum_{g \subseteq G'} (-1)^{b(G')-b(g)} X(g) \\ &= \sum_{G' \subseteq G} p^{b(G')}\, Q(X, G') \end{aligned} \tag{12}$$

where the expansion coefficient in the p series is

$$Q(X, G) \equiv \sum_{g \subseteq G} (-1)^{b(G)-b(g)} X(g) \tag{13}$$

Relations (12) and (13) hold for any $X(G')$, which may describe, e.g., connectivity properties of G'.

We can obtain an inverse of (13) if $X(G)$ is expressible as the trace over a product of edge factors, i.e.,

$$X(G) = \operatorname{Tr} \prod_{e \in G} h_{ij} \tag{14}$$

In this case it is easily verified[2] that

$$Q(X, G) = \mathrm{Tr} \prod_{e \in G} (h_{ij} - 1) \tag{15}$$

Writing in (14) $h_{ij} = 1 + (h_{ij} - 1)$ and again expanding the resulting expression using (8), we obtain the following inverse of (13):

$$X(G) = \sum_{G' \subseteq G} Q(X, G') \tag{16}$$

This is one example of the Möbius inversion,[13] which generally inverts summations over partially ordered sets[14] and follows as a consequence of the principle of inclusion and exclusion.[15]

3. HIGH-TEMPERATURE EXPANSIONS OF THE PARTITION FUNCTION

It is most natural to obtain expansions of the Potts partition function using the Mayer expansion. This is done by writing the edge Boltzmann factor in (3) as a sum of two terms and expanding using (8). This leads to different expansions when the Boltzmann factor is split differently.

The most often used expansion,[2,3] which forms the basis of relating the Potts model to the bond percolation problem, involves rewriting the partition function (3) as

$$Z_G = \mathrm{Tr} \prod_{e \in G} [1 + (e^K - 1)\, \delta_{\mathrm{Kr}}(\sigma_i, \sigma_j)] \tag{17}$$

Using (8) with $h_{ij} = (e^K - 1)\, \delta_{ij}$, one finds for the partition function (17) the simple form after taking the trace:

$$Z_G \equiv Z(q, e^K - 1) = \sum_{G' \subseteq G} (e^K - 1)^{b(G')} q^{n(G')} \tag{18}$$

where $n(G')$ is the number of components, including isolated points, in G'. This is a high-temperature expansion, since $e^K - 1 \to 0$ at high temperatures. The function $Z(x, y)$ is the dichromatic polynomial in graph theory [cf. ref. 8, (IX.1.12)] and the function $x^N Z(x, y/x)$ is known as the Whitney rank polynomial.[16,17] It is also known that the dichromatic polynomial, hence the Potts partition function, generates spanning trees[3] and forests[18] by taking appropriate $q = 0$ limits. See ref. 4 for a description of these connections.

[2] Expand (15) as in (8) and compare with (13). Note the extra minus signs.

One immediate consequence of (18) is, upon using (4),

$$P(q, G) = \sum_{G' \subseteq G} (-1)^{b(G')} q^{n(G')} \tag{19}$$

This is the Birkhoff[(19)] formula for the chromatic function, establishing the fact that $P(q, G)$ is a polynomial of q.

The quantity $(-1)^{b(G)} Z_G$ given by (17) is of the form of (15) with $h_{ij} = -(e^K - 1)\delta_{ij}$. It follows from (14) and (16) that the inverse of (18) is

$$(1 - e^K)^{b(G)} q^{n(G)} = \sum_{G' \subseteq G} (-1)^{b(G')} Z_{G'} \tag{20}$$

Equation (20) is a peculiar sum rule for the Potts partition function,[3] which does not appear to have been previously noted. In the special case of $K = -\infty$, (20) becomes, after introducing (4), the following inversion for the chromatic polynomial:[(16)]

$$q^{n(G)} = \sum_{G' \subseteq G} (-1)^{b(G')} P(q, G') \tag{21}$$

The partition function Z_G can now be written as a power series in p. Using (9) and (12) with $X = q^n$, $Q = (-1)^b P$, we obtain

$$\begin{aligned} Z_G &= e^{EK} \langle q^n \rangle_p \\ &= e^{EK} \sum_{G' \subseteq G} P(q, G')(-p)^{b(G')} \end{aligned} \tag{22}$$

where we have introduced (19) and

$$p = 1 - e^{-K} \tag{23}$$

Generally, any splitting of the edge Boltzmann factor (into a sum of two terms) other than that in (17) will lead to a different expansion equivalent to a resummation of the p series. A particularly useful splitting is to write

$$1 + (e^K - 1)\delta_{ij} = A[(1 - t) + qt\delta_{ij}] = A(1 + qtf_{ij}) \tag{24}$$

where

$$A = \frac{1}{q}(e^K + q - 1) \tag{25}$$

$$t = \frac{1 - e^{-K}}{1 + (q-1)e^{-K}} \tag{26}$$

$$f_{ij} = \delta_{ij} - q^{-1} = \frac{q-1}{q} \hat{\mathbf{s}}_i \cdot \hat{\mathbf{s}}_j \tag{27}$$

[3] If G is a connected graph such as a lattice, then $b(G) = E$, $n(G) = 1$.

For two-dimensional systems the variable t is the corresponding Boltzmann factor e^{-K} in the dual space, but more generally t is the thermal transmissivity arising in renormalization group treatments[20] of the Potts model. The substitution of the first identity in (24) into (17) now leads to the expression

$$\begin{aligned} Z_G &= A^E \langle q^{b+n} \rangle_t \\ &= q^N A^E \langle q^c \rangle_t \end{aligned} \tag{28}$$

Here, use has been made of the Euler relation

$$b(G') + n(G') = N + c(G') \tag{29}$$

where $c(G')$ is the number of independent circuits in G'. The advantage of introducing the t variable becomes apparent when we write Z_G as a power series in t. This is done by identifying q^c and F as, respectively, X and Q in (12) and (13). This leads to

$$Z_G = q^N A^E \sum_{G' \subseteq G} F(q, G')\, t^{b(G')} \tag{30}$$

where

$$F(q, G) \equiv \sum_{g \subseteq G} (-1)^{b(G) - b(g)} q^{c(g)} \tag{31}$$

The expansion coefficient $F(q, G)$ in the t series is precisely the flow polynomial occurring in graph theory.[7,8]

An alternate expression for the flow polynomial can now be formulated using the Potts model realization. Substituting the second identity of (24) into (17) and comparing the resulting expression with (30), we obtain the expression[4]

$$F(q, G') = q^{b(G') - N} \operatorname{Tr} \prod_{e \in G'} f_{ij} \tag{32}$$

valid for any $G' \subseteq G$. Now, (32) is of the form of (15) with $h_{ij} = 1 + q f_{ij}$; it follows that we can use (16) and (14) to obtain an inverse of (31). This leads to

$$\begin{aligned} \sum_{G' \subseteq G} F(q, G') &= q^{-N} \operatorname{Tr} \prod_{e \in G} (1 + q f_{ij}) \\ &= q^{-N} \operatorname{Tr} \prod_{e \in G} (q \delta_{ij}) \\ &= q^{-N + b(G) + n(G)} \\ &= q^{c(G)} \end{aligned} \tag{33}$$

[4] Direct evaluation of (32) using the first identity in (27) for f_{ij} again leads to (31). (Cf. ref. 6.)

If G is planar, then consider its dual D and the associated subgraph $D' \subseteq D$ complementing g; we then have $b(g)+b(D')=b(G)$, $c(g)=n(D')-1$ and, after introducing (19), (31) becomes

$$F(q, G)=q^{-1}P(q, D) \tag{34}$$

A general high-temperature expansion, which encompassess both the p and t expansions described above, is obtained by writing, in place of (24),

$$1+(e^K-1)\delta_{ij}=A_\mu[(1-t_\mu)+\mu t_\mu \delta_{ij}]=A_\mu[1+\mu t_\mu f_{ij}(\mu)] \tag{35}$$

where

$$A_\mu=(e^K+\mu-1)/\mu \tag{36}$$

$$t_\mu=(1-e^{-K})/[1+(\mu-1)e^{-K}] \tag{37}$$

$$f_{ij}(\mu)=\delta_{ij}-\mu^{-1} \tag{38}$$

and μ is a parameter which can be chosen at our disposal. By taking $\mu=1$ and $\mu=q$, e.g., (35) generates the p and t expansions, respectively.

In analogy to (28) and (30), we now have

$$\begin{aligned} Z_G &= \mu^N A_\mu^E \langle \mu^c (q/\mu)^n \rangle_{t_\mu} \\ &= \mu^N A_\mu^E \sum_{G' \subseteq G} F(\mu, q, G')\, t_\mu^{b(G')} \end{aligned} \tag{39}$$

where

$$F(\mu, q, G) \equiv \sum_{G' \subseteq G} (-1)^{b(G)-b(G')} \mu^{c(G')} \left(\frac{q}{\mu}\right)^{n(G')} \tag{40}$$

which becomes $(-1)^{b(G)} P(q, G)$ and $F(q, G)$, respectively, for $\mu=1$ and $\mu=q$. One can also write, as in (32),

$$F(\mu, q, G')=\mu^{b(G')-N}\,\mathrm{Tr} \prod_{e \in G'} f_{ij}(\mu) \tag{41}$$

from which one obtains the inverse of (40) by following (33):

$$\mu^{c(G)} \left(\frac{q}{\mu}\right)^{n(G)} = \sum_{G' \subseteq G} F(\mu, q, G') \tag{42}$$

Similarly, using arguments leading to (34), one derives for planar G the following duality relation:

$$\mu F(\mu, q, G)=(-1)^E \frac{q}{\mu} F\left(\frac{q}{\mu}, q, D\right) \tag{43}$$

The p expansion (22) of the Potts partition function is well known from its connection with the bond percolation.[2,3] Consideration of the t expansion (28) and (30) also has a long history. Nagle[21] considered the special case of $t = -(q-1)^{-1}$ in a numerical evaluation of the chromatic function for a lattice. Domb[22] analyzed the case of general t and explicitly evaluated what is equivalent to (31) for small star graphs. The expression (34) for planar graphs was observed by Wu,[4,23] who also expressed $F(q, G)$ in the form of a recursion relation of the Potts partition function. However, it was only recently that Essam and Tsallis[6] explicitly obtained (31) and pinpointed its connection with the flow polynomial in graph theory. The formulation in terms of the variable t_μ and the associated generalized coefficients $F(\mu, q, G)$ is due to de Magalhães and Essam.[9]

4. FLOW POLYNOMIAL

The flow polynomial (31) is known to possess a number of graph-theoretic properties. While many of these properties follow intuitively from the concept of "flow" of the polynomial (see description below), their derivations have so far appeared only as theorems in formal graph theory.[7,8] The formulation of the flow polynomial as expansion coefficients in the t expansion of the Potts partition function, particularly the representation (32) for $F(q, G)$, now provides alternate proof of these properties, which can be more easily visualized.

We first state some of the more important properties.[5]

1. $F(q, G) = 0$ if G has a vertex of degree one or an isthmus (an articulation edge).
2. $F(q, G)$ is topologically invariant.
3. If G has components (which may have articulation vertices in common), then $F(q, G)$ is equal to the product of the flow polynomials of individual components.
4. Contraction–deletion rule:

$$F(q, G) = F(q, G^\gamma) - F(q, G^\delta) \tag{44}$$

where G^γ is G with one edge contracted and G^δ is G with the same edge deleted.

[5] See ref. 8 for a complete list and formal proofs of all properties of the flow polynomial.

The above (and other) properties of $F(q, G)$ can be estalished by using the representation (32) for the flow polynomial and the readily verified identities

$$\mathrm{Tr}_j\, f_{ij} = 0 \tag{45}$$

$$\mathrm{Tr}_j\, f_{ij} f_{jk} = f_{ik} \tag{46}$$

where $f_{ij} = \delta_{ij} - q^{-1}$.

Consider first property 1. If G has a vertex of degree one, then, upon using (45), $F(q, G)$ vanishes identically by tracing over the spin variable of this spin. If G has an isthmus, we use the spin symmetry, which states that the trace over a cluster of spins except one is independent of the spin state of the untraced spin. Thus, we trace over all spins located in an isthmus except the one at the articulation point. Due to the spin symmetry, this gives rise to a common factor, irrespective of the spin state of the remaining spin. The remaining spin can therefore be treated as a vertex of degree one, and its trace now gives $F(q, G) = 0$.

To prove property 2, we see from (46) that a sequence of edges can be combined into a single one without affecting $F(q, G)$. Furthermore, the exponent $b(G') - N = c(G') - n(G')$ in (32) is also unchanged when a sequence of edges is combined. This establishes the fact that $F(q, G)$ is topologically invariant.

Property 3 is self-evident when the components are disjoint. When there are articulation vertices, decompose G into disjoint clusters by separating at the articulation points. The extra q factors thus introduced in (32) from the trace of the extra vertices (created in the decomposition process) cancel exactly with the q factors introduced by the increase of N, the number of vertices. Thus, the flow polynomial (32) is given by the product of those of its components as if they were disjoint. This establishes property 3.

Property 4 can be established as follows: Let the contracted and deleted edge be $(1, 2)$ and rewrite (32) as

$$F(q, G) = q^{b(G) - N}\, \mathrm{Tr}\left[f_{12} \prod_{e \neq (1,2)} f_{ij} \right] \tag{47}$$

Now both the contraction and deletion of a single edge decrease the total number of edges by 1, implying

$$b(G) - N = b(G^{\gamma}) - (N-1) = b(G^{\delta}) + 1 - N \tag{48}$$

Using (48) and by observation, we see[6]

$$F(q, G^{\gamma}) = F(q, G)|_{f_{12}=\delta_{12}} \tag{49}$$

$$F(q, G^{\delta}) = q^{-1} F(q, G)|_{f_{12}=1} \tag{50}$$

Property 4 is now established by substituting $f_{12} = \delta_{12} - q^{-1}$ into (47) and splitting (47) into two terms as in f_{12}.

For completeness, we now describe the graph-theoretic meaning of $F(q, G)$ in the context of a flow. Orient all adges of G. A given orientation of G defines an $N \times N$ incidence matrix whose elements are

$$\begin{aligned} D_{ij} = -D_{ji} &= 1 \quad \text{if edge } ij \text{ is directed from } i \text{ to } j \\ &= 0 \quad \text{if no edge connects } i \text{ and } j \end{aligned} \tag{51}$$

A flow on G is specified by assigning to each edge a number Φ_e satisfying the flow condition

$$\sum_j D_{ij} \Phi_{ij} = 0 \tag{52}$$

at all vertices i. If one visualizes G as a network with electric currents Φ_e flowing along its edges, then (52) is the mere statement of the first Kirchhoff law, that the net outgoing current is zero at all nodes. Clearly, the reversal of the orientation of one particular edge corresponds to the negation of the associated Φ_e, hence does not create a new flow configuration.

A mod-q flow is a flow specified by integral $\Phi_e = 0, 1, \ldots \pmod q$. A *proper* mod-$q$ flow is one for which *none* of the Φ_e is zero. The counting of the number of proper mod-q flows is a highly nontrivial problem. Tutte[(7,8)] showed that this number is given precisely by $F(q, G)$. The correctness of this counting can also be verified by applying the following argument:[(14)] First, (33) states that the total number of mod-q flows is $q^{c(G)}$.[7] To obtain the number of *proper* mod-q flows, we must subtract the number of flows with some zero Φ_e. We do this by applying the principle of inclusion and exclusion,[(15)] i.e., by deleting those flows with exactly one branch carrying zero current, including those with exactly two branches carrying zero current, etc. This leads to (31). See ref. 6 for a table of flow polynomials for all graphs with five or fewer independent circuits.

[6] In writing down (49) we use the fact that the contraction of an edge reduces the number of verices by 1.

[7] This fact is implicit in the Kirchhoff law, when one assigns loop currents to a network to describe its current configuration.

5. SPIN CORRELATION FUNCTIONS

In this section we show that correlation functions of the Potts model can also be represented by graph expansions.[9] To keep our presentation simple and concise, we confine it to essential results. See ref. 9 for more general discussions.

Consider first the m-spin correlation function $\Gamma_{12\cdots m}$ defined by (5), which is the probability that spins at sites 1, 2,..., m all point in the same direction α. By spin symmetry, this correlation is independent of α. It is then convenient to sum over α, which can be regarded as replacing $\hat{\mathbf{e}}_\alpha$ by a spin $\hat{\mathbf{s}}_0$. This leads to the consideration of a graph G^+ derived from G by adding an extra vertex numbered 0, the ghost vertex, connected to vertices 1, 2,..., m. Thus, upon introducing (24) and (27), (5) becomes

$$\begin{aligned}\Gamma_{12\cdots m} &= (qZ_G)^{-1} A^E \left(\frac{q}{q-1}\right)^m \mathrm{Tr}^+ \left[f_{10} \cdots f_{m0} \prod_{e \subseteq G} (1 + qtf_{ij}) \right] \\ &= (qZ_G)^{-1} A^E \left(\frac{q}{q-1}\right)^m \sum_{G' \subseteq G} (qt)^{b(G')} \mathrm{Tr}^+ \left[f_{10} \cdots f_{m0} \prod_{e \subseteq G'} f_{ij} \right]\end{aligned} \tag{53}$$

where we have used (8), and Tr^+ denotes that the trace is being taken over G^+.

Let

$$g' = G' \cup g_1 \cdots \cup g_m \tag{54}$$

where g_i is the edge linking vertices i and 0. Compared to G', g' has m more edges and one more vertex. Then, by (32), the flow polynomial on g' is

$$F(q, g') = q^{b(G') + m - (N+1)} \mathrm{Tr}^+ \left[f_{10} \cdots f_{m0} \prod_{e \subseteq G'} f_{ij} \right] \tag{55}$$

The substitution of (55) and (30) for Z_G into (53) now leads to the following expression for the m-spin correlation function:

$$\Gamma_{12 \cdot\cdot m} = \left(\frac{1}{q-1}\right)^m \frac{\sum_{G' \subseteq G} F(q, g') t^{b(G')}}{\sum_{G' \subseteq G} F(q, G') t^{b(G')}} \tag{56}$$

To obtain an expression similar to (56) for the partitioned correlation function $\Gamma_{P(m)}$ defined by (7), consider first a partitioned correlation without the restriction (ii) described in Section 2. That is, a partitioned correlation for which spin states in different blocks are not necessarily distinct. This "modified" partitioned correlation is given by

$$\Gamma^0_{P(m)} = \left\langle \prod_{B \in P} \prod_{i \in B} s_{i\alpha_B} \right\rangle^T \tag{57}$$

which is (7) with the prime over the second product removed. Following the same reasoning as in the above, it is straighforward to verify that $\Gamma^0_{P(m)}$ is also given by the rhs of (56), provided that a ghost vertex is introduced for each block of $P(m)$, and that the ith vertex is connected to its block ghost spin via the edge g_i.

It is not difficult to see that $\Gamma^0_{P(m)}$ can be written as a linear combination of $\Gamma_{P(m)}$. Write (57) as

$$\Gamma^0_{P(m)} = (Z_G)^{-1} \operatorname{Tr}\left[\left(\prod_{B \in P} \prod_{i \in B} s_{i\alpha_B}\right) e^{-\beta\mathscr{H}}\right] \tag{58}$$

and recall that the trace in (58) is taken without the restriction (ii), so that spin states of different blocks may or may not be equal. We can therefore expand this trace into a summation of traces for which spin states of different blocks are always distinct. A moment's reflection shows that this leads to the expression

$$\Gamma^0_{P(m)} = \sum_{P' \geqslant P} c(q, b')\, \Gamma_{P'(m)} \tag{59}$$

where the summation is taken over all block partitions $P'(m)$ of the m integers 1, 2,..., m such that every block of P is contained in a block of P', b' is the number of blocks in P', and $c(q, b')$ is the number of q colorings of the b' blocks such that they all bear different colors. This number is given by

$$c(q, b') \equiv q(q-1)\cdots(q-b'+1) \tag{60}$$

Now (59) is a sum over a partially ordered set of the partition $P(m)$, and therefore has a Möbius inverse. The inverse is[14]

$$\Gamma_{P(m)} = \frac{1}{c(q, b)} \sum_{P' \geqslant P} \mu(P, P')\, \Gamma^0_{P(m)} \tag{61}$$

with the Möbius function

$$\mu(P, P') = (-1)^{b-b'}(2!)^{a_3}(3!)^{a_4}\cdots[(m-1)!]^{a_m} \tag{62}$$

Here, a_i specifies the class of P', i.e., $a_i = 0, 1,..., b'$ is the number of blocks in P' (which are in the same block of P) having i elements.

ACKNOWLEDGMENTS

I am indebted to J. W. Essam for calling my attention to the importance of the t expansion. I have also benefited from discussions with A. C. N. de Magalhães, who explained to me the essence of her work with J. W. Essam. I would also like to thank M. Schick and D. Thouless for the hospitality kindly extended to me at the University of Washington, where this research was carried out. Work has been supported in part by National Science Foundation grants DMR-8219254, DMR-8613598, and DMR-8702596.

NOTE ADDED IN PROOF

After the completion of this work, Dr. de Magalhães has kindly pointed out to me that the relation (34) is an old result first obtained in W. T. Tutte, *Proc. Camb. Phil. Soc.* **43**:26 (1947).

REFERENCES

1. R. B. Potts, *Proc. Camb. Phil. Soc.* **48**:106 (1952).
2. P. W. Kasteleyn and C. M. Fortuin, *J. Phys. Soc. Jpn.* **26**(Suppl.):11 (1969).
3. C. M. Fortuin and P. W. Kasteleyn, *Physica* **57**:536 (1972).
4. F. Y. Wu, *Rev. Mod. Phys.* **54**:235 (1982).
5. F. Y. Wu, *J. Appl. Phys.* **55**:2421 (1984).
6. J. W. Essam and C. Tsallis, *J. Phys. A* **19**:409 (1986).
7. W. T. Tutte, *Can. J. Math.* **6**:80 (1954).
8. W. T. Tutte, *Encyclopedia of Mathematics and Its Aplications*, Vol. 21, *Graph Theory* (Addison-Wesley, Reading, Massachusetts, 1984), Chapter 9.
9. A. C. N. de Magalhães and J. W. Essam, *J. Phys. A* **19**:1655 (1986).
10. A. C. N. de Magalhães and J. W. Essam, *J. Phys. A* **21**:473 (1988).
11. A. C. N. de Magalhães and F. Y. Wu, unpublished.
12. J. E. Mayer and M. G. Mayer, *Statistical Mechanics* (McGraw-Hill, New York, 1940).
13. A. F. Möbius, *J. R. A. Math.* **9**:105 (1832).
14. G.-C. Rota, *J. Wahrsch.* **2**:340 (1966).
15. H. Whitney, *Bull. Am. Math. Soc.* **38**:572 (1932).
16. H. Whitney, *Ann. Math.* **38**:688 (1932).
17. N. Biggs, *Interaction Models* (Cambridge University Press, Cambridge, 1977).
18. M. J. Stephens, *Phys. Lett. A* **56**:149 (1976).
19. G. D. Birkhoff, *Ann. Math.* **14**:42 (1912).
20. C. Tsallis and S. V. F. Levy, *Phys. Rev. Lett.* **47**:950 (1981).
21. J. F. Nagle, *J. Comb. Theory B* **10**:42 (1971).
22. C. Domb, *J. Phys. A* **7**:1335 (1974).
23. F. Y. Wu, in *Studies in Foundations and Combinatorics*, G.-C. Rota, ed. (1978).

Printed by Catherine Press, Ltd., Tempelhof 41, B-8000 Brugge, Belgium

Ann. Inst. Fourier, Grenoble
49, 3 (1999), 1103–1114

ON THE ROOTED TUTTE POLYNOMIAL (*)

by F.Y. WU, C. KING and W.T. LU

1. The Tutte polynomial.

Consider a finite graph G with vertex set V and edge set E. A *spanning subgraph* $G'(S) \subseteq G$ is a subgraph of G containing all members of V and an edge set $S \subseteq E$. Let C be a set of q distinct colors. A *q-coloring* of G is a coloring of the vertices in V such that two vertices connected by an edge bear different colors. It is well-known that the number of q-colorings of G is given by the chromatic polynomial (see [1])

$$P(G;q) = \sum_{S \subseteq E} q^{p(S)}(-1)^{|S|}, \tag{1}$$

where $p(S)$ is the number of components in the spanning subgraph $G'(S)$. Alternately, we can regard (1) as generating colorings of components of spanning subgraphs of G with q colors with an edge weight -1.

As an extension of the chromatic polynomial, Tutte [2], [3], [4] introduced what is now known as the *Tutte polynomial*

$$Q(G;t,v) = \sum_{S \subseteq E} t^{p(S)} v^{|S|-|V|+p(S)}. \tag{2}$$

Indeed, one has the relation

$$P(G;q) = (-1)^{|V|} Q(G;-q,-1). \tag{3}$$

(*) This work is supported in part by the National Science Foundation grants DMR-9614170 (FYW and WTL) and DMS-9705779 (CK).
Keywords: Graph colorings – Rooted Tutte polynomial – Planar partition – Duality relation.
Math. classification: 05C30 – 05C15.

In view of (3), it is useful to write (2) as

$$Q(G;t,v) = v^{-|V|} \sum_{S \subseteq E} (vt)^{p(S)} v^{|S|}, \tag{4}$$

so that for $vt = q =$ positive integers, the Tutte polynomial (4) generates colorings of components of spanning subgraphs of G with q colors and edge weights v, instead of $v = -1$.

For planar G with dual graph G_D, it is well-known that the Tutte polynomial possesses the duality relation

$$v\, Q(G;t,v) = t\, Q(G_D;v,t), \tag{5}$$

a relation first observed by Whitney [5].

2. The rooted Tutte polynomial.

We extend the definition (4) to a rooted Tutte polynomial.

A vertex is rooted, or is a root, if it is colored with a prescribed (fixed) color. A graph is rooted if it contains rooted vertices.

Let R denote a set of n roots located at vertices $\{r_1, r_2, \ldots, r_n\}$. A *color configuration* is a map $x : R \mapsto C$, and as a convenient shorthand we write $x(r_i) = x_i$ for $i = 1, 2, \ldots, n$. A component of a spanning subgraph is *exterior* if it contains one or more roots, and is *interior* otherwise. An exterior component is *proper* if all roots in the component are of the same color. A spanning subgraph $G'(S)$ is proper if all its exterior components are proper. An edge set $S_x \subseteq E$ is proper if the spanning subgraph $G'(S_x)$ it generates is proper.

For a prescribed color configuration $\{x_1, x_2, \ldots, x_n\}$ of the n roots, we introduce in analogy to (4) the *rooted* Tutte polynomial[(1)]

$$Q_{x_1 x_2 \ldots x_n}(G;t,v) = v^{-|V|} \sum_{S_x \subseteq E} (vt)^{p_{\text{in}}(S_x)} v^{|S_x|}, \tag{6}$$

where the summation is taken over all proper edge sets S_x, and $p_{\text{in}}(S_x)$ is the number of interior components of $G'(S_x)$. Thus, as in (3), we have for

(1) Strictly speaking, it is the expression $v^{|V|} Q_{x_1 x_2 \ldots x_n}(G;t,v)$ which is a polynomial in v and t.

positive integral q the relation

$$(7)\quad (-1)^{|V|}Q_{x_1x_2\dots x_n}(G;-q,-1) = \text{the number of } q\text{-colorings of } G \text{ with color configuration } \{x_1,x_2,\dots,x_n\}.$$

Clearly, the expression (6) depends on how the n roots are partitioned into subsets of different colors, and the actual colors do not enter the picture.

The coloring configuration $\{x_1,x_2,\dots,x_n\}$ induces a partition X of R into blocks (subsets) such that all roots in one block are of one color, and colors of different blocks are different. Namely, two elements $r_i, r_j \in R$ belong to the same block of X if and only if they have the same prescribed color $x_i = x_j$.

Consider now the summation in (6). Let $G'(S)$ be any (not necessarily proper) spanning subgraph of G. The connected components of $G'(S)$ induce a partition on the set of vertices V of G. We get hence also a partition $\pi(S)$ on the set of rooted vertices R by restricting this partition to R. Clearly, the spanning subgraph $G'(S_x)$ is proper if and only if the partition $\pi(S_x)$ is a refinement of the partition X. It follow that we can rewrite (6) as

$$(8)\qquad Q_X(G;t,v) = \sum_{X' \preceq X} F_{X'}(G;t,v),$$

where

$$(9)\qquad F_{X'}(G;t,v) \equiv v^{-|V|} \sum_{\substack{S_x \subseteq E\\ \pi(S_x)=X'}} (vt)^{p_{\text{in}}(S_x)} v^{|S_x|}.$$

Here, we have abbreviated $Q_{x_1x_2\dots x_n}$ by Q_X, which is permitted since the actual colors do not enter the picture at this point. Also it is understood that G is now a rooted graph, with root set R.

The expression (8) assumes the form of a transformation of a partially ordered set. Its inverse is given by the Möbius inversion

$$(10)\qquad F_X(G;t,v) = \sum_{X'} \mu(X',X) Q_{X'}(G;t,v),$$

where (see [6])

$$(11)\qquad \mu(X',X) = \begin{cases} (-1)^{|X'|-|X|} \prod_{\text{blocks} \in X} (n_b(X')-1)!, & \text{if } X' \preceq X,\\ 0, & \text{otherwise,} \end{cases}$$

$n_b(X')$ being the number of blocks of X' that are contained in the block b of X. Note that for $n=1$ we have $|X| = |X'| = 1$, $p_{\text{in}}(S_x) = p(S_x) - 1$, and all edge sets $S \subseteq E$ are proper. Hence we have

$$F_X(G;t,v) = Q_X(G;t,v) = (vt)^{-1}Q(G;t,v), \quad n = 1. \tag{12}$$

This completes the definition and general description of the rooted Tutte polynomial for any graph G.

3. Planar graphs.

From here on we consider G being planar with the n roots residing around a single face of G. Without the loss of generality, we can choose the face to be the infinite face and order the roots in the sequence $\{r_1, r_2, \ldots, r_n, r_1\}$ as shown in Fig. 1.

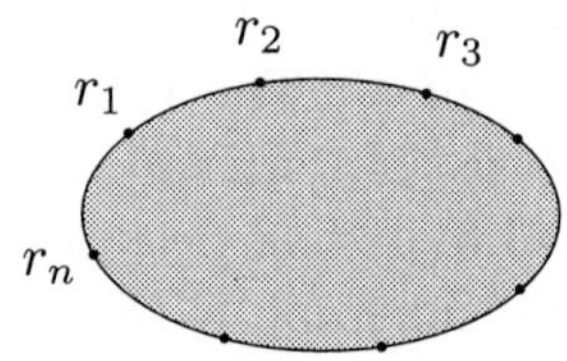

Figure 1. A planar graph G with n roots. The graph is denoted by the shaded region and the n roots by the black circles.

A partition X of the n roots is *non-planar* if two roots of one block separate two roots of another block in the cyclic sequence. Otherwise X is *planar*. For a given n, there are b_n partitions, where (see [9])

$$b_n = \sum_{m_\nu=0}^{\infty} \left[n!/\prod_{\nu=1}^{\infty}(\nu!)^{m_\nu} m_\nu!\right], \quad \sum_{\nu=1}^{\infty} \nu m_\nu = n, \tag{13}$$

and of the b_n partitions

$$c_n = \frac{(2n)!}{n!\,(n+1)!} \tag{14}$$

are planar (see [7], [8]). We shall adopt the convention of writing

$$X = \{ij, k\ell \ldots, \ldots\}$$

for colors $\{x_i = x_j,\ x_k = x_\ell = \cdots, \ldots\}$, with $\{ij\}, \{k\ell \ldots\}, \ldots$ each in order (see [8]). For example, two partitions for $n = 5$ are

$$\begin{aligned} X_1 &= \{123, 4, 5\}, \quad |X_1| = 3, \quad \text{planar}, \\ X_2 &= \{24, 351\}, \quad\ \ |X_2| = 2, \quad \text{non-planar}. \end{aligned} \tag{15}$$

Now if G is planar and X' is non-planar then by definition the summand in (9) is empty and one has $F_{X'}(G;t,v) = 0$. Thus we have

PROPOSITION 1. — *For planar G*

$$F_X(G;t,v) = 0, \quad \text{if } X \text{ is non-planar.} \tag{16}$$

This proposition was first established in [9] for the Potts model correlation function (see section 6) by considering its graphical expansion similar to the consideration given in the above.

As a consequence of Proposition 1 and the use of (10), we now have:

COROLLARY 1. — *Rooted Tutte polynomials associated with non-planar partitions can be written as linear combinations of the rooted Tutte polynomials associated with (refined) planar partitions.*

Corollary 1 leads to the sum-rule identities reported in [9] for the Potts correlation function. In the case of $n = 4$, for example, the identity

$$F_{\{13,24\}}(G;t,v) = 0$$

leads to the sum rule (see [9])

$$Q_{\{13,24\}}(G;t,v) = Q_{\{13,2,4\}}(G;t,v) + Q_{\{1,3,24\}}(G;t,v) - Q_{\{1,2,3,4\}}(G;t,v). \tag{17}$$

From here on we shall restrict our considerations to rooted Tutte polynomials associated with the c_n planar partitions only.

4. The graph G^*.

The rooted Tutte polynomial (6) possesses a duality relation for planar graphs, which relates the rooted Tutte polynomial on a graph G to that of a related graph G^*. Here we define G^*. Starting from a planar G, place an extra vertex f in the infinite face and connect it to each root of G by an edge. This gives a new graph G'', which has one more vertex than G and n additional edges. The dual graph of G'' is also planar, and it has a face F containing the extra vertex f. Now remove the n edges on the boundary of F, and the resulting graph is G^*.

It is readily seen that the graph G^* has

$$|V^*| = |V_D| + n - 1 \tag{18}$$

vertices where $|V_D|$ is the number of vertices of G_D, the dual of G, and there is a one-one correspondence between the edges of G and G^*. We denote the set of n vertices $\{r_1^*, r_2^*, \ldots, r_n^*\}$ of G^* surrounding the face F by R^*, with r_i^* residing between the two edges $\langle f, r_{i-1}\rangle$ and $\langle f, r_i\rangle$ of G'', where $r_0 = r_n$.

An example of a G and the related G^* for $n = 4$ is shown in Fig. 2.

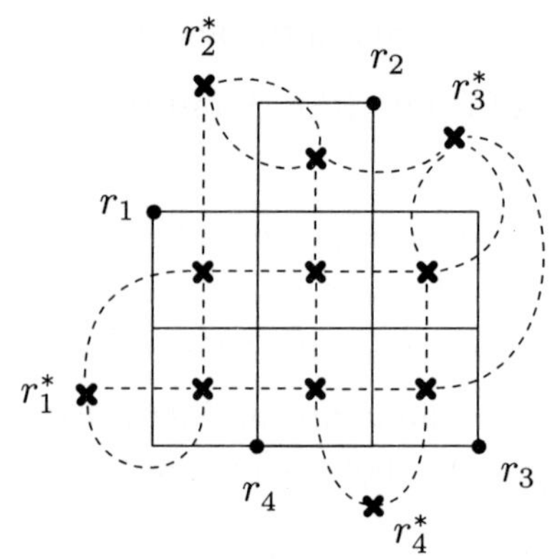

Figure 2. A graph G (solid lines) and the related graph G^* (broken lines) for $n = 4$. Black circles denote roots of G and crosses denote vertices of G^*.

Clearly, the relation of G to G^* is reciprocal, namely, we have

$$(G^*)^* = G.$$

Now each planar partition X of R induces a partition X^* of the set R^* (see [10]). In order to define X^*, for each block b of X we choose a point in the infinite face of G, and connect all roots r_i in b to this point by drawing new edges. Because X is planar, the points for the blocks can be chosen so that the edges of different blocks do not cross, and the resulting extended graph is still planar. So this process divides the infinite face into regions. The induced partition X^* is then described by the condition that all roots of R^* in one region are regarded as belonging to one block of X^*. Alternately, for another way of defining X^*, let P_{ij}, $1 \le i < j \le n$, be a partition of R into two blocks, the sets $\{r_i, \ldots, r_{j-1}\}$ and $\{r_j, \ldots, r_{i-1}\}$, where all numbers are modulo n. Then, the partition X^* induced by X is defined by the condition that the two roots r_i^* and r_j^* belong to the same block in X^* if and only if X is a refinement of P_{ij}.

If X induces X^*, we write

$$X \longrightarrow X^*.$$

Clearly, X^* is planar, and we have

$$|X| + |X^*| = n + 1. \tag{19}$$

For the planar partition X_1 in (15), for example, we have

$$X_1 = \{123, 4, 5\} \longrightarrow X_1^* = \{2, 3, 451\}, \quad |X_1| = |X_1^*| = 3. \tag{20}$$

However, as a result of our labelling convention, the color configuration of the partition $(X^*)^*$ further induced by X^* is a cyclic shift of that of X, namely,

$$\{x_1, x_2, \ldots, x_n\} \longrightarrow X^* \longrightarrow \{x_n, x_1, \ldots, x_{n-1}\}. \tag{21}$$

In the example above, for instance, we have

$$\{123, 4, 5\} \longrightarrow \{2, 3, 451\} \longrightarrow \{234, 5, 1\}. \tag{22}$$

Finally, there is a one-one correspondence between the edge sets E of G and E^* of G^*, an edge set $S \subseteq E$ defines a "complement" edge set $S^* \subseteq E^*$ by the condition that an edge is included in S^* if and only if its corresponding edge is not included in S. Clearly, we have

$$(S^*)^* = S.$$

5. The duality relation.

The rooted Tutte polynomial arises in statistical physics as the correlation function of the Potts model (see next section). In a recent paper [10] we have established a duality relation for the Potts correlation function for planar G. However, the proof of the duality relation given in [10] is cumbersome and not easily deciphered in graphical terms. Here we re-state the results as two propositions in the context of the rooted Tutte polynomial, and present direct graph-theoretical proofs of the propositions.

PROPOSITION 2. — *For planar G and G^* and the associated planar partitions $X \to X^*$, we have*

$$v^{|X|} F_X(G; t, v) = t^{|X^*|} F_{X^*}(G^*; v, t). \tag{23}$$

Proof. — Let S_x be a proper edge set on G. We have the Euler relation

$$|S_x| + |S_x^*| = |V| + |V_D| - 2 \tag{24}$$

and, after eliminating n and $|V_D|$ using (18), (19) and (24), the identity

$$|S_x| + |X| - |V| = |S_x^*| + |X^*| - |V^*|, \tag{25}$$

which holds for any proper edge set S_x. Note that we have also the fact

$$\pi(S_x) = X \quad \text{if and only if} \quad \pi(S_x^*) = X^*. \tag{26}$$

Let $c(S_x^*)$ the number of independent circuits in the spanning subgraph $G'(S_x^*)$. Then we have

$$p(S_x) = c(S_x^*) + |X|. \tag{27}$$

Also, starting from the $|V^*|$ isolated vertices on G^*, one constructs $G'(S_x^*)$ by drawing edges of S_x^* on G^* one at a time. Since each edge reduces the number of components by one except when the adding of an edge completes an independent circuit, one has also

$$p(S_x^*) = |V^*| - |S_x^*| + c(S_x^*). \tag{28}$$

Eliminating $c(S_x^*)$ using (27) and (28) and making use of the relations

$$\begin{cases} p(S_x) = p_{\text{in}}(S_x) + |X|, \\ p(S_x^*) = p_{\text{in}}(S_x^*) + |X^*|, \end{cases} \tag{29}$$

one obtains

$$p_{\text{in}}(S_x^*) = p_{\text{in}}(S_x) + |V^*| - |S_x^*| - |X^*|. \tag{30}$$

Proposition 2 now follows from the substitution of (30) into the right-hand side of (23) where, explicitly,

$$F_{X^*}(G^*; v, t) = t^{-|V^*|} \sum_{\substack{S_x^* \subseteq E^* \\ \pi(S_x^*) = X^*}} (vt)^{p_{\text{in}}(S_x^*)} t^{|S_x^*|}, \tag{31}$$

and the use of the identities (25) and (26). This completes the proof of Proposition 2.

Proposition 2 was first conjectured in [9] and established later in [10] in the context of Potts correlation functions (see next section) without the explicit reference to the polynomial form (9).

Remark. — For $n = 1$, the duality relation (23) for the rooted Tutte polynomial becomes the duality relation (5) for the Tutte polynomial. This is a consequence of (12).

PROPOSITION 3.

1) *The rooted Tutte polynomials associated with the c_n planar partitions for G and G^* are related by the duality transformation*

$$Q_X(G; t, v) = \sum_Y \boldsymbol{T}_n(X, Y) Q_Y(G^*; v, t), \tag{32}$$

where $\boldsymbol{T}_n$ is a $c_n \times c_n$ matrix with elements

$$\boldsymbol{T}_n(X, Y) = t^{n+1} \sum_{X' \preceq X} (vt)^{-|X'|} \mu(Y, Y'), \qquad X' \to Y'. \tag{33}$$

2) *The matrix $\boldsymbol{T}_n$ satisfies the identity*

$$[\boldsymbol{T}_n]^2(X, X') = \delta(x_1, x_2')\delta(x_2, x_3') \cdots \delta(x_n, x_1'). \tag{34}$$

Proof. — The transformation (32) follows by combining (8) and (10) with Proposition 2, and its uniqueness is ensured by the uniqueness of the Möbius inversion. The property (34) is a consequence of (21). □

Proposition 3 was first given in [10] in the context of the Potts correlation function (see next section). Explicit expression of $\boldsymbol{T}_n$ for $n = 2, 3, 4$ can be found in [10] and [11].

6. The Potts and the random cluster models.

It is well-known in statistical physics that the Tutte polynomial gives rise to the partition function of the Potts model (see [12]). In view of the prominent role played by the Potts model in many fields in physics, it is useful to review this equivalence and the further equivalence of the rooted Tutte polynomial with the Potts correlation function.

The q-state Potts model [13] is a spin model defined on a graph G. The spin model consists of $|V|$ spins placed at the vertices of G with each spin taking on q different states and interacting with spins connected by edges. Without going into details of the physics [12] which lead to the Potts model, it suffices for our purposes to define the Potts partition function

$$Z(G;q,v) \equiv \sum_{S \subseteq E} q^{p(S)} v^{|S|}, \tag{35}$$

the n-point partial partition function

$$Z_X(G;q,v) \equiv \sum_{S_x \subseteq E} q^{p_{\rm in}(S_x)} v^{|S_x|}, \tag{36}$$

and the n-point correlation function

$$P_n(G;x_1,x_2,\ldots,x_n) = P_n(G;X) \equiv Z_X(G;q,v)/Z(G;q,v), \tag{37}$$

where again, in analogy to notation in Sections 1 and 2, we have denoted the color configuration $\{x_1,x_2,\ldots,x_n\}$ by the associated partition X. More generally, for any real or complex q, the partition function (35) defines the random cluster model of Fortuin and Kasteleyn [14], which coincides with the Potts model for integral q.

Relating this to the Tutte polynomial, we now have

$$\begin{cases} Z(G;q,v) = v^{|V|}Q(G;t,v), \\ Z_X(G;q,v) = v^{|V|}Q_X(G;t,v), \\ P_n(G;X) = Q_X(G;t,v)/Q(G;t,v), \end{cases} \tag{38}$$

for $q = vt$. The duality relation (5) for the Tutte polynomial then implies the following duality relation for the Potts partition function [10], [13], [15]

$$v^{1-|V|}Z(G;q,v) = (v^*)^{1-|V_D|}Z(G_D;q,v^*), \tag{39}$$

where

$$vv^* = q. \tag{40}$$

One further defines the dual correlation function

$$P_n^*(G^*;X^*) \equiv q\, Z_{X^*}(G^*;q,v^*)/Z(G_D;q,v^*), \tag{41}$$

and also the functions A_X and B_{X^*} by

$$P_n(G;X) = \sum_{X' \preceq X} A_{X'}(G;q,v) \tag{42}$$

and

$$P_n^*(G^*;X^*) = \sum_{X^{*\prime} \preceq X^*} B_{X^{*\prime}}(G^*;q,v^*). \tag{43}$$

Then, Proposition 2 leads to the relation

$$A_X(G;q,v) = q^{-|X|} B_{X^*}(G^*;q,v^*), \quad X \to X^*. \tag{44}$$

which is the main result of [10].

7. Summary and discussions.

We have introduced the rooted Tutte polynomial (6) as a two-variable polynomial associated with a rooted graph and deduced a number of pertinent results.

Our first result is that the rooted Tutte polynomial assumes the form (8) of a partially order set for which the inverse can be uniquely determined. For planar graphs and all roots residing surrounding a single face, we showed that (Proposition 1) the inverse function vanishes for non-planar partitions of the roots. We further showed that the inverse function satisfies the duality relation (23) (Proposition 2) which, in turn, leads to the duality (32) for the rooted Tutte polynomial (Proposition 3). We also reviewed the connection of the Tutte and rooted Tutte polynomials with the Potts model in statistical physics.

Finally, we remark that results reported here have previously been obtained in [9] and [10] in the context of the Potts correlation function. Here, the results are reformulated as properties of the rooted Tutte polynomial and thereby permitting graph-theoretical proofs.

Noted added. — The polynomial $Q(G;x,y)$ is now commonly referred to as the *dichromatic polynomial*, and the dichromate

$$\chi(G;x,y) = (x-1)^{-1} Q(G;x-1,y-1)$$

is now commonly known as the *Tutte polynomial.* We are indebted to Professor W.T. Tutte for this remark.

Acknowledgments. — We are grateful to the referee for providing an independent proof of Proposition 1 and numerous suggested improvements on an earlier version of this paper.

BIBLIOGRAPHY

[1] G.D. BIRKHOFF, A determinant formula for the number of ways of coloring of a map, Ann. Math., 14 (1912), 42–46.

[2] W.T. TUTTE, A contribution to the theory of chromatic polynomials, Can. J. Math., 6 (1954), 80–91.

[3] W.T. TUTTE, On dichromatic polynomials, J. Comb. Theory, 2 (1967), 301–320.

[4] W.T. TUTTE, Graph Theory, in Encyclopedia of Mathematics and Its Applications, Vol. 21, Addison–Wesley, Reading, Massachusetts, 1984, Chap. 9.

[5] H. WHITNEY, The coloring of graphs, Ann. Math., 33 (1932), 688–718.

[6] See, for example, L.J. VAN LINT and R.M. WILSON, A course in combinatorics, Cambridge University Press, Cambridge, 1992, p. 301.

[7] H.N.V. TEMPERLEY and E.H. LIEB, Relations between the percolation and colouring problem and other graph–theoretical problems associated with regular planar lattice: some exact results for the percolation problem, Proc. Royal Soc. London A, 322 (1971), 251–280.

[8] W.T. TUTTE, The matrix of chromatic joins, J. Comb. Theory B, 57 (1993), 269–288.

[9] F.Y. WU and H.Y. HUANG, Sum rule identities and the duality relation for the Potts n–point boundary correlation function, Phys. Rev. Lett., 79 (1997), 4954–4957.

[10] W.T. LU and F.Y. WU, On the duality relation for correlation functions of the Potts model, J. Phys. A: Math. Gen., 31 (1998), 2823–2836.

[11] F.Y. WU, Duality relations for Potts correlation functions, Phys. Letters A, 228 (1997), 43–47.

[12] See, for example, F.Y. WU, The Potts Model, Rev. Mod. Phys., 54 (1982), 235–268.

[13] R.B. POTTS, Some generalized order–disorder transformations, Proc. Camb. Philos. Soc., 48 (1954), 106–109.

[14] C.M. FORTUIN and P.W. KASTELEYN, On the random–cluster model I. Introduction and relation to other models, Physica, 57 (1972), 536–564.

[15] F.Y. WU and Y.K. WANG, Duality transformation in a many–component spin model, J. Math. Phys., 17 (1976), 439–440.

F.Y. WU & W.T. LU,
Northeastern University
Department of Physics
Boston, Massachusetts 02115 (USA).
fywu@neu.edu
wtlu@albert.physics.neu.edu
&
C. KING,
Northeastern University
Department of Mathematics
Boston, Massachusetts 02115 (USA).
king@neu.edu

J. Phys. A: Math. Gen. **15** (1982) L395–L398. Printed in Great Britain

LETTER TO THE EDITOR

Random graphs and network communication

F Y Wu

Department of Physics, Northeastern University, Boston, Massachusetts 02115, USA

Received 26 April 1982

Abstract. The problem of random graphs, which arises in the analysis of network reliability in communication theory, is considered here as a bond percolation. A closed form expression is obtained for the cluster-size generating function from which the mean cluster size as well as the percolation probability are derived. In a network of $N\to\infty$ stations in which the communication between any two stations is intact with a probability α/N, it is found that for $\alpha\leqslant 1$ the network breaks into clusters of average size of $(1-\alpha)^{-1}$ stations and $\alpha/(1-\alpha)$ links, while for $\alpha>1$ there is a non-zero percolation probability.

A random graph is a collection of N vertices (sites) which are governed by a probability mechanism such that each pair of vertices is joined by an edge with a prescribed probability p, independent of the presence or absence of any other edges. If we regard the vertices as stations and the edges as communication links between the stations, then the random graphs simulate a communication network (see e.g. Welsh 1977). Writing $p=\alpha/N$ and α small, we expect the network to break down, even in the limit of $N\to\infty$, and decompose into isolated clusters of finite sizes which are not linked to one another via communication. But for α greater than a certain critical value α_c, a non-zero probability arises that a given station is linked with an infinite number of other stations.

The random graphs so defined also describe a bond percolation process (Welsh 1977), if the edges are regarded as occupying bonds. This consideration provides the possibility of an alternative approach to the problem of network reliability, a possibility which appears not to have been adequately examined. In this Letter we take up this consideration. We shall first formulate the percolation problem as a Potts model (Kasteleyn and Fortuin 1969), which is soluble in the limit of $N\to\infty$. Relevant information regarding the network reliability and random graphs is then derived from this solution.

We begin by writing down the Potts Hamiltonian relevant to the percolation problem. Since the bond percolation is long ranged in the sense that any two vertices can be connected, we consider a system of q-state Potts spins (for a review on the Potts model see Wu (1982)) having a similar long-range interaction. Thus, we consider the Hamiltonian $\mathscr{H}$ given by

$$\frac{-\mathscr{H}}{kT}=\frac{K}{N}\sum_{\langle ij\rangle}\delta(\sigma_i,\sigma_j)+\frac{M}{N}\sum_{\langle ij\rangle}\delta(\sigma_i,\sigma_j)\delta(\sigma_i,0)+L\sum_i\delta(\sigma_i,0), \tag{1}$$

where, in addition to the two-spin interactions K/N between *all* pairs $\langle ij\rangle$, there are also external fields M/N and L (cf equation (1.18) of Wu (1982)). These external fields are needed to generate quantities relevant to the cluster size, and will eventually be set to zero. In (1), $\sigma_i = 0, 1, \ldots, q-1$ refers to the spin state at the ith site, $i = 1, 2, \ldots, N$, and $\delta(\alpha, \beta)$ is the Kronecker delta function.

Consider the random graphs now regarded as configurations of the bond percolation. Following Kasteleyn and Fortuin (1969), we can establish that this bond percolation is generated by the Potts model (1). More specifically (cf equation (4.9) of Wu (1982)), let $Z(q; K, M, L)$ be the partition function of (1) and write

$$A(q; K, M, L) = (1/N) \ln Z(q; K, M, L); \tag{2}$$

then the cluster-size generating function, $G(L, L_1)$ for the percolation process is given by

$$G(L, L_1) \equiv \frac{1}{N} \left\langle \sum_c \exp(-Ls_c - L_1 b_c) \right\rangle = \left(\frac{\partial}{\partial q} A(q; K, M, L) \right)_{q=1} \tag{3}$$

with

$$e^{L_1} = (e^{(K+M)/N} - 1)/(e^{K/N} - 1). \tag{4}$$

Here the average $\langle \ \rangle$ is taken over all bond percolation configurations with the bond occupation probability

$$\alpha/N = 1 - e^{-(K+M)/N}, \tag{5}$$

the summation Σ_c in (3) is taken over all clusters of the percolation configuration and s_c, b_c denote, respectively, the numbers of sites and bonds in a cluster.

The cluster-size generating function $G(L, L_1)$ generates the various quantities of interest in the percolation problem. In particular, the percolation probability $P(\alpha)$ and the mean cluster size $S(\alpha)$ (of the finite cluster containing a given vertex) are given by (see e.g. Wu 1978)

$$P(\alpha) = 1 + M_{10}(\alpha), \tag{6}$$

$$\begin{aligned} S(\alpha) &= M_{20}(\alpha) \qquad \text{by site content,} \\ &= M_{11}(\alpha) \qquad \text{by bond content,} \end{aligned} \tag{7}$$

where

$$M_{rs}(\alpha) = \left(\frac{\partial^{r+s}}{\partial L^r \partial L_1^s} G(L, L_1) \right)_{L=L_1=0+}. \tag{8}$$

We next proceed to compute $G(L, L_1)$ by solving the Potts model (1). For N large, (4) and (5) give

$$K = \alpha\, e^{-L_1}, \qquad M = \alpha(1 - e^{-L_1}). \tag{9}$$

Also, in the limit of $N \to \infty$, Hamiltonian (1) is most conveniently dealt with by using a variational approach (Wu 1982).

Let x_i denote the fraction of spins that are in the spin state $i = 0, 1, \ldots, q-1$. We look for a solution with a long-range order in, say, the $i = 0$ spin state. To this end we write

$$x_0 = (1/q)[1 + (q-1)s], \qquad x_i = (1/q)(1-s), \qquad i \neq 0, \tag{10}$$

where $0 \leq s \leq 1$ is the order parameter. We then obtain from (1) and (2):

$$A(q;K,M,L)=\max_{0\leq s\leq 1}\left[\frac{K}{2q}[1+(q-1)s^2]+\frac{Mu^2}{2q^2}+\frac{Lu}{q}+\ln q-u\ln u\right.$$
$$\left.-\left(\frac{q-1}{q}\right)(1-s)\ln(1-s)\right] \tag{11}$$

where $u=1+(q-1)s$. Let s_0 be the value of the order parameter which maximises (11). Straightforward algebra leads to

$$A(q;K,M,L)=\frac{K}{2q}[(1-s_0)^2-qs_0^2]-\frac{M}{2}\left(s_0+\frac{1}{q}(1-s_0)\right)^2-\ln\left(\frac{1-s_0}{q}\right) \tag{12}$$

where s_0 is determined from

$$Ks_0+L+\frac{M}{q}[1+(q-1)s_0]=\ln\left(\frac{1+(q-1)s_0}{1-s_0}\right). \tag{13}$$

Substitution of (12) into (3) after using (9) now yields

$$G(L,L_1)=1-s_0-\tfrac{1}{2}\alpha\,\mathrm{e}^{-L_1}(1-s_0)^2 \tag{14}$$

where s_0 is determined from (13) at $q=1$, which now becomes, after introducing (9),

$$\alpha-\alpha\,\mathrm{e}^{-L_1}(1-s_0)+L+\ln(1-s_0)=0. \tag{15}$$

Finally, we obtain from (6), (7), and (14) and (15) the results

$$P(\alpha)=s_0, \tag{16}$$

$$\begin{aligned} S(\alpha)&=(1-s_0)/(1-\alpha+\alpha s_0) && \text{by site content,}\\ &=\alpha/(1-\alpha+\alpha s_0) && \text{by bond content,}\end{aligned} \tag{17}$$

where s_0 is determined from

$$\alpha s_0+\ln(1-s_0)=0. \tag{18}$$

Equations (16)–(18) are our main results. For $\alpha \leq \alpha_c = 1$, (18) has only one solution, namely, $s_0=0$, so that $P(\alpha)=0$ identically; the mean cluster size is then $(1-\alpha)^{-1}$ by site content and $\alpha/(1-\alpha)$ by bond content. For $\alpha>\alpha_c$, another solution $s_0>0$ arises which gives rise to a larger A (as seen from (11) in the limit of $q\to 1+$). Therefore, we should take this solution, and this leads to a non-zero percolation probability $P(\alpha)=s_0$. Near the threshold $\alpha_c=1$, (16), (17) and (18) give

$$P(\alpha)=2(\alpha-\alpha_c), \qquad S(\alpha)=|\alpha-\alpha_c|^{-1}, \tag{19}$$

leading to the classical percolation exponents $\beta=\gamma=\gamma'=1$. It is not surprising that we should obtain these 'mean field' exponents, since the expression (1) describes precisely a mean field Hamiltonian (Kac 1968) for the Potts model.

Erdös and Rényi (1960) have studied an equivalent graph problem in which the number of connecting edges is fixed at $\alpha N/2$, the average number of edges in the present problem. Using a purely probabilistic approach, they showed that the cluster structures of the random graphs exhibit a drastic change at $\alpha=1$. Therefore, our results are consistent with their findings. After the completion of this work I learned that A Coniglio has also considered this percolation problem using a different variational approach.

To summarise, we have considered the problem of network reliability using a Potts model approach. In a network of $N \to \infty$ stations, where the communication between any two stations is intact with a probability α/N, we have found that, for $\alpha \leqslant 1$, the network breaks into clusters of stations which have an average size of $(1-\alpha)^{-1}$ stations and $\alpha/(1-\alpha)$ communication lines. When $\alpha > 1$, there is a non-zero probability $P(\alpha)$, obtained from $\alpha P(\alpha)+\ln[1-P(\alpha)]=0$, that a given station maintains communication with an infinite number of other stations.

I wish to thank H E Stanley for the kind hospitality at the Center for Polymer Physics, Boston University, where this work was initiated. This research has been supported in part by grants from the National Science Foundation and the US Army Research Office.

References

Erdös P and Rényi A 1960 *Publ. Math. Inst. Hung. Acad. Sci.* **5** 17–60
Kac M 1968 in *Statistical Physics, Phase Transitions and Superfluidity* ed M Chrétien, E P Gross and S Deser (New York: Gordon and Breach) vol 1
Kasteleyn P W and Fortuin C M 1969 *J. Phys. Soc. Japan* **26** (Suppl.) 11–4
Welsh D J A 1977 *Sci. Prog., Oxf.* **64** 65–83
Wu F Y 1978 *J. Stat. Phys.* **18** 115–23
—— 1982 *Rev. Mod. Phys.* **54** 235–68

PERGAMON

Applied Mathematics Letters 13 (2000) 19–25

Applied Mathematics Letters

www.elsevier.nl/locate/aml

Spanning Trees on Hypercubic Lattices and Nonorientable Surfaces

W.-J. TZENG
Department of Physics, Tamkang University
Tamsui, Taipei, Taiwan 251, R.O.C.

F. Y. WU
National Center for Theoretical Sciences, Physics Division
P.O. Box 2-131, Hsinchu, Taiwan 300, R.O.C.
and
Department of Physics, Northeastern University
Boston, MA 02115, U.S.A.

(Received and accepted December 1999)

Communicated by F. Harary

Abstract—We consider the problem of enumerating spanning trees on lattices. Closed-form expressions are obtained for the spanning tree generating function for a hypercubic lattice in d dimensions under free, periodic, and a combination of free and periodic boundary conditions. Results are also obtained for a simple quartic net embedded on two nonorientable surfaces, a Möbius strip and the Klein bottle. Our results are based on the use of a formula expressing the spanning tree generating function in terms of the eigenvalues of an associated tree matrix. An elementary derivation of this formula is given. © 2000 Elsevier Science Ltd. All rights reserved.

Keywords—Spanning trees, Hypercubic lattices, Möbius strip, Klein bottle.

1. INTRODUCTION

The problem of enumerating spanning trees on a graph was first considered by Kirchhoff [1] in his analysis of electrical networks. Consider a graph $G = \{V, E\}$ consisting of a vertex set V and an edge set E. We shall assume that G is connected. A subset of edges $T \subset E$ is a spanning tree if it has $|V| - 1$ edges with at least one edge incident at each vertex. Therefore, T has no cycles. In ensuing discussions, we shall use T to also denote the spanning tree.

Number the vertices from 1 to $|V|$ and associate to the edge e_{ij} connecting vertices i and j a weight x_{ij}, with the convention of $x_{ii} = 0$. The enumeration of spanning trees concerns with the evaluation of the tree generating function

$$T(G; \{x_{ij}\}) = \sum_{T \subseteq E} \prod_{e_{ij} \in T} x_{ij}, \tag{1}$$

We are grateful to L. H. Kauffman for a useful conversation and to R. Shrock for calling our attention to references [2,3]. We thank T. K. Lee for the hospitality at the Center for Theoretical Sciences where this research is carried out. The work of F. Y. Wu is supported in part by NSF Grant DMR-9614170.

PII: S0893-9659(00)00071-9

where the summation is taken over all spanning trees T. Particularly, the number of spanning trees on G is obtained by setting $x_{ij} = 1$ as

$$N_{SPT}(G) = T(G; 1). \tag{2}$$

Considerations of spanning tree also arise in statistical physics [4] in the enumeration of close-packed dimers (perfect matchings) [5]. Using a similar consideration, for example, one of us [6] has evaluated the number of spanning trees for the simple quartic, triangular, and honeycomb lattices in the limit of $|V| \to \infty$. In this *letter*, we report new results on the evaluation of the generating function equation (1) for *finite* hypercubic lattices in arbitrary dimensions. Results are also obtained for a simple quartic net embedded on two nonorientable surfaces, the Möbius strip and the Klein bottle. As the main formula used in this letter is a relation expressing the tree generating function in terms of the eigenvalues of an associated tree matrix, for completeness we give an elementary derivation of this formula.

2. THE TREE MATRIX

For a given graph $G = \{V, E\}$ consider a $|V| \times |V|$ matrix $\mathbf{M}(G)$ with elements

$$M_{ij}(G) = \begin{cases} \sum_{k=1}^{|V|} x_{ik}, & i = j = 1, 2, \ldots, |V|, \\ -x_{ij}, & \text{if vertices } i,\ j,\ i \neq j, \text{ are connected by an edge,} \\ 0, & \text{otherwise.} \end{cases} \tag{3}$$

We shall refer to $\mathbf{M}(G)$ simply as the tree matrix. It is well known [7,8] that the tree generating function, equation (1), is given by the cofactor of any element of the tree matrix, and that the cofactor is the same for all elements. Namely, we have the identity

$$T(G; \{x_{ij}\}) = \text{the cofactor of } any \text{ element of the matrix } \mathbf{M}(G). \tag{4}$$

The tree generating function can also be expressed in terms of the eigenvalues of the tree matrix $\mathbf{M}(G)$ [2, p. 39]. We give here an elementary derivation of this result which we use in subsequent sections.

Let $\mathbf{M}(G)$ be the tree matrix of a graph $G = \{V, E\}$. Since the sum of all elements in a row of $\mathbf{M}(G)$ equals to zero, $\mathbf{M}(G)$ has 0 as an eigenvalue and, by definition, we have

$$\det |M_{ij}(G) - \lambda\delta_{ij}| = -\lambda F(\lambda), \tag{5}$$

where

$$F(\lambda) = \prod_{i=2}^{|V|} (\lambda_i - \lambda), \tag{6}$$

$\lambda_2, \lambda_3, \ldots, \lambda_{|V|}$ being the remaining eigenvalues.

Now the sum of all elements in a row of the determinant $|M_{ij}(G) - \lambda\delta_{ij}|$ is $-\lambda$. This permits us to replace the first column of $\det|M_{ij}(G) - \lambda\delta_{ij}|$ by a column of elements $-\lambda$ without affecting its value. Next we carry out a Laplace expansion of the resulting determinant along the modified column, obtaining

$$\det |M_{ij}(G) - \lambda\delta_{ij}| = -\lambda \sum_{i=1}^{|V|} C_{i1}(\lambda), \tag{7}$$

where $C_{i1}(\lambda)$ is the cofactor of the $(i1)^{\text{th}}$ element of the determinant. Combining equations (5)–(7), we are led to the identity

$$F(\lambda) = \sum_{i=1}^{|V|} C_{i1}(\lambda). \tag{8}$$

Now, $C_{i1}(0)$ is precisely the cofactor of the $(i1)^{\text{th}}$ element of $\mathbf{M}(G)$ which, by equation (4), is equal to the tree generating function $T(G; \{x_{ij}\})$. It follows that, after setting $\lambda = 0$ in equation (8), we obtain the expression

$$T(G; \{x_{ij}\}) = \frac{1}{|V|} \prod_{i=2}^{|V|} \lambda_i. \tag{9}$$

This result can also be deduced by considering the tree matrix of a graph obtained from G by adding an auxiliary vertex connected to all vertices with edges of weight x, followed by taking the limit of $x \to 0$ [9].

3. HYPERCUBIC LATTICES

We now deduce the closed-form expression for the tree generating function for a hypercubic lattice in d dimensions under various boundary conditions.

3.1. Free Boundary Conditions

THEOREM 1. *Let $\mathbf{Z}_d$ be a d-dimensional hypercubic lattice of size $N_1 \times N_2 \times \cdots \times N_d$ with edge weights x_i along the ith direction, $i = 1, 2, \ldots, d$. The tree generating function for $\mathbf{Z}_d$ is*

$$T(\mathbf{Z}_d; \{x_i\}) = \frac{2^{\mathcal{N}-1}}{\mathcal{N}} \prod_{\substack{n_1=0 \\ (n_1,\ldots,n_d) \neq (0,\ldots,0)}}^{N_1-1} \cdots \prod_{n_d=0}^{N_d-1} \left[\sum_{i=1}^{d} x_i \left(1 - \cos \frac{n_i \pi}{N_i}\right) \right], \tag{10}$$

where $\mathcal{N} = N_1 N_2 \ldots N_d$.

PROOF. The tree matrix of $\mathbf{Z}_d$ assumes the form of a linear combination of direct products of smaller matrices,

$$\begin{aligned} \mathbf{M}(\mathbf{Z}_d) = \sum_{i=1}^{d} x_i \, [2 I_{N_1} \otimes I_{N_2} \otimes \cdots \otimes I_{N_d} \\ - I_{N_1} \otimes \cdots \otimes I_{N_{i-1}} \otimes H_{N_i} \otimes I_{N_{i+1}} \otimes \cdots \otimes I_{N_d}], \end{aligned} \tag{11}$$

where I_N is an $N \times N$ identity matrix and H_N is the $N \times N$ tri-diagonal matrix

$$H_N = \begin{pmatrix} 1 & 1 & 0 & 0 & \ldots & 0 & 0 & 0 \\ 1 & 0 & 1 & 0 & \ldots & 0 & 0 & 0 \\ 0 & 1 & 0 & 1 & \ldots & 0 & 0 & 0 \\ \vdots & \vdots & \vdots & \vdots & \ddots & \vdots & \vdots & \vdots \\ 0 & 0 & 0 & 0 & \ldots & 1 & 0 & 1 \\ 0 & 0 & 0 & 0 & \ldots & 0 & 1 & 1 \end{pmatrix}. \tag{12}$$

It is readily verified that H_N is diagonalized by the similarity transformation

$$S_N H_N S_N^{-1} = \Lambda_N, \tag{13}$$

where S_N and S_N^{-1} are $N \times N$ matrices with elements

$$(S_N)_{mn} = \left(S_N^{-1}\right)_{nm} = \sqrt{\frac{2}{N}} \cos\left[(2n+1)\left(\frac{m\pi}{2N}\right)\right] + \left(\sqrt{\frac{1}{N}} - \sqrt{\frac{2}{N}}\right) \delta_{m,0}, \quad m, n = 0, 1, \ldots, N-1, \tag{14}$$

and Λ_N is an $N \times N$ diagonal matrix with diagonal elements

$$\lambda_n = 2\cos\frac{n\pi}{N}, \qquad n = 0, 1, \ldots, N-1. \tag{15}$$

Here $\delta_{m,n}$ is the Kronecker delta. It follows that $\mathbf{M}(\mathbf{Z}_d)$ is diagonalized by the similarity transformation

$$\mathbf{S}_{\mathcal{N}}\mathbf{M}\left(\mathbf{Z}_d\right)\mathbf{S}_{\mathcal{N}}^{-1} = \mathbf{\Lambda}_{\mathcal{N}}, \tag{16}$$

where

$$\mathbf{S}_{\mathcal{N}} = S_{N_1} \otimes S_{N_2} \otimes \cdots \otimes S_{N_d}, \tag{17}$$

and $\mathbf{\Lambda}_{\mathcal{N}}$ is an $\mathcal{N} \times \mathcal{N}$ diagonal matrix with diagonal elements

$$\lambda_{n_1,\ldots,n_d} = 2\sum_{i=1}^{d} x_i\left[1 - \cos\frac{n_i\pi}{N_i}\right], \qquad n_i = 0, 1, \ldots, N_i - 1. \tag{18}$$

Now, we have $\lambda_{n_1,\ldots,n_d} = 0$ for $n_1 = n_2 = \cdots = n_d = 0$. This establishes Theorem 1 after using equation (9). ∎

REMARK. The result equation (18) generalizes the $d = 2$ eigenvalues of $\mathbf{M}(\mathbf{Z}_2)$ for $x_i = 1$ reported in [2, p. 74].

3.2. Periodic Boundary Conditions

In applications in physics, one often requires periodic boundary conditions depicted by the condition that two "boundary" vertices at coordinates $(\ldots, n_i = 1, \ldots)$ and $(\ldots, n_i = N_i, \ldots)$, $i = 1, 2, \ldots, d$, are connected by an extra edge. This leads to a lattice $\mathbf{Z}_d^{\mathrm{Per}}$ which is a regular graph with degree $2d$ at all vertices. For $d = 2$, for example, $\mathbf{Z}_2^{\mathrm{Per}}$ can be regarded as being embedded on the surface of a torus.

THEOREM 2. *Let $\mathbf{Z}_d^{\mathrm{Per}}$ be a hypercubic lattice in d dimensions of size $N_1 \times N_2 \times \cdots \times N_d$ with edge weights x_i along the ith direction, $i = 1, 2, \ldots, d$ with periodic boundary conditions. The tree generating function for $\mathbf{Z}_d^{\mathrm{Per}}$ is*

$$T\left(\mathbf{Z}_d^{\mathrm{Per}}; \{x_i\}\right) = \frac{2^{\mathcal{N}-1}}{\mathcal{N}} \prod_{n_1=0}^{N_1-1} \cdots \prod_{n_d=0}^{N_d-1} \left[\sum_{i=1}^{d} x_i\left(1 - \cos\frac{2n_i\pi}{N_i}\right)\right], \quad (n_1, \ldots, n_d) \neq (0, \ldots, 0). \tag{19}$$

PROOF. The tree matrix assumes the form

$$\mathbf{M}\left(\mathbf{Z}_d^{\mathrm{Per}}\right) = \sum_{i=1}^{d} x_i\left[2I_{N_1} \otimes I_{N_2} \otimes \cdots \otimes I_{N_d} - I_{N_1} \otimes \cdots \otimes I_{N_{i-1}} \otimes G_{N_i} \otimes I_{N_{i+1}} \otimes \cdots \otimes I_{N_d}\right], \tag{20}$$

where G_N is the $N \times N$ cyclic matrix

$$G_N = \begin{pmatrix} 0 & 1 & 0 & 0 & \ldots & 0 & 0 & 1 \\ 1 & 0 & 1 & 0 & \ldots & 0 & 0 & 0 \\ 0 & 1 & 0 & 1 & \ldots & 0 & 0 & 0 \\ \vdots & \vdots & \vdots & \vdots & \ddots & \vdots & \vdots & \vdots \\ 0 & 0 & 0 & 0 & \ldots & 1 & 0 & 1 \\ 1 & 0 & 0 & 0 & \ldots & 0 & 1 & 0 \end{pmatrix}. \tag{21}$$

As in equation (16), the matrix $\mathbf{M}(\mathbf{Z}_d^{\text{Per}})$ can be diagonalized by a similarity transformation generated by

$$\mathbf{R}_{\mathcal{N}} = R_{N_1} \otimes R_{N_2} \otimes \cdots \otimes R_{N_d}, \tag{22}$$

where R_N is an $N \times N$ matrix with elements

$$(R_N)_{nm} = \left(R_N^{-1}\right)^*_{mn} = N^{-1/2} e^{i2\pi mn/N}, \tag{23}$$

where * denotes the complex conjugate, yielding eigenvalues of G_N as

$$\lambda_n = 2\cos\frac{2n\pi}{N}, \qquad n = 0, 1, \ldots, N-1. \tag{24}$$

This establishes Theorem 2 after using equation (9). ∎

3.3. Periodic Boundary Conditions Along $m \le d$ Directions

COROLLARY. *Let $\mathbf{Z}_d^{\text{Per}(m)}$ be a hypercubic lattice in d dimensions of size $N_1 \times N_2 \times \cdots \times N_d$ with periodic boundary conditions in directions $1, 2, \ldots, m \le d$ and free boundaries in the remaining $d - m$ directions. The tree generating function is*

$$\begin{aligned} T\left(\mathbf{Z}_d^{\text{Per}(m)}; \{x_i\}\right) = \frac{2^{\mathcal{N}-1}}{\mathcal{N}} \prod_{n_1=0}^{N_1-1} \cdots \prod_{n_d=0}^{N_d-1} & \left[\sum_{i=1}^{m} x_i \left(1 - \cos\frac{2n_i\pi}{N_i}\right)\right. \\ & \left. + \sum_{i=m+1}^{d} x_i \left(1 - \cos\frac{n_i\pi}{N_i}\right)\right], \qquad (n_1, \ldots, n_d) \ne (0, \ldots, 0). \end{aligned} \tag{25}$$

4. THE MÖBIUS STRIP AND THE KLEIN BOTTLE

Due to the interplay with the conformal field theory [10], it is of current interest in statistical physics to study lattice systems on nonorientable surfaces [11,12]. Here, we consider two such surfaces, the Möbius strip and the Klein bottle, and obtain the respective tree generating functions.

4.1. The Möbius Strip

THEOREM 3. *Let $\mathbf{Z}_2^{\text{Mob}}$ be an $M \times N$ simple quartic net embedded on a Möbius strip forming a Möbius net of width M and twisted in the direction N, with edge weights x_1 and x_2 along directions M and N, respectively. The tree generating function for $\mathbf{Z}_2^{\text{Mob}}$ is*

$$\begin{aligned} T\left(\mathbf{Z}_2^{\text{Mob}}; \{x_1, x_2\}\right) = \frac{2^{MN-1}}{MN} \prod_{m=0}^{M-1} \prod_{n=0}^{N-1} & \left[x_1 \left(1 - \cos\frac{m\pi}{M}\right)\right. \\ & \left. + x_2 \left(1 - \cos\frac{4n+1-(-1)^m}{2N}\pi\right)\right], \qquad (m, n) \ne (0, 0). \end{aligned} \tag{26}$$

PROOF. Specifically, let the the two vertices at coordinates $\{m, 1\}$ and $\{M - m, N\}$, $m = 1, 2, \ldots, M$ be connected with a lattice edge of weight x_2. Then the tree matrix assumes the form

$$\mathbf{M}\left(\mathbf{Z}_2^{\text{Mob}}\right) = 2(x_1 + x_2) I_M \otimes I_N - x_1 H_M \otimes I_N - x_2 \left[I_M \otimes F_N + J_M \otimes K_N\right], \tag{27}$$

where

$$F_N = \begin{pmatrix} 0 & 1 & 0 & 0 & \dots & 0 & 0 & 0 \\ 1 & 0 & 1 & 0 & \dots & 0 & 0 & 0 \\ 0 & 1 & 0 & 1 & \dots & 0 & 0 & 0 \\ \vdots & \vdots & \vdots & \vdots & \ddots & \vdots & \vdots & \vdots \\ 0 & 0 & 0 & 0 & \dots & 1 & 0 & 1 \\ 0 & 0 & 0 & 0 & \dots & 0 & 1 & 0 \end{pmatrix}, \qquad J_M = \begin{pmatrix} 0 & 0 & \dots & 0 & 0 & 1 \\ 0 & 0 & \dots & 0 & 1 & 0 \\ 0 & 0 & \dots & 1 & 0 & 0 \\ \vdots & \vdots & \ddots & \vdots & \vdots & \vdots \\ 0 & 1 & \dots & 0 & 0 & 0 \\ 1 & 0 & \dots & 0 & 0 & 0 \end{pmatrix},$$

$$K_N = \begin{pmatrix} 0 & 0 & 0 & \dots & 0 & 1 \\ 0 & 0 & 0 & \dots & 0 & 0 \\ 0 & 0 & 0 & \dots & 0 & 0 \\ \vdots & \vdots & \vdots & \ddots & \vdots & \vdots \\ 0 & 0 & 0 & \dots & 0 & 0 \\ 1 & 0 & 0 & \dots & 0 & 0 \end{pmatrix}.$$

Since H_M and J_M commute, they can be simultaneously diagonalized by applying the similarity transformation equation (13). The transformed matrix $\mathbf{S}_{\mathcal{N}}\mathbf{M}(\mathbf{Z}_2^{\text{Mob}})\mathbf{S}_{\mathcal{N}}^{-1}$ is block diagonal with $N \times N$ blocks

$$2\left(x_1 - x_1 \cos\frac{m\pi}{M} + x_2\right) I_N - x_2\left(F_N + (-1)^m K_N\right), \qquad m = 0, 1, \dots, M-1. \tag{28}$$

Now, the eigenvalues of $G_N = F_N + K_N$ and $F_N - K_N$ are, respectively, $2\cos[2(n+1)\pi/N]$ and $2\cos[(2n+1)\pi/N]$, $n = 0, 1, \dots, N-1$. Theorem 3 is established by combining these results with equation (9). ∎

REMARK. For $M = 2$ and $x_1 = x_2 = 1$, equation (26) gives the number of spanning trees on a $2 \times N$ Möbius ladder as

$$\begin{aligned} N_{SPT} &= \frac{1}{2N} \prod_{j=1}^{2N-1} \left[3 - (-1)^j - 2\cos\frac{j\pi}{N}\right] \\ &= \frac{N}{2}\left[2 + \left(2+\sqrt{3}\right)^N + \left(2-\sqrt{3}\right)^N\right]. \end{aligned} \tag{29}$$

These two equivalent expressions have previously been given by [2, p. 218] and by Guy and Harary [3], respectively.

4.2. The Klein Bottle

The embedding of an $M \times N$ simple quartic net on a Klein bottle is accomplished by further imposing a periodic boundary condition to $\mathbf{Z}_2^{\text{Mob}}$ in the M direction, namely, by connecting vertices of $\mathbf{Z}_2^{\text{Mob}}$ at coordinates $\{1, n\}$ and $\{M, n\}$, $n = 1, 2, \dots, N$ with an edge of weight x_1. This leads to a lattice $\mathbf{Z}_2^{\text{Klein}}$ of the topology of a Klein bottle.

THEOREM 4. *The tree generating function for* $\mathbf{Z}_2^{\text{Klein}}$ *(described in the above) is*

$$\begin{aligned} T\left(\mathbf{Z}_2^{\text{Klein}}; \{x_1, x_2\}\right) &= \frac{2^{MN-1}}{MN}\left[\prod_{n=1}^{N-1} x_2\left(1 - \cos\frac{2n\pi}{N}\right)\right] \\ &\times \prod_{m=1}^{[M-1/2]} \prod_{n=0}^{2N-1} \left[x_1\left(1 - \cos\frac{2m\pi}{M}\right) + x_2\left(1 - \cos\frac{n\pi}{N}\right)\right] \\ &\times \begin{cases} \displaystyle\prod_{n=0}^{N-1} \left[2x_1 - x_2\left(1 - \cos\frac{(2n+1)\pi}{N}\right)\right], & \text{for } M \text{ even}, \\ 1, & \text{for } M \text{ odd}, \end{cases} \end{aligned} \tag{30}$$

where $[n]$ *is the integral part of* n.

PROOF. The tree matrix of $\mathbf{Z}_2^{\text{Klein}}$ assumes the form

$$\mathrm{M}\left(\mathbf{Z}_2^{\text{Klein}}\right) = 2\,(x_1+x_2)\,I_M \otimes I_N - x_1 G_M \otimes I_N - x_2\left[I_M \otimes F_N + J_M \otimes K_N\right]. \tag{31}$$

To obtain its eigenvalues, we first apply the similarity transformation generated by R_M in the M subspace. While this diagonalizes G_M with eigenvalues $2\cos(2m\pi/M)$, $m = 0, 1, \dots, M-1$, it transforms the tree matrix $\mathrm{M}(\mathbf{Z}_2^{\text{Klein}})$ into

$$\begin{pmatrix} A_0+B_0 & 0 & 0 & \dots & 0 & 0 & 0 \\ 0 & A_1 & 0 & \dots & 0 & 0 & B_1 \\ 0 & 0 & A_2 & \dots & 0 & B_2 & 0 \\ \vdots & \vdots & \vdots & \ddots & \vdots & \vdots & \\ 0 & 0 & B_{-2} & \dots & 0 & A_{-2} & 0 \\ 0 & B_{-1} & 0 & \dots & 0 & 0 & A_{-1} \end{pmatrix}, \tag{32}$$

where A_m and B_m are $N \times N$ matrices given by

$$\begin{aligned} A_m &= 2\left[x_1 + x_2 - x_1 \cos\frac{2m\pi}{M}\right] I_N - x_2 F_N, \\ B_m &= -e^{2\pi i m/M} x_2 K_N, \qquad m = 0, 1, \dots, M-1. \end{aligned}$$

The matrix equation (32) is block diagonal with blocks $A_0 + B_0, \left(\begin{smallmatrix} A_m & B_m \\ B_{-m} & A_{-m} \end{smallmatrix}\right)$, $m = 1, 2, \dots, [(M-1)/2]$ and, for $m =$ even, $A_{M/2} + B_{M/2}$. The eigenvalues of individual blocks can be deduced from those of $F_N \pm K_N$. We are led to the theorem after using equation (9). ∎

REFERENCES

1. G. Kirchhoff, Über die Auflösung der Gleichungen, auf welche man bei der Untersuchung der linearen Verteilung galvanischer Ströme geführt wird, *Ann. Phys. und Chemie.* **72**, 497–508, (1847).
2. D.M. Cvetković, M. Doob and H. Sachs, *Spectra of Graphs—Theory and Applications*, Academic Press, New York, (1979).
3. R.K. Guy and F. Harary, On the Möbius ladders, The University of Calgary Research Report, No. 2, (November 1966); J. Sedlácek, On the Skeleton of a Graph or Digraph, *Combinatorial Structures and Applications*, Gordon and Breach, New York, (1970).
4. H.N.V. Temperley, On the mutual cancellation of cluster integrals in Mayer's fugacity series, *Proc. Phys. Soc.* **83**, 3–16, (1964).
5. H.N.V Temperley, *Combinatorics: Proceedings of the British Combinatorial Conference*, Lecture Notes Series #13, London Math. Soc., (1974).
6. F.Y. Wu, Number of spanning trees on a lattice, *J. Phys. A* **10**, L113–L115, (1977).
7. R.L. Brooks, C.A.B. Smith, A.H. Stone and W.T. Tutte, Dissection of a rectangle into squares, *Duke Math. J.* **7**, 312–340, (1940).
8. F. Harary, *Graph Theory*, Addison-Wesley, Reading, MA, (1969).
9. R.W. Kenyon, J.G. Propp and D.B. Wilson, Trees and Matchings, LANL preprint math.CO/9903025.
10. H.W.J. Blöte, J.C. Cardy and M.P. Nightingale, Conformal invariance, the central charge, and universal finite-size amplitudes at criticality, *Phys. Rev. Lett.* **56**, 742–745, (1986).
11. W.T. Lu and F.Y. Wu, Dimer statistics on the Möbius strip and the Klein bottle, *Phys. Lett.* **A259**, 108–114, (1999).
12. N. Biggs and R. Shrock, $T = 0$ partition functions for Potts antiferromagnets on square lattice strips with (twisted) periodic boundary conditions, *J. Phys. A* (to appear).

J. Phys. A: Math. Gen., **13** (1980) 629–636. Printed in Great Britain

On the triangular Potts model with two- and three-site interactions

F Y Wu† and K Y Lin

Department of Physics, Northeastern University, Boston, Massachusetts 02115, USA

Received 26 April 1979

Abstract. The equivalence of the triangular Potts model having two- and three-site interactions with a 20-vertex Kelland model is rederived using a graphical method. The conjectured critical point of this Potts model is shown to agree with the known results in two instances.

1. Introduction

The Potts model (Potts 1952) has remained to this date one of the most intriguing lattice statistical models of phase transitions. While its exact solution is not yet known, significant progress has been made in recent years in exact analyses of its properties. The breakthrough came in 1971 when Temperley and Lieb (1971) established a remarkable equivalence of the nearest-neighbour Potts model on the square lattice with an ice-rule model, a fact that made possible the exact determination of its critical properties (Baxter 1973). These considerations have recently been extended to the Potts model with two- and three-site interactions (Baxter *et al* 1978). In these analyses an operator method has been used to establish the equivalence of the Potts model with an ice-rule model. A simpler and more direct graphical analysis for proving this equivalence was later developed by Baxter *et al* (1976) for the pure two-site problem. In view of the usefulness and richness of the new results of Baxter *et al* (1978), it appears desirable to extend the graphical approach to models with two- and three-site interactions. This is the subject matter of the present paper.

We shall proceed in a way which differs slightly from that of Baxter *et al* (1976). We define, in § 2, a five-vertex model on the triangular lattice, and show in § 3 that this vertex model is equivalent to the Potts model under consideration. A simple symmetry of the vertex model then leads to a duality relation for the Potts model, which, in turn, determines the Potts critical point. This conjectured critical point is shown to reduce to the known exact results in two instances. In § 4 we show that the five-vertex model is also equivalent to a Kelland (1974) model. It follows that the Potts model with two- and three-site interactions is equivalent to an ice-rule Kelland model, thus rederiving the result obtained by Baxter *et al* (1978).

† Work supported in part by the National Science Foundation.

0305-4470/80/020629+08$01.00

2. Five-vertex model on the triangular lattice

Consider a triangular lattice $\mathscr{L}'$ of N sites. Cover all edges of $\mathscr{L}'$ with bonds and join the ends of bonds so that the bonds form non-crossing paths. A typical joining of the bonds is shown in figure 1. Note that the bonds form closed, non-intersecting polygons. The six bonds incident at a vertex can join in only five distinct ways. These five configurations are shown in figure 2.

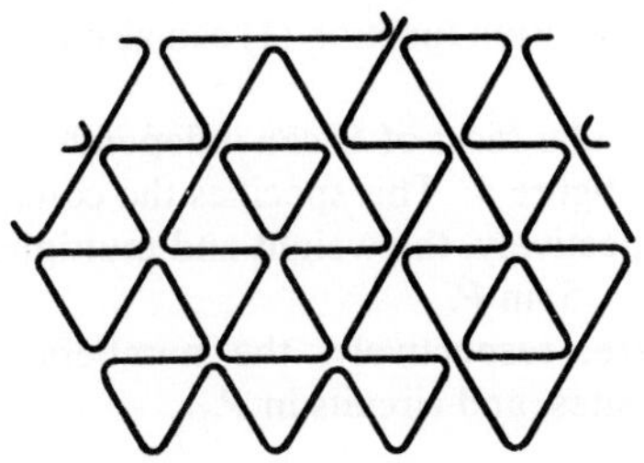

Figure 1. A typical bond graph on $\mathscr{L}'$. The bonds form closed, non-intersecting polygons.

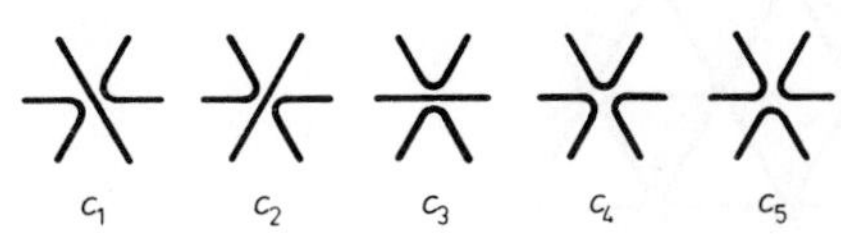

Figure 2. Vertex configurations of the five-vertex model.

Associate weights c_i, $i = 1, 2, \ldots, 5$, with these five configurations as shown in figure 2. Further, with each polygon on $\mathscr{L}'$ we associate a weight z. The partition generating function for this five-vertex model is defined to be

$$Z_{12345} \equiv Z(z; c_1, c_2, c_3, c_4, c_5)$$
$$= \sum_p z^p \prod_{i=1}^{5} c_i^{n_i} \qquad (1)$$

where the summation is taken over the 5^N polygonal configurations, or bond joinings, on $\mathscr{L}'$, n_i is the number of vertices of type i satisfying

$$n_1 + n_2 + n_3 + n_4 + n_5 = N \qquad (2)$$

and p is the number of polygons. The partition function (1) possesses the obvious 60° rotational symmetry

$$Z_{12345} = Z_{31254} = Z_{23145} = Z_{12354} = Z_{31245} = Z_{23154}. \qquad (3)$$

It follows that Z is invariant under the cyclic permutations of the indices 1, 2, 3, and/or 4, 5.

3. Reduction to a Potts model

We now show that the five-vertex model (1) is equivalent to the Potts model with two- and three-site interactions considered by Baxter *et al* (1978).

There are two kinds of faces in the triangular lattice $\mathscr{L}'$, namely, the up-pointing and down-pointing triangles. Following Baxter *et al* (1976), we shade one kind of the faces, say the down-pointing triangles, and regard such shaded areas as 'land', and the remaining unshaded areas as 'water'. Then, as shown in figure 3, a typical polygonal configuration P will consist of *connected* lands surrounded by water.

Next, we place a site at the middle of each of the N shaded triangles, and join as shown in figure 3 the two or three neighbouring sites whose lands are connected. The connecting lines are either boomerang- or Y-shaped. Consider now the triangular lattice $\mathscr{L}$ formed by these N sites. The partition function (1) can also be interpreted as defined on $\mathscr{L}$ as follows.

(1) Each of the N up-pointing triangular faces of $\mathscr{L}$ can independently take one of the five configurations shown in figure 4. This specifies the configuration P.

(2) The numbers c_i and n_i are, respectively, the weight and multiplicity of the ith vertex configuration, $i = 1, 2, \ldots, 5$, in P.

(3) $p = C + S$, where C and S are, respectively, the numbers of connected components, including isolated sites, and circuits in P.

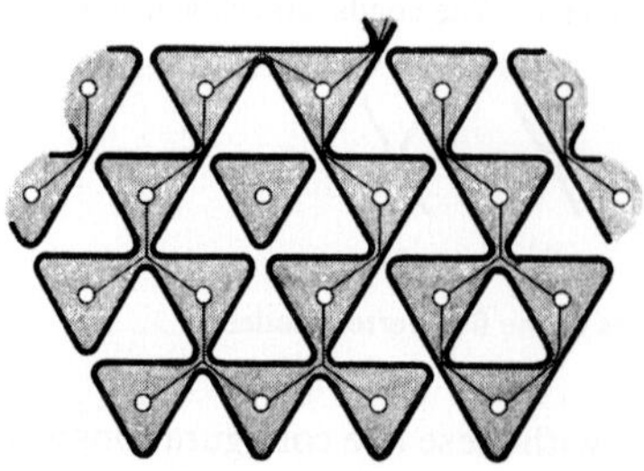

Figure 3. The same bond graph as in figure 1. The down-pointing triangular faces are shaded showing connected lands surrounded by water. The circles form a triangular lattice $\mathscr{L}$.

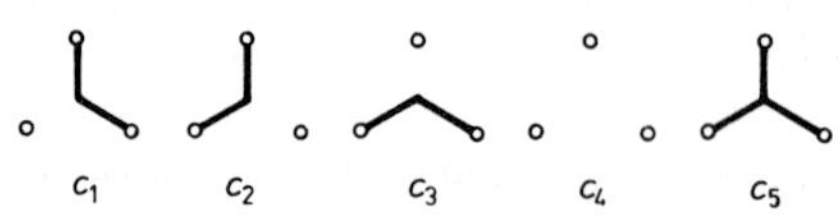

Figure 4. The five possible bond configurations for the up-pointing triangular faces of $\mathscr{L}$.

Here, use has been made of the fact that, for a given P, each closed polygon on $\mathscr{L}'$ is the outside perimeter of either a circuit or a connected component of the associated configuration on $\mathscr{L}$.

Consider a q-state Potts model on $\mathscr{L}$ whose interactions consist of two-site interactions $\epsilon_1, \epsilon_2, \epsilon_3$ and three-site interactions ϵ among every three sites surrounding an up-pointing (triangular) face. This is shown in figure 5. The Hamiltonian now reads

$$\mathscr{H} = \sum_{\Delta} E_{abc} \tag{4}$$

where the summation is over all up-pointing faces of $\mathscr{L}$ and

$$E_{abc} = -(\epsilon_1 \delta_{bc} + \epsilon_2 \delta_{ca} + \epsilon_3 \delta_{ab} + \epsilon \delta_{abc}). \tag{5}$$

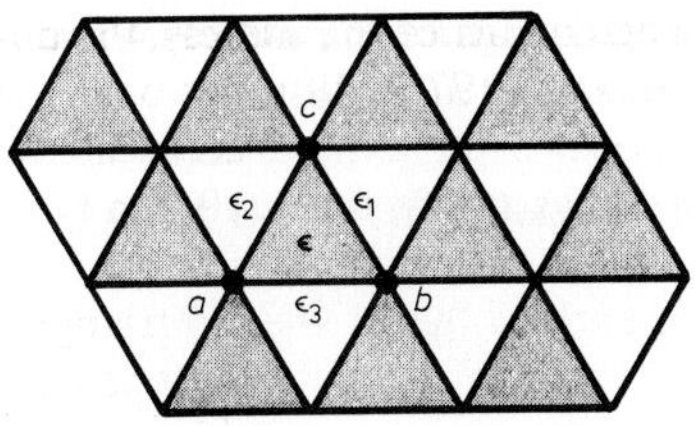

Figure 5. The Potts model on $\mathcal{L}$ with two- and three-site interactions at each shaded triangle.

Here, $\delta_{ab} = \delta_{\text{Kr}}(\xi_a, \xi_b)$, $\delta_{abc} = \delta_{ab}\delta_{bc}$, and $\xi_a = 1, 2, \ldots, q$ refer to the spin state at the site a.

Following Baxter *et al* (1978), we write

$$\exp(-\beta E_{abc}) = 1 + f_1\delta_{bc} + f_2\delta_{ca} + f_3\delta_{ab} + y\delta_{abc} \tag{6}$$

where

$$\begin{aligned} f_i &= \exp(\beta\epsilon_i) - 1 \\ g &= \exp(\beta\epsilon) - 1 \\ y &= f_1f_2 + f_2f_3 + f_3f_1 + f_1f_2f_3 + g(1+f_1)(1+f_2)(1+f_3) \end{aligned} \tag{7}$$

and $\beta = 1/kT$. The partition function of the Potts model is

$$Z_{\text{Potts}}(q; f_1, f_2, f_3, y) = \sum \prod_{\Delta} (1 + f_1\delta_{bc} + f_2\delta_{ca} + f_3\delta_{ab} + y\delta_{abc}) \tag{8}$$

where the summation is over the q^N spin states. The product is taken over the N shaded triangles shown in figure 5.

Expand the product in (8). A natural graphical representation of the expansion is as follows. To each factor $f_i\delta_{ab}$ associate a boomerang-shaped bond connecting the sites a and b, and to each factor $y\delta_{abc}$ associate a Y-shaped bond connecting the sites a, b and c. Since these are precisely the configurations shown in figure 4, we can write, as in (1),

$$Z_{\text{Potts}} = \sum_P q^C f_1^{n_1} f_2^{n_2} f_3^{n_3} y^{n_5} \tag{9}$$

where the summation is taken over the 5^N configurations P on $\mathcal{L}$. Also, since f_i connects two sites and each Y connects three sites, we have the Euler relation

$$N + S = C + n_1 + n_2 + n_3 + 2n_5. \tag{10}$$

Eliminating N and S from (2), (10) and the relation $p = C + S$, we obtain

$$C = \tfrac{1}{2}(p + n_4 - n_5). \tag{11}$$

Substituting (11) into (9) and comparing with (1), we arrive at the identity

$$Z_{\text{Potts}}(q; f_1, f_2, f_3, y) = Z(\sqrt{q}; f_1, f_2, f_3, \sqrt{q}, y/\sqrt{q}). \tag{12}$$

This states that the Potts model (4) is equivalent to the five-vertex model (1), a result we set out to prove. Note that there is no loss of generality in taking $z = c_4$ in (1), since Z is homogeneous in c_i.

The invariance of Z_{12345} under the interchange of the indices 4 and 5 now implies the following duality relation for Z_{Potts} (Baxter *et al* 1978):

$$Z_{\text{Potts}}(q; f_1, f_2, f_3, y) = (y/q)^N Z_{\text{Potts}}(q; f_1', f_2', f_3', y'), \tag{13}$$

where

$$f_i' = qf_i/y \qquad y' = q^2/y. \tag{14}$$

The transformation (13) maps the partition function at a temperature $T > T_c$ into one at another temperature $T < T_c$, and vice versa, where T_c is determined by the fixed point

$$y = q \tag{15}$$

of the transformation. In the isotropic case ($\epsilon_1 = \epsilon_2 = \epsilon_3$) (15) reads

$$3f^2 + f^3 + g(1+f)^3 = q \tag{16}$$

and we plot (16) in figure 6 to give T_c as a function of $\alpha \equiv \epsilon/\epsilon_1$. Along the $\alpha = 0$ axis the T_c in figure 6 is known to be exact and unique (Hintermann *et al* 1978). If, for $\alpha \neq 0$, one assumes the transition also to be unique, then the critical point is given by (16). We expect a similar uniqueness argument to lead to the critical condition (15) in the general anisotropic case. Indeed, the general Potts model (4) is exactly soluble for $q = 2$. In this case the state ξ_a may be described by the Ising variables $\sigma_a = \pm 1$ and we write $\delta_{ab} = \frac{1}{2}(1 + \sigma_a \sigma_b)$. The Potts model is then exactly equivalent to a triangular Ising model whose interactions are $J_i = \frac{1}{2}\epsilon_i + \frac{1}{4}\epsilon$. From the known solution of the triangular Ising model (Houtappel 1950), one verifies that its critical condition is indeed (15) in the region $\epsilon + \epsilon_i + \epsilon_j \geq 0$, $i \neq j$, or

$$(1+g)(1+f_i)(1+f_j) \geq 1 \qquad i \neq j. \tag{17}$$

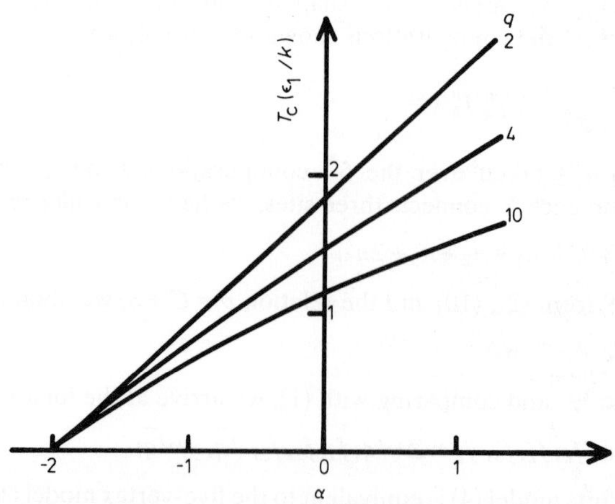

Figure 6. The transition temperature T_c in the isotropic case, T_c in units of ϵ_1/k and $\alpha = \epsilon/\epsilon_1$. The straight line $\alpha + 2 = T_c \ln 3$ for $q = 2$ is exact.

Since the Ising critical condition is different from (15) outside the region (17), the validity of (15) will generally be limited. It appears safe, however, to expect (15) to hold at least for positive g and f_i. We note that (15) indeed reduces to the exact results of Hintermann *et al* (1978) for $g=0$ and $f_i \geqslant 0$.

4. Equivalence with an ice-rule model

In this section we show that the five-vertex model (1) is also equivalent to an ice-rule model, thereby deriving the equivalence of the latter with the Potts model.

Consider the partition generating function (1) for the five-vertex model. Write

$$z = t^3 + t^{-3} \tag{18}$$

and expand the factor $(t^3+t^{-3})^p$ in (1). Following Baxter *et al* (1976), a natural graphical representation of this expansion is to direct the polygons in P and associate the weights t^3 and t^{-3} to the directed polygons. As shown in figure 7, let t^3 (t^{-3}) be the weight of a clockwisely (counterclockwisely) directed polygon. The polygonal weights $t^{\pm 3}$ can also be associated with the vertices with the following rule (Baxter *et al* 1976): each directed line turning an angle θ to the right (left) carries a weight $t^{3\theta/2\pi}$ ($t^{-3\theta/2\pi}$). This leads us to consider a vertex problem on $\mathscr{L}'$ whose edges are directed. Since there are always three arrows out and three arrows in at each vertex, we are led to the Kelland (1974) model, namely, the 20-vertex ice-rule model on $\mathscr{L}'$. Collecting the weights of those vertices having the same arrow arrangement, we obtain, as shown in figure 8, the following equivalence:

$$Z_{12345} = Z_{\text{Kelland}}(u_i, u_i'). \tag{19}$$

Here Z_{Kelland} is the partition function of the Kelland model. The vertex weights of the Kelland model are obtained from figure 8:

$$\begin{aligned} &u_1 = c_2 \qquad u_2 = c_1 \qquad u_3 = c_3 \\ &u_4 = c_1+c_2+c_3+c_4t^3+c_5t^{-3} \qquad u_5 = c_3t^2+c_5t^{-1} \\ &u_6 = c_3t^{-2}+c_4t \qquad u_7 = c_2t^2+c_4t^{-1} \\ &u_8 = c_2t^{-2}+c_5t \qquad u_9 = c_1t^2+c_5t^{-1} \\ &u_{10} = c_1t^{-2}+c_4t \qquad u_i' = u_i(t \to t^{-1}). \end{aligned} \tag{20}$$

The configuration of u_i' is the same as that of u_i with all arrows reversed.

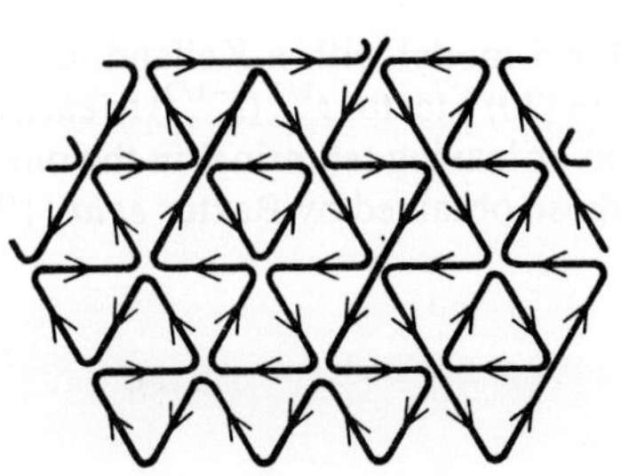

Figure 7. A typical directed polygonal configuration on $\mathscr{L}'$. Each polygon can be directed either clockwisely or counterclockwisely carrying respective weights t^3 and t^{-3}.

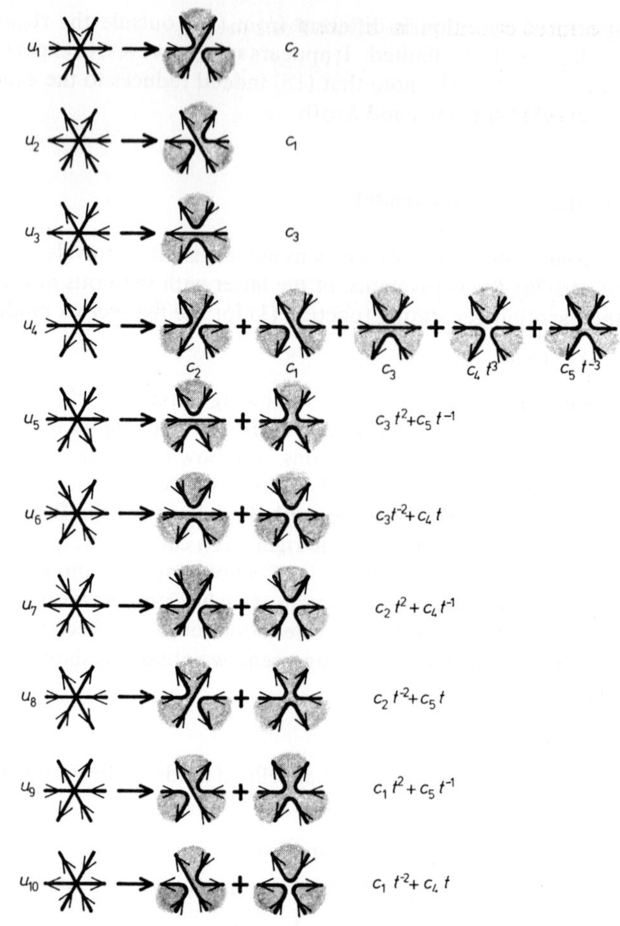

Figure 8. Vertex weights and vertex decompositions of a Kelland model.

The equivalence (20) is valid for general c_i and t, or z. Specialising to the Potts model for which, from (12),

$$z=\sqrt{q} \qquad c_1=f_1 \qquad c_2=f_2 \qquad c_3=f_3 \qquad c_4=\sqrt{q} \qquad c_5=y/\sqrt{q}, \tag{21}$$

(19) leads to an equivalence of the Potts model with a Kelland model. If, without changing Z_{Kelland}, we further introduce in (20) a factor $t^{1/2}$ ($t^{-1/2}$) to each arrow entering (leaving) a vertex in the three-direction or leaving (entering) in the one-direction, the resulting u_i and u_i' reduce exactly to those obtained by Baxter *et al* (1978). We have thus rederived their result.

5. Summary

We have established from a graphical consideration the equivalence of the triangular Potts model (4) with a Kelland model whose parameters u_i and u_i' are given by (20) and

(21). The conjectured critical point of this Potts model,

$$f_1f_2+f_2f_3+f_3f_1+f_1f_2f_3+g(1+f_1)(1+f_2)(1+f_3)=q, \tag{22}$$

agrees with the exact results for $q=2$ and/or for $g=0$.

References

Baxter R J 1973 *J. Phys. C: Solid St. Phys.* **6** L445–8
Baxter R J, Kelland S B and Wu F Y 1976 *J. Phys. A: Math. Gen.* **9** 397–406
Baxter R J, Temperley H N V and Ashley S E 1978 *Proc. R. Soc.* A **358** 535–59
Hintermann A, Kunz H and Wu F Y 1978 *J. Statist. Phys.* **19** 623–32
Houtappel R M F 1950 *Physica* **16** 425–55
Kelland S B 1974 *Aust. J. Phys.* **27** 813–29
Potts R B 1952 *Proc. Camb. Phil. Soc.* **48** 106–9
Temperley H N V and Lieb E H 1971 *Proc. R. Soc.* A **322** 251–80

Reprinted from JOURNAL OF STATISTICAL PHYSICS Vol. 42, Nos. 5/6, March 1986
Printed in Belgium

Nonintersecting String Model and Graphical Approach: Equivalence with a Potts Model

J. H. H. Perk[1] and F. Y. Wu[2]

Received August 8, 1985

Using a graphical method we establish the exact equivalence of the partition function of a q-state nonintersecting string (NIS) model on an arbitrary planar, even-valenced, lattice with that of a q^2-state Potts model on a related lattice. The NIS model considered in this paper is one in which the vertex weights are expressible as sums of those of basic vertex types, and the resulting Potts model generally has multispin interactions. For the square and Kagomé lattices this leads to the equivalence of a staggered NIS model with Potts models with anisotropic pair interactions, indicating that these NIS models have a first-order transition for $q > 2$. For the triangular lattice the NIS model turns out to be the five-vertex model of Wu and Lin and it relates to a Potts model with two- and three-site interactions. The most general model we discuss is an oriented NIS model which contains the six-vertex model and the NIS models of Stroganov and Schultz as special cases.

KEY WORDS: Nonintersecting string model; Potts model; vertex model; graphical approach.

1. INTRODUCTION

Great progress has been made in recent years in solving lattice models in statistical physics.[(1)] Many of the solved problems can be formulated as vertex models in which the system is described by assigning states to the lattice edges and Boltzmann weight factors to the vertices dependent on the incident states. For many of the earlier solved models the edges can be in one of two states (colors) and the configurations can be described in terms

[1] Institute for Theoretical Physics, State University of New York at Stony Brook, Stony Brook, New York 11794-3840.

[2] Department of Physics, Northeastern University, Boston, Massachusetts 02115.

0022-4715/86/0300-0727$05.00/0

of strings of conserved color on the lattice edges. Included are the ice-rule models,[2,3] the eight-vertex model,[4] and the (critical) Potts model.[5,6]

Recently, there has been increasing interest in considering string models with more than two colors.[7] One of the earliest investigations is by Stroganov,[8] who considered a 3-state nonintersecting string model, a model in which the states are described by strings of conserved colors that do *not* intersect, and obtained its solution in two special cases. Stroganov's result was generalized to an arbitrary number, $q \geqslant 2$, of states by Schultz[9]; and Perk and Schultz[7,10,11] further extended the general $q \geqslant 3$ solution to $q+1$ distinct cases. (The details of the analysis together with the consideration of some additional complex cases can be found in Ref. 12.) These investigations, which have been carried out using the commuting transfer matrices approach and the matrix inversion trick, lead, quite surprisingly, to a bulk partition function identical to that of the critical Potts model (or the six-vertex model). There has been no direct simple proof of this mystifying fact which Baxter[13] referred to as "weak equivalence". In general there is only the heuristic matrix inversion argument, except for one case for which a Bethe Ansatz could be carried out.[11],[3]

In this and a forthcoming paper[14] we shall report on further exact results on the general q-state vertex problem. We shall use a graphical approach which permits us to discuss vertex models on arbitrary planar lattices. We shall establish new equivalences between lattice-statistical models and resolve, among other things, in simple graphical terms the problem concerning the weak equivalence observed above.

In this paper we start defining the general q-state vertex model on a square lattice. We shall show how it can be formulated, equivalently, as an interaction-around-a-face model. This equivalence establishes a connection between two types of lattice-statistical problems, which are often considered in different contexts. In Section 2 we shall also define the nonintersecting string (NIS) model. The particular NIS model considered in this paper is a "separable" one in which the vertex weights can be written as sums of those of basic types. In Section 3 we shall consider such a q-state NIS model on an arbitrary planar lattice of valence 4, and show that it is equivalent to a q^2-state Potts model. This equivalence can be extended to an oriented NIS model in which edges of certain colors also carry arrows. This model contains the ice-rule model as a special case when all edges are oriented. In Section 4 we shall consider a q-state NIS model on an

[3] After the completion of this research we received a preprint from T. T. Truong, who has given a proof of this weak equivalence through the consideration of the model of A. B. Zamolodchikov and M. I. Monastyrskiĭ, Zh. Eksp. Teor. Fiz. *77*: 325 (1979) [Sov. Phys. JETP *50*: 167 (1979)].

arbitrary even-valenced lattice, and show that it is also equivalent to a q^2-state Potts model, although now in general with pair as well as multispin interactions. In Section 5 we shall apply these results to regular lattices and deduce critical properties of the separable NIS model from the known properties of the Potts model.

In a later paper[14] we shall study the Baxter–Yang relation for the general NIS model. We shall verify that it is also satisfied by our general oriented NIS model for a suitable parametrization. We shall discuss implications of this, including a graphical derivation of the inversion relation and the solution of the solvable NIS models.

2. GENERAL VERTEX MODEL

2.1. Definition

In this section we consider a square lattice $\mathscr{L}$ of N sites with periodic boundary conditions. Each lattice edge of $\mathscr{L}$ can be in one of q distinct states (colors) which are specified by an edge (string) variable $\mu = 1, 2,..., q$. A vertex weight $\omega_i(\lambda, \mu, \alpha, \beta)$ is assigned to the ith vertex whose four incident edges are in respective states λ, μ, α, and β. Then, in the most general case, we have q^4 distinct vertex weights, and a q^4-vertex model. Particularly for $q=2$, this becomes the 16-vertex model.[15] We wish to compute the per site partition function

$$\kappa = \lim_{N \to \infty} Z^{1/N} \tag{1}$$

Here Z is the partition function given by

$$Z = \sum \prod_{i=1}^{N} \omega_i(\lambda, \mu, \alpha, \beta) \tag{2}$$

where the summation is taken over all $2N$ edge configurations of the lattice and the product is taken over all N vertex weights.

2.2. Equivalence with an IRF Model

It has become customary to study lattice models utilizing the interaction-around-a-face (IRF) language for which states are assigned to lattice faces, rather than edges.[1] To establish a connection with our considerations we shall now show that the IRF model and the vertex model formulations can be seen as entirely equivalent. Specifically, we shall show that a q-state vertex model can always be transcribed into a q^2-state IRF model defined on the *same* lattice, and that, conversely, any q-state IRF model can be reformulated as a q^2-state vertex problem. Specific 1–1 mappings are given in Figs. 1a and 1b. In Fig. 1a we assign to each edge between faces with states a and b the state $\overline{ab} \equiv (a-1)\,q + b$; to all vertices with a configuration inconsistent with this assignment we give a weight $\omega = 0$; if the configuration is consistent we identify the weights, i.e., $\omega(\overline{ad}, \overline{bc}, \overline{ab}, \overline{dc}) \equiv W(a, b, c, d)$. In Fig. 1b we assign to each face a state

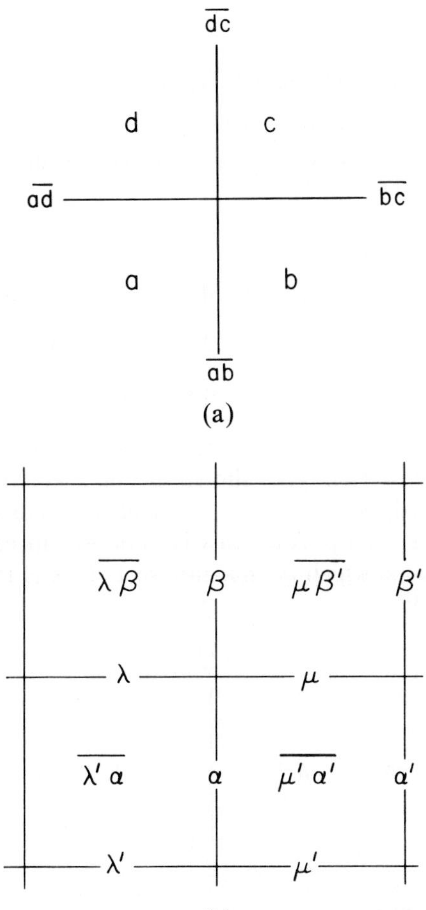

Fig. 1. (a) Configuration with Boltzmann weight $W(a, b, c, d)$ of a q-state IRF model, $a, b, c, d = 1, 2,..., q$, and the corresponding q^2-state vertex model configuration.
(b) Configuration with Boltzmann weight $\omega(\lambda, \mu, \alpha, \beta)$ of a q-state vertex model, $\lambda, \mu, \alpha, \beta = 1, 2,..., q$, and the corresponding q^2-state IRF-model configuration.

made up from the states to the right of and below the face; we assign weight $W=\omega$ or 0 depending on whether the IRF model configuration is consistent or inconsistent (following the prescription of Fig. 1b) with a vertex-model configuration.

We should note that there are, in specific models, other interesting mappings between vertex and IRF models. As examples, we mention the relation between the eight-vertex model and the Ising model with two- and four-spin interactions[15–17] and between the hard-hexagon model and a 3-state vertex model.[11]

2.3. Nonintersecting String (NIS) Model

The most general q-state model defined by (2) is a q^4-vertex model. Studies in the past,[7–12] however, have focused primarily on a subclass when the vertex weights satisfy

$$\begin{aligned} \omega_i(\lambda, \mu, \alpha, \beta) &= w^d_{\rho\rho}, && \text{if} \quad \lambda=\mu=\alpha=\beta=\rho \\ &= w^r_{\rho\sigma}, && \text{if} \quad \rho=\lambda=\alpha\neq\beta=\mu=\sigma \\ &= w^l_{\rho\sigma}, && \text{if} \quad \rho=\mu=\alpha\neq\lambda=\beta=\sigma \\ &= 0, && \text{otherwise} \end{aligned} \tag{3}$$

where the indices λ, μ, α, β are positioned as shown in Fig. 1b. In other words, only the three vertex types shown in Fig. 2 are allowed. If one now traces along lattice edges of the same color always making 90° turns, one eventually completes a loop. After this is done for all edges, the lattice is decomposed into loops which do not intersect. This is the nonintersecting string (NIS) model.[11]

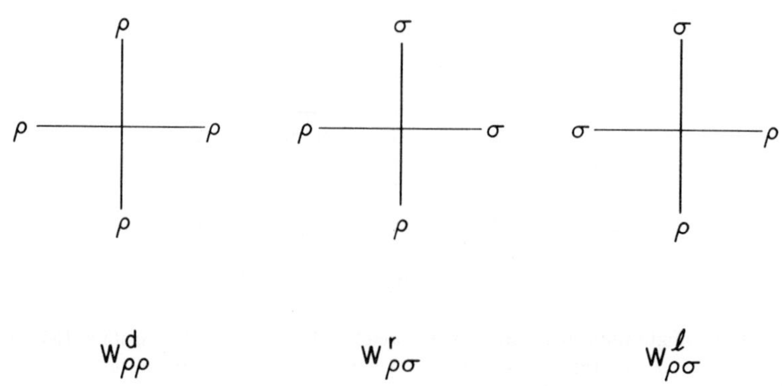

Fig. 2. The three allowed vertex types in the NIS model on a square lattice.

Generally, the restriction (3) permits $q+2q(q-1)=q(2q-1)$ distinct vertex configurations, and the q-state NIS model becomes a $q(2q-1)$-vertex model. This leads to, for $q=2$, a 6-vertex model which can be directly mapped into a staggered ice-rule model.[4] The $q=3$ problem was considered by Stroganov,[(8)] who found two soluble cases. The most general solution is by Perk and Schultz[(7,10,11)] who solved the NIS model (3) in $q+1$ distinct cases for arbitrary $q \geqslant 3$. It was found that, in all soluble cases, the solution is the same as that of the critical Potts model. In the next section we show, more generally, a special NIS model can always be formulated as a Potts model, and that this can be done for any planar lattice with arbitrary Potts interactions. Particularly, the critical Potts model on the square lattice which is exactly soluble, leads to one of the previously solved cases, and this explains the weak equivalence mentioned above.

The definition of the NIS model can be extended to any lattice which has even valences at all sites. In the general NIS model only those vertex configurations which can be decomposed into non-intersecting trajectories are allowed. Globally, the lattice is decomposed into loops of given colors, which do not intersect. Explicit examples of allowed vertex configurations will be given later for the case of valence 6.

3. EQUIVALENCE OF A NIS MODEL WITH A POTTS MODEL: ARBITRARY LATTICE OF VALENCE 4

3.1. NIS Model on a Surrounding Lattice

In this section we consider a NIS model on an arbitrary planar lattice $\mathscr{L}'$ of valence 4, which does not have to be regular. The NIS model is a $q(2q-1)$-vertex model defined by the vertex types shown in Fig. 2. Since the set of faces of an even-valenced lattice is bipartite, it is convenient to shade every other face of $\mathscr{L}'$, so that the pairs of two edges having the same label (color) will either separate or join two shaded areas at a given vertex. Then we consider a NIS model with site-dependent vertex weights w_i, $i=1, 2,\ldots, N$,

$$
\begin{aligned}
w_i &= A_i + B_i, && \text{if all 4 edges have the same color} \\
&= A_i, && \text{if the shaded areas are joined} \\
&= B_i, && \text{if the shaded areas are separated}
\end{aligned} \tag{4}
$$

[4] The mapping can be carried out by following the prescription given by Fig. 5 of Ref. 15. The resulting ice-rule model has the weights $(w^l_{12}, w^l_{21}, w^r_{21}, w^r_{12}, w^d_{22}, w^d_{11})$ and $(w^l_{21}, w^l_{12}, w^r_{12}, w^r_{21}, w^d_{11}, w^d_{22})$ alternately, on the two sublattices I and II of the square lattice, and is soluble if $w^l_{12}=w^l_{21}$, $w^r_{12}=w^r_{21}$, $w^d_{11}=w^d_{22}$.

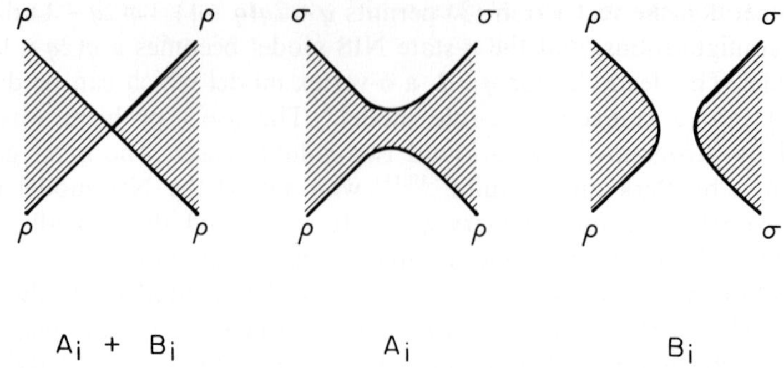

Fig. 3. Vertex weights of a NIS model on an arbitrary lattice of valence 4 ($\rho \neq \sigma$).

These situations are shown in Fig. 3. Note that in the case of a square lattice the weights w^r (and w^l) in (3) are equal to A_i and B_i alternately as i ranges from sublattice I to sublattice II, and the NIS model is therefore "staggered." Note that in (4) and Fig. 3 we can regard the vertices with weights A_i and B_i as two basic types. Then the weight of the vertex with 4 edges having the same color can be written as the sum of those of the basic types, corresponding to the two ways the vertex configuration can be decomposed. In this sense the vertex weight given by (4) is separable.

The four-coordinated lattice $\mathscr{L}'$ can be regarded as the surrounding graph (lattice) of another lattice $\mathscr{L}$ (or $\mathscr{L}_D$, the dual of $\mathscr{L}$) whose sites reside in the shaded (or unshaded) faces of $\mathscr{L}'$.[18] For planar $\mathscr{L}$ we need to pay special attention at the boundary. The boundary sites of $\mathscr{L}$ (or $\mathscr{L}_D$) are closed in by introducing "external" sites for $\mathscr{L}'$ and connecting them by straight edges. Readers are referred to Ref. 18 for examples of explicit constructions. In particular, $\mathscr{L}$ is a simple square lattice if $\mathscr{L}'$ is simple square, and $\mathscr{L}$ is either triangular (or hexagonal, the dual of triangular) if $\mathscr{L}'$ is the Kagomé lattice. At the boundary we require the edge colors be conserved at all external sites so that $\mathscr{L}'$ can again be decomposed into nonintersecting loops. While this requirement imposes a severe constraint on the vertex types that may occur at the boundary, it will not affect the bulk partition function (1) as long as the vertex weights (4) are all positive. Finally, the external sites always carry weights 1, independent of the color of the two incident edges.

3.2. Equivalence with a Potts Model

Our main result in this section is the equivalence of the q-state NIS model (4), defined on $\mathscr{L}'$, with a q^2-state Potts model defined on $\mathscr{L}$

(or $\mathscr{L}_D$). It is clear that the N sites of $\mathscr{L}'$ coincide with the N edges of $\mathscr{L}$ (or $\mathscr{L}_D$). Let the numbers of sites of $\mathscr{L}$ and $\mathscr{L}_D$ be M and M_D, respectively. The equivalence is then stated in the following theorem:

Theorem:

$$\begin{aligned} Z_{\text{NIS}}(q) &= \left[\prod_{i=1}^{N} B_i\right] q^{-M} Z_{\text{Potts}}(q^2) \\ &= \left[\prod_{i=1}^{N} A_i\right] q^{-M_D} Z_{\text{Potts}}^{(D)}(q^2) \end{aligned} \tag{5}$$

where $Z_{\text{NIS}}(q)$ is the partition function (2) of the q-state NIS model (4), $Z_{\text{Potts}}(q^2)$ is the partition function of a q^2-state Potts model on $\mathscr{L}$ defined in the standard way[19] with edge-dependent Boltzmann weights

$$e^{K_i} = 1 + qA_i/B_i, \qquad i = 1, 2,..., N \tag{6}$$

and $Z_{\text{Potts}}^{(D)}(q^2)$ is the partition function of a q^2-state Potts model on $\mathscr{L}_D$ with Boltzmann factors

$$e^{K_i^{(D)}} = 1 + qB_i/A_i, \qquad i = 1, 2,..., N \tag{7}$$

Proof. To prove the theorem we observe that the particular separable form of the vertex weights (4) permits us to write the partition function (2) as a sum over all nonintersecting polygonal decompositions P of $\mathscr{L}'$. The summation is linear in A_i and B_i, for all i, with $A_i(B_i)$ appearing in terms where shaded areas at the ith vertex are joined (separated). The partition function takes the form

$$Z_{\text{NIS}}(q) = \sum_P q^{p(P)} \prod_i W_i(P) \tag{8}$$

where $p(P)$ is the number of polygons (loops) in P, each of which can be colored independently in q different ways. Here $W_i(P)$ is the weight of the ith site in P, equal to $A_i(B_i)$ if the shaded areas at the site are joined (separated). The expansion (8) is the key expression which leads to an equivalence to a Potts model. The theorem is proved since the partition function of a Potts model can also be written as a sum over the same polygonal decompositions.[18] The theorem now follows from a direct comparison of (8) with Eq. (9) of Ref. 18. ∎

Corollary. The NIS model (4) is also equivalent to an ice-rule model on $\mathscr{L}'$. The ice-rule model has the following vertex weights:

$$\begin{aligned} \omega_1, \omega_2, \omega_3, \omega_4, \omega_5, \omega_6 = {} & z^{\alpha-\gamma}B_i,\ z^{\gamma-\alpha}B_i,\ z^{\beta-\delta}A_i,\ z^{\delta-\beta}A_i, \\ & z^{-\beta-\delta}B_i + z^{\alpha+\gamma}A_i,\ z^{\beta+\delta}B_i + z^{-\alpha-\gamma}A_i \end{aligned} \tag{9}$$

for internal sites, and the weight

$$\omega = z^{\alpha} \tag{10}$$

for external sites. Here, α, β, γ, and δ are the relative angles spanned by the incident edges at a given site as defined in Figs. 4 and 5, the six vertex weights (9) are numbered as in Ref. 18, and

$$z^{2\pi} + z^{-2\pi} = q \tag{11}$$

The proof that the two models have identical partition functions parallels that given in Ref. 18, and will not be repeated here.

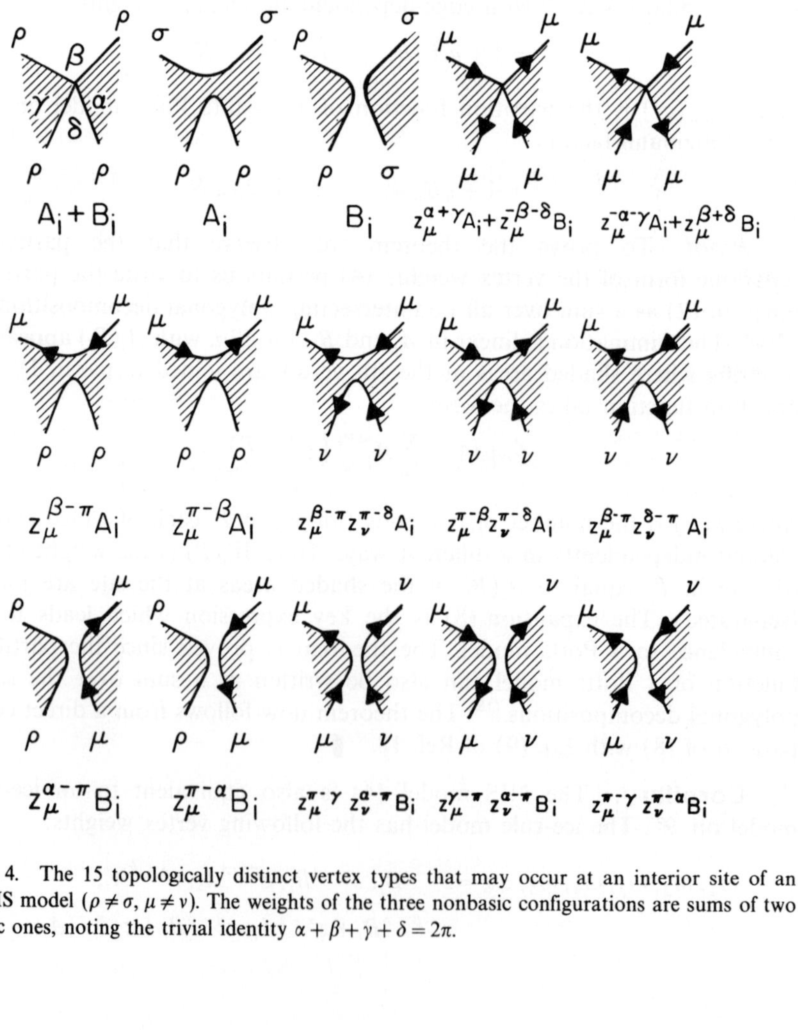

Fig. 4. The 15 topologically distinct vertex types that may occur at an interior site of an ONIS model ($\rho \neq \sigma$, $\mu \neq \nu$). The weights of the three nonbasic configurations are sums of two basic ones, noting the trivial identity $\alpha + \beta + \gamma + \delta = 2\pi$.

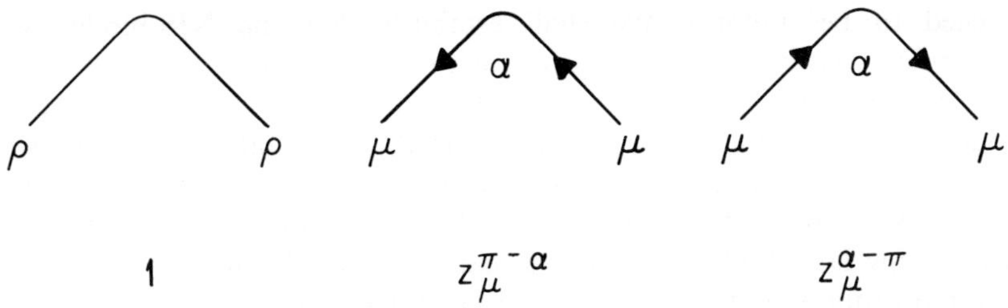

Fig. 5. The three vertex types that may occur at an external site of an ONIS model.

3.3. An Oriented NIS Model

Our proof of the theorem in the preceding section indicates that the equivalence of the NIS model (4) with a Potts model is the consequence of a local property. This observation permits an extension of the equivalence to an oriented NIS (ONIS) model.

Consider a vertex model in which the lattice edges can be colored in q_1 distinct colors and, in addition, can be colored as well as oriented (in either direction along the edge) in q_2 colors, with restriction that the numbers of in and out arrows of a given color at a vertex are the same. Again, the allowed configurations are those decomposable into altogether 15 topologically distinct vertex types that may occur at an internal site and three types that may occur at an external site. These vertex types together with their vertex weights are shown in Figs. 4 and 5. We note that the case $q_2=0$ is the (nonoriented) NIS model, and that the case $q_1=0$, $q_2=1$ is the ice-rule model. Following the same argument as in deriving (5), we can equal the partition function $Z_{\mathrm{ONIS}}(q_1, q_2)$ of this ONIS model to the right-hand side of (5) or (8), provided that we take

$$q=q_1+\sum_{\mu=1}^{q_2}(z_\mu^{2\pi}+z_\mu^{-2\pi}) \tag{12}$$

Therefore, the ONIS model defined in Figs. 4 and 5 is equivalent to a q^2-state Potts model and to a q-state (nonoriented) NIS model.

4. EQUIVALENCE OF A NIS MODEL WITH A POTTS MODEL: ARBITRARY LATTICES OF EVEN VALENCE

In this section we consider more generally a NIS model on an arbitrary planar lattice $\mathscr{L}'$ which is even valenced, i.e., the valence v_i is even for all sites i. The lattice does not have to be regular, nor does the valence

v_i need to be uniform. We shall establish that the NIS model with appropriate vertex weights is equivalent to a q^2-state Potts model.

As in Section 3, we shade every other lattice face of $\mathscr{L}'$, and the Potts model is defined on a related lattice $\mathscr{L}$ (or $\mathscr{L}_D$, the dual of $\mathscr{L}$) whose sites reside in the shaded (or unshaded) faces of $\mathscr{L}'$. The lattice $\mathscr{L}'$, which was previously introduced by one of us,[(19)] now serves the same role as a surrounding lattice. However, as we shall see, the related Potts model will have multispin as well as pair interactions for $v_i \geqslant 6$.

As before, we close in the boundary Potts spins by introducing new external sites with valence 2. The external sites conserve edge colors and always carry the weight 1. The appropriate assignment of vertex weights for internal sites is best illustrated by considering the case $v_i = 6$. In analogy to Fig. 3 where, for $v_i = 4$, two basic vertex types, namely, those associated with weights A_i and B_i, may occur at a vertex in a polygonal decomposition P, there are now five basic vertex types. These are the vertices with weights $C_1, C_2, ..., C_5$ in Fig. 6. The weights of other NIS vertices with four or more edges having the same color are then written as appropriate linear combinations of all possible NIS decompositions of the vertex. There are altogether 15 distinct NIS vertex configurations, which are shown in Fig. 6 together with the associated vertex weights. We next write the partition function $Z_{\mathrm{NIS}}(q)$ in a graphical expansion in terms of polygonal decompositions P. This leads to the expression (8), but now with

$$\begin{aligned} W_i(P) &= \{A_i, B_i\}, \qquad && \text{if} \quad v_i = 4 \\ &= \{C_1, C_2, ..., C_5\}, \qquad && \text{if} \quad v_i = 6 \end{aligned} \tag{13}$$

If $v_i = 6$ for all i, then the NIS model is precisely the five-vertex model considered by Wu and Lin.[(20)] In fact, in this case Wu and Lin have shown that the NIS partition function (8), (13) is precisely the partition function of a q^2-state Potts model[5] which has two-site interactions K_1, K_2, and K_3, and a three-site interaction K for every three Potts spins surrounding a site of $\mathscr{L}'$, provided that we take

$$\begin{aligned} C_n &= e^{K_n} - 1, \qquad n = 1, 2, 3 \\ C_4 &= q \\ C_5 &= (e^{K + K_1 + K_2 + K_3} - e^{K_1} - e^{K_2} - e^{K_3} + 2)/q \end{aligned} \tag{14}$$

For details we refer to Ref. 20. If the lattice $\mathscr{L}'$ has mixed valences 4 and 6, then the NIS model is equivalent to a q^2-state Potts model using either (6)

[5] Compare (13) with Eq. (1) of Ref. 20 and use Eq. (12) of Ref. 20 after replacing q by q^2.

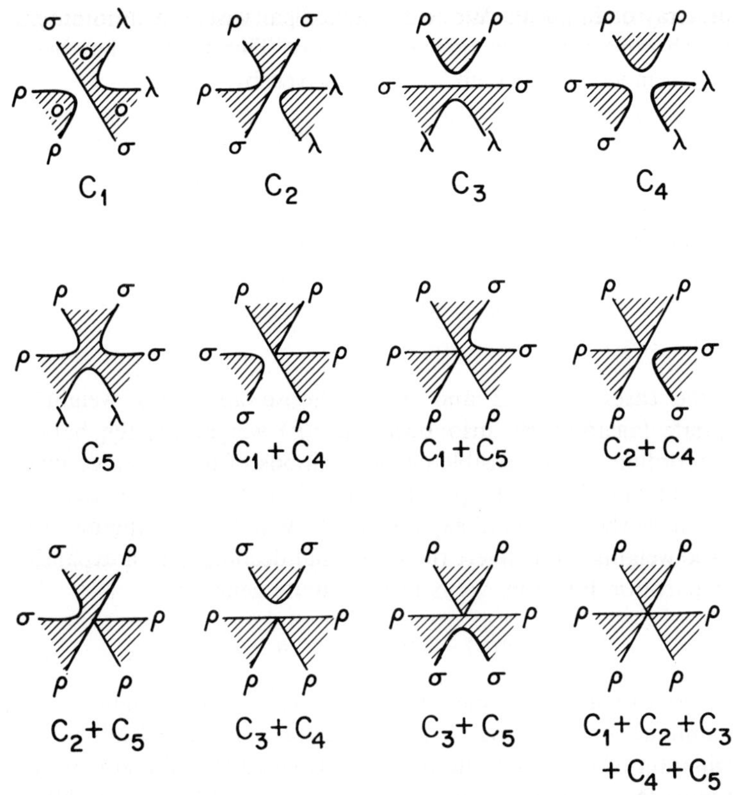

Fig. 6. NIS vertex configurations at the ith vertex of valence 6 ($\rho \neq \sigma \neq \lambda \neq \rho$, except ρ and λ may be equal in diagrams with weights C_1, C_2, and C_3). The small circles, in the first diagram, indicate the positions of the Potts spins on lattice $\mathscr{L}$.

or (14), as required by the valence, for obtaining the Potts interactions surrounding the ith site of $\mathscr{L}'$. The exact equivalence between the two partition functions is still given by (5), identifying $A_i \equiv C_5$ and $B_i \equiv C_4$ for sites of valence 6.[6]

The picture for general even valence v_i is now clear. At the ith site of $\mathscr{L}'$, there are $f(n)$ pair- and multispin Potts interactions and $g(n)$ basic NIS vertex types, where $n = v_i/2$. We have already seen that $f(2) = g(2) - 1 = 1$ and $f(3) = g(3) - 1 = 4$. Since only $g(n) - 1$ of the $g(n)$ basic vertex weights are independent, this leads to a unique determination of the Potts interactions from a given set of basic vertex weights and vice versa, for $n = 2$, 3. More generally, $g(n)$ is the number of distinct ways that

[6] This result, which holds for finite lattices, is more general than that given in Ref. 20, which assumes the thermodynamic limit.

a linear array of $2n$ points can be connected pairwisely by n nonintersecting lines which remain on one side of the array. This number has been computed by Temperley and Lieb,[21] and is given by

$$g(n)=\frac{1}{n+1}\binom{2n}{n} \tag{15}$$

The number of distinct Potts interactions is

$$f(n)=\sum_{r=2}^{n}\binom{n}{r}=2^n-n-1 \tag{16}$$

As in the cases of $n=2$ and 3 we choose all vertex weights to be appropriate linear combinations of the $g(n)$ weights of the basic vertex types, according to the possible decompositions of the vertex configuration (into basic types). Then, the partition function of the NIS model is written in the form of the polygonal expansion (8), with $W_i(P)$ ranging over these $g(n)$ basic weights. To convert the polygonal decomposition expansion into a Potts partition function, we note that, quite generally,

$$g(n)-1>f(n), \qquad n\geqslant 4 \tag{17}$$

so that we can always equate the $g(n)$ basic vertex weights with Potts Boltzmann factors involving $f(n)$ Potts interactions. This leads to a unique determination of the Potts interactions, provided that the $g(n)$ weights are constrained for $n\geqslant 4$. So we have an equivalence of the q-state NIS model on an even-valenced lattice with a q^2-state Potts model. This equivalence can again be generalized to an ONIS model with appropriate vertex weights.

5. EXACT SOLUTIONS FOR THE SEPARABLE NIS MODEL

In the preceding sections we have established the equivalence of a q-state NIS model (on an arbitrary planar even-valenced lattice) with a q^2-state Potts model. This equivalence makes it possible to deduce properties of the NIS model from known solutions of the Potts model for regular lattices.

While the NIS model is defined for integer values of q, the particular model considered in this paper, for which the vertex weights are separable, permits a natural continuation of the partition function, through the polygonal expansion (8), to noninteger values of q. We can now discuss its critical properties.

5.1. Square Lattice

Consider a q-state staggered NIS model on a square lattice whose weights are given by (4) and Fig. 3 with

$$\begin{aligned}\{A_i, B_i\} &= \{A_1, B_1\}, \qquad i \in \text{sublattice I} \\ &= \{A_2, B_2\}, \qquad i \in \text{sublattice II}\end{aligned} \tag{18}$$

or, in terms of the vertex weights (3) and Fig. 2,

$$\begin{aligned}\{w^r_{\rho\sigma}, w^l_{\rho\sigma}\} &= \{A_1, B_1\}, \qquad i \in \text{sublattice I} \\ &= \{B_2, A_2\}, \qquad i \in \text{sublattice II} \\ w^d_{\rho\rho} &= A_i + B_i\end{aligned} \tag{19}$$

Theorem (5) now relates the partition function of the NIS model with that of an anisotropic q^2-state Potts model with interactions (6) or (7).

From known properties of the Potts model[19,22] we have the following:

(i) The NIS model is exactly soluble for $q = \sqrt{2}$ for which it becomes an Ising model.

(ii) The NIS model exhibits a continuous transition for $1 < q \leqslant 2$, with the critical exponents varying continuously with the value of q.

(iii) The NIS model exhibits a first-order transition for $q > 2$ accompanied by a nonzero latent heat which can be computed.

The anisotropic Potts model is exactly soluble[1,5] at its critical point

$$A_1 A_2 = B_1 B_2 \tag{20}$$

If we divide all vertex weights on sublattice I and II by A_1 and B_2, respectively, then the NIS model has uniform weights and becomes the soluble case NIS1 considered by Perk and Schultz.[11] Thus, we have a simple proof that the NIS1 model free energy of Ref. 11 is identical to that of the critical Potts model. This is the weak equivalence referred to in Ref. 13. Explicit expressions of the critical free energy will be discussed in a later paper.[14]

5.2. Kagomé Lattice

Consider a q-state NIS model on the Kagomé lattice with vertex weights given by (4) and Fig. 3, with

$$\begin{aligned}\{A_i, B_i\} &= \{A_1, B_1\}, &\quad i \in \text{sublattice I}\\ &= \{A_2, B_2\}, &\quad i \in \text{sublattice II}\\ &= \{A_3, B_3\}, &\quad i \in \text{sublattice III}\end{aligned} \tag{21}$$

Then Theorem (5) relates its partition function to that of an anisotropic q^2-state Potts model on a triangular (honeycomb) lattice with interactions (6) and (7). The same general conclusions on its critical properties can now be drawn as in the case of the square lattice. The NIS model is exactly soluble at its critical point

$$q + \frac{B_1}{A_1} + \frac{B_2}{A_2} + \frac{B_3}{A_3} = \frac{B_1 B_2 B_3}{A_1 A_2 A_3} \tag{22}$$

obtained from the corresponding Potts critical point.[6,23] The solution has been given in Ref. 6.

5.3. Triangular Lattice

Consider a q-state NIS model on a triangular lattice with vertex weights given by Fig. 6. Our analysis in Section 4 relates its partition function to that of a triangular Potts model with two- and three-spin interactions for every three spins surrounding an up-pointing triangle. The Potts interactions K_1, K_2, K_3, and K are uniquely determined from (14). The NIS model is self-dual and is critical at its self-dual point[6,24]

$$C_4 = C_5 \tag{23}$$

6. SUMMARY

We have established the exact equivalence of the partition function of a separable q-state NIS model on any even-valenced planar lattice with that of a q^2-state Potts model on a related lattice. The equivalence also holds for the partition function of a generalized separable q-state NIS model for which the lattice edges can be oriented. The resulting Potts model generally has multispin interactions. Critical properties of the NIS model are derived from this equivalence and the known properties of the Potts model. This includes a previously solved NIS model on a regular square lattice, now identified as the exact solution of the Potts model at the critical point.

ACKNOWLEDGMENT

This work was supported in part by NSF grants Nos. DMR-8206390 and DMR-8219254.

REFERENCES

1. R. J. Baxter, *Exactly Solved Models in Statistical Mechanics* (Academic, London, 1982).
2. E. H. Lieb, *Phys. Rev. Lett.* **18**:692, 1046 (1967); *Phys. Rev. Lett.* **19**:108 (1967); B. Sutherland, *Phys. Rev. Lett.* **19**:103 (1967); C. P. Yang, *Phys. Rev. Lett.* **19**:586 (1967); B. Sutherland, C. N. Yang, and C. P. Yang, *Phys. Rev. Lett.* **19**:588 (1967).
3. S. B. Kelland, *Aust. J. Phys.* **27**:813 (1974).
4. R. J. Baxter, *Ann. Phys. (N.Y.)* **70**:193 (1972).
5. R. J. Baxter, *J. Phys. C* **6**:L445 (1973).
6. R. J. Baxter, H. N. V. Temperley, and S. E. Ashley, *Proc. R. Soc. London Ser. A* **358**:535 (1978).
7. J. H. H. Perk and C. L. Schultz, in *Non-Linear Integrable Systems—Classical Theory and Quantum Theory*, Proc. RIMS Symposium, M. Jimbo and T. Miwa, eds. (World Scientific, Singapore, 1983).
8. Yu. G. Stroganov, *Phys. Lett.* **74A**:116 (1979).
9. C. L. Schultz, *Phys. Rev. Lett.* **46**:629 (1981).
10. J. H. H. Perk and C. L. Schultz, *Phys. Lett.* **84A**:407 (1981).
11. J. H. H. Perk and C. L. Schultz, *Physica* **122A**:50, (1983).
12. C. L. Schultz, Ph. D. dissertation (State University of New York, Stony Brook, 1982).
13. R. J. Baxter, *J. Stat. Phys.* **28**:1 (1982).
14. J. H. H. Perk and F. Y. Wu, *Physica A* (in press).
15. E. H. Lieb and F. Y. Wu, in *Critical Phenomena and Phase Transitions*, Vol. I, C. Domb and M. S. Green, eds. (Academic, London, 1972), p. 331.
16. F. Y. Wu, *Phys. Rev.* **B4**:2312 (1971).
17. L. P. Kadanoff and F. J. Wegner, *Phys. Rev.* **B4**:3989 (1971).
18. R. J. Baxter, S. B. Kelland, and F. Y. Wu, *J. Phys.* **A9**:397 (1976).
19. F. Y. Wu, *Rev. Mod. Phys.* **54**:235 (1982).
20. F. Y. Wu and K. Y. Lin, *J. Phys.* **A13**:629 (1980).
21. H. N. V. Temperley and E. H. Lieb, *Proc. R. Soc. London Ser. A* **322**:251 (1971).
22. F. Y. Wu, *J. Appl. Phys.* **55**:2421 (1984).
23. A. Hintermann, H. Kunz, and F. Y. Wu, *J. Stat. Phys.* **19**:623 (1978).
24. F. Y. Wu and R. K. P. Zia, *J. Phys.* **A14**:721 (1981).

Printed by the St. Catherine Press Ltd., Tempelhof 41, Bruges, Belgium

9. Knot Invariants

Knot theory and statistical mechanics

F. Y. Wu

Department of Physics, Northeastern University, Boston, Massachusetts 02115

This is a tutorial review on knot invariants and their construction using the method of statistical mechanics. We begin with brief reviews of the elements of knot theory and relevant results in statistical mechanics. We then show how knot invariants, including those discovered recently, can be obtained by applying techniques used in solving lattice models in lattice statistics. Our approach is based on the consideration of solvable models with strictly local Boltzmann weights. The presentation, which is self-contained and elementary, is intended for a general readership. A table of polynomial invariants for knot and links containing up to six crossings is included in the Appendix.

CONTENTS

I. INTRODUCTION

An exciting development occurring recently in the mathematical theory of knots and links is the discovery of new knot invariants (Freyd *et al.*, 1985; Jones, 1985; Akutsu and Wadati, 1987a; Kauffman, 1990) and their connection with statistical mechanics (Kauffman, 1987a; Jones, 1989). Particularly, the newly discovered connection with statistical mechanics has permitted a simple and direct formulation of knot invariants, a longstanding fundamental problem in knot theory. (For an account of these developments readable by nonexperts see Jones, 1990a).

Knots and links are loops of strings possessing threading and knotting properties that are topological in nature. To characterize these topological properties algebraically, mathematicians have found that certain polynomials can be used. These polynomials are knot invariants. Traditionally, each of the polynomial invariants was discovered and constructed under different circumstances, often requiring lengthy and tedious analyses. While knot invariants can now be analyzed using braid groups and understood within the framework of quantum groups (see, for example, Reshetikhin and Turaev, 1991), the statistical mechanical approach remains, in contrast, simple and elementary. Despite its simplicity and usefulness, however, the connection of knot invariants with statistical mechanics has remained largely a topic unfamiliar to physicists, including many in the community of statistical mechanics where the main idea of the new approach is rooted. To be sure, a number of monographs and articles addressing the connection of knot theory with physics have appeared recently (e.g., Kauffman, 1988a, 1988b, 1991; Turaev, 1988; Jones, 1989; Wadati *et al.*, 1989; Yang and Ge, 1989), but they have been written mostly for mathematicians and cover aspects that may not be familiar to all physicists. There exists an apparent gap of communication between these two communities.

The purpose of this review is to introduce the recent advances in knot theory to physicists, explain what knot invariants are about, and show how they can be derived and understood from the point of view of statistical physics. To accomplish this, it is necessary to reformulate and rework a large body of existing and known results in knot theory and recast them within the framework of conventional statistical mechanics. The scope of this review is therefore necessarily limited, and confined only to the stated purpose. It is not our intent to review knot theory, nor do we intend to explore the braid group and the associated algebraic approach. We also do not discuss the role played by knot invariants in topological field theory (Witten, 1989a, 1989b, 1990), which could be the topic of a treatise by itself.

This review assumes no prior knowledge of knot theory and statistical mechanics and is therefore self-contained and suitable for a general readership. We shall cite and refer to original references as they arise in the course of our presentation, but no attempt will be made at a complete literature survey. Readers are referred to two recent books by Kohno (1991) and Kauffman (1991), which contain complete listings of the relevant literature in mathematics.

The organization of this review is as follows: In Sec. II we present elements of knot theory and introduce knot invariants and their traditional definition in terms of the Skein relation. The basic idea of approaching knot invariants using statistical mechanical methods is explained in Sec. III. This is followed by a review of relevant solvable, or integrable, vertex models in Sec. IV. These results are used in Sec. V to obtain knot invariants. We next show in Sec. VI that interaction-round-a-face (IRF) models can always be formulated as vertex models and use this formulation to derive knot invariants from IRF models. In Sec. VII we consider the construction of knot invariants from edge-interaction models, the spin models with pure two-spin interactions. For completeness we include in the Appendix a table of knot invariants for prime knots and links containing six or fewer crossings.

II. THEORY OF KNOTS

A. Definitions

We begin with a description of some terms in knot theory. A knot is the embedding of a circle, or a nonintersecting loop, in three-dimensional space. A link is a collection of two or more knots, or components, which may or may not be physically intertwined. In this paper we use the term *knot* loosely to denote either a single-component knot or a link. A knot is *oriented* if its loops are directed; otherwise, the knot is *unoriented*. While knots are unoriented to begin with, it is often convenient to direct the loops and consider oriented knots. Starting from a given unoriented knot consisting of n components, one has generally 2^n versions of oriented knots.

Knots can be projected onto a plane and thus represented by planar diagrams. We shall always have planar diagrams, or projections, in mind when we speak of knots. Diagrams of some simple knots are shown in Fig. 1. It can be seen that when using planar diagrams to represent knots we need to break one of the two lines crossing at an intersection to indicate their relative positionings in three-dimensional space. For oriented knots this leads to two kinds of line crossings, denoted by the signs + and −, as shown in Fig. 2, where the + and − crossings are related by a 90° rotation.

An intrinsic property, the *writhe* $w(K)$, of a knot K is defined by first orienting the knot and then computing

$$w(K)=n_+-n_- , \tag{2.1}$$

where n_+ and n_- are, respectively, the numbers of + and − crossings in the knot. For single-component knots the writhe $w(K)$ is uniquely defined, independent of the line orientation chosen. For example, the writhe of the right-handed trefoil in Fig. 1(c) remains 3 if the orientation is reversed. For knots with two or more components, the writhe will generally depend on the relative orientations of the components.

An *unknot* is a knot represented by a circle. Two knots are equivalent if they can be transformed into each other by a continuous deformation of the lines. Thus the knot in Fig. 1(b) is equivalent to an unknot and, in fact, all single-component knots with one or two crossings are equivalent to an unknot. The simplest nontrivial knot is the trefoil with three crossings. There are a total of 12 965 distinct single-component knots with 13 or fewer crossings, excluding mirror images, for which a complete

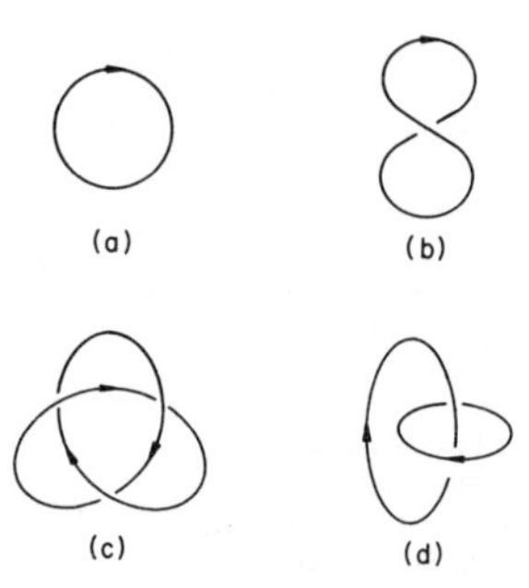

FIG. 1. Examples of planar representations of oriented knots: (a) and (b), unknot; (c) right-handed trefoil; (d) Hopf link.

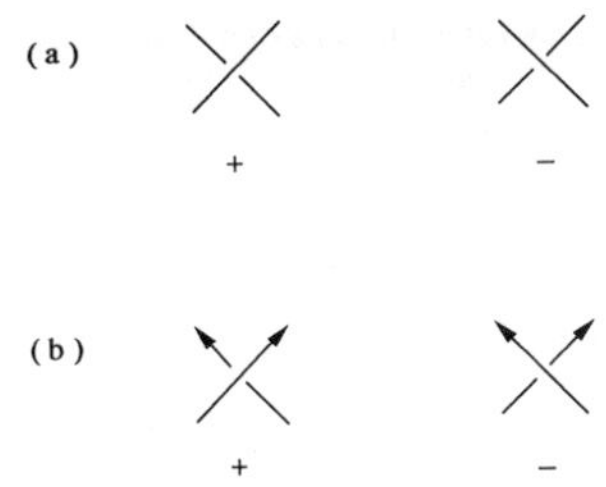

FIG. 2. Two kinds of line crossings: (a) unoriented knots, (b) oriented knots.

tabulation exists (Thistlethwaite, 1985).

While it is relatively easy to verify by inspection that the two knots in Figs. 1(a) and 1(b) are indeed equivalent, it is generally difficult to identify the equivalence of knots without doing some tedious checking. The problem lies in the fact that two equivalent knots may have very different planar projections. This leads to the need for characterizing knots algebraically. The idea is to associate algebraic functions with knots such that equivalent knots possess the same identical function. Such algebraic functions are the topological invariants of knots, or simply *knot invariants.* It should be mentioned that, despite recent progress, the construction of knot invariants that are different for all distinct knots still remains an open problem.

B. Reidemeister moves

As a first step in constructing knot invariants, one needs to understand the process of deforming knots. To this end it was shown by Reidemeister (1948) that all deformations of knots (in three-dimensional space) can be broken down into sequences of three basic types of line moves (in the two-dimensional projection), the Reidemeister moves. Thus it is sufficient to consider each of the three Reidemeister moves individually. We describe these moves for oriented and unoriented knots separately.

1. Unoriented knots

The three types of Reidemeister moves for unoriented knots are shown in Fig. 3; all other moves can be obtained from those shown by applying a rotation and/or a reflection. The algebraic function associated with a knot is said to be of an invariant of *ambient isotopy* if it is invariant under all I, II, and III types of moves, and of *regular isotopy* if invariant under the type-II and type-III moves only.

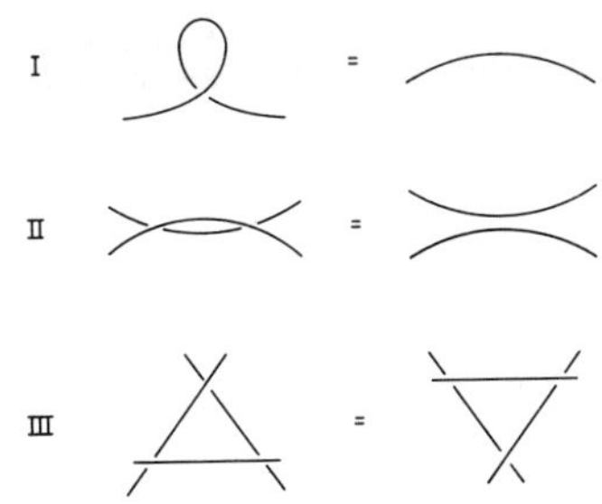

FIG. 3. Three types of Reidemeister moves for unoriented knots.

2. Oriented knots

The consideration of Reidemeister moves is more involved for oriented knots, since the lines are directed, thus breaking some symmetry. However, the basic topology of the moves remains unaltered. For type-I and type-II moves, it can be seen that there exists two independent moves of each type. These are the moves shown in Fig. 4; all other type-I and type-II moves can be obtained from those shown by applying a rotation and/or a reflection. For example, a reflection of the move IIA about a horizontal axis yields a move given by the same diagram but with the crossings + and − interchanged.

The type-III moves for oriented knots require more attention. Basically, there exist two distinct kinds of line orientations, types IIIA and IIIB, shown in Figs. 5 and 6, which differ in the way that the three lines are oriented. In each kind of line orientation, there further exist six distinct possible moves, and all other type-III moves are related to those shown by rotations and/or reflections. Note that the Reidemeister moves shown are those dictated by legitimate line moves (in three-dimensional

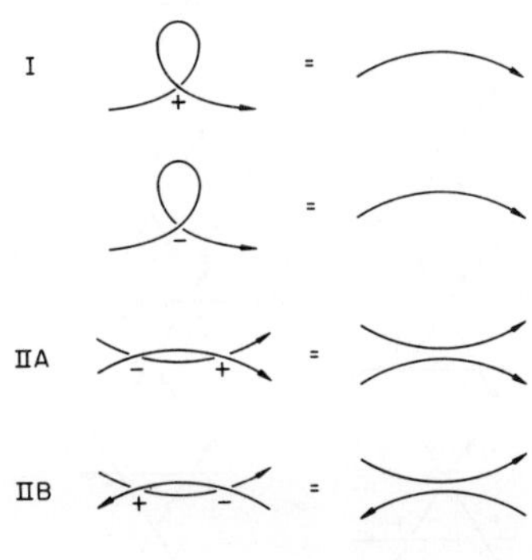

FIG. 4. Type-I and type-II Reidemeister moves for oriented knots. In move IIA the two lines point into the same half-plane, and in move IIB the lines point in opposite directions.

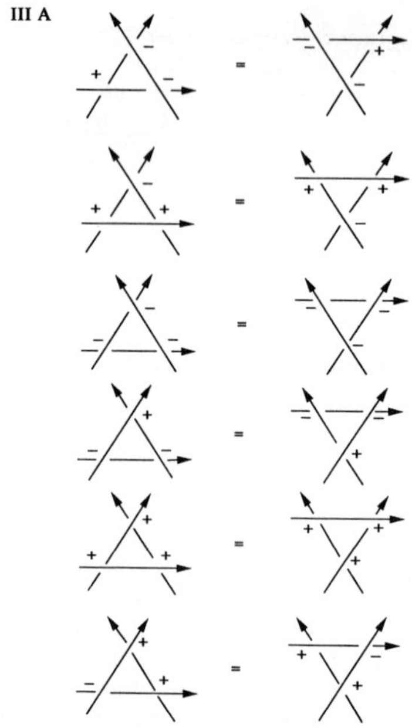

FIG. 5. Type-IIIA moves for oriented knots. The three lines are oriented to point into the same half-plane.

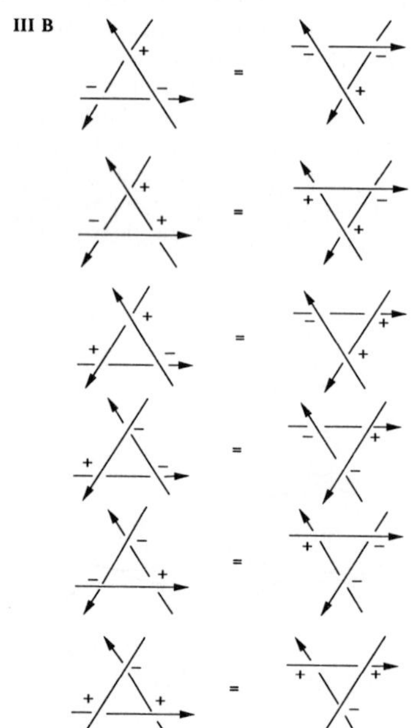

FIG. 6. Type-IIIB moves for oriented knots. The three lines are oriented to form a net circulation.

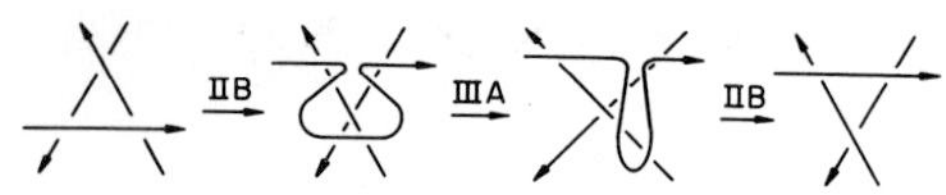

FIG. 7. A type-IIIB move as deduced from type-IIB and one of the six type-IIIA moves.

space), and therefore are not necessarily all independent. In fact, it suffices to consider only the type-I and type-II moves shown in Fig. 4 and any one of the six IIIA moves. All other moves, including the six-type-IIIB, follow as a consequence (Turaev, 1988). We include in Fig. 7 an example of how the type-IIIB move shown in the second line of Fig. 6 can be deduced by using the IIB and one of the IIIA moves (Kauffman, 1991). For comparison, we show in Fig. 8 configurations that cannot be disentangled by line moves. Note that configurations in Fig. 8 complement those in Figs. 4–6, so that altogether they give rise to all possible kinds of crossings that two and three lines can form.

As in the case of unoriented knots, the term ambient isotopy refers to invariance under all types (I, II, and III) of moves, and regular isotopy to invariance under type-II and type-III moves only. It is clear that the writhe $w(K)$ of a knot given by Eq. (2.1) is regular isotopy invariant, and that, under type-I moves, it changes by 1.

C. The Skein relation

Skein relations are recursion relations relating the invariants of knots whose diagrams are identical except that the connectivity of lines in a small region embedded in the knot is different. The most common Skein relations are described below.

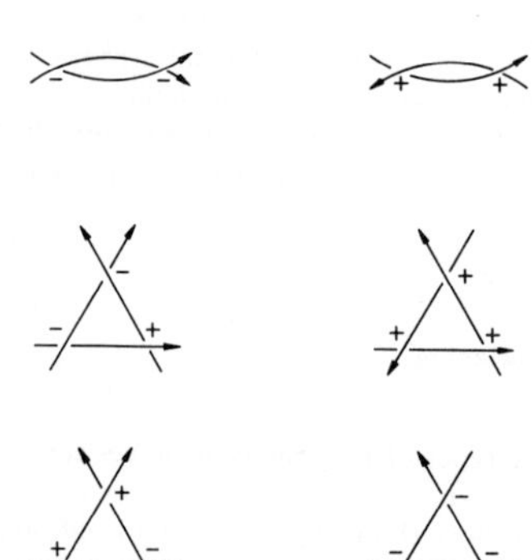

FIG. 8. Configurations that cannot be disentangled by line moves.

FIG. 9. The three line configurations considered in the Skein relation for oriented knots.

1. Oriented knots

For oriented knots there are two kinds of line crossings denoted by the signs + and − shown in Fig. 2 and again in Fig. 9. If a plus-type line crossing in a given knot is switched to a minus type, we arrive at, of course, a different knot. Furthermore, if the crossing is spliced so that it is replaced by a configuration denoted by 0 in Fig. 9, one obtains yet a third knot. We denote the three knots, respectively, by L_+, L_-, and L_0 and their associated invariants P_{L_+}, P_{L_-}, and P_{L_0}. The simplest Skein relation is then a recursion relation connecting these three P functions,

$$xP_{L_+}(x,y,z)+yP_{L_-}(x,y,z)=zP_{L_0}(x,y,z) , \qquad (2.2)$$

where x, y, z are variables of the invariant.

The knot containing the configuration L_0 is simpler in the sense that it contains one less line crossing than the other two. Then, by applying the Skein relation Eq. (2.2) and Reidemeister moves repeatedly, one eventually equates the P function of any knot to a product of two factors: a Laurent polynomial of homogeneous degree zero in variables x, y, and z, and $P_{\text{unknot}}(x,y,z)$, the P function of an unknot which, without loss of generality, can be taken to be $P_{\text{unknot}}(x,y,z)=1$. Thus the P function is a Laurent polynomial of degree zero in x, y, z.

For example, applying Eq. (2.2) to the three knots shown in the first row in Fig. 10, one obtains

$$x\cdot 1+y\cdot 1=zP_{2l}(x,y,z) , \qquad (2.2a)$$

and thus the P function for two unlinked loops is $P_{2l}(x,y,z)=(x+y)/z$. Applying the Skein relation to the three knots in the second row in Fig. 10, one obtains

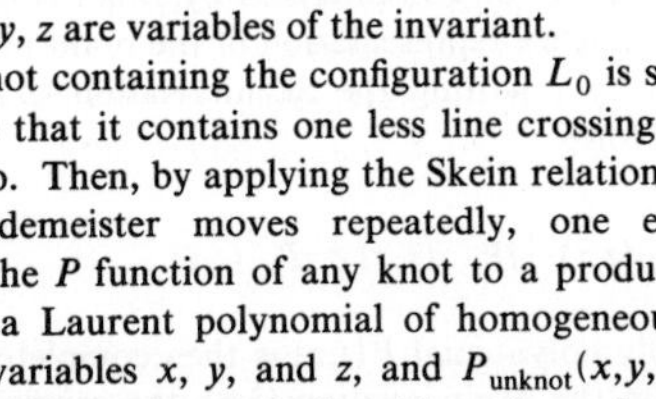

FIG. 10. Examples of knots connected by the Skein relation.

$$xP_{\text{Hopf}}(x,y,z)+y\cdot\frac{1}{z}(x+y)=z\cdot 1 , \qquad (2.2b)$$

and hence the P function for the Hopf link is $P_{\text{Hopf}}=(z^2-xy-y^2)/xz$. From the third row of Fig. 10 one has

$$xP_{\text{RT}}(x,y,z)+y\cdot 1=z\cdot\frac{1}{xz}(z^2-xy-y^2) , \qquad (2.2c)$$

leading to the P function $P_{\text{RT}}(x,y,z)=(z^2-2xy-y^2)/x^2$ for the right-handed trefoil. One may also verify that the same results are always reached, independent of the order in which the Skein relation is applied to line crossings.

For the P function to be a true knot invariant, we need to ascertain its existence and uniqueness. That is, the P function so obtained is independent of the order in which the Skein relation is applied to line crossings for *arbitrary* knots. This is the crux of this approach. Indeed, the fact that the Skein relation Eq. (2.2) with general x, y, and z actually defines a knot invariant, the Homfly polynomial, was not recognized until very recently (Freyd *et al.* 1985), and only after its validity in two special instances became known (Conway, 1970; Jones, 1985).

The P function of the mirror image of a knot is obtained by interchanging x and y, since reflections interchange the crossings + and −. For example, the mirror image of the right-handed trefoil of Fig. 1(c) produces a left-handed trefoil with the invariant $P_{\text{LT}}(x,y,z)=(z^2-2xy-x^2)/y^2$. It is also clear that the reversal of all arrows does not change the ± types of crossings and therefore leaves the P function unchanged. But the reversal of arrows in one component of a link generally leads to a different P function.

2. Unoriented knots

Skein relations can be similarly written down for unoriented knots. However, as the four lines at a crossing can now be connected in four different ways, denoted by +, −, 0, and ∞, as shown in Fig. 11, the Skein relation for unoriented knots relates four functions P_{D_+}, P_{D_-}, P_{D_0}, and P_{D_∞}, where D_+ is the knot with the configuration +, etc. An example is the Skein relation for the Kauffman polynomial (Kauffman, 1990) given in Sec. II.D.5.

3. Other Skein relations

More generally, one can define Skein relations relating knots that differ in a small disk containing other types of

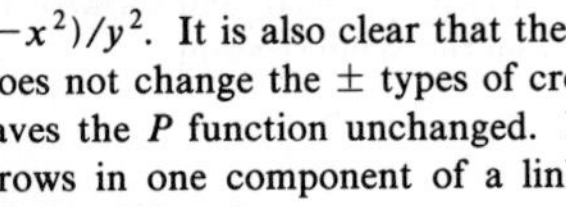

FIG. 11. The four line configurations considered in the Skein relation for unoriented knots.

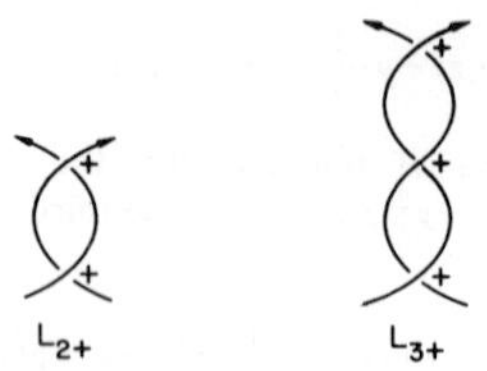

FIG. 12. Line configurations for the general Skein relation.

configurations, such as those denoted by L_{2+}, L_{3+} in Fig. 12. The Akutsu-Wadati polynomial satisfies Skein relations relating precisely these configurations.

D. Polynomial invariants

In this section we give the traditional definition of several known knot invariants using the Skein relation.

1. The Alexander-Conway polynomial

Alexander (1928) discovered the first knot invariant $\Delta(t)$ for oriented knots. His derivation is a combinatorial one. Years later Conway (1970) obtained an invariant $\nabla(z)$, the Conway polynomial, and showed that it can be determined from the Skein relation

$$\nabla_{L_+}(z)-\nabla_{L_-}(z)=z\nabla_{L_0}(z) \tag{2.3}$$

in conjunction with the conditions that $\nabla(z)$ be ambient isotopy invariant and

$$\nabla_{\text{unknot}}(z)=1 \; . \tag{2.4}$$

He further showed that $\nabla(z)$ is related to the Alexander polynomial $\Delta(t)$ through the relation

$$\nabla(z)=\Delta(t), \quad z=\frac{1}{\sqrt{t}}-\sqrt{t} \; . \tag{2.5}$$

It is clear that the Skein relation (2.3) is a special case of Eq. (2.2) with $x=1$, $y=-1$. Alexander polynomials of single-component knots are polynomials of z^2 (Lickorish and Millett, 1987) and are therefore symmetric in t and $1/t$. It also follows from Eq. (2.3) that the Alexander-Conway polynomial of a knot containing unlinked components vanishes identically. Until the discovery of the Jones polynomial in 1984, the Alexander-Conway polynomial had remained as the single known and useful polynomial invariant for decades.

2. The Jones polynomial

As alluded to earlier in Sec. I, recent advances in knot theory were brought about by Jones' discovery several years ago (Jones, 1985) of a new polynomial invariant $V(t)$, now known as the Jones polynomial. While Jones originally obtained $V(t)$ by analyzing the braid-group representation of knots using von Neumann algebras, he also pointed out that $V(t)$ satisfies the Skein relation[1]

$$\frac{1}{t}V_{L_+}(t)-tV_{L_-}(t)=\left[\sqrt{t}-\frac{1}{\sqrt{t}}\right]V_{L_0}(t) \tag{2.6}$$

and can be determined from it by further requiring ambient isotopy invariance and

$$V_{\text{unknot}}(t)=1 \; . \tag{2.7}$$

3. The Homfly polynomial

The similarity between the Jones and Alexander polynomials in terms of the Skein relation is very appealing. Within a few months after the announcement of Jones' result, four groups of researchers (Freyd *et al.*, 1985; see also Hoste, 1986; Lickorish and Millett, 1987) independently extended Jones's result to arrive at a new invariant. They showed that the Skein relation (2.2) with general x,y,z indeed defines a two-variable invariant, the Homfly polynomial.[2] This new invariant, which has since been rederived and analyzed by Jones (1987) using the Hecke algebra representation of the braid group, can be defined by rewriting the Skein relation (2.2) in the equivalent form

$$\frac{1}{t}P_{L_+}(t,z)-tP_{L_-}(t,z)=zP_{L_0}(t,z) \; . \tag{2.8}$$

The Homfly polynomial $P(t,z)$ is then completely determined from the Skein relation (2.8), the requirement of ambient isotopy, and the condition

$$P_{\text{unknot}}(t,z)=1 \; . \tag{2.9}$$

We have the relations

$$\nabla(z)=P(1,z), \quad V(t)=P(t,\sqrt{t}-1/\sqrt{t}\,) \; . \tag{2.10}$$

4. The Akutsu-Wadati polynomial

The Akutsu-Wadati polynomial is an example of a new knot invariant derived from exactly solvable models in statistical mechanics (Akutsu and Wadati, 1987a). For each $N=2,3,\ldots,$ the Akutsu-Wadati polynomial $A^{(N)}(t)$ is a Laurent polynomial in t satisfying ambient

[1]The definition of $V(t)$ adopted here is the same as that used in Jones (1987) and Kauffman (1991), and differs from that in Jones (1985) and Freyd *et al.* (1985) by the change of $t\to 1/t$.

[2]Named after the initials of the six coauthors of Freyd *et al.* (1985). A fifth group of researchers also obtained the same results. However, their announcement (Przytycki and Traczyk, 1987) arrived late due to slow mail (from Poland) and was not included in the joint paper.

isotopy invariance. For $N=2$ the Akutsu-Wadati polynomial coincides with the Jones polynomial, that is, we have

$$A^{(2)}(t)=V(t)\,, \tag{2.11}$$

which satisfies the Skein relation (2.6). For general N, however, the Akutsu-Wadati polynomial satisfies a Skein relation connecting knots differing in a small disk containing configurations L_0, L_-, and L_{n+}, with $n=1,2,\ldots,N-1$. For $N=3$, for example, the Skein relation is

$$A^{(3)}_{L_{2+}}(t)=t(1-t^2+t^3)A^{(3)}_{L_+}(t)+t^2(t^2-t^3+t^5)\times A^{(3)}_{L_0}(t)-t^8A^{(3)}_{L_-}(t)\,. \tag{2.12}$$

5. The Kauffman polynomial

The Kauffman polynomial (Kauffman, 1990) is a two-variable invariant of regular isotopy for unoriented knots. That is, it is invariant only under type-II and type-III Reidemeister moves. The Kauffman polynomial $L(\alpha,z)$ is defined by the Skein relation

$$L_{D_+}(\alpha,z)+L_{D_-}(\alpha z)=z[L_{D_0}(\alpha,z)+L_{D_\infty}(\alpha,z)]\,, \tag{2.13}$$

where $+$, $-$, 0, and ∞ are configurations shown in Fig. 11, and L_{D_+}, L_{D_-}, L_{D_0}, L_{D_∞} are the Kauffman polynomials of four knots D_+, D_-, D_0, D_∞ that are identical except that a small disk containing a single line crossing is replaced by the respective configurations $+$, $-$, 0, and ∞. In addition, the Kauffman polynomial is required to satisfy regular isotopy and the conditions

$$L_{\text{unknot}}(\alpha,z)=1\,, \tag{2.14}$$

$$L_{G_\pm}(\alpha,z)=\alpha^{\mp1}L_{G_0}(\alpha,z)\,, \tag{2.15}$$

Here, G_0, G_+, G_- are configurations shown in Fig. 13, and $L_{G_0}(\alpha,z)$, $L_{G_+}(\alpha,z)$, and $L_{G_-}(\alpha,z)$ are the Kauffman polynomials of three knots that are identical except that one disk containing two incident lines is replaced by G_0, G_+, and G_-, respectively.

For our purposes it is convenient to consider the Dubrovnik version[3] of the Kauffman polynomial. The Dubrovnik version of the Kauffman polynomial, $Q(\alpha,z)$ is defined by the Skein relation

$$Q_{D_+}(\alpha,z)-Q_{D_-}(\alpha,z)=z[Q_{D_0}(\alpha,z)-Q_{D_\infty}(\alpha,z)] \tag{2.13a}$$

and subject to the conditions

FIG. 13. The three line configurations considered in Eq. (2.15).

$$Q_{\text{unknot}}(\alpha,z)=1\,, \tag{2.14a}$$

$$Q_{G_\pm}(\alpha,z)=\alpha^{\mp1}Q_{G_0}(\alpha,z)\,. \tag{2.15a}$$

It is related to $L(\alpha,z)$ by the relation (Lickorish, in Kauffman, 1990)

$$Q(\alpha,z)=-(-1)^{c(K)}i^{-w(K)}L(i\alpha,-iz)\,, \tag{2.16}$$

where $i=\sqrt{-1}$, $c(K)$ is the number of components of the knot K, and the writhe $w(K)$ is given by Eq. (2.1). Since the reversal of the orientation of one component of a link induces a change of writhe $\Delta w(K)=4n$, n being an integer, Eq. (2.16) is actually independent of the orientation chosen.

E. The semioriented invariant

Given an invariant of regular isotopy for unoriented knots, we can always use it to construct an invariant of ambient isotopy for oriented knots (Kauffman, 1988a). We state this result as a theorem:[4]

Theorem II.E. If $L(\alpha)$ is a polynomial of regular isotopy for an unoriented knot K satisfying Eqs. (2.14) and (2.15), then

$$F(\alpha)=\alpha^{-w(K)}L(\alpha) \tag{2.17}$$

is an invariant of ambient isotopy for an oriented knot derived from K. Here, $w(K)$ is the writhe [Eq. (2.1)] of the oriented knot.

The proof of the theorem follows from the facts that both $w(K)$ and $L(\alpha)$ are regular isotopy invariants, i.e., invariant under Reidemeister moves II and III, and that the factor $\alpha^{-w(K)}$ in Eq. (2.17) cancels precisely those powers of α induced under Reidemeister moves I, to render $F(\alpha)$ ambient invariant.[5]

As examples, applying Theorem II.E to the bracket polynomial (Kauffman, 1987a; see Sec. V.B.2 below), one obtains the Jones polynomial, and applying it to the Kauffman polynomial, one obtains the F polynomial

$$F(\alpha,z)=\alpha^{-w(K)}L(\alpha,z)\,, \tag{2.18}$$

which is a two-variable polynomial of ambient isotopy

[3]Discovered by Kauffman is 1985 while visiting the city of Dubrovnik of the former Yugoslavia.

[4]Theorems are numbered by the sections in which they appear.

[5]Note that attaching orientations to configurations $G_\pm$ in Fig. 13 leads to the respective configurations $L_\mp$ for oriented knots, independent of the orienting direction chosen.

for oriented knots (Kauffman, 1990). A table of the F polynomials can be found in Lickorish and Millett (1988).

As a corollary, Theorem II.E implies that, when the orientation of one component of a link is reversed, the net results to the invariant $F(\alpha)$ is the introduction of an overall factor $\alpha^{-\Delta w(K)}$, where $\Delta w(K)$ is the change of writhe and is always in the form of $\Delta w(K)=4n$, n being an integer. For the Jones polynomial, for examples, we have $\alpha=-t^{-3/4}$ (Sec. V.B.2). Then the reversal of the orientation of one component introduces a factor t^{3n}, a fact verified by checking the list given in the Appendix. Since reversals of line orientations induce only changes of an overall factor, invariants given by Eqs. (2.17) and (2.18), including the Jones polynomial $V(t)$, have been termed "semioriented" (Lickorish, 1988; Lickorish and Millett, 1988).

III. LATTICE MODELS AND KNOT INVARIANTS

Lattice models are mathematical models of physical systems defined on lattices. While in the real world one deals with regular lattices of infinite size, many results on lattice models also hold for arbitrary finite lattices. It is these latter results that are useful in knot theory.

In lattice models one is interested in the computation of a partition function

$$Z=\sum W \tag{3.1}$$

where the summation is taken over all spin (or edge) states, and W is a Boltzmann factor defined for each configuration of spin (or edge) states. The Boltzmann factors are usually local in nature, that is, they can be decomposed into products of factors, each of which depends on states of few spins (edges) located in the immediate neighborhood. In statistical mechanics one further computes thermodynamic properties by taking derivatives of the partition function for infinite lattices. In knot theory, however, one deals mostly with partition functions.

The strategy of deriving knot invariants using statistical mechanics is the following: For each given knot, one constructs a two-dimensional lattice. One then seeks to construct lattice models on the lattice such that the partition function is identical for lattices constructed from equivalent knots. Then, by definition, the partition function is a knot invariant.

There are generally two different kinds of lattice models. If one places spins at lattice sites and introduces interactions among spins around an elementary cell of the lattice, one is led to spin models. This includes the special case of edge-interaction models for which only pair interactions are present. When there are multisite and/or hard-core interactions, the spin models are also known as interaction-round-a-face (IRF) models. Alternatively, if one places spins on lattice edges and associates weights with vertices according to the spin states of the incident edges, then one has vertex models. Vertex and IRF models are closely related and can always be transformed into each other (Perk and Wu, 1986a). For applications in knot theory, however, we shall see that it is convenient to begin with vertex models.

Historically, spin models originated from studies of the Ising model of ferromagnetism (Ising, 1925; Onsager, 1944). The study of vertex models was initiated in 1967 following Lieb's pioneering work on the exact determination of the residue entropy for square ice (Lieb, 1967a, 1967d), culminating in Baxter's exact solution of the two-state eight-vertex model (Baxter, 1971, 1972). The two-state vertex models have since been generalized to general q states (Kulish and Sklyanin, 1980, 1982; Schultz, 1981). The IRF model, a term coined by Baxter (1980), is another generalization of the eight-vertex model along a somewhat different route. A summary of early progress in lattice models can be found in the review by Lieb and Wu (1972) and the book by Baxter (1980). More recent results, particularly those on general q-state vertex and IRF models applicable to knot theory, are scattered through the literature.

The connection between knot theory and statistical mechanics was first noted by Jones (1985). In his derivation of the Jones polynomial, Jones noticed the resemblance of the von Neumann algebra used by him to the algebra occurring in the Temperley-Lieb formulation of the Potts model (Temperley and Lieb, 1971). The direct connection between the two seemingly unrelated fields came to light in 1986, when Kauffman (1987a) produced a remarkably simple derivation of the Jones polynomial using the bracket polynomial, a diagrammatic formulation which also arose in the consideration of the nonintersecting string model (Perk and Wu, 1986a) (see Sec. V.B.2 below). Soon thereafter, Jones worked out a derivation of the Homfly polynomial using a vertex-model approach. His derivation, while unpublished at the time, became widely known[6] and was extended by Turaev (1988) to the Kauffman polynomial. The connection of knot theory with statistical mechanics was formalized and further extended to include spin models by Jones (1989). Particularly, Jones introduced angle dependences to vertex models characterized by local weights.

The approach presented in this review follows closely that of Jones (1989). In particular, we consider vertex models with strictly local weights through the introduction of piecewise-linear lattices. We further establish that the IRF-model approach to knot invariants can be deduced as a special instance of the vertex-model formulation, thus simplifying the task of its derivation.

Finally, we point out the essence of the statistical mechanical approach. The statistical mechanical approach to knot invariants is based on the integrability of lattice models. Since we are seeking lattice models whose partition functions are invariant under Reidemeister moves, the main idea is that the partition function of integrable models (in the infinite-rapidity limit) naturally

[6]See example 1.16 in Jones (1989).

fulfills the requirement of Reidemeister moves IIIA (see Sec. V.A.1 below), a point not readily seen in the braid-group approach (Jones, 1990b). In addition, Reidemeister move IIA is satisfied as a consequence of unitarity. It is then a relatively simple matter to require the invariance of the partition function only under Reidemeister moves I and IIB.

Our main results are summarized in Theorems V.A.1 (for vertex models), VI.F (for IRF models), and VII.A (for models with pure two-spin interactions).

IV. VERTEX MODELS

A. Formulation

Consider a finite lattice $\mathcal{L}$ of N sites (vertices), E edges, and arbitrary shape. For our purposes we shall confine ourselves to lattices with a uniform coordination number and without free edges, i.e., every edge terminates at two vertices. An example of one such lattice is shown in Fig. 14.

Place spins on lattice edges, and let each spin independently take on q distinct values, or states. It is often convenient to associate colors with spin states so that one may regard edges as being colored. Then the partition function Z in Eq. (2.1) generates q^E edge colorings of $\mathcal{L}$. In the case of $q=2$, for example, one may regard the edges as having two colors, and thus one is led to consider two-state vertex models that have been analyzed extensively (for reviews see Lieb and Wu, 1972, and Baxter, 1980).

In vertex models the Boltzmann factor in Eq. (3.1) is taken to be a product of individual vertex weights, and the partition function reads

$$Z_{\text{vertex}}(\omega)=\sum_{\{\text{edge states}\}}\prod_{i=1}^{N}\omega_i\ , \tag{4.1}$$

where ω_i, the vertex weight of the ith vertex, is a function of the spin states of its four incident edges.

Since we have arbitrary lattices in mind, in which vertices can assume arbitrary orientations, a local frame of reference is needed to properly define the weights. This can be provided by directing lattice edges such that each vertex is formed by the crossing of two directed lines. For example, the lattice in Fig. 14 can be directed as shown in Fig. 15. We can now write the vertex weight as

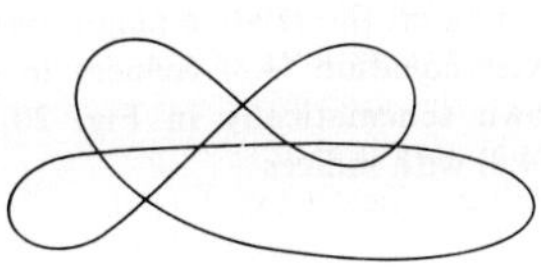

FIG. 14. A finite lattice of coordination number four. The lattice contains 6 vertices and 12 edges.

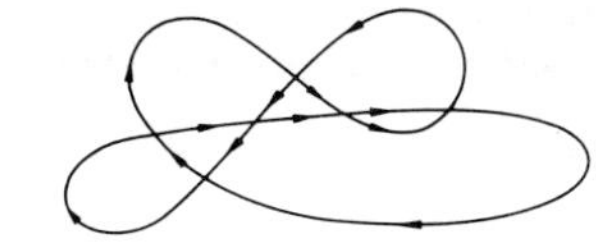

FIG. 15. A directed lattice.

$$\omega_i=\omega_i(a,b|x,y)=\omega_i\begin{bmatrix} y & b \\ a & x\end{bmatrix}, \tag{4.2}$$

where a, b, x, y are numerical numbers denoting the spin states of the four incident edges of a vertex as arranged in Fig. 16. We shall assume the indices $\{a,b,x,y\}\in\mathcal{J}$, where $\mathcal{J}$ is a set of q numerical values distributed *symmetrically* about zero.

In the most general case, Eq. (4.2) gives rise to q^4 distinct vertex weights and a q^4-vertex model. For $q=2$, for example, this becomes the 16-vertex model (Lieb and Wu, 1972). Several special case cases of the $q=2$ problem have been considered in the past; these include the six-vertex (Lieb, 1967a, 1967b, 1967c; Sutherland, 1967) and the eight-vertex (Fan and Wu, 1970; Baxter, 1971, 1972) models.

B. The Yang-Baxter equation

The q^4-vertex model is *integrable* if the q^4 vertex weights satisfy a condition known as the *Yang-Baxter equation*. In practice, integrability of lattice models often leads to closed-form solutions of the partition function and other physical quantities such as correlation functions. For our purposes, however, it suffices to consider only solutions of the Yang-Baxter equation, which, as we shall see, lead naturally to the realization of type-IIIA Reidemeister moves. As alluded to earlier in Sec. III, this is the key to the statistical mechanical derivation of knot invariants.

Consider two clusters of lattice edges containing three lattice sites represented by the upward-pointing and downward-pointing triangles in Fig. 17. The Yang-Baxter equation is the condition on the vertex weights such that the partition functions of these two small lattices are identical for any given states $\{a,b,c,d,e,f\}$. This implies that one may replace an upward-pointing triangle that is part of a lattice by a downward-pointing one, and vice versa, without affecting the overall partition function. Algebraically, this condition reads

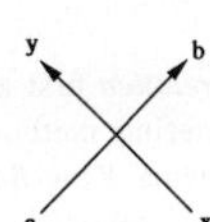

FIG. 16. The orientation of a vertex.

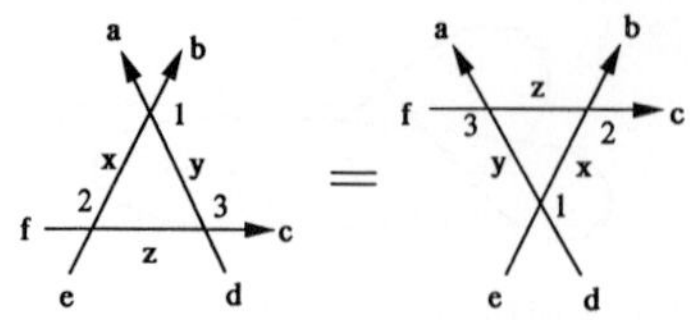

FIG. 17. The Yang-Baxter equation for vertex models.

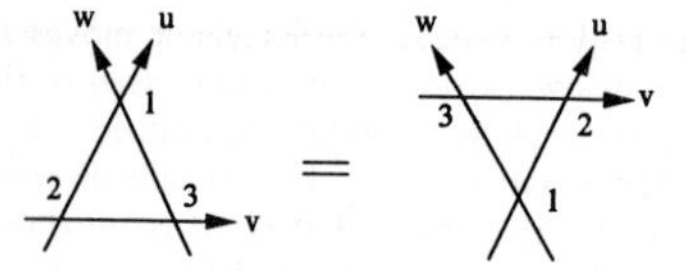

FIG. 18. Association of rapidities u, v, w, with the lines.

$$\sum_{x,y,z\in\mathcal{J}} \omega_1(x,b|y,a)\omega_2(f,z|e,x)\omega_3(z,c|d,y)=\sum_{x,y,z,\in\mathcal{J}} \omega_1(e,x|d,y)\omega_2(z,c|x,b)\omega_3(f,z|y,a)$$

$$\text{for all } a,b,c,d,e,f\in\mathcal{J} . \quad (4.3)$$

Here, we have allowed vertex weights to be different at the three sites.

Historically, the Yang-Baxter equation arose for $q=2$ as the factorizability condition in the Bethe-ansatz approach to the one-dimensional delta-function gas [McGuire (1964) for the Bose gas considered by Lieb and Liniger (1963); Gaudin (1967) and Yang (1967) for the Fermi gas] and as the star-triangle relation in solvable two-dimensional models in statistical mechanics [Onsager (1944) for the Ising model; Baxter (1972) for the eight-vertex model; Baxter (1978) for the general Z-invariant model]. The Yang-Baxter equation, a term introduced by the Faddeev school,[7] also arises in the theory of the factorized S matrix in quantum field theory (Zamolodchikov, 1979), for which the vertex weight ω is known as the R matrix. For general q, a problem first studied by Kulish and Sklyanin (1980, 1982) and Schultz (1981), Eq. (4.3) is a set of q^6 equations with $3\times q^4$ unknowns and is highly overdetermined.

The most general solution of the Yang-Baxter equation is not yet known, but families of solutions, including many of the special solutions found by brute force, can be constructed by using finite-dimensional representations of simple Lie algebras (Bazhanov, 1985; Jimbo, 1986) connected with quantum groups (Drinfel'd, 1986). These solutions are parametrized by assigning line variables, or rapidities (spectral parameters), u, v, w to the three lines, as shown in Fig. 18, so that one has

$$\begin{aligned}\omega_1(a,b|x,y)&=\omega(a,b|x,y|u-w)\\&=\omega\begin{bmatrix}y & b\\ a & x\end{bmatrix}(u-w) ,\\ \omega_2(a,b|x,y)&=\omega(a,b|x,y|v-u) ,\\ \omega_3(a,b|x,y)&=\omega(a,b|x,y|v-w) .\end{aligned} \quad (4.4)$$

[7]The term *Baxter-Yang relation* first appeared in a review on the quantum inverse scattering method by Takhtadzhan and Faddeev (1979), and the name *Yang-Baxter equation* was used thereafter by the Faddeev school. A useful collection of relevant reprints on the Yang-Baxter equation can be found in Jimbo (1989).

That is, vertex weights depend on a parameter that is the difference of the two rapidities of the two lines crossing at each vertex. Furthermore, it can be shown (Perk and Wu, 1986b) that the decoupling (initial) condition

$$\omega(a,b|x,y|0)=\delta_{ay}\delta_{bx} \quad (4.5a)$$

usually satisfied by solutions of the Yang-Baxter equation leads to the unitarity condition

$$\sum_{b,y\in\mathcal{J}} \omega(a,b|x,y|v-u)\omega(y,z|b,c|u-v)=\delta_{ac}\delta_{xz} , \quad (4.5b)$$

a situation shown in Fig. 19. Here, the Kronecker delta indicates that there is a contribution only when the two lines have identical indices. Solutions of the Yang-Baxter equation useful in constructing knot invariants are those with trigonometric parametrizations, usually the degenerate critical manifolds of more general soluble families with elliptic function parametrizations. This leads, as we shall see, to various generalizations of the two-state six-vertex models solved by Lieb (1967c, 1967d) to general q states.

The infinite rapidity limit. In pursuing realizations of knots as vertex models, we need two kinds of vertex weights for the $+$ and $-$ types of crossings. This need can be fulfilled by taking the infinite-rapidity limit $u\to\infty$, $v\to\infty$, $w\to\infty$ and writing[8]

$$\omega_\pm\begin{bmatrix}c & d\\ a & b\end{bmatrix}\equiv\lim_{u\to\pm\infty}\omega\begin{bmatrix}c & d\\ a & b\end{bmatrix}(u) . \quad (4.6)$$

Here it is understood that the right-hand side of Eq. (4.6) has been divided by a divergent factor, such as $\sinh u$ or $e^{\beta|u|}$ where β is a constant, such that only the leading weights contribute. The less divergent weights, if any, vanish in this limit.

Then, depending on the relative magnitudes of u, v, w, the Yang-Baxter equation (4.3) reduces to six different equations shown schematically in Fig. 20. These are given by Eq. (4.3) with indices

[8]The subscripts $\pm$ and the argument u of a vertex weight ω serve to remind us that ω is a solution of the Yang-Baxter equation.

$$\{1,2,3\}=\{--+\},\{-++\},\{---\},\{+--\},\{+++\},\{+-+\} . \quad (4.7)$$

It is intriguing to note that the six configurations specified by Eq. (4.7) coincide precisely with those in Fig. 5 representing the six possible Reidemeister moves IIIA. Similarly, the unitarity relation Eq. (4.5b) reduces to

$$\sum_{b,y\in\mathcal{J}} \omega_{\pm}(a,b|x,y)\omega_{\mp}(y,z|b,c)=\delta_{ac}\delta_{xz} , \quad (4.8)$$

represented by configurations coinciding with that of Reidemeister move IIA shown in Fig. 4. Conversely starting from a given $\omega_{\pm}\equiv\omega(\pm\infty)$ obtained from, say, braid-group analysis, one may seek to reconstruct the weight $\omega(u)$. This inverse process is termed *Baxterization* (Jones, 1990b).

C. Enhanced vertex models

It often happens that vertex weights occurring in a vertex model contain factors depending explicitly on angles between the incident lattice edges, a local parameter that may vary from vertex to vertex. It further transpires that one often regroups these local factors according to global loops, a technique first used in an analysis of the Potts model by Baxter *et al.* (1976) for arbitrary two-dimensional lattices. It is then convenient to replace curved edges, such as those shown in Fig. 15, by zigzag lines. This leads to the consideration of *piecewise-linear* lattices $\mathcal{L}^*$. For example, the conversion of the oriented lattice in Fig. 15 into one that is piecewise linear is shown in Fig. 21. Note that the conversion creates new vertices of degree 2.

Consider next an *enhanced* vertex model on $\mathcal{L}^*$ derived from the vertex model on $\mathcal{L}$ by associated angle dependences with vertex weights. For vertices of degree two, shown in Figs. 22(a) and 22(b), we associate vertex weights

$$\begin{aligned}\omega^*(a)&=\lambda^{a\theta/2\pi} \text{ if the line turns an angle } \theta \text{ to the left ,}\\ &=\lambda^{-a\theta/2\pi} \text{ if the line turns an angle } \theta \text{ to the right ,}\end{aligned} \quad (4.9)$$

where a is the state variable, and λ a variable yet to be specified.[9] The product of the vertex weights $\omega^*(a)$ along the zigzag path shown in Fig. 23 is given by

$$\prod_{\text{path}} \omega^*(a)=\lambda^{a(-\theta_1+\theta_2+\cdots)}=\lambda^{a\theta/2\pi} , \quad (4.10)$$

where θ is the angle between the final and initial directions of the path. Thus one always obtains the same product, independent of the way that the curved edges are linearized. In addition, the creation of vertices of degree 2 leads to the consideration of lattices in the shape of a ring. Since the product of vertex weights along a ring is

$$\begin{aligned}\prod_{\text{closed path}} \omega^*(a)&=\lambda^{a} \text{ arrows in counterclockwise direction ,}\\ &=\lambda^{-a} \text{ arrows in clockwise direction ,}\end{aligned} \quad (4.11)$$

the partition function of a ring,

$$Z_{\text{ring}}(\omega^*)=\sum_{a\in\mathcal{J}} \lambda^{a} , \quad (4.12)$$

is independent of the arrow direction for $\mathcal{J}$ symmetric about zero.

In the same spirit, we modify all other vertex weights by multiplying them by a factor to yield the angle-dependent weights

$$\omega^*(a,d|b,c|u)\equiv\omega^*\begin{bmatrix} c & d\\ a & b\end{bmatrix}(u) =\lambda^{(a+c-b-d)\theta/4\pi}\omega(a,d|b,c|u) \quad (4.13a)$$

and the infinite-rapidity limits

$$\omega_{\pm}^*(a,d|b,c)=\lambda^{(a+c-b-d)\theta/4\pi}\omega_{\pm}(a,d|b,c) , \quad (4.13b)$$

where a, b, c, d are arranged as shown in Fig. 22(c), and θ is the angle between the two incoming (or outgoing) arrows.[10] Explicitly, the partition function of the enhanced

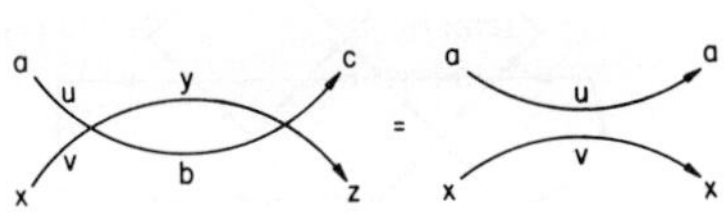

FIG. 19. The unitarity condition for vertex weights.

[9]Since λ is as yet unspecified, we may write λ^a as λ^{h_a} for some function h_a indexing the lattice edge. Then the discussions of this section and Sec. IV.D below can be carried through, provided that we replace the condition $a+b=c+d$ in Eq. (4.16) by $h_a+h_b=h_c+h_d$, and the factor $a-d$ in the exponent in Eq. (4.18) by h_a-h_d. This generalization proves to be useful when vertex-model results are applied to IRF models in Sec. VII.

[10]We shall assume that all vertices are formed by the crossing of two straight lines, so that $0<\theta<\pi$.

vertex model is

$$Z_{\text{vertex}}(\omega^*)=\sum_{\{\text{edge states}\}}\prod \omega^*(a,b|x,y|u)\prod \omega^*(a)\ , \tag{4.14a}$$

and, in the infinite-rapidity limit,

$$Z_{\text{vertex}}(\omega_\pm^*)=\sum_{\{\text{edge states}\}}\prod \omega_\pm^*(a,b|x,y)\prod \omega^*(a)\ , \tag{4.14b}$$

where the two products are taken over vertices of degrees 4 and 2, respectively.

For the enhanced vertex model to be useful in knot theory, we require the *enhanced* Yang-Baxter equation

$$\sum_{x,y,z\in\mathcal{I}}\omega^*(x,b|y,a|u-w)\omega^*(f,z|e,x|v-u)\omega^*(z,c|d,y|v-w)$$
$$=\sum_{x,y,z,\in\mathcal{I}}\omega^*(e,x|d,y|u-w)\omega^*(z,c|x,b|v-u)\omega^*(f,z|y,a|v-w) \tag{4.15}$$

to hold. This leads us to consider charge-conserving models.

D. Charge-conserving vertex models

In most of our applications we shall have

$$\omega\begin{bmatrix} c & d \\ a & b\end{bmatrix}(u)=0 \quad \text{unless } a+b=c+d\ . \tag{4.16}$$

If we regard the functions a, b, c, d as defining charges with edges, then the total charge of incoming/outgoing arrows is conserved, and we refer to Eq. (4.16) as the condition of *charge conservation.* In charge-conserving models the angle-dependent weights, Eqs. (4.13a) and (4.13b), are, respectively,

$$\omega^*(a,d|b,c|u)=\omega^*\begin{bmatrix} c & d \\ a & b\end{bmatrix}(u)$$
$$=\lambda^{(a-d)\theta/2\pi}\omega(a,d|b,c|u)\ , \tag{4.17a}$$

$$\omega_\pm^*(a,d|b,c)=\omega_\pm^*\begin{bmatrix} c & d \\ a & b\end{bmatrix}$$
$$=\lambda^{(a-d)\theta/2\pi}\omega_\pm\begin{bmatrix} c & d \\ a & b\end{bmatrix}\ . \tag{4.17b}$$

Using the identity $\theta_3=\theta_1+\theta_2$, where θ_i is the angle between the two incoming arrows at site i in Fig. 18, one can readily verify that the vertex weight Eq. (4.17a) is a solution of the enhanced Yang-Baxter equation (4.15), provided that $\omega(a,d,|b,c|u)$ is a solution of the Yang-Baxter equation (4.3). It follows that Eq. (4.17b) is the solution of Eq. (4.15) in the infinite-rapidity limit. Along the same lines, since the rapidity differences also satisfy $u_3=u_1+u_2$, where u is the difference of the two rapidities at site i,

$$\tilde{\omega}(a,d,|b,c|u)\equiv e^{\beta(a-d)u}\omega(a,d,|b,c|u)\ , \tag{4.18}$$

where β is arbitrary, is also a solution of the Yang-Baxter equation. This is a "symmetry-breaking" transformation, which provides to be useful in latter applications. We shall leave open the possibility of introducing this

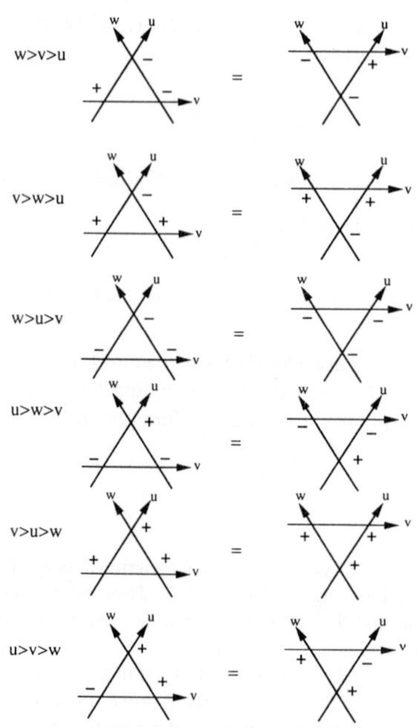

FIG. 20. The Yang-Baxter equation in the infinite-rapidity limit.

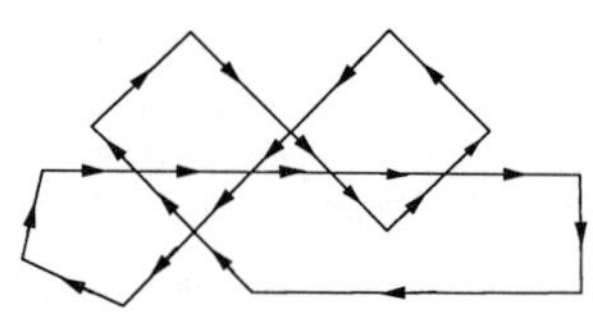

FIG. 21. A directed piecewise-linear lattice.

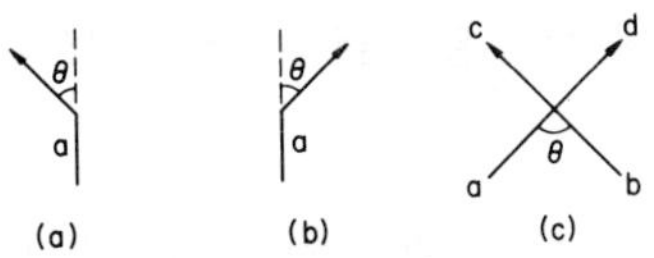

FIG. 22. Vertices in piecewise-linear lattices.

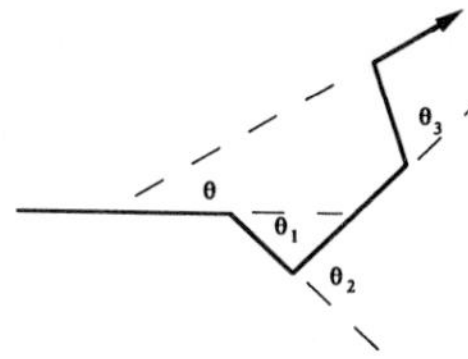

FIG. 23. A directed path.

symmetry breaking and use ω to denote either ω or $\tilde{\omega}$, whichever is needed in the application.

E. Integrable vertex models

We now present examples of integrable vertex models known in statistical mechanics.

1. The spin-conserving model

Schultz (1981) and Perk and Schultz (1981, 1983) have carried out a systematic study of solutions of the Yang-Baxter equation in some special cases. The first case is a $q(2q-1)$-vertex model generalizing the $(q=2)$ six-vertex ice-type models (Lieb, 1967a, 1967b, 1967c). In this vertex model all weights vanish except those associated with the $q(2q-1)$ configurations shown in Fig. 24. If one identifies edge variables as spins, then the incoming/outgoing spins are conserved,[11] that is, we have either

$$\{a=c,\ b=d\} \quad \text{or} \quad \{a=d,\ b=c\} . \tag{4.19}$$

Thus the spin-conserving model satisfies charge conservation, $a+b=c+d$. For $q=2$, the condition (4.19) is equivalent to the ice rule (Wu, 1967, 1968) leading to the six-vertex models solved by Lieb (1967a, 1967b, 1967c).

Let W_{aa}, S_{ab}, and T_{ab}, $a\neq b$, be the vertex weights shown in Fig. 24. Then we can write

$$\omega\begin{bmatrix} c & d \\ a & b \end{bmatrix}(u)=W_{aa}(u)\delta_{abcd}+S_{ab}(u)(\delta_{ad}\delta_{bc}-\delta_{abcd}) + T_{ab}(u)(\delta_{ac}\delta_{bd}-\delta_{abcd}) , \tag{4.20}$$

where δ_{ab} is the Kronecker delta function and

$$\begin{aligned} \delta_{abcd}&=1 \quad \text{if } a=b=c=d , \\ &=0 \quad \text{otherwise} . \end{aligned} \tag{4.21}$$

Perk and Schultz (1983) found that, within the family of Eq. (4.20), the solution of the Yang-Baxter equation is, excluding an overall normalization,[12]

$$\begin{aligned} W_{aa}(u)&=\sinh(\eta+\epsilon_a u) , \\ S_{ab}(u)&=\sinh u , \\ T_{ab}(u)&=e^{-u[\mathrm{sgn}(a-b)]}\sinh\eta , \end{aligned} \tag{4.22}$$

where $\epsilon_a=1$ or -1, and η is arbitrary. In the infinite-rapidity limit, Eqs. (4.20) and (4.22) become

$$\omega_{\pm}\begin{bmatrix} c & d \\ a & b \end{bmatrix}=A_{\pm}(\epsilon_a e^{\pm\epsilon_a\eta}\delta_{abcd}+(\delta_{ad}\delta_{bc}-\delta_{abcd}) \pm(e^{\eta}-e^{-\eta})\theta[\pm(b-a)]\delta_{ac}\delta_{bd}) , \tag{4.23}$$

where we have divided the weight, Eq. (4.20), by $\sinh u$ and introduced normalization factor $A_{\pm}$ and the step function

$$\begin{aligned} \theta(b-a)&=1 \quad \text{if } b>a , \\ &=0 \quad \text{if } b\leq a . \end{aligned} \tag{4.24}$$

It is instructive to write out Eq. (4.23) explicitly. Excluding the normalization factor $A_{\pm}$, we have

$$\begin{aligned} \omega_{\pm}\begin{bmatrix} a & a \\ a & a \end{bmatrix}&=\epsilon_a e^{\pm\epsilon_a\eta} , \\ \omega_{\pm}\begin{bmatrix} b & a \\ a & b \end{bmatrix}&=1, \quad a\neq b , \\ \omega_{+}\begin{bmatrix} a & b \\ a & b \end{bmatrix}&=e^{\eta}-e^{-\eta}, \quad a<b , \\ \omega_{-}\begin{bmatrix} a & b \\ a & b \end{bmatrix}&=-(e^{\eta}-e^{-\eta}), \quad a>b , \\ \omega_{\pm}\begin{bmatrix} c & d \\ a & b \end{bmatrix}&=0, \quad \text{otherwise} . \end{aligned} \tag{4.25}$$

As we shall see, this spin-conserving model leads to the Homfly polynomial.

2. The N-state vertex model

Another family of integrable charge-conserving models is the N-state vertex model (Sogo *et al.*, 1983). Let the edge variables take on N distinct values,

---

[11]The term *spin conservation* has been used by different authors in different contexts. Here we use it to refer to models in which the states of the incoming and outgoing arrows are strictly the same.

[12]Here we have omitted some multiplication factors corresponding to gauge transformations and external fields that do not concern us.

$$\mathcal{J}=\left\{-\left[\frac{N-1}{2}\right],-\left[\frac{N-3}{2}\right],\ldots,\left[\frac{N-3}{2}\right],\left[\frac{N-1}{2}\right]\right\}, \tag{4.26}$$

for fixed $N=2,3,\ldots$. Then the N-state vertex model is specified by

$$a+b=c+d=l,\quad |l|\leq N-1\,, \tag{4.27}$$

that is, charge-conserving weights with the total charge $l\leq N-1$. This yields a total of $2\sum_{l=0,1,2,\ldots}^{N-1} l^2+N^2 = N(2N^2+1)/3$ vertex configurations, leading to the 6-vertex (ice-type) models for $N=2$ and the 19-vertex model (Zamolodchikov and Fateev, 1980) for $N=3$. For example, the $N=2$ solution is (Akutsu and Wadati, 1987b)

$$\begin{aligned}
&\omega\begin{bmatrix}\frac{1}{2} & \frac{1}{2}\\ \frac{1}{2} & \frac{1}{2}\end{bmatrix}(u)=\omega\begin{bmatrix}-\frac{1}{2} & -\frac{1}{2}\\ -\frac{1}{2} & -\frac{1}{2}\end{bmatrix}(u)=\sinh(\eta-u)\,,\\
&\omega\begin{bmatrix}\frac{1}{2} & -\frac{1}{2}\\ -\frac{1}{2} & \frac{1}{2}\end{bmatrix}(u)=\omega\begin{bmatrix}-\frac{1}{2} & \frac{1}{2}\\ \frac{1}{2} & -\frac{1}{2}\end{bmatrix}(u)=\sinh u\,,\\
&\omega\begin{bmatrix}\frac{1}{2} & -\frac{1}{2}\\ \frac{1}{2} & -\frac{1}{2}\end{bmatrix}(u)=e^{-u}\sinh\eta\,,\\
&\omega\begin{bmatrix}-\frac{1}{2} & \frac{1}{2}\\ -\frac{1}{2} & \frac{1}{2}\end{bmatrix}(u)=e^{u}\sinh\eta\,,
\end{aligned} \tag{4.28}$$

where the vertex weights are those including the symmetry-breaking factor in Eq. (4.18) with $\beta=1$. We observe that this solution is identical to that of the spin-conserving model given by Eqs. (4.20) and (4.22) with $q=2$, $\epsilon_a=-1$. But the general N-state vertex model is different from the spin-conserving model. Explicit expressions of vertex weights for $N=3,4$ can be found in Sogo *et al.* (1983) and Akutsu and Wadati (1987b). The general N model can be constructed using, for example, a fusion procedure (Kulish and Sklyanin, 1982b) according to the ideas of Kulish *et al.* (1981).

An expression for vertex weights $\omega_\pm$, the infinity-rapidity limit of $w(u)$, for general N has been given by Jones (1989). This expression, which includes the symmetry-breaking factor in Eq. (4.18) with $\beta=1$ and reduces to our $\epsilon_a=1$ solution in the case of $N=2$, is

$$\omega_+\begin{bmatrix}c & d\\ a & b\end{bmatrix}(\eta)=e^{\eta[(ab+cd)+k(k-1)/2]}\times\Gamma_{b,k}^{N}(\eta)[\theta(a-b)+\delta_{ab}]\,, \tag{4.29}$$

where $k=a-d$,

$$\Gamma_{b,k}^{N}(\eta)=\prod_{n=0}^{k-1}\left[\frac{2\sinh[(b+n+\frac{1}{2}N+\frac{1}{2})\eta]\sinh[(b+n-\frac{1}{2}N-\frac{1}{2})\eta]}{-\sinh[(n+1)\eta]}\right], \tag{4.30}$$

with empty product being 1, and

$$\omega_-\begin{bmatrix}c & d\\ a & b\end{bmatrix}(\eta)=\omega_+\begin{bmatrix}d & c\\ b & a\end{bmatrix}(-\eta)\,. \tag{4.31}$$

This vertex model leads to the Akutsu-Wadati polynomial.

3. The nonintersecting-string model

Another family of solutions considered by Perk and Schultz (1983) is the $q(2q-1)$-vertex model with vertex weights shown in Fig. 25. These are vertex weights characterized by either

$$\{a=c,\ b=d\}\ \text{or}\ \{a=b,\ c=d\}\,. \tag{4.32}$$

Thus the vertex weight is

$$\omega\begin{bmatrix}c & d\\ a & b\end{bmatrix}(u)=W_{aa}(u)\delta_{abcd}+S_{ab}(u)(\delta_{ab}\delta_{cd}-\delta_{abcd})+T_{ab}(u)(\delta_{ac}\delta_{bd}-\delta_{abcd})\,. \tag{4.33}$$

If at each vertex we connect edges in the same states, then the partition function (4.1) generates polygons on $\mathcal{L}$ that do not intersect. Hence this is a model of nonintersecting strings. For $q=2$, this reduces to a six-vertex

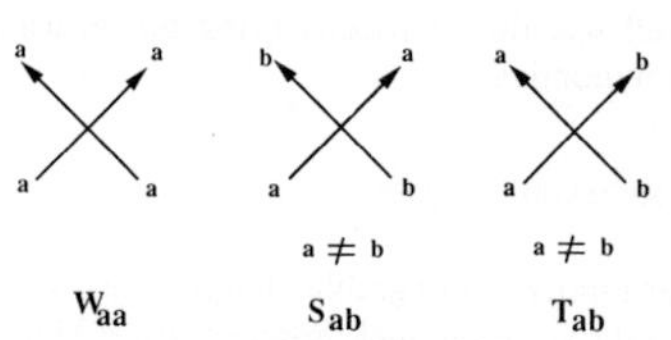

FIG. 24. Vertex weights of the spin-conserving vertex model.

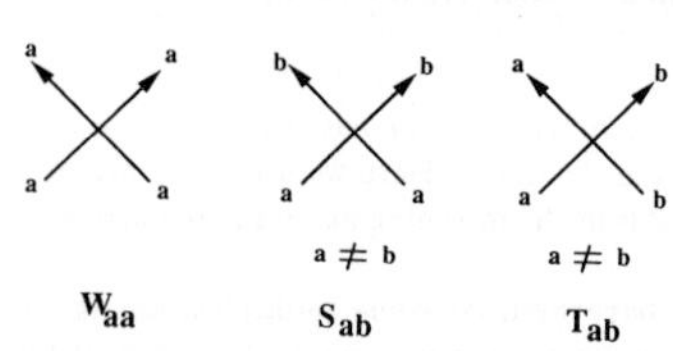

FIG. 25. Vertex weights of the nonintersecting-string model.

model discussed by Lieb and Wu (1972). Two models for $q=3$ have been considered by Stroganov (1979). Note that the nonintersecting-string model does not obey the charge-conserving condition,[13] Eq. (4.16), and consequently we take $\lambda=1$ so that all vertices of degree two in $\mathcal{L}^*$ have the weight

$$\omega(a)=1, \quad \text{for all } a \ . \tag{4.34}$$

Perk and Schultz (1981, 1983) found that there exist exactly $q+1$ distinct integrable models in the form of Eq. (4.33). One of the integrable cases, termed the separable model by Perk and Wu (1986a), is

$$\begin{aligned} W_{aa}(u)&=\sinh u+\sinh(\eta-u) \ , \\ S_{ab}(u)&=\sinh u \ , \\ T_{ab}(u)&=\sinh(\eta-u) \ . \end{aligned} \tag{4.35}$$

In the infinite-rapidity limit, we obtain from Eq. (4.33)

$$\omega_{\pm}\begin{bmatrix} c & d \\ a & b \end{bmatrix}=A_{\pm}[\delta_{ab}\delta_{cd}-e^{\mp\eta}\delta_{ac}\delta_{bd}] \ , \tag{4.36}$$

where we have again divided Eq. (4.33) by sinh$u$ and included a normalization factor $A_{\pm}$. Explicitly, excluding the normalization factor $A_{\pm}$, we have

$$\begin{aligned} \omega_{\pm}\begin{bmatrix} a & a \\ a & a \end{bmatrix}&=1-e^{\pm\eta} \ , \\ \omega_{\pm}\begin{bmatrix} a & b \\ a & b \end{bmatrix}&=-e^{\pm\eta}, \quad a\neq b \ , \\ \omega_{\pm}\begin{bmatrix} a & a \\ b & b \end{bmatrix}&=1, \quad a\neq b \ , \\ \omega_{\pm}\begin{bmatrix} c & d \\ a & b \end{bmatrix}&=0, \quad \text{otherwise} \ . \end{aligned} \tag{4.37}$$

As we shall see, this vertex model leads to the Jones polynomial (Lipson, 1992; Wu, 1992b).

V. KNOT INVARIANTS FROM VERTEX MODELS

A. Oriented knots

1. Formulation

Starting from a given oriented knot, one constructs a directed lattice $\mathcal{L}$ and the associated piecewise-linear lattice $\mathcal{L}^*$ by regarding lines of the knot as lattice edges and line crossings as lattice sites (vertices). This leads naturally to two types of vertices, + and −, corresponding to the two kinds of line crossings + and −. For example, from a trefoil one constructs the directed lattice in Fig. 26 and the piecewise-linear lattice in Fig. 27, both having three + crossings.

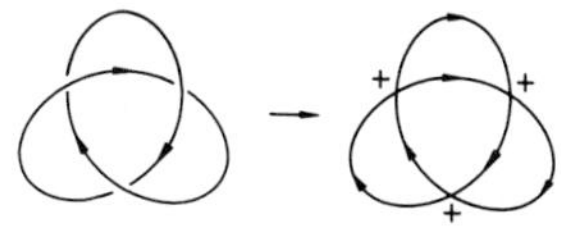

FIG. 26. Construction of a lattice from a knot.

We next seek to construct an enhanced vertex model on $\mathcal{L}^*$ with correspondingly two different kinds of vertex weights $\omega_{\pm}^*$, such that its partition function $Z(\omega_{\pm}^*)$ is a knot invariant. That is, we require $Z(\omega_{\pm}^*)$ to remain invariant under Reidemeister moves of the lattice edges. To accomplish this, we use vertex weights $\omega_{\pm}^*$ derived from the enhanced Yang-Baxter equation (4.15). Indeed, as remarked after Eq. (4.7), configurations of the Yang-Baxter equation in the infinite-rapidity limits coincide precisely with those of type-IIIA Reidemeister moves. As a result, the partition function $Z(\omega_{\pm}^*)$ is *by definition* invariant under type-IIIA moves. We therefore need only examine its invariance under Reidemeister moves I and II (moves IIIB follow as a consequence). Note that the use of the infinite-rapidity limit, Eq. (4.6), a crucial step whose meaning is not well understood in the braid-group approach (Witten, 1989b; Jones, 1990b), now emerges naturally as a condition for ensuring invariance under Reidemeister moves IIIA.

The invariance of $Z(\omega_{\pm}^*)$ under Reidemeister moves I, shown in Fig. 28, reads

$$\sum_{a\in\mathcal{J}}\lambda^{a(2\pi-\theta)/2\pi}\omega_{\pm}^*(a,b|x,a)=\lambda^{-b\theta/2\pi}\delta_{bx} \quad (\text{I}) \ , \tag{5.1}$$

where we have used the identity $\theta_1+\theta_2+\theta_3=2\pi-\theta$. Similarly, consideration of the invariance of $Z(\omega_{\pm}^*)$ under Reidemeister moves IIA and IIB, shown in Figs. 29 and 30, respectively, leads to the conditions

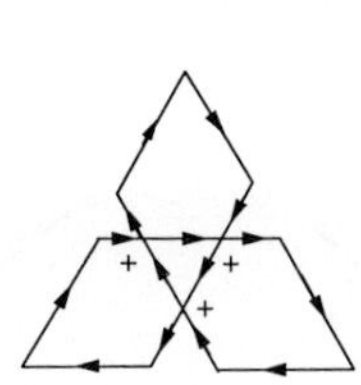

FIG. 27. The piecewise-linear lattice constructed from the lattice in Fig. 26.

[13]However, by applying a staircase-type transformation generalizing the one used by Fan and Wu (1970) for the eight-vertex model, one can view the nonintersecting-string model as a checkerboard spin-conserving model. I am indebted to J. H. H. Perk for this remark.

$$\sum_{b,y\in\mathcal{J}} \lambda^{(b-y)\theta/2\pi}\omega^*_{\pm}(a,b|x,y)\omega^*_{\mp}(y,z|b,c) = \lambda^{(a-x)\theta/2\pi}\delta_{ac}\delta_{xz} \quad \text{(IIA)} , \tag{5.2}$$

$$\sum_{b,y\in\mathcal{J}} \lambda^{(y+b)(\pi-\theta)/2\pi}\omega^*_{\pm}(y,x|a,b)\omega^*_{\mp}(b,c|z,y) = \lambda^{(a+x)(\pi-\theta)/2\pi}\delta_{xz}\delta_{ac} \quad \text{(IIB)} . \tag{5.3}$$

For completeness, although it is redundant, we write down the requirement imposed by Reidemeister moves IIIB. Using the labelings shown in Fig. 31, we have

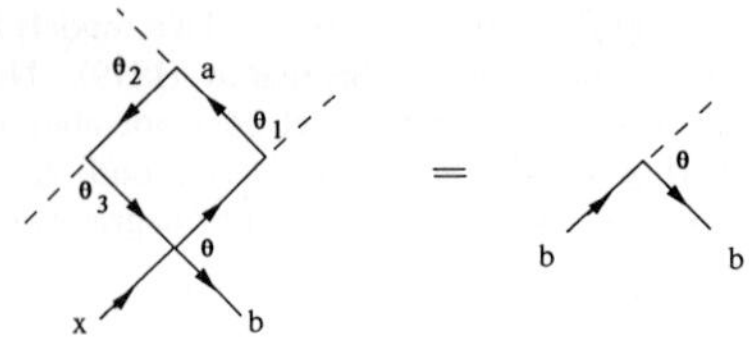

FIG. 28. Labelings for Reidemeister moves I.

$$\sum_{x,y,z\in\mathcal{J}} \omega^*_1(y,a|b,x)\omega^*_2(x,c|f,z)\omega^*_3(z,c|d,y) = \sum_{x,y,z\in\mathcal{J}} \omega^*_1(d,y|x,e)\omega^*_2(b,c|z,c)\omega^*_3(f,z|y,a) \quad \text{(IIIB)} , \tag{5.4}$$

$$\{1,2,3\}=\{+--\},\{+-+\},\{++-\},\{-+-\},\{--+\},\{-++\} . \tag{5.5}$$

For charge-conserving models, we use Eqs. (4.17a) and (4.17b) and obtain from Eqs. (5.1)–(5.3) the equivalent conditions

$$\sum_{a\in\mathcal{J}} \lambda^{a}\omega_{\pm}(a,b|x,a)=\delta_{bx}, \quad \text{(I)} , \tag{5.1a}$$

$$\sum_{b,y\in\mathcal{J}} \omega_{\pm}(a,b|x,y)\omega_{\mp}(y,z|b,c)=\delta_{ac}\delta_{xz} \quad \text{(IIA)} , \tag{5.2a}$$

$$\sum_{b,y\in\mathcal{J}} \lambda^{b-a}\omega_{\pm}(y,x|a,b)\omega_{\mp}(b,c|z,y)=\delta_{xz}\delta_{ac} \quad \text{(IIB)} . \tag{5.3a}$$

Here, $\omega_{\pm}(a,b,|c,d)$ are defined by Eqs. (4.6) and (4.4) and are deduced from the solution of the Yang-Baxter equation (4.3). These are the *fundamental* conditions, which do not refer to enhanced weights. Note that they do not depend on the angle θ and the condition (5.2a) coincides with the unitarity relation, Eq. (4.8).

We now collect our main results and state them as a theorem:

Theorem V.A.1. For each oriented knot construct a directed lattice $\mathcal{L}$ and the associated piecewise-linear lattice $\mathcal{L}^*$. Then the partition function (4.14b), with vertex weights $\omega^*(a)$ given by Eq. (4.9) and $\omega^*_{\pm}$ by the infinite-rapidity limit of the solution of the enhanced Yang-Baxter equation (4.15), is a knot invariant, provided that Eqs. (5.1)–(5.3) hold.

For charge-conserving models, $\omega^*_{\pm}$ is given by Eq. (4.17b), and Eqs. (5.1)–(5.3) reduce to the fundamental conditions (5.1a)–(5.3a).

Skein relation. With knot invariants formulated as partition functions, Skein relations relating knot invariants can be formulated in terms of vertex weights. For the Homfly polynomial, for example, it is readily verified that the Skein relation (2.8) is equivalent to the following relation among the enhanced vertex weights:

$$t^{-1}\omega^*_{+}\begin{bmatrix} c & d \\ a & b \end{bmatrix} - t\omega^*_{-}\begin{bmatrix} c & d \\ a & b \end{bmatrix} = z\lambda^{(a-d)\theta/2\pi}\delta_{ac}\delta_{bd} . \tag{5.6}$$

For charge-conserving models Eq. (5.6) reduces to

$$t^{-1}\omega_{+}\begin{bmatrix} c & d \\ a & b \end{bmatrix} - t\omega_{-}\begin{bmatrix} c & d \\ a & b \end{bmatrix} = z\delta_{ac}\delta_{bd} . \tag{5.6a}$$

Similar relations can be written down for other Skein relations.

We now apply our formulation to obtain knot invariants.

2. The Homfly polynomial

We now show that the q-state spin-conserving model described in Sec. IV.E.1 (Perk and Schultz, 1981, 1983) generates the Homfly polynomial (Jones, 1989). The vertex weights $\omega^*_{\pm}$ of the spin-conserving model are given by

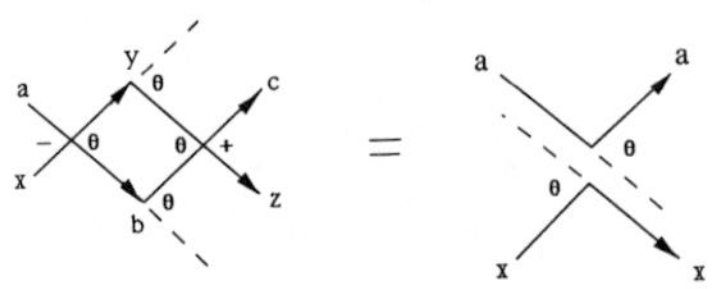

FIG. 29. Labelings for Reidemeister moves IIA.

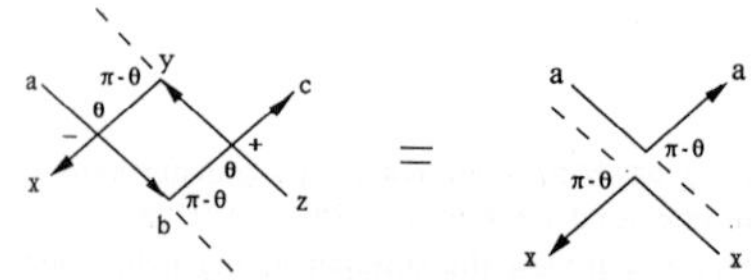

FIG. 30. Labelings for Reidemeister moves IIB.

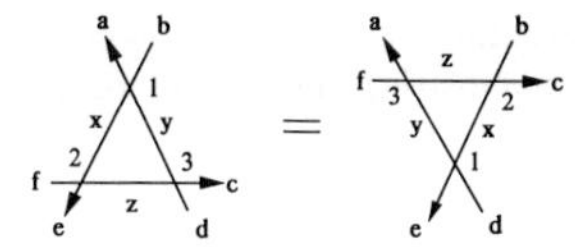

FIG. 31. Labelings for Reidemeister moves IIIB.

Eqs. (4.20) and (4.23) with[14]

$$\epsilon_a=-1 \quad \text{for all } a\ , \tag{5.7}$$

and

$$\mathcal{J}=\{-(q-1),-(q-3),\ldots,(q-3),(q-1)\}\ , \tag{5.8}$$

containing q integers with intervals of 2 in between.

The partition function is invariant under Reidemeister moves IIIA by construction. To satisfy invariance under Reidemeister moves I, we substitute the vertex weight (4.23) with $\epsilon_a=-1$ in Eq. (5.1a) and obtain

$$A_\pm\left[-e^{\mp\eta}\lambda^b\pm(e^\eta-e^{-\eta})\sum_{a\in\mathcal{J}}\theta[\pm(b-a)]\lambda^a\right]=1,\quad b\in\mathcal{J}\ . \tag{5.9}$$

Using the identity

$$\sum_{a\in\mathcal{J}}\theta[\pm(b-a)]\lambda^a=(\lambda^{\mp(q-1)}-\lambda^b)/(1-\lambda^{\pm 2})\ , \tag{5.10}$$

we deduce from Eq. (5.9) the condition

$$A_\pm e^{\pm\eta}\left[\lambda^b+\frac{(1-e^{-2\eta})(e^{\mp(q+1)\eta}-\lambda^b)}{1-\lambda^2}\right]=1\ , \tag{5.11}$$

which is satisfied by taking

$$\lambda=e^\eta,\quad A_\pm=-e^{\pm q\eta}\ . \tag{5.12}$$

Similarly, substituting Eq. (4.23) with $\epsilon_a=-1$ in Eqs. (5.2a) and (5.3a) required by Reidemeister moves IIA and IIB, we verify that they are also satisfied with the choice of Eq. (5.12). For example, to verify Eq. (5.2a), which is the same as the unitarity relation Eq. (4.8), one substitutes Eq. (5.12), which equates the left-hand side of Eq. (5.2a) to

$$\begin{aligned}\text{LHS}=\sum_{b,y\in\mathcal{J}}&[-e^{-\eta}\delta_{abxy}+(\delta_{ab}\delta_{xy}-\delta_{abxy})+(e^\eta-e^{-\eta})\theta(b-a)\delta_{ay}\delta_{bx}]\\&\times[-e^\eta\delta_{yzbc}+(\delta_{ba}\delta_{yz}-\delta_{bcyz})=(e^\eta-e^{-\eta})\theta(y-z)\delta_{yc}\delta_{bz}]\ .\end{aligned} \tag{5.13}$$

Expanding the fist square bracket in Eq. (5.13) and carrying out the summations term by term, one obtains

$$\begin{aligned}\text{LHS}&=-e^{-\eta}[(-e^\eta\delta_{acxz})+0+0]+[0+(\delta_{ac}\delta_{xz}-\delta_{acxz})+0]+(e^\eta-e^{-\eta})[0+0+0]\\&=\delta_{ac}\delta_{xz}\ .\end{aligned} \tag{5.14}$$

This establishes Eq. (5.2a). In a similar fashion and using Eq. (5.10) in conjunction with the identity

$$\sum_{b\in\mathcal{J}}\theta(a-b)\theta(b-c)\lambda^b=(\lambda^{c+2}-\lambda^a)/(1-\lambda^2)\ , \tag{5.15}$$

we verify that Eq. (5.3a) is satisfied. It can also be checked, although this is not necessary, that the condition (5.4) required by Reidemeister moves IIIB is also satisfied.

Combining Eqs. (5.12) with (4.17b) and (4.23), we arrive at the following explicit expression for the angle-dependent vertex weight:

$$\omega^*_\pm\begin{bmatrix}c & d\\ a & b\end{bmatrix}=e^{\pm q\eta}((e^{\pm\eta}+1)\delta_{abcd}-\delta_{ad}\delta_{bc}\mp(e^\eta-e^{-\eta})e^{(a-d)\eta\theta/2\pi}\theta[\pm(b-a)]\delta_{ac}\delta_{bd})\ . \tag{5.16}$$

Here θ is the angle between the two incoming arrows at the vertex, and $\theta(a)$ is the step function defined by Eq. (4.24). The partition function $Z(\omega^*_\pm)$ of this vertex model generates knot invariant.[15]

To see that this knot invariant is the Homfly polynomial, we need to establish the Skein relation (5.6). Indeed, using the identity

$$\theta(a-b)+\theta(b-a)+\delta_{ab}=1\ , \tag{5.17}$$

one verifies that Eq. (5.6) is satisfied by the $\omega^*_\pm$ Eq. (5.16) by identifying

$$t=e^{q\eta},\quad z=e^{-\eta}-e^\eta\ . \tag{5.18}$$

Furthermore, as discussed in Sec. II.C.1, the Skein relation expresses the partition function $Z(q,e^\eta)$ of the enhanced spin-conserving vertex model as the product of

[14]The choice of $\epsilon_a=1$, which leads to $\lambda=e^{-\eta}$ and $A_\pm=e^{\mp q\eta}$, also yields the Homfly polynomial.

[15]It is instructive to verify that the vertex weight (5.16) does not disentangle the configurations shown in Fig. 8. Thus the vertex weight (5.16) is exactly what is required of knot invariances, no more and no less.

two factors: a Laurent polynomial $P(t,z)$ in t and z and the partition function of a ring, $Z_{\rm ring}=\sinh q\eta/\sinh\eta$, deduced from Eq. (4.12). Thus the Laurent polynomial

$$P(t,z)=\left[\frac{\sinh\eta}{\sinh q\eta}\right]Z_{\rm vertex}(q,e^{\eta}) \tag{5.19}$$

satisfies the normalization condition (2.9) and hence is the Homfly polynomial for integral q. By analytically continuing Eq. (5.19) to all values of q, we establish the existence and uniqueness of the Homfly polynomial $P(t,z)$ for general t and z. This completes the construction of the Homfly polynomial.

3. The Jones polynomial

We have seen in Sec. II.D.3 that the Jones polynomial $V(t)$ is obtained from the Homfly polynomial $P(t,z)$ by taking $z=\sqrt{t}-1/\sqrt{t}$, indicating that the Jones polynomial is constructed from the $q=2$ spin-conserving vertex model. In view of its fundamental importance, we give here another construction of the Jones polynomial using the nonintersecting-string model of Sec. IV.E.3 (Lipson, 1992; Wu, 1992b). This construction is direct, as there is no need of introducing piecewise-linear lattices nor the writhe; it also expresses the Jones polynomial directly as a Potts model partition function (Wu, 1992b).

In the nonintersecting-string model we have $\lambda=1$, so that there is no angle dependence in vertex weights. As before, the condition for Reidemeister moves IIIA is automatically satisfied by the vertex weight (4.36). Substituting this weight in Eq. (5.1a) with $\lambda=1$, we obtain

$$A_{\pm}=-e^{\pm 2\eta}, \quad q=e^{\eta}+e^{-\eta} . \tag{5.20}$$

Explicitly, the vertex weight Eq. (4.36) is now

$$\omega_{\pm}\begin{bmatrix} c & d \\ a & b\end{bmatrix}=-e^{\pm 2\eta}\delta_{ab}\delta_{cd}+e^{\pm\eta}\delta_{ac}\delta_{bd} . \tag{5.21}$$

It can be checked that Eqs. (5.2a) and (5.3a) are now satisfied by Eq. (5.21). Hence, the partition function $Z(q,e^{\eta})$ of the nonintersecting-string model with weight (5.21) is a knot invariant. To identify this knot invariant as the Jones polynomial, we obtain from Eq. (5.21)

$$e^{-2\eta}\omega_{+}^{*}\begin{bmatrix} c & d \\ a & b\end{bmatrix}-e^{2\eta}\omega_{-}^{*}\begin{bmatrix} c & d \\ a & b\end{bmatrix}$$
$$=(-e^{\eta}+e^{-\eta})e^{(a-d)\theta/2\pi}\delta_{ac}\delta_{bd} . \tag{5.22}$$

This leads to the Skein relation Eq. (2.6) for $V(t)$ upon identifying $e^{\eta}=-\sqrt{t}$. Furthermore, $Z(q,e^{\eta})$ is proportional to $Z_{\rm ring}(q,e^{\eta})=q$. It follows that the Laurent polynomial

$$V(t)=q^{-1}Z_{\rm vertex}(q,e^{\eta}) \tag{5.23}$$

is the Jones polynomial when one sets

$$q=-(\sqrt{t}+1/\sqrt{t}), \quad e^{\eta}=-\sqrt{t} . \tag{5.24}$$

Another construction of the Jones polynomial based on the nonintersecting-string model will be given in Sec. V.B.2 below.

4. The Alexander-Conway polynomial

The Alexander-Conway polynomial $\nabla(z)$ is obtained from the Homfly polynomial $P(t,z)$ by taking $t=1$. According to Eq. (5.18), this corresponds to taking $q=0$ in our derivation of $P(t,z)$. We shall therefore assume that we have analytically continued Eq. (5.19) to permit us to take the $q\to 0$ limit. This is very much similar to the $q\to 0$ limit of the Potts model, which generates percolations (Fortuin and Kasteleyn, 1972).

Alternatively, $\nabla(z)$ can also be constructed from a two-state vertex model (Kauffman, 1991). This is done by considering two-strand knots which convert to lattices possessing two open lattice edges. It can then be shown that the partition function of a $q=2$ spin-conserving vertex model with weights given by Eqs. (4.23) and (4.17b), with $q=2$, $\{a,b,c,d\}=\pm 1$, $\epsilon_{\pm 1}=\pm 1$, and $\lambda=\sqrt{-1}$, gives rise to $\nabla(e^{\eta}-e^{-\eta})$ for two-strand knots. Readers are referred to Kauffman (1991) for details of this analysis.

5. The Akutsu-Wadati polynomial

In a similar fashion the angle-dependent vertex weight (4.17b) with $\lambda=e^{2\eta}$ and $\omega_{\pm}$ given by Eqs. (4.29)–(4.31) for the N-state vertex model can be used to derive knot invariants. This leads to the Akutsu-Wadati polynomial $A^{(N)}(t)$ (Akutsu and Wadati, 1987a, 1987b). Expressions for the $N=3$ Akutsu-Wadati polynomial have been obtained, and tabulated, for knots of closed three-braids (Akutsu *et al.*, 1987). The extension to two-variable polynomials has been made (Deguchi *et al.*, 1988), and Ge *et al.* (1989) have also given an explicit derivation of the $N=3,4$ polynomials. The Akutsu-Wadati polynomial satisfies the general Skein relation relating knots with configurations L_{-}, L_{0}, L_{+}, and L_{n+}, $n=2,3,\ldots,N-1$, and is more powerful than the Jones polynomial in differentiating knots. For example, the two knots found by Birman (1985) to possess an identical Jones polynomial can be distinguished using the Akutsu-Wadati polynomial (Akutsu *et al.*, 1987).

B. Unoriented knots

1. Formulation

Polynomial invariants for unoriented knots can be constructed by following the same route as that for oriented knots. For each knot one constructs an *unoriented* lattice $\mathcal{L}$. Consider a vertex model on $\mathcal{L}$ and require the partition function to remain invariant under all Reidemeister moves of lattice edges. The partition function is then a knot invariant.

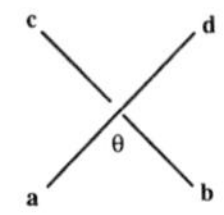

FIG. 32. Labelings of a vertex for unoriented knots.

For unoriented knots, however, there is only one type of line crossing, and hence the partition function has a uniform vertex weight. We label the vertex edges as in Fig. 32 and write the vertex weight as

$$\omega(a,d|b,c)=\omega\begin{bmatrix}c & d\\ a & b\end{bmatrix}=\omega\begin{bmatrix}b & a\\ d & c\end{bmatrix}. \tag{5.25}$$

If we further label the Reidemeister moves as shown in Fig. 33, assuming regular isotopy and Eq. (2.15) for type-I moves, we can read off from Fig. 33 conditions imposed by Reidemeister moves. This leads to

$$\sum_{c\in\mathcal{J}}\omega(a,c|b,c)=\alpha^{-1}\delta_{bx}\ , \tag{5.26a}$$

$$\sum_{c\in\mathcal{J}}\omega(c,b|a,c)=\alpha\delta_{ab}\ , \tag{5.26b}$$

$$\sum_{b,y\in\mathcal{J}}\omega(a,b|x,y)\omega(b,c|y,z)=\delta_{ac}\delta_{xz}\ , \tag{5.26c}$$

$$\sum_{xyz\in\mathcal{J}}\omega(x,b|y,a)\omega(f,z|e,x)\omega(z,c|d,y) = \sum_{xyz\in\mathcal{J}}\omega(e,x|d,y)\omega(z,c|x,b)\omega(f,z|y,a)\ , \tag{5.26d}$$

where, as before, $\mathcal{J}$ is a set of q integers. We now state the main result as a theorem:

Theorem V.B.1. For a given knot we construct an unoriented lattice $\mathcal{L}$. The partition function $Z(\omega)$ [Eq. (4.1)] of a vertex model on $\mathcal{L}$ with vertex weight as given in Eq. (5.25) and satisfying Eq. (2.15) under type-I Reidemeister moves is an invariant of regular isotopy for unoriented knots, provided that Eqs. (5.26a)–(5.26d) hold.

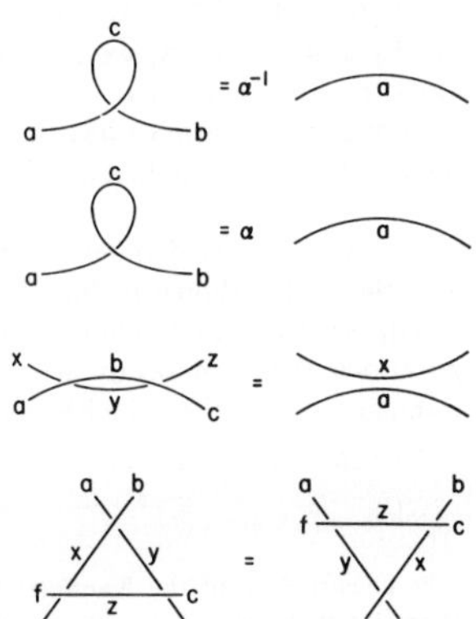

FIG. 33. Labelings for Reidemeister moves.

Collorary V.B.1. The function $\alpha^{-w(K)}Z(\omega)$ is a semioriented invariant for oriented knots.

2. The bracket polynomial and ice-type vertex models

As an example, we construct an invariant using the nonintersecting-string model of Sec. IV.E.3, which leads to the Jones polynomial.

For the vertex shown in Fig. 32, we assign the vertex weight given by Eqs. (4.33) and (4.35), namely,

$$\omega\begin{bmatrix}c & d\\ a & b\end{bmatrix}=A\delta_{ac}\delta_{bd}+B\delta_{ab}\delta_{cd}\ , \tag{5.27}$$

where $A\equiv\sinh(\eta-u)$, $B\equiv\sinh u$. Note, however, that we regard A and B as two independent parameters and apply the vertex weight (5.27) to all vertices.

Perk and Wu (1986a) pointed out that the particular form of the weight given in Eq. (5.27) permits one to write the partition function $Z_{\rm NIS}$ as a generating function of nonintersecting polygonal decompositions P of $\mathcal{L}$. Indeed, by substituting Eq. (5.27) into Eq. (4.1) and summing over all edge states, one finds $Z_{\rm NIS}$ in the form of a polynomial in q, A, and B (Perk and Wu, 1986a),

$$Z_{\rm NIS}(q,A,B)=\sum_P q^{p(P)}\prod_{i=1}^N W_i(P)\ , \tag{5.28}$$

where $p(P)$ is the number of polygons (loops) in P, and $W_i(P)$ is the weight of the ith site in P, equal to either A or B. Since the lattice has at least one loop, $Z_{\rm NIS}$ is divisible by q.

The polynomial

$$P(q,A,B)=q^{-1}Z_{\rm NIS}(q,A,B) \tag{5.29}$$

was discovered independently by Kauffman and named the *bracket polynomial* of a *state model* (Kauffman, 1987a). Clearly, in this picture, the state model is characterized by *nonlocal* Boltzmann weights. In a remarkable piece of pioneering insight connecting knot theory with statistical mechanics, Kauffman (1987a) showed that the bracket polynomial can be used to provide a simple derivation of the Jones polynomial (see also Wu, 1992a). Perk and Wu (1986a), Truong (1986), and Kauffman (1988b) have also shown that the bracket polynomial is completely equivalent to a q^2-state Potts model partition function, a fundamental connection relating the Potts model with the Jones polynomial. Kauffman (1988b) went further and reformulated the Potts model in terms of a formalism of alternating link diagrams.

It is straightforward to verify that Eqs. (5.26a)–(5.26d) are satisfied by taking

$$B=A^{-1},\quad q=-(A^2+A^{-2}),\quad \alpha=-A^3\ . \tag{5.30}$$

It follows from Theorem V.B.1 that the one-variable function

$$f(A)=P(-A^2-A^{-2},A,A^{-1})\,, \tag{5.31}$$

which is normalized to $f_{\rm ring}(A)=1$, is an invariant for unoriented knots. This is the Kauffman bracket invariant. Furthermore, by Corollary V.B.1, the function

$$V(t)=(-t^{-3/4})^{-w}f(t^{-1/4}) \tag{5.32}$$

is an invariant of ambient isotopy for oriented knots, where we have written $A=t^{-1/4}$. To identify $V(t)$ as the Jones polynomial, one verifies the identity

$$\frac{1}{t}\left[\alpha^{-1}\omega\begin{bmatrix} c & d\\ a & b\end{bmatrix}\right]-t\left[\alpha\omega\begin{bmatrix} d & b\\ c & a\end{bmatrix}\right] =\left[\sqrt{t}-\frac{1}{\sqrt{t}}\right]\delta_{ac}\delta_{bd}\,. \tag{5.33}$$

This shows that $V(t)$ satisfies the Skein relation (2.6) and hence is the Jones polynomial.

The noninteracting-string model can be further generalized by associating line orientations. This leads to the oriented nonintersecting-string (ONIS) and generalized ice-type models (Perk and Wu, 1986a). In the ONIS model the lattice edges can be colored in q_1 distinct colors and, in addition, colored as well as oriented in q_2 colors, with the restriction that the numbers of in and out arrows of a given color at a vertex must be the same (the ice rule). This permits one to consider the piecewise-linear lattice $\mathcal{L}^*$ and introduce, for vertices of degree 2, weights in the form of Eq. (4.9) with λ_μ, $\mu=1,2,\ldots,q_2$ replacing λ for each of the q_2 colors. For a model with separable weights one finds the partition function again given by Eq. (5.28), but with

$$q=q_1+\sum_{\mu=1}^{q_2}(\lambda_\mu+\lambda_\mu^{-1})\,. \tag{5.34}$$

The case of $q_1=0$, $q_2=1$ leads to the usual ice-rule model (Temperley and Lieb, 1971; Baxter *et al.*, 1976), a correspondence that has also been discussed by Kauffman (1988b).

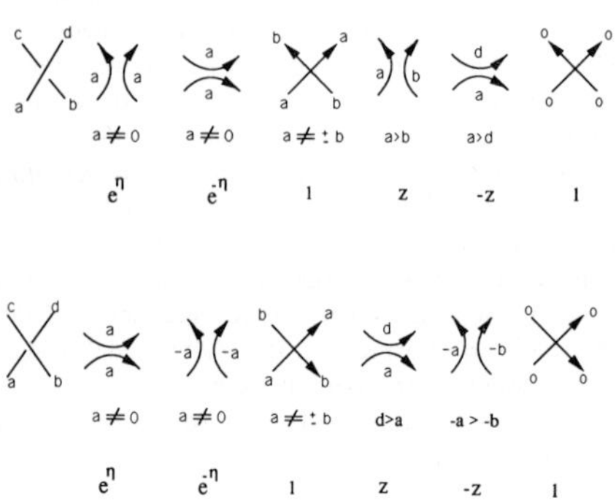

FIG. 34. Vertex configurations and weights for the Kauffman polynomial. Configurations in the two rows are related by a 90° rotation.

FIG. 35. Orientation and sign convention.

3. The Kauffman polynomial

The construction of the Kauffman polynomial (Turaev, 1988) requires special attention. The following is essentially a reformulation of the diagrammatic analysis (of the Turaev construction) due to Kauffman (1991), modified by considering a vertex model with local weights on piecewise-linear lattices.

To begin with consider a $(q+1)$-state vertex model with edge variables $\{a,b,\ldots,x,y,\ldots\}$ taking on $q+1$ numerical values contained in the set

$$\mathcal{J}_P=\{\mathcal{J},0\}\,, \tag{5.35}$$

Here $\mathcal{J}$ is the set of q numerical values given by Eq. (5.8). For our purposes we shall consider $q=2,4,6,\ldots$ so that $\mathcal{J}$ does not contain the value zero.[16]

Again, one looks for vertex weights that are solutions of the Yang-Baxter equation (5.26d), so that the Reidemeister move III is automatically satisfied. To obtain the Kauffman polynomial one uses a representation of simple Lie algebras $A_{q-1}^{(1)}$, giving rise to nonvanishing vertex weights of the form (Turaev, 1988)

$$\omega\begin{bmatrix} a & b\\ a & b\end{bmatrix},\quad \omega\begin{bmatrix} b & -b\\ a & -a\end{bmatrix},\quad \text{or }\omega\begin{bmatrix} a & b\\ b & a\end{bmatrix}. \tag{5.36}$$

However, to write down the explicit expression of ω we need first to orient and decompose $\mathcal{L}$.

Connect at each vertex the edge indexed a with that indexed $\pm a$, and b with $\pm b$, a process that is unique and that leads to one of the six configurations shown in the first row of Fig. 34. This process decomposes $\mathcal{L}$ into disconnected components, each of which contains an edge in states $\pm a$. [Components may cross each other via bridges, however, due to the presence of the third vertex weight in Eqs. (5.36) and (5.37).] To uniquely specify each component by a single index, we now *orient* components and adopt the convention that

(i) the negation of an edge variable has the same effect as reversing the orientation, a situation shown in Fig. 35, and

(ii) a component has the same index as its upward-pointing edges at the vertex shown in Fig. 32, assuming the latter vertex edges are oriented to point upward.

Thus configurations that can occur at a vertex are those shown in the first row of Fig. 34 with weights

[16]For q = odd the construction of the Kauffman polynomial still holds, but there are then two zeros in the set $\mathcal{J}_P$, and one needs to distinguish them carefully in Eq. (5.37) and Fig. 34 below.

$$\omega\begin{bmatrix} a & a \\ a & a \end{bmatrix} = e^{\eta}, \quad a \neq 0 ,$$

$$\omega\begin{bmatrix} -a & a \\ a & -a \end{bmatrix} = e^{-\eta}, \quad a \neq 0 ,$$

$$\omega\begin{bmatrix} b & a \\ a & b \end{bmatrix} = 1, \quad a \neq \pm b ,$$

$$\omega\begin{bmatrix} a & b \\ a & b \end{bmatrix} = z, \quad a > b , \tag{5.37}$$

$$\omega\begin{bmatrix} -d & d \\ a & -a \end{bmatrix} = -z, \quad a > d ,$$

$$\omega\begin{bmatrix} 0 & 0 \\ 0 & 0 \end{bmatrix} = 1 ,$$

$$\omega\begin{bmatrix} c & d \\ a & b \end{bmatrix} = 0, \quad \text{otherwise} ,$$

where η is arbitrary and

$$z = e^{\eta} - e^{-\eta} . \tag{5.38}$$

Note that edges with state zero also form connected components. For later use we show in the second row of Fig. 34 the same configurations rotated 90° clockwise, where we have adopted the sign and orientation convention and negated some edge variables.

The weight equation (5.37) can be summarized as

$$\begin{aligned}\omega\begin{bmatrix} c & d \\ a & b \end{bmatrix} &= e^{\eta}\delta_{abcd}(1-\delta_{a0}) + e^{-\eta}\delta_{a,-b-c,d}(1-\delta_{a0}) \\ &\quad + \delta_{ad}\delta_{bc}(1-\delta_{ab})(1-\delta_{a,-b}) + z\delta_{ac}\delta_{bd}\theta(a-b) \\ &\quad - z\delta_{b,-a}\delta_{c,-d}\theta(a-d) + \delta_{abcd0} \\ &= \omega\begin{bmatrix} -b & -a \\ -d & -c \end{bmatrix} .\end{aligned} \tag{5.39}$$

Note that the symmetry of the vertex weight indicated in the last line is different from that given in Eq. (5.25). However, due to the sign and orientation convention, the symmetry shown ensures its consistency and does not affect the overall partition function.

Substituting Eq. (5.39) into the partition function equations (4.1) and (4.2), we can write the partition function as

$$Z_{\text{vertex}}(\omega) = \sum_{a \in c} \sum_{i} \omega_i , \tag{5.40}$$

where the summation is taken over all possible decompositions of $\mathcal{L}$ into oriented components c, each of which is now indexed by a single edge variable a.

We next introduce the piecewise-linear lattice $\mathcal{L}^*$ with angle-dependent vertex weights. For vertices of degree two, the weights $\omega^*(a)$ are those given previously in Eq. (4.9). For other vertices, we require that the new weight ω^* satisfy the Yang-Baxter equation. If we color components of $\mathcal{L}$ by different colors, then as seen in Fig. 34 the incoming/outgoing colors are conserved at each vertex. This color conservation, which is a special case of charge conservation in the sense that charges (colors) remain unchanged, permits us to introduce angle factors as in Eq. (4.17b) for each term in Eq. (5.39), leading to the new weight

$$\begin{aligned}\omega^*\begin{bmatrix} c & d \\ a & b \end{bmatrix}(\theta) &= e^{\eta}\delta_{abcd}(1-\delta_{a0}) + e^{-\eta}\delta_{a,-b,-c,d}(1-\delta_{a0}) + \delta_{ad}\delta_{bc}(1-\delta_{ab})(1-\delta_{a,-b}) + z\lambda^{(a-d)\theta/2\pi}\delta_{ac}\delta_{bd}\theta(a-d) \\ &\quad - z\lambda^{(d-a)(\pi-\theta)/2\pi}\delta_{a,-b}\delta_{c,-d}\theta(a-d) + \delta_{abcd0} \\ &= \omega^*\begin{bmatrix} -b & -a \\ -d & -c \end{bmatrix} .\end{aligned} \tag{5.41}$$

As a consequence of color conservation, the weight ω^* now satisfies the Yang-Baxter equation.[17] The partition function $Z(\omega^*)$ with angle-dependent weights is now invariant under Reidemeister moves IIIA of the lattice edges.

The expression of ω^* differs from that of ω only in the appearance of angle-dependent factors in the fourth and fifth terms. In the latter (fifth) term we can write $\lambda^{(d-a)(\pi-\theta)/2\pi} = \lambda^{(d-a)/2}\lambda^{(a-d)\theta/2\pi}$, giving rise to a factor $\lambda^{(a-d)/2}$ noted in another context (Kauffman, 1991). Here this factor arises naturally as a consequence of the requirement that ω^* satisfy the Yang-Baxter equation.

We now choose λ so that $Z(\omega^*)$ is invariant under the two distinct Reidemeister moves I shown in Fig. 36. Adopting line orientations as shown, we obtain from Eq. (2.15) the conditions

[17]This fact can also be seen by noting that ω^* can be generated from ω by separating the angle-dependent factor into factors $\lambda^{\pm a\theta/2\pi}$ and associating them separately with the two paths of different colors passing through a vertex. The desired property can then be established by using the property Eq. (4.10).

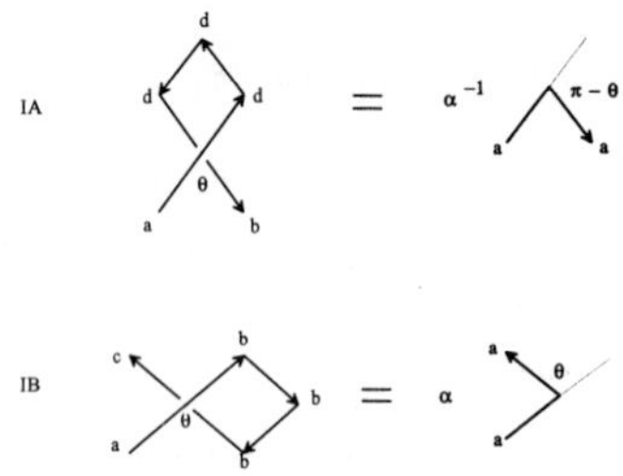

FIG. 36. Labelings for Reidemeister moves I.

$$\sum_{d\in\mathcal{J}_p}\lambda^{d(\pi+\theta)/2\pi}\omega^*\begin{bmatrix}-d & d\\ a & -b\end{bmatrix}(\theta)=\alpha^{-1}\lambda^{-a(\pi-\theta)/2\pi}\delta_{a,-b}\quad(\text{IA})\,, \tag{5.42}$$

$$\sum_{b\in\mathcal{J}_p}\lambda^{-b(2\pi-\theta)/2\pi}\omega^*\begin{bmatrix}c & b\\ a & b\end{bmatrix}(\theta)=\alpha\lambda^{a\theta/2\pi}\delta_{ac}\quad(\text{IB})\,, \tag{5.43}$$

where, as in Eq. (5.1), we have included weights of the three vertices of degree 2.

When we substitute Eq. (5.41) into Eq. (5.43) and use the identity

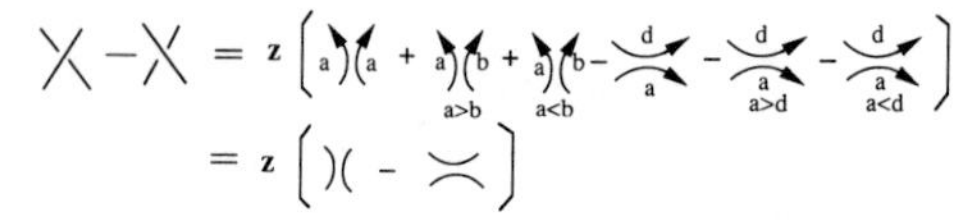

FIG. 37. Skein relation for the Kauffman polynomial.

$$\sum_{d\in\mathcal{J}_p}\lambda^{-d}\theta(a-d)=(1-\delta_{a0})\theta(a)+\left[\frac{\lambda^{q-1}-\lambda^{-a}}{1-\lambda^{-2}}\right]+\left[\frac{1-\lambda^{-1}}{1-\lambda^{-2}}\right]\delta_{a0}\,, \tag{5.44}$$

it is straightforward to show that Eq. (5.43) is satisfied if we take

$$\lambda=e^{\eta},\quad \alpha=e^{-q\eta}\,. \tag{5.45}$$

In a similar manner one shows that Eq. (5.42) is satisfied. One also establishes that conditions imposed by Reidemeister moves II (and III) are all satisfied by the vertex weight (5.41), details of which we omit. It follows that the partition function $Z(\omega^*)$ defines a knot invariant.

To identify this knot invariant as the Kauffman polynomial, we need to show that the partition function $Z(\omega^*)$ satisfies the Skein relation (2.13) or (2.13a). Now the vertex configurations and weights of a minus-type crossing are given in the second row in Fig. 34 (for which the "upward-pointing" direction is pointing towards the right). By taking the difference of the two weights in Fig. 34 and making use of the identity (5.17), one obtains

$$\omega^*\begin{bmatrix}c & d\\ a & b\end{bmatrix}(\theta)=\omega^*\begin{bmatrix}d & -b\\ -c & a\end{bmatrix}(\pi-\theta)=z[\lambda^{(a-d)\theta/2\pi}\delta_{ac}\delta_{bd}-\lambda^{(d-a)(\pi-\theta)/2\pi}\delta_{a,-b}\delta_{c,-d}]\,. \tag{5.46}$$

Here the negation of b and c in the second expression in Eq. (5.46) is due to our orientation convention.

Inserting this expression into $Z(\omega^*)$ written in the form of Eq. (5.40), a procedure shown schematically in Fig. 37, one arrives at the identity

$$Z_{D_+}(\omega^*)-Z_{D_-}(\omega^*)=z[Z_{D_0}(\omega^*)-Z_{D_\infty}(\omega^*)]\,, \tag{5.47}$$

which is precisely the Skein relation (2.13a) for the Dubrovnik version of the Kauffman polynomial.

As before, recursive applications of the Skein relation eventually equate $Z(\omega^*)$ to the product of two factors, a Laurent polynomial in α and z, and the partition function of a ring, now given by

$$\begin{aligned}Z_{\text{ring}}(\alpha,z)&=\sum_{a\in\mathcal{J}_p}\lambda^{\pm a}=1+\frac{\sinh q\eta}{\sinh\eta}\\&=1+(\alpha-\alpha^{-1})/z\,.\end{aligned} \tag{5.48}$$

It follows that the Laurent polynomial

$$Q(\alpha,z)=Z_{\text{vertex}}(\omega^*)/Z_{\text{ring}}(\alpha,z)\,, \tag{5.49}$$

normalized to $Q_{\text{unknot}}(\alpha,z)=1$, is the Dubrovnik version of the Kauffman polynomial. By analytically continuing $Z_{\text{vertex}}(e^{q\eta},e^{\eta}-e^{-\eta})$ to all q, we finally establish the existence and uniqueness of $Q(\alpha,z)$ for arbitrary α and z. This completes the construction of the Kauffman polynomial.

VI. KNOT INVARIANTS FROM IRF MODELS

A. The IRF model

Consider a directed lattice $\mathcal{L}$ of N sites, arbitrary shape, and a uniform coordination number 4. Place spins inside the faces of $\mathcal{L}$ as shown in Fig. 38, where the spin locations are indicated by solid circles. Let the spins take on values, or spin states, designated by variables $\{a,b,\ldots\}\in\mathcal{J}$, where $\mathcal{J}$ is a set of q integers. Let the

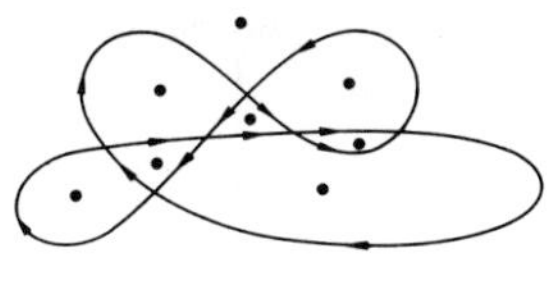

FIG. 38. A directed lattice for the interaction-round-a-face (IRF) model. Spins are denoted by solid circles.

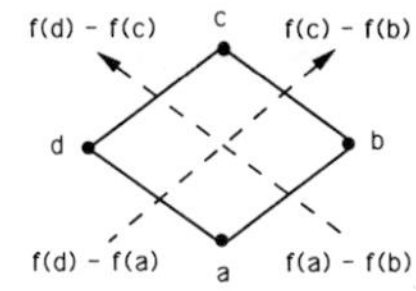

FIG. 39. The four interacting spins in the IRF model. Edge indices are defined as in Fig. 42.

four spins surrounding a site of $\mathcal{L}$ interact via a Boltzmann weight $B(a,b,c,d)$, where spins a, b, c, d are arranged as shown in Fig. 39. In the figure we have drawn the edges of $\mathcal{L}$ as broken lines and connected the four spins along the edges of $\mathcal{L}_D$, the dual of $\mathcal{L}$, to indicate the "domain" of the interaction.

If one regards spin states $a,b,\ldots$ as defining heights, then an overall spin configuration describes a height assignment of faces of $\mathcal{L}$. This is then a solid-on-solid (SOS) model describing the interface of two solids. The overall Boltzmann factor W is a product of individual Boltzmann weights B, and the partition function (3.1) reads

$$Z_{\mathrm{IRF}}(B)=\sum_{\{\mathrm{heights}\}}\prod_{i=1}^{N}B_i(a,b,c,d) \ . \tag{6.1}$$

Here the product is taken over all vertices of $\mathcal{L}$ or, equivalently, all faces $\mathcal{L}_D$, including the exterior (infinite) one. This defines an interaction-round-a-face (IRF) model (Baxter, 1980).

Generally there can be q^4 different Boltzmann weights $B(a,b,c,d)$. But in practice one considers IRF models for which $B(a,b,c,d)$ vanishes unless the heights of two neighboring (adjacent) faces are related in a specific way. For example, the *restricted* eight-vertex SOS model solved by Andrews, Baxter, and Forrester (1984), the ABF model, is an SOS model with *q finite* and for which the difference of two adjacent heights is always 1. Particularly, the $q=\infty$ version is the *unrestricted* eight-vertex SOS model. Such rules are conveniently represented by line graphs in which heights are represented by numbered dots and allowed adjacent heights by line connections.[18] For example, the unrestricted eight-vertex SOS model is described by the graph shown in Fig. 40, and the ABF model is described by graph A_q in Fig. 41.

Generally, there is a one-to-one correspondence between line graphs and certain IRF models (Akutsu *et al.*, 1988), a consideration leading to hierarchies of integrable models (Date *et al.*, 1986; see also Akutsu *et al.*, 1986a, 1986b; Kuniba *et al.*, 1986a–1986e; Pearce and Seaton, 1988). In particular, there exists an integrable IRF mole for each Dynkin diagram of simply-laced classical or affine Lie algebras of the A, D, E series (Pasquier, 1987a; Jimbo *et al.*, 1988), examples of which are shown in Fig. 41. The IRF model corresponding to A_n is the ABF model; the model corresponding to D_n has been solved by Pasquier (1987b), and the cyclic eight-vertex SOS model (Baxter, 1973a, 1973b) corresponding to $A_n^{(1)}$ has been solved by Pearce and Seaton (1989).

[18]The connecting lines will be directed in the case of IRF models with chiral Boltzmann weights.

B. Equivalence with charge-conserving vertex models

The construction of knot invariants from IRF models is most conveniently done via the equivalence of IRF models with a charge-conserving vertex model. We first elucidate this equivalence (Akutsu *et al.*, 1988; Jones, 1989; see also Kadanoff and Wegner, 1971 and Wu, 1971).

Consider an IRF model with the partition function (6.1). Consider further the partition function $Z_{\mathrm{IRF}}^{(a)}(B)$ defined by Eq. (6.1) with the height of one face, say, the exterior, fixed at a. Then Eq. (6.1) can be written as

$$Z_{\mathrm{IRF}}(B)=\sum_{a\in\mathcal{J}}Z_{\mathrm{IRF}}^{(a)}(B) \ . \tag{6.2}$$

To each height a we assign a value $f(a)$ where the function f is one-to-one; to each directed edge we assign an index

$$h_{ab}=f(a)-f(b) \ , \tag{6.3}$$

where a is the height to the left, and b to the right, of the edge, as shown in Fig. 42. An example of f is $f(a)=a$; but more generally the function f can be chosen at our discretion. A height configuration is now mapped into an edge indexing. Clearly, as can be seen from Fig. 39, the edge indexing satisfies the charge conservation condition, Eq. (4.16), as generalized in footnote 9. Conversely, each charge-conserving edge indexing in the form of Eq. (6.3) is mapped into a height configuration, provided that the height a, or the function $f(a)$, of the exterior face is given. This leads to the equivalence

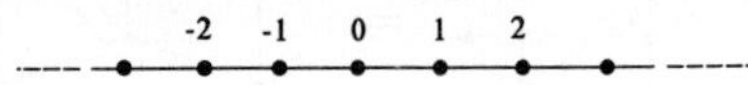

FIG. 40. Line graph for the Andrews, Baxter, and Forrester (ABF) model.

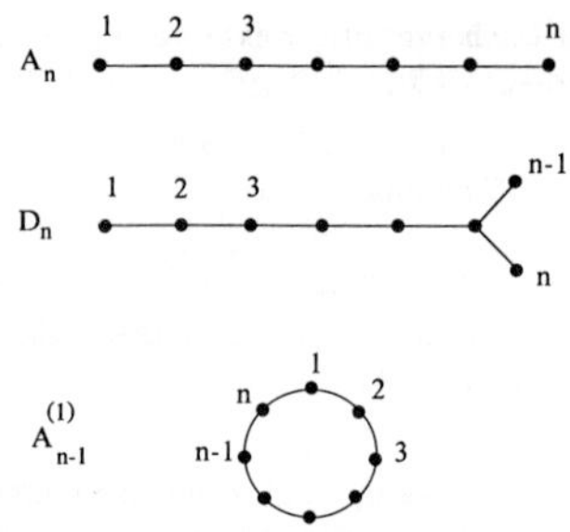

FIG. 41. Dynkin diagrams of Lie algebras.

$$Z^{(a)}_{\rm IRF}(B)=Z^{f(a)}_{\rm vertex}(w)\ , \tag{6.4}$$

where $Z^{f(a)}_{\rm vertex}(w)$ is the vertex-model partition function (4.1), with edges indexed by h_{ab}, and the function f of the exterior face fixed at $f(a)$. Explicitly, we have the equivalence

$$\omega\begin{bmatrix} f(d)-f(c) & f(c)-f(b) \\ f(d)-f(a) & f(a)-f(b)\end{bmatrix}=B(a,b,c,d)\ . \tag{6.5}$$

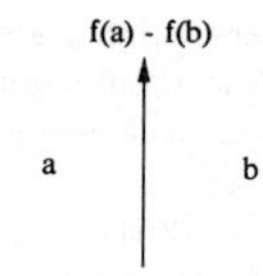

FIG. 42. Convention of lattice edge indexing.

C. The Yang-Baxter equation

An IRF model is *integrable* if its Boltzmann weight $B(a,b,c,d)$ satisfies a Yang-Baxter equation. The Yang-Baxter equation can now be written down from the equivalence with a vertex model, by assuming appropriate edge indexings in Eqs. (4.3) and (4.4). To completely describe the Yang-Baxter equation, one needs further to specify the factor $f(a)$ associated with the exterior face. It is then more convenient to write down the Yang-Baxter equation directly in terms of the IRF-model Boltzmann weights $B(a,b,c,d)$. As may be surmised from Fig. 43, this is equivalent to considering a cluster of seven spins with interactions arranged in two different ways, as shown, and requiring the partition functions of the two clusters to be identical for any given spin states $\{a,b,c,d,e,f\}$. The Yang-Baxter equation in IRF language then reads (Baxter, 1980)

$$\sum_{g\in\mathcal{J}} B(g,c,b,a|u-w)B(f,e,g,a|v-u)B(e,d,c,g|v-w)$$
$$=\sum_{g\in\mathcal{J}} B(e,d,g,f|u-w)B(g,d,c,b|v-w)B(f,g,b,a|v-w)\quad\text{for all } a,b,c,d,e,f\in\mathcal{J}\ . \tag{6.6}$$

The unitarity condition, Eq. (4.5b), now reads, after changing edge indexings,

$$\sum_{c\in\mathcal{J}} B(a,b,c,d|u-v)B(c,b,d,e|v-u)=\delta_{ae}\ , \tag{6.7}$$

which we show graphically in Fig. 44. In analogy to Eq. (4.18) for the vertex model, one verifies that the Boltzmann weight

$$\tilde{B}(a,b,c,d|u)=e^{\beta(f_a+f_c-f_b-f_d)u}B(a,b,c,d|u) \tag{6.8}$$

is also a solution of Eq. (6.6) for any β. We shall leave open, the possibility of using this symmetry-breaking Boltzmann weight, and use B to denote either B or $\tilde{B}$, whichever arises in applications. In the infinite-rapidity limit, we have

$$B_{\pm}(a,b,c,d)=\lim_{u\to\pm\infty} B(a,b,c,d|u)\ , \tag{6.9}$$

where, as before, the right-hand side of Eq. (6.9) has been divided by the leading diverging Boltzmann weight.

D. Integrable IRF models

We now present examples of integrable IRF models.

1. The unrestricted eight-vertex SOS model

The unrestricted eight-vertex SOS model, the $q=\infty$ ABF model, is characterized by the line graph of Fig. 40.

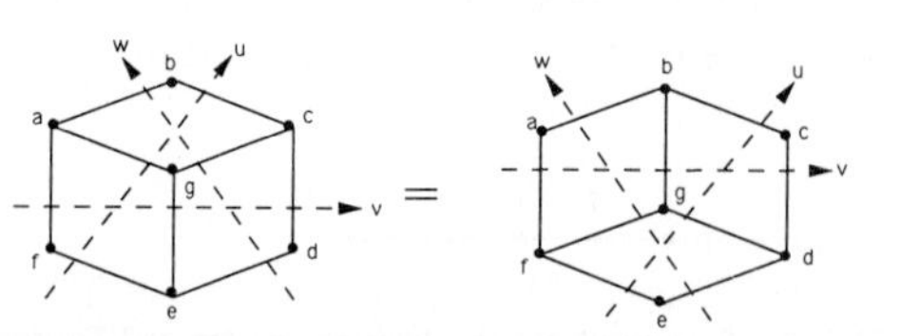

FIG. 43. The Yang-Baxter equation for IRF models.

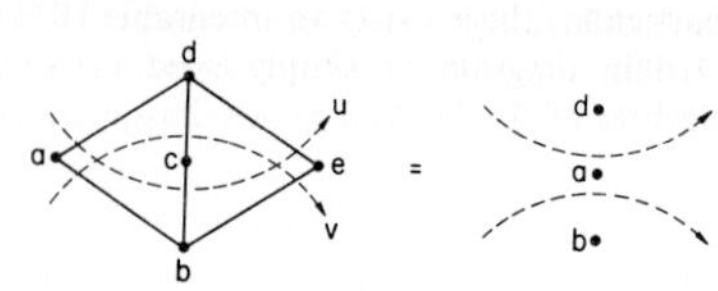

FIG. 44. The unitarity condition for Boltzmann weights.

In this model, adjacent heights always differ by 1, and there are six contributing configurations, as shown in Fig. 45. It is also clear that we need only consider the partition function $Z^{(a)}(B)$.

Boltzmann weights of integrable IRF models are given in terms of elliptical theta functions. At criticality, however, they reduce to hyperbolic functions. In the case of the $q=\infty$ ABF model they can be written in the form

$$\begin{aligned} &B_1=B_2=1\ ,\\ &B_3=B_4=\sinh u/\sinh(\eta-u)\ .\\ &B_5=e^{u}\sinh\eta/\sinh(\eta-u)\ ,\\ &B_6=e^{-u}\sinh\eta/\sinh(\eta-u)\ , \end{aligned} \tag{6.10}$$

where u is the rapidity and η is arbitrary, and we have included the symmetry-breaking factor in Eq. (6.8) with $\beta=1/2$ and $f(a)=a$.

The Boltzmann weights of Eq. (6.10) can be rewritten as

$$\begin{aligned} &B(a,b,c,d|u)\\ &\quad=\delta_{ac}+\delta_{bd}\left[\frac{\sinh u}{\sinh(\eta-u)}\right]e^{[(a+c)/2-b]\eta}\\ &\quad=0,\quad (a-b)(b-c)(c-d)(a-d)\neq\pm1\ , \end{aligned} \tag{6.11}$$

where a,b,c,d are integers. Taking the infinite-rapidity limit, we obtain

$$\begin{aligned} B_\pm(a,b,c,d)&=A_\pm[\delta_{ac}-\delta_{bd}e^{[(a+c)/2-b\pm1]\eta}]\\ &=0,\quad (a-b)(b-c)(c-d)(a-d)\neq\pm1\ , \end{aligned} \tag{6.12}$$

where we have included a normalization factor $A_\pm$.

2. The cyclic SOS model

The q-state cyclic SOS model (Pearce and Seaton, 1988, 1989) is characterized by the Dynkin diagram $A_q^{(1)}$ of Fig. 41. The contributing configurations are also those shown in Fig. 45, but now with indices $a,b,\ldots,\ \mathrm{mod}(q)$. The critical vertex weights are again those given by Eqs. (6.10) and (6.11), but with

$$\eta=i2\pi s/q,\quad s=1,2,\ldots,q-1\ . \tag{6.13}$$

Since the q states are cyclic, the partition function is independent of the height of the exterior face, and we have $Z_{\mathrm{IRF}}(B)=qZ_{\mathrm{IRF}}(a)(B)$.

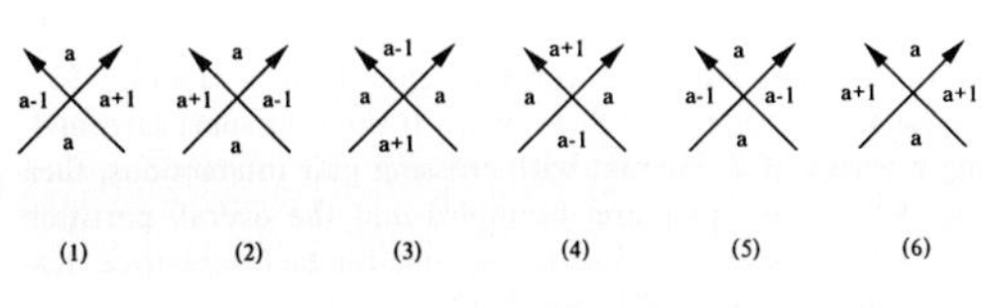

FIG. 45. Configurations of the ABF and cyclic solid-on-solid (SOS) models.

E. Enhanced IRF models

Analogous to the discussions in Sec. IV.C, we introduce the piecewise-linear lattice $\mathcal{L}^*$ and *enhanced* IRF models on $\mathcal{L}^*$. The enhanced IRF model has angle-dependent Boltzmann weights

$$B^*(a,b,c,d|u)=\lambda^{(h_{da}-h_{cb})\theta/2\pi}B(a,b,c,d|u)\ , \tag{6.14}$$

where θ is the angle of the two edges bordering the face indexed a, and, for vertices of degree 2 on $\mathcal{L}^*$,

$$\begin{aligned} B^*(a,b)&=\lambda^{h_{ab}\theta/2\pi}\quad\text{if the line turns an angle }\theta\text{ to the left}\\ &=\lambda^{-h_{ab}\theta/2\pi}\quad\text{if the line turns an angle }\theta\text{ to the right}\\ &=0\quad\text{if adjacent heights } a \text{ and } b \text{ are forbidden.} \end{aligned} \tag{6.15}$$

Here the arrangement of a and b is the same as in Fig. 42. This enhanced IRF model now maps into an enhanced vertex model with vertex weights as in Eqs. (4.17a) and (4.17b) and the replacement of a by $f(a)$.

The partition function of the enhanced IRF model is now

$$Z_{\mathrm{IRF}}(B^*)=\sum_{\{\text{heights}\}}\prod B^*(a,b,c,d,|u)\prod B^*(a,b)\ , \tag{6.16}$$

and, in the infinite-rapidity limit,

$$Z_{\mathrm{IRF}}(B_\pm^*)=\sum_{\{\text{heights}\}}\prod B_\pm^*(a,b,c,d)\prod B^*(a,b)\ , \tag{6.17}$$

where

$$B_\pm^*(a,b,c,d)=\lambda^{(h_{da}-h_{cb})\theta/2\pi}B_\pm(a,b,c,d)\ . \tag{6.18}$$

The creation of vertices of degree two leads to the consideration of lattices in the form of a ring. We shall assume that the integer set $\mathcal{J}$ and the function f have been chosen such that the partition function of a ring,

$$Z_{\mathrm{ring}}(B^*)=\sum_{\{a,b\}\in\mathcal{J}}\lambda^{\pm h_{ab}}=\Lambda\ , \tag{6.19}$$

is a constant. Here the summation is taken over heights a and b, consistent with the adjacency requirement.

F. Construction of knot invariants

We now construct knot invariants from IRF models. From a given knot we consider an integrable IRF model

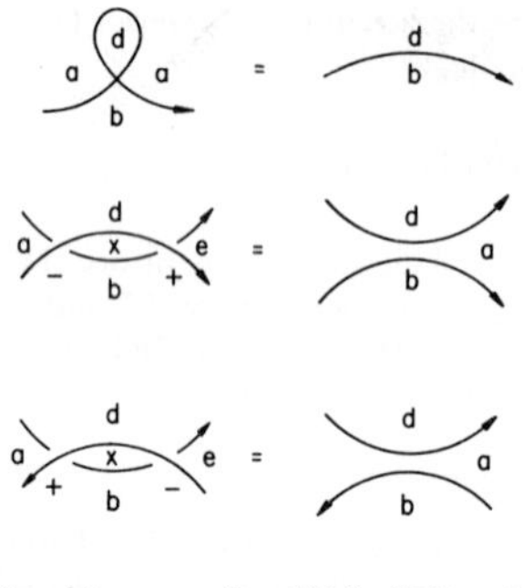

FIG. 46. Reidemeister moves I and II for IRF models.

and its equivalent enhanced vertex model. We can then use Theorem V.A.1, and, since the equivalent vertex model is charge conserving, we need only consider conditions (5.1a)–(5.3a). Recasting these conditions for Reidemeister moves I and II in terms of Boltzmann weights $B_{\pm}(a,b,c,d)$, a process we show in Fig. 46, we obtain

$$\sum_{d\in\mathcal{I}} \lambda^{f(d)-f(a)}B_{\pm}(a,b,a,d)=1, \quad \text{for all } a,b. \quad (\mathrm{I}), \tag{6.20a}$$

$$\sum_{x\in\mathcal{I}} B_{\pm}(a,b,x,d)B_{\mp}(x,b,e,d)=\delta_{ae} \quad (\mathrm{IIA}), \tag{6.20b}$$

$$\sum_{x\in\mathcal{I}} \lambda^{f(a)+f(x)-f(b)-f(d)}B_{\pm}(d,a,b,x) \times B_{\mp}(b,e,d,x)=\delta_{ae} \quad (\mathrm{IIB}). \tag{6.20c}$$

These conditions have been obtained by Akutsu *et al.* (1988). Note that, as in the case of vertex models, Eq. (6.20b) is a consequence of the unitarity condition, Eq. (6.7).

We now state our results on IRF models as a theorem:

Theorem VI.F. For each oriented knot we construct a directed lattice $\mathcal{L}$ and the associated piecewise-linear lattice $\mathcal{L}^*$. Then the partition function (6.17) of an enhanced IRF model with Boltzmann weights (6.15) and (6.18) is a knot invariant, provided that Eqs. (6.20a)–(6.20c) hold and that the partition function of an unknot is Eq. (6.19).

G. Examples

We now apply Theorem VI.F to the ABF and cyclic SOS models, both of which lead to the Jones polynomial (Akutsu and Wadati, 1988).

Using the Boltzmann weight given by Eq. (6.12), we find that Eq. (6.20a) is satisfied by choosing

$$f(a)=a, \quad \lambda=e^{\eta}, \quad A_{\pm}=e^{\mp\eta}. \tag{6.21}$$

With these choices, one readily verifies that both Eqs. (6.20b) and (6.20c) hold and that the partition function of a ring, Eq. (6.19), is $\Lambda=e^{\eta}+\epsilon^{-\eta}$ for the ABF model and $\Lambda=q(e^{\eta}+\epsilon^{-\eta})$ for the cyclic SOS model. It follows that $Z(B^*_{\pm})$ is a knot invariant.

To identify this invariant as the Jones polynomial, we obtain from Eqs. (6.12), (6.14), and (6.21) the identity

$$e^{-2\eta}B^*_{+}(a,b,c,d)-e^{2\eta}B^*_{-}(a,b,c,d) = (-e^{\eta}+\epsilon^{-\eta})e^{(d-a)\theta/2\pi}e^{-(a-b)\theta/2\pi} \times\delta_{d-a,d-c}\delta_{a-b,c-b}\,. \tag{6.22}$$

This is precisely Eq. (5.22) leading to the Skein relation (2.6) for the Jones polynomial $V(t)$ after identifying $e^{\eta}=-\sqrt{t}$. This establishes that

$$V(t)=\Lambda^{-1}Z(B^*_{\pm})\,. \tag{6.23}$$

By considering multicomponent spins, Akutsu *et al.* (1989) have shown that the Homfly and Kauffman polynomials can also be constructed from IRF models.

VII. KNOT INVARIANTS FROM EDGE-INTERACTION MODELS

A. Formulation

In our discussion of constructing knot invariants from IRF models, we have not inquired about explicit realizations of the Boltzmann weight $B(a,b,c,d)$. In this section we consider the realization of B by explicitly introducing two-spin interactions. While it is possible to do this by further specializing our results on IRF models, it is more convenient to take advantage of the simplicity of the interaction and proceed directly. This direct approach also eliminates the need for introducing the piecewise-linear lattice $\mathcal{L}^*$ and the associated enhanced lattice models. This leads to the consideration of edge-interaction models.

Starting from a given knot consisting of N line crossings, we construct an *unoriented* lattice $\mathcal{L}$ of N sites, while disregarding the line orientations. In the simplest case we consider a spin model whose spins reside in one set of the bipartite faces of $\mathcal{L}$ forming a lattice $\mathcal{L}'$.[19] To help us visualize, it is convenient to shade faces of $\mathcal{L}$ containing spins, a device first introduced by Baxter *et al.* (1976) in an analysis of the Potts model for arbitrary planar lattices.[20] An example of a lattice $\mathcal{L}$ with shaded

---

[19]It is also possible to consider spin models (Jones, 1989) whose spins reside in all faces of $\mathcal{L}$. If the four spins surrounding a vertex of $\mathcal{L}$ interact with crossing pair interactions, then the two sets of spins are decoupled and the overall partition function becomes a product of two, one for each sublattice (Kadanoff and Wegner, 1971; Wu, 1971).

[20]The designations of $\mathcal{L}$ and $\mathcal{L}'$ here are interchanged from that in Baxter *et al.* (1976).

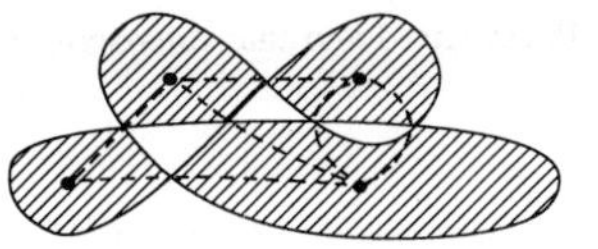

FIG. 47. Example of a lattice for a spin model with pure pair interactions. The solid circles denote spins and the dashed lines denote lattice edges and interactions.

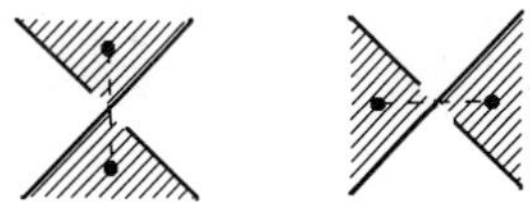

FIG. 48. Two kinds of interactions in the spin model. The interaction is of type + (−) if one finds the shaded area on the left (right) upon leaving the vertex along an edge that is an "overpass."

faces is shown in Fig. 47. The lattice $\mathcal{L}$ is the surrounding, or the covering, lattice of $\mathcal{L}'$.

Let the spins interact with two-spin interactions placed across lattice sites of $\mathcal{L}$ (and along lattice edges of $\mathcal{L}'$) as indicated by the dashed lines in Fig. 47. Then, depending on the relative positionings of the shaded faces with respect to the line crossing, we assign two kinds of interactions, + and −, as shown in Fig. 48.[21]

We let the Boltzmann factors be $W_{\pm}(a,b)$, and, for simplicity, we assume symmetric interactions, i.e.,

$$W_{\pm}(a,b)=W_{\pm}(b,a) . \tag{7.1}$$

As in the case of the IRF model, we assume that spin variables $a,b,\ldots$ take on q integral values in the set $\mathcal{J}$. The partition function (3.1) now reads

$$Z(W_{\pm})=q^{-N/2}\sum_{\text{spin states}}\prod W_{\pm}(a,b) , \tag{7.2}$$

where the product is over all interacting spin pairs in $\mathcal{L}'$, and we have introduced to each spin summation[22] a factor $q^{-1/2}$. The partition function of a single spin corresponding to an unknot is then

$$Z_{\text{single spin}}=q^{-1/2}\sum_{a\in\mathcal{J}}1=\sqrt{q} . \tag{7.3}$$

We require the partition function $Z(W_{\pm})$ to be an invariant of regular isotopy under Reidemeister moves of lattice edges. Taking into account all possible face shadings, this leads to the independent moves shown in Figs. 49 and 50. Figure 49 shows the four independent Reidemeister moves I of regular isotopy derived by shading faces of the two type-I moves shown in Fig. 13 and Eq. (2.15). Similarly, Fig. 50 contains independent Reidemeister moves of types II and III derived by shading faces of the corresponding moves in Fig. 3. Explicitly, the conditions are

$$\frac{1}{\sqrt{q}}\sum_{b\in\mathcal{J}}W_{\pm}(a,b)=\alpha^{\mp1} , \tag{7.4}$$

$$W_{\pm}(a,a)=\alpha^{\pm1} , \tag{7.5}$$

for Reidemeister moves I, and

$$\frac{1}{\sqrt{q}}\sum_{b\in\mathcal{J}}W_{-}(a,b)W_{+}(b,c)=\sqrt{q}\,\delta_{ac} , \tag{7.6}$$

$$W_{+}(a,b)W_{-}(a,b)=1 , \tag{7.7}$$

$$\frac{1}{\sqrt{q}}\sum_{d\in\mathcal{J}}W_{\pm}(a,d)W_{\mp}(b,d)W_{\pm}(c,d)=W_{\mp}(a,b)W_{\mp}(b,c)W_{\pm}(c,a) \tag{7.8}$$

for Reidemeister moves II and III.[23] Conditions (7.4)–(7.8) can be more conveniently represented by linear graphs on $\mathcal{L}'$, as shown in Fig. 51.

Note that according to Eq. (7.2) there is a factor $q^{-1/2}$ for each shaded area; this leads to the compensating factors occurring in the left-hand side of Eqs. (7.4), (7.6), and (7.8). Furthermore, conditions (7.4)–(7.8) are not all independent. Setting $b=c$ in Eq. (7.8), for example, one obtains Eq. (7.4) after using Eqs. (7.5) and (7.7). The condition (7.8) is the Yang-Baxter equation, which is a generalization of the star-triangle equation for the Ising model (Onsager, 1944).

We now state the main result as a theorem:

Theorem VII.A. For each knot we construct an unoriented lattice $\mathcal{L}$ and a q-state spin model with spins occupying every other face of $\mathcal{L}$, with its partition function $Z(W_{\pm})$ given by Eqs. (7.2). Then $q^{-1/2}Z(W_{\pm})$ is an invariant of regular isotopy for unoriented knots satisfying Eqs. (2.14) and (2.15), provided that Eqs. (7.4)–(7.8) hold.

Corollary VII.A. The function $\alpha^{-w(K)}Z(W_{\pm})/\sqrt{q}$ is an invariant of ambient isotopy for oriented knots.

Finally, we remark that since the faces of $\mathcal{L}$, or the lattice $\mathcal{L}'$, are bipartite, there exist two choices for shading the faces, and hence two ways of constructing the spin model. However, these two choices lead to the same invariant (Jones, 1989).

[21]It should be noted that the + and − types of vertices in this context are different from the + and − types of crossings introduced in Sec. II.

[22]More generally one introduces a factor $\tau^{-1/2}$ for each summation. Then setting $a=c$ in Eq. (7.6) and using Eq. (7.7), one obtains $\tau=q$.

[23]The condition imposed by Reidemeister moves III must now be checked, since we are not basing our derivation on solutions of the Yang-Baxter equation.

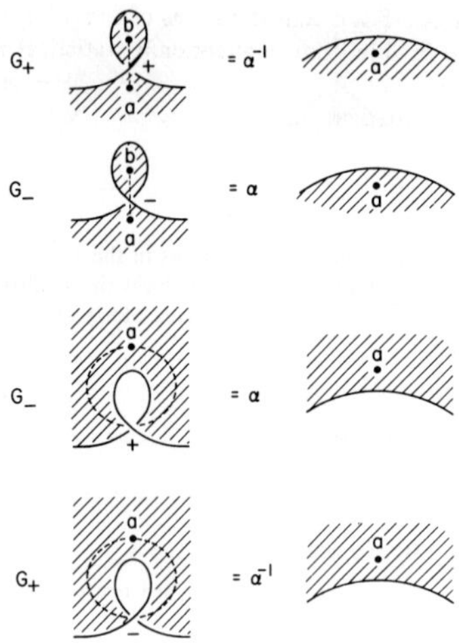

FIG. 49. Type-I Reidemeister moves.

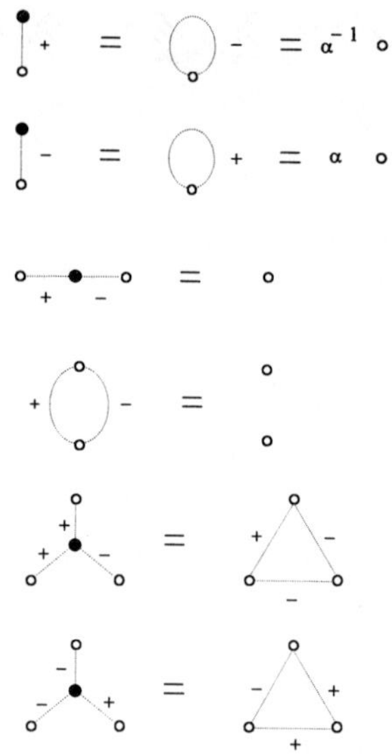

FIG. 51. Equivalent representations of Reidemeister moves. Open circles are rooted denoting fixed spin states; solid circles denote spin states under summations.

B. Example

As an example of the formulation, we show that the Potts model leads to the Jones polynomial (Kauffman, 1988b).

The Potts model (Potts, 1952; for a review see Wu, 1982) is characterized by the two-spin Boltzmann factor

$$W_{\pm}(a,b)=A_{\pm}e^{K_{\pm}\delta_{ab}}$$
$$=A_{\pm}(1+v_{\pm}\delta_{ab}) , \tag{7.9}$$

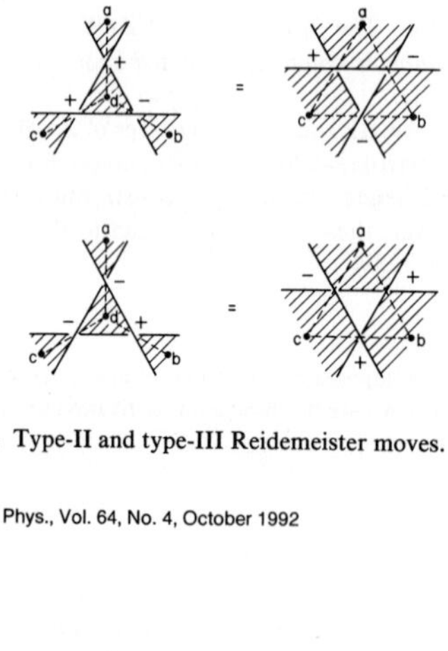

FIG. 50. Type-II and type-III Reidemeister moves.

where

$$v_{\pm}=e^{K_{\pm}}-1 . \tag{7.10}$$

Then the substitution of Eq. (7.9) into Eq. (7.7) leads to

$$A_{+}A_{-}=1, \quad v_{+}+v_{-}+v_{+}v_{-}=0 . \tag{7.11}$$

The second relation in Eq. (7.11) corresponds to $K_{+}=-K_{-}$. Similarly, Eq. (7.6) leads, after using Eq. (7.11), to

$$q=v_{+}v_{-} , \tag{7.12}$$

and Eq. (7.8) leads to

$$A_{\pm}^{2}=\sqrt{q}\,/v_{\pm} . \tag{7.13}$$

Finally, it can be checked that both Eqs. (7.4) and (7.5) are satisfied if one takes

$$\alpha=A_{+}(1+v_{+})=[A_{-}(1+v_{-})]^{-1} . \tag{7.14}$$

It is readily verified that Eqs. (7.11)–(7.14) are satisfied by writing

$$\begin{aligned} &t\equiv -e^{-K_{+}}=-e^{K_{-}} , \\ &v_{\pm}=-(1+t^{\mp 1}) , \\ &A_{\pm}=t^{\pm 1/4} , \\ &q=t+2+1/t , \\ &\alpha=-t^{-3/4} . \end{aligned} \tag{7.15}$$

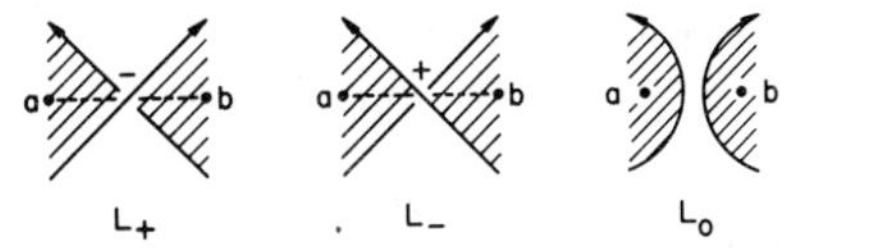

FIG. 52. Skein relation configurations. Note that the W_+ (W_-) interaction corresponds to the L_- (L_+) crossing.

Then, by Theorem VII.A, $Z(W_\pm)$ is an invariant for unoriented knots, and, by Corollary VII.A, $P(t)=(-t^{-3/4})^{-w(K)}Z(W_\pm)$ is an invariant for oriented knots. To identify $P(t)$ as the Jones polynomial other than a normalization factor, we consider the three configurations shown in Fig. 52. A moment's reflection shows that $Z(W_\pm)$ satisfies the Skein relation (2.6), provided that we have [compare with Eq. (5.33)]

$$\frac{1}{t}[\alpha^{-1}W_-(a,b)]-t[\alpha W_+(a,b)]=\sqrt{t}-\frac{1}{\sqrt{t}}\ . \tag{7.16}$$

Indeed, using Eq. (7.15) one verifies that Eq. (7.16) is an identity. Now $P(t)$ has $P_{\mathrm{unknot}}(t)=\sqrt{q}$ as a factor. We thus conclude that

$$V(t)=q^{-1/2}(-t^{-3/4})^{w(K)}Z(W_\pm) \tag{7.17}$$

is a Laurent polynomial normalized to $V_{\mathrm{unknot}}(t)=1$ and is thus the Jones polynomial.

For further examples of invariants derived from spin models with pure two-spin interactions, see Jones (1989).

VIII. SUMMARY

We have presented the formulation of knot invariants using the method of two-dimensional models in statistical mechanics. The underlying theme of the statistical mechanical approach is the construction of lattice models on lattices deduced from planar projections of knots, with the requirement that the partition function remain invariant under Reidemeister moves of lattice edges. When this is done, the partition function is a knot invariant.

The requirement of invariance under Reidemeister moves leads naturally to the consideration of integrable lattice models. It is shown that the integrability of a lattice model leads to invariance under two of the required Reidemeister moves, namely, IIIA and IIA. Then the job is done if the remaining Reidemeister moves, I and IIB, are also realized.

The main results using vertex and IRF models are summarized in Theorems V.A.1 and VI.F, respectively. The construction of knot invariants can also be carried out using spin models with pure two-spin interactions. This leads to Theorem VII.A and the semioriented invariants.

Finally, we emphasize that the approach presented in this review utilizes lattice models whose Boltzmann weights are strictly *local,* without reference to global considerations. Such considerations of local Boltzmann factors are in line with conventional statistical mechanics. With this perspective in mind, we have presented a genuine statistical mechanical approach to knot invariants.

ACKNOWLEDGMENTS

I am grateful to C. King for a critical reading of the manuscript and for comments and suggestions that have greatly improved the clarity of the presentation. I am also indebted to J. H. H. Perk for critical and helpful comments and for calling my attention to relevant references. I would like to thank L. H. Kauffman for sending me a copy of his book (Kauffman, 1991) prior to publication, and V. F. R. Jones for comments. The knot table of Fig. 53 is produced from computer graphics designed by D. Rolfsen and R. Scharein; I am grateful to D. Rolfsen for providing a copy of the figure for our use. This work is supported in part by the National Science Foundation Grant DMR-9015489.

APPENDIX: TABLE OF KNOT INVARIANTS

Traditionally, knots are classified according to the minimum number of crossings in a planar projection. Prime knots and links with up to thirteen crossings have been tabulated in Thistlethwaite (1985). Here we include in Fig. 53 graphs of prime knots and links with up to six

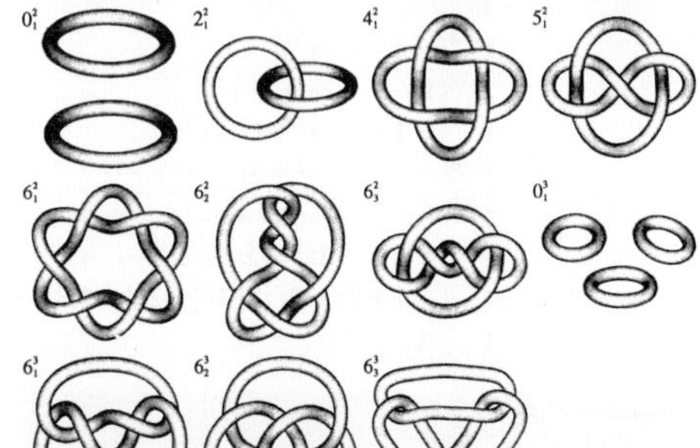

FIG. 53. Planar projections of prime knots and links with six or fewer crossings.

crossings. We also include a table of the associated polynomial invariants. The knot notation of 6_2^3, for example, denotes the second three-component knot (link) with six crossings. In the case of links for which there exist more than one orientation, only those generating distinct invariants are given. They are specified by the subscript $i=1,2$ in $[\ ]_i$. Our convention of specifying the subscript is that if $w_i(K)$ is the writhe of the oriented knot $[K]_i$, $i=1,2$, then $w_2(K)>w_1(K)$.

1. The Alexander-Conway polynomial

The Alexander-Conway polynomial $\Delta(t)=\nabla(z)$, $z=\sqrt{t}-1/\sqrt{t}$, is defined in Sec. II.D.1. Further listings of the Alexander polynomial can be found in Burde and Zieschang (1985) and Rolfsen (1976).

0_1	1	
3_1	$t^{-1}(1-t+t^2)$	$=1+z^2$
4_1	$t^{-1}(-1+3t-t^2)$	$=1-z^2$
5_1	$t^{-2}(1-t+t^2-t^3+t^4)$	$=1+3z^2+z^4$
5_2	$t^{-1}(2-3t+2t^2)$	$=1+2z^2$
6_1	$t^{-1}(-2+5t-2t^2)$	$=1-2z^2$
6_2	$t^{-2}(-1+3t-3t^2+3t^3-t^4)$	$=1-z^2-z^4$
6_3	$t^{-2}(1-3t+5t^2-3t^3-t^4)$	$=1+z^2+z^4$

Alexander polynomials for links with two or more components vanish identically.

2. The Jones polynomial

The Jones polynomial $V(t)$ listed below is defined in Sec. II.D.2 and is the same as in Jones (1987). Further listings of the Jones polynomial for single-component knots can be found in Jones (1985, 1987). Note, however, definitions of $V(t)$ in Jones (1985) and Jones (1987) are related by $t\to t^{-1}$, and expressions in Jones (1985) contain several misprints.[24]

0_1	1
3_1	$t^{-4}(-1+t+t^3)$
4_1	$t^{-2}(1-t+t^2-t^3+t^4)$
5_1	$t^{-7}(-1+t-t^2+t^3+t^5)$
5_2	$t^{-6}(-1+t-t^2+2t^3-t^4+t^5)$
6_1	$t^{-4}(1-t+t^2-2t^3+2t^4-t^5+t^6)$
6_2	$t^{-5}(1-2t+2t^2-2t^3+2t^4-t^5+t^6)$
6_3	$t^{-3}(-1+2t-2t^2+3t^3-2t^4+2t^5-t^6)$
0_1^2	$t^{-1/2}(-1-t)$
2_1^2	$t^{-5/2}(-1-t^2)$
$[4_1^2]_1$	$t^{-11/2}(-1+t-t^2-t^4)$
$[4_1^2]_2$	$\sqrt{t}(-1+t-t^2-t^4)$
5_1^2	$t^{-7/2}(1-2t+t^2+2t^3+t^4-t^5)$
$[6_1^2]_1$	$t^{-17/2}(-1+t-t^2+t^3-t^4-t^6)$
$[6_1^2]_2$	$\sqrt{t}(-1+t-t^2+t^3-t^4-t^6)$
6_2^2	$t^{3/2}(-1+t-2t^2-2t^3+2t^4+t^5-t^6)$
$[6_3^2]_1$	$t^{-15/2}(-1+2t-2t^2+2t^3-3t^4+t^5-t^6)$
$[6_3^2]_2$	$t^{-3/2}(-1+2t-2t^2+2t^3-3t^4+t^5-t^6)$
0_1^3	$t^{-1}(1+2t+t^2)$
$[6_1^3]_1$	$t^{-1}(1-t+3t^2-t^3+3t^4-2t^5+t^6)$
$[6_1^3]_2$	$t^{-7}(1-t+3t^2-t^3+3t^4-2t^5+t^6)$
6_2^3	$t^{-3}(-1+3t-2t^2+4t^3-2t^4+3t^5-t^6)$
$[6_3^3]_1$	$t^{-4}(1+t^2+2t^4)$
$[6_3^3]_2$	$t^2(1+t^2+2t^4)$

[24]Specifically, the expression for 6_1 in Jones (1985) is in error (but correct in Jones, 1987), and expressions for the links 4_1^2 (the second expression), 5_1^2, 6_2^2, 6_3^2, 6_1^3, and 6_3^3 are given in the variable t^{-1}, instead of t. The expressions for 6_1^2 and 6_2^3 given in Jones (1985) are correct.

3. The Homfly polynomial

The Homfly polynomial $P(t,z)$ given below is defined in Sec. II.D.3 and computed from the list of $P(l,m)$ given by Lickorish and Millett (1987, 1988), by substituting with $l=it^{-1}$, $m=iz$. Setting $z=\sqrt{t}-1/\sqrt{t}$ in the expressions below we recover the Jones polynomial, and setting $t=1$ we recover the Alexander-Conway polynomial.

0_1	1
3_1	$t^{-4}(-1+2t^2+z^2t^2)$
4_1	$t^{-2}(1-t^2+t^4-t^2z^2)$
5_1	$t^{-6}[-2+3t^2+z^2(-1+4t^2)+t^2z^4]$
5_2	$t^{-6}[-1+t^2+t^4+z^2t^2(1+t^2)]$
6_1	$t^{-4}[1-t^2+t^6-t^2(1+t^2)z^2]$
6_2	$t^{-4}[1-2t^2+2t^4+(1-3t^2+t^4)z^2-t^2z^4]$
6_3	$t^{-2}(-1+3t^2-t^4)(1+z^2)+z^4$
0_1^2	$(zt)^{-1}(1-t^2)$
2_1^2	$(zt^3)^{-1}(1-t^2)-zt^{-1}$
$[4_1^2]_1$	$(zt^5)^{-1}(1-t^2)-3zt^{-3}(1-t^2)-z^3t^{-3}$
$[4_1^2]_2$	$(zt)^{-1}(1-t^2)-zt^{-3}(1-t^2)^2+z^3t^{-1}$
5_1^2	$(zt)^{-1}(1-t^2)-zt^{-3}(1-t^2)^2+z^3t^{-1}$
6_1^2	$(zt^7)^{-1}(1-t^2)+3zt^{-7}(1-2t^2)+z^3t^{-7}(1-5t^2)$ $-z^5t^{-5}$
6_2^2	$t^5z^{-1}(1-t^2)+zt^3(2+2t^2-t^4)+z^3t^3(1+t^2)$
$[6_3^2]_1$	$(t^5z)^{-1}(1-t^2)+zt^{-7}(1-t^2-2t^4)-z^3t^{-5}(1+t^2)$
$[6_3^2]_2$	$t^3z^{-1}(1-t^2)+zt^{-1}(1-t^2+2t^4)-tz^3$
0_1^3	$(zt)^{-1}(1-2t^2+t^4)$
$[6_1^3]_1$	$(1-t^2)z^{-2}+(1-3t^2+2t^4)+(1-3t^2+t^4)z^2$ $-t^2z^4$
$[6_1^3]_2$	$t^{-8}(1-t^2)^2z^{-2}+3t^{-6}(-1+t^2)+t^{-4}(2+t^2)z^2$
6_2^3	$t^{-2}(1-t^2)^2z^{-2}-t^{-2}(1-t^2)^2z^2+z^4$
$[6_3^3]_1$	$t^{-4}(1-t^2)^2z^{-2}+t^{-4}(1-3t^2+2t^4)-t^{-2}z^2$
$[6_3^3]_2$	$t^4(1-t^2)^2z^{-2}+3t^4(1-t^2)+t^4(4-t^2)z^4+t^4z^4$

4. The three-state Akutsu-Wadati polynomial

The N-state Akutsu-Wadati polynomial $A^{(N)}(t)$ is defined in Sec. V.A.5. The following list of $A^{(3)}(t)$ for knots of closed three-braids is taken from Akutsu *et al.* (1987).

3_1	$t^2(1+t^3-t^5+t^6-t^7-t^8+t^9)$
4_1	$t^{-6}(1-t-t^2+2t^3-t^4-t^5+3t^6-t^7-t^8+2t^9-t^{10}-t^{11}+t^{12})$
5_1	$t^4(1+t^3-t^5+t^6-t^8+t^9-2t^{11}+t^{12}-t^{14}+t^{15})$
5_2	$t^2(1-t+3t^3-2t^4-t^5+4t^6-3t^7-t^8+3t^9-2t^{10}-t^{11}+2t^{12}-t^{13}-t^{14}+t^{15})$
6_2	$t^{-4}(1-t-t^2+3t^3-t^4-3t^5+5t^6-t^7-5t^8+6t^9-6t^{11}+6t^{12}-5t^{14}+4t^{15}-2t^{17}+t^{18})$
6_3	$t^{-9}(1-2t-t^2+5t^3-4t^4-3t^5+9t^6-5t^7-5t^8+11t^9-5t^{10}-5t^{11}$ $+9t^{12}-3t^{13}-4t^{14}+5t^{15}-t^{16}-2t^{17}+t^{18})$

5. The Kauffman polynomial—the Dubrovnik version

The Kauffman polynomial $L(\alpha,z)$ is defined in Sec. II.D.5. In the following we list $Q(\alpha,z)$, the Dubrovnik version of the Kauffman polynomial, computed from the list of $L(\alpha,z)$ given by Kauffman (1987b) and using Eq. (2.16).

3_1	1
3_1	$(2\alpha-\alpha^{-1})+(1-\alpha^{-2})z+(\alpha-\alpha^{-1})z^2$
4_1	$(\alpha^2-1+\alpha^{-2})+(-\alpha+\alpha^{-1})z+(\alpha^2-2+\alpha^{-2})z^2+(-\alpha+\alpha^{-1})z^3$
5_1	$(3\alpha-2\alpha^{-1})+(2-\alpha^{-2}-\alpha^{-4})z+(4\alpha-3\alpha^{-1}-\alpha^{-3})z^2+(1-\alpha^{-2})z^3+(\alpha-\alpha^{-1})z^4$
5_2	$(-\alpha+\alpha^{-1}+\alpha^{-3})+(-2+2\alpha^{-2})z+(-2\alpha+\alpha^{-1}+\alpha^{-3})z^2+(\alpha^2-2+\alpha^{-2})z^3+(-\alpha+\alpha^{-1})z^4$
6_1	$(-\alpha^2+1-\alpha^{-4})+2(\alpha-\alpha^{-1})z+(-3\alpha^2+4-\alpha^{-4})z^2$ $+(-3\alpha+2\alpha^{-1}+\alpha^{-3})z^3+(\alpha^2-2+\alpha^{-2})z^4+(-\alpha+\alpha^{-1})z^5$
6_2	$(-2\alpha^2+2+\alpha^{-2})+(\alpha^{-1}-\alpha^{-3})z+(-3\alpha^2+6-2\alpha^{-2}-\alpha^{-4})z^2$ $+(-2\alpha+2\alpha^{-3})z^3+(\alpha^2-3+2\alpha^{-2})z^4+(-\alpha+\alpha^{-1})z^5$
6_3	$(-\alpha^2+3-\alpha^{-2})+(\alpha^3-2\alpha+2\alpha^{-1}-\alpha^{-3})z+(-3\alpha^2+6-3\alpha^{-2})z^2$ $+(\alpha^3-\alpha+\alpha^{-1}-\alpha^{-3})z^3+(-2\alpha^2+4-2\alpha^{-2})z^4+(\alpha-\alpha^{-1})z^5$
0_1^2	$1+(\alpha-\alpha^{-1})z^{-1}$
0_1^3	$[1+(\alpha-\alpha^{-1})z^{-1}]^2$

REFERENCES

Akutsu, Y., T. Deguchi, and M. Wadati, 1987, "Exactly solved models and new link polynomials. II. Link polynomials for closed 3-braids," J. Phys. Soc. Jpn. **56**, 3464–3479.

Akutsu, Y., T. Deguchi, and M. Wadati, 1988, "Exactly solved models and new link polynomials. IV. IRF models," J. Phys. Soc. Jpn. **57**, 1173–1185.

Akutsu, Y., T. Deguchi, and M. Wadati, 1989, "The Yang-Baxter relation: a new tool for knot theory," in *Braid Group, Knot Theory, and Statistical Mechanics,* edited by C. N. Yang and M. L. Ge (World Scientific, Singapore), pp. 151–200.

Akutsu, Y., A. Kuniba, and M. Wadati, 1986a, "Exactly solvable IRF models. II. S_N-generalizations," J. Phys. Soc. Jpn. **55**, 1466–1474.

Akutsu, Y., A. Kuniba, and M. Wadati, 1986b, "Exactly solvable IRF models. III. A new hierarchy of solvable models," J. Phys. Soc. Jpn. **55**, 1466–1474.

Akutsu, Y., and M. Wadati, 1987a, "Knot invariants and the critical statistical systems," J. Phys. Soc. Jpn. **56**, 839–842.

Akutsu, Y., and M. Wadati, 1987b, "Exactly solvable models and new link polynomials. I. N-state vertex models," J. Phys. Soc. Jpn. **56**, 3039–3051.

Alexander, J. W., 1928, "Topological invariants of knots and knots," Trans. Am. Math. Soc. **30**, 275–306.

Andrews, G. E., R. J. Baxter, and P. J. Forrester, 1984, "Eight-vertex SOS model and generalized Rogers-Ramanujan-type identities," J. Stat. Mech. **35**, 193–266.

Baxter, R. J., 1971, "Eight-vertex model in lattice statistics," Phys. Rev. Lett. **26**, 832-833.

Baxter, R. J., 1972, "Partition function of the eight-vertex lattice model," Ann. Phys. (N.Y.) **70**, 193–228.

Baxter, R. J., 1973a, "Eight-vertex model in lattice statistics and one-dimensional anisotropic Heisenberg chain. I. Some fundamental eigenvectors," Ann. Phys. (N.Y.) **76**, 1–24.

Baxter, R. J., 1973b "Eight-vertex model in lattice statistics and one-dimensional anisotropic Heisenberg chain. II. Equivalence to a general ice-type lattice model," Ann. Phys. (N.Y.) **76**, 25–47.

Baxter, R. J., 1978, "Solvable eight-vertex model on an arbitrary planar lattice," Philos. Trans. R. Soc. London **289**, 315–346.

Baxter, R. J., 1980, "Exactly solved models," in *Fundamental Problems in Statistical Mechanics V,* edited by E. G. D. Cohen (North-Holland-Amsterdam), pp. 109–141.

Baxter, R. J., 1982, *Exactly Solved Models in Statistical Mechanics* (Academic, New York).

Baxter, R. J., S. B. Kelland, and F. Y. Wu, 1976, "Equivalence of the Potts model or Whitney polynomial with an ice-type model," J. Phys. A **9**, 397–406.

Bazhanov, V. V., 1985, "Trigonometric solutions of the star-triangle equation and classical Lie algebras," Phys. Lett. B **159**, 321–324.

Birman, J. S., 1985, "On the Jones polynomial of closed 3-braids," Invent. Math. **81**, 287–294.

Burde, G., and H. Zieschang, 1985, *Knots* (Walter de Gruyter, New York).

Conway, J. H., 1970, "An enumeration of knots and links and some of their algebraic properties," in *Computational Problems in Abstract Algebra,* edited by J. Leech (Pergamon, New York), pp. 329–358.

Date, E., M. Jimbo, T. Miwa, and M. Okado, 1986, "Fusion of the eight-vertex SOS model," Lett. Math. Phys. **12**, 209–215.

Deguchi, T., Y. Akutsu, and M. Wadati, 1988, "Exactly solvable models and new link polynomials. III. Two-variable topological invariants," Phys. Soc. Jpn. **57**, 757–776.

Drinfel'd, V. G., 1986, "Quantum groups," in *Proceedings of the International Congress of Mathematicians,* Berkeley, edited by A. M. Gleason (Academic, New York), pp. 798–820.

Fan, C., and F. Y. Wu, 1970, "General lattice statistical model of phase transitions," Phys. Rev. B **2**, 723–733.

Fortuin, C. M., and P. W. Kasteleyn, 1972, "On the random-cluster model I. Introduction and relation to other models," Physica **57**, 536–564.

Freyd, P., D. Yetter, J. Hoste, W. B. R. Lickorish, K. C. Millett, and A. Oceanau, 1985, "A new polynomial invariant of knots and links," Bull. Am. Math. Soc. **12**, 239–246.

Gaudin, M., 1967, "Un système à une dimension de fermions en interaction," Phys. Lett. A **24**, 55-56.

Ge, M. L., L. Y. Wang, K. Xue, and Y. S. Wu, 1989, "Akutsu-Wadati polynomials from Feynman-Kauffman diagrams," in *Braid Group, Knot Theory, and Statistical Mechanics,* edited by C. N. Yang and M. L. Ge (World Scientific, Singapore), pp. 201–237.

Hoste, J., 1986, "A polynomial invariant for knots and links," Pacific J. Math. **124**, 295–320.

Ising, E., 1925, "Beitrag zur theorie des ferromagnetismus," Z. Phys. **31**, 253–258.

Jimbo, M., 1986, "Quantum R matrix for the generalized Toda system," Commun. Math. Phys. **102**, 537–547.

Jimbo, M., 1989, *Yang-Baxter Equation in Integrable Systems* (World Scientific, Singapore).

Jimbo, M., T. Miwa, and M. Okado, 1988, "Solvable lattice models related to the vector representation of classical simple Lie algebras," Commun. Math. Phys. **116**, 507–525.

Jones, V. F. R., 1985, "A polynomial invariant for links via von Neumann algebras," Bull. Am. Math. Soc. **12**, 103–112.

Jones, V. F. R., 1987, "Hecke algebra representations of braid groups and link polynomials," Ann. Math. **126**, 103–112.

Jones, V. F. R., 1989, "On knot invariants related to some statistical mechanical models," Pacific J. Math. **137**, 311–334.

Jones, V. F. R., 1990a, "Knot theory and statistical mechanics," Sci. Am. November, 98–103.

Jones, V. F. R., 1990b, "Baxterization," Int. J. Mod. Phys. B **4**, 701–713.

Kadanoff, L. P., and F. J. Wegner, 1971, "Some critical properties of the eight-vertex model," Phys. Rev. B **4**, 3989–3993.

Kauffman, L. H., 1987a, "State models and the Jones polynomial," Topology **26**, 395–407.

Kauffman, L. H., 1987b, *On Knots* (Princeton University, Princeton, NJ).

Kauffman, L. H., 1988a, "New invariants in the theory of knots," Am. Math. Monthly **95**, 195–242.

Kauffman, L. H., 1988b, "Statistical mechanics and the Jones polynomial," Contemp. Math. **78**, 263–312.

Kauffman, L. H., 1990, "An invariant of regular isotopy," Trans. Am. Math. Soc. **318**, 417–471.

Kauffman, L. H., 1991, *Knots and Physics* (World Scientific, Singapore).

Kohno, T., 1991, *New Developments in the Theory of Knots* (World Scientific, Singapore).

Kulish, P. P., N. Y. Reshetkhin, and E. K. Sklyanin, 1981, "Yang-Baxter equation and representation theory. I," Lett. Math. Phys. **5**, 393–403.

Kulish, P. P., and E. K. Sklyanin, 1980, "On the solution of the Yang-Baxter equation," Zap. Nauchn. Semin. Lenningr. Otd. Mat. Inst. Steklova **95**, 129–160 (in Russian).

Kulish, P. P., and E. K. Sklyanin, 1982, "Solutions of the Yang-Baxter equation," J. Sov. Math. **19**, 1596-1620.

Kulish, P. P., and E. K. Sklyanin, 1982b, "Quantum spectral transform method. Recent developments," in *Integrable Quantum Field Theories,* edited by J. Hietarinta and C. Montonen, Lecture Notes in Physics Vol. 151 (Springer, Berlin), pp. 61–119.

Kuniba, A., Y. Akutsu, and M. Wadati, 1986a, "Exactly solvable IRF models. I. A three-state model," J. Phys. Soc. Jpn. **55**, 1092–1101.

Kuniba, A., Y. Akutsu, and M. Wadati, 1986b, "Exactly solvable IRF models. IV. Generalized Rogers-Ramanujan identities and a solvable hierarchy," J. Phys. Soc. Jpn. **55**, 2166–2176.

Kuniba, A., Y. Akutsu, and M. Wadati, 1986c, "Exactly solvable IRF models. V. A further new hierarchy," J. Phys. Soc. Jpn. **55**, 2605–2617.

Kuniba, A., Y. Akutsu, and M. Wadati 1986d, "The Gordon-generalization hierarchy of exactly solvable IRF models," J. Phys. Soc. Jpn. **55**, 3338–3353.

Kuniba, A., Y. Akutsu, and M. Wadati, 1986e, "Inhomogeneous eight-vertex SOS model and solvable IRF hierarchies," J. Phys. Soc. Jpn. **55**, 2907–2910.

Lickorish, W. B. R., 1988, "Polynomials for links," Bull. London Math. Soc. **20**, 558–588.

Lickorish, W. B. R., and K. C. Millett, 1987, "A polynomial invariant of oriented links," Topology **26**, 107–141.

Lickorish, W. B. R., and K. C. Millett, 1988, "The new polynomial invariants of knots and links," Math. Magazine **61**, 3–23.

Lieb, E. H., 1967a, "Exact solution of the problem of the entropy of square ice," Phys. Rev. Lett. **18**, 692–694.

Lieb, E. H., 1967b, "Exact solution of the F model of an antiferroelectric," Phys. Rev. Lett. **18**, 1046–1048.

Lieb, E. H., 1967c, "Exact solution of the two-dimensional Slater KDP model of a ferroelectric," Phys. Rev. Lett. **19**, 108–110.

Lieb, E. H., 1967d, "Residue entropy of square ice," Phys. Rev. **162**, 162–171.

Lieb, E. H., and W. Liniger, 1963, "Exact analysis of an interacting Bose gas. I. The general solution and the ground state," Phys. Rev. **130**, 1605–1616.

Lieb, E. H., and F. Y. Wu, 1972, "Two-dimensional ferroelectric models", in *Phase Transitions and Critical Phenomena, Vol. I,* edited by C. Domb and M. S. Green (Academic, New York), pp. 331–490.

Lipson, A. S., 1992, "Some more states models for link invariants," Pacific J. Math. **152**, 337–346.

McGuire, J. B., 1964, "Studies of exactly solvable one-dimensional N-body problems," J. Math. Phys. **5**, 622–636.

Onsager, L., 1944, "Crystal statistics I. A two-dimensional model with an order-disorder transition," Phys. Rev. **65**, 117–149.

Pasquier, V., 1987a, "Two-dimensional critical systems labelled by Dynkin diagrams," Nucl. Phys. B **285**, [FS19], 162–172.

Pasquier, V., 1987b, "Exact solubility of the D_n series," J. Phys. A **20**, L217–L220.

Pearce, P., and K. A. Seaton, 1988, "Solvable hierarchy of cyclic solid-on-solid lattice models," Phys. Rev. Lett. **60**, 1347–1350.

Pearce, P., and K. A. Seaton, 1989, "Exact solution of cyclic solid-on-solid lattice models," Ann. Phys. (N.Y.) **193**, 326–366.

Perk, J. H. H., and C. L. Schultz, 1981, "New families of commuting transfer matrices in q-state vertex models," Phys. Lett. A **84**, 407–410.

Perk, J. H. H., and C. L. Schultz, 1983, "Families of commuting transfer matrices in q-state vertex models," in *Nonlinear Integrable Systems—Classical and Quantum Theory,* edited by M. Jimbo and T. Miwa (World Scientific, Singapore), pp. 137–152.

Perk, J. H. H., and F. Y. Wu, 1986a, "Nonintersecting string model and graphical approach: Equivalence with a Potts model," J. Stat. Phys. **42**, 727–742.

Perk J. H. H., and F. Y. Wu, 1986b, "Graphical approach to the nonintersecting string model: Star-triangle equation, inversion relation, and exact solution," Physica A **138**, 100–124.

Potts, R. B., 1952, "Some generalized order-disorder transformations," Proc. Cambridge Philos. Soc. **48**, 106–109.

Przytycki, J. H., and P. Traczyk, 1987, "Invariants of links of Conway type," Kobe J. Math. **4**, 115–139.

Reidemeister, K., 1948, *Knotentheorie* (Chelsea, New York); English translation, edited by L. F. Boron, C. D. Christenson, and B. A. Smith, *Knot Theory* (BCS Associates, Moscow, Idaho, 1983).

Reshetikhin, N.Y., and V. Turaev, 1991, "Invariants of three-manifolds via link polynomials and quantum groups," Invent. Math. **103**, 547–597.

Rolfsen, D., 1976, *Knots and Links* (Publish or Perish, Berkeley).

Schultz, D. L., 1981, "Solvable q-state models in lattice statistics and quantum field theory," Phys. Rev. Lett. **46**, 629–632.

Sogo, K., Y. Akutsu, and A. Takayuki, 1983, "New factorized S-matrix and its application to exactly solvable q-state model. I," Prog. Theor. Phys. **70**, 730-738.

Stroganov, Y. G., 1979, "A new calculation method for partition functions in some lattice models," Phys. Lett. A **74**, 116–118.

Sutherland, B., 1967, "Exact solution of a two-dimensional model for hydrogen-bonded crystals," Phys. Rev. Lett. **19**, 103–104.

Takhtadzhan, L. A., and L. D. Faddeev, 1979, "The quantum inverse problem method and the *XYZ* Heisenberg model," Uspekhi Mat. Nauk **34**(5), 13–63 (in Russian) [Russian Math. Surveys **34**(5), 11–68 (1979)].

Temperley, H. N. V., and E. H. Lieb, 1971, "Relations between the percolation and colouring problem and other graph-theoretical problems associated with regular planar lattice: some exact results for the percolation problem," Proc. R. Soc. London A **322**, 251–280.

Thistlethwaite, M. B., 1985, "Knot tabulations and related topics," in *Aspects of Topology,* edited by I. M. James and E. H. Kronheimer, London Mathematical Society Lecture Notes 93 (Cambridge University Press, London), pp. 1–76.

Truong, T. T., 1986, "Structure properties of a $Z(N^2)$-spin model and its equivalent $Z(N)$-vertex model," J. Stat. Phys. **42**, 349–379.

Turaev, V. G., 1988, "The Yang-Baxter equation and invariants of links," Invent. Math. **92**, 527–553.

Wadati, M., T. Deguchi, and Y. Akutsu, 1989, "Exactly solvable models and knot theory," Phys. Rep. **180**, 247–332.

Witten, E., 1989a, "Quantum field theory and the Jones polynomial," Commun. Math. Phys. **121**, 351–399

Witten, E., 1989b, "Gauge theories and integrable lattice models," Nucl. Phys. B **322**, 629–697.

Witten, E., 1990, "Gauge theories, vertex models, and quantum groups," Nucl. Phys. B **330**, 285–346.

Wu, F. Y., 1967, "Exactly soluble model of the ferroelectric phase transition in two dimensions," Phys. Rev. Lett. **18**, 605–607.

Wu, F. Y., 1968, "Remarks on the modified potassium dihydrogen phosphate model of a ferroelectric," Phys. Rev. **168**, 539–543.

Wu, F. Y., 1971, "Ising model with four-spin interactions," Phys. Rev. B **4**, 2312-2314.

Wu, F. Y., 1982, "The Potts Model," Rev. Mod. Phys. **54**, 235–268.

Wu, F. Y., 1992a, "Knot theory and statistical mechanics," in *Computer-Aided Statistical Physics,* AIP Conf. Proc. No. 248 (American Institute of Physics, New York), pp. 3–11.

Wu, F. Y., 1992b, "Jones polynomial as a Potts model partition function," J. Knot Theory and Its Ramifications **1**, 47–57.

Yang, C. N., 1967, "Some exact results for the many-body problem in one dimension with repulsive delta-function interaction," Phys. Rev. Lett. **19**, 1312-1314.

Yang, C. N., and M. L. Ge, 1989, *Braid Group, Knot Theory, and Statistical Mechanics* (World Scientific, Singapore).

Zamolodchikov, A. B., 1979, "Z_4-symmetric factorized S-matrix in two space-time dimensions," Commun. Math. Phys. **69**, 165–178.

Zamolodchikov, A. B., and V. A. Fateev, 1980, "A model factorized S-matrix and an integrable spin-1 Heisenberg chain," Sov. J. Nucl. **32**, 298–303.

VOLUME 72, NUMBER 25 PHYSICAL REVIEW LETTERS 20 JUNE 1994

New Link Invariant from the Chiral Potts Model

F. Y. Wu and P. Pant
Department of Physics, Northeastern University, Boston, Massachusetts 02115

C. King
Department of Mathematics, Northeastern University, Boston, Massachusetts 02115
(Received 31 January 1994)

A new link invariant is obtained using the exactly solvable chiral Potts model and a generalized Gaussian summation identity. The new link invariant is characterized by roots of unity and does not appear to belong to the usual quantum group family of invariants. The invariant is given in terms of a matrix associated with the link diagram, and can be readily written down for any given link.

PACS numbers: 05.50.+q, 02.90.+p

It is rare that important advances in different branches of science are found to be closely related. One example of such a happening is the recent discovery of the connection between exactly solvable models in statistical mechanics and the generation of knot and link invariants in mathematics. Link invariants are algebraic quantities associated with embeddings of circles in R^3, which are topologically invariant. In 1985 Jones [1] discovered a new invariant, the Jones polynomial, and noticed some relationship with the Potts model. It was soon shown that the Jones polynomial can be derived from statistical mechanical models [2], and that statistical mechanical considerations can further be used to generate new link invariants [3,4]. Several reviews now exist elucidating this connection [5–7], and related recent development on spin models and link invariants can be found in [8–11].

In this Letter we first briefly review a formulation which generates link invariants for *oriented* links from spin models with chiral interactions. We then apply the formulation to the recently solved chiral Potts model [12], and obtain a link invariant characterized by roots of unity and a form which is very different from those previously known. In particular, it does not seem to belong to the usual quantum group family of invariants. While link invariants arising from chiral Potts models have previously been analyzed from other perspectives [13,14], this is the first time that these invariants are explicitly evaluated.

Consider an oriented link K with a planar projection given by a directed graph $\mathcal{L}$. We shall assume $\mathcal{L}$ to be connected. Consider an N-state spin model with spins residing in alternate faces of $\mathcal{L}$ and interactions spanning across the line crossings. The spins form a graph G with vertices designating spins and edges the spin interactions. It is convenient to shade the faces containing spins [4]. Then, depending on the relative positioning of the shaded faces with respect to the line orientations and crossings, there exist four distinct types of line intersections, and hence four types of spin interactions. These situations are shown in Fig. 1. We write the four Boltzmann weights as

$$u_\pm(a-b),\quad \bar{u}_\pm(a-b)\,, \tag{1}$$

where $a,b=1,2,\ldots,N$ denote the spin states. Here, we allow the possibility that the interactions are chiral in the sense that $u_\pm(a)\neq u_\pm(-a)$.

Following the standard formulation [4,5], the partition function $Z(u_\pm,\bar{u}_\pm)$ of the spin model will be a topological invariant, provided that the Boltzmann weights satisfy certain conditions imposed by Reidemeister moves [15]. In enumerating the Reidemeister moves, however, one must consider all possible face shadings and line crossings for the same line movement. This leads to the possibilities shown in Fig. 2, from which one reads off the following conditions:

$$\bar{u}_+(0)=1\,, \tag{2a}$$

$$\frac{1}{\sqrt{N}}\sum_{b=0}^{N-1} u_+(a-b)=1\,, \tag{2b}$$

$$u_+(a-b)u_-(a-b)=1\,, \tag{2c}$$

$$\frac{1}{N}\sum_{b=0}^{N-1} \bar{u}_+(a-b)\bar{u}_-(b-c)=\delta_{ac}\,, \tag{2d}$$

$$\bar{u}_+(a-b)\bar{u}_-(b-a)=1\,, \tag{2e}$$

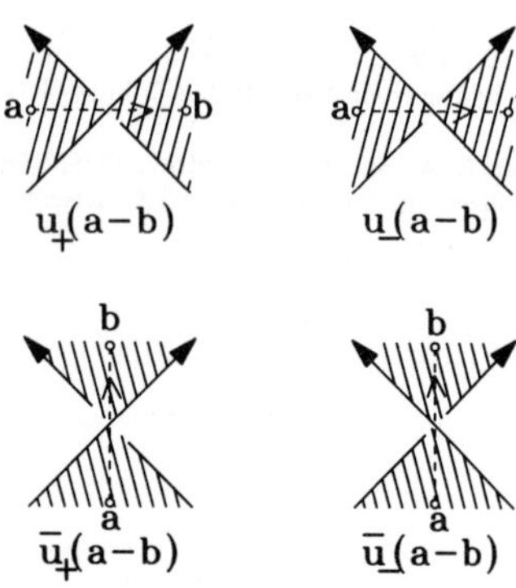

FIG. 1. The four different kinds of line intersections and face shadings that can occur at a vertex.

0031-9007/94/72(25)/3937(4)$06.00

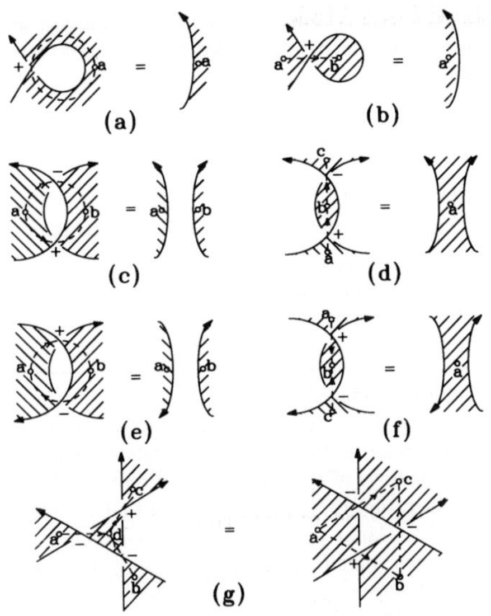

FIG. 2. Reidemeister moves for oriented knots with two different kinds of face shadings.

$$\frac{1}{N}\sum_{b=0}^{N-1} u_+(a-b)u_-(c-b)=\delta_{ac}\,, \tag{2f}$$

$$\frac{1}{\sqrt{N}}\sum_{d=0}^{N-1} u_-(a-d)\bar{u}_-(b-d)\bar{u}_+(d-c) =\bar{u}_-(b-c)u_-(a-c)u_+(a-b)\,. \tag{2g}$$

Provided that conditions (2a)–(2f) are met, the quantity

$$I_K(N)\equiv N^{-(S+1)/2}Z(u_\pm,\bar{u}_\pm)\,, \tag{3}$$

where S is the number of spins (shaded faces) in $\mathcal{L}$, is an invariant for the link K [5,16]. Note that the normalization of (3) is $I_{\text{unknot}}=1$.

The self-dual chiral Potts model.—The N-state chiral Potts model is a spin model whose Boltzmann weights $W(n)$ and $\overline{W}(n)$ are N periodic, namely, they satisfy $W(n)=W(n+N)$, $\overline{W}(n)=\overline{W}(n+N)$. In the integrable self-dual case [17] the Boltzmann weights are related through the Fourier transform

$$\overline{W}_{pq}(n)=\frac{1}{\sqrt{N}}\sum_{n=1}^{N}\omega^{mn}W_{pq}(m)\,, \tag{4}$$

where $\omega=e^{2\pi i/N}$, and the weights are parametrized by associating line rapidities a_p,b_p [12]. Explicitly, one writes

$$g_{pq}(n)\equiv\frac{W_{pq}(n)}{W_{pq}(0)}=\prod_{j=1}^{n}\frac{b_q-a_p\omega^j}{b_p-a_q\omega^j}\,,$$
$$\bar{g}_{pq}(n)\equiv\frac{\overline{W}_{pq}(n)}{\overline{W}_{pq}(0)}=\prod_{j=1}^{n}\frac{\omega a_p-a_q\omega^j}{b_q-b_p\omega^j}\,. \tag{5}$$

The N periodicity then requires $a_p^N+b_p^N=0$, or, equivalently [18],

$$a_p=tb_p,\quad t=\omega^{-1/2}\,. \tag{6}$$

A crucial step in generating knot invariants from exactly soluble models is to specialize the weights by taking a certain "infinite" rapidity limit. For our purposes we define the limits

$$u_\pm(n)=A_\pm\lim_{b_p/b_q\to\pm\infty}g_{pq}(n)\,,$$
$$\bar{u}_\pm(n)=B_\pm\lim_{b_p/b_q\to\pm\infty}\bar{g}_{pq}(n)\,. \tag{7}$$

It can be verified that conditions (2a)–(2g) are all satisfied, provided that we take $A_\pm=e^{\pm i(N-1)\pi/4}$, $B_\pm=1$. Thus, one obtains the Boltzmann weights

$$u_\pm(n)=(-1)^n e^{\pm i(N-1)\pi/4}\omega^{\pm n^2/2}\,,$$
$$\bar{u}_\pm(n)=(-1)^n\omega^{\mp n^2/2}\,, \tag{8}$$

which satisfy all requirements imposed by Reidemeister moves [18]. The substitution of (8) into (3) now yields the desired invariant $I_K(N)$ for the link K.

Evaluation of knot invariants.—We can rewrite the invariant (3) in a form suitable for evaluation. To each link K we associate matrices $\mathbf{Q}$ and $\mathbf{M}$ as follows. Let G be the graph associated to the spins. We assign a number $\epsilon_l=\pm1$ to each edge in G according to the following rules: If the associated line crossing is of type $\bar{u}_+$ or u_-, assign $\epsilon_l=+1$; if the crossing is of type $\bar{u}_-$ or u_+, assign $\epsilon_l=-1$ [4,5]. Then $\mathbf{Q}$ is the $S\times S$ symmetric matrix with elements

$$Q_{ij}=\sum_{l=\langle i,j\rangle}\epsilon_l,\quad i\neq j,\quad Q_{ii}=-\sum_{k(\neq i)}Q_{ik}\,, \tag{9}$$

where the summation in the first line is over *all* edges l connecting the ith and jth vertices; $\mathbf{M}$ is the $(S-1)\times(S-1)$ cofactor matrix of $\mathbf{Q}$ obtained by deleting, say, the Sth row and column from $\mathbf{Q}$. Since the sum of each row or column vanishes, matrix $\mathbf{Q}$ has the property that all cofactors are equal and generating spanning trees of the graph G [19]. Let $\mathbf{n}=(n_1,\ldots,n_S)$ be a vector whose component $n_i=0,1,\ldots,N-1$, $i=1,\ldots,S$, denotes the state of the ith spin. Further introduce a vector $\mathbf{z}$ with components $z_i=Q_{ii}/2$, $i=1,\ldots,S$. Then the link invariant (3) when substituted with (8) can be written compactly as

$$I_K(N)=N^{-(S+1)/2}e^{\pi i(N-1)\zeta(K)/4}\sum_{n_1,\ldots,n_S=0}^{N-1}\exp\left[\frac{\pi i}{N}\mathbf{n}\cdot(\mathbf{Qn})+2\pi i\mathbf{n}\cdot\mathbf{z}\right]\,, \tag{10}$$

where $\zeta(K)$ is the number of u_+ weights minus the number of u_- weights in K.

VOLUME 72, NUMBER 25 PHYSICAL REVIEW LETTERS 20 JUNE 1994

Because of the N periodicity of Boltzmann weights, there is no loss of generality to fix the state of, say, the Sth spin at $n_S=0$. Then the index n_S disappears completely in (10) and the summation over n_S is replaced by an overall factor N. Then (10) can be rewritten as

$$I_K(N)=N^{-(S-1)/2}e^{\pi i(N-1)\zeta(K)/4}\sum_{n_1,\ldots,n_{S-1}=0}^{N-1}\exp\left[\frac{\pi i}{N}\mathbf{n}\cdot(\mathbf{Mn})+2\pi i\mathbf{n}\cdot\mathbf{y}\right], \tag{11}$$

where $\mathbf{n}$ and $\mathbf{y}$ are $(S-1)$-dimensional vectors, and the summation is $(S-1)$-fold.

The expression (11) can be evaluated using a generalized Gaussian summation identity [20] which converts the role of N from a summation limit to an explicit parameter. This leads to

$$I_K(N)=e^{\pi i(N-1)\zeta(K)/4}e^{\pi i\eta(\mathbf{M})/4}\frac{1}{\sqrt{D}}\sum_{\mathbf{n}\in\Delta}\exp\{-\pi iN(\mathbf{n}+\mathbf{y})\cdot[\mathbf{M}^{-1}(\mathbf{n}+\mathbf{y})]\}, \tag{12}$$

where $\mathbf{n}=(n_1,\ldots,n_{S-1})$, $\mathbf{y}=(y_1,\ldots,y_{S-1})$, $y_i=Q_{ii}/2$, $D=|\det\mathbf{M}|$, Δ is the fundamental domain (unit cell) of the lattice formed by the collection of vectors $\mathbf{Mn}$, and $\eta(\mathbf{M})$ is the signature of $\mathbf{M}$, namely, the number of positive eigenvalues minus the number of negative eigenvalues. This completes the evaluation of $I_K(N)$.

The invariants evaluated using (12) for selected knots and links are given in Table I. A complete table of invariants for links with 8 and less crossings will be given elsewhere [21]. Generally, the matrix $\mathbf{Q}$, and hence $\mathbf{M}$, can be read off by inspection from the shadings of G. For the right-handed trefoil, e.g., one has either

$$\mathbf{M}=\begin{pmatrix}-2 & 1\\ 1 & -2\end{pmatrix},$$

$\zeta(K)=0$, $\eta(\mathbf{M})=-2$, $\Delta=(0,0),(0,-1),(-1,0)$ (if one shades three faces with $S=3$), or $M=3$, $\zeta(K)=3$, $\eta(M)=1$, $\Delta=0,1,2$ (if one shades two faces with $S=2$). In either case, one is led to the expression given in the table. It is clear from (8) that the invariant for the mirror image of the link K is the complex conjugate $I_K^*(N)$. It is also clear that, since $\mathbf{M}$ is independent of line orientations, the reversal of the orientation of individual components in a link introduces an overall factor $e^{i(N-1)\pi\Delta\zeta/4}$, where $\Delta\zeta$ is the induced change of $\zeta(K)$.

TABLE I. Selected knot and link invariants $I_K(N)$. [$w(K)$ is the number of + crossings minus the number of − crossings.]

K	$w(K)$	$I_K(N)$
3_1	$+3$	$-i(1+2e^{2\pi iN/3})/\sqrt{3}$
4_1	0	$(1+2e^{-2\pi iN/5}+2e^{2\pi iN/5})/\sqrt{5}$
5_1	$+5$	$-(1+2e^{-4\pi iN/5}+2e^{4\pi iN/5})/\sqrt{5}$
5_2	$+5$	$-i(1+2e^{2\pi iN/7}+2e^{4\pi iN/7}+2e^{8\pi iN/7})/\sqrt{7}$
4_1^2	-4	$e^{3i\pi/4}(1+2e^{\pi iN/4}+e^{\pi iN})/2$
	$+4$	$e^{-i\pi/4}(1+2e^{\pi iN/4}+e^{\pi iN})/2$
5_1^2	$+1$	$e^{-i\pi/4}(1+e^{5\pi iN/8}+e^{\pi iN/2}+e^{13\pi iN/8})/\sqrt{2}$
6_1^2	$+6$	$e^{-i\pi/4}(1+2e^{\pi iN/6}+2e^{2\pi iN/3}+e^{3\pi iN/2})/\sqrt{6}$
7_2	$+7$	$-i(1+2e^{2\pi iN/11}+2e^{6\pi iN/11}+2e^{8\pi iN/11}+2e^{10\pi iN/11}+2e^{18\pi iN/11})/\sqrt{11}$
7_7	$+1$	$(1+4e^{2\pi iN/21}+4e^{8\pi iN/21}+2e^{2\pi iN/3}+2e^{6\pi iN/7}+2e^{10\pi iN/7}+4e^{32\pi iN/21}+2e^{12\pi iN/7})/\sqrt{21}$

Uniqueness of the invariant.—For the invariant to be unique, we require that $I_K(N)$ be independent of the choice of shadings of the faces of $\mathcal{L}$. From a general duality consideration [22] it can be shown that the partition function of a planar spin model with Boltzmann weight $u_{ij}(n)$ between spins i and j is unchanged when evaluated in the dual space, provided that the dual Boltzmann weight across the same edge is taken to be

$$u_{ij}^{(D)}(m)=\frac{1}{\sqrt{N}}\sum_{n=0}^{N-1}\omega^{mn}u_{ij}(n). \tag{13}$$

For the self-dual model satisfying (4), we obtain from (13)

$$u_{\pm}^{(D)}(n)=\bar{u}_{\pm}(-n),\quad \bar{u}_{\pm}^{(D)}(n)=u_{\pm}(n). \tag{14}$$

Referring to Fig. 1, the right-hand sides of (14) are precisely those weights that we would have used, had we shaded the faces the other way. Thus the partition function, hence $I_K(N)$, is independent of the choice of face shadings, and the invariant is unique [16]. The proof of uniqueness can be extended to disconnected links $\mathcal{L}$ [21].

Skein relation.—The link invariant given by (12) satisfies a Skein relation. To see this, suppose two shaded regions are separated by a crossing of type u_+. Call the invariant I_1. Replace the crossing by k consecutive twists, all of type u_+. Then the factor $u_+(a-b)$ in the partition function is replaced by $[u_+(a-b)]^k$, giving the invariant I_k. Since $u_+(a-b)$ is a root of unity, it satisfies a polynomial equation (of order at most N) with coefficients independent of $a-b$. Therefore the invariants $\{I_k\}$ satisfy a linear equation, which is a Skein relation.

Since the number of terms in the Skein relation grows with N, it is not very helpful for evaluating the invariants. However, the Skein relation shows that $I_K(2)=(-1)^{c(K)+1}V(-i)$ where $V(t)$ is the Jones polynomial and $c(K)$ is the number of components in the link K, and $I_K(3)$ is the HOMFLY polynomial $P(e^{2\pi i/3},-i)$ as well as $(-1)^{c(K)+1}V(e^{-\pi i/e})$ (see [5] for notation).

A prerequisite for our result (12) to hold is that the matrix $\mathbf{M}$ is nonsingular ($D\neq 0$). The matrix $\mathbf{M}$, which is a cofactor matrix of $\mathbf{Q}$, also arises in the evaluation of the conductivity of an electrical network represented by G. It can be shown [23] that the condition of $D\neq 0$ is equivalent to the statement that the conductance between any two nodes does not vanish. Now the rules of combining and rearranging conductances are the same as those of the Reidemeister moves. If the network can be "split" into disjoint networks by combining and rearranging conductances (allowing conductances of both signs), then it must follow that $D=0$. Indeed, by using the fact that **Q** generates the spanning tree polynomial for G, one can establish the identity $D=|V(-1)|$. Now $V(-1)=\Delta(-1)$, where $\Delta(t)$ is the Alexander polynomial for which it is known $\Delta(-1)=0$ for split links. It follows that $D=0$ for split links. The converse, however, is false; there are unsplit links with $D=0$ [24].

We would like to thank Helen Au-Yang for pointing out the resemblance of properties of the chiral Potts model weights with those of the Reidemeister moves, which has led to this investigation. We thank Vaughan F. R. Jones for calling our attention to Ref. [8], and Pierre de la Harpe and Eiichi Bannai for sending us relevant reprints and preprints prior to publication. Work by one of us (F.Y.W.) has been supported in part by National Science Foundation Grants No. DMR-9313648 and No. INT-9207261.

Note added.—After submission of this manuscript, it has come to our attention that the invariant $I_K(N)$ has previously been defined (using a modified Goeritz matrix) in a form equivalent to (11) but without explicit evaluation [25].

[1] V. F. R. Jones, Bull. Am. Math. Soc. **12**, 103 (1985).
[2] L. H. Kauffman, Topology **26**, 395 (1987).
[3] Y. Akutsu and M. Wadati, J. Phys. Soc. Jpn. **56**, 3039 (1987).
[4] V. F. R. Jones, Pacific J. Math. **137**, 311 (1989).
[5] F. Y. Wu, Rev. Mod. Phys. **64**, 1099 (1992).
[6] L. H. Kauffman, *Knots and Physics* (World Scientific, Singapore, 1991).
[7] W. Wadati, T. Deguchi, and Y. Akutsu, Phys. Rep. **180**, 247 (1989).
[8] F. Jaeger, Geom. Dedicata **44**, 23 (1992).
[9] P. de la Harpe and V. F. R. Jones, J. Comb. Theory B **57**, 207 (1993).
[10] E. Bannai and E. Bannai, in *Memoirs of the Faculty of Science* [Kyushu University, A **47**, 397 (1993)].
[11] P. de la Harpe, Pacific J. Math. **162**, 57 (1994).
[12] R. J. Baxter, J. H. H. Perk, and H. Au-Yang, Phys. Lett. A **128**, 138 (1988).
[13] D. Goldschmidt and V. F. R. Jones, Geom. Dedicata **31**, 165 (1989); V. F. R. Jones, Commun. Math. Phys. **125**, 459 (1989).
[14] E. Date, M. Jimbo, K. Miki, and T. Miwa, Pacific J. Math. **154**, 37 (1992).
[15] K. Reidemeister, *Knotentheorie* (Chelsea, New York, 1948).
[16] In case $\mathcal{L}$ has several connected components, the expression for the invariant is $I_K(N)=N^{(c_D-S-2)/2}Z(u\pm,\bar{u}\pm)$ where c_D is the number of components of the graph given by the other (dual) choice of shading. For connected $\mathcal{L}$ we have $c_D=1$. We thank Alex Suciu for help on this point.
[17] V. A. Fateev and A. B. Zamolodchikov, Phys. Lett. **92A**, 37 (1982).
[18] More generally, one may write $t=\omega^{l-1/2}$, where l is any integer, and carry the analysis through, obtaining genuinely chiral and l-dependent vertex weights. However, it can be shown that the resulting knot invariant is always independent of l. Anticipating this result, we present only the result with $l=0$.
[19] F. Harary, *Graph Theory* (Addison-Wesley, New York, 1971).
[20] C. L. Siegel, Nachr. der Akad. Wiss. Göttingen Math.-Phys. Klasse **1**, 1 (1960). The identity equates (11) and (12) for integral $N>0, M_{ij}$, and any **y** satisfying $N(2y_i+M_{ii})$ = even integers.
[21] F. Y. Wu, P. Pant, and C. King (to be published).
[22] F. Y. Wu and Y. K. Wang, J. Math. Phys. **17**, 439 (1976).
[23] G. Kirchhoff, Ann. Phys. Chem. **72**, 497 (1847).
[24] Links 8^3_{10} and 8^4_3 have $D=0$. We thank Alex Suciu for calling our attention to the link 8^4_3.
[25] T. Kobayashi, H. Murakami, and J. Murakami, Proc. Japan Acad. **64A**, 235 (1988).

10. Other Topics

INSTITUTE OF PHYSICS PUBLISHING JOURNAL OF PHYSICS A: MATHEMATICAL AND GENERAL

J. Phys. A: Math. Gen. **37** (2004) 6653–6673 PII: S0305-4470(04)78587-X

Theory of resistor networks: the two-point resistance

F Y Wu

Department of Physics, Northeastern University Boston, MA 02115, USA

Received 30 March 2004, in final form 18 May 2004
Published 16 June 2004
Online at stacks.iop.org/JPhysA/37/6653
doi:10.1088/0305-4470/37/26/004

Abstract
The resistance between two arbitrary nodes in a resistor network is obtained in terms of the eigenvalues and eigenfunctions of the Laplacian matrix associated with the network. Explicit formulae for two-point resistances are deduced for regular lattices in one, two and three dimensions under various boundary conditions including that of a Möbius strip and a Klein bottle. The emphasis is on lattices of finite sizes. We also deduce summation and product identities which can be used to analyse large-size expansions in two and higher dimensions.

PACS numbers: 01.55.+b, 02.10.Yn

1. Introduction

A classic problem in electric circuit theory studied by numerous authors over many years is the computation of the resistance between two nodes in a resistor network (for a list of relevant references up to 2000 see, e.g., [1]). Besides being a central problem in electric circuit theory, the computation of resistances is also relevant to a wide range of problems ranging from random walks (see [2] and [3][1], and discussions below), the theory of harmonic functions [4], first-passage processes [5], to lattice Green's functions [6]. The connection with these problems originates from the fact that electrical potentials on a grid are governed by the same difference equations as those occurring in the other problems. For this reason, the resistance problem is often studied from the point of view of solving the difference equations, which is most conveniently carried out for infinite networks. In the case of Green's function approach, for example, past efforts [1, 7] have been focused mainly on infinite lattices. Little attention has been paid to *finite* networks, even though the latter are those occurring in real life. In this paper, we take up this problem and present a general formulation for computing two-point resistances in finite networks. Particularly, we show that known results for infinite networks are recovered by taking the infinite-size limit. In later papers we plan to study effects of lattice defects and carry out finite-size analyses.

[1] I am indebted to S Redner for calling my attention to this reference.

The study of electric networks was formulated by Kirchhoff [8] more than 150 years ago as an instance of a linear analysis. Our starting point is along the same line by considering the Laplacian matrix associated with a network. The Laplacian is a matrix whose off-diagonal entries are the conductances connecting pairs of nodes. Just as in graph theory where everything about a graph is described by its adjacency matrix (whose elements is 1 if two vertices are connected and 0 otherwise), everything about an electric network is described by its Laplacian. Indeed, in section 2 below we shall derive an expression for the two-point resistance between two arbitrary nodes in terms of the eigenvalues and eigenvectors of the Laplacian matrix[2]. In the following sections our formulation is applied to networks of one-dimensional and two-dimensional lattices under various boundary conditions including those embedded on a Möbius strip and a Klein bottle, and lattices in higher spatial dimensions. We also deduce summation and product identities which can be used to reduce the computational labour, and analyse large-size expansions in two and higher dimensions.

Let $\mathcal{L}$ be a resistor network consisting of $\mathcal{N}$ nodes numbered $i = 1, 2, \ldots, \mathcal{N}$. Let $r_{ij} = r_{ji}$ be the resistance of the resistor connecting nodes i and j. Hence, the conductance is

$$c_{ij} = r_{ij}^{-1} = c_{ji}$$

so that $c_{ij} = 0$ (as in an adjacency matrix) if there is no resistor connecting i and j.

Denote the electric potential at the ith node by V_i and the net current flowing *into* the network at the ith node by I_i (which is zero if the ith node is not connected to the external world). Since there exist no sinks or sources of current including the external world, we have the constraint

$$\sum_{i=1}^{\mathcal{N}} I_i = 0. \tag{1}$$

The Kirchhoff law states

$$\sum_{j=1}^{\mathcal{N}}{}' c_{ij}(V_i - V_j) = I_i \qquad i = 1, 2, \ldots, \mathcal{N} \tag{2}$$

where the prime denotes the omission of the term $j = i$. Explicitly, equation (2) reads

$$\mathbf{L}\vec{V} = \vec{I} \tag{3}$$

where

$$\mathbf{L} = \begin{pmatrix} c_1 & -c_{12} & \ldots & -c_{1\mathcal{N}} \\ -c_{21} & c_2 & \ldots & -c_{2\mathcal{N}} \\ \vdots & \vdots & \ddots & \vdots \\ -c_{\mathcal{N}1} & -c_{\mathcal{N}2} & \cdots & c_{\mathcal{N}} \end{pmatrix} \tag{4}$$

is the Laplacian matrix of $\mathcal{L}$ with

$$c_i \equiv \sum_{j=1}^{\mathcal{N}}{}' c_{ij}. \tag{5}$$

Here $\vec{V}$ and $\vec{I}$ are $\mathcal{N}$-vectors whose components are V_i and I_i respectively. We note in passing that if all nonzero resistances are equal to 1, then the off-diagonal elements of the matrix $-\mathbf{L}$ are precisely those of the adjacency matrix of $\mathcal{L}$.

[2] A study of random walks using matrices can be found in [3].

The Laplacian matrix $\mathbf{L}$ of the network $\mathcal{L}$ is also known as the Kirchhoff matrix, or simply the tree matrix; the latter name derived from the fact that all cofactors of $\mathbf{L}$ are equal, and equal to the spanning tree generating function for $\mathcal{L}$ [9].

To compute the resistance $R_{\alpha\beta}$ between two nodes α and β, we connect α and β to the two terminals of an external battery and measure the current I going through the battery while no other nodes are connected to external sources. Let the potentials at the two nodes be, respectively, V_α and V_β. Then, the desired resistance is the ratio

$$R_{\alpha\beta} = \frac{V_\alpha - V_\beta}{I}. \tag{6}$$

The computation of the two-point resistance $R_{\alpha\beta}$ is now reduced to solving (3) for V_α and V_β with the current given by

$$I_i = I(\delta_{i\alpha} - \delta_{i\beta}). \tag{7}$$

A probabilistic interpretation. The two-point resistance has a probabilistic interpretation. Consider a random walker walking on the network $\mathcal{L}$ with the probability

$$p_{i\to j} = c_{ij}/c_i \tag{8}$$

of hopping from node i to node j, where $p_{i\to j}$ can be different from $p_{j\to i}$ (and thus the matrix $\mathbf{L}$ is not necessarily symmetric). Let $P(\alpha;\beta)$ be the probability that the walker starting from node α will reach node β before returning to α, which is the probability of first passage. Then one has the relation [2, 5]

$$P(\alpha;\beta) = \frac{1}{c_\alpha R_{\alpha\beta}} \tag{9}$$

where c_α is defined in (5) and $R_{\alpha\beta}$ is obtained from (6) by solving (3) using the Laplacian defined by (8). If all resistances are 1, then (9) becomes

$$P(\alpha;\beta) = \frac{1}{z_\alpha R_{\alpha\beta}} \tag{10}$$

where z_α is the coordination number, or the number of nodes connected to, of the node α.

2. The two-point resistance: a theorem

Let Ψ_i and λ_i be the eigenvectors and eigenvalues of $\mathbf{L}$, namely,

$$\mathbf{L}\Psi_i = \lambda_i \Psi_i \qquad i = 1, 2, \ldots, \mathcal{N}.$$

Let $\psi_{i\alpha}, \alpha = 1, 2, \ldots, \mathcal{N}$ be the components of Ψ_i. Since $\mathbf{L}$ is Hermitian, the Ψ_i can be taken to be orthonormal satisfying

$$(\Psi_i^*, \Psi_j) = \sum_\alpha \psi_{i\alpha}^* \psi_{j\alpha} = \delta_{ij}.$$

Now the sum of all columns (or rows) of $\mathbf{L}$ is identically zero, so one of the eigenvalues of $\mathbf{L}$ is zero. It is readily verified that the zero eigenvalue $\lambda_1 = 0$ has the eigenvector

$$\psi_{1\alpha} = 1/\sqrt{\mathcal{N}} \qquad \alpha = 1, 2, \ldots, \mathcal{N}.$$

We now state our main result as a theorem.

Theorem. *Consider a resistance network whose Laplacian has nonzero eigenvalues λ_i with orthonormal eigenvectors $\Psi_i = (\psi_{i1}, \psi_{i2}, \ldots, \psi_{i\mathcal{N}})$, $i = 2, 3, \ldots, \mathcal{N}$. Then the resistance between nodes α and β is given by*

$$R_{\alpha\beta} = \sum_{i=2}^{\mathcal{N}} \frac{1}{\lambda_i} |\psi_{i\alpha} - \psi_{i\beta}|^2. \tag{11}$$

Proof. We proceed to solve equation (3) by introducing the inverse of the Laplacian $\mathbf{L}$, or Green's function [6]. However, since one of the eigenvalues of $\mathbf{L}$ is zero, the inverse of $\mathbf{L}$ must be considered with care.

We circumvent this problem by adding a small term $\epsilon\mathbf{I}$ to the Laplacian, where $\mathbf{I}$ is the $\mathcal{N} \times \mathcal{N}$ identity matrix, and set $\epsilon = 0$ at the end. The modified Laplacian

$$\mathbf{L}(\epsilon) = \mathbf{L} + \epsilon\mathbf{I}$$

is of the same form $\mathbf{L}$ except that the diagonal elements c_i are replaced by $c_i + \epsilon$. It is clear that $\mathbf{L}(\epsilon)$ has eigenvalues $\lambda_i + \epsilon$ and is diagonalized by the same unitary transformation which diagonalizes $\mathbf{L}$.

The inverse of $\mathbf{L}(\epsilon)$, which is Green's function, is now well defined and we write

$$\mathbf{G}(\epsilon) = \mathbf{L}^{-1}(\epsilon).$$

Rewrite (3) as $\mathbf{L}(\epsilon)\vec{V}(\epsilon) = \vec{I}$ and multiply from the left by $\mathbf{G}(\epsilon)$ to obtain $\vec{V}(\epsilon) = \mathbf{G}(\epsilon)\vec{I}$. Explicitly, this reads

$$V_i(\epsilon) = \sum_{j=1}^{\mathcal{N}} G_{ij}(\epsilon) I_j \qquad i = 1, 2, \ldots, \mathcal{N} \tag{12}$$

where $G_{ij}(\epsilon)$ are the ijth elements of the matrix $\mathbf{G}(\epsilon)$.

We now compute Green's function $G_{ij}(\epsilon)$.

Let $\mathbf{U}$ be the unitary matrix which diagonalizes $\mathbf{L}(\epsilon)$ and $\mathbf{L}$, namely,

$$\mathbf{U}^\dagger\mathbf{L}\mathbf{U} = \Lambda \qquad \mathbf{U}^\dagger\mathbf{L}(\epsilon)\mathbf{U} = \Lambda(\epsilon). \tag{13}$$

It is readily verified that elements of $\mathbf{U}$ are $U_{ij} = \psi_{ji}$, and Λ and $\Lambda(\epsilon)$ are, respectively, diagonal matrices with elements $\lambda_i\delta_{ij}$ and $(\lambda_i + \epsilon)\delta_{ij}$.

The inverse of the second equation in (13) is

$$\mathbf{U}^\dagger\mathbf{G}(\epsilon)\mathbf{U} = \Lambda^{-1}(\epsilon)$$

where $\Lambda^{-1}(\epsilon)$ has elements $(\lambda_i + \epsilon)^{-1}\delta_{ij}$. It follows that we have

$$\mathbf{G}(\epsilon) = \mathbf{U}\Lambda^{-1}(\epsilon)\mathbf{U}^\dagger$$

or, explicitly,

$$\begin{aligned} G_{\alpha\beta}(\epsilon) &= \sum_{i=1}^{\mathcal{N}} U_{\alpha i} \left(\frac{1}{\lambda_i + \epsilon}\right) U^*_{\beta i} \\ &= \frac{1}{\mathcal{N}\epsilon} + g_{\alpha\beta}(\epsilon) \end{aligned} \tag{14}$$

where

$$g_{\alpha\beta}(\epsilon) = \sum_{i=2}^{\mathcal{N}} \frac{\psi_{i\alpha}\psi^*_{i\beta}}{\lambda_i + \epsilon}. \tag{15}$$

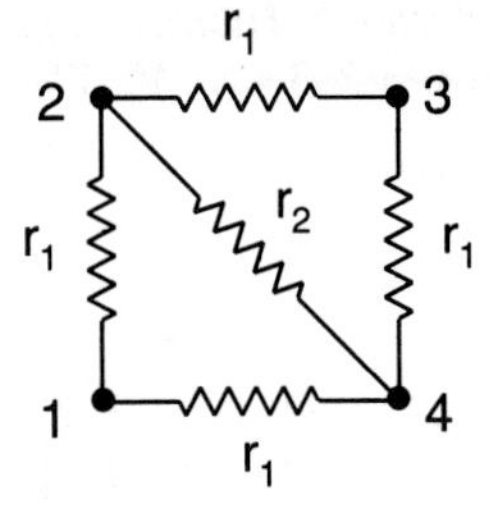

Figure 1. A network of four nodes.

The substitution of (14) into (12) now yields, after making use of the constraint (1),

$$V_i(\epsilon) = \sum_{j=1}^{\mathcal{N}} g_{ij}(\epsilon) I_j.$$

It is now safe to take the $\epsilon \to 0$ limit in $g_{ij}(\epsilon)$ to obtain

$$V_i = \sum_{j=1}^{\mathcal{N}} g_{ij}(0) I_j. \tag{16}$$

Finally, by combining (6), (7) with (16) we obtain

$$R_{\alpha\beta} = g_{\alpha\alpha}(0) + g_{\beta\beta}(0) - g_{\alpha\beta}(0) - g_{\beta\alpha}(0)$$

which becomes (11) after introducing (15). □

The usefulness of (11) is illustrated by the following examples.

Example 1. Consider the 4-node network shown in figure 1 with the Laplacian

$$\mathbf{L} = \begin{pmatrix} 2c_1 & -c_1 & 0 & -c_1 \\ -c_1 & 2c_1 + c_2 & -c_1 & -c_2 \\ 0 & -c_1 & 2c_1 & -c_1 \\ -c_1 & -c_2 & -c_1 & 2c_1 + c_2 \end{pmatrix}$$

where $c_1 = 1/r_1, c_2 = 1/r_2$. The nonzero eigenvalues of $\mathbf{L}$ and their orthonormal eigenvectors are

$$\lambda_2 = 4c_1 \qquad \Psi_2 = \tfrac{1}{2}(1, -1, 1, -1)$$
$$\lambda_3 = 2c_1 \qquad \Psi_3 = \tfrac{1}{\sqrt{2}}(-1, 0, 1, 0)$$
$$\lambda_4 = 2(c_1 + c_2) \qquad \Psi_4 = \tfrac{1}{\sqrt{2}}(0, -1, 0, 1).$$

Using (11), we obtain

$$R_{13} = \frac{1}{\lambda_2}(\psi_{21} - \psi_{23})^2 + \frac{1}{\lambda_3}(\psi_{31} - \psi_{33})^2 + \frac{1}{\lambda_4}(\psi_{41} - \psi_{43})^2 = r_1$$
$$R_{12} = \frac{1}{\lambda_2}(\psi_{21} - \psi_{22})^2 + \frac{1}{\lambda_3}(\psi_{31} - \psi_{32})^2 + \frac{1}{\lambda_4}(\psi_{41} - \psi_{42})^2 = \frac{r_1(2r_1 + 3r_2)}{4(r_1 + r_2)}.$$

Example 2. A complete graph is a network in which every node is connected to every other node. We consider $\mathcal{N}(\mathcal{N} - 1)/2$ resistors of equal resistance r embedded on a complete graph

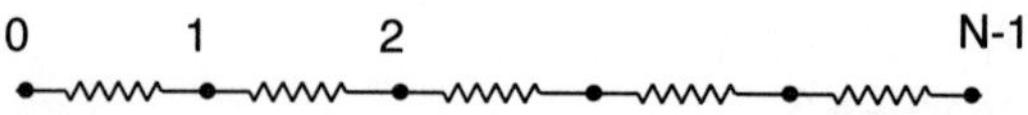

Figure 2. A one-dimensional network of $\mathcal{N}$ nodes with free ends.

of $\mathcal{N}$ nodes. The Laplacian is therefore

$$\mathbf{L}^{\text{complete graph}} = r^{-1}\begin{pmatrix} \mathcal{N}-1 & -1 & \cdots & -1 & -1 \\ -1 & \mathcal{N}-1 & \cdots & -1 & -1 \\ \vdots & \vdots & \ddots & \vdots & \vdots \\ -1 & -1 & \cdots & \mathcal{N}-1 & -1 \\ -1 & -1 & \cdots & -1 & \mathcal{N}-1 \end{pmatrix}. \tag{17}$$

It is readily verified that the Laplacian (17) has eigenvalues $\lambda_0 = 0$ and $\lambda_n = \mathcal{N}r^{-1}, n = 1, 2, \ldots, \mathcal{N}-1$, with corresponding eigenvectors Ψ_n having components

$$\psi_{n\alpha} = \frac{1}{\sqrt{\mathcal{N}}}\exp(\mathrm{i}2\pi n\alpha/\mathcal{N}) \qquad n, \alpha = 0, 1, \ldots, \mathcal{N}-1.$$

It follows from (11) that the resistance between any two nodes α and β is

$$R_{\alpha,\beta} = r\sum_{n=1}^{\mathcal{N}}\frac{|\psi_{n\alpha}-\psi_{n\beta}|^2}{\mathcal{N}} = \frac{2}{\mathcal{N}}r. \tag{18}$$

In subsequent sections we consider applications of (11) to regular lattices.

3. One-dimensional lattice

It is instructive to first consider the one-dimensional case of a linear array of resistors, and to demonstrate that our formulation produces results implied by simple applications of Ohm's law. We consider free and periodic boundary conditions separately.

3.1. Free boundary condition

Consider $N-1$ resistors of resistance r each connected in series forming a chain of N nodes numbered $0, 1, 2, \ldots, N-1$ as shown in figure 2, where each of the two end nodes connects to only one interior node. This is the Neumann (or the free) boundary condition.

The Laplacian (4) assumes the form

$$\mathbf{L}^{\text{free}}_{\{N\times 1\}} = r^{-1}\mathbf{T}^{\text{free}}_N$$

where $\mathbf{T}^{\text{free}}_N$ is the $N \times N$ matrix

$$\mathbf{T}^{\text{free}}_N = \begin{pmatrix} 1 & -1 & 0 & \cdots & 0 & 0 & 0 \\ -1 & 2 & -1 & \cdots & 0 & 0 & 0 \\ \vdots & \vdots & \vdots & \ddots & \vdots & \vdots & \vdots \\ 0 & 0 & 0 & \cdots & -1 & 2 & -1 \\ 0 & 0 & 0 & \cdots & 0 & -1 & 1 \end{pmatrix}. \tag{19}$$

The eigenvalues and eigenvectors of $\mathbf{T}_N$ can be readily computed (see, for example, [10]), and are found to be

$$\lambda_n = 2(1-\cos\phi_n) \qquad n = 0, 1, \ldots, N-1$$

$$\psi_{nx}^{(N)} = \frac{1}{\sqrt{N}} \qquad n = 0, \text{ for all } x \tag{20}$$
$$= \sqrt{\frac{2}{N}} \cos((x + 1/2)\phi_n) \qquad n = 1, 2, \ldots, N-1, \text{ for all } x$$

where $\phi_n = n\pi/N$. Thus, using (11), the resistance between nodes x_1 and x_2 is

$$R_{\{N\times 1\}}^{\text{free}}(x_1, x_2) = \frac{r}{N} \sum_{n=1}^{N-1} \frac{\left[\cos\left(x_1 + \frac{1}{2}\right)\phi_n - \cos\left(x_2 + \frac{1}{2}\right)\phi_n\right]^2}{1 - \cos\phi_n}$$
$$= r\left[F_N(x_1 + x_2 + 1) + F_N(x_1 - x_2) - \frac{1}{2}F_N(2x_1 + 1) - \frac{1}{2}F_N(2x_2 + 1)\right] \tag{21}$$

where

$$F_N(\ell) = \frac{1}{N} \sum_{n=1}^{N-1} \frac{1 - \cos(\ell\phi_n)}{1 - \cos\phi_n}. \tag{22}$$

Note that without loss of generality we can take $0 \leqslant \ell < 2N$.

The function $F_N(\ell)$ can be evaluated by taking the limit $\lambda \to 0$ of the function $I_1(0) - I_1(\ell)$ evaluated in (61) below. It is, however, instructive to evaluate $F_N(\ell)$ directly. To do this we consider the real part of the summation

$$T_N(\ell) \equiv \frac{1}{N} \sum_{n=1}^{N-1} \frac{1 - \exp(\mathrm{i}\ell\phi_n)}{1 - \exp(\mathrm{i}\phi_n)}.$$

First, writing out the real part of the summand, we obtain

$$\operatorname{Re} T_N(\ell) = \frac{1}{N} \sum_{n=1}^{N-1} \frac{1 - \cos\phi_n - \cos\ell\phi_n + \cos(\ell - 1)\phi_n}{2(1 - \cos\phi_n)}$$
$$= \frac{1}{2}[F_N(1) + F_N(\ell) - F_N(\ell - 1)]. \tag{23}$$

Second, we expand the summand to obtain

$$T_N(\ell) = \frac{1}{N} \sum_{n=1}^{N-1} \sum_{\ell'=0}^{\ell-1} \exp(\mathrm{i}\pi n\ell'/N)$$

and carry out the summation over n. The term $\ell' = 0$ yields $F_N(1)$ and terms $\ell' \geqslant 1$ can be explicitly summed, leading to

$$T_N(\ell) = F_N(1) + \frac{1}{N} \sum_{\ell'=1}^{\ell-1} \left[\frac{1 - (-1)^{\ell'}}{1 - \exp(\mathrm{i}\pi\ell'/N)} - 1\right] \qquad \ell < 2N. \tag{24}$$

We now evaluate the real part of $T_N(\ell)$ given by (24). Using the identity

$$\operatorname{Re}\left(\frac{1}{1 - \mathrm{e}^{\mathrm{i}\theta}}\right) = \frac{1}{2} \qquad 0 < \theta < 2\pi \tag{25}$$

we find

$$\operatorname{Re} T_N(\ell) = F_N(1) - \frac{1}{4N}[2\ell - 3 - (-1)^\ell]. \tag{26}$$

Equating (23) with (26) and noting $F_N(1) = 1 - 1/N$, we are led to the recursion relation

$$F_N(\ell) - F_N(\ell - 1) = 1 - \frac{1}{2N}[2\ell - 1 - (-1)^\ell]$$

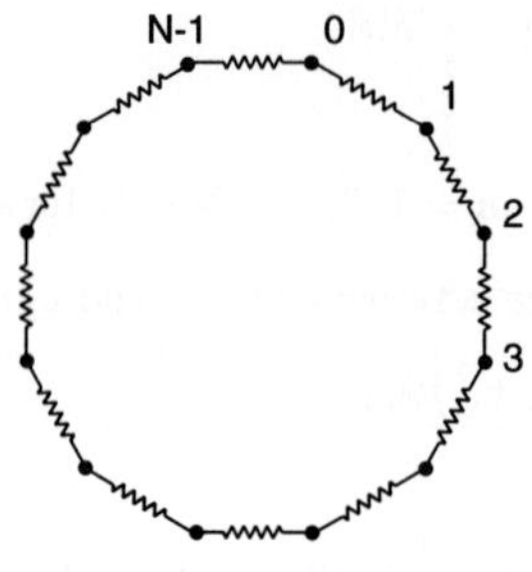

Figure 3. A one-dimensional network of $\mathcal{N}$ nodes with periodic boundary conditions.

which can be solved (cf section 10) to yield

$$F_N(\ell) = |\ell| - \frac{1}{N}\left(\frac{\ell^2+|\ell|}{2} - \left[\frac{|\ell|}{2}\right]\right) \tag{27}$$

where $[x]$ denotes the integral part of x. The substitution of (27) into (21) now gives the answer

$$R^{\text{free}}_{\{N\times 1\}}(x_1, x_2) = r|x_1 - x_2| \tag{28}$$

which is the expected expression obtainable by a simple application of Ohm's law.

3.2. Periodic boundary conditions

Consider next periodic boundary conditions for which nodes 0 and $N-1$ are also connected as shown in figure 3. The Laplacian (4) of the lattice is therefore $\mathbf{L}^{\text{per}}_{\{N\times 1\}} = r^{-1}\mathbf{T}^{\text{per}}_N$ where

$$\mathbf{T}^{\text{per}}_N = \begin{pmatrix} 2 & -1 & 0 & \cdots & 0 & 0 & -1 \\ -1 & 2 & -1 & \cdots & 0 & 0 & 0 \\ \vdots & \vdots & \vdots & \ddots & \vdots & \vdots & \vdots \\ 0 & 0 & 0 & \cdots & -1 & 2 & -1 \\ -1 & 0 & 0 & \cdots & 0 & -1 & 2 \end{pmatrix}. \tag{29}$$

The eigenvalues and eigenfunctions of $\mathbf{T}^{\text{per}}_N$ are well known and are, respectively,

$$\begin{aligned} \lambda_n &= 2(1-\cos 2\phi_n) \\ \psi^{\text{per}}_{nx} &= \frac{1}{\sqrt{N}}\exp(\mathrm{i}2x\phi_n) \qquad n, x = 0, 1, \ldots, N-1. \end{aligned} \tag{30}$$

Substituting (30) into (11), we obtain the resistance between nodes x_1 and x_2

$$\begin{aligned} R^{\text{per}}_{\{N\times 1\}}(x_1, x_2) &= \frac{r}{N}\sum_{n=1}^{N-1} \frac{|\exp(\mathrm{i}2x_1\phi_n) - \exp(\mathrm{i}2x_2\phi_n)|^2}{2(1-\cos 2\phi_n)} \\ &= rG_N(x_1 - x_2) \end{aligned} \tag{31}$$

where

$$G_N(\ell) = \frac{1}{N}\sum_{n=1}^{N-1}\frac{1-\cos(2\ell\phi_n)}{1-\cos 2\phi_n} = G_N(|\ell|).$$

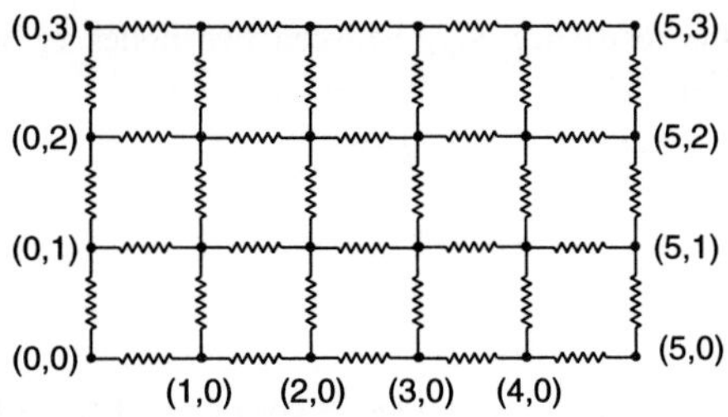

Figure 4. A 5×4 rectangular network.

Again, the function $G_N(\ell)$ can be evaluated by taking the limit $\lambda \to 0$ of $I_2(0) - I_2(\ell)$ given in (62) below. Alternately, it can be evaluated directly as in section 3.1 above by considering the real part of the summation

$$\frac{1}{N}\sum_{n=1}^{N-1}\frac{1-\exp(\mathrm{i}2\ell\phi_n)}{1-\exp(\mathrm{i}2\phi_n)}$$

in two different ways. This leads to the recursion relation

$$G_N(\ell) - G_N(\ell-1) = 1 - \frac{1}{N}(2\ell-1)$$

which can be solved to yield

$$G_N(\ell) = |\ell| - \ell^2/N. \tag{32}$$

It follows that we have

$$R^{\mathrm{per}}_{\{N\times 1\}}(x_1, x_2) = r|x_1 - x_2|\left[1 - \frac{|x_1 - x_2|}{N}\right]. \tag{33}$$

Expression (33) is the expected resistance of two resistors $|x_1 - x_2|r$ and $(N - |x_1 - x_2|)r$ connected in parallel as in a ring.

4. Two-dimensional network: free boundaries

Consider a rectangular network of resistors connected in an array of $M \times N$ nodes forming a network with free boundaries as shown in figure 4.

Number the nodes by coordinates $\{m, n\}$, $0 \leqslant m \leqslant M-1$, $0 \leqslant n \leqslant N-1$ and denote the resistances along the two principal directions by r and s. The Laplacian is therefore

$$\mathbf{L}^{\mathrm{free}}_{\{M\times N\}} = r^{-1}\mathbf{T}^{\mathrm{free}}_M \otimes \mathbf{I}_N + s^{-1}\mathbf{I}_M \otimes \mathbf{T}^{\mathrm{free}}_N \tag{34}$$

where $\otimes$ denotes direct matrix products and $\mathbf{T}^{\mathrm{free}}_N$ is given by (19). The Laplacian can be diagonalized in the two subspaces separately, yielding eigenvalues

$$\lambda_{(m,n)} = 2r^{-1}(1-\cos\theta_m) + 2s^{-1}(1-\cos\phi_n) \tag{35}$$

and eigenvectors

$$\psi^{\mathrm{free}}_{(m,n);(x,y)} = \psi^{(M)}_{mx}\psi^{(N)}_{ny}. \tag{36}$$

It then follows from (11) that the resistance $R_{\rm free}$ between two nodes $\mathbf{r}_1 = (x_1, y_1)$ and $\mathbf{r}_2 = (x_2, y_2)$ is

$$R^{\rm free}_{\{M\times N\}}(\mathbf{r}_1,\mathbf{r}_2) = \sum_{m=0}^{M-1}\sum_{n=0}^{N-1}{}_{(m,n)\neq(0,0)} \frac{\left|\psi^{\rm free}_{(m,n);(x_1,y_1)} - \psi^{\rm free}_{(m,n);(x_2,y_2)}\right|^2}{\lambda_{(m,n)}}$$
$$= \frac{r}{N}|x_1-x_2| + \frac{s}{M}|y_1-y_2| + \frac{2}{MN}$$
$$\times \sum_{m=1}^{M-1}\sum_{n=1}^{N-1} \frac{\left[\cos\left(x_1+\frac{1}{2}\right)\theta_m \cos\left(y_1+\frac{1}{2}\right)\phi_n - \cos\left(x_2+\frac{1}{2}\right)\theta_m \cos\left(y_2+\frac{1}{2}\right)\phi_n\right]^2}{r^{-1}(1-\cos\theta_m)+s^{-1}(1-\cos\phi_n)} \tag{37}$$

where

$$\theta_m = \frac{m\pi}{M} \qquad \phi_n = \frac{n\pi}{N}.$$

Here, use has been made of (28) for summing over the $m = 0$ and $n = 0$ terms. As expected, the resulting expression (37) now depends on the coordinates x_1, y_1, x_2, y_2 of the two nodes explicitly.

The usefulness of (37) is best illustrated by applications. Several examples are now given.

Example 3. For $M = 5, N = 4, r = s$, the resistance between nodes $\{0, 0\}$ and $\{3, 3\}$ is computed from (37) as

$$R^{\rm free}_{\{5\times 4\}}(\{0,0\},\{3,3\}) = \left(\frac{3}{4}+\frac{3}{5}+\frac{9877\,231}{27\,600\,540}\right) r$$
$$= (1.707\,86\ldots)r. \tag{38}$$

Example 4. For $M = N = 4$, we find

$$R^{\rm free}_{\{4\times 4\}}(\{0,0\},\{3,3\}) = \frac{(r+s)(r^2+5rs+s^2)(3r^2+7rs+3s^2)}{2(2r^2+4rs+s^2)(r^2+4rs+2s^2)}. \tag{39}$$

Example 5. We evaluate the resistance between two nodes in the interior of a large lattice. Consider, for definiteness, both M, N = odd (for other parities the result (40) below is the same) and compute the resistance between two nodes in the centre region,

$$\mathbf{r}_1 = (x_1, y_1) = \left(\frac{M-1}{2}+p_1, \frac{N-1}{2}+q_1\right)$$
$$\mathbf{r}_2 = (x_2, y_2) = \left(\frac{M-1}{2}+p_2, \frac{N-1}{2}+q_2\right)$$

where $p_i, q_i \ll M, N$ are integers. The numerator of the summand in (37) becomes

$$\left[\cos\left(\frac{m\pi}{2}+p_1\theta_m\right)\cos\left(\frac{n\pi}{2}+q_1\phi_n\right) - \cos\left(\frac{m\pi}{2}+p_2\theta_m\right)\cos\left(\frac{n\pi}{2}+q_2\phi_n\right)\right]^2$$
$$= (\cos p_1\theta_m \cos q_1\phi_n - \cos p_2\theta_m \cos q_2\phi_n)^2 \qquad m, n = \text{even}$$
$$= (\sin p_1\theta_m \sin q_1\phi_n - \sin p_2\theta_m \sin q_2\phi_n)^2 \qquad m, n = \text{odd}$$
$$= (\sin p_1\theta_m \cos q_1\phi_n - \sin p_2\theta_m \cos q_2\phi_n)^2 \qquad m = \text{odd}, n = \text{even}$$
$$= (\cos p_1\theta_m \sin q_1\phi_n - \cos p_2\theta_m \sin q_2\phi_n)^2 \qquad m = \text{even}, n = \text{odd}$$

and the summation in (37) breaks into four parts. In the $M, N \to \infty$ limit the summations can be replaced by integrals. After some reduction we arrive at the expression

$$R_\infty(\mathbf{r}_1,\mathbf{r}_2) = \frac{1}{(2\pi)^2}\int_0^{2\pi} d\phi \int_0^{2\pi} d\theta \left(\frac{1-\cos(x_1-x_2)\theta\cos(y_1-y_2)\phi}{r^{-1}(1-\cos\theta)+s^{-1}(1-\cos\phi)}\right). \tag{40}$$

Our expression (40) agrees with the known expression obtained previously [1]. It can be verified that expression (40) holds between any two nodes in the lattice, provided that the two nodes are far from the boundaries.

5. Two-dimensional network: periodic boundary conditions

We next consider an $M \times N$ network with periodic boundary conditions. The Laplacian in this case is

$$\mathbf{L}^{\text{per}}_{\{M\times N\}} = r^{-1}\mathbf{T}^{\text{per}}_M \otimes \mathbf{I}_N + s^{-1}\mathbf{I}_M \otimes \mathbf{T}^{\text{per}}_N \tag{41}$$

where $\mathbf{T}^{\text{per}}_N$ is given by (29). The Laplacian (41) can again be diagonalized in the two subspaces separately, yielding eigenvalues and eigenvectors

$$\begin{aligned} \lambda_{(m,n)} &= 2r^{-1}(1-\cos 2\theta_m) + 2s^{-s}(1-\cos 2\phi_n) \\ \psi_{(m,n);(x,y)} &= \frac{1}{\sqrt{MN}} \exp(\mathrm{i}2x\theta_m)\exp(\mathrm{i}2y\phi_n). \end{aligned} \tag{42}$$

This leads to the resistance between nodes $\mathbf{r}_1 = (x_1, y_1)$ and $\mathbf{r}_2 = (x_2, y_2)$,

$$\begin{aligned} R^{\text{per}}_{\{M\times N\}}(\mathbf{r}_1,\mathbf{r}_2) &= \sum_{m=0}^{M-1}\sum_{n=0 \, (m,n)\neq(0,0)}^{N-1} \frac{|\psi_{(m,n);(x_1,y_1)} - \psi_{(m,n);(x_2,y_2)}|^2}{\lambda_{(m,n)}} \\ &= \frac{r}{N}\left[|x_1-x_2| - \frac{(x_1-x_2)^2}{M}\right] + \frac{s}{M}\left[|y_1-y_2| - \frac{(y_1-y_2)^2}{N}\right] \\ &\quad + \frac{1}{MN}\sum_{m=1}^{M}\sum_{n=1}^{N} \frac{1-\cos[2(x_1-x_2)\theta_m + 2(y_1-y_2)\phi_n]}{r^{-1}(1-\cos 2\theta_m) + s^{-1}(1-\cos 2\phi_n)} \end{aligned} \tag{43}$$

where the two terms in the second line are given by (33). It is clear that the result depends only on the differences $|x_1 - x_2|$ and $|y_1 - y_2|$, as it should under periodic boundary conditions.

Example 6. Using (43) the resistance between nodes $\{0, 0\}$ and $\{3, 3\}$ on a 5×4 periodic lattice with $r = s$ is

$$\begin{aligned} R^{\text{per}}_{\{5\times 4\}}(\{0,0\},\{3,3\}) &= \left(\frac{3}{10} + \frac{3}{20} + \frac{1799}{7790}\right) r \\ &= (0.680\,937\ldots)r. \end{aligned} \tag{44}$$

This is to be compared to the value $(1.707\,863\ldots)r$ for free boundary conditions given in example 3. It can also be verified that the resistance between nodes $\{0, 0\}$ and $\{2, 1\}$ is also given by (44) as it must for a periodic lattice.

In the limit of $M, N \to \infty$ with $|\mathbf{r}_1 - \mathbf{r}_2|$ finite, (43) becomes

$$\begin{aligned} R_\infty(\mathbf{r}_1,\mathbf{r}_2) &= \frac{1}{(2\pi)^2}\int_0^{2\pi} \mathrm{d}\phi \int_0^{2\pi} \mathrm{d}\theta \frac{1-\cos[(x_1-x_2)\theta + (y_1-y_2)\phi]}{r^{-1}(1-\cos\theta) + s^{-1}(1-\cos\phi)} \\ &= \frac{1}{(2\pi)^2}\int_0^{2\pi} \mathrm{d}\phi \int_0^{2\pi} \mathrm{d}\theta \frac{1-\cos(x_1-x_2)\theta\cos(y_1-y_2)\phi}{r^{-1}(1-\cos\theta) + s^{-1}(1-\cos\phi)} \end{aligned} \tag{45}$$

which agrees with (40).

6. Cylindrical boundary conditions

Consider an $M \times N$ resistor network embedded on a cylinder with periodic boundary in the direction of M and free boundaries in the direction of N.

The Laplacian is

$$\mathbf{L}^{\mathrm{cyl}}_{\{M\times N\}} = r^{-1}\mathbf{T}^{\mathrm{per}}_M \otimes \mathbf{I}_N + s^{-1}\mathbf{I}_M \otimes \mathbf{T}^{\mathrm{free}}_N$$

which can again be diagonalized in the two subspaces separately. This gives the eigenvalues and eigenvectors

$$\lambda_{(m,n)} = 2r^{-1}(1-\cos 2\theta_m) + 2s^{-1}(1-\cos\phi_n)$$

$$\psi^{\mathrm{cyl}}_{(m,n);(x,y)} = \frac{1}{\sqrt{M}}\exp(\mathrm{i}2x\theta_m)\psi^{(N)}_{ny}.$$

It follows that the resistance R_{free} between nodes $\mathbf{r}_1 = (x_1, y_1)$ and $\mathbf{r}_2 = (x_2, y_2)$ is

$$\begin{aligned}
R^{\mathrm{cyl}}_{\{M\times N\}}(\mathbf{r}_1,\mathbf{r}_2) &= \sum_{m=0}^{M-1}\sum_{n=0\,(m,n)\neq(0,0)}^{N-1} \frac{\left|\psi^{\mathrm{cyl}}_{(m,n);(x_1,y_1)} - \psi^{\mathrm{cyl}}_{(m,n);(x_2,y_2)}\right|^2}{\lambda_{(m,n)}} \\
&= \frac{r}{N}\left[|x_1-x_2| - \frac{(x_1-x_2)^2}{M}\right] + \frac{s}{M}|y_1-y_2| \\
&\quad + \frac{2}{MN}\sum_{m=1}^{M-1}\sum_{n=1}^{N-1}\left(\frac{C_1^2 + C_2^2 - 2C_1C_2\cos 2(x_1-x_2)\theta_m}{r^{-1}(1-\cos 2\theta_m) + s^{-1}(1-\cos\phi_n)}\right)
\end{aligned}$$

where

$$C_1 = \cos\left(y_1 + \tfrac{1}{2}\right)\phi_n \qquad C_2 = \cos\left(y_2 + \tfrac{1}{2}\right)\phi_n. \tag{46}$$

It can be verified that in the $M, N \to \infty$ limit (46) leads to the same expression (40) for two interior nodes in an infinite lattice.

Example 7. The resistance between nodes $\{0, 0\}$ and $\{3, 3\}$ on a 5×4 cylindrical lattice with $r = s$ is computed to be

$$\begin{aligned}
R^{\mathrm{cyl}}_{\{5\times 4\}}(\{0,0\},\{3,3\}) &= \left(\frac{3}{10} + \frac{3}{5} + \frac{5023}{8835}\right)r \\
&= (1.468\,53\ldots)r.
\end{aligned} \tag{47}$$

This is compared to the values of $(1.707\,86\cdots)r$ for free boundary conditions and $(0.680\,937\ldots)r$ for periodic boundary conditions.

7. Möbius strip

We next consider an $M \times N$ resistor lattice embedded on a Möbius strip of width N and length M, which is a rectangular strip connected at two ends after a 180° twist of one of the two ends of the strip. The schematic figure of a Möbius strip is shown in figure 5(a). The Laplacian for this lattice assumes the form

$$\mathbf{L}^{\mathrm{Mob}}_{\{M\times N\}} = r^{-1}[\mathbf{H}_M \otimes \mathbf{I}_N - \mathbf{K}_M \otimes \mathbf{J}_N] + s^{-1}\mathbf{I}_M \otimes \mathbf{T}^{\mathrm{free}}_N \tag{48}$$

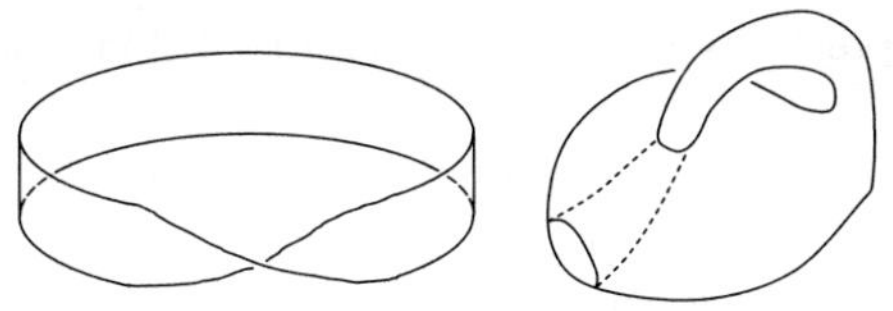

Figure 5. (*a*) The schematic plot of an Möbius strip. (*b*) The schematic plot of a Klein bottle.

where

$$\mathbf{K}_N = \begin{pmatrix} 0 & 0 & \cdots & 0 & 1 \\ 0 & 0 & \cdots & 0 & 0 \\ \vdots & \vdots & \ddots & \vdots & \vdots \\ 0 & 0 & \cdots & 0 & 0 \\ 1 & 0 & \cdots & 0 & 0 \end{pmatrix} \qquad \mathbf{J}_N = \begin{pmatrix} 0 & 0 & \cdots & 0 & 1 \\ 0 & 0 & \cdots & 1 & 0 \\ \vdots & \vdots & \ddots & \vdots & \vdots \\ 0 & 1 & \cdots & 0 & 0 \\ 1 & 0 & \cdots & 0 & 0 \end{pmatrix}$$

$$\mathbf{H}_N = \mathbf{T}_N^{\text{per}} + \mathbf{K}_N = \begin{pmatrix} 2 & -1 & \cdots & 0 & 0 \\ -1 & 2 & \cdots & 0 & 0 \\ \vdots & \vdots & \ddots & \vdots & \vdots \\ 0 & 0 & \cdots & 2 & -1 \\ 0 & 0 & \cdots & -1 & 2 \end{pmatrix}$$

are $N \times N$ matrices. Now $\mathbf{T}_N^{\text{free}}$ and $\mathbf{J}_N$ commute so they can be replaced by their respective eigenvalues $2(1 - \cos\phi_n)$ and $(-1)^n$ and we need only diagonalize an $M \times M$ matrix. This leads to the following eigenvalues and eigenvectors of the Laplacian (48) [10][3]:

$$\begin{aligned} \lambda_{(m,n)} &= 2r^{-1}\cos\left[(4m + 1 - (-1)^n)\frac{\pi}{2M}\right] + 2s^{-1}\left(1 - \cos\frac{n\pi}{N}\right) \\ \psi^{\text{Mob}}_{(m,n);(x,y)} &= \frac{1}{\sqrt{M}}\exp\left[\mathrm{i}(4m + 1 - (-1)^n)\frac{x\pi}{2M}\right]\psi^{(N)}_{ny} \end{aligned} \tag{49}$$

where $\psi^{(N)}_{ny}$ is given in (20). Substituting these expressions into (11) and after a little reduction, we obtain

$$\begin{aligned} R^{\text{Mob}}_{\{M\times N\}}(\mathbf{r}_1, \mathbf{r}_2) &= \sum_{m=0}^{M-1}\sum_{n=0}^{N-1}{}_{(m,n)\neq(0,0)} \frac{\left|\psi^{\text{Mob}}_{(m,n);(x_1,y_1)} - \psi^{\text{Mob}}_{(m,n);(x_2,y_2)}\right|^2}{\lambda_{(m,n)}} \\ &= \frac{r}{N}\left[|x_1 - x_2| - \frac{(x_1 - x_2)^2}{M}\right] \\ &\quad + \frac{1}{MN}\sum_{m=0}^{M-1}\sum_{n=1}^{N-1}\frac{C_1^2 + C_2^2 - 2C_1C_2\cos\left[(x_1 - x_2)(4m + 1 - (-1)^n)\frac{\pi}{2M}\right]}{r^{-1}\left[1 - \cos(4m + 1 - (-1)^n)\frac{\pi}{2M}\right] + s^{-1}(1 - \cos\phi_n)} \end{aligned} \tag{50}$$

where C_1 and C_2 have been given in (46).

Example 8. The 2×2 Möbius strip is a complete graph of $\mathcal{N} = 4$ nodes. For $r = s$ expression (50) gives a resistance $r/2$ between any two nodes which agrees with (18).

[3] I am indebted to W-J Tzeng for working out (49) and (53).

Example 9. The resistance between nodes (0, 0) and (3, 3) on a 5×4 Möbius strip with $r = s$ is computed from (50) as

$$R^{\rm Mob}_{\{5\times 4\}}(\{0,0\},\{3,3\}) = \left(\frac{3}{10} + \frac{1609}{2698}\right) r$$
$$= (0.896\,367\ldots)r. \tag{51}$$

This is to be compared to the corresponding values for the same network under other boundary conditions in examples 3, 6 and 7.

8. Klein bottle

A Klein bottle is a Möbius strip with a periodic boundary condition imposed in the other direction, a schematic diagram of which is shown in figure 5(*b*). We consider an $M \times N$ resistor grid embedded on a Klein bottle.

Let the network have a twisted boundary condition in the direction of the length M and a periodic boundary condition in the direction of the width N. Then, analogous to (48), the Laplacian of the network assumes the form

$$\mathbf{L}^{\rm Klein}_{\{M\times N\}} = r^{-1}[\mathbf{H_M} \otimes \mathbf{I}_N - \mathbf{K}_M \otimes \mathbf{J}_N] + s^{-1}\mathbf{I}_M \otimes \mathbf{T}^{\rm per}_N. \tag{52}$$

Now the matrices $\mathbf{J}_N$ and $\mathbf{T}^{\rm per}_N$ commute so they can be replaced by their respective eigenvalues ± 1 and $2(1 - \cos 2\phi_n)$ in (52) and one need only diagonalize an $M \times M$ matrix. This leads to the following eigenvalues and eigenvectors for $\mathbf{L}^{\rm Klein}_{\{M\times N\}}$ [10] (see footnote 3):

$$\lambda_{(m,n)}(\tau) = 2r^{-1}\left[1 - \cos\left((2m+\tau)\frac{\pi}{M}\right)\right] + 2s^{-1}\left(1 - \cos\frac{2n\pi}{N}\right)$$
$$\psi^{\rm Klein}_{(m,n);(x,y)} = \frac{1}{\sqrt{M}}\exp\left[\mathrm{i}(2m+\tau)\frac{x\pi}{M}\right]\psi^{(N)\dagger}_{ny} \tag{53}$$

where

$$\tau = \tau_n = 0 \qquad n = 0, 1, \ldots, \left[\frac{N-1}{2}\right]$$
$$= 1 \qquad n = \left[\frac{N+1}{2}\right], \ldots, N-1$$

and

$$\psi^{(N)\dagger}_{ny} = \frac{1}{\sqrt{N}} \qquad n = 0$$
$$= \sqrt{\frac{2}{N}}\cos\left[(2y+1)\frac{n\pi}{N}\right] \qquad n = 1, 2, \ldots, \left[\frac{N-1}{2}\right]$$
$$= \frac{1}{\sqrt{N}}(-1)^y \qquad n = \frac{N}{2} \qquad \text{for even } N \text{ only}$$
$$= \sqrt{\frac{2}{N}}\sin\left[(2y+1)\frac{n\pi}{N}\right] \qquad n = \left[\frac{N}{2}\right] + 1, \ldots, N-1.$$

Substituting these expressions into (11), separating out the summation for $n = 0$, and making use of the identity

$$\sin\left[(2y+1)\left(\frac{N}{2} + n\right)\frac{\pi}{N}\right] = (-1)^y \cos\left[(2y+1)\frac{n\pi}{N}\right]$$

we obtain after some reduction

$$
\begin{aligned}
R^{\text{Klein}}_{\{M\times N\}}(\mathbf{r}_1,\mathbf{r}_2) &= \sum_{m=0}^{M-1}\sum_{n=0}^{N-1}{}_{(m,n)\neq(0,0)} \frac{\left|\psi^{\text{Klein}}_{(m,n);(x_1,y_1)} - \psi^{\text{Klein}}_{(m,n);(x_2,y_2)}\right|^2}{\lambda_{(m,n)}(\tau_n)} \\
&= \frac{r}{N}\left[|x_1-x_2| - \frac{(x_1-x_2)^2}{M}\right] \\
&\quad + \sum_{m=0}^{M-1}\sum_{n=1}^{N-1} \frac{1}{\lambda_{(m,n)}(\tau_n)}\left|\psi^{\text{Klein}}_{(m,n);(x_1,y_1)} - \psi^{\text{Klein}}_{(m,n);(x_2,y_2)}\right|^2 \\
&= \frac{r}{N}\left[|x_1-x_2| - \frac{(x_1-x_2)^2}{M}\right] + \Delta_N + \frac{2}{MN}\sum_{\tau=0}^{1}\sum_{m=0}^{M-1}\sum_{n=1}^{[\frac{N-1}{2}]} \\
&\quad \times \frac{C_1^2 + C_2^2 - 2(-1)^{(y_1-y_2)\tau}C_1C_2\cos[2(x_1-x_2)\Theta_m(\tau)]}{\lambda_{(m,n)}(\tau)}
\end{aligned} \tag{54}
$$

where

$$
\begin{aligned}
\Theta_m(\tau) &= \left(m+\frac{\tau}{2}\right)\frac{\pi}{M} \\
\Delta_N &= \frac{2}{MN}\sum_{m=0}^{M-1}\frac{1-(-1)^{y_1-y_2}\cos[2(x_1-x_2)\Theta_m(1)]}{\lambda_{(m,N/2)}(1)} \qquad N = \text{even} \\
&= 0 \qquad N = \text{odd}
\end{aligned} \tag{55}
$$

and $C_i = \cos[(y_i+1/2)n\pi/N]$, $i = 1, 2$, as defined in (46).

Example 10. The resistance between nodes (0, 0) and (3, 3) on a 5×4 (N = even) Klein bottle with $r = s$ is computed from (54) as

$$
\begin{aligned}
R^{\text{Klein}}_{\{5\times4\}}(\{0,0\},\{3,3\}) &= \left(\frac{3}{10}+\frac{5}{58}+\frac{56}{209}\right)r \\
&= (0.654\,149\cdots)r
\end{aligned} \tag{56}
$$

where the three terms in the first line are from the evaluation of corresponding terms in (54). The result is to be compared to the corresponding value for the same 5×4 network under the Möbius boundary condition considered in example 9, which is the Klein bottle without periodic boundary connections.

9. Higher dimensional lattices

The two-point resistance can be computed using (11) for lattices in any spatial dimensionality under various boundary conditions. To illustrate, we give the result for an $M\times N\times L$ cubic lattice with free boundary conditions.

Number the nodes by $\{m,n,\ell\}$, $0\leqslant m\leqslant M-1$, $0\leqslant n\leqslant N-1$, $0\leqslant \ell\leqslant L-1$, and let the resistances along the principal axes be, respectively, r, s and t. The Laplacian then assumes the form

$$
\mathbf{L}^{\text{free}}_{\{M\times N\times L\}} = r^{-1}\mathbf{T}^{\text{free}}_M\otimes\mathbf{I}_N\otimes\mathbf{I}_L + s^{-1}\mathbf{I}_M\otimes\mathbf{T}^{\text{free}}_N\otimes\mathbf{I}_L + t^{-1}\mathbf{I}_M\otimes\mathbf{I}_N\otimes\mathbf{T}^{\text{free}}_L
$$

where $\mathbf{T}^{\text{free}}_N$ is given by (19). The Laplacian can be diagonalized in the three subspaces separately, yielding eigenvalues

$$
\lambda_{(m,n,\ell)} = 2r^{-1}(1-\cos\theta_m) + 2s^{-1}(1-\cos\phi_n) + 2t^{-1}(1-\cos\alpha_\ell) \tag{57}
$$

and eigenvectors

$$\psi^{\text{free}}_{(m,n,\ell);(x,y,z)} = \psi^{(M)}_{mx}\psi^{(N)}_{ny}\psi^{(L)}_{\ell z}$$

where $\psi^{(M)}_{mx}$ is given by (20) and $\alpha_\ell = \ell\pi/L$. It then follows from (11) that the resistance R_{free} between two nodes $\mathbf{r}_1 = (x_1, y_1, z_1)$ and $\mathbf{r}_2 = (x_2, y_2, z_2)$ is

$$R^{\text{free}}_{\{M\times N\times L\}}(\mathbf{r}_1, \mathbf{r}_2) = \sum_{m=0}^{M-1}\sum_{n=0}^{N-1}\sum_{\ell=0 \, (m,n,\ell)\neq(0,0,0)}^{L-1} \lambda^{-1}_{(m,n,\ell)} \left|\psi^{\text{free}}_{(m,n,\ell);(x_1,y_1,z_1)} - \psi^{\text{free}}_{(m,n,\ell);(x_2,y_2,z_2)}\right|^2. \tag{58}$$

The summation can be broken down as

$$\begin{aligned} R^{\text{free}}_{\{M\times N\times L\}}(\mathbf{r}_1, \mathbf{r}_2) &= \sum_{m=1}^{M-1}\sum_{n=1}^{N-1}\sum_{\ell=1}^{L-1} \frac{\left|\psi^{\text{free}}_{(m,n,\ell);(x_1,y_1,z_1)} - \psi^{\text{free}}_{(m,n,\ell);(x_2,y_2,z_2)}\right|^2}{\lambda_{(m,n,\ell)}} \\ &\quad + \frac{1}{L} R^{\text{free}}_{\{M\times N\}}(\{x_1, y_1\}, \{x_2, y_2\}) + \frac{1}{M} R^{\text{free}}_{\{N\times L\}}(\{y_1, z_1\}, \{y_2, z_2\}) \\ &\quad + \frac{1}{N} R^{\text{free}}_{\{L\times M\}}(\{z_1, x_1\}, \{z_2, x_2\}) - \frac{1}{MN} R^{\text{free}}_{\{L\times 1\}}(x_1, x_2) \\ &\quad - \frac{1}{NL} R^{\text{free}}_{\{M\times 1\}}(y_1, y_2) - \frac{1}{LM} R^{\text{free}}_{\{N\times 1\}}(z_1, z_2). \end{aligned} \tag{59}$$

All terms in (59) have previously been computed except the summation in the first line.

Example 11. The resistance between the nodes (0, 0, 0) and (3, 3, 3) in a $5 \times 5 \times 4$ lattice with free boundaries and $r = s = t$ is computed from (59) as

$$\begin{aligned} R^{\text{free}}_{\{5\times 5\times 4\}}(\{0, 0, 0\}; \{3, 3, 3\}) &= \left(\frac{327\,687\,658\,482\,872}{352\,468\,567\,489\,225}\right) r \\ &= (0.929\,693\ldots)r. \end{aligned} \tag{60}$$

Example 12. The resistance between two interior nodes $\mathbf{r}_1$ and $\mathbf{r}_2$ can be worked out as in example 5. The result is

$$\begin{aligned} R_\infty(\mathbf{r}_1, \mathbf{r}_2) &= \frac{1}{(2\pi)^3}\int_0^{2\pi} \mathrm{d}\phi \int_0^{2\pi} \mathrm{d}\theta \int_0^{2\pi} \mathrm{d}\alpha \\ &\quad \times \left(\frac{1 - \cos(x_1 - x_2)\theta \cos(y_1 - y_2)\phi \cos(z_1 - z_2)\alpha}{r^{-1}(1 - \cos\theta) + s^{-1}(1 - \cos\phi) + t^{-1}(1 - \cos\alpha)}\right) \end{aligned}$$

which is the known result [1].

10. Summation and product identities

The reduction of the two-point resistances for one-dimensional lattices to the simple and familiar expressions of (28) and (33) is facilitated by the use of the summation identities (27) and (32). In this section, we extend the consideration and deduce generalizations of these identities which can be used to reduce the computational labour for lattice sums as well as analyse large-size expansions in two and higher dimensions.

We state two new lattice sum identities as a proposition.

Proposition. *Define*

$$I_\alpha(\ell) = \frac{1}{N}\sum_{n=0}^{N-1} \frac{\cos\left(\alpha\ell\frac{n\pi}{N}\right)}{\cosh\lambda - \cos\left(\alpha\frac{n\pi}{N}\right)} \qquad \alpha = 1, 2.$$

Then the following identities hold for λ real and $N = 1, 2, \ldots$:

$$I_1(\ell) = \frac{\cosh(N-\ell)\lambda}{(\sinh\lambda)\sinh(N\lambda)} + \frac{1}{N}\left[\frac{1}{\sinh^2\lambda} + \frac{1-(-1)^\ell}{4\cosh^2(\lambda/2)}\right] \qquad 0 \leqslant \ell < 2N \tag{61}$$

$$I_2(\ell) = \frac{\cosh\left(\frac{N}{2}-\ell\right)\lambda}{(\sinh\lambda)\sinh(N\lambda/2)} \qquad 0 \leqslant \ell < N. \tag{62}$$

Remarks.

(1) It is clear that without the loss of generality we can restrict ℓ to the ranges indicated.
(2) For $\ell = 0$ and $\lambda \to 0$, $I_1(0)$ leads to (27) and $I_2(0)$ leads to (32).
(3) In the $N \to \infty$ limit both (61) and (62) become the integral[4]

$$\frac{1}{\pi}\int_0^\pi \frac{\cos(\ell\theta)}{\cosh\lambda - \cos\theta}\,d\theta = \frac{e^{-\ell|\lambda|}}{\sinh|\lambda|} \qquad \ell \geqslant 0. \tag{63}$$

(4) Set $\ell = 0$ in (61), multiplying by $\sinh\lambda$ and integrating over λ, we obtain the product identity [11]

$$\prod_{n=0}^{N-1}\left(\cosh\lambda - \cos\frac{n\pi}{N}\right) = (\sinh N\lambda)\tanh(\lambda/2). \tag{64}$$

(5) Set $\ell = 0$ in (62), multiplying by $\sinh\lambda$ and integrating over λ, we obtain the product identity

$$\prod_{n=0}^{N-1}\left(\cosh\lambda - \cos\frac{2n\pi}{N}\right) = \sinh^2(N\lambda/2). \tag{65}$$

Proof of the proposition. It is convenient to introduce the notation

$$S_\alpha(\ell) = \frac{1}{N}\sum_{n=0}^{N-1}\frac{\cos(\ell\theta_n)}{1+a^2-2a\cos\theta_n} \qquad a < 1 \quad \alpha = 1, 2 \tag{66}$$

so that

$$I_\alpha(\ell) = 2aS_\alpha(\ell) \qquad a = e^{-|\lambda|}. \tag{67}$$

It is readily seen that we have the identity

$$S_\alpha(1) = \frac{1}{2a}[(1+a^2)S_\alpha(0) - 1]. \tag{68}$$

Proof of (61). First we evaluate $S_1(0)$ by carrying out the following summation, where Re denotes the real part, in two different ways. First we have

$$\begin{aligned}
\mathrm{Re}\,\frac{1}{N}\sum_{n=0}^{N-1}\frac{1}{1-a\,e^{i\theta_n}} &= \mathrm{Re}\,\frac{1}{N}\sum_{n=0}^{N-1}\frac{1-a\,e^{-i\theta_n}}{|1-a\,e^{i\theta_n}|^2} \\
&= \frac{1}{N}\sum_{n=0}^{N-1}\frac{1-a\cos\theta_n}{1+a^2-2a\cos\theta_n} \\
&= S_1(0) - aS_1(1) \\
&= \frac{1}{2}[1+(1-a^2)S_1(0)].
\end{aligned} \tag{69}$$

[4] This integral is equivalent to the integral (A6) of [1], where it is evaluated using the method of residues.

Secondly, by expanding the summand we have

$$\mathrm{Re}\,\frac{1}{N}\sum_{n=0}^{N-1}\frac{1}{1-a\,\mathrm{e}^{\mathrm{i}\theta_n}}=\mathrm{Re}\,\frac{1}{N}\sum_{n=0}^{N-1}\sum_{\ell=0}^{\infty}a^{\ell}\,\exp(\mathrm{i}\ell n\pi/N)$$

and carry out the summation over n for fixed ℓ. It is clear that all $\ell=$ even terms vanish except those with $\ell=2mN$, $m=0,1,2,\ldots$ which yield $\sum_{m=0}^{\infty}a^{2mN}=1/(1-a^{2N})$. For $\ell=\mathrm{odd}=2m+1$, $m=0,1,2,\ldots$ we have

$$\mathrm{Re}\sum_{n=0}^{N-1}\exp(\mathrm{i}(2m+1)n\pi/N)=\mathrm{Re}\,\frac{1-(-1)^{2m+1}}{1-\exp(\mathrm{i}(2m+1)\pi/N)}=1$$

after making use of (25). So the summation over $\ell=$ odd terms yields $N^{-1}\sum_{m=0}^{\infty}a^{2m+1}=a/N(1-a^2)$, and we have

$$\mathrm{Re}\sum_{n=0}^{N-1}\frac{1}{1-a\,\mathrm{e}^{\mathrm{i}\theta_n}}=\frac{1}{1-a^{2N}}+\frac{a}{N(1-a^2)}. \tag{70}$$

Equating (69) with (70) we obtain

$$S_1(0)=\frac{1}{1-a^2}\left[\left(\frac{1+a^{2N}}{1-a^{2N}}\right)+\frac{2a}{N(1-a^2)}\right]. \tag{71}$$

To evaluate $S_1(\ell)$ for general ℓ, we consider the summation

$$\begin{aligned}\mathrm{Re}\,\frac{1}{N}\sum_{n=0}^{N-1}\frac{1-(a\,\mathrm{e}^{\mathrm{i}\theta_n})^{\ell}}{1-a\,\mathrm{e}^{\mathrm{i}\theta_n}}&=\mathrm{Re}\,\frac{1}{N}\sum_{n=0}^{N-1}\frac{(1-a^{\ell}\,\mathrm{e}^{\mathrm{i}\ell\theta_n})(1-a\,\mathrm{e}^{-\mathrm{i}\theta_n})}{|1-a\,\mathrm{e}^{\mathrm{i}\theta_n}|^2}\\&=S_1(0)-aS_1(1)-a^{\ell}S_1(\ell)+a^{\ell+1}S_1(\ell-1)\end{aligned} \tag{72}$$

where the second line is obtained by writing out the real part of the summand as in (69). On the other hand, by expanding the summand we have

$$\begin{aligned}\mathrm{Re}\,\frac{1}{N}\sum_{n=0}^{N-1}\frac{1-(a\,\mathrm{e}^{\mathrm{i}\theta_n})^{\ell}}{1-a\,\mathrm{e}^{\mathrm{i}\theta_n}}&=\mathrm{Re}\,\frac{1}{N}\sum_{n=0}^{N-1}\sum_{m=0}^{\ell-1}a^m\exp(\mathrm{i}\pi mn/N)\\&=1+\mathrm{Re}\,\frac{1}{N}\sum_{m=1}^{\ell-1}a^m\left(\frac{1-(-1)^m}{1-\exp(\mathrm{i}\pi m/N)}\right)\\&=1+\frac{a(1-a^{\ell})}{N(1-a^2)}\qquad \ell=\mathrm{even}<2N\\&=1+\frac{a(1-a^{\ell-1})}{N(1-a^2)}\qquad \ell=\mathrm{odd}<2N\end{aligned} \tag{73}$$

where again we have used (25).

Equating (73) with (72) and using (68) and (71), we obtain the recursion relation

$$S_N(\ell)-aS_N(\ell-1)=Aa^{-\ell}+B_{\ell} \tag{74}$$

where

$$A=\frac{a^{2N}}{1-a^{2N}}\qquad B_{\ell}=\frac{a^{(1+(-1)^{\ell})/2}}{N(1-a^2)}. \tag{75}$$

The recursion relation (74) can be solved by standard means. Define the generating function

$$G_{\alpha}(t)=\sum_{\ell=0}^{\infty}S_{\alpha}(\ell)t^{\ell}\qquad \alpha=1,2. \tag{76}$$

Multiplying (74) by t^ℓ and sum over ℓ, we obtain

$$(1-at)G_1(t) - S_1(0) = \frac{Aa^{-1}t}{1-a^{-1}t} + \frac{t+at^2}{N(1-a^2)(1-t^2)}. \tag{77}$$

This leads to

$$\begin{aligned} G_1(t) &= \frac{1}{1-at}\left[S_1(0) + \frac{Aa^{-1}t}{1-a^{-1}t} + \frac{t+at^2}{N(1-a^2)(1-t^2)}\right] \\ &= \frac{1}{(1-a^2)(1-a^{2N})}\left[\frac{1}{1-at} + \frac{a^{2N}}{1-a^{-1}t}\right] \\ &\quad + \frac{1}{2N(1-a)^2(1-t)} - \frac{1}{2N(1+a)^2(1+t)} \end{aligned}$$

from which one obtains

$$\begin{aligned} S_1(\ell) &= \frac{a^\ell + a^{2N-\ell}}{(1-a^2)(1-a^{2N})} + \frac{1}{2N(1-a)^2} - \frac{(-1)^\ell}{2N(1+a)^2} \\ &= \frac{a^\ell + a^{2N-\ell}}{(1-a^2)(1-a^{2N})} + \frac{1}{2N}\left[\frac{4a}{(1-a^2)^2} + \frac{1-(-1)^\ell}{(1+a)^2}\right]. \end{aligned} \tag{78}$$

It follows that using $I_1(\ell) = 2aS_1(\ell)$ we obtain (61) after setting $a = \mathrm{e}^{-|\lambda|}$. □

Proof of (62). Again, we first evaluate $S_2(0)$ by carrying out the summation

$$\mathrm{Re}\,\frac{1}{N}\sum_{n=0}^{N-1}\frac{1}{1-a\,\mathrm{e}^{\mathrm{i}2\theta_n}} \qquad a<1 \tag{79}$$

in two different ways. First as in (69) we have

$$\mathrm{Re}\,\frac{1}{N}\sum_{n=0}^{N-1}\frac{1}{1-a\,\mathrm{e}^{\mathrm{i}2\theta_n}} = \frac{1}{2}[1+(1-a^2)S_2(0)] \tag{80}$$

where $S_2(\ell)$ is defined in (66). Secondly, by expanding the summand we have

$$\frac{1}{N}\sum_{n=0}^{N-1}\frac{1}{1-a\,\mathrm{e}^{\mathrm{i}2\theta_n}} = \frac{1}{N}\sum_{n=0}^{N-1}\sum_{\ell=0}^{\infty} a^\ell \exp(\mathrm{i}2\ell n\pi/N) = \frac{1}{1-a^N} \tag{81}$$

where by carrying out the summation over n for fixed ℓ all terms in (70) vanish except those with $\ell = mN, m = 0, 1, 2, \ldots$ Equating (81) with (80) we obtain

$$S_2(0) = \frac{1}{1-a^2}\left(\frac{1+a^N}{1-a^N}\right) \tag{82}$$

and from (68)

$$S_2(1) = \frac{1}{1-a^N}.$$

We consider next the summation

$$\mathrm{Re}\,\frac{1}{N}\sum_{n=0}^{N-1}\frac{1-(a\,\mathrm{e}^{\mathrm{i}2\theta_n})^\ell}{1-a\,\mathrm{e}^{\mathrm{i}2\theta_n}} \qquad a<1. \tag{83}$$

Evaluating the real part of the summand directly as in (72), we obtain

$$\mathrm{Re}\,\frac{1}{N}\sum_{n=0}^{N-1}\frac{1-(a\,\mathrm{e}^{\mathrm{i}2\theta_n})^\ell}{1-a\,\mathrm{e}^{\mathrm{i}2\theta_n}} = S_2(0) - aS_2(1) - a^\ell S_2(\ell) + a^{\ell+1}S_2(\ell-1). \tag{84}$$

Secondly, expanding the summand in (83) we obtain

$$\frac{1}{N}\sum_{n=0}^{N-1}\frac{1-(a\exp(\mathrm{i}2\theta_n)^\ell)}{1-a\exp(\mathrm{i}2\theta_n)} = \frac{1}{N}\sum_{n=0}^{N-1}\sum_{m=0}^{\ell-1} a^m \exp(\mathrm{i}2\pi mn/N)$$
$$= \frac{1}{N}\left[N + \sum_{m=1}^{\ell-1}\frac{1-\exp(\mathrm{i}2m\pi)}{1-\exp(\mathrm{i}2m\pi/N)}\right]$$
$$= 1 \qquad m < \ell \leqslant N. \tag{85}$$

Equating (85) and (84) and making use of (82) for $S_2(0)$, we obtain

$$S_2(\ell) - aS_2(\ell-1) = \frac{a^{N-\ell}}{1-a^N}. \tag{86}$$

The recursion relation (86) can be solved as in the above. Define the generating function $G_2(t)$ by (76). We find

$$G_2(t) = \frac{1}{1-at}\left[S_2(0) + \frac{a^{N-1}t}{(1-a^N)(1-a^{-1}t)}\right]$$
$$= \frac{1}{(1-a^2)(1-a^{2N})}\left[\frac{1}{1-at} + \frac{a^N}{1-a^{-1}t}\right] \tag{87}$$

from which one reads off

$$S_2(\ell) = \frac{a^\ell + a^{N-\ell}}{(1-a^2)(1-a^{2N})}. \tag{88}$$

Using the relation $I_2(\ell) = 2aS_2(\ell)$ with $a = \mathrm{e}^{-|\lambda|}$, we obtain (62). □

Acknowledgments

I would like to thank D H Lee for discussions and the hospitality at Berkeley where this work was initiated. I am grateful to W-J Tzeng for a critical reading of the manuscript and help in clarifying the Möbius strip and Klein bottle analyses, and to W T Lu for assistance in preparing the graphs. Work is supported in part by NSF Grant no DMR-9980440.

References

[1] Cserti J 2000 Application of the lattice Green's function for calculating the resistance of an infinite network of resistors *Am. J. Phys.* **68** 896–906 (*Preprint* cond-mat/9909120)

[2] Doyle P G and Snell J L 1984 *Random Walks and Electric Networks* (*The Carus Mathematical Monograph* series 22) (Washington, DC: The Mathematical Association of America)) pp 83–149 (*Preprint* math.PR/0001057)

[3] Lovász L 1996 *Random Walks on Graphs: A Survey, in Combinatorics, Paul Erdöis Eighty* vol 2, ed D Miklós, V T Sós and T Szónyi (Budepest: János Bolyai Mathematical Society) pp 353–98 (at http://research.microsoft.com/~lovasz/ as a survey paper)

[4] van der Pol B 1959 The finite-difference analogy of the periodic wave equation and the potential equation *Probability and Related Topics in Physical Sciences* (Lectures in Applied Mathematics vol 1) ed M Kac (London: Interscience) pp 237–57

[5] Redner S 2001 *A Guide to First-Passage Processes* (Cambridge: Cambridge University Press)

[6] Katsura S, Morita T, Inawashiro S, Horiguchi T and Abe Y 1971 Lattice Green's function: introduction *J. Math. Phys.* **12** 892–5

[7] Cserti J, Dávid G and Attila Piróth 2002 Perturbation of infinite networks of resistors *Am. J. Phys.* **70** 153–9 (*Preprint* cond-mat/0107362)

[8] Kirchhoff G 1847 Über die Auflösung der Gleichungen, auf welche man bei der Untersuchung der linearen Verteilung galvanischer Ströme geführt wird *Ann. Phys. Chem.* **72** 497–508

[9] See, for example, Harary F 1969 *Graph Theory* (Reading, MA: Addison-Wesley)

[10] Tzeng W-J and Wu F Y 2000 Spanning trees on hypercubic lattices and non-orientable surfaces *Appl. Math. Lett.* **13** (7) 19–25 (*Preprint* cond-mat/0001408)

[11] This product identity has previously been given in Grashteyn I S and Ryzhik I M 1965 *Table of Integrals, Summations and Products* (New York: Academic) 1.396.1

CHINESE JOURNAL OF PHYSICS VOL. 9, NO. 1 APRIL, 1971

On the Eigenvalues of Orbital Angular Momentum

DAVID M. KAPLAN

Department of Physics, Virginia Polytechnic Institute
Blacksburg, Viginia 24061, U.S.A.

and

F. Y. WU (伍法岳)+

Department of Physics, Northeastern University at
Boston, Boston, Massachusetts 02115, U.S.A.

(Received 30 April 1971)

Using only the elementary commutation relations in quantum mechanics, it is shown that the eigenvalues of $L_z = xp_y - yp_x$ are integers.

The eigenvalue problem for the orbital angular momentum operator

$$\boldsymbol{L} = \boldsymbol{r} \times \boldsymbol{p} \tag{1}$$

has been one of the least satisfactorily discussed topics in elementary quantum mechanics. In the discussions found in most of the textbooks,[1] one usully starts from the commutation relations implied by (1) and derives the result that the eigenvalues of L_z can only be half-integers $(0, \pm\frac{1}{2}, \pm 1, \pm\frac{3}{2} \cdots)$. This result is obtained by purely abstract considerations without any need for the use of fuction spaces. It is sufficient to simply use an abstract Hilbert space without demanding any specific realization of the space. At this point the problem of he elimination of the $\frac{1}{2}$ integral eigenvalues arises. This is usually done by gong outside the abstract Hilbert space framework and realizing (1) as an operator in a function space. Then with the help of some further restrictions, such as the single-valuedness requirement on the eigenfunction in the Schrodinger represntation,[1] one rules out the half-odd integral values $(\pm\frac{1}{2}, \pm\frac{3}{2}, \cdots)$. The necessity for inclusion of a physical constraint and the explicit use of a particular representation in the discussion of an eigenvalue problem has caused some uneasy feelings among physicists and has been a subject of considerable debate for many

---

+ Supported in part by National Science Foundation Grant, Nos. GP-9041 and GP-25306.

(1) See, for example, E. Merzbacher, *Quantum Mechanics* (John Wiley & Sons, Inc., New York, 1963), pp. 359 and 174.

years.[2-6] While the restriction can certainly be formulated in a variety of seemingly harmless statements,[7] it is nevertheless annoying to have the necessity for introducing such conditions. Several years ago Buchdahl[8] and Louck[9] gave independent derivations of the eigenvalues of L_z without using any requirement. While they both recognized the fact that the correct eigenvalues are implied by the particular form (1) of the orbital angular momentum operator, their arguments do not take the most elegant form. Besides being rather lengthy and quite indirect, their derivations involve the use of particular representations for the orbital angular momentum operator.[10] Shortly thereafter, Merzbacher[11] pointed out the connection between the two-dimensional harmonic oscillator and the angular momentum in three dimensions which provides, for the first time, a direct derivation of the correct eigenvalues.[12] However, as far as we know, this proof has never been adopted in any textbook of quantum mechanics, presumably because the ingenious trick involved is not an everyday tool familiar to all students.

We wish to present in this note another proof which seems to us to be more direct and simpler in structure and, therefore, more suitable for classroom presentations. The proof is abstract in structure depending only on the *form* of the operator L_z (i.e. that it is built in a specific way out of the operators $\underset{\sim}{r}$ and $\underset{\sim}{p}$) and on the fact we are (as in the general angular momentum theoy) working in an abstract Hilbert space.

First let us write L_z as

$$\begin{aligned} L_z &= xp_y - yp_x \\ &= C^+C - (A^+A + B^+B) \end{aligned} \tag{2}$$

where

$$A = \frac{1}{\sqrt{2}}(p_x - ix)$$

(2) W. Pauli, Helv. Phys. Acta **12**, 147 (1939).

(3) D. Bohm, *Quantum Theory* (Prentice-Hall, Inc., Englewood Cliffs. New Jersey, 1951), pp. 389–390.

(4) J. M. Blatt and V. F. Weisskopf, *Theoretical Nuclear Physics* (John Wiley & Sons, Inc., New York, 1952), pp. 783 and 787.

(5) E. Merzbacher, Am. J. Phys. **30**, 237 (1962).

(6) M. L. Whipman, Am. J. Phys. **34**, 656 (1966).

(7) For example, the comparison with experiments is considered in Ref. 3. The condition of the absence of source and sink for the probability current is mentioned in Ref. 4 and discussed in detail in Ref. 6. References to other considerations can also be found in Ref. 6.

(8) H. Buchdahl, Am. J. Phys. **30**, 829 (1962).

(9) J. D. Louck, Am. J. Phys. **31**, 378 (1963),

(10) An alternate derivation was also given by Louck (Ref. 9) for the operator (1) in the four-dimensional Cartesian space.

(11) E. Merzbacher, Am. J. Phys. **31**, 549 (1963).

(12) There also exist group-theoretical arguments which lead to the correct result. See, for example, J. Schwinger in *Quantum Theory of Angular Momentum*, L. C. Biedenharn and H. Van Dam, Eds. (Academic Press Inc., New York, 1965) and J. M. Lévy-Leblond, Am. J. Phys. **35**, 444 (1967).

$$B = \frac{1}{\sqrt{2}}(p_y - iy)$$

$$C = B + iA.$$

The following relations can then be readily established by using the commutation relations between $\boldsymbol{r}$ and $\boldsymbol{p}$ $(=1)$.

$$[A, A^+] = 1 \tag{3}$$

$$[B, B^+] = 1 \tag{4}$$

$$[C, C^+] = 2 \tag{5}$$

$$[A^+A, B^+B] = 0 \tag{6}$$

$$[C^+C, A^+A + B^+B] = 0. \tag{7}$$

The proof is based on the following results well-known to all students of quantum mechanics[13] which we now present as two lemmas. If A and B are two operators in a Hilbert space, then;

Lemma 1.

The commutation relation $[A, A^+] = \lambda$ implies that the eigenvalues of A^+A are $0, \lambda, 2\lambda, 3\lambda, \ldots.$

Lemma 2.

If A, B commute, then the eigenvalues of $A+B$ (or $A-B$) are some sums (or differences) of the eigenvalues of A and B.

From (3), (4) and Lemma 1, the eigenvalues of A^+A and B^+B are $0, 1, 2, \ldots.$ Hence by (6) and Lemma 2, the eigenalues of A^+A+B^+B are 0. 1,. 2, ... Sim ilarly from (5) and Lemma 1, the eigenvalues of C^+C are $0, 2, 4, \ldots.$ Hence from (7), Lemma 2, and the eigenvalues of A^+A+B^+B just deduced, the eigenvalves of L_z can only have positive of negative integral values including zero. This completes the proof.[14]

To summarize, we have shown that the operator L_z defined in a Hilbert space has integral eigenvalues only. The proof does not use any additional condition usually needed in the Schrodinger representation.

(13) Lemma 1 is proved in almost any elementary textbook in quantum mechanics. See, for example, pp. 349–351 of Ref. 1. Lemma 2 follows from the fact that commuting operators have simultaneous eigenvectors.

(14) Technically speaking, our proof only rules out the non-integral eigenvalues. But this is the desired result.

J. Phys. A: Math. Gen. **20** (1987) L299–L306. Printed in the UK

LETTER TO THE EDITOR

The vicious neighbour problem

R Tao and F Y Wu†

Department of Physics, Northeastern University, Boston, MA 02115, USA

Received 18 November 1986

Abstract. We compute the probability that a person will survive a shootout. The shootout involves N persons randomly placed in a d-dimensional space, each firing a single shot and killing his nearest neighbour with a probability p. We present a formulation which gives $P_N(p)$, the probability that a given person will survive, as a polynomial of p containing a finite number of terms. The coefficients appearing in the polynomial are explicitly evaluated for $d=1$ and $d=2$ in the limit of $N \to \infty$ to yield exact expressions for $P_\infty(p)$. In particular, $P_\infty(1)$ gives the probability that a given particle is *not* the nearest neighbour of any other particle in a classical ideal gas, and we further determine $P_\infty(1)$ for $d=3, 4$ and 5 using Monte Carlo simulations.

Consider N persons placed randomly in a bounded d-dimensional space. At a given instance, each person shoots, and kills, his nearest neighbour (called vicious neighbours) with a probability p. What is the fraction of persons who will survive the shootout in the limit of $N \to \infty$ and neglecting boundary corrections?

This problem of vicious neighbours, first posed by Abilock (1967) for $p=1$, has remained unsolved for almost two decades. The $d=2$ version of the $p=1$ problem re-appeared recently as a puzzle for which a prize was posted (Morris 1986, 1987). In this letter we present a solution to the general p problem for any spatial dimension d. More precisely, we present a formulation which gives $P_N(p)$, the fraction of persons who will survive the shootout, as a finite polynomial in p. We further show that coefficients of the polynomial are given in terms of finite-dimensional integrals in the limit of $N \to \infty$. For $d=1, 2$ these integrals are relatively simple and are explicitly evaluated to yield exact expressions for $P_\infty(p)$. For three and higher dimensions we compute $P_\infty(1)$ using independent Monte Carlo simulations.

We first summarise our findings for $p=1$, the problem originally proposed by Abilock (1967),

$$\begin{aligned} P_\infty(1) &= \tfrac{1}{4} && \text{for } d=1 \\ &= 0.284\,051\ldots && \text{for } d=2 \\ &= 0.303\ldots && \text{for } d=3 \text{ (Monte Carlo result)} \\ &= 0.318\ldots && \text{for } d=4 \text{ (Monte Carlo result)} \\ &= 0.328\ldots && \text{for } d=5 \text{ (Monte Carlo result).} \end{aligned} \tag{1}$$

Explicit expressions for $P_\infty(p)$ for $d=1$ and $d=2$ are given by (14) and (39).

† Work supported in part by NSF Grant DMR-8219254.

0305-4470/87/050299+08$02.50

It is convenient to regard the N persons as being particles in a many-body system. Then $P_N(p)$ is the probability that a given particle will survive the shootout, averaged over all particle configurations. As an example of possible application, $P_\infty(1)$ now gives the probability that a given particle is *not* the nearest neighbour of any other particle in a classical ideal gas. Our goal is to compute the thermodynamic limit

$$P_\infty(p) = \lim_{N\to\infty} P_N(p). \tag{2}$$

Number the particles from 0 to $N-1$ and consider the survival of particle 0. Each particle (other than 0) can be in one of two 'states': that it either kills, or does not kill, particle 0. Regard the occurrence of these two states as a probabilistic event and denote the probability that n particles, numbered $j_1, j_2, \ldots j_n$, *all* shoot (and kill) particle 0, regardless of the states of the other $N-n-1$ particles, by

$$p(j_1, j_2, \ldots, j_n) = p^n w(j_1, j_2, \ldots, j_n) \qquad n = 1, 2, \ldots, N-1 \tag{3}$$

where $w(j_i, j_2, \ldots, j_n)$ is the probability that the n particles $j_1, j_2, \ldots, j_n$ will find 0 as their common nearest neighbour. Then as a consequence of an identity in probability theory (Whitney 1932) we can express $P_N(p)$, the probability that all $N-1$ particles are in one state (of not killing 0), as a linear combination of $p(j_1, j_2, \ldots j_n)$, the probability that the n particles $j_1, j_2, \ldots, j_n$ are in the other state (*all* killing 0), as follows:

$$\begin{aligned} P_N(p) = 1 - \sum_{j=1}^{N-1} p(j) + \sum_{1\le j_i<j_2\le N-1} p(j_1, j_2) + \ldots \\ + \sum_{1\le j_1<\ldots<j_n\le N-1} (-1)^n p(j_1, \ldots, j_n) + \ldots \\ + (-1)^{N-1} p(1, 2, \ldots, N-1). \end{aligned} \tag{4}$$

Since all particles 1, 2, ..., $N-1$ are equivalent in the consideration of the survival of particle 0, we can write (4) as

$$P_N(p) = 1 + \sum_{n=1}^{N-1} C_n(-p)^n \tag{5}$$

where

$$C_n = \binom{N-1}{n} w(1, 2, \ldots, n). \tag{6}$$

The intriguing fact which allows the problem to be exactly soluble is that

$$w(1, 2, \ldots, n) = 0 \qquad \text{for } n > n_d \tag{7}$$

where n_d is a finite integer whose value depends on the spatial dimension d. That is, no more than n_d particles can simultaneously find particle 0 as their common nearest neighbour. It is easy to see this for $d=1$ since, for particles arranged on a line, there can be at most two particles having particle 0 as their nearest neighbours. Thus, we have $n_1 = 2$. In the case of $d=2$ we assume there exist n particles, numbered $1, 2, \ldots, n$, all having particle 0 as their nearest neighbours (cf figure 1). For particles 1 and 2 to have particle 0 as their nearest neighbour, we must have $r_{12} > r_{01}$ and r_{02}, where r_{ij} is the distance between particles i and j, and, consequently, $\theta_1 > \pi/3$, where θ_1 is the angle between $\boldsymbol{r}_1$ and $\boldsymbol{r}_2$. Similarly we find $\theta_i > \pi/3$, $i = 2, 3, \ldots, m$, for the other $n-1$ angles. The sum rule $\Sigma_{j=1}^n \theta_j = 2\pi$ now implies that $n \le 5$ and hence $n_2 = 5$.

Generally, the integer n_d for $d \ge 2$ is bounded by the maximum number of d-dimensional regular $(d+1)$-polyhedra that can be fitted together such that they all

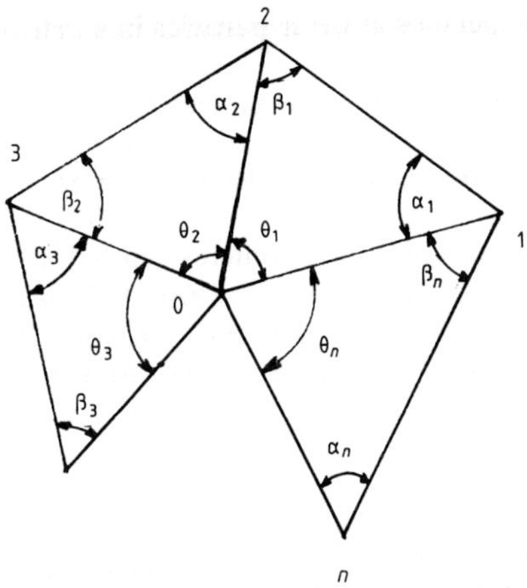

Figure 1. Configuration showing that n particles have particle 0 as their common nearest neighbour.

have the origin as a common vertex and there is still room for the polyhedra to rotate slightly about the origin without spoiling the fit. In three dimensions one can fit at most 22 regular tetrahedra at the origin without exhausting the whole solid angle 4π (Coxeter 1969). It follows that n_3 cannot be greater than 22.

Up to this point we have regarded N finite and have not considered the fact that the region confining the N particles is bounded. Let Ω be the volume of the region. We shall take the thermodynamic limit $N\to\infty$, $\Omega\to\infty$ with the density $\rho=N/\Omega$ held constant, a limit we denote by $N\to\infty$ for brevity. While there is no intrinsic length in the problem, so that the final result is expected to be independent of ρ, the introduction of the density ρ is a convenient tool which enables us to take the limit appropriately.

It is relatively easy to see that

$$\lim_{N\to\infty} C_1 \equiv (N-1)w(1) = 1 \qquad \text{for all } d. \tag{8}$$

This is so since $w(1)$, the probability that r_{01} is the shortest among the $N-1$ distances r_{i1}, $i=0,2,3,\ldots,N-1$, is $1/(N-1)$ after the boundary corrections are ignored.

Consider next the evaluation of $C_2=\binom{N-1}{2}w(1,2)$, where $w(1,2)$ is the probability that both particles 1 and 2 have particle 0 as their nearest neighbours. For this to happen we must have $r_1, r_2<r_{12}$ and, in addition, $r_1<r_{i1}$, $r_2<r_{i2}$, for $i=3,4,\ldots,N-1$. Let $S_2(r_1, r_2, \Omega)$ be the volume common to Ω and the union of two spheres centred at $\boldsymbol{r}_2$ and $\boldsymbol{r}_3$ with respective radii r_2 and r_3 (thus both passing through the origin). Then, since $N-3$ particles must stay outside S_2, we have

$$C_2=\frac{(N-1)(N-2)}{2!}\int_{r_1,r_2<r_{12}}\frac{\mathrm{d}\boldsymbol{r}_1}{\Omega}\frac{\mathrm{d}\boldsymbol{r}_2}{\Omega}\left(1-\frac{S_2(\boldsymbol{r}_1,\boldsymbol{r}_2,\Omega)}{\Omega}\right)^{N-3}. \tag{9}$$

Taking the thermodynamic limit now leads to

$$\begin{aligned} c_2 \equiv \lim_{N\to\infty} C_2 &= \frac{1}{2!}\rho^2\int_{r_1,r_2<r_{12}}\mathrm{d}\boldsymbol{r}_1\,\mathrm{d}\boldsymbol{r}_2\exp[-\rho V_2(\boldsymbol{r}_1,\boldsymbol{r}_2)] \\ &= \frac{1}{2!}\int_{r_1,r_2<r_{12}}\mathrm{d}\boldsymbol{r}_1\,\mathrm{d}\boldsymbol{r}_2\exp[-V_2(\boldsymbol{r}_1,\boldsymbol{r}_2)] \end{aligned} \tag{10}$$

where $V_2(\boldsymbol{r}_1, \boldsymbol{r}_2)$ is the volume occupied by the two aforementioned spheres, a situation shown in figure 2 for $d=2$.

Proceeding in the same fashion we obtain, quite generally,

$$c_n \equiv \lim_{N\to\infty} C_n = \frac{1}{n!}\int_{r_i<r_{ij}} \mathrm{d}\boldsymbol{r}_1 \ldots \mathrm{d}\boldsymbol{r}_n \exp[-V_n(\boldsymbol{r}_1, \ldots, \boldsymbol{r}_n)] \tag{11}$$

where $V_n(\boldsymbol{r}_1, \ldots, \boldsymbol{r}_n)$ is the volume occupied by n spheres centred at $\boldsymbol{r}_1, \ldots, \boldsymbol{r}_n$ and intersecting at the origin. Here, the restrictions $r_i < r_{ij}$, $i \neq j = 1, 2, \ldots, n$ ensure that particle 0 is the common nearest neighbour of particles $i = 1, 2, \ldots, n$. Finally, the survival probability $P_\infty(p)$ is obtained by combining (2) and (5) as

$$P_\infty(p) = 1 - p + c_2p^2 - c_3p^3 + \ldots + c_{n_d}(-p)^{n_d} \tag{12}$$

where c_n is given by (11).

In one dimension we have $n_1 = 2$ and hence, from (10) and (12),

$$P_\infty(p) = 1 - p + \tfrac{1}{2}p^2 \int_{x_1, x_2 < x_{12}} \mathrm{d}x_1\, \mathrm{d}x_2 \exp[-V_2(x_1, x_2)] \tag{13}$$

where $x_{12} = |x_1 - x_2|$ and the two integrations range from $-\infty$ to ∞. Now the restriction $x_1, x_2 < x_{12}$ implies that x_1 and x_2 must have opposite signs and, therefore, $V_2(x_1, x_2) = 2(|x_1| + |x_2|)$. The integrations in (13) are thus simply carried out, yielding

$$P_\infty(p) = 1 - p + 2\times\tfrac{1}{2}p^2 \int_{-\infty}^{0} \exp(-2|x_1|)\, \mathrm{d}x_1 \int_0^\infty \exp(-2|x_2|)\, \mathrm{d}x_2 = 1 - p + \tfrac{1}{4}p^2 \tag{14}$$

which, for $p = 1$, reduces to $\frac{1}{4}$, the result quoted in (1).

We now evaluate (12) in two dimensions, where $n_2 = 5$, term by term by explicitly reducing the coefficients c_i into quadratures.

The integrand in (11) is invariant under permutations of the n vectors $\boldsymbol{r}_1, \boldsymbol{r}_2, \ldots, \boldsymbol{r}_n$. This permits us to focus on a particular arrangement of the vectors, such as the one shown in figure 1, which occurs $(n-1)!$ times. The key to the reduction of (11) lies in the observation that, for r_i fixed, the constraints r_i, $r_{i+1} < r_{i,i+1}$ effectively restrict $t_{i+1} \equiv r_{i+1}/r_i$ to range from $t_{\min}(\theta_i)$ to $t_{\max}(\theta_i)$, where θ_i is the angle between the vectors $\boldsymbol{r}_i$ and $\boldsymbol{r}_{i+1}$, and with

$$\{t_{\min}(\theta_i), t_{\max}(\theta_i)\} = \begin{cases} \{2\cos\theta_i, (2\cos\theta_i)^{-1}\} & \text{if } \pi/3 < \theta_i < \pi/2 \\ \{0, \infty\} & \text{if } \pi/2 < \theta_i < 3\pi/2. \end{cases} \tag{15}$$

It is therefore important to treat the cases of $\theta_i < \frac{1}{2}\pi$ and $\theta_i > \frac{1}{2}\pi$ separately.

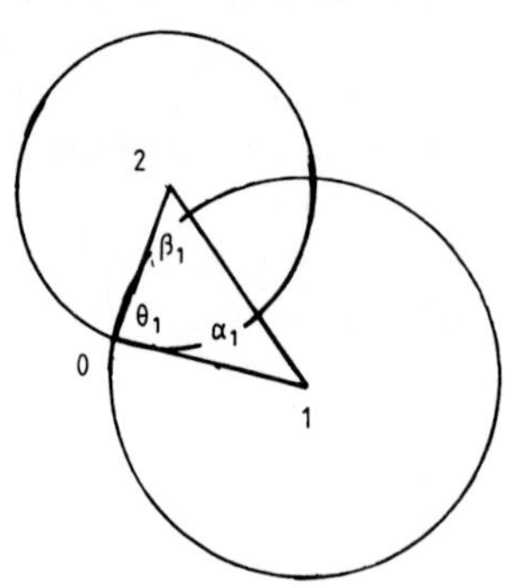

Figure 2. Configuration showing the area occupied by two circles.

Consider first the integrations in (11) over the directions of $\boldsymbol{r}_1, \boldsymbol{r}_2, \ldots, \boldsymbol{r}_n$. We decompose the phase space of these angular integrations into regions according to whether each of the n angles θ_i is greater or less than $\frac{1}{2}\pi$. This decomposition is facilitated by assigning a two-valued variable σ_i to the angle θ_i such that $\sigma_i = 0$ if $\theta_i > \frac{1}{2}\pi$ and $\sigma_i = 1$ if $\theta_i < \frac{1}{2}\pi$. Let $I(\sigma_1, \sigma_2, \ldots, \sigma_n)$ be the contribution to c_n with $\{\theta_1, \ldots, \theta_n\}$ in the range specified by $\sigma_1, \sigma_2, \ldots, \sigma_n$. Clearly, I is invariant under cyclic permutations of its arguments, i.e. $I(\sigma_1, \ldots, \sigma_n) = I(\sigma_2, \ldots, \sigma_n, \sigma_1)$, etc. After taking this degeneracy into consideration, we find

$$
\begin{aligned}
c_2 &= I(0,0) + 2I(1,0) \\
c_3 &= I(0,0,0) + 3I(1,0,0) + 3I(1,1,0) \\
c_4 &= 4I(1,0,0,0) + 4I(1,1,0,0) + 2I(1,0,1,0) + 4I(1,1,1,0) \\
c_5 &= 5I(1,1,1,1,0) + I(1,1,1,1,1)
\end{aligned}
\tag{16}
$$

where we have used the fact that each θ_i must range between $\frac{1}{3}\pi$ and $\frac{5}{3}\pi$ and thus, e.g., there can be only two ways to fit five angles in the case of $n = 5$.

Consider next the n radial integrations over $\mathrm{d}r_1, \ldots, \mathrm{d}r_n$. Using the fact that the volume $V_n(\boldsymbol{r}_1, \ldots, \boldsymbol{r}_n)$ is homogeneous and quadratic in $r_1, \ldots, r_n$, we can write for each term in (16)

$$V_n = r_1^2 V_n(t_2, t_3, \ldots, t_n; \theta_1, \theta_2, \ldots, \theta_{n-1}) \tag{17}$$

where $t_i \equiv r_i / r_{i-1}$. Thus, after introducing the variables t_i, $i = 2, \ldots, n$, into the integrand, the integration over r_1 can be carried out, yielding a factor

$$2\pi \int_0^\infty r_1^{2n-1}\, \mathrm{d}r_1 \exp[-r_1^2 V_n(t_2, t_3, \ldots, t_n; \theta_1, \theta_2, \ldots, \theta_{n-1})] = \pi(n-1)!(V_n)^{-n}. \tag{18}$$

Thus we find, for c_2,

$$I(0,0) = \pi \int_{\pi/2}^{3\pi/2} \mathrm{d}\theta_1 \int_0^\infty t_2\, \mathrm{d}t_2 [V_2(t_2; \theta_1)]^{-2} = 0.258\,572\,168\ldots \tag{19}$$

$$I(1,0) = \pi \int_{\pi/3}^{\pi/2} \mathrm{d}\theta_1 \int_{2\cos\theta_1}^{(2\cos\theta_1)^{-1}} t_2\, \mathrm{d}t_2 [V_2(t_2; \theta_1)]^{-2} = 0.028\,880\,652\ldots \tag{20}$$

where

$$V_2(t_2; \theta_1) = z_1 + t_2^2 z_2 \tag{21}$$

and

$$
\begin{aligned}
z_1 &= \pi - \alpha_1 + \tfrac{1}{2}\sin(2\alpha_1) \\
z_2 &= \pi - \beta_2 + \tfrac{1}{2}\sin(2\beta_2)
\end{aligned}
\tag{22}
$$

α_1 and β_2 being the angles shown in figure 2 and given by, with $i = 1$,

$$
\begin{aligned}
\sin\alpha_i &= t_{i+1} \sin\theta_i (1 + t_{i+1}^2 - 2t_{i+1}\cos\theta_i)^{-1/2} \\
\sin\beta_{i+1} &= \sin\theta_i (1 + t_{i+1}^2 - 2t_{i+1}\cos\theta_i)^{-1/2}.
\end{aligned}
\tag{23}
$$

Substitution of (19) and (20) into (16) now gives

$$c_2 = 0.316\,3335\ldots. \tag{24}$$

In a similar fashion we find, for c_3,

$$I(0,0,0)=\frac{2\pi}{3}\int_{\pi/2}^{\pi}\mathrm{d}\theta_1\int_{\pi/2}^{3\pi/2-\theta_1}\mathrm{d}\theta_2\int_0^{\infty}t_2^3\,\mathrm{d}t_2\int_0^{\infty}t_3\,\mathrm{d}t_3[V_3]^{-3}=0.011\,207\,724\ldots \tag{25}$$

$$I(1,0,0)=\frac{2\pi}{3}\int_{\pi/3}^{\pi/2}\mathrm{d}\theta_1\int_{\pi/2}^{3\pi/2-\theta_1}\mathrm{d}\theta_2\int_{2\cos\theta_1}^{(2\cos\theta_1)^{-1}}t_2^3\,\mathrm{d}t_2\int_0^{\infty}t_3\,\mathrm{d}t_3[V_3]^{-3}$$
$$=0.005\,621\,972\ldots \tag{26}$$

$$I(1,1,0)=\frac{2\pi}{3}\int_{\pi/3}^{\pi/2}\mathrm{d}\theta_1\int_{\pi/3}^{\pi/2}\mathrm{d}\theta_2\int_{2\cos\theta_1}^{(2\cos\theta_1)^{-1}}t_2^3\,\mathrm{d}t_2\int_{2\cos\theta_2}^{(2\cos\theta_2)^{-1}}t_3\,\mathrm{d}t_3[V_3]^{-3}$$
$$=0.001\,168\,842\ldots \tag{27}$$

where, quite generally,

$$V_n=z_1+t_2^2z_2+\ldots+(t_2\ldots t_n)^2z_n \qquad (n=3,4,5) \tag{28}$$

$$z_i=\pi-\alpha_i+\tfrac{1}{2}\sin 2\alpha_i-\beta_{i-1}+\tfrac{1}{2}\sin 2\beta_{i-1} \qquad (i=1,2,\ldots,n). \tag{29}$$

Here the angles α_i and β_i, shown in figure 1, are related to the integration variables through (23), with $\theta_n=2\pi-\theta_1-\theta_2-\ldots-\theta_{n-1}$, and subject to the constraints $\beta_0\equiv\beta_n$. Special care must be taken for $n=3$, a situation shown in figure 3, for which we must set $\alpha_2=\beta_2=0$ if $\theta_2>\pi$ and $\alpha_3=\beta_3=0$ if $\theta_1+\theta_2<\pi$.

Substitution of (25)-(27) into (16) now gives

$$c_3=0.032\,9390\ldots. \tag{30}$$

Similarly, for c_4 we find

$$I(1,0,0,0)=\frac{3\pi}{2}\int_{\pi/3}^{\pi/2}\mathrm{d}\theta_1\int_{\pi/2}^{\pi-\theta_1}\mathrm{d}\theta_2\int_{\pi/2}^{3\pi/2-\theta_1-\theta_2}\mathrm{d}\theta_3$$
$$\times\int_{2\cos\theta_1}^{(2\cos\theta_1)^{-1}}t_2^5\,\mathrm{d}t_2\int_0^{\infty}t_3^3\,\mathrm{d}t_3\int_0^{\infty}t_4\,\mathrm{d}t_4[V_4]^{-4} \tag{31}$$

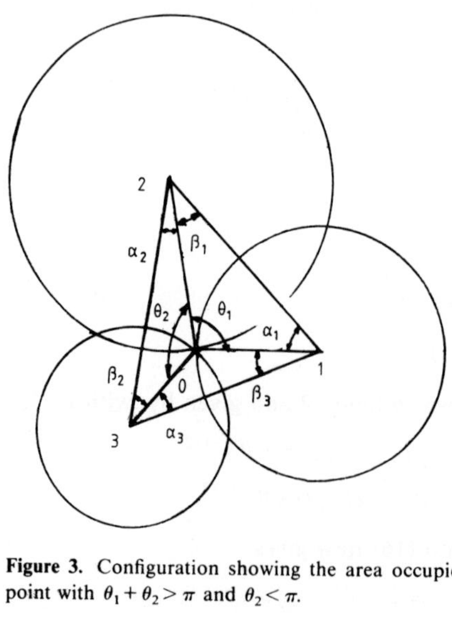

Figure 3. Configuration showing the area occupied by three circles intersecting at one point with $\theta_1+\theta_2>\pi$ and $\theta_2<\pi$.

$$I(1,1,0,0)=\frac{3\pi}{2}\int_{\pi/3}^{\pi/2}\mathrm{d}\theta_1\int_{\pi/3}^{\pi/2}\mathrm{d}\theta_2\int_{\pi/2}^{3\pi/2-\theta_1-\theta_2}\mathrm{d}\theta_3\int_{2\cos\theta_1}^{(2\cos\theta_1)^{-1}}t_2^5\,\mathrm{d}t_2$$
$$\times\int_{2\cos\theta_2}^{(2\cos\theta_2)^{-1}}t_3^3\,\mathrm{d}t_3\int_0^\infty t_4\,\mathrm{d}t_4[V_4]^{-4} \tag{32}$$

$$I(1,0,1,0)=\frac{3\pi}{2}\int_{\pi/3}^{\pi/2}\mathrm{d}\theta_1\int_{\pi/2}^{7\pi/6-\theta_1}\mathrm{d}\theta_2\int_{\pi/3}^{\min\{3\pi/2-\theta_1-\theta_2,\pi/2\}}\mathrm{d}\theta_3$$
$$\times\int_{2\cos\theta_1}^{(2\cos\theta_1)^{-1}}t_2^5\,\mathrm{d}t_2\int_0^\infty t_3^3\,\mathrm{d}t_3\int_{2\cos\theta_3}^{(2\cos\theta_3)^{-1}}t_4\,\mathrm{d}t_4[V_4]^{-4} \tag{33}$$

$$I(1,1,1,0)=\frac{3\pi}{2}\int_{\pi/3}^{\pi/2}\mathrm{d}\theta_1\int_{\pi/3}^{\pi/2}\mathrm{d}\theta_2\int_{\pi/3}^{\pi/2}\mathrm{d}\theta_3$$
$$\times\int_{2\cos\theta_1}^{(2\cos\theta_1)^{-1}}t_2^5\,\mathrm{d}t_2\int_{2\cos\theta_2}^{(2\cos\theta_2)^{-1}}t_3^3\,\mathrm{d}t_3\int_{2\cos\theta_3}^{(2\cos\theta_3)^{-1}}\mathrm{d}t_4[V_4]^{-4}. \tag{34}$$

Numerical evaluation of (31)-(34) yields

$$c_4=0.000\,6575\ldots. \tag{35}$$

In the evaluation of c_5 we have

$$I(1,1,1,1,0)=\frac{24\pi}{5}\int_{\pi/3}^{\pi/2}\mathrm{d}\theta_1\int_{\pi/3}^{\pi/2}\mathrm{d}\theta_2\int_{\pi/3}^{\pi/2}\mathrm{d}\theta_3\int_{\min\{\pi/3,3\pi/2-\theta_1-\theta_2-\theta_3\}}^{3\pi/2-\theta_1-\theta_2-\theta_3}\mathrm{d}\theta_4$$
$$\times\int_{2\cos\theta_1}^{(2\cos\theta_1)^{-1}}t_2^7\,\mathrm{d}t_2\int_{2\cos\theta_2}^{(2\cos\theta_2)^{-1}}t_3^5\,\mathrm{d}t_3\int_{2\cos\theta_3}^{(2\cos\theta_3)^{-1}}t_4^3\,\mathrm{d}t_4$$
$$\times\int_{2\cos\theta_4}^{(2\cos\theta_4)^{-1}}t_5\,\mathrm{d}t_5[V_5]^{-5} \tag{36}$$

$$I(1,1,1,1,1)=\frac{24\pi}{5}\int_{\pi/3}^{\pi/2}\mathrm{d}\theta_1\int_{\pi/3}^{\pi/2}\mathrm{d}\theta_2\int_{\pi/3}^{\pi/2}\mathrm{d}\theta_3$$
$$\times\int_{\min\{\max\{\pi/3,3\pi/2-\theta_1-\theta_2-\theta_3\},5\pi/3-\theta_1-\theta_2-\theta_3\}}^{\min\{\pi/2,5\pi/3-\theta_1-\theta_2-\theta_3\}}\mathrm{d}\theta_4\int_{2\cos\theta_1}^{(2\cos\theta_1)^{-1}}t_2^7\,\mathrm{d}t_2$$
$$\times\int_{2\cos\theta_2}^{(2\cos\theta_2)^{-1}}t_3^5\,\mathrm{d}t_3\int_{2\cos\theta_3}^{(2\cos\theta_3)^{-1}}t_4^3\,\mathrm{d}t_4\int_{\max\{2\cos\theta_4,2\cos\theta_5\}}^{\min\{(2\cos\theta_4)^{-1},(2\cos\theta_5)^{-1}\}}t_5\,\mathrm{d}t_5[V_5]^{-5}. \tag{37}$$

Numerical evaluation of (36) and (37) gives

$$c_5=0.000\,0010\ldots. \tag{38}$$

Finally, upon combining (12), (24), (30), (35) and (38), we obtain

$$P_\infty(p)=1-p+0.316\,3335p^2-0.032\,9390p^3+0.000\,6575p^4-0.000\,0010p^5 \tag{39}$$

which, for $p=1$, reduces to 0.284 051 ..., the result quoted in (1).

The evaluation of $P_\infty(p)$ given by (12) can, in principle, be carried for any d. For $d=3$, for example, we replace circles by spheres in the above consideration and it is necessary to evaluate 21 terms at most in (12), each of which is a multidimensional integral. However, these integrals are fairly complicated and, instead, we have used independent Monte Carlo simulations to obtain estimates of $P_\infty(1)$. Simulations on a VAX computer for systems consisting of up to 10 000 particles yield the results in (1). To check the accuracy of our simulations, we applied the same procedure to the $d=2$ system and obtained the number $P_\infty(1)=0.284\pm0.003$, in excellent agreement with the exact result (39).

Note added. After the submission of this letter, Veit Elser and Friend Kierstead Jr have called our attention to the known fact that $n_3=12$. Dr Elser also provided upper bounds on n_d for d up to 24.

References

Abilock R 1967 *Am. Math. Mon.* **74** 720
Coxeter H S M 1969 *Introduction to Geometry* (New York: Wiley)
Morris S 1986 *Omni* **8** no 4, 113
—— 1987 *Omni* **9** to appear
Whitney H 1932 *Bull. Am. Math. Soc.* **38** 572–9

VOLUME 76, NUMBER 2 PHYSICAL REVIEW LETTERS 8 JANUARY 1996

Directed Compact Lattice Animals, Restricted Partitions of an Integer, and the Infinite-State Potts Model

F. Y. Wu, G. Rollet, and H. Y. Huang
Department of Physics, Northeastern University, Boston, Massachusetts 02115

J. M. Maillard
Laboratoire de Physique Théorique et Hautes Energies, Tour 16, 1er étage, 4 place Jussieu, 75252 Paris Cedex, France

Chin-Kun Hu and Chi-Ning Chen
Institute of Physics, Academia Sinica, Nankang, Taipei, Taiwan 11529, Republic of China
(Received 14 April 1995)

We consider a directed compact site lattice animal problem on the d-dimensional hypercubic lattice, and establish its equivalence with (i) the infinite-state Potts model and (ii) the enumeration of $(d-1)$-dimensional restricted partitions of an integer. The directed compact lattice animal problem is solved exactly in $d = 2,3$ using known solutions of the enumeration problem. The maximum number of lattice animals of size n grows as $\exp(cn^{(d-1)/d})$. Also, the infinite-state Potts model solution leads to a conjectured limiting form for the generating function of restricted partitions for $d > 3$, the latter an unsolved problem in number theory.

PACS numbers: 05.50.+q

An intriguing aspect of lattice statistics is that seemingly totally different problems are sometimes related to each other, and that the solution of one problem can often be used to solve other outstanding unsolved problems. An example is the $d = 2$ directed lattice site animals solved by Dhar [1] who used Baxter's exact solution of a hard-square lattice gas model [2,3] to deduce its solution. In this Letter we consider a directed *compact* site lattice animal problem in d dimensions, and show that it is related to (i) the infinite-state Potts model in d dimensions and (ii) the enumeration of $(d-1)$-dimensional restricted partitions of an integer. The known solutions of restricted partitions in two and three dimensions [4,5] now solve the corresponding compact lattice animal problems, and, similarly, the established solution of the infinite-state Potts model [6] leads to a conjectured limiting form for the generating function of restricted partitions for $d > 3$, which is an outstanding unsolved problem in number theory. For clarity of presentation, we present details of discussions for $d = 2$. Considerations in higher dimensions are similar, and relevant results will be given.

Directed compact lattice animals and restricted partitions of an integer.—Starting from the origin $\{1,1\}$ of an $L_1 \times L_2$ simple quartic lattice $\mathcal{L}$ whose columns and rows are numbered, respectively, by $i = 1,\ldots,L_1$ and $j = 1,\ldots,L_2$, a site animal grows in the directions of increasing i and j. In contrast to the previously considered directed animal problem [1] for which a site $\{i,j\}$ can be occupied when either the site $\{i-1,j\}$ or the site $\{i,j-1\}$ is occupied, we introduce a more restricted growth rule. Our rule is that a site $\{i,j\}$ can be occupied only when both $\{i-1,j\}$ and $\{i,j-1\}$ are occupied. (When applying the growth rule, sites with coordinates $i = 0$ or $j = 0$ are regarded as being occupied.) Compared to the usual directed lattice animals [1], the present model generates compact animals since it excludes configurations with unoccupied interior sites. In addition, we keep L_1, L_2 finite, so that there exists a maximum animal size of L_1L_2.

Let $A_n(L_1,L_2)$ be the number of distinct n-site compact animals that can grow on $\mathcal{L}$. In considering animal problems, one is primarily interested in finding the asymptotic behavior $A_n(L_1,L_2)$ for large n. It is clear that by keeping L_1, L_2 finite the asymptotic behavior will depend on the relative magnitudes of n, L_1, L_2. The study of enumerations is facilitated by the use of generating functions. In the present case we define the generating function

$$G(L_1,L_2;t) = 1 + \sum_{n=1}^{L_1L_2} A_n(L_1,L_2)t^n. \tag{1}$$

For example, the generating function for the 3×3 lattice is

$$G(3,3;t) = 1 + t + 2t^2 + 3t^3 + 3t^4 + 3t^5 + 3t^6 + 2t^7 + t^8 + t^9. \tag{2}$$

We observe that $A_n(L_1,L_2)$ reaches a maximum at $n \sim L_1L_2/2$.

Let h_i, $i = 1,2,\ldots,L_1$, be the number of occupied sites in the ith column of $\mathcal{L}$. One observes that our growth rule implies the restriction

$$L_2 \ge h_1 \ge h_2 \ge \cdots \ge h_{L_1} \ge 0. \tag{3}$$

In addition, one has the (one-dimensional) sum rule

$$\sum_{i=1}^{L_1} h_i = n, \tag{4}$$

where n is the animal size. It is convenient to regard (4) as specifying the partitions of a positive integer n into

0031-9007/96/76(2)/173(4)$06.00

VOLUME 76, NUMBER 2 PHYSICAL REVIEW LETTERS 8 JANUARY 1996

sums of integers h_i, and the condition (3) ensures that all partitions are distinct. Then $A_n(L_1, L_2)$ is precisely the number of distinct ways that an integer n is partitioned into at most L_1 parts with each part less than or equal to L_2. This leads to a classic restricted *partitio numerorum* problem dating back to Gauss [4]. Particularly, the generating function (1) can be evaluated in a closed form [5,7]

$$G(L_1, L_2; t) = (t)_{L_1+L_2}/(t)_{L_1}(t)_{L_2}, \tag{5}$$

where $(t)_L \equiv \prod_{p=1}^{L}(1 - t^p)$. Note that, despite its appearance, all zeros in the denominator are canceled and (5) is a true polynomial in t as shown in (2). The generating function (1) is known as the Gaussian polynomial or the "q coefficient."

There are $L_1L_2 + 1$ terms in (1) whose coefficients satisfy the sum rule

$$1 + \sum_{n=1}^{L_1L_2} A_n(L_1, L_2) = \binom{L_1 + L_2}{L_1} \tag{6}$$

and the symmetry

$$G(L_1, L_2; t) = t^{L_1L_2} G(L_1, L_2; t^{-1}). \tag{7}$$

While these two properties are relatively easy to prove [5], the Gaussian polynomial possesses a unimodal property, namely, $A_{n-1}(L_1, L_2) < A_n(L_1, L_2)$ for $n \leq L_1L_2/2$, which is very deep. A combinatorial proof of this unimodal property appeared only very recently [8].

The Gaussian polynomial can be inverted by the Cauchy integral to yield

$$A_n(L_1, L_2) = \frac{1}{2\pi i} \oint \frac{1}{z^{n+1}} G(L_1, L_2; z)\, dz, \tag{8}$$

where the integration is taken over a contour inside $|z| = 1$, enclosing the origin. The asymptotic behavior of $A_n(L_1, L_2)$ for large n can be deduced by applying saddle point analyses to (8). For $n < \min\{L_1, L_2\}$, the rows and columns of $\mathcal{L}$ are never fully filled so that the partition of n is actually without restrictions. Then, the classic analysis of (8) by Rademacher [9] with $G(L_1, L_2; z)$ effectively replaced by the Eulerian product $(t)_\infty^{-1}$ yields the celebrated Hardy-Ramanujan [10] result

$$A_n(L_1, L_2) \sim \frac{1}{4n\sqrt{3}} \exp\left(\pi\sqrt{\frac{2n}{3}}\right), \quad n < \min\{L_1, L_2\}. \tag{9}$$

Clearly, the asymptotic behavior of $A_n(L_1, L_2)$ changes as n increases, and the partition of n becomes more restricted. When $A_n(L_1, L_2)$ assumes its maximum value at $n = L_1L_2/2$ (the unimodal property), the leading contribution can be obtained by observing that the left-hand side of (6) consists of $L_1L_2 + 1$ *positive* terms of which the largest term is of the order of $e^{c\sqrt{n}}$, where c is a constant. It follows that to the leading order the largest term is well approximated by the sum $\binom{L_1+L_2}{L_1}$. This leads to the asymptotic behavior

$$A_n(L_1, L_2) \propto e^{c(\alpha)\sqrt{n}}, \quad n \sim L_1L_2/2, \tag{10}$$

where

$$c(\alpha) = \sqrt{2}\left[\sqrt{\frac{1}{\alpha}}\ln(1+\alpha) + \sqrt{\alpha}\ln\left(1+\frac{1}{\alpha}\right)\right] = c(\alpha^{-1}),$$
$$\alpha = L_1/L_2. \tag{11}$$

Assuming a Gaussian distribution for $A_n(L, L)$ near its center $n = L^2/2$ ($\alpha = 1$), it can be shown [11] that

$$A_n(L, L) \sim \left(\frac{\sqrt{3}}{2\pi n}\right) 2^{2\sqrt{2n}}, \quad n \sim L^2/2. \tag{12}$$

It therefore appears safe to conclude that the asymptotic behavior of $A_n(L_1, L_2)$ assumes the universal form of $n^{-1}e^{c\sqrt{n}}$, where c is a constant which decreases with increasing n. The initial value of c is $\pi\sqrt{2/3} = 2.5651\ldots$ for $n < \{L_1, L_2\}$. Its value decreases to $c(\alpha) < c(1) = 2\sqrt{2}\ln 2 = 1.9605\ldots$ for $n \sim L_1L_2/2$ and eventually to 0 for $n \sim L_1L_2$ when the lattice is fully occupied. This is to be compared with the asymptotic behavior $3^n n^{-3/2}$ of the usual directed animals [1].

Equivalence with an infinite-state Potts model.— Consider the standard Potts model with reduced nearest-neighbor interactions K on an $L_1 \times L_2$ simple quartic lattice with the special boundary conditions shown in Fig. 1. It has been recently conjectured [12] that zeros of the Potts partition function on this (self-dual) lattice lie on the unit circle $|x| = 1$ in the $\mathrm{Re}(x) > 0$ half plane for all L_1 and L_2, where $x = (e^K - 1)/q^{1/2}$. As a by-product of our analysis, we shall establish this conjecture in the $q = \infty$ limit.

The high-temperature expansion of the Potts partition function assumes the form [13]

$$Z_{L_1,L_2}(q, x) = \sum_{\text{bond config}} x^b q^{n+b/2}, \tag{13}$$

where the summation is taken over all $2^{2L_1L_2}$ bond-covering configurations of the lattice; b and n are, respectively, the numbers of bonds and connected clusters (including isolated points) of each configuration. In the large q limit, the leading terms in (13) are of the order of $q^{L_1L_2+1}$. This factor can be achieved by taking, for example, the fully covered bond configuration of $n = 1, b = 2L_1L_2$ with the weight $q^{L_1L_2+1}x^{2L_1L_2}$. It is then convenient to introduce the reduced ($q = \infty$) Potts partition function

$$Y_{L_1,L_2}(x) \equiv \lim_{g\to\infty} q^{-(L_1L_2+1)} Z_{L_1,L_2}(q, x). \tag{14}$$

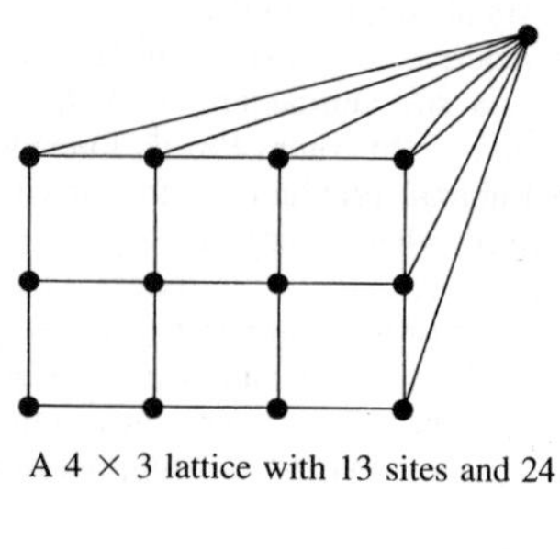

FIG. 1. A 4×3 lattice with 13 sites and 24 edges.

We now establish the identity

$$\begin{aligned} Y_{L_1,L_2}(x) &= x^{2L_1L_2}G(L_1,L_2;x^{-2}) \\ &= G(L_1,L_2;x^2). \end{aligned} \quad (15)$$

To prove (15), we consider the generation of Y_{L_1,L_2} from a systematic removal of bonds starting from the fully covered configuration. Generally, to hold the number $n + b/2$ constant, the minimum one can do is to decrease b by 2 while increasing n by 1. Thus, one always looks for sites connected to exactly two neighboring sites. Starting from the fully covered configuration, one observes from Fig. 1 that there is only one such site, namely, the site {1,1} at the lower-left corner, which is connected to the two sites {1,2} and {2,1}. Removing the two bonds connecting to {1,1}, one creates a configuration of $n = 2$ and $b = 2L_1L_2 - 2$ with the weight $x^b = x^{2L_1L_2}x^{-2}$. We regard the now isolated site {1,1} as a one-site animal.

Repeating this procedure, one next looks for the one-site animal configuration sites which are connected to exactly two neighboring sites. There are now two such sites, namely, {1,2} and {2,1}. By removing the two bonds connected to either of the two sites, one finds the next term in the reduced partition function having $n = 3$, $b = 2L_1L_2 - 4$ and the weight $2x^b = 2x^{2L_1L_2}x^{-4}$. The resulting configurations now have two isolated sites which can be regarded as two-site animals. Continuing in this fashion, it is recognized that the process of creating isolating sites (by removing two bonds at a time) follows precisely our rule of growing directed animals on $\mathcal{L}$. It follows that we have established the first line of (15). The second line of (15) now follows from (7). It should be pointed out that our proof of (15) works equally well for the Potts model with anisotropic reduced interactions K_1 and K_2. The reduced partition function is again given by (15) but with the replacement of x^2 by x_1x_2, where $x_i = (e^{K_i} - 1)/\sqrt{q}$. We have also established that all zeros of $Z_{L_1,L_2}(\infty, x)$ are on the unit circle $|x| = 1$, verifying a conjecture of [12] in the $q = \infty$ limit.

Since all zeros of the Gaussian polynomial are on the unit circle $|x| = 1$, one can introduce a per-site reduced free energy for the $q = \infty$ Potts model as [12,14]

$$\begin{aligned} f(x^2) &\equiv \lim_{L_1,L_2\to\infty} (L_1L_2)^{-1}\ln G(L_1,L_2;x^2) \\ &= \int_{-\pi}^{\pi} g(\theta)\ln(e^{i\theta} - x^2)\,d\theta, \end{aligned} \quad (16)$$

where $L_1L_2g(\theta)$ is the density of zeros of $G(L_1,L_2;x^2)$ on the unit circle in the complex x^2 plane. To determine $g(\theta)$, we note that the zeros of $(t)_L = \prod_{p=1}^{L}(1 - t^p)$ are at $e^{i\theta_{\ell p}}$, where

$$\theta_{\ell p} = 2\pi\ell/p, \quad p = 1,2,\ldots,L; \quad \ell = 1,2,\ldots,p. \quad (17)$$

This implies that, as p ranges from 1 to L, the number of zeros on an arc of the unit circle $|t| = 1$ between the real axis and any angle θ is equal to $\theta L^2/4\pi$, the area of the right triangle with perpendicular sides L and $\theta L/2\pi$. It follows that the density of the zeros of $(t)_L$ on the circle $|t| = 1$ is a constant equal to

$$H_L(\theta) = L^2/4\pi. \quad (18)$$

Consequently, from (5), the density of zeros of $G(L_1,L_2;t)$ in the complex-t plane is also a constant and equal to $H_{L_1+L_2}(\theta) - H_{L_1}(\theta) - H_{L_2}(\theta) = L_1L_2/2\pi$. This leads to $g(\theta) = 1/2\pi$, and the integral (16) can be evaluated, yielding

$$f(x^2) = \begin{cases} \ln|x^2|, & |x| > 1, \\ 0, & |x| < 1, \end{cases} \quad (19)$$

confirming the known first-order transition of the infinite-state Potts model [6,15].

Results in d dimensions.—The above consideration can be extended to d dimensions [16]. Define directed compact lattice animals which grow from the origin of a d-dimensional hypercubic lattice $\mathcal{L}$ of size $L_1 \times \cdots \times L_d$ in the d-positive directions subject to the constraint that a site $\{i_1, i_2, \ldots, i_d\}$ can be occupied only when the d sites $\{i_1, i_2, \ldots, i_s - 1, \ldots, i_d\}$, $s = 1,2,\ldots,d$ are all occupied. Let $A_n(L_1,L_2,\ldots,L_d)$ be the number of directed compact animals of size n. Then $A_n(L_1,L_2,\ldots,L_d)$ is the number of distinct partitions of a positive integer n into sums of nonnegative integers $m(n_1,n_2,\ldots,n_{d-1})$ associated with vertices $\{n_1,n_2,\ldots,n_{d-1}\}$ of a $(d-1)$-dimensional hypercubic lattice, or, explicitly,

$$n = \sum_{n_1=1}^{L_1}\sum_{n_2=1}^{L_2}\cdots\sum_{n_{d-1}=1}^{L_{d-1}} m(n_1,\ldots,n_{d-1}), \quad m(n_1,\ldots,n_{d-1}) > 0, \quad (20)$$

such that

$$0 \le m(n_1', n_2', \ldots, n_{d-1}') \le m(n_1,n_2,\ldots,n_{d-1}) \le L_d, \quad (21)$$

whenever $n_1 \le n_1', n_2 \le n_2', \ldots, n_{d-1} \le n_{d-1}'$. This defines a $(d-1)$-dimensional restricted partition [5].

In a similar fashion one defines the generating function

$$G(L_1,L_2,\ldots,L_d;t) = 1 + \sum_{n=1}^{L_1L_2\cdots L_d} A_n(L_1,L_2,\ldots,L_d)t^n, \quad (22)$$

and, analogous to (15), establishes [16] that the generating function (22) is precisely the reduced partition function of the infinite-state Potts model [17] on $\mathcal{L}$, provided that one identifies $t = x^d$ and $x = (e^K - 1)/q^{1/d}$.

But explicit expressions of the generating function (22) are known only for $d = 2$ and $d = 3$. For $d = 2$ it is given by (5), and for $d = 3$ it is [5,7]

$$G(L_1,L_2,L_3;t) = \frac{[t]_{L_1+L_2+L_3-1}[t]_{L_1-1}[t]_{L_2-1}[t]_{L_3-1}}{[t]_{L_1+L_2-1}[t]_{L_2+L_3-1}[t]_{L_1+L_3-1}}, \quad (23)$$

VOLUME 76, NUMBER 2 PHYSICAL REVIEW LETTERS 8 JANUARY 1996

where

$$[t]_L \equiv \prod_{p=1}^{L} (t)_p, \quad (t)_L \equiv \prod_{p=1}^{L} (1 - t^p). \tag{24}$$

We observe from (23) that zeros of $G(L_1, L_2, L_3; t)$ are on the unit circle $|t| = 1$ with a uniform density $L_1L_2L_3/2\pi$, leading again to the per-site reduced free energy (19) with x^2 replaced by x^3 in agreement with the known solution [6]. In addition, the asymptotic behavior of the largest $A_n(L_1, L_2, L_3)$, which we expect as in $d = 2$ to occur at $n \sim L_1L_2L_3/2$ and is the same as that of $G(L_1, L_2, L_3; 1)$, is [18]

$$A_n(L_1, L_2, L_3) \propto \exp[c(\alpha_1, \alpha_2, \alpha_3)n^{2/3}], \quad n \sim L_1L_2L_3/2, \tag{25}$$

where

$$c(\alpha_1, \alpha_2, \alpha_3) = 2^{-1/3}\left[\left(\frac{\alpha_1}{\alpha_2}\right)^{1/3} + \left(\frac{\alpha_2}{\alpha_3}\right)^{1/3} + \left(\frac{\alpha_3}{\alpha_1}\right)^{1/3}\right]t(\alpha_1, \alpha_2, \alpha_3),$$

$$t(\alpha_1, \alpha_2, \alpha_3) = (x_1x_2 + x_2x_3 + x_3x_1)^{-1} \sum_{i=1}^{3} [x_i^2 \ln x_i - (1 - x_i)^2 \ln(1 - x_i)], \tag{26}$$

with $x_i = (1 + \alpha_j + 1/\alpha_k)^{-1}, \alpha_i = L_j/L_k, i, j, k$ in cyclic order of 1,2,3. Particularly, for $L_1 = L_2 = L_3 = L$, one has $c(1, 1, 1) = 2^{2/3}(9 \ln\sqrt{3} - 3 \ln 4) = 1.245\,907$. Expressions (10) and (25) suggest the asymptotic behavior

$$A_n(L_1, L_2, \dots, L_d) \propto \exp(cn^{(d-1)/d}), \quad n \sim L_1L_2 \cdots L_d/2 \tag{27}$$

for general d. However, the problem of $(d-1)$-dimensional restricted partitions of a positive integer for $d > 3$ is an outstanding unsolved problem in number theory. In fact, it can be verified by considering a $2 \times 2 \times 2 \times 2$ lattice that zeros of the generating function (22) are no longer on the unit circle. On the other hand, the $q = \infty$ Potts model is known to have a first-order transition at $x \equiv (e^K - 1)/q^{1/d} = 1$ [6]. Our results in $d = 2, 3$ then suggest that the generating function (22) can be evaluated in the thermodynamic limit as

$$\lim_{L_1,\dots,L_d \to \infty} (L_1 \cdots L_d)^{-1} \ln G(L_1, \dots, L_d; t) = \begin{cases} \ln|t|, & |t| > 1, \\ 0, & |t| < 1. \end{cases} \tag{28}$$

We conjecture that (27) and (28) hold for any $d > 1$.

Finally, we remark that in deducing (28) we have assumed the special boundary condition [17] and interchanged the $q \to \infty$ and the thermodynamic limits. While the interchange of the two limits is a subtle matter, it can be explicitly verified in the $d = 1$ solution that the two limits indeed commute under the boundary conditions of [17].

We would like to thank P. Flajolet for illuminating discussions and for providing the estimate (13) for $A_n(L, L)$. This work is supported in part by CNRS and by NSF Grants No. DMR-9313648, No. INT-9113701, and No. INT-9207261. The work by C.K.H. and C.N.C. is supported by the National Science Council Grants No. NSC 84-2112-M-001-93Y and No. 84-0501-I-001-037-1312, and G.R. acknowledges the support of a Lavoisier grant from the Ministère des Affaires Etrangères.

[1] D. Dhar, Phys. Rev. Lett. **49**, 959 (1982).
[2] R.J. Baxter, J. Phys. A **13**, L61 (1980).
[3] A.M.W. Verhagen, J. Stat. Phys. **15**, 219 (1976).
[4] C.F. Gauss and Werke, Königliche Gesellschaft der Wissenschaften, (Göttingen, Germany, 1870), Vol. 2.
[5] For properties of restricted partitions, see, for example, G.E. Andrews, *Theory of Partitions,* edited by G.-C. Rota, in Encyclopedia of Mathematics and Its Applications (Addison-Wesley, Reading, MA, 1976), Chaps. 3 and 11.
[6] P.A. Pearce and R.B. Griffiths, J. Phys. A **13**, 2143 (1980).
[7] P.A. MacMahon, *Combinatory Analysis* (Cambridge, Cambridge, England, 1916), Vol. 2.
[8] K.M. O'Hara, J. Comb. Theory A **53**, 29 (1990).
[9] H. Rademacher, Proc. London Math. Soc. **43**, 241 (1937).
[10] G.H. Hardy and S. Ramanujan, Proc. London Math. Soc. (2) **17**, 75 (1918).
[11] P. Flajolet (private communication).
[12] C.N. Chen, C.K. Hu, and F.Y. Wu, preceding Letter, Phys. Rev. Lett. **76**, 169 (1996).
[13] For a review on the Potts model and its physical relevance, see F.Y. Wu, Rev. Mod. Phys. **54**, 235 (1982).
[14] M.E. Fisher, in *Lecture Notes in Theoretical Physics,* edited by W.E. Brittin (University of Colorado Press, Boulder, 1965), Vol. 7c.
[15] R.J. Baxter, J. Phys. C **6**, L445 (1973).
[16] Details can be found in F.Y. Wu (to be published).
[17] The hypercubic lattice Potts lattice $\mathcal{L}$ assumes the boundary condition that an extra site is introduced which connects by an edge to *every* site in the d hyperplanes intersecting at the point $\{L_1, L_2, \dots, L_d\}$.
[18] V. Elser, J. Phys. A **17**, 1509 (1984).

International Journal of Modern Physics B, Vol. 11, Nos. 1 & 2 (1997) 121–126

THE INFINITE-STATE POTTS MODEL AND SOLID PARTITIONS OF AN INTEGER

H. Y. HUANG and F. Y. WU
Department of Physics and
Center for Interdisciplinary Research in Complex Systems
Northeastern University, Boston, Massachusetts 02115, USA

It has been established that the infinite-state Potts model in d dimensions generates restricted partitions of integers in $d-1$ dimensions, the latter a well-known intractable problem in number theory for $d > 3$. Here we consider the $d = 4$ problem. We consider a Potts model on an $L \times M \times N \times P$ hypercubic lattice whose partition function $G_{LMNP}(t)$ generates restricted solid partitions on an $L \times M \times N$ lattice with each part no greater than P. Closed-form expressions are obtained for $G_{222P}(t)$ and we evaluated its zeroes in the complex t plane for different values of P. On the basis of our numerical results we conjecture that all zeroes of the enumeration generating function $G_{LMNP}(t)$ lie on the unit circle $|t| = 1$ in the limit that any of the indices L, M, N, P becomes infinite.

1. Introduction

It has been recently established[1] that the q-state Potts model in the $q \to \infty$ limit is intimately related to the problem of partitions of integers in number theory. Specifically, it was shown[1,2] that the d dimensional Potts model[3,4] in the infinite-state limit generates $(d-1)$-dimensional restricted partitions of integers.[5] Using this equivalence and the known solutions of the enumeration problem[6] for $d = 2, 3$, the infinite-state Potts model is solved[1] on certain finite lattices in $d = 2, 3$. But the solution for the partition enumeration problem is open for $d > 3$. Here, we investigate this open problem for $d = 4$ by making use of the Potts equivalence. Specifically, we study zeroes of the enumeration generating function of restricted solid partitions, and show that their distribution approaches a unit circle as the size of the partitioned parts increases.

The consideration of zeroes of the partition function plays an important role in the analysis of phase transitions in statistical mechanics.[8,9] However, the precise location of the zero distribution are known only in a very few instances. This includes the Ising lattice gas whose partition function zeroes lie on a unit circle in the complex fugacity plane.[8] For the enumeration problem in $d = 2, 3$ alluded to in the above, one finds that the zeroes of the generating function also lie on a unit circle.[2] But for $d > 3$ the zeroes of the generating function computed for small lattices are found to scatter, and their distribution does not appear to follow a regular pattern. On the other hand, the related Potts model has been solved,[7] and the solution is consistent with the assumption that, in the thermodynamic limit (of infinite lattices), all partition function zeroes lie on a unit circle. This suggests that a fruitful approach to the enumeration problem is to look into the zeroes of the generating function. This is the topic of this investigation.

We evaluate the generating function of restricted solid partitions on finite lattices and study the location of its zeroes. Our main finding, which is suggested by the Potts counterpart, is that the zeroes approach a unit circle as the size of the partitioned parts increases. This leads us to conjecture that the zeroes of the enumeration generating function of restricted solid partitions lie on a circle, when the size of partitioned parts, or equivalently the lattice size, becomes infinite.

2. The Potts model and restricted partitions

Restricted solid partitions can be generated by considering a Potts model. Consider a Potts model on a four-dimensional hypercubic lattice of size $L \times M \times N \times P$. The lattice sites are specified by coordinates i, j, k, p, where $1 \le i \le L$, $1 \le j \le M$, $1 \le k \le N$, and $1 \le p \le P$. Introduce an extra site which is connected by edges to every site in the hyperplanes $i = L$, $j = M$, $k = N$ and $p = P$. The resulting lattice contains $LMNP + 1$ vertices and $4LMNP$ edges.

The high-temperature expansion of the Potts partition function assumes the form[4]

$$Z_{LMNP}(q,x) = \sum_{\text{bond config.}} x^b q^{n_c + b/4}, \tag{1}$$

where $x = (e^K - 1)/q^{1/4}$, b and n_c are, respectively, the numbers of bonds and connected clusters, including isolated points. In the large q limit the leading terms in (1) are of the order of q^{LMNP+1}. One introduces the reduced partition function

$$G_{LMNP}(x^4) \equiv \lim_{q \to \infty} q^{-(LMNP+1)} Z_{LMNP}(q,x), \tag{2}$$

which is a polynomial of degree $LMNP$ in x^4.

It has been shown[1] that the reduced partition function $G_{LMNP}(t)$ is precisely the generating function of restricted solid partitions of a positive integer into a sum of parts on an $L \times M \times N$ cubic lattice, with each part no greater than P. The generating function for the solid partition is defined by

$$G_{LMNP}(t) = 1 + \sum_{n=1}^{LMNP} A_n(L,M,N)\, t^n, \tag{3}$$

where $A_n(L, M, N)$ is the number of distinct ways that a positive integer n is partitioned into the sum of nonnegative integers $m(i, j, k)$,

$$n = \sum_{i=1}^{L} \sum_{j=1}^{M} \sum_{k=1}^{N} m(i,j,k), \tag{4}$$

subject to

$$0 \le m(i,j,k) \le P, \tag{5}$$

and

$$m(i,j,k) \le \{m(i-1,j,k), m(i,j-1,k), m(i,j,k-1)\}. \tag{6}$$

We point out that, despite the apparent asymmetric footing, $G_{LMNP}(t)$ is actually symmetric in the 4 indices L, M, N and P, a fact which is obvious from the Potts equivalence.[1]

The explicit expression of $G_{LMNP}(t)$ for general $\{L, M, N, P\}$ is not known. However, for $L = M = N = 2$ MacMahon[6] has obtained a closed-form expression given by

$$G_{222P}(t) = \sum_{i=0}^{4} L_i \frac{(t)_{P+8-i}}{(t)_8 (t)_{P-i}}, \tag{7}$$

with

$$\begin{aligned} L_0 &= 1 \\ L_1 &= 2t^2 + 2t^3 + 3t^4 + 2t^5 + 2t^6 \\ L_2 &= t^5 + 3t^6 + 4t^7 + 8t^8 + 4t^9 + 3t^{10} + t^{11} \\ L_3 &= 2t^{10} + 2t^{11} + 3t^{12} + 2t^{13} + 2t^{14} \\ L_4 &= t^{16}, \end{aligned} \tag{8}$$

and

$$(t)_m = \prod_{p=1}^{m} (1 - t^p), \qquad m \ge 1. \tag{9}$$

Before we proceed further, we first cast (7) into an alternate form which is more suggestive.

For $d = 2$ and 3 the partition generating functions for similarly defined line and planar partitions assume the form[6]

$$G_{LM}(t) = \frac{(t)_{L+M}}{(t)_L (t)_M}, \tag{10}$$

$$G_{LMN}(t) = \frac{[t]_{L+M+N}[t]_L[t]_M[t]_N}{[t]_{L+M}[t]_{M+N}[t]_{N+L}}, \tag{11}$$

where

$$[t]_m = \prod_{p=1}^{m-1} (t)_p, \qquad m \ge 2. \tag{12}$$

The expression which straightforwardly generalizes (10) and (11) to $d = 4$ is

$$G^{(0)}_{LMNP}(t) = \frac{\{t\}_{L+M+N+P}\{t\}_{L+M}\{t\}_{L+N}\{t\}_{M+N}\{t\}_{L+P}\{t\}_{M+P}\{t\}_{N+P}}{\{t\}_{L+M+N}\{t\}_{L+M+P}\{t\}_{L+N+P}\{t\}_{M+N+P}\{t\}_L\{t\}_M\{t\}_N\{t\}_P}, \tag{13}$$

with

$$\{t\}_m = \prod_{p=2}^{m-1} [t]_p, \qquad m \geq 3. \tag{14}$$

Note that the $LMNP + 1$ zeroes of $G^{(0)}_{LMNP}(t)$ lie on the unit circle $|t| = 1$. This suggests one to write

$$G_{LMNP}(t) = G^{(0)}_{LMNP}(t) + C_{LMNP}(t), \tag{15}$$

where $C_{LMNP}(t)$ is a "correction" to the straightforward extension (13). For $L = M = N = 2$ we find that (7) can indeed be be rewritten in the form of (15) with

$$C_{222P}(t) = -\left(\frac{t^6(t+1)^2(t^4 - 2t^3 + t^2 - 2t + 1)}{t^2 + t + 1}\right)\left(\frac{(t)_{P+6}}{(t)_8(t)_{P-2}}\right). \tag{16}$$

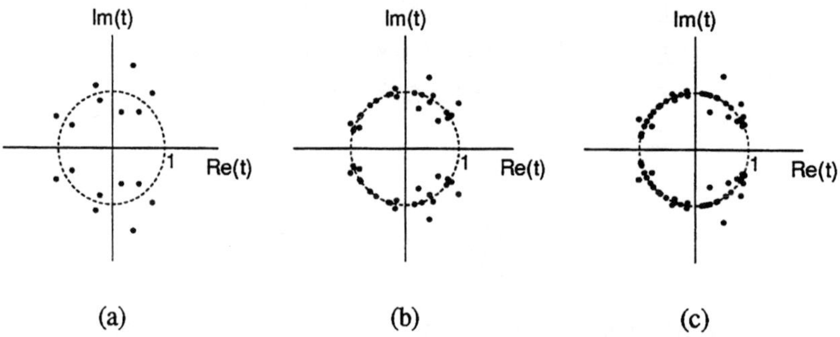

Fig. 1. Zeroes of G_{222P} for (a) $P = 2$, (b) $P = 6$, and (c) $P = 10$.

3. Zeroes of the generating function

We now use (7), or equivalently (15), to evaluate the zeroes of $G_{222P}(t)$. Selected results for $P = 2, 6, 10$ are shown in Fig. 1. It is seen that, while for $P = 2$ none of the zeroes lie on the unit circle $|t| = 1$, more zeroes are found on the unit circle as P increases. To measure quantitatively the deviation of the zeroes from the unit circle, we have computed d_1, the average distance, and d_2, the root-mean-square distance, of the zeroes from the unit circle given by

$$\begin{aligned} d_1 &= \frac{1}{LMNP} \sum_{i=1}^{LMNP} (|t_i| - 1) \\ d_2 &= \frac{1}{LMNP} \sqrt{\sum_{i=1}^{LMNP} (|t_i| - 1)^2}. \end{aligned} \tag{17}$$

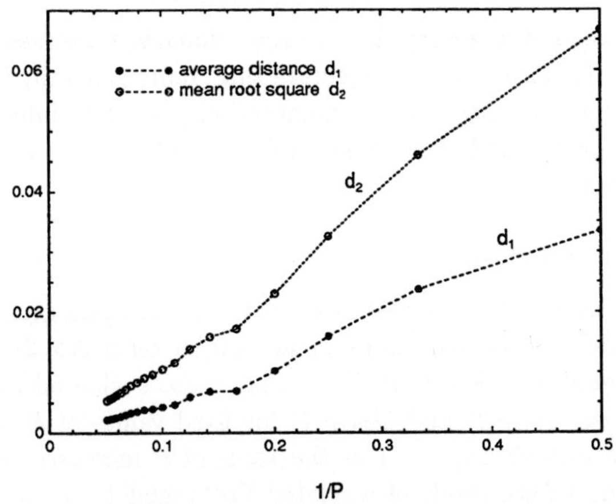

Fig. 2. Distance to the unit circle $|t| = 1$ for $G_{222P}(t)$.

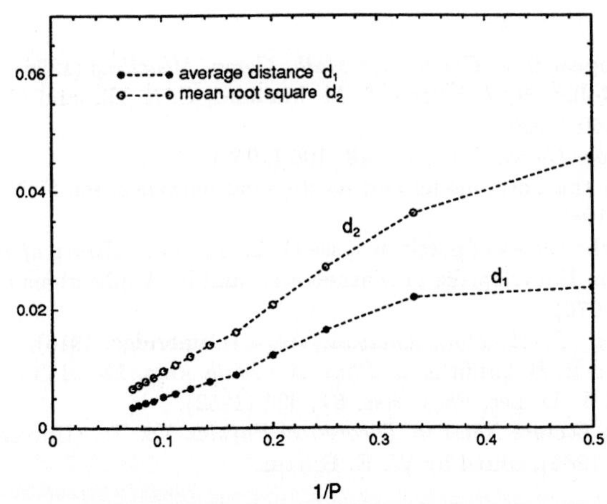

Fig. 3. Distance of zeroes to the unit circle $|t| = 1$ for $G_{322P}(t)$.

Plots of d_1 and d_2 are shown in Fig. 2. It is clear that both d_1 and d_2 extrapolate to zero in the limit of $P \to \infty$.

We have also generated explicitly the generating function $G_{322P}(t)$ for $P = 3$ to 14, and evaluated their zeroes. The results are shown in in Fig. 3. Again, it is seen that both d_1 and d_2 extrapolate to zero, namely, all zeroes lie on the unit circle, as $P \to \infty$. This leads us to conjecture that, quite generally and making use of the indices symmetry,[1] zeroes of the enumeration generating function $G_{LMNP}(t)$ of restricted solid partitions lie on a unit circle in the limit that any of the indices $\{L, M, N, P\}$ becomes infinite.

4. Summary and conclusion

We have obtained closed-form expressions for the enumeration generation function $G_{222P}(t)$ for restricted solid partitions of an integer on a $2 \times 2 \times 2$ lattice into parts which are equal to or less than P. We have also evaluated the zeroes of the generating functions $G_{222P}(t)$ and $G_{322P}(t)$ for fixed values of P, and found that they approach the unit circle $|t| = 1$ as the value of P increases. On the basis of our finding and the known result of a related Potts model, we conjecture that all zeroes of the enumeration generating function $G_{LMNP}(t)$ lie on a unit circle in the limit that any of the indices $\{L, M, N, P\}$ becomes infinite.

Acknowledgements

The work is supported in part by a Northeastern University RSDF grant and by National Science Foundation grants DMR-9313648 and DMR-9614170.

References

1. F. Y. Wu, to appear in *J. Comp. and Math. Comp. Modelling* (1996).
2. F. Y. Wu, G. Rollet, H. Y. Huang, J. M. Maillard, C. K. Hu, and C. N. Chen, *Phys. Rev. Lett.* **76**, 169 (1996).
3. R. B. Potts, *Proc. Camb. Phil. Soc.* **48**, 106 (1952).
4. For a review on the Potts model and its physical relevance, see F. Y. Wu, *Rev. Mod. Phys.* **54**, 235 (1982).
5. For a review of the theory of partitions, see G. E. Andrews, *Theory of Partitions*, edited by G.-C. Rota, in Encyclopedia of Mathematics and Its Applications (Addison-Wesley, Reading, MA, 1976).
6. P. A. MacMahon, *Combinatory Analysis*, Vol. 2 (Cambridge, 1916).
7. P. A. Pearce and R. B. Griffiths, *J. Phys. A : Math. Gen.* **13**, 2143 (1980).
8. C. N. Yang and T. D. Lee, *Phys. Rev.* **87**, 404 (1952).
9. M. E. Fisher, in *Lecture Notes in Theoretical Physics*, Vol. 7c, (University of Colorado Press, Boulder, 1965), edited by W. E. Brittin.

VOLUME 20, NUMBER 25 PHYSICAL REVIEW LETTERS 17 JUNE 1968

ABSENCE OF MOTT TRANSITION IN AN EXACT SOLUTION OF THE SHORT-RANGE, ONE-BAND MODEL IN ONE DIMENSION

Elliott H. Lieb* and F. Y. Wu
Department of Physics, Northeastern University, Boston, Massachusetts
(Received 22 April 1968)

The short-range, one-band model for electron correlations in a narrow energy band is solved exactly in the one-dimensional case. The ground-state energy, wave function, and the chemical potentials are obtained, and it is found that the ground state exhibits no conductor-insulator transition as the correlation strength is increased.

The correlation effect of electrons in a partially filled energy band has been a subject of interest for many years.[1-4] A realistic model which takes this correlation into consideration, and which is hopefully amenable to mathematical treatment, is the short-range, one-band model considered by a number of authors.[2-5] In this model, one pictures the electrons in a narrow energy band hopping between the Wannier states of neighboring lattice sites, with a repulsive interaction energy between two electrons of opposite spins occupying the same lattice site. The central problems of interest have been (a) the possible existence of a "Mott transition" between conducting and insulating states as the strength of the interaction is increased, and (b) the magnetic nature (ferromagnetic or antiferromagnetic) of the ground state. While previous treatments of this model have always been approximate, we have succeeded in solving the model exactly in the one-dimensional case. Our exact result shows that the Mott transition does occur in the ground state of the one-dimensional model. Furthermore, a general theorem of Lieb and Mattis[6] on one-dimensional systems tells us that the ground state is necessarily antiferromagnetic.

It may be argued that the absence of a Mott transition in one dimension is irrelevant for the study of real three-dimensional systems because of the folkloristic dictum that there are never any phase transitions in one dimension with short-range interactions. In actual fact, the dictum is only true for nonzero temperature; the ground state is another matter. Generally speaking, when a Hamiltonian is considered to be a function of some parameter, U (which in our case is the electron-electron repulsion), singularities with respect to U usually do appear in the ground-state wave function, energy, polarizability, etc., even in one dimension. A good example of this is the one-dimensional Heisenberg chain (to which the present model is very close) which, when considered as a function of the anisotropy parameter, does have two singularities in the ground state and, presumably, no singularities for nonzero temperatures.[7,8]

Consider a crystal (one-, two-, or three-dimensional) of N_a lattice sites with a total of $N \leq 2N_a$ electrons. We suppose that the electrons can hop between the Wannier states of neighboring lattice sites, and that each site is capable of accommodating two electrons of opposite spins, with an interaction energy $U>0$. The Hamiltonian to consider is then[2-5]

$$H=T\sum_{\langle ij\rangle}\sum_{\sigma}c_{i\sigma}^{\dagger}c_{j\sigma}+U\sum_{i}c_{i\uparrow}^{\dagger}c_{i\uparrow}c_{i\downarrow}^{\dagger}c_{i\downarrow}, \tag{1}$$

where $c_{i\sigma}^{\dagger}, c_{i\sigma}$ are, respectively, the creation and annihilation operators for an electron of spin σ in the Wannier state at the ith lattice site, and the sum

$$\sum_{\langle ij\rangle}$$

is restricted to nearest-neighbor sites.

First of all, it can be shown that the energy spectrum of H is invariant under the replacement of T by $-T$.[9] Therefore, for simplicity we shall take, in appropriate units, $T=-1$. Since the numbers M of down-spin electrons and M' of up-spin electrons are good quantum numbers $(M+M'=N)$, we may designate the ground-state energy of H by $E(M,M';U)$. It is then easy to derive the following relations [by considering holes instead of particles in (1)]:

$$\begin{aligned}E(M,M';U)&=-(N_a-M-M')U\\&\quad+E(N_a-M,N_a-M';U)\\&=MU+E(M,N_a-M';-U)\\&=M'U+E(N_a-M,M';-U).\end{aligned} \tag{2}$$

Without loss of generality, therefore, we may take

$$S_z\equiv\tfrac{1}{2}(N-2M)\geq 0 \text{ and } N\leq N_a$$

(less than half-filled band).

VOLUME 20, NUMBER 25 PHYSICAL REVIEW LETTERS 17 JUNE 1968

It can similarly be shown that the maximum energy $G(M,M';U)$ is related to the ground-state energy by

$$G(M,M';U)=M'U-E(N_a-M,M';U).$$

Therefore, a knowledge of the ground-state energies also tells us about the maximum energies.

For a one-dimensional system, the lattice sites can be numbered consecutively from 1 to N_a. Let $f(x_1,x_2,\cdots,x_M,x_{M+1},\cdots,x_N)$ represent the amplitude in ψ for which the down spins are at the sites $x_1,\cdots,x_M$, and the up spins at $x_{M+1},\cdots,x_N$. Then the eigenvalue equation $H\psi=E\psi$ leads to

$$-\sum_{i=1}^{N}\sum_{s=\pm1} f(x_1,\cdots,x_i+s,\cdots,x_N) + U\sum_{i<j}\delta(x_i-x_j)f(x_1\cdots x_N) = Ef(x_1\cdots x_N), \tag{3}$$

where it is understood that we require a solution of the form

$$f(x_1,x_2,\cdots,x_M|x_{M+1},x_{M+2},\cdots,x_N) \tag{4}$$

which is antisymmetric in the first M and the last $N-M$ variables.

In each region defined by $1\le x_{Q1}\le x_{Q2}\le\cdots x_{QN}\le N$, we make the following Ansatz for f:

$$f(x_1,\cdots,x_M|x_{M+1},\cdots,x_N) = \sum_P [Q,P]\exp(i\sum_{j=1}^{N} k_{Pj}x_{Qj}), \tag{5}$$

where $P=(P1,P2,\cdots,PN)$ and $Q=(Q1,Q2,\cdots,QN)$ are two permutations of the numbers $(1,2,\cdots,N)$, $\{k_1,k_2,\cdots k_N\}$ is a set of N unequal real numbers, and $[Q,P]$ is a set of $N!\times N!$ coefficients to be determined.

The coefficients $[Q,P]$ are not all independent. The condition of single valuedness (or continuity) of f and the requirement that (5) be a solution of (3) lead to the following:

$$E=-2\sum_{j=1}^{N}\cos k_j \tag{6}$$

and, for all Q and P, the coefficients $[Q,P]$ must be chosen to satisfy the relations

$$[Q,P]=Y_{nm}^{ab}[Q,P']. \tag{7}$$

In (7), Y_{nm}^{ab} is an operator defined by

$$Y_{nm}^{ab}=\frac{-\frac{1}{2}iU}{\sin k_n-\sin k_m+\frac{1}{2}iU}+\frac{\sin k_n-\sin k_m}{\sin k_n-\sin k_m+\frac{1}{2}iU}P^{ab}, \tag{8}$$

where, for $j=i+1$,

$Qi=a=Q'j$, $Qj=b=Q'i$,

$Qk=Q'k$ for all $k\ne i,j$;

$Pi=m=P'j$, $Pj=n=P'i$,

$Pk=P'k$ for all $k\ne i,j$;

and P^{ab} is an operator which exchanges $Qi=a$ and $Qj=b$.

It is fortunate that the Ansatz (5) and the algebraic consistency conditions (7) and (8) have, in essence, appeared before in the study of the one-dimensional delta-function gas for particles in a continuum. The first solution of that problem was for bosons (symmetric f) by Lieb and Liniger[10] but this case is not relevant here, besides which the consistency conditions there are trivial to solve. The two-component fermion case was solved by McGuire[11] for $M=1$, but again (7) is trivial because of translational invariance. The next development was the solution of the case $M=2$ by Flicker and Lieb[12] by an inelegant algebraic method which could not be easily generalized. However, the case $M=2$ is the simplest one which displays the full difficulty of the problem. Shortly thereafter, Gaudin[13] published the solution of the general-M problem. The method of his brilliant solution did not appear for some time and is now available as his thesis.[14] In the meantime, Yang[15] also discovered the method of solution (essentially the same as Gaudin's) and published it with considerable detail. Here, we have followed Yang's notation with slight modification.

The important point is that our Eqs. (7) and (8) are the same as for the continuum gas except for the replacement of k by $\sin k$ in the latter. This has no effect on the beautiful algebraic analysis which finally leads to the following conditions which determine the set $\{k_{1,2}, \cdots, k_N\}$:

$$N_a k_j = 2\pi I_j + \sum_{\beta=1}^{M} \theta(2\sin k_j - 2\Lambda_\beta),\ j=1,2,\cdots,N, \tag{9}$$

where the Λ's are a set of real numbers related to the k's through

$$-\sum_{j=1}^{N} \theta(2\Lambda_\alpha - 2\sin k_j) = 2\pi J_\alpha - \sum_{\beta=1}^{M} \theta(\Lambda_\alpha - \Lambda_\beta),\quad \alpha = 1,2,\cdots,M, \tag{10}$$

$$\theta(p) \equiv -2\tan^{-1}(2p/U),\quad -\pi \leqslant \theta < \pi, \tag{11}$$

and I_j = integers (or half-odd integers) for M = even (or odd), J_α = integers (or half-odd integers) for M' = odd (or even). An immediate consequence is

$$\sum_{j=1}^{N} k_j = \frac{1}{N_a}\left(\sum_j I_j + \sum_\alpha J_\alpha\right). \tag{12}$$

For the ground state, J_α and I_j are consecutive integers (or half-odd integers) centered around the origin and satisfying $\sum_j k_j = 0$.

In the limit of $N\to\infty$, $N_a\to\infty$, $M\to\infty$ with the ratios N/N_a, M/N_a kept finite, the real numbers k and Λ are distributed continuously between $-Q$ and $Q \leqslant \pi$ and $-B$ and $B \leqslant \infty$, with density functions $\rho(k)$ and $\sigma(\Lambda)$, respectively. Equations (9) and (10) then lead to the coupled integral equations for the distribution function $\rho(k)$ and $\sigma(\Lambda)$:

$$2\pi\rho(k) = 1 + \cos k \int_{-B}^{B} \frac{8U\sigma(\Lambda)d\Lambda}{U^2 + 16(\sin k - \Lambda)^2}, \tag{13}$$

$$\int_{-Q}^{Q} \frac{8U\rho(k)dk}{U^2+16(\Lambda - \sin k)^2} = 2\pi\sigma(\Lambda) + \int_{-B}^{B} \frac{4U\sigma(\Lambda')d\Lambda'}{U^2+4(\Lambda-\Lambda')^2}, \tag{14}$$

where Q and B are determined by the conditions

$$\int_{-Q}^{Q} \rho(k)dk = N/N_a, \tag{15}$$

$$\int_{-B}^{B} \sigma(\Lambda)d\Lambda = M/N_a. \tag{16}$$

The ground-state energy (6) now becomes

$$E = -2N_a \int_{-Q}^{Q} \rho(k)\cos k\, dk. \tag{17}$$

We have established the following:

(a) Equations (13)-(16) have a unique solution which is positive for all allowed B and Q.

(b) M/N is a monotonically increasing function of B reaching a maximum of $\frac{1}{2}$ at $B=\infty$. This is the antiferromagnetic case, $S_z = 0$, and corresponds to the absolute ground state.

(c) N/N_a is a monotonically increasing function of Q, reaching a maximum of 1 (half-filled band) at $Q=\pi$.

For $B=\infty$ and $Q=\pi$, (13)-(16) can be solved in closed form by Fourier transforms with the result

$$\sigma(\Lambda) = (2\pi)^{-1}\int_0^\infty \operatorname{sech}(\tfrac{1}{4}\omega U) \times \cos(\omega\Lambda) J_0(\omega)d\omega, \tag{18}$$

$$\rho(k) = (2\pi)^{-1} + \pi^{-1}\cos k \times \int_0^\infty \frac{\cos(\omega \sin k) J_0(\omega)d\omega}{1+\exp(\frac{1}{2}\omega U)}, \tag{19}$$

$$E \equiv E(\tfrac{1}{2}N_a, \tfrac{1}{2}N_a; U) = -4N_a \int_0^\infty \frac{J_0(\omega)J_1(\omega)d\omega}{\omega[1+\exp(\frac{1}{2}\omega U)]}, \tag{20}$$

where J_0 and J_1 are Bessel functions.

To investigate whether the ground state is conducting or insulating, we compute the chemical

VOLUME 20, NUMBER 25 PHYSICAL REVIEW LETTERS 17 JUNE 1968

potentials μ_+ and μ_- as defined in a forthcoming paper by Mattis[16]:

$$\mu_+ \equiv E(M+1, M; U) - E(M, M; U),$$
$$\mu_- \equiv E(M, M; U) - E(M-1, M; U). \quad (21)$$

If μ_+ and μ_- are equal, the system has the property of a conductor. If, on the other hand, we find $\mu_+ > \mu_-$, then the system shares the property of an insulator. We can compute μ_- directly from (9) and (10) by replacing $M \to M-1$ and $N \to N-1$, while letting all the k's, Λ's, and their distribution functions change slightly. The procedure is quite similar to the calculation of the excitation spectrum of the continuum gas.[10] If $N < \frac{1}{2}N_a$, we can compute μ_+ in the same way and thereby find that $\mu_+ = \mu_-$ for all U. If, however, N is exactly $\frac{1}{2}N_a$, then we must compute μ_+ by using the first line of (2) which tells us that

$$\mu_+ = U - \mu_- \text{ (half-filled band).} \quad (22)$$

The calculation of μ_- can be done in closed form for a half-filled band with the result

$$\mu_- - 2 = -4\int_0^\infty \frac{J_1(\omega)d\omega}{\omega[1+\exp(\frac{1}{2}\omega U)]}$$
$$= -4 \sum_{n=1}^{\infty} (-1)^n [(1 + \tfrac{1}{4}n^2U^2)^{\frac{1}{2}} - \tfrac{1}{2}nU]. \quad (23)$$

It can be established from (22) and (23) that, indeed, $\mu_+ > \mu_-$ for $U > 0$, and

$$\lim_{U\to 0} \mu_\pm = 0.$$

Therefore, we conclude that the ground state for a half-filled band is insulating for any nonzero U, and conducting for $U = 0$. That is, there is no Mott transition for nonzero U. This absence of a Mott transition is also reflected by the fact that the ground-state energy and the ground-state wave function are analytic in U on the real axis (except at the origin).

We have also investigated the excitation spectrum $E(p)$ for a given total momentum $\sum_j k_j = p$ and a given value of S_z. Just as in the case of a continuum gas for which the spectrum can be regarded as consisting of several elementary excitations,[10,15] we find three types of excitations: (I) a "hole" state in the Λ distribution, (II) a "hole" state in the k distribution, and (III) a "particle" state in the k distribution. While the $S_z = 0$ spin-wave state may have any of these three types of spectra, the $S_z = 1$ spin-wave state is always associated with the type-I spectrum. The type-I excitation has the lowest energy and is characterized by a double periodicity similar to that of an antiferromagnetic chain.[7] In the limit $U \to 0$, it goes over to $E(p) = |\sin p|$, while the type-II and -III spectra have the identical limiting form $E(p) = |2\sin(\frac{1}{2}p)|$. Detailed discussions of these matters will be given elsewhere.

We are grateful to Dr. D. C. Mattis for helpful advice and suggestions and E. L. would like to thank Dr. J. Zittartz for interesting him in the problem.

---

*Work partially supported by National Science Foundation Grant No. GP-6851.

[1]J. H. Van Vleck, Rev. Mod. Phys. 25, 220 (1953).

[2]M. C. Gutzwiller, Phys. Rev. Letters 10, 159 (1963), and Phys. Rev. 134, A923 (1964), and 137, A1726 (1965).

[3]J. Hubbard, Proc. Roy. Soc. (London), Ser. A 276, 238 (1963), and 277, 237 (1964).

[4]G. Kemeny, Ann. Phys. (N.Y.) 32, 69, 404 (1965).

[5]See also the review article by C. Herring, in Magnetism, edited by G. T. Rado and H. Suhl (Academic Press, Inc., New York, 1966), Vol. IV, Chap. 10.

[6]E. Lieb and D. C. Mattis, Phys. Rev. 125, 164 (1962). The proof of the theorem given in this reference can be adopted to our Hamiltonian (1).

[7]J. des Cloizeaux and M. Gaudin, J. Math. Phys. 7, 1384 (1966).

[8]C. N. Yang and C. P. Yang, Phys. Rev. 150, 321, 327 (1966).

[9]The proof assumes A and B sublattices and uses the unitary transformation

$$\exp[i\pi \sum_\sigma \sum_{i\in A} c_{i\sigma}^\dagger c_{i\sigma}]$$

which changes $c_{i\sigma}$ to $-c_{i\sigma}$ and $c_{i\sigma}^\dagger$ to $-c_{i\sigma}^\dagger$ for $i \in A$. This transformation does not change the number operator.

[10]E. Lieb and W. Liniger, Phys. Rev. 130, 1605 (1963); E. Lieb, Phys. Rev. 130, 1616 (1963).

[11]J. B. McGuire, J. Math. Phys. 6, 432 (1965).

[12]M. Flicker, thesis, Yeshiva University, 1966 (unpublished); M. Flicker and E. Lieb, Phys. Rev. 161, 179 (1967).

[13]M. Gaudin, Phys. Letters 24A, 55 (1967).

[14]M. Gaudin, thesis, University of Paris, 1967 (unpublished).

[15]C. N. Yang, Phys. Rev. Letters 19, 1312 (1967).

[16]D. Mattis, to be published.

Available online at www.sciencedirect.com
SCIENCE @ DIRECT®
ELSEVIER
Physica A 321 (2003) 1–27
PHYSICA A
www.elsevier.com/locate/physa

The one-dimensional Hubbard model: a reminiscence☆

Elliott H. Lieb[a], F.Y. Wu[b,*]

[a] *Departments of Mathematics and Physics, Princeton University, Princeton, NJ 08544, USA*
[b] *Department of Physics, Northeastern University, Boston, MA 02115, USA*

Abstract

In 1968 we published the solution of the ground state energy and wave function of the one-dimensional Hubbard model, and we also showed that there is no Mott transition in this model. Details of the analysis have never been published, however. As the Hubbard model has become increasingly important in condensed matter physics, relating to topics such as the theory of high-T_c superconductivity, it is appropriate to revisit the one-dimensional model and to recall here some details of the solution.

PACS: 71.10.Fd; 71.27.+a; 75.10.Lp

Keywords: Hubbard model; One dimension; Exact solution

1. Introduction

In a previous paper [1] we reported the solution of the one-dimensional (1D) Hubbard model, showing the absence of the Mott transition in its ground state, but the letter format of the paper did not permit the presentation of all the details of the analysis. Over the years the Hubbard model [2,3] has become more important, for it plays an essential role in several topics in condensed matter physics, including 1D conductors and high-T_c superconductivity. It also plays a role in the chemistry of aromatic compounds (e.g., Benzene [4,5]). Several books [6–9] now exist in which the 1D Hubbard

☆ This article may be reproduced in its entirety for non-commercial purposes.

* Corresponding author.
E-mail address: fywu@neu.edu (F.Y. Wu).

0378-4371/03/$ - see front matter © 2002 E.H. Lieb and F.Y. Wu. Published by Elsevier Science B.V. All rights reserved.
doi:10.1016/S0378-4371(02)01785-5

model is analyzed, and numerous papers have been written on properties of the model. [1] Almost invariably these publications are based upon results of [1], including the absence of a Mott transition, but without derivation. While other rigorous results on higher dimensional Hubbard models exist, and some of these are reviewed in Refs. [10,11], the ID model stands as the only Hubbard model whose ground state can be found exactly. It has been brought to our attention that it would be useful to students and researchers if some details of the solution could be made available. Here, taking the opportunity of the symposium, StatPhys-Taiwan 2002, which takes place in the year when both of the authors turn 70, we revisit the 1D Hubbard model and present some details of the 34-year old solution.

While our paper [1] contained significant results about the excitation spectrum, it was mainly concerned with the integral equations for the ground state and we concentrate on those equations here. The new, unpublished results are contained in Sections 5–7.

Consider a crystal of N_a lattice sites with a total of N itinerant electrons hopping between the Wannier states of neighboring lattice sites, and that each site is capable of accommodating two electrons of opposite spin, with an interaction energy $U > 0$, which mimics a screened Coulomb repulsion among electrons. The Hubbard model [3] is described by the Hamiltonian

$$\mathscr{H} = T\sum_{\langle ij\rangle}\sum_{\sigma} c_{i\sigma}^{\dagger}c_{j\sigma} + U\sum_{i} n_{i\uparrow}n_{i\downarrow}\,, \tag{1}$$

where $c_{i\sigma}^{\dagger}$ and $c_{i\sigma}$ are, respectively, the creation and annihilation operators for an electron of spin σ in the Wannier state at the ith lattice site and $n_{i\sigma}=c_{i\sigma}^{\dagger}c_{i\sigma}$ is the occupation number operator. The summation $\langle ij\rangle$ is over nearest neighbors, and one often considers (as we do here) periodic boundary conditions, which means that $\langle ij\rangle$ includes a term coupling opposite edges of the lattice. We are interested in the ground state solution of the Schrödinger equation $\mathscr{H}|\psi\rangle = E|\psi\rangle$.

For bipartite lattices (i.e., lattices in which the set of sites can be divided into two subsets, A and B, such that there is no hopping between A sites or between B sites), such as the 1D chain, the unitary transformation $V^{\dagger}\mathscr{H}V$ leaves $\mathscr{H}$ unchanged except for the replacement of T by $-T$. Here $V = \exp[i\pi\sum_{i\in A}(n_{i\uparrow} + n_{i\downarrow})]$, with A being one of the two sublattices. Without loss of generality we can, therefore, take $T = -1$. In any event, bipartite or not, we can renormalize U by redefining U to be $U/|T|$. Henceforth, the value of T in (1) is -1 and U is positive and fixed. The dependence of the Hamiltonian and the energy on U will not be noted explicitly.

The commutation relations

$$\left[\sum_i n_{i\uparrow}, \mathscr{H}\right] = \left[\sum_i n_{i\downarrow}, \mathscr{H}\right] = 0$$

[1] For example, there have been over 500 citations to Ref. [1] in papers published in the *Physical Review* and in *Physical Review Letters* alone from 1968 to 2002. Most of the papers on the one-dimensional Hubbard model can be traced from the *PROLA* link of the American Physical Society web page.

imply that the numbers of down-spin electrons M and up-spin electrons M' are good quantum numbers. Therefore we characterize the eigenstates by M and M', and write the Schrödinger equation as

$$\mathscr{H}|M,M'\rangle = E(M,M')\,|M,M'\rangle \ . \tag{2}$$

Naturally, for any fixed choice of M, M' there will generally be many solutions to (2), so that $|M,M'\rangle$ and $E(M,M')$ denote only generic eigenvectors and eigenvalues. Furthermore, by considering particles as holes, and vice versa, namely, introducing fermion operators

$$d_{i\sigma} = c_{i\sigma}^{\dagger}, \qquad d_{i\sigma}^{\dagger} = c_{i\sigma}$$

and the relation $n_{i\sigma} = 1 - d_{i\sigma}^{\dagger} d_{i\sigma}$, we obtain the identity

$$E(M,M') = -(N_a - N)U + E(N_a - M, N_a - M') \ , \tag{3}$$

where

$$N = M + M'$$

is the total number of electrons. Since $N \geqslant N_a$ if, and only if, $(N_a - M) + (N_a - M') \leqslant N_a$, we can restrict our considerations to

$$N \leqslant N_a \ ,$$

namely, the case of at most a "half-filled band". In addition, owing to the spin-up and spin-down symmetry, we need only consider

$$M \leqslant M' \ .$$

2. The 1D model

We now consider the 1D model, and write $|M,M'\rangle$ as a linear combination of states with electrons at specific sites. Number the lattice sites by $1, 2, \ldots, N_a$ and, since we want to use periodic boundary conditions, we require N_a to be an even integer in order to retain the bipartite structure. For later use it is convenient also to require that $N_a = 2 \times$ (odd integer) in order to be able to have $M = M' = N_a/2$ with M odd. For the 1D model the sum in (1) over $\langle ij \rangle$ is really a sum over $1 \leqslant i \leqslant N_a$, $j = i+1$, plus $1 \leqslant j \leqslant N_a$, $i = j+1$ with $N_a + 1 \equiv 1$.

Let $|x_1, \ldots, x_N\rangle$ denote the state in which the down-spin electrons are located at sites $x_1, x_2, \ldots, x_M$ and the up-spin electrons are at sites $x_{M+1}, \ldots, x_N$. The eigenstate is now written as

$$|M,M'\rangle = \sum_{1 \leqslant x_i \leqslant N_a} f(x_1, \ldots, x_N)|x_1, \ldots, x_N\rangle \ , \tag{4}$$

where the summation is over all $x_1, \ldots, x_N$ from 1 to N_a, and $f(x_1, \ldots, x_N)$ is the amplitude of the state $|x_1, \ldots, x_N\rangle$.

It is convenient to denote the N-tuple $x_1, x_2, \ldots, x_N$ simply by X.

By substituting (4) into the Schrödinger equation (2), we obtain (recall $T = -1$)

$$-\sum_{i=1}^{N} [f(x_1, \ldots, x_i + 1, \ldots, x_N) + f(x_1, \ldots, x_i - 1, \ldots, x_N)]$$

$$+ U \left[\sum_{i<j} \delta(x_i - x_j) \right] f(x_1, \ldots, x_N) = E f(x_1, \ldots, x_N) , \tag{5}$$

where δ is the Kronecker delta function. We must solve (5) for f and E, with the understanding that site 0 is the same as site N_a and site $N_a + 1$ is the same as site 1 (the periodic boundary condition). Eq. (5) is the 'first quantized' version of the Schrödinger equation (2). It must be satisfied for all $1 \leqslant x_i \leqslant N_a$, with $1 \leqslant i \leqslant N$.

As electrons are governed by Fermi–Dirac statistics, we require that $f(X)$ be antisymmetric in its first M and last M' variables *separately*. This antisymmetry also ensures that $f = 0$ if any two x's in the same set are equal, which implies that the only delta-function term in (5) that are relevant are the ones with $i \leqslant M$ and $j > M$. This is consistent with the definition of $\mathscr{H}$ in (1), in which the only interaction is between up- and down-spin electrons.

The antisymmetry allows us to reinterpret (5) in the following alternative way. Define the region R to be the following subset of all possible values of X (note the $<$ signs):

$$R = \left\{ X : \begin{pmatrix} 1 \leqslant x_1 < x_2 < \cdots < x_M \leqslant N_a \\ 1 \leqslant x_{M+1} < x_{M+2} < \cdots < x_N \leqslant N_a \end{pmatrix} \right\} . \tag{6}$$

In R any of the first M x_i's can be equal to any of the last M', with an interaction energy nU, where n is the number of overlaps of the first set with the second.

The antisymmetry of f tells us that f is completely determined by its values in the subset R, together with the requirement that $f = 0$ if any two x's in the same set are equal (e.g., $x_1 = x_2$). Therefore, it suffices to satisfy the Schrödinger equation (5) when X on the right-hand side of (5) is only in R, together with the additional fact that we set $f = 0$ on the left-hand side of (5) if $x_i \pm 1$ takes us out of R, e.g., if $x_1 + 1 = x_2$. (*Warning:* With this interpretation, Eq. (5) then becomes a self-contained equation in R alone and one should not ask it to be valid if $X \notin R$.)

There is one annoying point about restricting attention to R in (5). When $x_1 = 1$ the left-hand side of (5) asks for the value of f for $x_1 = N_a$, which takes us outside R. Using the antisymmetry we conclude that

$$f(N_a, x_2, \ldots, x_N) = (-1)^{M-1} f(x_2, \ldots, x_M, N_a, x_{M+1}, \ldots, x_N) \tag{7}$$

with similar relations holding for $x_M = 1$, $x_{M+1} = N_a$ or $x_N = 1$. Eq. (7) and its three analogues reflect the "periodic boundary conditions" and, with its use, (5) becomes a self-contained equation on R alone.

We now come to the main reason for introducing R. *Let us assume that M and M' are both odd integers.* Then $(-1)^{M-1} = (-1)^{M'-1} = 1$ and we claim that: *For all U, the ground state of our Hamiltonian satisfies*

(1) There is only one ground state and
(2) $f(X)$ is a strictly positive function in R.

To prove (2) we think of (5) as an equation in R, as explained before. We note that all the off-diagonal terms in the Hamiltonian (thought of as a matrix $\widehat{\mathscr{H}}$ from $L^2(R)$ to $L^2(R)$) are non-positive (this is where we use the fact that $(-1)^{M-1} = (-1)^{M'-1} = 1$). If E_0 is the ground state energy and if $f(X)$ is a ground state eigenfunction (in R), which can be assumed to be real, then, by the variational principle, the function $g(X) = |f(X)|$ (in R) has an energy at least as low as that of f, i.e., $\langle g \,|\, g\rangle = \langle f \,|\, f\rangle$ and $\langle g \,|\widehat{\mathscr{H}}|\, g\rangle \leqslant \langle f \,|\widehat{\mathscr{H}}|\, f\rangle$ since $|f(X)|\,\widehat{\mathscr{H}}(X,Y)\,|f(Y)| \leqslant f(X)\widehat{\mathscr{H}}(X,Y)f(Y)$ for every X,Y. Hence g must be a ground state as well (since it cannot have a lower energy than E_0, by the definition of E_0). Therefore, $g(X)$ must satisfy (5) with the same E_0. Moreover, we see from (5) that $g(X)$ is strictly positive for *every* $X \in R$ (because if $g(Y) = 0$ for some $Y \in R$ then $g(Z) = 0$ for every Z that differs from Y by one 'hop'; tracing this backward, $g(X) = 0$ for every $X \in R$).

Returning now to f, let us assume the contrary of (2), namely, $f(X) > 0$ for some $X \in R$, and $f(Y) \leqslant 0$ for some $Y \in R$. We observe that since $h = g - f$ must also be a ground state (because sums of ground states are ground states, although not necessarily normalized), we have a ground state (namely h) that is non-negative and non-zero, but not strictly positive; this contradicts the fact, which we have just proved, that every non-negative ground state must be strictly positive. Thus, (2) is proved.

A similar argument proves (1). If f and f' are two linearly independent ground states then the state given by $k(X) = f(X) + cf'(X)$ is also a ground state and, for suitable c, $k(X)=0$ for some $X \in R$, but k cannot be identically zero. Then $|k|$ will be a non-negative ground state that is not strictly positive, and this contradicts statement (2).

The uniqueness statement (1) is important for the following reason. Suppose that we know the ground state for some particular value of U (e.g., $U = \infty$) and suppose we have a U-dependent solution to (5) in some interval of U values (e.g., $(0,\infty)$) with an energy $E(U)$ such that: (a) $E(\infty)$ is the known ground state energy and (b) $E(U)$ is continuous on the interval. Then $E(U)$ is necessarily the ground state energy in that interval. If not, the curve $E(U)$ would have to cross the ground state curve (which is always continuous), at which point there would be a degeneracy—which is impossible according to (1).

Items (1) and (2) can be used in two main applications. The first is the proof of the fact that when M and M' are odd the ground state belongs to total spin S equal to $|M - M'|/2$ and not to some higher S value. The proof is the same as in Ref. [12]. In Ref. [12] this property was shown to hold for *all* values of M and M', but for an open chain instead of a closed chain. In the thermodynamic limit this distinction is not important.

The second main application of these items (1) and (2) is a proof that the state we construct below using the Bethe Ansatz really is the ground state. This possibility is addressed at the end of Section 3 where we outline a strategy for such a proof. Unfortunately, we are unable to carry it out and we leave it as an open problem.

We also mention a theorem [13], which states that *the ground state is unique for* $M = M' = N_a/2$ *and* N_a= *even* (*the half-filled band*). There is no requirement for $M = M'$ to be odd.

3. The Bethe Ansatz

The Bethe Ansatz was invented [14] to solve the Heisenberg spin model, which is essentially a model of lattice bosons. The boson gas in the continuum with a positive delta function interaction and with positive density in the thermodynamic limit was first treated in Ref. [15]. McGuire [16] was the first to realize that the method could be extended to continuum fermions with a delta function interaction for $M = 1$. (The case $M = 0$ is trivial.) The first real mathematical difficulty comes with $M = 2$ and this was finally solved in Ref. [17]. The solution was inelegant and not transparent, but was a precursor to the full solution for general M by Gaudin [18] and Yang [19].

We now forget about the region R and focus, instead, on the fundamental regions (note the $\leqslant$ signs)

$$R_Q = \{X\colon 1 \leqslant x_{Q1} \leqslant \cdots \leqslant x_{QN} \leqslant N_a\} . \tag{8}$$

Here $Q = \{Q1, Q2, \ldots, QN\}$ is the permutation that maps the ordered set $\{1, 2, \ldots, N\}$ into $\{Q1, Q2, \ldots, QN\}$. There are $N!$ permutations and corresponding regions R_Q. The union of these regions is the full configuration space. These regions are disjoint *except* for their boundaries (i.e., points where $x_{Qi} = x_{Q(i+1)}$).

Let $k_1 < k_2 < \cdots < k_N$ be a set of *unequal*, *ordered* and *real* numbers in the interval $-\pi < k \leqslant \pi$, and let $[Q,P]$ be a set of $N! \times N!$ coefficients indexed by a pair of permutations Q, P, all yet to be determined.

When $X \in R_Q$ we write the function $f(X)$ as (the Bethe Ansatz)

$$f(X) = f_Q(x_1, \ldots, x_N) = \sum_P [Q,P] \exp[\mathrm{i}(k_{P1}x_{Q1} + \cdots + k_{PN}x_{QN})] . \tag{9}$$

In order for (9) to represent a function on the whole configuration space it is essential that the definitions (9) agree on the intersections of different R_Q's. This will impose conditions on the $[Q,P]$'s.

Choose some integer $1 \leqslant i < N$ and let $j = i+1$. Let P, P' be two permutations such that $Pi = P'j$ and $Pj = P'i$, but otherwise $Pm = P'm$ for $m \neq i, j$. Similarly, let Q, Q' be a pair with the same property (for this same choice of i) but otherwise P, P' and Q, Q' are unrelated.

The common boundary between R_Q and $R_{Q'}$ is the set in which $x_{Qi} = x_{Qj}$. In order to have $f_Q = f_{Q'}$ on this boundary it is sufficient to require that

$$[Q,P] + [Q,P'] = [Q',P] + [Q',P'] . \tag{10}$$

The reason that this suffices is that on this boundary we have $x_{Qi} = x_{Qj}$ and $k_{Pi} + k_{Pj} = k_{P'i} + k_{P'j}$. Thus, (10) expresses the fact that the exponential factor $\exp[\mathrm{i}(k_{Pi}x_{Qi} + k_{Pj}x_{Qj})] = \exp[\mathrm{i}(k_{P'i}x_{Qi} + k_{P'j}x_{Qj})]$ is the same for Q and Q', and for all values of the other x_m's.

Next we substitute the Ansatz (9) into (5). If $|x_i - x_j| > 1$ for all i,j then, clearly, we have

$$E = E(M, M') = -2\sum_{j=1}^{N} \cos k_j \ . \tag{11}$$

We next choose the coefficients $[Q,P]$ to make (11) hold generally—even if it is not possible to have $|x_i - x_j| > 1$ for all i,j when, for example, the number of electrons exceeds $N_a/2$. The requirement that (11) holds will impose further conditions on $[Q,P]$ similar to (10).

Sufficient conditions are obtained by setting $x_{Qi} = x_{Qj}$ on the right-hand side of (5) and requiring the exponential factors with x_{Qi} and x_{Qj} alone to satisfy (5). In other words, we require that

$$\begin{aligned}
&[Q,P]\mathrm{e}^{-\mathrm{i}k_{Pj}} + [Q,P']\mathrm{e}^{-\mathrm{i}k_{P'j}} + [Q,P]\mathrm{e}^{+\mathrm{i}k_{Pi}} + [Q,P']\mathrm{e}^{+\mathrm{i}k_{P'i}} \\
&\quad = [Q',P]\mathrm{e}^{-\mathrm{i}k_{Pi}} + [Q',P']\mathrm{e}^{-\mathrm{i}k_{P'i}} + [Q',P]\mathrm{e}^{+\mathrm{i}k_{Pj}} + [Q',P']\mathrm{e}^{+\mathrm{i}k_{P'j}} \\
&\quad\quad + U([Q,P] + [Q,P']) \ .
\end{aligned} \tag{12}$$

If we combine (12) with (10) and recall that $k_{Pj} = k_{P'i}$, etc., we obtain

$$\begin{aligned}
[Q,P] = &\frac{-\mathrm{i}U/2}{\sin k_{Pi} - \sin k_{Pj} + \mathrm{i}U/2}[Q,P'] \\
&+ \frac{\sin k_{Pi} - \sin k_{Pj}}{\sin k_{Pi} - \sin k_{Pj} + \mathrm{i}U/2}[Q',P'] \ .
\end{aligned} \tag{13}$$

It would seem that we have to solve both (13) *and* (10) for the $(N!)^2$ coefficients $[Q,P]$, and for each $1 \leqslant i \leqslant N-1$. Nevertheless, (13) alone is sufficient because it implies (10). To see this, add (13), as given, to (13) with $[Q',P]$ on the left side. Since $Q'' = Q$, the result is (10). Our goal, then, is to solve (13) for the coefficients $[Q,P]$ such that the amplitude f has the required symmetry.

These equations have been solved in Refs. [18,19], as we stated before, and we shall not repeat the derivation here. In these papers the function $\sin k$ appearing in (13) is replaced by k, which reflects the fact that Refs. [18,19] deal with the continuum and we are working on a lattice. This makes no difference as far as the algebra leading to Eqs. (14) below is concerned, but it makes a big difference for constructing a proof that these equations have a solution (the reason being that the sine function is not one-to-one).

The algebraic analysis in Refs. [18,19] leads to the following set of $N+M$ equations for the N ordered, real, unequal k's. (Recall that $M \leqslant M'$.) They involve an additional set of M ordered, unequal real numbers $\Lambda_1 < \Lambda_2 < \cdots < \Lambda_M$.

$$e^{ik_jN_a} = \prod_{\beta=1}^{M} \frac{i\sin k_j - i\Lambda_\beta - U/4}{i\sin k_j - i\Lambda_\beta + U/4}, \quad j = 1,2,\ldots,N$$

$$\prod_{j=1}^{N} \frac{i\sin k_j - i\Lambda_\alpha - U/4}{i\sin k_j - i\Lambda_\alpha + U/4} = -\prod_{\beta=1}^{M} \frac{-i\Lambda_\beta + i\Lambda_\alpha + U/2}{-i\Lambda_\beta + i\Lambda_\alpha - U/2}, \quad \alpha = 1,2,\ldots,M . \tag{14}$$

We remark that an explicit expression for the wave function $f(X)$ has been given by Woynarovich [20, part 1, Eqs. (2.5)–(2.9)].

These equations can be cast in a more transparent form (in which we now really make use of the fact that the k's and Λ's are ordered) by defining

$$\theta(p) = -2\tan^{-1}\left(\frac{2p}{U}\right), \quad -\pi \leqslant \theta \leqslant \pi .$$

Then, taking the logarithm of (14), we obtain two sets of equations

$$N_a k_j = 2\pi I_j + \sum_{\beta=1}^{M} \theta(2\sin k_j - 2\Lambda_\beta), \quad j = 1,2,\ldots,N , \tag{15}$$

$$\sum_{j=1}^{N} \theta(2\sin k_j - 2\Lambda_\alpha) = 2\pi J_\alpha - \sum_{\beta=1}^{M} \theta(\Lambda_\alpha - \Lambda_\beta), \quad \alpha = 1,2,\ldots,M , \tag{16}$$

where I_j is an integer (half-odd integer) if M is even (odd), while J_α is an integer (half-odd integer) if M' is odd (even).

It is noteworthy that in the $U \to \infty$ limit the two sets of equations essentially decouple. The Λ's are proportional to U in this limit, but the sum in (15) becomes independent of j. In particular, when the Λ's are balanced (i.e., for every Λ there is a $-\Lambda$) as in our case, then this sum equals zero.

From (15) and (16) we have the identity

$$k_{total} \equiv \sum_{j=1}^{N} k_j = \frac{2\pi}{N_a}\left(\sum_{j=1}^{N} I_j + \sum_{\alpha=1}^{M} J_\alpha\right) . \tag{17}$$

For the ground state, with $N = 2\times$ (odd integer) and $M = N/2 =$ odd, we make the choice of the I_j and J_α that agrees with the correct values in the case $U = \infty$, namely

$$I_j = j - (N+1)/2, \quad J_\alpha = \alpha - (M+1)/2 . \tag{18}$$

We are not able to prove the existence of solutions to (15) and (16) that are real and increasing in the index j and α. In the next section, however, we show that the $N \to \infty$ limit of (15) and (16) has a solution, and in Section 6 we obtain the solution explicitly for $N/2M = N/N_a = 1$. This leaves little doubt that (15) and (16) can be solved as well, at least for large N.

Assuming that $M = M' = N/2$ is odd, the solution is presumably unique with the given values of I_j and J_α and belongs to total spin $S = 0$.

Assuming that the solution exists, we would still need a few more facts (which we have not proved) in order to prove that the Bethe Ansatz gives the ground state:

(a) prove that the wave function (9) is not identically zero,
(b) prove that the wave function (9) is a continuous function of U.

From the uniqueness of the ground state proved in Section 2, and the fact that solution (9) coincides with the exact solution for $U = \infty$ (in which case $f_Q(x)$ is a Slater determinant of plane waves with wavenumbers $k_j = 2\pi I_j/N_a$), (a) and (b) now establish that wave function (9) must be the ground state for all U.

Remark. Assuming that the Bethe Ansatz gives the ground state for a given $M \leqslant M'$ then, as remarked at the end of Section 2 (and assuming M and M' to be odd) the value of the total spin in this state is $S = (M' - M)/2$. Thus, the solution to the Bethe Ansatz we have been looking at is a highest weight state of $SU(2)$, i.e., a state annihilated by spin raising operators.

4. The ground state

For the ground state $I_j = I(k_j)$ and $J_\alpha = J(\Lambda_\alpha)$ are consecutive integers or half-odd integers centered around the origin. As stated in Section 3, each k_j lies in $[-\pi, \pi]$ (since $k_j \to k_j + 2\pi n$ defines the same wave function). In the limit of N_a, N, M, $M' \to \infty$ with their ratios kept fixed, the real numbers k and Λ are distributed between $-Q$ and $Q \leqslant \pi$ and $-B$ and $B \leqslant \infty$ for some $0 < Q \leqslant \pi$ and $0 < B \leqslant \infty$. In a small interval $\mathrm{d}k$ the number of k values, and hence the number of j values in (15), is $N_a\rho(k)\,\mathrm{d}k$, where ρ is a density function to be determined. Likewise, in a small interval $\mathrm{d}\Lambda$ the number of Λ values and α values in (16) is $N_a\sigma(\Lambda)\,\mathrm{d}\Lambda$. An alternative point of view is to think of $I(k)$ as a function of the variable k. Then $I(k + \mathrm{d}k) - I(k)$ counts the number of k values between k and $k + \mathrm{d}k$ so we have $\mathrm{d}I(k)/\mathrm{d}k = N_a\rho(k)$. A similar remark holds for $J(\Lambda)$.

The density functions $\rho(k)$ and $\sigma(\Lambda)$ satisfy the obvious normalization

$$\int_{-Q}^{Q} \rho(k)\,\mathrm{d}k = N/N_a, \qquad \int_{-B}^{B} \sigma(\Lambda)\,\mathrm{d}\Lambda = M/N_a\,. \tag{19}$$

By subtracting (15) with j from (15) with $j+N_a\rho(k)\,\mathrm{d}k$, and taking the limit $N_a \to \infty$ we obtain (20) below. Likewise, subtracting (16) with α from (16) with $\alpha+N_a\sigma(\Lambda)\,\mathrm{d}\Lambda$, and taking the limit $N_a \to \infty$ we obtain (21). An alternative point of view is to take the derivatives of (15) and (16) with respect to k_j and Λ_α, respectively, set $\mathrm{d}I/\mathrm{d}k = N_a\rho(k)$, $\mathrm{d}J/\mathrm{d}\Lambda = N_a\sigma(\Lambda)$, and take the $N_a \to \infty$ limit. In either case we obtain

$$1 = 2\pi\rho(k) + 2\cos k \int_{-B}^{B} \mathrm{d}\Lambda\,\sigma(\Lambda)\theta'(2\sin k - 2\Lambda)\,, \tag{20}$$

$$-2\int_{-Q}^{Q} \mathrm{d}k\,\rho(k)\theta'(2\sin k - 2\Lambda)$$

$$= 2\pi\sigma(\Lambda) - \int_{-B}^{B} \mathrm{d}\Lambda'\sigma(\Lambda')\theta'(\Lambda - \Lambda') \tag{21}$$

or, equivalently,

$$\boxed{\rho(k) = \frac{1}{2\pi} + \cos k \int_{-B}^{B} K(\sin k - \Lambda)\sigma(\Lambda)\,\mathrm{d}\Lambda\,,} \tag{22}$$

$$\boxed{\sigma(\Lambda) = \int_{-Q}^{Q} K(\sin k - \Lambda)\rho(k)\,\mathrm{d}k - \int_{-B}^{B} K^2(\Lambda - \Lambda')\sigma(\Lambda')\,\mathrm{d}\Lambda'\,,} \tag{23}$$

where

$$K(\Lambda - \Lambda') = -\frac{1}{\pi}\,\theta'(2\Lambda - 2\Lambda') = \frac{1}{2\pi}\left[\frac{8U}{U^2 + 16(\Lambda - \Lambda')^2}\right],$$

$$K^2(\Lambda - \Lambda') = -\frac{1}{2\pi}\,\theta'(\Lambda - \Lambda') = \frac{1}{2\pi}\left[\frac{4U}{U^2 + 4(\Lambda - \Lambda')^2}\right]$$

$$= \int_{-\infty}^{\infty} K(\Lambda - x)K(x - \Lambda')\,\mathrm{d}x\,.$$

Note that K^2 is the square of K in the sense of operator products. Note also that (22) and (23) are to be satisfied only for $|k| \leqslant Q$ and $|\Lambda| \leqslant B$. Outside these intervals ρ and σ are not uniquely defined, but we *can* and will define them by the right-hand sides of (22) and (23).

The following Fourier transforms will be used in later discussions:

$$\int_{-\infty}^{\infty} \mathrm{e}^{\mathrm{i}\omega\Lambda}K(\Lambda)\,\mathrm{d}\Lambda = \mathrm{e}^{-U|\omega|/4}, \qquad \int_{-\infty}^{\infty} \mathrm{e}^{\mathrm{i}\omega\Lambda}K^2(\Lambda)\,\mathrm{d}\Lambda = \mathrm{e}^{-U|\omega|/2}\,. \tag{24}$$

The ground state energy (11) now reads

$$E(M, M') = -2N_a \int_{-Q}^{Q} \rho(k)\cos k\,\mathrm{d}k\,, \tag{25}$$

where $\rho(k)$ is to be determined together with $\sigma(\Lambda)$ from the coupled integral equations (22) and (23) subject to the normalizations (19).

5. Analysis of the integral equations

In this section, we shall prove that Eqs. (22) and (23) have unique solutions for each given $0 < Q \leqslant \pi$ and $0 < B \leqslant \infty$ and that the solutions are positive and have certain monotonicity properties. These properties guarantee that the normalization conditions (19) uniquely determine values of Q, B for each given value of N when $M = M' = N/2$ (in this case we have $B = \infty$). However, we have not proved uniqueness of Q, B when $M \neq M'$ (although we believe there is uniqueness). But this does not matter for the absolute ground state since, as remarked earlier, the ground state has $S = 0$ (in the thermodynamic limit) and so we are allowed to take $S^z = 0$. For $M \neq M'$, we have remarked earlier that the solution probably has $S = |M' - M|/2$ and is the ground state for $S = |M' - M|/2$.

An important first step is to overcome the annoying fact (which is relevant for $Q > \pi/2$) that $\sin k$ is not a monotonic function of k in $[-\pi, \pi]$. To do this we note that $(\cos k) K(\sin k - \Lambda)$ is an odd function of $k - \pi/2$ (for each Λ) and hence $\rho(k) - 1/2\pi$ also has this property. On the other hand, $K(\sin k - \Lambda)$ appearing in (23) is an even function of $k - \pi/2$. As a result $\rho(k)$ appearing in the first term on the right-hand side of (23) can be replaced by $1/2\pi$ in the intervals $Q' < k < Q$ and $-Q < k < -Q'$, where $Q' = \pi - Q$. Thus, when $Q > \pi/2$, we can rewrite the $[Q', Q]$ portion of the first integral in (23) as

$$\int_{Q'}^{Q} K(\sin k - \Lambda)\rho(k)\,dk = \int_{Q'}^{Q} K(\sin k - \Lambda)\frac{1}{2\pi}\,dk$$
$$= \frac{2}{2\pi}\int_{Q'}^{\pi/2} K(\sin k - \Lambda)\,dk\,.$$

A similar thing can be done for the $[-Q, -Q']$ portion and for the corresponding portions of (19).

The integrals over k now extend at most over the interval $[-\pi/2, \pi/2]$, in which $\sin k$ is monotonic.

We are now in a position to change variables as follows. For $-1 \leqslant x \leqslant 1$ let

$$t(x) = \frac{1}{2\pi}(1 - x^2)^{-1/2}, \quad f(x) = (1 - x^2)^{-1/2}\rho(\sin^{-1} x)\,. \tag{26}$$

In case $Q < \pi/2$, $\rho(\sin^{-1} x)$ is defined only for $\sin x \leqslant Q$, but we shall soon see (after (28)) how to extend the definition of f in this case. We define the step functions for all real x by

$$\begin{aligned} B(x) &= 1, \quad |x| < B, && = 0, \quad \text{otherwise} \\ A(x) &= 1, \quad |x| < a, && = 0, \quad \text{otherwise} \\ D(x) &= H(Q), \quad a < |x| < 1, && = 0, \quad \text{otherwise}\,, \end{aligned} \tag{27}$$

where $a = \sin Q = \sin Q'$ and where

$$H(Q) = 0 \quad \text{if } Q \leqslant \pi/2, \qquad H(Q) = 2 \quad \text{if } Q > \pi/2\,.$$

The integral equations (22) and (23) become

$$f(x) = t(x) + \int_{-\infty}^{\infty} K(x - x')B(x')\sigma(x')\,dx', \quad |x| \leqslant a\,, \tag{28}$$

$$\sigma(x) = \int_{-\infty}^{\infty} K(x-x')A(x')f(x')\,dx' + \int_{-\infty}^{\infty} K(x-x')D(x')t(x')\,dx'$$
$$- \int_{-\infty}^{\infty} K^2(x-x')B(x')\sigma(x')\,dx', \quad |x| < B . \tag{29}$$

Although these equations have to be solved in the stated intervals we can use their right-hand sides to define their left-hand sides for *all* real x. We define $t(x) \equiv 0$ for $|x| > 1$. It is obvious that the extended equations have (unique) solutions if and only if the original ones do. Henceforth, we shall understand the functions f and σ to be defined for all real x.

These equations read, in operator form

$$f = t + \hat{K}\hat{B}\sigma , \tag{30}$$

$$\sigma = \hat{K}\hat{A}f + \hat{K}\hat{D}t - \hat{K}^2\hat{B}\sigma , \tag{31}$$

where $\hat{K}$ is convolution with K and $\hat{A}, \hat{B}, \hat{D}$ are the multiplication operators corresponding to A, B, D (and which are also projections since $\hat{A}^2 = \hat{A}$, etc.).

In view of the normalization requirements (19), the space of functions to be considered is, obviously, $L^1([-a,a])$ for f and $L^1([-B,B])$ for σ. (L^p is the pth power integrable functions and L^∞ is the bounded functions.) Since $K(x)$ is in $L^1(\mathbb{R}) \cap L^\infty(\mathbb{R})$, it is a simple consequence of Young's inequality that the four integrals in (28) and (29) are automatically in $L^1(\mathbb{R}) \cap L^\infty(\mathbb{R})$ when $f \in L^1([-a,a])$ and $\sigma \in L^1([-B,B])$. In particular, the integrals are in $L^2(\mathbb{R})$, which allows us to define the operators in (30), (31) as bounded operators on $L^2(\mathbb{R})$. In addition, t is in $L^1(\mathbb{R})$, but not in $L^2(\mathbb{R})$. To summarize, solutions in which f and σ are in $L^1(\mathbb{R})$ automatically have the property that $f - t$ and σ are both in $L^p(\mathbb{R})$ for all $1 \leqslant p \leqslant \infty$.

Theorem 1 (Uniqueness). *The solutions $f(x)$ and $\sigma(x)$ are unique and positive for all real x.*

Remark. The uniqueness implies that f and σ are even functions of x (because the pair $f(-x), \sigma(-x)$ is also a solution). The theorem implies (from the definition (26)) that $\sigma(\Lambda) > 0$ for all real Λ and it implies that $\rho(k) > 0$ for all $|k| \leqslant \pi/2$. It does *not* imply that $\rho(k)$, defined by the right-hand side of (22), is non-negative for all $|k| > \pi/2$. We shall prove this positivity, however, in Lemma 3. Note that the positivity of ρ is equivalent to the statement that $f(x) < 2t(x)$ for all $|x| \leqslant 1$ because, from (22) and the evenness or σ, $\rho(\pi - k) = (1/\pi) - \rho(k)$.

Proof. By substituting (30) into (31) and rearranging slightly we obtain

$$(1 + \hat{K}^2)\sigma = \hat{K}(\hat{A} + \hat{D})t + \hat{K}^2(1 - \hat{B})\sigma + \hat{K}\hat{A}\hat{K}\hat{B}\sigma . \tag{32}$$

Since $\hat{K}^2$ is positive definite, $1 + \hat{K}^2$ has an inverse $1/(1 + \hat{K}^2)$, which we can apply to both sides of (32). The convolution operator

$$\hat{R} = \hat{K}(1 + \hat{K}^2)^{-1} \tag{33}$$

has a Fourier transform $\frac{1}{2}\mathrm{sech}(\omega/4)$. The inverse Fourier transform is proportional to $\mathrm{sech}(2\pi x)$ (see (55) below), which is positive. In other words, $\hat{R}$ is not only a positive operator, it also has a positive integral kernel.

We can rewrite (32) as

$$(1-\hat{W})\sigma = \hat{R}(\hat{A}+\hat{D})t \equiv \xi \tag{34}$$

with

$$\hat{W} = \hat{R}\hat{K}(1-\hat{B}) + \hat{R}\hat{A}\hat{K}\hat{B} = \hat{R}[\hat{K} - (1-\hat{A})\hat{K}\hat{B}] . \tag{35}$$

The middle expression shows that the integral kernel of $\hat{W}$ is positive.

Clearly, $\xi > 0$ as a function and $\xi \in L^1(\mathbb{R}) \cap L^2(\mathbb{R})$. Also, $\hat{W}$ has a positive integral kernel. We note that $\|\hat{R}\| = 1/2$ on $L^2(\mathbb{R})$ since $y/(1+y) \leqslant 1/2$ for $y \geqslant 0$. Also, $\|\hat{K}\| = 1$, and $\|1-\hat{B}\| = 1$, $\|\hat{A}\| = 1$, $\|\hat{B}\| = 1$. In fact, it is easy to check that $\|\hat{R}\hat{A}\hat{K}\hat{B}\| < 1/2$. From this we conclude that $\|\hat{W}\| < 1$ on $L^2(\mathbb{R})$ and thus $1-\hat{W}$ has an inverse (as a map from $L^2(\mathbb{R}) \to L^2(\mathbb{R})$).

Therefore, we can solve (32) by iteration:

$$\sigma = (1 + \hat{W} + \hat{W}^2 + \hat{W}^3 + \cdots)\xi . \tag{36}$$

This is a strongly convergent series in $L^2(\mathbb{R})$ and hence (36) solves (32) in $L^2(\mathbb{R})$. It is the unique solution because the homogeneous equation $(1-\hat{W})\phi = 0$ has no solution. Moreover, since each term is a positive function, we conclude that σ is a positive function as well. □

Lemma 1 (Monotonicity in B). *When B increases with Q fixed, $\sigma(x)$ decreases pointwise for all $x \in \mathbb{R}$.*

Proof. Since $1-\hat{A}$ is fixed and positive, we see from the right-hand side of (35) that the integral kernel of $\hat{W}$ is monotone decreasing in $\hat{B}$. The lemma then follows from the representation (36). □

Lemma 2 (Monotonicity in B). *When B increases with Q fixed, $f(x)$ increases pointwise for all $x \in \mathbb{R}$. This implies, in particular, that $\rho(k)$ increases for all $|k| \leqslant \pi/2$ and decreases for all $\pi/2 \leqslant |k| \leqslant \pi$.*

Proof. Consider Eq. (32) for the case $A = 0$. Theorem 1 and Lemma 1 hold in this case, of course. We also note that their proofs do not depend on any particular fact about the function Dt, other than the fact that it is a non-negative function. From these observations we learn that the solution to the equation

$$(1 + \hat{K}^2\hat{B})S = \hat{K}g \tag{37}$$

has the property, for all $x \in R$, that $S(x) \geqslant 0$ and that $S(x)$ is a non-increasing function of B, provided only that $g(x) \geqslant 0$ for all $x \in \mathbb{R}$.

Another way to say this is that the integral kernel of $\hat{V} = (1+\hat{K}^2\hat{B})^{-1}\hat{K}$ is positive and is a pointwise monotone decreasing function of B.

Now let us rewrite (37) as

$$\hat{K}^2(1+\hat{K}^2)^{-1}g = \hat{U}g + \hat{K}(1+\hat{K}^2)^{-1}(1-\hat{B})S \tag{38}$$

with

$$\hat{U} = \hat{K}\hat{B}\hat{V} = \hat{K}\hat{B}(1+\hat{K}^2\hat{B})^{-1}\hat{K}\,. \tag{39}$$

The operator $\hat{U}$ has a positive integral kernel since $\hat{V}$, $\hat{K}$, and $\hat{B}$ have one. As B increases the second term on the right-hand side of (38) decreases pointwise (because $(1-\hat{B})$ decreases as a kernel and S decreases, as we have just proved). The left-hand side of (38) is independent of B and, therefore, the first term on the right-hand side of (38) must increase pointwise. Since this holds for arbitrary positive g, we conclude that the integral kernel of $\hat{U}$, in contrast to that of $\hat{V}$, is a pointwise *increasing* function of B.

Having established the monotonicity property of $\hat{U}$ let us return to f, which we can write (from (30)) as

$$f = (1+\hat{U}\hat{D})t + \hat{U}\hat{A}f \tag{40}$$

$$=[1+\hat{U}\hat{A}+(\hat{U}\hat{A})^2+\cdots](1+\hat{U}\hat{D})t\,. \tag{41}$$

The series in (41) is strongly convergent (since $\|\hat{A}\| = 1$ and $\|\hat{U}\| \leqslant 1/2$) and thus defines the solution to (40). Since $\hat{U}$ is monotone in $\hat{B}$, (41) tells us that f is also pointwise monotone, as claimed.

Eq. (26) tells us that $\rho(k)$ is increasing in B for $|k| \leqslant \pi/2$ and is decreasing in B for $\pi/2 \leqslant |k| \leqslant \pi$. □

Theorem 2 (Monotonicity in B). *When B increases with Q fixed, N/N_a and M/N increase. When $B=\infty$, we have $2M = N$, and when $B<\infty$ we have $2M < N$ (for all Q).*

Proof. The integral for N/N_a in (19) can be written as $\int_{-\infty}^{\infty}[A\rho+(1/2\pi)D]$, and this is monotone increasing in B since ρ is monotone for $|k| \leqslant \pi/2$ and $A(k)=0$ for $|k| > \pi/2$.

If we integrate (23) from $\Lambda=-\infty$ to ∞, and use the fact that $\int K = 1$ from (24), we obtain

$$\frac{N}{N_a} = \int_{-Q}^{Q}\rho(k)\,\mathrm{d}k = \int_{-\infty}^{\infty}\sigma(\Lambda)\,\mathrm{d}\Lambda + \int_{-B}^{B}\sigma(\Lambda)\,\mathrm{d}\Lambda \tag{42}$$

which becomes, after making use of the normalization (19)

$$1 = 2\frac{M}{N} + \frac{N_a}{N}\left[\int_{-\infty}^{-B} + \int_{B}^{\infty}\right]\sigma(\Lambda)\,\mathrm{d}\Lambda\,. \tag{43}$$

Now the integrals in (43) decrease as B increases by Lemma 1 and converge to 0 as $B \to \infty$, while N/N_a increases, as we have just proved. It follows that M/N increases monotonically with B, reaching $M/N = 1/2$ at $B = \infty$. If $B < \infty$ then $M/N < 1/2$ since σ is a strictly positive function. □

We turn now to the dependence of σ, ρ on Q, with fixed B. First, Lemma 3 (which was promised in the remark after Theorem 1) is needed.

Lemma 3 (Positivity of ρ). *For all* $B \leqslant \infty$, *all* $Q \leqslant \pi$, *and all* $|k| \leqslant \pi$, *we have* $\rho(k) > 0$.

Proof. As mentioned in the Remark after Theorem 1, the positivity of ρ is equivalent to the statement that $f(x) < 2\, t(x)$ for all $|x| \leqslant 1$. We shall prove $f(x) < 2t(x)$ here.

Owing to the monotonicity in B of f (Lemma 2) it suffices to prove the lemma for $B = \infty$, which we assume now. We see from (41) that for any given value of a the worst case is $Q > \pi/2$, whence $H(Q) = 2$ and $D > 0$. We assume this also.

For the purpose of this proof (only) we denote the dependence of $f(x)$ on a by $f_a(x)$.

We first consider the case $a=0$, corresponding to $Q=\pi$. Let us borrow some information from the next section, where we actually solve the equations for $B = \infty$, $Q = \pi$ and discover (Lemma 5) that $f(x) < 2t(x)$ for $|x| \leqslant 1$ (for $U > 0$).

We see from (40) or (41) that f_a is continuous in a and differentiable in a for $0 < a < 1$ (indeed, it is real analytic). Also, since the kernel $K(x-y)$ is smooth in (x, y) and $t(x)$ is smooth in $x \in (-1, 1)$, it is easy to see that f_a is smooth, too, for $x \in (-1, 1)$. Eq. (28) defines $f_a(x)$ pointwise for all x and $f_a(x)$ is jointly continuous in a, x.

In detail, (40) reads

$$f_a(x) = t(x) + 2\left[\int_{-1}^{-a} + \int_a^1\right] U(x,x')t(x')\,\mathrm{d}x' + \int_{-a}^{a} U(x,x')f_a(x')\,\mathrm{d}x' \,. \tag{44}$$

Take the derivative with respect to a and set $h_a(x) = \partial f_a(x)/\partial a$. Observe that $\hat{U}$ does not depend on a. We obtain

$$h_a(x) = U(x,a)[f_a(a) - 2t(a)] + U(x,-a)[f_a(-a) - 2t(-a)] + \int_{-a}^{a} U(x,x')h_a(x')\,\mathrm{d}x' \,. \tag{45}$$

(This equation makes sense because $f_a(x)$ is jointly continuous in x, a and $t(x)$ is continuous for $|x| < 1$. Recall that f and t are even functions of x. Note that U here is the kernel of (39) with $B = \infty$, i.e., $\hat{U} = \hat{K}^2(1 + \hat{K}^2)^{-1}$, which is self-adjoint and positive as an operator and positive as a kernel.)

Eq. (45) can be iterated in the same manner as (41) (since $\|\hat{U}\| = 1/2$)

$$h_a(x) = [\hat{U} + \hat{U}\hat{A}\hat{U} + \hat{U}\hat{A}\hat{U}\hat{A}\hat{U} + \cdots](x,a)F(a) \equiv T(x,a)F(a) \,, \tag{46}$$

where $[\cdot](x,a)$ denotes the integral kernel of $\hat{T}=[\cdot]$, and where $F(a)=f_a(a)-2t(a)$ is a number. As an operator. $\hat{T}$ is self-adjoint and positive.

Now $\hat{U}$ has a positive kernel and thus $T(x,a)\geqslant 0$, so $h_a(x)<0$ for all x if and only if $F(a)<0$. We have already noted that $F(0)<0$.

We can integrate (46) to obtain

$$f_a(x)=f_0(x)+\int_0^a h_{a'}(x)\,\mathrm{d}a'=f_0(x)+\int_0^a T(x,a')[f_{a'}(x)-2t(a')]\,\mathrm{d}a' . \quad (47)$$

If we subtract $2t(a)$ from this and set $x=a$, we obtain

$$F(a)=G(a)+\int_0^a T(a,a')F(a')\,\mathrm{d}a' , \quad (48)$$

where $G(a)=f_0(a)-2t(a)<0$. Another way to state (48) is $F=G+\hat{T}\hat{A}F$.

Eq. (48) implies that $F(a)<0$ for all a, as desired. There are two ways to see this. One way is to note that $\hat{T}$ is monotone increasing in a (as an operator and as a kernel), so $\hat{T}\leqslant\hat{U}+\hat{U}^2+\cdots=\hat{K}^2<1$. Therefore, (48) can be iterated as $F=[1+\hat{T}\hat{A}+\hat{T}\hat{A}\hat{T}\hat{A}+\cdots]G$, and this is negative. The second way is to note that $f_a(a)$ (and hence $F(a)$) is continuous in a. Let a^* be the smallest a for which $F(a)=0$. Then, from (48), $0=F(a^*)=G(a^*)+\int_0^{a^*}T(a^*,a')F(a')\,\mathrm{d}a'<0$, which is a contradiction.

From $F(a)<0$ we can deduce that $f_a(x)-2t(x)<0$ for all $|x|\leqslant 1$. Simply subtract $2t(x)$ from both sides of (47). Then $f_a(x)-2t(x)=\{f_0(x)-2t(x)\}+(\hat{T}\hat{A}F)(x)$. The first term $\{\}<0$ by Lemma 5, which we prove in Section 6 below, and the second term is <0 (since $F<0$). □

Lemma 4 (Monotonicity in Q). *Consider the dependence of the solution to* (30), (31) *on the parameter* $0\leqslant a\leqslant 1$ *for fixed* $B\leqslant\infty$. *For* $Q\leqslant\pi/2$ (*i.e.*, $H(Q)=D=0$), *both* f *and* σ *increase pointwise as a increases. For* $Q>\pi/2$ (*i.e.*, $H(Q)=2, Dt=2(1-A)t$), *both* f *and* σ *decrease pointwise as a increases.*

If, instead of the dependence on a, we consider the dependence on $0\leqslant Q\leqslant\pi$ *of* $\rho(k)$ (*which is defined by* (22) *for all* $|k|\leqslant\pi$) *and of* $\sigma(\Lambda)$ (*which is defined by* (23) *for all real* Λ), *then, as* Q *increases*

$$\rho(k) \textit{ increases for } 0\leqslant|k|<\pi/2 \textit{ and decreases for } \pi/2\leqslant|k|\leqslant\pi$$

$$\sigma(\Lambda) \textit{ increases for all real } \Lambda . \quad (49)$$

Proof. Concerning the monotonicities stated in the second part of the lemma, (49), we note that as Q goes from 0 to $\pi/2$, *a increases* from 0 to 1, but when Q goes from $\pi/2$ to 0, *a decreases* from 1 to 0. Moreover, $H(Q)=0$ in the first case and $H(Q)=2$ in the second case. This observation shows that the first part of the lemma implies the statement about σ in (49). The statement about ρ in (49) also follows, if we take note of the $\cos k$ factor in (49).

We now turn to the first part of the lemma. The easy case is $Q \leqslant \pi/2$ or $H(Q)=0$. Then (41) does not have the $\hat{U}\hat{D}t$ term and, since $\hat{U}$ has a positive kernel and since $\hat{A}$ has a kernel that increases with a, we see immediately that f increases with a. Likewise, from (34), (35), we see that $\hat{W}$ and ξ increase with a and, from (36), we see that σ increases.

For $Q > \pi/2$ or $H(Q) = 2$, we proceed as in the proof of Lemma 3 by defining $h_a(x) = \partial f_a(x)/\partial a$ and proceeding to (46) (but with $\hat{U}$ given by (39)). This time we know that $F(a) < 0$ (by Lemma 3) and hence $h_a(x) < 0$, as claimed. The monotonicity of $\sigma(x)$ follows by differentiating (29) with respect to a. Then $(\partial\sigma(x)/\partial a)=(\hat{V}\hat{A}h_a)(x)+V(x,a)F(a)$, where $V(x,y)$ is the kernel of $\hat{V}$, which is positive, as noted in the proof of Lemma 2. □

Theorem 3 (Monotonicity in Q). *When Q increases with fixed B, N/N_a and M/N_a increase. When $Q=\pi$, $N/N_a = 1$ (for all B), while $N/N_a < 1$ if $Q < \pi$.*

Proof. From (42), $N/N_a = 2\int_{-B}^{B}\sigma + 2\int_{B}^{\infty}\sigma$ and this increases with Q by (49). Also, by (42), $N/N_a = \int_{-Q}^{Q}\rho$. When $Q=\pi$, we see from (22) that $\int_{-Q}^{Q}\rho = \int_{-\pi}^{\pi}(1/2\pi) = 1$, so $N/N_a = 1$. To show that $N/N_a < 1$ when $Q < \pi$ we use the monotonicity of σ with respect to B (Lemma 2) and Q (Lemma 3) (with $\sigma_0(\Lambda)=$ the value of $\sigma(\Lambda)$ for $B=\infty$, $Q=\pi$) to conclude that $N/N_a \leqslant 2\int_{-B}^{B}\sigma_0+2\int_{B}^{\infty}\sigma_0=\int_{-\infty}^{\infty}\sigma_0=1-2\int_{B}^{\infty}\sigma_0 < 1$, since σ_0 is a strictly positive function.

Finally, from (42) we have that $M/N_a=\int_{-B}^{B}\sigma$, and this increases with Q by (49). □

6. Solution for the half-filled band

In the case of a half-filled band, we have $N=N_a$, $M=M'=N/2$ and, from Theorems 2 and 3, $Q=\pi$, $B=\infty$. In this case the integral equations (22) and (23) can be solved. We use the notation $\rho_0(k)$ and $\sigma_0(\Lambda)$ for these solutions.

Substituting (22) into (23) where, as explained earlier, we use $\rho_0(k)=1/2\pi$ in the first term on the right-hand side of (23). Then the integral equation (23) involves only $\sigma_0(\Lambda)$ and can be solved by Fourier transform. Using equations (24) it is straightforward to obtain the solution for σ_0 and its Fourier transform $\hat{\sigma}_0$ as

$$\hat{\sigma}_0(\omega) = \int_{-\infty}^{\infty} e^{i\omega\Lambda}\sigma_0(\Lambda)\,d\Lambda = \frac{J_0(\omega)}{2\cosh(U\omega/4)}, \tag{50}$$

$$\boxed{\sigma_0(\Lambda) = \frac{1}{2\pi}\int_0^{\infty}\frac{J_0(\omega)\cos(\omega\Lambda)}{\cosh(\omega U/4)}\,d\omega\,,} \tag{51}$$

where

$$J_0(\omega) = \frac{2}{\pi}\int_0^{\pi/2}\cos(\omega\cos\theta)\,d\theta = \frac{1}{\pi}\int_0^{\pi}\cos(\omega\sin\theta)\,d\theta \tag{52}$$

is the zeroth order Bessel function.

Next we substitute (51) into (22) and this leads (with (24)) to

$$\boxed{\rho_0(k) = \frac{1}{2\pi} + \frac{\cos k}{\pi} \int_0^\infty \frac{\cos(\omega \sin k) J_0(\omega)}{1 + e^{\omega U/2}} \, d\omega .} \tag{53}$$

The substitution of (53) into (25) finally yields the ground state energy, E_0, of the half-filled band as

$$\boxed{E_0\left(\frac{N_a}{2}, \frac{N_a}{2}\right) = -4N_a \int_0^\infty \frac{J_0(\omega) J_1(\omega)}{\omega(1 + e^{\omega U/2})} \, d\omega ,} \tag{54}$$

where $J_1(\omega) = \pi^{-1} \int_0^\pi \sin(\omega \sin p) \sin p \, dp = \omega \pi^{-1} \int_0^\pi \cos(\omega \sin p) \cos^2 p \, dp$ is the Bessel function of order one.

Remarks. (A) When there is no interaction ($U=0$), $\hat{K}$ is a δ-function; we can evaluate (51) and (53) as

$$\sigma_0(\Lambda) = \frac{1}{2\pi\sqrt{1-\Lambda^2}}, \quad |\Lambda| \leqslant 1; \qquad = 0, \text{ otherwise} ,$$

$$\rho_0(k) = \frac{1}{\pi}, \quad |k| \leqslant \frac{\pi}{2}; \qquad = 0, \text{ otherwise} .$$

This formula for $\rho_0(k)$ agrees with what is expected for an ideal Fermi gas.

(B) The $U \to \infty$ limit is peculiar. From (50) we see that $\hat{\sigma}_0(0) = \frac{1}{2}$, so $\int \sigma_0 = \frac{1}{2}$, but from (51) we see that $\sigma_0(\Lambda) \to 0$ in this limit, uniformly in Λ. On the other hand $\rho_0(k) \to \frac{1}{2}\pi$, for all $|k| \leqslant \pi$, which is what one would expect on the basis of the fact that this 'hard core' gas becomes, in effect, a one-component ideal Fermi gas of $N = N_a$ particles.

We now derive alternative, more revealing expressions for σ_0, ρ_0.

For $\sigma_0(\Lambda)$ we substitute the integral representation (52) for J_0 into (51) and recall the Fourier cosine transform (for $\alpha > 0$)

$$\int_0^\infty \frac{\cos(\omega x)}{\cosh(\omega \alpha)} \, d\omega = \left(\frac{\pi}{2\alpha}\right) \frac{1}{\cosh(\pi x / 2\alpha)} . \tag{55}$$

Then, using $2 \cos a \cos b = \cos(a-b) + \cos(a+b)$ we obtain

$$\boxed{\sigma_0(\Lambda) = \frac{1}{\pi U} \int_0^\pi \frac{d\theta}{\cosh[2\pi(\Lambda + \cos\theta)/U]} > 0 .} \tag{56}$$

An alternate integral representation can be derived similarly for $\rho_0(k)$, but the derivation and the result is more complicated. We substitute $(1 + e^x)^{-1} = \sum_{n=1}^\infty (-1)^n \exp[-nx]$, with $x = \omega U/2$, into (53) and make use of the identity (with $\alpha = -is \pm c$ in the notation of Gradshteyn and Ryzhik [21, 6.611.1])

$$2\int_0^\infty e^{-c\omega}J_0(\omega)\cos(\omega s)\,d\omega = [(-c-is)^2+1]^{-1/2}+[(c-is)^2+1]^{-1/2} \tag{57}$$

for $c>0$, and where the square roots $[\quad]^{-1/2}$ in (57) are taken to have a positive real part. This leads to

$$\rho_0(k) = \frac{1}{2\pi}+\frac{\cos k}{2\pi}\sum_{n=1}^{\infty}(-1)^{n+1}\{[(-nU/2-i\sin k)^2+1]^{-1/2} + [(nU/2-i\sin k)^2+1]^{-1/2}\}\,. \tag{58}$$

We can rewrite the sum of the two terms in (58) as a single sum from $n=-\infty$ to ∞, after making a correction for the $n=0$ term (which equals $\cos k/|\cos k|$ for $k\neq\pi/2$). We obtain

$$\begin{aligned}\rho_0(k) &= \frac{1}{2\pi}\left[1+\frac{\cos k}{|\cos k|}\right]-\frac{\cos k}{2\pi}\sum_{n=-\infty}^{\infty}(-1)^n[(nU/2-i\sin k)^2+1]^{-1/2}\\ &= \frac{1}{2\pi}\left[1+\frac{\cos k}{|\cos k|}\right]-\frac{\cos k}{2\pi}\frac{1}{2\pi i}\int_C\frac{dz}{\sqrt{(zU/2-i\sin k)^2+1}}\frac{\pi}{\sin(\pi z)}\,.\end{aligned} \tag{59}$$

The contour C encompasses the real axis, i.e., it runs to the right just below the real axis and to the left just above the real axis.

The integrand has two branch points $y_\pm$ on the imaginary axis, where $y_\pm=(2i/U)\times(\sin k\pm 1)$. In order to have the correct sign of the square root in the integrand we define the branch cuts of the square root to extend along the imaginary axis from y_+ to $+\infty$ and from y_- to $-\infty$. We then deform the upper half of the contour C into a contour that runs along both sides of the upper branch cut and in two quarter circles of large radius down to the real axis. In a similar fashion we deform the lower half of C along the lower cut. As the radius of the quarter circles goes to ∞ this gives rise to the following expression:

$$\boxed{\rho_0(k) = \frac{1}{2\pi}\left[1+\frac{\cos k}{|\cos k|}\right]-\frac{\cos k}{2\pi U}[I_-(k)+I_+(k)]>0\,,} \tag{60}$$

where

$$I_\pm(k) = \int_{1\pm\sin k}^{\infty}\frac{d\alpha}{\sinh(2\pi\alpha/U)\sqrt{(\alpha\mp\sin k)^2-1}}\,. \tag{61}$$

By introducing the variable $\alpha=\cosh x\pm\sin k$ we finally obtain the simple expression

$$I_\pm(k) = \int_0^\infty\frac{dx}{\sinh\{(2\pi/U)(\cosh x\pm\sin k)\}}\,. \tag{62}$$

As a consequence of expressions (60) and (62) for ρ_0, we have the crucial bound needed as input at the end of the proof of Lemma 3:

Lemma 5 (ρ bounds). *When $B=\infty$, $Q=\pi$, and $U>0$*

$$1/2\pi < \rho_0(k) < 1/\pi \quad for\ 0 \leqslant |k| < \pi/2\,,$$

$$0 < \rho_0(k) < 1/2\pi \quad for\ \pi/2 < |k| \leqslant \pi\,. \tag{63}$$

Equivalently, $f_0(x) < 2t(x)$ for all $|x| \leqslant 1$.

Proof. When $\pi/2 < |k| \leqslant \pi$ and $\cos k < 0$ the first term [ ] in (60) is zero while the second term is positive (since $I_\pm(k) > 0$). On the other hand, when $0 \leqslant |k| \leqslant \pi/2$, Theorem 1 shows that $\rho_0(k) > 0$. Thus, we conclude that $\rho_0(k) > 0$ for all $|k| \leqslant \pi$.

From (22) and the positivity of σ_0 we conclude that $\rho_0(k) < 1/2\pi$ when $\pi/2 < |k| \leqslant \pi$. From the positivity of $\rho_0(k)$ when $\pi/2 \leqslant |k| \leqslant \pi$ we conclude that the integral in (22) is less than $1/2\pi$ for all values of $0 \leqslant \sin k < 1$ and, therefore, $1/2\pi < \rho_0(k) < 1/\pi$ for $0 \leqslant |k| < \pi/2$. □

7. Absence of a Mott transition

A system of itinerant electrons exhibits a Mott transition if it undergoes a conducting–insulating transition when an interaction parameter is varied. In the Hubbard model one inquires whether a Mott transition occurs at some critical $U_c > 0$. Here we show that there exists no Mott transition in the 1D Hubbard model for all $U > 0$.

Our strategy is to compute the chemical potential μ_+ (resp. μ_-) for adding (resp. removing) one electron. The system is conducting if $\mu_+ = \mu_-$ and insulating if $\mu_+ > \mu_-$.

In the thermodynamic limit we can define μ by $\mu = \mathrm{d}E(N)/\mathrm{d}N$, where $E(N)$ denotes the ground state energy with $M = M' = N/2$. As we already remarked, this choice gives the ground state energy for all U, at least in the thermodynamic limit.

The thermodynamic limit is given by the solution of the integral equations, which we analyzed in Section 5. In this limit one cannot distinguish the odd and even cases (i.e., $M = M' = N/2$ if N is even or $M = M' - 1 = (N-1)/2$ if N is odd.) and one simply has $M/N = 1/2$ in the limit $N_a \to \infty$. In this case Theorem 2 says that we must have $B = \infty$. Then only Q is a variable and Theorem 3 says that Q is uniquely determined by N provided $N \leqslant N_a$.

In the thermodynamic limit we know, by general arguments, that $E(N)$ has the form $E(N) = N_a e(N/N_a)$ and e is a convex function of N/N_a. It is contained in (25) when $N/N_a \leqslant 1$. A convex function has right and left derivatives at every point and, therefore, $\mu_+ =$ right derivative and $\mu_- =$ left derivative are well defined. Convexity implies that $\mu_- \leqslant \mu_+$.

For less than a half-filled band it is clear that $\mu_+ = \mu_-$ since $E(M,M)$ is smooth in $M = N/2$ for $N \leqslant N_a$. The chemical potential cannot make any jumps in this region. But, for $N > N_a$ we have to use hole–particle symmetry as discussed in Section 1 to calculate $E(N)$. The derivatives of $E(N)$, namely μ_+ and μ_-, can now be different above and below the half-filling point $N=N_a$ and this gives rise to the possibility of having an insulator. We learn from (3) that

$$\mu_+ + \mu_- = U \tag{64}$$

and hence $\mu_+ > \mu_-$ if $\mu_- < U/2$.

We calculate μ_- in two ways, and arrive at the same conclusion

$$\boxed{\mu_-(U) = 2 - 4\int_0^\infty \frac{J_1(\omega)}{\omega[1+\exp(\omega U/2)]}\,\mathrm{d}\omega\,.} \tag{65}$$

The first way is to calculate μ_- from the integral equations by doing perturbation theory at the half-filling point analyzed in Section 6. This is a 'thermodynamic' or 'macroscopic' definition of μ_- and it is given in Section 7.1 below. (From now on μ_- means the value at the half-filling point.)

In Section 7.2 we calculate μ_- 'microscopically' by analyzing the Bethe Ansatz directly with $N=N_a-4$ electrons. Not surprisingly, we find the same value of μ_-. This was the method we originally employed to arrive at [1, Eq. (23)].

Before proceeding to the derivations of (65), we first show that (65) implies $\mu_- < U/2$ for every $U>0$. To see that $\mu_- < U/2$ we observe that

$$\mu_-(0) = 2 - 2\int_0^\infty \frac{J_1(\omega)}{\omega}\,\mathrm{d}\omega = 0\,, \tag{66}$$

$$u'_-(0) = \frac{1}{2}\int_0^\infty J_1(\omega)\,\mathrm{d}\omega = \frac{1}{2}\,. \tag{67}$$

Then $\mu_- < U/2$ holds if $\mu''_-(U) < 0$, which we turn to next. Here, (66) is in [21, 6.561.17] and (67) is in [21, 6.511.1].

Expanding the denominator in the integrand of (88) and integrating term by term, we obtain

$$\mu_-(U) = 2 - 4\sum_1^\infty (-1)^n \left[\sqrt{1+\frac{n^2U^2}{4}} - \frac{nU}{2}\right]$$

using which one obtains

$$\begin{aligned}\mu''_-(U) &= 2\sum_{-\infty}^{\infty}(-1)^n \frac{n^2}{(1+n^2U^2/4)^{3/2}}\\ &= \frac{2}{2\pi i}\int_C \frac{z^2}{(1+U^2z^2/4)^{3/2}}\,\frac{\pi}{\sin\pi z}\,\mathrm{d}z\,,\end{aligned} \tag{68}$$

where we have again replaced the summation by a contour integral with the contour C encompassing the real axis. The integrand in (68) is analytic except at the poles on the real axis and along two branch cuts on the imaginary axis. This allows us to deform the path to coincide the imaginary axis, thereby picking up contributions from the cuts. This yields

$$\mu''_-(U) = -\frac{32}{U^3}\int_1^\infty \frac{y^2}{(y^2-1)^{3/2}}\frac{1}{\sinh(2\pi y/U)}\,\mathrm{d}y < 0 \quad \text{for all } U > 0\,. \tag{69}$$

Thus, we have established $\mu_+(U) > \mu_-(U)$, and hence the 1D Hubbard model is insulating for all $U > 0$. *There is no conducting–insulating transition in the ground state of the 1D Hubbard model* (except at $U = 0$).

7.1. Chemical potential from the integral equations

As noted, we take $B = \infty$ and $Q < \pi$. In fact we take $Q = \pi - a$ with a small. (In the notation of Section 5, $a = \sin Q$, but to leading order in a, $\sin Q = \pi - Q$ and we need not distinguish the two numbers.) Our goal is to calculate δE, the change in E using (25) and δN, the change in N using (19); μ_- is the quotient of the two numbers.

As before, we use the notation $\rho(k)$ for the density at $Q = \pi - a$ and $\rho_0(k)$ for the density at $Q = \pi$, as given in (53), (60).

We start with N. As explained earlier, $\rho - 1/2\pi$ is odd around $\pi/2$ so, from (19),

$$\begin{aligned}\frac{N}{N_a} &= \int_{-Q}^{Q}\rho = 2\int_0^Q \rho = 2\int_0^a \rho + 2\int_a^{\pi-a}\frac{1}{2\pi} \\ &= 2\int_0^a \rho + \frac{1}{\pi}(\pi - 2a) \approx 1 + 2a\left(\rho_0(0) - \frac{1}{\pi}\right)\,.\end{aligned} \tag{70}$$

In the last expression we used the fact (and will use it again) that ρ is continuous in k and a (as we see from (41)); therefore, we can replace $\int_0^a \rho$ by $a\rho_0(0)$ to leading order in a. We learn from (70) that $\delta N/N_a = 2a(\rho_0(0) - 1/\pi) < 0$.

The calculation of δE is harder. From (25)

$$\begin{aligned}\frac{E}{N_a} &= -4\int_0^Q \rho\cos k = -4\int_0^a \rho\cos k - 4\int_a^{\pi-a}\rho\cos k \\ &\approx -4a\rho_0(0) - 4\int_a^{\pi-a}\left(\rho - \frac{1}{2\pi}\right)\cos k - \frac{2}{\pi}\int_a^{\pi-a}\cos k \\ &= -4a\,\rho_0(0) - 8\int_a^{\pi/2}\left(\rho - \frac{1}{2\pi}\right)\cos k \\ &= -4a\rho_0(0) + \frac{8}{2\pi}(1 - \sin a) - 8\int_0^{\pi/2}\rho\cos k + 8\int_0^a \rho\cos k \\ &\approx +4a\rho_0(0) - \frac{4a}{\pi} - 8\int_0^{\pi/2}\delta\rho\cos k + \frac{4}{\pi} - 8\int_0^{\pi/2}\rho_0\cos k\,,\end{aligned} \tag{71}$$

where $\delta\rho \equiv \rho - \rho_0$. The last two terms in (71) are the energy of the half-filled band, $N = N_a$.

Our next task is to compute $\delta\rho$ to leading order in a. It is more convenient to deal with the function $\delta f \equiv f - f_0$ and to note (from (26)) that $\int_0^{\pi/2} \rho(k)\cos k\, dk = \int_0^1 f(x)\sqrt{1-x^2}\, dx$. We turn to (41) and find, to leading order, that

$$f \approx (1+2\hat{U})t - \hat{U}\hat{A}t + 2\hat{U}\hat{A}\hat{U}\,t = f_0 + \hat{U}\hat{A}f_0 - 2\hat{U}\hat{A}\,t \tag{72}$$

with $f_0 = (1+2\hat{U})t$. We note that $\hat{U} = \hat{K}^2(1+\hat{K}^2)^{-1}$ since $B = \infty$ (see (39)) and has a kernel which we will call $u(x-y)$. If g is continuous near 0 (in our case $g = f_0$ or $g = t$) then $(\hat{U}\hat{A}\,g)(x) = \int_{-a}^{a} u(x-y)g(y)\,dy \approx 2au(x-0)g(0)$ to leading order in a. We also note from (26) that $f_0(0) = \rho_0(0)$. Therefore,

$$\int_0^{\pi/2} \delta\rho \cos k \approx a\left[\rho_0(0) - \frac{1}{\pi}\right] \int_{-1}^{1} \sqrt{1-x^2}\, u(x)\, dx\,. \tag{73}$$

The integral in (73) is most easily evaluated using Fourier transforms and Plancherel's theorem,

$$\int_{-1}^{1} \sqrt{1-x^2}e^{i\omega x}\, dx = 2\int_0^1 \cos(\omega x)\sqrt{1-x^2}\, dx$$

$$= 2\int_0^{\pi/2} \cos(\omega \sin\theta)\cos^2\theta\, d\theta = \frac{\pi}{\omega} J_1(\omega) \tag{74}$$

and from (24)

$$\int_{-\infty}^{\infty} u(x)e^{i\omega x}\, dx = [1 + e^{|\omega U/2|}]^{-1}\,. \tag{75}$$

By combining these transforms we can evaluate δE from (71).

$$\frac{\delta E}{N_a} = 2a\left[\rho_0(0) - \frac{1}{\pi}\right]\left[2 - 4\int_0^{\infty} \frac{J_1(\omega)}{\omega[1+\exp(\omega U/2)]}\, d\omega\right]\,. \tag{76}$$

By dividing (76) by (70) we obtain (65).

7.2. Chemical potential from the Bethe Ansatz

The evaluation of the chemical potentials from the Bethe Ansatz is reminiscent of the calculation of the excitation spectrum of the 1D delta-function Bose gas solved by one of us [15]. We consider the case of a half-filled band. To use our results in the previous sections, which hold for M, M' odd, we calculate μ_- by removing 4 electrons, 2 with spin up and 2 with spin down, from a half-filled band. This induces the changes

$$N \to N - 4 = N_a - 4, \quad M \to M - 2 = N_a/2 - 2\,. \tag{77}$$

Eqs. (15) and (16) determining the new k' and Λ' now read

$$N_a k'_j = 2\pi I'_j + \sum_{\beta=2}^{M-1} \theta(2\sin k'_j - 2\Lambda'_\beta), \quad j = 3,4,\ldots,N-2\,, \tag{78}$$

$$\sum_{j=3}^{N-2} \theta(2\sin k'_j - 2\Lambda'_\alpha) = 2\pi J'_\alpha - \sum_{\beta=2}^{M-1} \theta(\Lambda'_\alpha - \Lambda'_\beta), \quad \alpha = 2,\ldots,M-1\,. \tag{79}$$

Under the changes (77), the values of I' and J' are the same as those of I and J, namely,

$$I'_j = I_j, \quad j = 3,4,\ldots,N-2\,,$$

$$J'_\alpha = J_\alpha, \quad \alpha = 2,3,\ldots.M-1\,,$$

so they are centered around the origin with $k'_{total} = k_{total}$.

The removal of four electrons causes the values of k and Λ to shift by small amounts, and we write

$$k'_j = k_j + \frac{1}{N_a} w(k_j), \quad \Lambda'_\alpha = \Lambda_a + \frac{1}{N_a} u(\Lambda_a)\,.$$

By taking the differences of (78) and (15), and (79) and (16), and keeping the leading terms, one obtains

$$w(k_j) = \frac{1}{N_a} \sum_{\beta=2}^{M-1} \theta'(2\sin k_j - 2\Lambda_\beta)[2\cos k_j w(k_j) - 2u(\Lambda_\beta)]\,, \tag{80}$$

$$4\theta(2\Lambda_\alpha) - \frac{1}{N_a} \sum_{j=3}^{N-2} \theta'(2\Lambda_\alpha - 2\sin k_j)[2u(\Lambda_\alpha) - 2\cos k_j w(k_j)]$$
$$= -\frac{1}{N_a} \sum_{\beta=2}^{M-1} \theta'(\Lambda_\alpha - \Lambda_\beta)[u(\Lambda_\alpha) - u(\Lambda_\beta)]\,. \tag{81}$$

In deriving these equations we have used facts from our analysis of the integral equations, namely that when $M = M'$, $-\Lambda_1 = \Lambda_M \approx \infty$ (i.e., $=\infty$ in the limit $N_a \to \infty$) and that when $N = N_a$, $-k_1 = k_N \approx -k_2 = k_{N-1} \approx \pi$ as $N_a \to \infty$. Without using these facts there would be extra terms in (80) and (81), e.g., $\theta(2\sin k_j - 2\Lambda_1) + \theta(2\sin k_j - 2\Lambda_M)$, which is ≈ 0 because $-\Lambda_1 = \Lambda_M \approx \infty$.

By replacing the sums by integrals and making use of (20) and (21), we are led to the coupled integral equations

$$r(k) = \int_{-\infty}^{\infty} K(\sin k - \Lambda) s(\Lambda)\, d\Lambda \tag{82}$$

$$4\theta(2\Lambda) + 2\pi s(\Lambda) - \int_{-\pi}^{\pi} K(\sin k - \Lambda) r(k) \cos k \, dk$$

$$= - \int_{-\infty}^{\infty} K^2(\Lambda - \Lambda') s(\Lambda') \, d\Lambda' , \tag{83}$$

where

$$r(k) = w(k)\rho_0(k), \quad s(\Lambda) = u(\Lambda)\sigma_0(\Lambda) . \tag{84}$$

Eqs. (82) and (83) can be solved as follows. Note that the third term on the left-hand side of (83) vanishes identically after substituting (82) for $r(k)$. Next introduce the Fourier transforms (24) and

$$\int_{-\infty}^{\infty} e^{i\omega\Lambda} \theta(2\Lambda) \, d\Lambda = - \left(\frac{2\pi i}{\omega}\right) e^{-|\omega|U/4} , \tag{85}$$

and we obtain from (83)

$$s(\Lambda) = \frac{2}{\pi} \int_0^{\infty} \frac{\sin \omega\Lambda}{\omega \cosh(\omega U/4)} \, d\omega . \tag{86}$$

Thus, from (82)

$$r(k) = \frac{4}{\pi} \int_0^{\infty} \frac{\sin(\omega \sin k) \, d\omega}{\omega(1 + e^{\omega U/2})} . \tag{87}$$

Note that we have $r(-k) = -r(k)$ and $s(-\Lambda) = -s(\Lambda)$.

The chemical potential μ_- for a half-filled band is now computed to be

$$\begin{aligned}
\mu_-(U) &= \frac{1}{4} \left[E\left(\frac{N_a}{2}, \frac{N_a}{2}\right) - E\left(\frac{N_a}{2} - 2, \frac{N_a}{2} - 2\right) \right] \\
&= \frac{1}{4} \left[-2 \sum_{j=1}^{N} \cos k_j + 2 \sum_{j=3}^{N-2} \cos k'_j \right] \\
&= \frac{1}{4} \left[-2(-1 - 1 - 1 - 1) + 2 \sum_{j=3}^{N-2} (\cos k'_j - \cos k_j) \right] \\
&= 2 - \frac{1}{2} \int_{-\pi}^{\pi} r(k) \sin k \, dk \\
&= 2 - 4 \int_0^{\infty} \frac{J_1(\omega)}{\omega(1 + e^{\omega U/2})} \, d\omega
\end{aligned} \tag{88}$$

which agrees with (65).

8. Conclusion

We have presented details of the analysis of ground state properties of the 1D Hubbard model previously reported in Ref. [1]. Particularly, the analyses of the integral equations and of the absence of a Mott transition presented here have not heretofore appeared in print.

It is important to note that in order to establish that our solution is indeed the true ground state of the 1D Hubbard model, it is necessary to establish the existence of ordered real solutions to the Bethe Ansatz equations (14) and, assuming the solution exists, proofs of (a) and (b) as listed at the end of Section 3. The fulfillment of these steps remains as an open problem.

Acknowledgements

We are indebted to Daniel Mattis for encouraging us to investigate the jump in the chemical potential as an indicator of the insulator–conductor transition. We also thank Helen Au-Yang and Jacques Perk for helpful discussions. FYW would like to thank Dung-Hai Lee for the hospitality at the University of California at Berkeley and Ting-Kuo Lee for the hospitality at the National Center for Theoretical Sciences, Taiwan, where part of this work was carried out. Work has been supported in part by NSF grants PHY-0139984, DMR-9980440 and DMR-9971503.

References

[1] E.H. Lieb, F.Y. Wu, Phys. Rev. Lett. 20 (1968) 1445–1448, Erratum; E.H. Lieb, F.Y. Wu, Phys. Rev. Lett. 21 (1968) 192.

[2] M. Gutzwiller, Phys. Rev. Lett. 10 (1963) 159–162.

[3] J. Hubbard, Proc. R. Soc. London A 276 (1963) 238–257; J. Hubbard, Proc. R. Soc. London A 277 (1964) 237–259.

[4] O.J. Heilman, E.H. Lieb, Trans. N.Y. Acad. Sci. 33 (1970) 116–149; O.J. Heilman, E.H. Lieb, Ann. N.Y. Acad. Sci. 172 (1971) 583–617.

[5] E.H. Lieb, B. Nachtergaele, Phys. Rev. B 51 (1995) 4777–4791.

[6] M. Gaudin, La Fonction d'onde de Bethe, Masson, Paris, 1983.

[7] Z.N.C. Ha, Quantum Many-Body Systems in One Dimension, World Scientific, Singapore, 1996.

[8] A. Montorsi, The Hubbard Model, World Scientific, Singapore, 1992.

[9] M. Takahashi, Thermodynamics of One-dimensional Solvable Models, Cambridge University Press, London, 1999.

[10] E.H. Lieb, in: D. Iagolnitzer (Ed.), Proceedings of the XIth International Congress of Mathematical Physics, Paris, 1994, International Press, 1995, pp. 392–412.

[11] H. Tasaki, J. Phys. Cond. Matt. 10 (1998) 4353–4378.

[12] E.H. Lieb, D.C. Mattis, Phys. Rev. 125 (1962) 164–172.

[13] E.H. Lieb, Phys. Rev. Lett. 62 (1989) 1201–1204, Errata; E.H. Lieb, Phys. Rev. Lett. 62 (1989) 1927.

[14] H.A. Bethe, Zeits. f. Physik 71 (1931) 205–226 (Engl. trans. in D.C. Mattis, The Many-Body Problem, World Scientific, Singapore, 1993).

[15] E.H. Lieb, W. Liniger, Phys. Rev. 130 (1963) 1605–1616; E.H. Lieb, Exact analysis of an interacting bose gas, II. The excitation spectrum, Phys. Rev. 130 (1963) 1616–1624.

[16] J.B. McGuire, J. Math. Phys. 6 (1965) 432;
J.B. McGuire, Attractive potential, J. Math. Phys. 7 (1966) 123.

[17] E.H. Lieb, M. Flicker, Phys. Rev. 161 (1967) 179–188.

[18] M. Gaudin, Phys. Lett. 24 A (1967) 55–56. See also Thesis, University of Paris, 1967, which is now in book form as Travaux de Michel Gaudin: Modèles exactement résolus, Les Editions de Physique, Paris, Cambridge, USA, 1995.

[19] C.N. Yang, Phys. Rev. Lett. 19 (1967) 1312–1314.

[20] F. Woynarovich, J. Phys. C 15 (1982) 85–96, See also;
F. Woynarovich, J. Phys. C 15 (1982) 97–109;
F. Woynarovich, J. Phys. C 16 (1983) 5293–5304;
F. Woynarovich, J. Phys. C 16 (1983) 6593–6604.

[21] I.S. Gradshteyn, I.M. Ryzhik, Tables of Integrals, Series and Products, Academic Press, San Diego, 2000.

Journal of Statistical Physics, Vol. 119, Nos. 3/4, May 2005 (© 2005)
DOI: 10.1007/s10955-004-2112-z

Book Review: *Lectures on the Kinetic Theory of Gases, Non-equilibrium Thermodynamics and Statistical Theories*

Lectures on the Kinetic Theory of Gases, Non-equilibrium Thermodynamics and Statistical Theories. Ta-You Wu, National Tsing Hua University Press, Hsin Chu, Taiwan. $50.00 (226 pp.), ISBN 957-02-8205-3, Email: thup@my.nthu.edu.tw

The lecture notes by Ta-You Wu on the kinetic theory of gases, non-equilibrium thermodynamics and statistical theories, have recently been published by the National Tsing Hua University Press, Taiwan. Professor Ta-You Wu (1907–2000), a prominent researcher, writer, educator, and science administrator in China, Canada, the U.S., and Taiwan, was the third Chinese physicist to receive a Ph.D. in theoretical physics. One year after obtaining his Ph.D. from the University of Michigan, he returned to China in 1934 where he taught throughout the difficult wartime years. After the war he was the head of the Theoretical Division of the Physics Institute of the National Research Council of Canada and taught at SUNY Buffalo until his retirement in 1978. Later he moved to Taiwan and served as President of the Academia Sinica in Taipei from 1983 to 1994. Starting in the early 1960's, he single-handedly developed from scratch a scientific research program in Taiwan which became the National Science Council, the counterpart of NSF now with an annual budget about one tenth of that of the NSF's. In China, Taiwan, and among Chinese physicists, Professor Wu is widely known as the teacher of T.D. Lee and C.N. Yang, Nobel laureates of 1957, during their student years.

Professor Ta-You Wu is also known for his prolific writings in theoretical physics. His authoritative monograph, *Vibrational Spectra and Structure of Polyatomic Molecules*, written under the most difficult conditions during the war is well-known. Equally important are his eight volumes of lecture notes on theoretical physics. Educated under the influence of S.A. Goudsmit and G.E. Uhlenbeck of the Michigan (and Dutch)

School, Wu closely followed the development of modern physics at the time. While most of his lecture notes are not readily accessible to students and researchers in the West, it is very fortunate that the last set of Wu's lecture notes, delivered by Professor Wu at the ripe age of 87, is now being published as a book. The book, *Lectures on the Kinetic Theory of Gases, Non-equilibrium Thermodynamics and Statistical Theories*, records expanded lecture notes delivered by Wu in the Spring of 1994. In these lectures, Professor Wu presented in his unique style of clarity and simplicity, the formulation and development of kinetic theory, statistical physics, and non-equilibrium thermodynamics. The lectures cover a large part of the theory of non-equilibrium statistical thermodynamics, and examine the fundamental problem of the irreversible direction of time. The lectures are brief (223 pages), but are complete in the sense that the derivations of central results from clearly stated assumptions are given in full detail. The strength of the book lies in the five chapters, Chapters III–VII (133 pages), on kinetic theory and non-equilibrium statistical thermodynamics which contain materials not readily found in standard textbooks.

After an introductory statement of purpose for the lectures, professor Ta-You Wu discusses the laws of thermodynamics giving particular emphasis to a precise definition of the law of increasing entropy. This involves a clear separation of entropy as a sum of entropies of the system and of the surroundings. As is typical of the lecture style of professor Ta-You Wu, the treatment is short yet careful and complete.

Chapters III and IV of the lectures discuss the Boltzmann equation. It is shown how the law of increasing entropy is a consequence of probability assumptions implicit in the kinetic equation. Particular care is taken to describe properly how the conservation laws enter into the collision operators, and the resulting connection is made between collision operator properties and the macroscopic limit of fluid mechanics in gases. The justification of the Boltzmann equation proceeds along the lines set out by Bogoliubov which are derived in detail. The Frieman-Sandri theory of the Boltzmann equation is also discussed. The BBGKY hierarchy of equations must be terminated in order to obtain a closed kinetic theory (e.g. the generalized Boltzmann equation). Several termination procedures are discussed leading to closed kinetic theories for dilute gases. Professor Ta-You Wu again faces the problem that the BBGKY hierarchy is time reversal symmetric yet the resulting kinetic theories must choose an "arrow of time". The problem of dynamically and spontaneously breaking time reversal symmetry has been present starting from the pioneering statistical thermodynamic work of Boltzmann and Gibbs. But Professor Wu made it clear in his discussions where this symmetry breaking enters. In Chapter

V the Boltzmann equation is applied to the Vlasov-Landau theory of a dilute plasma, which is discussed from a physical kinetics viewpoint.

Professor Ta-You Wu goes on in Chapter VI to discuss the general irreversible processes in condensed matter from the viewpoint of linear transport coefficient matrices along the lines set out by Onsager. The symmetry properties of the matrices are derived on the basis of thermal fluctuations and microscopic reversibility. In all cases where a reasonable irreversible kinetic model has been useful in describing an approach to equilibrium, the final ensemble equilibrium probability distribution turns out to be either the micro-canonical distribution of Boltzmann or the canonical distribution of Gibbs. These are equivalent for large systems. While these results have not been rigorously derived from microscopic dynamics (no such derivation presently exists), it is argued that the results are eminently reasonable, and that the results can be and have been born out experimentally. The Einstein theory of Brownian motion is discussed from such a viewpoint, and Professor Wu then discusses several simply solvable models in thermal equilibrium. Methods of describing equilibrium fluctuations are also discussed. Both classical and quantum statistical thermodynamic canonical distributions are covered in a clear and concise manner.

In Chapter VII, Professor Ta-You Wu returns yet again to the problem which has haunted many other distinguished researchers, including R. Kubo and L.D. Landau, on the foundations of statistical physics: from whence comes the "arrow of time"? Each so-called derivation of irreversible kinetic model contains at least one point at which a statistical assumption chooses for the theorist a time direction. Landau was convinced that the derivation involved the irreversibility of quantum measurements but even Landau here admitted that he had no proof of such a conjecture. Here, the style of professor Ta-You Wu's lectures is to provide the mathematics, where it is available, to make the underlying assumptions explicit. Where no mathematical proofs are available, the qualitative discussions remain clear.

In summary, the lectures of Professor Ta-You Wu will prove to be very useful to students and researchers. The central and fundamental concepts of physical kinetics are more than adequately discussed, and those parts of the theory not yet understood are presented in a manner inviting the reader to contribute to their solution. While fewer topic are covered than may be found in, say, the treatise *Physical Kinetics* by L.D. Landau and E.M. Lifshitz, the simple yet elegant detailed discussions make the lectures a delight to read. As remarked by Professor T.D. Lee in his *Introduction* at the beginning of the book: "reading these lecture notes is an experience that will make you closer to the Master and to Nature". This is a book that must be read by anyone who is serious about learning the foundations of equilibrium and irreversible statistical thermodynamics.

A. Widom and F.Y. Wu
Department of Physics
Northeastern University
Boston, Massachusetts 02115
U.S.A.
E-mail: fywu@neu.edu

In Memorial of Shang-Keng Ma

A talk given at the 50th statistical mechanics meeting at Rutgers University, December 15, 1983

I speak with great sorrow and deep grief of the loss of our friend, colleague, and coworker, Shang-Keng Ma. Shang-Keng was born in 1940 in Chungking, the World War II capital of China. He came to this country at the age of eighteen and entered U. C. Berkeley to study physics. He obtained his B. S. degree there in 1962 with the honor of the "most promising senior," and continued on to earn a Ph.D. in 1966 at Berkeley under the direction of Professor Kenneth Watson. His Ph.D. work was in many-body theory, and it was natural that he did his postdoctoral work with Professor Keith Brueckner at U. C. San Diego, where he eventually became a full professor in 1975.

His early work reflected much of his Ph.D. training. His first publication was with Chia-Wei Woo, who also was a postdoctoral research associate working with Brueckner at the same time. They collaborated on two papers on the charged Bose gas, obtaining the same results using two entirely different approaches, one using Green's functions and the other using correlated basis wave functions. Ma subsequently worked on various problems in different fields, including the electron gas, fermion liquids, and quantum electrodynamics, all with the flavor of Green's functions. During those early years, Chia-Wei once related to me that Shang-Keng had confided to him that he could not do anything without Green's functions. But that was soon to change.

During the period of 1969-72, Shang-Keng continued to work in both condensed matter as well as high energy physics, often bridging the two, producing papers with titles such as *Singularities in Forward Multi-Particle Scattering Amplitudes* and *The S-Matrix Interpretation of Higher Virial Coefficients.* In 1972 his interest shifted to the then rapidly emerging area of renormalization group theory. To learn the development first-hand from the originators, he took a leave from La Jolla and spent a few months at Cornell with Ken Wilson and Michael Fisher. Soon thereafter he produced a number of important and influential papers on the subject, among them, the $1/N$ and $1/n$ expansions, and the first review article on renormalization group. Since then, he worked in diverse areas of critical phenomena and statistical physics, including critical dynamics of ferromagnet and spin glasses, magnetic chains, and the study of the Boltzmann equation.

In 1976 Shang-Keng introduced the idea of Monte Carlo renormalization group, an approach that has now become fashionable. His most recent contribution with Joseph Imry was on random systems, on the change of the critical dimensionality of spin systems due to the presence of random fields. He also had formulated a new way of considering entropy in dynamical systems. These works are full of physical intuition and new ideas, and are very different from his earlier Green's function calculations. It is clear that Shang-Keng was just at the beginning of making an impact in many areas of statistical physics. He visited many institutions to pursue his ideas, including Cornell, the Institute for Advanced Studies at Princeton, Berkeley, Saclay, Harvard, National Tsing Hua University of Taiwan, and the IBM Watson Research Center.

Shang-Keng's work is characterized by a unique style of elegance and profound thinking. He was, as Leo Kadanoff remarked to me, a deep thinker, not just a calculator. He wrote two books. His first book *Modern Theory of Critical Phenomena*, was published by Benjamin and has been translated into Russian. His most recent book *Statistical Mechanics* was written in Chinese. In this book which was published earlier this year, statistical mechanics is presented in an unconventional way reflecting his unique style and way of thinking. The book was intended to students of all fields and is very readable. Fortunately for readers in the West, it is now available in an English edition.

Shang-Keng was a dedicated teacher and researcher, and a devoted father and husband. He was also talented in many areas outside physics. His greatest past-time was reading Chinese classics. He was a regular contributor of articles to newspapers and magazines in Taiwan. One of his unfinished works on his desk was a novel on cancer patients written in Chinese. He was a student of oil painting for many years, and he enjoyed and sang Chinese operas and played the ancient Chinese musical instrument "tseng" very well.

Although he was not a smoker, Shang-Keng was found to have lung cancer in May 1982 shortly after returning home from a sabbatical leave in Taiwan. While he worked hard in Taiwan including finishing his second book, the hard work took an apparent toll since by then it was too late for treatment. Doctors soon gave up on him and he gave up on the doctors in return. In order to lead a normal family life, especially with his children, Shang-Keng chose to stay at home and work as usual, despite all the pain he had to suffer. It was a courageous fight from the very beginning. He mentioned on the phone the pain that kept him awake at night, but he did not give up. He continued to teach and do research until two weeks before Thanksgiving, when the doctor brought the worst news after a blood test. But he was confined to bed only in his last four days. The abstract of his last paper *Entropy of Polymer Chains Moving in a Two Dimensional Square Lattice* was finished one week before his death. By that time, he was unable to read and had to rely on Claudia, his wife, to read the text for corrections. He passed away in his home, in the early hours of Thanksgiving Day, leaving Claudia and two sons, Tian-Shan and Tian-Mo, ages three and fifteen months.

Last night, I spoke to Claudia and asked her if there was anything that Shang-Keng would have wanted to tell us, his friends, colleagues, and coworkers, on this occasion. After a pause, she said that Shang-Keng had told her that he would like to be remembered as an ordinary person. Yes, just an ordinary person.

There is an old Chinese saying which says "the truly greatness is being ordinary." With this quote I would like to close, and hope we all remember our friend and colleague, Shang-Keng Ma, as the ordinary person who worked so hard and contributed so much in physics.

International Journal of Modern Physics B
Vol. 22, No. 12 (2008) 1899–1909

World Scientific
www.worldscientific.com

PROFESSOR C. N. YANG AND STATISTICAL MECHANICS

F. Y. WU*

Department of Physics, Northeastern University,
Boston, Massachusetts 02115, USA
**fywu@neu.edu*

Received 10 January 2008

Professor Chen Ning Yang has made seminal and influential contributions in many different areas in theoretical physics. This talk focuses on his contributions in statistical mechanics, a field in which Professor Yang has held a continual interest for over sixty years. His Master's thesis was on the theory of binary alloys with multi-site interactions, some 30 years before others studied the problem. Likewise, his other works opened the door and led to subsequent developments in many areas of modern day statistical mechanics and mathematical physics. He made seminal contributions in a wide array of topics, ranging from the fundamental theory of phase transitions, the Ising model, Heisenberg spin chains, lattice models, and the Yang–Baxter equation, to the emergence of Yangian in quantum groups. These topics and their ramifications will be discussed in this talk.

Keywords: Phase transition; Ising and lattice models; Yang–Baxter equation.

1. Introduction

Statistical mechanics is the subfield of physics that deals with systems consisting of large numbers of particles. It provides a framework for relating the macroscopic properties of a system, such as the occurrence of phase transitions, to microscopic properties of individual atoms and molecules.

The theory of statistical mechanics was founded by Gibbs (1834–1903), who based his considerations on the earlier works of Boltzmann (1844–1906) and Maxwell (1831–1879). By the end of the 19th century, classical mechanics was fully developed and successfully applied to rigid body motions. However, after it was recognized that ordinary materials consist of 10^{23} molecules, it soon became apparent that the application of traditional classical mechanics is fruitless for explaining physical phenomena on the basis of molecular considerations. To overcome this difficulty, Gibbs proposed a statistical theory for computing the bulk properties of real materials.

*This paper will also appear in the proceedings of the Conference in Honour of C. N. Yang's 85th Birthday, to be published by World Scientific and NTU.

Statistical mechanics as proposed by Gibbs applies to all physical systems regardless of their macroscopic states. But in the early years, there had been doubts about whether it could fully explain physical phenomena such as phase transitions. In 1937, Mayer[1] developed the method of cluster expansions for analyzing the statistical mechanics of a many-particle system, which worked well for systems in the gas phase. This offered some hope of explaining phase transitions, and the Mayer theory subsequently became the main frontier of statistical mechanical research. This was unfortunate in hindsight since, as Yang and Lee would later show (see Sec. 4), the grand partition function used in the Mayer theory cannot be extended into the condensed phase, and hence it does not settle the question it set out to answer.

This was the stage and status of statistical mechanics in the late 1930's when Professor C. N. Yang entered college.

2. A Quasi-chemical Mean-field Model of Phase Transition

In 1938, Yang entered the National Southwest Associate University, a university formed jointly by National Tsing Hua University, National Peking University and Nankai University during the Japanese invasion, in Kunming, China. As an undergraduate student, Yang attended seminars given by J. S. (Zhuxi) Wang, who had recently returned from Cambridge, England, where he had studied the theory of phase transitions under R. H. Fowler. These lectures brought C. N. Yang in contact with the Mayer theory and other latest developments in statistical mechanics.[2–4]

After obtaining his B.S. degree in 1942, Yang continued to work on an M.S. degree in 1942–1944, and he chose to work in statistical mechanics under the direction of J. S. Wang. His Master's thesis included a study of phase transitions using a quasi-chemical method of analysis, and led to the publication of his first paper.[5]

In this paper, Yang generalized the quasi-chemical theory of Fowler and Guggenheim[6] of phase transitions in a binary alloy to encompass 4-site interactions. The idea of introducing multi-site interactions to a statistical mechanical model was novel and new. In contrast, the first mention of a lattice model with multi-site interactions was by myself[7] and by Kadanoff and Wegner[8] in 1972 — that the 8-vertex model solved by Baxter[9] is also an Ising model with 4-site interactions. Thus, Yang's quasi-chemical analysis of a binary alloy, an Ising model in disguise, predated the important study of a similar nature by Baxter in modern-day statistical mechanics by three decades!

3. Spontaneous Magnetization of the Ising Model

The two-dimensional Ising model was solved by Onsager in 1944.[10] In a legendary footnote of a conference discussion, Onsager[11] announced without proof a formula of the spontaneous magnetization of the two-dimensional Ising model with nearest-neighbor interactions K,

$$I = (1 - \sinh^{-4} 2K)^{1/8} . \quad (1)$$

Onsager never published his derivation since, as related by him later, he had made use of some unproven results on Toeplitz determinants, which he did not feel comfortable to put in print. Since the subject matter was close to his Master's thesis, Yang had studied the Onsager paper extensively and attempted to derive Eq. (1). But the Onsager paper was full of twists and turns, offering very few clues to the computation of the spontaneous magnetization.[12]

A simplified version of the Onsager solution by Kauffman[13] appeared in 1949. With the new insight to Onsager's solution, Yang immediately realized that the spontaneous magnetization I can be computed as an off-diagonal matrix element of Onsager's transfer matrix. This started Yang on the most difficult and longest calculation of his career.[12]

After almost 6 months of hard work off and on, Yang eventually succeeded in deriving the expression (1), and published the details in 1952.[14] Several times during the course of the work, the calculation stalled and Yang almost gave up, only to have it picked up again days later with the discovery of new tricks or twists.[12] It was a most formidable *tour de force* algebraic calculation in the history of statistical mechanics.

3.1. *Universality of the critical exponent β*

At Yang's suggestion, C. H. Chang[15] extended Yang's analysis of the spontaneous magnetization to the Ising model with anisotropic interactions K_1 and K_2, obtaining the expression

$$I = (1 - \sinh^{-2} 2K_1 \sinh^{-2} 2K_2)^{1/8} . \quad (2)$$

This expression exhibits the same critical exponent $\beta = 1/8$ as in the isotropic case, and marked the first ever recognition of universality of critical exponents, a fundamental principle of critical phenomena proposed by Griffiths 20 years later.[16]

3.2. *An integral equation*

A key step in Yang's evaluation of the spontaneous magnetization is the solution of an integral equation (Eq. (84) in Ref. 13) whose kernel is a product of 4 factors — I, II, III, and IV. Yang pioneered the use of Fredholm integral equations in the theory of exactly solved models (see also Sec. 7.1). This particular kernel and similar ones have been used later by others, as they also occurred in various forms in studies of the susceptibility[17] and the n-spin correlation function of the Ising model.[18–20]

4. Fundamental Theory of Phase Transitions

As described above, the frontier of statistical mechanics in the 1930's focused on the Mayer theory and the question of whether the theory was applicable to all phases of matter. Being thoroughly versed in the Mayer theory as well as the Ising lattice gas, Yang investigated this question in collaboration with T. D. Lee. Their investigation resulted in two fundamental papers on the theory of phase transitions.[21,22]

In the first paper,[21] Yang and Lee examined the question of whether the cluster expansion in the Mayer theory can apply to both the gas and condensed phases. This led them to examine the convergence of the grand partition function series in the thermodynamic limit, a question that had not been previously investigated closely. To see whether a single equation of state can describe different phases, they looked at zeroes of the grand partition function in the complex fugacity plane, again a consideration that revolutionized the study of phase transitions. Since an analytic function is defined by its zeroes, under this picture, the onset of phase transitions is signified by the pinching of zeroes on the real axis. This shows that the Mayer cluster expansion, while working well in the gas phase, cannot be analytically continued, and hence does not apply in the condensed phase. It also rules out any possibility in describing different phases of matter by a single equation of state.

In the second paper,[22] Lee and Yang applied the principles formulated in the first paper to the example of an Ising lattice gas. By using the spontaneous magnetization result (1), they deduced the exact two-phase region of the liquid-gas transition. This established without question that the Gibbs statistical mechanics holds in all phases of matter. The analysis also led to the discovery of the remarkable Yang–Lee circle theorem, which states that zeroes of the grand partition function of a ferromagnetic Ising lattice gas always lie on a unit circle.

These two papers have profoundly influenced modern-day statistical mechanics, as described in the following:

4.1. *The existence of the thermodynamic limit*

Real physical systems typically consist of $N \sim 10^{23}$ particles confined in a volume V. In applying Gibbs statistical mechanics to real systems, one takes the thermodynamic (bulk) limit $N, V \to \infty$ with N/V held constant, and implicitly assumes that such a limit exists. But in their study of phase transitions,[21] Yang and Lee demonstrated the necessity of a closer examination of this assumption. This insight initiated a host of rigorous studies of a similar nature.

The first comprehensive study was by Fisher[23] who, on the basis of earlier works of van Hove[24] and Groeneveld,[25] established in 1964 the existence of the bulk free energy for systems with short-range interactions. For Coulomb systems with long-range interactions, the situation is more subtle, and Lebowitz and Lieb established the bulk limit by making use of the Gauss law unique to Coulomb systems.[26] The existence of the bulk free energy for dipole-dipole interactions was subsequently established by Griffiths.[27] These rigorous studies led to a series of later studies on the fundamental question of the stability of matter.[28]

4.2. *The Yang–Lee circle theorem and beyond*

The consideration of Yang–Lee zeroes of the Ising model opened a new window in statistical mechanics and mathematical physics. The study of Yang–Lee zero loci has been extended to Ising models of arbitrary spins,[29] to vertex models,[30] and to numerous other spin systems.

While the Yang–Lee circle theorem concerns zeroes of the grand partition function, in 1964, Fisher[31] proposed to consider zeroes of the partition function, and demonstrated that they also lie on circles. The Fisher argument has since been made rigorous with the density of zeroes explicitly computed by Lu and myself.[32,33] The partition function zero consideration has also been extended to the Potts model by numerous authors.[34]

The concept of considering zeroes has also proven to be useful in mathematical physics. A well-known intractable problem in combinatorics is the problem of solid partitions of an integer.[35] But a study of the zeroes of its generating function by Huang and myself[36] shows that they tend towards a unit circle as the integer becomes larger. Zeroes of the Jones polynomial in knot theory have also been computed, and found to tend towards the unit circle as the number of nodes increases.[37] These findings appear to point to some unifying truth lurking beneath the surface of many unsolved problems in mathematics and mathematical physics.

5. The Quantization of Magnetic Flux

During a visit to Stanford University in 1961, Yang was asked by W. M. Fairbank whether or not the quantization of magnetic flux, if found, would be a new physical principle. The question arose at a time when Fairbank and B. S. Deaver were in the middle of an experiment investigating the possibility of magnetic flux quantization in superconducting rings. Yang, in collaboration with N. Byers, began to ponder over the question.[38,39]

By the time Deaver and Fairbank[40] successfully concluded from their experiment that the magnetic flux is indeed quantized, Byers and Yang[41] have also reached the conclusion that the quantization result did not indicate a new property. Rather, it can be deduced from usual quantum statistical mechanics. This was the "first true understanding of flux quantization".[42]

6. The Off-Diagonal Long-Range Order

The physical phenomena of superfluidity and superconductivity have been among the least-understood macroscopic quantum phenomena occurring in nature. The practical and standard explanation has been based on bosonic considerations: the

Bose condensation in superfluidity and Cooper pairs in the BCS theory of superconductivity. But there had been no understanding of a fundamental nature in substance. That was the question Yang pondered on in the early 1960's.[43]

In 1962, Yang published a paper[44] with the title *Concept of off-diagonal long-range order and the quantum phases of liquid helium and of superconductors*, which crystallized his thoughts on the essence of superfluidity and superconductivity. While the long-range order in the condensed phase in a real system can be understood, and computed, as the diagonal element of the two-particle density matrix, Yang proposed in this paper that the quantum phases of superfluidity and superconductivity are manifestations of a long-range order in off-diagonal elements of the density matrix. Again, this line of thinking and interpretation was totally new, and the paper has remained to be one that Yang has "always been fond of".[43]

7. The Heisenberg Spin Chain and the 6-vertex Model

After the publication of the paper on the long-range off-diagonal order, Yang experimented using the Bethe ansatz in constructing a Hamiltonian which can actually produce the off-diagonal long-range order.[45] Instead, this effort led to ground-breaking works on the Heisenberg spin chain, the 6-vertex model, and the one-dimensional delta function gas described below.

7.1. *The Heisenberg spin chain*

In a series of definitive papers in collaboration with C. P. Yang, Yang studied the one-dimensional Heisenberg spin chain with the Hamiltonian

$$H = -\frac{1}{2}\sum(\sigma_x\sigma_x' + \sigma_y\sigma_y' + \Delta\sigma_z\sigma_z')\,. \qquad (3)$$

Special cases of the Hamiltonian had been considered before by others. But Yang and Yang analyzed the Bethe ansatz solution of the eigenvalue equation of (3) with complete mathematical rigor, including a rigorous analysis of a Fredholm integral equation arising in the theory in the full range of Δ. The ground state energy is found to be singular at $\Delta = \pm 1$. Furthermore, this series of papers has become important, as it formed the basis of ensuing studies of the 6-vertex model, the one-dimensional delta function gas and numerous other related problems.

7.2. *The 6-vertex model*

In 1967, Lieb[48] solved the residual entropy problem of square ice, a prototype of the two-dimensional 6-vertex model, using the method of Bethe ansatz. Subsequently, the solution was extended to 6-vertex models in the absence of an external field.[49,50] These solutions share the characteristics that they are all based on Bethe ansatz analyses involving real momentum k.

In the same year 1967, Yang, Sutherland and C. P. Yang[51] published a solution of the general 6-vertex model in the presence of external fields, in which they used the Bethe ansatz with complex momentum k. But the Sutherland–Yang–Yang paper did not provide details of the solution. This led others to fill in the gap in ensuing years, often with analyses starting from scratch, to understand the thermodynamics. Thus, the $\Delta < 1$ case was studied by Nolden,[52] the $\Delta \geq 1$ case by Shore and Bukman,[53,54] and the case $|\Delta| = \infty$ by myself in collaboration with Huang *et al.*[55] The case of $|\Delta| = \infty$ is of particular interest, since it is also a 5-vertex model as well as a honeycomb lattice dimer model with a nonzero dimer-dimer interaction. It is the only known soluble interacting close-packed dimer model.

8. One-Dimensional Delta Function Gases

8.1. *The Bose gas*

The first successful application of the Bethe ansatz to a many-body problem was the one-dimensional delta function Bose gas solved by Lieb and Liniger.[56,57] Subsequently, by extending considerations to include all excitations, Yang and C. P. Yang deduced the thermodynamics of the Bose gas.[58] Their theoretical prediction has recently been found to agree very well with experiments on a one-dimensional Bose gas trapped on an atom chip.[59]

8.2. *The Fermi gas*

The study of the delta function Fermi gas was more subtle. In a seminal work having profound and influential impacts in many-body theory, statistical mechanics and mathematical physics, Yang in 1967 produced the full solution of the delta function Fermi gas.[60] The solution was obtained as a result of the combined use of group theory and the nested Bethe ansatz, a repeated use of the Bethe ansatz devised by Yang.

One very important ramification of the Fermi gas work is the exact solution of the ground state of the one-dimensional Hubbard model obtained by Lieb and myself.[61–63] The solution of the Hubbard model is similar to that of the delta function gas except with the replacement of the momentum k by $\sin k$ in the Bethe ansatz solution. Due to its relevance in high T_c superconductivity, the Lieb–Wu solution has since led to a torrent of further works on the one-dimensional Hubbard model.[64]

9. The Yang–Baxter Equation

The two most important integrable models in statistical mechanics are the delta function Fermi gas solved by Yang[60] and the 8-vertex model solved by Baxter.[9,65] The key to the solubility of the delta function gas is an operator relation[66] of the S-matrix,

$$Y_{jk}{}^{ab} Y_{ik}{}^{bc} Y_{ij}{}^{ab} = Y_{ij}{}^{bc} Y_{ik}{}^{ab} Y_{jk}{}^{bc} , \tag{4}$$

and for the 8-vertex model, the key is a relation[67] of the 8-vertex operator,

$$U_{i+1}(u)U_i(u+v)U_{i+1}(v) = U_i(v)U_{i+1}(u+v)U_i(u)\,. \tag{5}$$

Noting the similarity of the two relations and realizing that they are fundamentally the same, in a paper on the 8-vertex model, Takhtadzhan and Faddeev[68] called it the Baxter–Yang relation. Similar relations also arise in other quantum and lattice models. These relations have since been referred to as the *Yang–Baxter equation.*[69,70]

The Yang–Baxter equation is an internal consistency condition among parameters in a quantum or lattice model, and can usually be written down by considering a star-triangle relation.[69,70] The solution of the Yang–Baxter equation, if found, often aids in solving the model itself. The Yang–Baxter equation has been shown to play a central role in connecting many subfields in mathematics, statistical mechanics and mathematical physics.[71]

9.1. *Knot invariants*

One example of the role played by the Yang–Baxter equation in mathematics is the construction of knot (link) invariants. Knot invariants are algebraic quantities, often in polynomial forms, which preserve topological properties of three-dimensional knots. In the absence of definite prescriptions, very few knot invariants were known for decades. The situation changed dramatically after the discovery of the Jones polynomial by Jones in 1985,[72] and the subsequent revelation that knot invariants can be constructed from lattice models in statistical mechanics.[73]

The key to constructing knot invariant from statistical mechanics is the Yang–Baxter equation. Essentially, from each lattice model whose Yang–Baxter equation possesses a solution, one constructs a knot invariant. One example is the Jones polynomial, which can be constructed from a solution of the Yang–Baxter equation of the Potts model, even though the solution is in an unphysical regime.[74] Other examples are described in a 1992 review on knot theory and statistical mechanics by myself.[75]

9.2. *The Yangian*

In 1985, Drinfeld[76] showed that there exists a Hopf algebra (quantum group) over $SL(n)$ associated with the Yang–Baxter equation (4) after the operator Y is expanded into a series. Since Yang found the first rational solution of the expanded equation, he named the Hopf algebra the *Yangian* in honor of Yang.[76]

Hamiltonians with the Yangian symmetry include, among others, the one-dimensional Hubbard model, the delta function Fermi gas, the Haldane–Shastry model,[77] and the Lipatov model.[78] The Yangian algebra is of increasing importance in quantum groups, and has been used very recently in a formulation of quantum entangled states.[79]

10. Conclusion

In this talk, I have summarized the contributions made by Professor Chen Ning Yang in statistical mechanics. It goes without saying that it is not possible to cover all aspects of Professor Yang's work in this field, and undoubtedly, there are omissions. But it is clear from what is presented, however limited, that Professor C. N. Yang has made immense contributions to this relatively young field of theoretical physics.

A well-known treatise in statistical mechanics is the 20-volume *Phase Transitions and Critical Phenomena* published in 1972–2002.[80,81] The series covers almost every subject matter of traditional statistical mechanics. The first chapter of Volume 1 is an introductory note by Professor Yang, in which he assessed the status of the field and remarked about possible future directions of statistical mechanics.

In the conclusion he wrote:

> *One of the great intellectual challenges for the next few decades is the question of brain organization.*

As research in biophysics and brain memory functioning has mushroomed into a major field in recent years, this is an extraordinary prophecy and a testament to the insight and foresight of Professor Chen Ning Yang.

Acknowledgments

I would like to thank Dr. K. K. Phua for inviting me to the Symposium. I am grateful to M.-L. Ge and J. H. H. Perk for inputs in the preparation of the talk, and to J. H. H. Perk for a critical reading of the manuscript.

References

1. J. E. Mayer, *J. Chem. Phys.* **5**, 67 (1937).
2. C. N. Yang, in *Selected Papers (1945–1980) with Commentary* (World Scientific, Singapore, 2005).
3. C. N. Yang, *Int. J. Mod. Phys. B* **2**, 1325 (1988).
4. T. C. Chiang, *Biography of Chen-Ning Yang: The Beauty of Gauge and Symmetry* (in Chinese) (Tian Hsia Yuan Jian Publishing Co., Taipei, 2002).
5. C. N. Yang, *J. Chem. Phys.* **13**, 66 (1943).
6. R. H. Fowler and E. A. Guggenheim, *Proc. Roy. Soc. A* **174**, 187 (1940).
7. F. Y. Wu, *Phys. Rev. B* **4**, 2312 (1971).
8. L. P. Kadanoff and F. Wegner, *Phys. Rev. B* **4**, 3989 (1972).
9. R. J. Baxter, *Phys. Rev. Lett.* **26**, 832 (1971).
10. L. Onsager, *Phys. Rev.* **65**, 117 (1944),
11. L. Onsager, *Nuovo Cimento* **6**(Suppl.), 261 (1949).
12. Ref. 2, p. 12.
13. B. Kauffman, *Phys. Rev.* **76**, 1232 (1949).
14. C. N. Yang, *Phys. Rev.* **85**, 808 (1952).
15. C. H. Chang, *Phys. Rev.* **88**, 1422 (1952).
16. R. B. Griffiths, *Phys. Rev. Lett.* **24**, 1479 (1970).

17. E. Barouch, B. M. McCoy and T. T. Wu, *Phys. Rev. Lett.* **31**, 1409 (1973).
18. B. M. McCoy, C. A. Tracy and T. T. Wu, *Phys. Rev. Lett.* **38**, 793 (1973).
19. D. B. Abraham, *Commun. Math. Phys.* **59**, 17 (1978).
20. D. B. Abraham, *Commun. Math. Phys.* **60**, 205 (1978).
21. C. N. Yang and T. D. Lee, *Phys. Rev.* **87**, 404 (1952).
22. T. D. Lee and C. N. Yang, *Phys. Rev.* **87**, 410 (1952).
23. M. E. Fisher, *Arch. Rat. Mech. Anal.* **17**, 377 (1964).
24. L. van Hove, *Physica* **15**, 951 (1949).
25. J. Groeneveld, *Phys. Lett.* **3**, 50 (1962).
26. J. L. Lebowitz and E. H. Lieb, *Phys. Rev. Lett.* **22**, 631 (1969).
27. R. B. Griffiths, *Phys. Rev.* **176**, 655 (1968).
28. E. H. Lieb, *Rev. Mod. Phys.* **48**, 553 (1976).
29. R. B. Griffiths, *J. Math. Phys.* **10**, 1559 (1969).
30. M. Suzuki and M. E. Fisher, *J. Math. Phys.* **12**, 235 (1971).
31. M. E. Fisher, in *Lecture Notes in Theoretical Physics*, Vol. 7c, ed. W. E. Brittin (University of Colorado Press, Boulder, 1965).
32. W. T. Lu and F. Y. Wu, *Physica A* **258**, 157 (1998).
33. W. T. Lu and F. Y. Wu, *J. Stat. Phys.* **102**, 953 (2001).
34. See, for example, C. N. Chen, C. K. Hu and F. Y. Wu, *Phys. Rev. Lett.* **76**, 169 (1996).
35. P. A. MacMahon, *Combinatory Analysis*, Vol. 2 (Cambridge University Press, United Kingdom, 1916).
36. H. Y. Huang and F. Y. Wu, *Int. J. Mod. Phys. B* **11**, 121 (1997).
37. F. Y. Wu and J. Wang, *Physica A* **296**, 483 (2001).
38. Ref. 2, pp. 49–50.
39. Ref. 3, p. 1328.
40. B. S. Deaver and W. M. Fairbank, *Phys. Rev. Lett.* **7**, 43 (1961).
41. N. Byers and C. N. Yang, *Phys. Rev. Lett.* **7**, 46 (1961).
42. Ref. 3, p. 1328.
43. Ref. 2, p. 54.
44. C. N. Yang, *Rev. Mod. Phys.* **34**, 694 (1962).
45. Ref. 2, p. 63.
46. C. N. Yang and C. P. Yang, *Phys. Rev.* **150**, 321, 327 (1966).
47. C. N. Yang and C. P. Yang, *Phys. Rev.* **151**, 258 (1966).
48. E. H. Lieb, *Phys. Rev. Lett.* **18**, 692 (1967).
49. E. H. Lieb, *Phys. Rev. Lett.* **18**, 1046 (1967).
50. E. H. Lieb, *Phys. Rev. Lett.* **19**, 588 (1967).
51. B. Sutherland, C. N. Yang and C. P. Yang, *Phys. Rev. Lett.* **19**, 588 (1967).
52. I. Nolden, *J. Stat. Phys.* **67**, 155 (1992).
53. J. D. Shore and D. J. Bukman, *Phys. Rev. Lett.* **72**, 604 (1994).
54. D. J. Bukman and J. D. Shore, *J. Stat. Phys.* **78**, 1227 (1995).
55. H. Y. Huang, F. Y. Wu, H. Kunz and D. Kim, *Physica A* **228**, 1 (1996).
56. E. H. Lieb and W. Liniger, *Phys. Rev.* **130**, 1605 (1963).
57. E. H. Lieb, *Phys. Rev.* **130**, 1616 (1963).
58. C. N. Yang and C. P. Yang, *J. Math. Phys.* **10**, 1315 (1969).
59. A. H. van Amerongen, J. J. P. van Es, P. Wicke, K. V. Kheruntsyan and N. J. van Druten, *Phys. Rev. Lett.* **100**, 090402 (2008).
60. C. N. Yang, *Phys. Rev. Lett.* **19**, 1312 (1967).
61. E. H. Lieb and F. Y. Wu, *Phys. Rev. Lett.* **20**, 1445 (1968).
62. E. H. Lieb and F. Y. Wu, *Phys. Rev. Lett.* **21**, 192 (1968).
63. E. H. Lieb and F. Y. Wu, *Physica A* **321**, 1 (2003).

64. See, for example, F. H. L. Essler, H. Frahm, F. Göhmann, A. Klümper and V. E. Korepin, *The One-dimensional Hubbard Model* (Cambridge University Press, United Kingdom, 2005).
65. R. J. Baxter, *Exactly Solved Models* (Academic Press, London, 1980).
66. Equation (8) in Ref. 60.
67. Equation (10.4.31) in Ref. 65.
68. L. A. Takhtadzhan and L. D. Faddeev, *Russian Math. Surveys* **34**(5), 11 (1979).
69. J. H. H. Perk and H. Au-Yang, Yang–Baxter equations, in *Encyclopedia of Mathematical Physics*, eds. J.-P. Francoise, G. L. Naber and S. T. Tsou (Oxford, Elsevier, 2006).
70. J. H. H. Perk and H. Au-Yang, Yang–Baxter equations, arXiv: math-ph/0606053.
71. C. N. Yang and M.-L. Ge, *Int. J. Mod. Phys.* **20**, 2223 (2006).
72. V. F. R. Jones, *Bull. Am. Math. Soc.* **12**, 103 (1985).
73. L. H. Kauffman, *Topology* **26**, 395 (1987).
74. L. H. Kauffman, *Contemp. Math.* **78**, 263 (1988).
75. F. Y. Wu, *Rev. Mod. Phys.* **64**, 1099 (1992).
76. V. G. Drinfeld, *Soviet Math. Dokl.* **32**(1), 254 (1985).
77. F. D. M. Haldane, in *Proceedings of 16th Taniguchi Symposium on Condensed Matter Physics*, eds. O. Okiji and N. Kawakami (Springer, Berlin, 1994).
78. L. Dolan, C. R. Nappi and E. Witten, *J. High Energy Phys.* **10**, 017 (2003).
79. C. M. Bai, M.-L. Ge and X. Kang, *Proc. Conference in Honor of C. N. Yang's 85th Birthday* (World Scientific, Singapore, 2008), (to be published).
80. C. Domb and M. S. Green (eds.), *Phase Transitions and Critical Phenomena*, Vols. 1–6 (Academic Press, New York, 1972–2002).
81. C. Domb and J. L. Lebowitz (eds.), *Phase Transitions and Critical Phenomena*, Vols. 7–20 (Academic Press, New York, 1972–2002).

APPENDIX

A Challenge in Enumerative Combinatorics:
The Graph of Contribution of Professor Fa-Yueh Wu
Review of F. Y. Wu's Research by J.-M. Maillard

CHINESE JOURNAL OF PHYSICS VOL. 40, NO. 4 AUGUST 2002

A Challenge in Enumerative Combinatorics: The Graph of Contributions* of Professor Fa-Yueh Wu

J.-M. Maillard†

LPTHE, Université de Paris VI,
Tour 16, 1er étage, boîte 126, 4 Place Jussieu,
F-75252 Paris Cedex 05, France

(Received May 8, 2002)

We will try to sketch Professor F. Y. Wu's contributions in lattice statistical mechanics, solid state physics, graph theory, enumerative combinatorics and so many other domains of physics and mathematics. We will recall F. Y. Wu's most important and well-known classic results, and we will also sketch his most recent research dedicated to the connections of lattice statistical mechanical models with deep problems in pure mathematics. Since it is hard to provide an exhaustive list of all his contributions, to give some representation of F. Y. Wu's "mental connectivity", we will concentrate on the interrelations between the various results he has obtained in so many different domains of physics and mathematics. Along the way we will also try to understand Wu's motivations and his favorite concepts, tools and ideas.

PACS. 05.50.+q – Lattice theory and statistics; Ising problems.

I. Introduction

The publish-or-perish period of science could soon be seen as a golden age: our brave new world now celebrates the triumph of Enron's financial and accounting creativity. Sadly science is now also, increasingly, considered from an accountant's viewpoint. In this respect, if one takes this "modern" point of view, Professor F. Y. Wu's contribution[1] is clearly a very good return on investment: he has given more than 270 talks in meetings or conferences, published over 200 papers and monographs in refereed journals, and had many students. He has also published in, or is the editor[2] of, many books [21, 31, 71, 122, 138, 157, 171, 178, 179, 196].

Professor Wu was trained in theoretical condensed matter physics [3, 4, 19, 20, 27, 35, 108], but he is now seen as a mathematical physicist who is a leading expert in mathematical modeling of phase transition phenomena occurring in complex systems. Wu's research includes

[1] Professor F. Y. Wu is presently the Matthews University Distinguished Professor at Northeastern University. He is a fellow of the American Physical Society and a permanent member of the Chinese Physical Society (Taipei). His research has been supported by the National Science Foundation since 1968, a rare accomplishment by itself in an environment of declining research support in the U.S., and he currently serves as the editor of three professional journals: the Physica A, International Journal of Modern Physics B and the Modern Physics Letters B.

[2] For instance, Ref. [180] contains the proceedings of the conference on "Exactly Soluble Models in Statistical Mechanics: Historical Perspectives and Current Status", held at Northeastern University in March 1996 – the first ever international conference to deal exclusively with this topic. The proceedings reflect the broad range of interest in exactly soluble models as well as the diverse fields in physics and mathematics that they connect.

both theoretical studies and practical applications[3]. Among his recent researches he has studied connections of statistical mechanical models with deep problems in pure mathematics. This includes the generation of knot and link invariants from soluble models of statistical mechanics and the study of the long-standing unsolved mathematical problem of multidimensional partitions of integers in number theory using a Potts model approach.

Professor Wu's contributions to lattice statistical mechanics have been mostly in the area of exactly solvable lattice models. While integrable models have continued to occupy a prominent place in his work (such as the exact solution of two- and three-dimensional spin models and interacting dimer systems), his work has ranged over a wide variety of problems including exact lattice statistics in two and three dimensions, graph theory and combinatorics, to mention just a few. His work in many-body theory [3, 4, 7, 8, 15, 22, 28, 36, 66], especially those on liquid helium [2, 3, 6, 25, 26], has also been influential for many years.

F. Y. Wu joined the faculty of Northeastern University to work with Elliott Lieb in 1967, and in 1968 they published a joint paper[4] on the ground state of the Hubbard model [11] which has since become a classic. The Baxter-Wu model [45, 49] is also, clearly, an important milestone in the history of integrable lattice models.

F. Y. Wu has published several very important reviews of lattice statistical mechanics. First, Lieb and Wu wrote a monograph in 1970 on vertex models which became the fundamental reference in the field for decades [31]. Wu's 1982 review on the Potts model is another classic [89]. At more than one hundred citations per year ever since it was published, it is one of the most cited papers in physics[5]. In 1992 F. Y. Wu published yet another extremely well-received review on knot theory and its connection with lattice statistical mechanics [154]. In addition, in 1981, F. Y. Wu and Z. R. Yang published a series of expository papers on critical phenomena written in Chinese [84] - [88]. This review is well-known to Chinese researchers.

I-1. The choice of presentation: a challenge in enumerative combinatorics

An intriguing aspect of lattice statistics is that seemingly totally different problems are sometimes related to each other, and that the solution of one problem can often lead to solving other outstanding unsolved problems. At first sight, most of the work of F. Y. Wu could be said to correspond to *exact results in lattice statistical mechanics*, but because of the relations between seemingly totally different problems it can equivalently be seen, and sometimes be explicitly presented, as exact results in various domains of mathematical physics or mathematics: sometimes exact results in *graph theory*, sometimes in *enumerative combinatorics*, sometimes in *knot theory*, sometimes in *number theory*, etc. Wu's "intellectual walk" goes from vertex models to circle theorems or duality relations, from dimers to Ising models and back, from percolations or animal problems to Potts models, from Potts models to the Whitney-Tutte Polynomials, to polychromatic

[3] He has considered, for instance, the modeling of physical adsorption and applied it to describe processes used in chemical and environmental engineering [148, 175]. He has even published one experimental paper on slow neutron detectors [5].

[4] This paper has become prominent in the theory of high-T_c superconductors. P. W. Anderson even attributed to this paper as "predicting" the existence of quarks in his *Physics Today* (October, 1997) article on the centennial of the discovery of electrons.

[5] There was once a study published in 1984 (E. Garfield, Current Comments **48**, 3 (1984)) on citations in physics for the year of 1982. It reports that in 1982, the year this Potts review was published, it was the fifth most-cited paper among papers published in all of physics.

polynomials or to knot theory, from results, or conjectures, on critical manifolds[6] to Yang-Baxter integrability, perhaps on the way revisiting duality or Lee-Yang zeros, etc., etc. The simple listing of Professor Wu's results and contributions, and the inter-relations between these results and the associated concepts and tools, is by itself a *challenge in enumerative combinatorics.*

Actually it is impossible to describe Wu's contributions *linearly*, in a sequence of sections in a review paper like this, or even with a website-like "*tree* organization" of paragraphs. F. Y. Wu's contributions really correspond to a quite large "graph" of concepts, results, tools and models, with many "intellectual loops". The only possible "linear" and exhaustive description of Wu's contributions is his list of publications.

We have therefore chosen to give his exhaustive list of publications at the end of this paper. No other references are given.

We have chosen to keep the notation F.Y. Wu used in his publications[7], and *not to normalize them*, so that the reader who wants to see more and goes back to the cited publications will immediately be able to recover the equations and notations.

Obviously, we will not try to provide an exhaustive description of Wu's contributions but, rather, to provide some considered well-suited specific "morceaux choisis[8]", *comments* on some of his results, some hints of the kind of concepts he likes to work with, and try to explain why his results are important, fruitful and stimulating for anyone who works in lattice statistical mechanics or in mathematical physics.

II. Even before vertex models: the exact solution of the Hubbard model

Elliott H. Lieb and F. Y. Wu published in 1968 a joint paper on the ground state of the Hubbard model [11] which has since become a classic, and served as a cornerstone in the theory of high-T_c superconductors. An important question there corresponds to the spin-charge decoupling, which is exact and explicit in one-dimensional models: is the spin-charge decoupling a characteristic of one dimension? Is it possible that some "trace" of spin-charge decoupling remains for quantum two-dimensional models which are supposedly related to high-T_c superconductors?

Let us describe briefly the classic Lieb-Wu solution of the Hubbard model. One assumes that the electrons can hop between the Wannier states of neighboring lattice sites and that each site is capable of accommodating two electrons of opposite spins with an interaction energy $U > 0$. The corresponding Hamiltonian reads:

$$H = T \sum_{<ij>} \sum_{\sigma} c^{\dagger}_{i\sigma} c_{j\sigma} + U \sum_{i} c^{\dagger}_{i\uparrow}\, c_{i\uparrow}\, c^{\dagger}_{i\downarrow}\, c_{i\downarrow},$$

[6] The critical manifolds deduced or conjectured by F. Y. Wu are mostly algebraic varieties and not simple differentiable or analytical manifolds.

[7] The price paid is, for instance, that the spin edge Boltzmann weights will sometimes be denoted e^{K_1}, e^{K_2}, e^{K_3}, e^{K_4}, or a, b, c, d, or x_1, x_2, x_3, x_4, and the vertex Boltzmann weights $\omega_1, \omega_2, \cdots$ or $a, b, c, d, a', b', c', d'$. This corresponds to the spectrum of notations used in the lattice statistical mechanics literature. These different notations were often introduced when one faced large polynomial expressions and the e^{K_i} or $e^{-\beta \cdot J_i}$ notations for Boltzmann weights would be painful.

[8] I apologize, in advance, for the fact that these "morceaux choisis" are obviously biased by my personal taste for effective birational algebraic geometry in lattice statistical mechanics.

where $c^{\dagger}_{i\sigma}$ and $c_{i\sigma}$ are the creation and annihilation operators for an electron of spin σ in the Wannier state at the i-th lattice site and the first sum is taken over nearest neighbor sites. Denoting $f(x_1, x_2, \cdots, x_M; x_{M+1}, \cdots, x_N)$ the amplitude of the wavefunction for which the down spins are located at sites $x_1, x_2, \cdots, x_M$ and the up spins are located at sites $x_{M+1}, \cdots, x_N$. The eigenvalue equation $H\psi = E\psi$ leads to:

$$\begin{aligned} -\sum_{i=1}^{N}\sum_{s=\pm 1} & f(x_1, x_2, \cdots, x_i + s, \cdots, x_N) \\ +U\sum_{i<j} & \delta(x_i - x_j) f(x_1, x_2, \cdots, x_N) = E\, f(x_1, x_2, \cdots, x_N), \end{aligned} \tag{1}$$

where $f(x_1, x_2, \cdots, x_N)$ is antisymmetric in the first M and the last $N-M$ variables (separately). Let μ_+ (resp. μ_-) denote the chemical potential of adding (resp. removing) one electron. In the half-filled band one has $\mu_+ = U - \mu_-$, and the calculation of μ_- can be done in closed form with the result:

$$\mu_- = 2 - 4\int_0^{\infty} \frac{J_1(\omega)\cdot d\omega}{\omega\cdot(1 + \exp(\omega U/2))}, \tag{2}$$

where J_1 is the Bessel function. It can be established from (2) and $\mu_+ = U - \mu_-$ that $\mu_+ > \mu_-$ for $U > 0$. In other words, the ground state for a half-filled band is insulating for any nonzero U, and conducting for $U = 0$. Equivalently, there is no Mott transition for nonzero U, *i.e.*, the ground state is analytic in U on the real axis except at the origin.

III. Vertex models

The distinction between vertex models and spin models is traditional in lattice statistical mechanics, but there are "bridges" between these two sets of lattice models [78]. Roughly speaking one can say that F. Y. Wu first obtained results on vertex models [13, 14] (five-vertex models [9, 10], free-fermion vertex models [50], dimer models seen as vertex models, ...) and then obtained results on spin models (Ising model with second-neighbor Interactions [12], the Baxter-Wu model [45, 49], Potts model, ...), introducing more and more graph theoretical approaches, up to looping the loop with knot theory, which is, in fact, closely related to vertex models and to Potts models! As far as vertex models are concerned, we will first sketch the approach given in his monograph with Lieb (section (III-1)), in a second step we will sketch his free-fermion results (section (III-2-1)) closely followed by his dimer results (section (III-3)), and, then, we will discuss some miscellaneous results he obtained on five-, six- and eight-vertex models (section (III-4)).

III-1. Two-dimensional ferroelectric models

Elliott Lieb and F. Y. Wu wrote a monograph on vertex models in 1970, entitled "Two-dimensional Ferroelectric Models", which became a fundamental reference in the field for decades [31]. This monograph gives the best introduction to the sixteen-vertex model, which is a fundamental model in lattice statistical mechanics. Unfortunately it is not known well enough, even to many specialists of lattice models, that it contains the most general eight-vertex model, most of the (Yang-Baxter) integrable vertex models (the symmetric eight-vertex model, various free-fermion models, the asymmetric free-fermion model, the asymmetric and symmetric six-vertex model, the five-vertex models, three-coloring of square maps, and others) and also fundamental *non-integrable* models such as, for instance, the Ising model in a magnetic field. In particular the monograph mentions explicitly the weak-graph duality (see section (V) below) on the sixteen-vertex model (see page 457 of [31]):

$$\begin{aligned} \omega_1^* &= \frac{1}{4}\cdot\sum_{i=1}^{16}\omega_i, \qquad \omega_2^* = \frac{1}{4}\cdot\Big(\sum_{i=1}^{8}\omega_i - \sum_{i=9}^{16}\omega_i\Big) \\ \omega_3^* &= 14\cdot\Big(\sum_{i=1}^{4}\omega_i - \sum_{i=5}^{8}\omega_i + (\omega_{10}+\omega_{12}+\omega_{14}+\omega_{16}) \\ &\quad -(\omega_9+\omega_{11}+\omega_{13}+\omega_{15})\omega_i\Big), \cdots \end{aligned} \tag{3}$$

The 154 pages of this monograph are still, by today's standard, an extremely valuable document for any specialist of lattice models. Beyond the taxonomy of ferro and ferrielectric models (ice model, KDP [9, 18], modified KDP [41], F model [13], modified F model [38, 75, 80], F model with a staggered field, ...), this monograph remains extremely modern and valuable from a technical viewpoint.

Among the exactly soluble models (the bread-and-butter of F. Y. Wu) was one that, for a long time, was a "sleeper", namely, Bethe's 1931 solution of the ground state energy and elementary excitations of the one-dimensional quantum-mechanical spin-$\frac{1}{2}$ Heisenberg model of antiferromagnetism. We will see below a large set of results from the Lieb-Wu monograph on *vertex models*, in particular the six-vertex model.

The monograph gives an extremely lucid exposition of the Bethe ansatz for the six-vertex model. The Bethe ansatz is analyzed and explained in the most general framework (with horizontal and vertical fields) and it is a must-read anyone who wants to work seriously on the coordinate Bethe ansatz. It is certainly much more interesting and deeper than so many subsequent papers that have revisited, at nauseum, the Bethe ansatz of the symmetric six-vertex model, re-styling this simple Bethe ansatz with a conformal resp. quantum group, resp. knot theory, resp. ... framework. The analysis of the conditions for the transfer matrix T of the most general sixteen-vertex model to have a non-trivial "linear operator" (1D quantum Hamiltonian) that commutes[9] with T (pages 367 to 373) are probably one of the first pages any student who wants to study integrable lattice models should read.

The monograph makes crystal clear the fact that the Bethe ansatz is related to the conservation of a certain charge. This can be seen from the fact that most of the analysis (from page 374 to page 444) relies on the use (page 363 equation (81)) of the variable $y = 1 - 2\,n/N$, which in spin language is the average z-component of the spin per vertical bond, namely, $y =< S_z > /N$ for a square lattice of size $N \times M$, where n denotes the number of down arrows and N the number of vertical bonds in a row.

We use the same notation as in Lieb-Wu. In particular, let us introduce the horizontal and vertical fields H and V, respectively. The partition function per site in the thermodynamic limit is:

[9] Which is the most obvious manifestation of the Yang-Baxter integrability.

$$\lim_{N\to\infty} \frac{1}{N} \cdot \ln(\Lambda) = \max_{-1\cdot\ y\cdot\ +1} [z(y) + V \cdot y], \tag{4}$$

where Λ denotes the largest eigenvalue of the transfer matrix. The monograph details a large set of situations. Let us consider here the regime

$$\Delta \equiv (\omega_1\omega_2 + \omega_3\omega_4 - \omega_5\omega_6)/2\sqrt{\omega_1\omega_2\omega_3\omega_4} < -1, \tag{5}$$

and introduce the variable:

$$e^{\theta_0} = \frac{1+\eta e^{\lambda}}{e^{\lambda}+\eta}, \qquad 0\cdot\ \theta_0 \cdot\ \lambda \qquad \text{where:}$$

$$\eta = e^{|K_1|} = \left(\frac{\omega_1\omega_2}{\omega_3\omega_4}\right)^{1/2} \qquad \text{or:} \qquad \left(\frac{\omega_3\omega_4}{\omega_1\omega_2}\right)^{1/2}.$$

When $\Delta < -1$, $z(y)$ reads:

$$z(y) = -K_2 + \max(0, -K_1) + \frac{1}{4\pi} \cdot \int_{-b}^{+b} R(\alpha) \cdot C(\alpha) \cdot d\alpha,$$

$$\text{where:} \qquad C(\alpha) = ln\left(\frac{\cosh(2\lambda-\theta_0) - \cos(\alpha)}{\cosh(\theta_0) - \cos(\alpha)}\right),$$

and the (normalized) density[10] $R(\alpha)$ satisfies the Bethe-ansatz integral equation with the kernel $K(\alpha)$:

$$R(\alpha) = \frac{\sinh(\lambda)}{\cosh(\lambda) - \cos(\alpha)} - \int_{-b}^{+b} K(\alpha-\beta) \cdot R(\beta)\, d\beta \tag{6}$$

$$\text{with:} \qquad 2\pi \cdot K(\alpha-\beta) = \frac{\sinh(2\lambda)}{\cosh(2\lambda) - \cos(\alpha-\beta)}.$$

The integral equation (6) is nothing but the well-known Yang-Yang Bethe ansatz integral equation on the density $\rho(q)$:

$$1 = 2\pi \cdot \rho(p) - \int_{-Q}^{+Q} \frac{d\theta(p,q)}{dp} \rho(q) \cdot dq \quad \text{with:} \quad Q = \frac{\pi \cdot (1-y)}{2}.$$

The range b of the new variable α in the integral relation (6) can be deduced from the definition of the density $R(\alpha)$:

$$\pi \cdot (1-y) = \int_{-b}^{+b} R(\alpha)\, d\alpha = \int_{-Q}^{+Q} \rho(q) \cdot dq.$$

When $y = 0$, the integral attains its maximum range and one can solve (6) by using a Fourier series of a Fourier transform. One thus gets $R(\alpha)$ as a simple dn elliptic function. Not surprisingly one can also calculate all the derivatives of $z(y)$ at $y = 0$. One can thus expand $z(y)$ namely, write $z(y) = z(0) - z'(0) \cdot y + z'''(0) \cdot y^3/6 + \cdots$. To first order in y one obtains $z'(0) = -\Xi(\lambda - \theta_0)$, where the function Ξ is related to the Jacobian elliptic function nd:

[10] One has $R(\alpha) \cdot d\alpha = 2\pi\rho(p) \cdot dp$.

$$\Xi(\phi) = \ln\left(\frac{\cosh((\lambda+\phi)/2)}{\cosh((\lambda-\phi)/2)}\right) - \frac{\phi}{2} - \sum_{n=1}^{\infty} \frac{(-1)^n \cdot e^{-2n\lambda} \cdot \sinh(n\phi)}{n \cdot \cosh(n\lambda)}.$$

The function $\Xi(\phi)$ also satisfies the nice involutive functional relations[11]:

$$\Xi(\phi) = -\Xi(-\phi), \quad \Xi(\lambda+\phi)\Xi(\lambda-\phi), \quad \Xi(4\lambda+\phi) = \Xi(\phi).$$

Let us consider the thermodynamic properties of the model when $H = 0$ and $V \neq 0$. From (4) one sees that the thermodynamic properties depend on the optimal choice of y given by:

$$z'(y) = -V.$$

When lowering the temperature the slope of $z(y)$ corresponding to the transition sticks at $y \simeq 0$, and one thus has (see page 425 of [31]) an antiferroelectric transition occurring at $T_c(V)$ given by:

$$V = \Xi(\lambda - \theta_0). \tag{7}$$

This gives a beautiful example of a *transcendental critical manifold* which reduces, in some domain of the parameters (low temperatures), to a transcendental equation (7) and *not to an algebraic one*, as one is used to seeing in exactly solvable models. One thus has a *transcendental critical manifold* for a vertex model for which one can actually write down the exact Bethe ansatz (see equation (6)). Writing a closed simple formula for the solution is not possible, but one can certainly find numerical solutions on a computer. Should we say that the model is exactly solvable but not "computable"? We will revisit these questions of the algebraicity of the critical manifold versus integrability in other sections of this paper with other critical manifold conjectures, or results, of F. Y. Wu (see for instance sections (VI-3), (IV-1) below). For those who have a "naive" point of view on the character of critical manifolds[12], example (7) shows that a model having a Bethe ansatz can have a *transcendental* critical manifold.

The Lieb-Wu review provides wonderful pieces of analytical work (analysis in one complex variable, see for instance pages 410-411 and the analysis of the analytic structure of the F model or the temperature Riemann structure for the free energy of the F model). One finds a festival of one complex variable analytical tools (the Maclaurin formula, tools for the evaluations of asymptotic behaviors, path integration, etc.).

Many more results can be found in the monograph (the three-color problem, the hard square model, the F model on the triangular lattice, three coloring of the edges of the hexagonal lattice ...). Let us mention, in particular, the six-vertex model with *site-dependent* weights (which can be considered as the first example of a Z-invariant model). Let us introduce $\omega_j(I, J)$ where $j = 1$,

[11] In agreement with the inversion relations on the model.

[12] With, for instance, a prejudice of algebraicity of the critical manifolds of "solvable" models: all examples known in the literature are polynomial expressions in well-suited variables e^{K_i}. These include, for instance, the critical varieties of the anisotropic Ising, or Potts, models on square, and triangular lattices, or the critical varieties of the Baxter model. For non-integrable models the common wisdom is, probably, that critical manifolds are always analytic, or may be differentiable, and the algebraicity of the critical manifolds is ruled out by the non-integrability. This is also a naive point of view: see (36) in section (VI-1).

$\cdots$,6, are the six possible Boltzmann factors of the vertex in row I and column J, and let us require that the algebraic invariant Δ be independent of I, J:

$$\Delta = \frac{\omega_1(I,J)\omega_2(I,J) + \omega_3(I,J)\omega_4(I,J) - \omega_5(I,J)\omega_6(I,J)}{2 \cdot (\omega_1(I,J)\omega_2(I,J)\omega_3(I,J)\omega_4(I,J))^{1/2}}. \tag{8}$$

Up to a multiplicative factor $\Box_{I,J}$, a (rational) parametrization of these invariance conditions (8) is:

$$\begin{aligned}
&\omega_1(I,J) = (1 - t \cdot p_{I,J}) \cdot \alpha_{I,J}\beta_{I,J}, \\
&\omega_2(I,J) = (1 - t \cdot p_{I,J}) \cdot \frac{1}{\alpha_{I,J}\beta_{I,J}}, \\
&\omega_3(I,J) = (p_{I,J} - t) \cdot \frac{\alpha_{I,J}}{\beta_{I,J}}, \quad \omega_4(I,J) = (p_{I,J} - t) \cdot \frac{\beta_{I,J}}{\alpha_{I,J}}, \\
&\omega_5(I,J) = \Big(\frac{1}{t} - t\Big) \cdot p_{I,J} \cdot \gamma_{I,J}, \quad \omega_6(I,J) = (1 - t^2) \cdot \frac{1}{\gamma_{I,J}}.
\end{aligned} \tag{9}$$

Baxter's Z-invariance condition for integrability requires that the $p_{I,J}$'s are actually products of a (spectral) parameter depending on the row and another parameter depending on the column: $p_{I,J} = \rho_I \cdot \sigma_J$. We will see in section (7) when sketching the correspondence between the standard scalar Potts model and a staggered asymmetric six-vertex model, that these product conditions, $p_{I,J} = \rho_I\sigma_J$, actually correspond in the case of the checkerboard Potts model to criticality, or to the vanishing conditions of a staggering field H_{stag} from a Lee-Yang zeros viewpoint: $|z| = e^{H_{\text{stag}}} = 1$.

Provided that the $p_{I,J} = \rho_I \cdot \sigma_J$ integrability conditions are satisfied, the partition function, with parametrization (9), can be expressed as a multiplicative closed formula:

$$\begin{aligned}
&Z = 2 \prod_{I=1}^{M} \prod_{I=1}^{M} \frac{\sqrt{\omega_5(I,J)\omega_6(I,J)}}{1 - t^2} \cdot F(\rho_I\sigma_J) \cdot F\Big(\frac{1}{\rho_I\sigma_J}\Big), \\
&\text{where:} \qquad F(z) = \prod_{m=1}^{\infty} \frac{1 - t^{4m-1}z}{1 - t^{4m+1}z}.
\end{aligned} \tag{10}$$

III-2. Vertex models: free fermions

Another classic work of F. Y. Wu is his 1970 paper with C. Fan in which they coined the term the *free-fermion model* [16]. This work was later extended to its checkerboard version during one of Wu's visits to Taiwan [50, 52]. In the following we shall arrange the homogeneous vertex weights in a matrix R, whose size and form vary according to the number of edge states and the coordination number of the lattice. Typical examples we will consider are the 2D square and triangular lattices shown below:

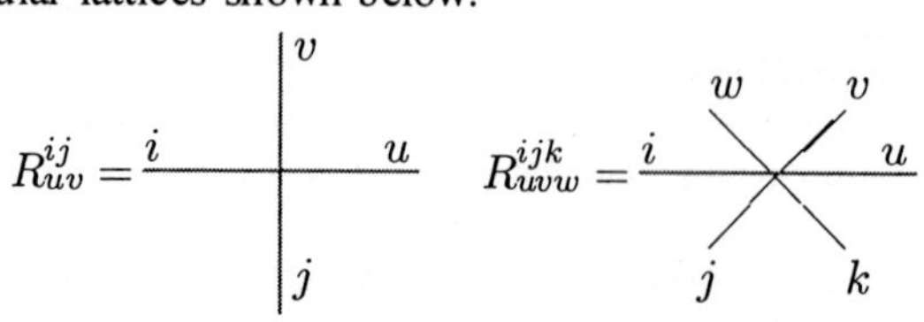

2D square 2D triangular

III-2-1. Free-fermion asymmetric eight-vertex model

C. Fan and F. Y. Wu obtained many free-fermion results [12, 16]. The free energy of the most general free-fermion model on a square lattice evaluated by Fan and Wu reads:

$$f = \frac{1}{16\pi^2}\int_0^{2\pi} d\theta\, d\phi\, \ln\Big(2a + 2b\cos\theta + 2c\cos\phi + 2d\cos(\theta-\phi) + 2\,e\,\cos(\theta+\phi)\Big),$$

where:

$$a = \frac{1}{2}\cdot(\omega_1^2+\omega_2^2+\omega_3^2+\omega_4^2), \qquad b = \omega_1\omega_3 - \omega_2\omega_4,$$
$$c = \omega_1\omega_4 - \omega_2\omega_3, \qquad d = \omega_3\omega_4 - \omega_7\omega_8, \qquad e = \omega_3\omega_4 - \omega_5\omega_6,$$

provided that the free fermion condition:

$$\omega_1\omega_2 + \omega_3\omega_4 = \omega_5\omega_6 + \omega_7\omega_8 \tag{11}$$

is satisfied.

Let us revisit some of their results from an inversion relation viewpoint. Renaming the vertex weights as $a = \omega_1$, $a' = \omega_2$, $b = \omega_3$, $b' = \omega_4$, $c = \omega_5$, $c' = \omega_6$, $d = \omega_7$, $d' = \omega_8$, the matrix R of the eight-vertex model is then:

$$R = \begin{pmatrix} a & 0 & 0 & d' \\ 0 & b & c' & 0 \\ 0 & c & b' & 0 \\ d & 0 & 0 & a' \end{pmatrix}. \tag{12}$$

A matrix of the form (12) can be brought, by a similarity transformation, to a block-diagonal form:

$$R = \begin{pmatrix} R_1 & 0 \\ 0 & R_2 \end{pmatrix}, \quad \text{with} \quad R_1 = \begin{pmatrix} a & d' \\ d & a' \end{pmatrix} \quad \text{and} \quad R_2 = \begin{pmatrix} b & c' \\ c & b' \end{pmatrix}.$$

If one introduces $\delta_1 = aa' - dd'$ and $\delta_2 = bb' - cc'$, the determinants of the two blocks, then the (homogeneous) matrix inverse I (namely $R \to \det(R)\cdot R^{-1}$) reads:

$$\begin{aligned}(a, a', d, d') &\to (a'\cdot\delta_2,\, a\cdot\delta_2, -d\cdot\delta_2, -d'\cdot\delta_2)\\ (b, b', c, c') &\to (b'\cdot\delta_1,\, b\cdot\delta_1, -c\cdot\delta_1, -c'\cdot\delta_1).\end{aligned} \tag{13}$$

It is straightforward to see that the free-fermion condition (11) is $\delta_1 = -\delta_2$ which has the effect of *linearizing* the inversion (13) into an involution given by:

$$a \leftrightarrow a', \quad b \leftrightarrow -b', \quad (d, d') \to (-d, d'), \quad b \leftrightarrow -b', \quad (c, c') \to (c, c').$$

The group generated by the two inversion relations of the model is then realized by permutations of the entries mixing with sign changes, and its orbits are thus *finite*. The *finiteness* condition of the group is a common feature of all free-fermion models.

III-2-2. Free-fermion for the 32-vertex model on a triangular lattice

We next consider the free-fermion conditions of J. E. Sacco and F. Y. Wu [53] for the 32-vertex model on a triangular lattice. Using the same notation as in [53], we have:

$$R = \begin{bmatrix} f_0 & 0 & 0 & f_{23} & 0 & f_{13} & f_{12} & 0 \\ 0 & f_{36} & f_{26} & 0 & f_{16} & 0 & 0 & \bar{f}_{45} \\ 0 & f_{35} & f_{25} & 0 & f_{15} & 0 & 0 & \bar{f}_{46} \\ f_{56} & 0 & 0 & \bar{f}_{14} & 0 & \bar{f}_{24} & \bar{f}_{34} & 0 \\ 0 & f_{34} & f_{24} & 0 & f_{14} & 0 & 0 & \bar{f}_{56} \\ f_{46} & 0 & 0 & \bar{f}_{15} & 0 & \bar{f}_{25} & \bar{f}_{35} & 0 \\ f_{45} & 0 & 0 & \bar{f}_{16} & 0 & \bar{f}_{26} & \bar{f}_{36} & 0 \\ 0 & \bar{f}_{12} & \bar{f}_{13} & 0 & \bar{f}_{23} & 0 & 0 & \bar{f}_0 \end{bmatrix}. \tag{14}$$

By permuting rows and columns, this matrix can be brought into the block diagonal form:

$$R = \begin{pmatrix} R_1 & 0 \\ 0 & R_2 \end{pmatrix}, \qquad \text{with:} \tag{15}$$

$$R_1 = \begin{bmatrix} f_0 & f_{13} & f_{12} & f_{23} \\ f_{46} & \bar{f}_{25} & \bar{f}_{35} & \bar{f}_{15} \\ f_{45} & \bar{f}_{26} & \bar{f}_{36} & \bar{f}_{16} \\ f_{56} & \bar{f}_{24} & \bar{f}_{34} & \bar{f}_{14} \end{bmatrix}, \quad R_2 = \begin{bmatrix} f_{14} & f_{34} & f_{24} & \bar{f}_{56} \\ f_{16} & f_{36} & f_{26} & \bar{f}_{45} \\ f_{15} & f_{35} & f_{25} & \bar{f}_{46} \\ \bar{f}_{23} & \bar{f}_{12} & \bar{f}_{13} & \bar{f}_0 \end{bmatrix}. \tag{16}$$

The inverse I, written polynomially (homogeneous matrix inverse), is now a transformation of degree 7. If one introduces the two determinants, $\Delta_1 = \det(R_1)$ and $\Delta_2 = \det(R_2)$, then each term in the expression of $I(R)$ is a product of a degree three minor, taken within a block, times the determinant of the other block. This inverse I clearly singles out one of the three directions of the triangular lattice. These three involutions do not commute and generate a quite large infinite discrete group $\Gamma_{\rm triang}$ (see also section (IV-1) below).

The free-fermion conditions of Sacco and Wu [53] read:

$$\begin{aligned} f_0 f_{ijkl} &= f_{ij} f_{kl} - f_{ik} f_{jl} + f_{il} f_{jk}, \qquad \forall\, i, j, k, l = 1, \dots, 6 \\ f_0 \bar{f}_0 &= f_{12}\bar{f}_{12} - f_{13}\bar{f}_{13} + f_{14}\bar{f}_{14} - f_{15}\bar{f}_{15} + f_{16}\bar{f}_{16}, \end{aligned} \tag{17}$$

which we denote by $\mathcal{V}$. What is remarkable is that, not only is the rational variety $\mathcal{V}$ globally invariant under $\Gamma_{\rm triang}$, but *again* the realization of this (generically very large infinite discrete group) $\Gamma_{\rm triang}$ on this variety becomes *finite.* This comes about from the degeneration of I into a mixture of sign changes and permutations of the entries, as in the preceding subsection.

Remark. The ordinary *matrix product of three matrices* (14) *solutions* of (17), is *another solution*! In other words, if R_α, R_β, $R_\gamma \in \mathcal{V}$, then $R_\alpha \cdot R_\beta \cdot R_\gamma \in \mathcal{V}$, while $R_\alpha \cdot R_\beta \notin \mathcal{V}$. This was also the case for involutions of (11) in the case of the square lattice, but the mechanism is more subtle here as the conditions (17) imply $\Delta_1 = \Delta_2$.

III-3. Dimers and spanning trees

Before Fan and Wu's free-fermion vertex models, the Onsager solution of the two-dimensional Ising model was clearly the first free-fermion model ever solved. There were also several approaches to the two-dimensional Ising model that did not use the transfer matrix formalism; the most interesting one is perhaps the mapping of the problem onto a dimer-covering problem on a slightly more complicated lattice. The dimer problem was first solved by Temperley-Fisher and Kasteleyn. Kasteleyn found out how to treat the most general planar graph.

The dimer problem has a life of its own and has generated since many followup works, not only in statistical mechanics, but also in combinatorial theory. In this regard, Wu has provided a large number of new results [37, 173, 184, 194, 207, 205], including applications to condensed matter physics, as well as in pure combinatorial analysis. In addition, Wu has obtained new results on the spanning tree problem [198, 200], a problem intimately related to the dimer problem through a bijection due to Temperley. In the following we shall describe some of the contributions in this area.

III-3-1. Revisiting dimers: the honeycomb lattice

The dimer model on the honeycomb lattice was first solved by Kasteleyn, but he never published the solution, except for hinting at the existence of a transition. This deficiency was made up by Wu in a 1968 paper [10] in which he presented details of the analysis for the honeycomb lattice, and applied the results to describe the physics of a modified KDP model.

III-3-2. Revisiting dimers: Interacting dimers in 2 and 3 dimensions

Almost 30 years after the publication of the solution for the dimers on the honeycomb lattice [10], Wu and his co-workers made two important extensions of the earlier Kasteleyn solution. In the first, H. Y. Huang, F. Y. Wu, H. Kunz, and D. Kim [173] considered the case where the dimers have nearest-neighbor interaction. This model turns out to be identical to the most general five-vertex model, a degenerate case of the six-vertex model which requires a special Bethe ansatz analysis. The resulting phase diagram of this five-vertex model is very complicated and the analysis extremely lengthy.

In the second work H. Y. Huang, V. Popkov and F. Y. Wu [177, 184] introduced, and solved, a three-dimensional model consisting of layered honeycomb dimer lattices, as described in the preceding subsection, but with a specific layer-layer interaction. Again, the phase diagram is very complicated. It is noted that this model is the only solvable three-dimensional lattice model with *physical* Boltzmann weights (the Baxter solution of the 3D Zamolodchikov model has negative weights). However, the layered dimer model, while having strictly positive weights, describes dimer configurations in which the dimers are confined in planes. As a consequence the critical behavior is essentially two-dimensional.

III-3-3. Revisiting dimers: a continuous-line model

F. Y. Wu and H. Y. Huang [158] have further used a dimer mapping to solve a continuous-line lattice model in three and higher dimensions. They have also applied it to model a type-II superconductor [160. In three dimensions, the model is a special case of an $O(n)$ model on a finite $L_1 \times L_2 \times L_3$ cubic lattice with periodic boundary conditions with the partition function:

$$Z(n) = \sum_{closed\ polygons} n^l \cdot z^b,$$

where the summation is taken over all closed non-intersecting polygonal configurations, l is the number of polygons, and b is the number of edges of each configuration. They considered the $n = -1$ special case, which they showed to be in one-to-one correspondence with a dimer problem whose partition function can be evaluated as a *Pfaffian*. The result for a finite lattice is:

$$Z(-1) = \prod_{n_1=1}^{L_1} \prod_{n_2=1}^{L_2} \prod_{n_3=1}^{L_3} \left| 1 + \sum_{i=1}^{3} z \cdot e^{2\pi n_i / N_i} \right| .$$

In the thermodynamic limit, this leads to the per-site free energy:

$$f = \frac{1}{(2\pi)^3} \cdot \int_0^{2\pi} d\theta_1 \int_0^{2\pi} d\theta_2 \int_0^{2\pi} d\theta_3 \, \ln \left| 1 + \sum_{i=1}^{3} z \cdot e^{2\pi n_i / N_i} \right| .$$

The phase diagram is rich and quite non-trivial.

However, it must be said that this exactly soluble three-dimensional $\mathrm{O}(-1)$ model describes line configurations running only in a *preferred direction* and, secondly,[13] the Boltzmann weights can be negative.

III-3-4. Revisiting dimers: nonorientable surfaces

More recently W. T. Lu and F. Y. Wu initiated studies on dimers and Ising models on *nonorientable* surfaces [194, 203, 205]. For dimers on an $M \times N$ net, embedded on non-orientable surfaces, they solved *both* the Möbius strip and the Klein bottle problems for all sizes M and N and obtained the dimer generating function $Z_{M,N}$ as:

$$\frac{Z_{M,N}}{z^{M\,N/2}} = \mathrm{Re}\Big((1-i) \cdot \prod_{m=1}^{(M+1)/2} \prod_{n=1}^{N} \Big(2i(-1)^{M/2+m+1} \sin\Big(\frac{(4\,n-1)\pi}{2N} \Big) + 2X_m \Big) \Big),$$

where Re denotes the real part, z_v and z_h are the dimer weights in the vertical and horizontal directions, respectively, and X_m is given for the Möbius strip and the Klein bottle respectively by:

$$X_m = \frac{z_v}{z_h} \cdot \cos\Big(\frac{m\pi}{M+1} \Big), \qquad X_m = \frac{z_v}{z_h} \cdot \cos\Big(\frac{(2m-1)\pi}{M} \Big).$$

In paper [205] they also obtained an *extension of the Stanley-Propp reciprocity theorem* for dimers[14]. Inspired by this work, there is now much activity in this area. There is also very much current interest in finite-size corrections and conformal field theories on more complicated surfaces (higher genus pretzels, ...).

III-3-5. Dimers on a square lattice with a boundary defect

In a very recent paper [207], fittingly dedicated to the 70th birthday of Michael Fisher, who first solved the dimer problem for the square lattice, W. J. Tzeng and F. Y. Wu obtained the dimer

---

[13] But we are used to this after R. J. Baxter's solution of the 3D Zamolodchikov model.

[14] A subject matter of pertinent interest to mathematicians.

generating function for the square lattice with one corner (or some other boundary site) of the lattice missing. In this work they made use of a bijection between the dimer and spanning tree configurations due to Temperley (and extended by Wu in his unpublished 1976 lecture notes as well as well as more recently by Kenyon, Propp and Wilson). They also carried out a finite-size analyses which lead to a logarithmic correction term in the large-size expansion for the vacancy problem with free boundary conditions. They found a central charge $c = -2$ for the vacancy problem, to be compared with $c = -1$ when there is no vacancy. This central charge $c = -2$ is in contradiction with the prediction of a naive conformal field theory.

III-3-6. Spanning trees

As mentioned above, the problem of spanning trees in graph theory is intimately related to the dimer problem, and it is not surprising that Wu found his way to spanning trees. In 1977 he published a paper [62] on the counting of spanning trees on two-dimensional lattices using the equivalence with a Potts model. Very recently he refined the tools by using a result in algebraic graph theory, which he and W. J. Tzeng rederived using elementary means. Tzeng and Wu enumerated spanning trees for general d-dimensional lattices as well as non-orientable surfaces [198]. Applying these results to general graphs and regular lattices, R. Shrock and F. Y. Wu [200] published a lengthy paper in which they established new theorems on spanning trees as well as enumerating spanning trees for a large number of regular lattices in the thermodynamic limit.

III-4. Miscellaneous results on vertex models

In this section we describe an arbitrary choice of miscellaneous results obtained by F. Y. Wu on vertex models.

III-4.1. Boundary conditions

The six-vertex model is known to be a boundary condition dependent model. However H. J. Brascamp, H. Kunz and F. Y. Wu [43] established, for the first time, that, at sufficiently low temperatures or sufficiently high fields, the six-vertex models with either periodic or free boundary conditions are equivalent.

III-4-2. The eight-vertex model in a field

A simple result due to F. Y. Wu [105] is that a very general staggered eight-vertex model in the Ising language (as introduced by Kadanoff and Wegner and by F. Y. Wu [78]) with the special Yang-Lee magnetic field $i\pi kT/2$, is equivalent to Baxter's symmetric eight-vertex model and hence is soluble. This result is remarkable, since the general eight-vertex model without this field is not known to be soluble.

III-4-3. The eight-vertex model on the honeycomb lattice

F. Y. Wu has always been keen in providing results for the *honeycomb* lattice [48, 130]. One interesting result is that he has established the exact equivalence of the eight-vertex model on the honeycomb lattice with an Ising model in a nonzero magnetic field [48, 130]. The equivalence also leads to exact analysis of the Blume-Emery-Griffiths model for the honeycomb lattice (for details see section (IV-3) below).

In pursuit of applications of these results, P. Pant and J. H. Barry and Wu [181, 182] obtained exact results for a model of a ternary polymer mixture which is equivalent to an eight-vertex model. A model ternary polymer mixture was considered with bi- and tri-functional monomers and a solvent placed on the sites of a honeycomb lattice. Using the equivalence with an eight-vertex model which further maps the problem into an Ising model in a nonzero magnetic field, an exact analysis of the model was carried out. The phase boundary of the three-phase equilibrium polymerization regime was determined exactly.

Comment: These kinds of results are particularly interesting when one realizes that concepts and structures corresponding to the Yang-Baxter integrability do not exist at first sight. This leads to the natural question: *how to construct a Yang-Baxter relation for the honeycomb lattice*?

III-4-4. Exact critical line of a vertex model in 3 dimensions

Wu [46] has introduced a vertex model in three dimensions with real vertex weights, and determined its exact first-order phase transition line by mapping it to an Ising model in a field. It also exhibits a critical point. This is one of the very few lattice statistical models for which exact results can be deduced in higher-than-two dimensions.

IV. Spin models: Ising models and other models

We will consider in this section F. Y. Wu's results on *spin models*, mostly *Ising* models. Due to its importance the Potts models will be treated separately in section (7).

The distinction between vertex models and spin models (more generally Interaction Round a Face (IRF) models) is an important one in lattice statistical mechanics. However Wu showed in paper [78] an equivalence between an Ising model with a vertex model. Similarly in paper [114] Wu and K. Y. Lin studied the Ising model on the Union Jack lattice, showing it to be a free-fermion model. Many of the free-fermion results on the vertex models in sections (III-2-1) and (III-3) can also be re-styled as free-fermion Ising models.

As far as the Union Jack lattice is concerned Wu has also obtained the spontaneous magnetization of the three-spin Ising model [51]. It is, in fact, obtained in terms of the magnetic and ferroelectric orderings of the eight-vertex model, or, equivalently, the spontaneous magnetization and polarization of the eight-vertex model. It was found that the two sublattices possess different critical exponents.

An important development in the history of lattice models is the analysis of the phase diagram of the Ashkin-Teller model on the square lattice by F. Y. Wu and K. Y. Lin [47]. The Ashkin-Teller model is another example of spin models for which the traditional distinction of lattice statistical mechanics between spin and vertex models is irrelevant. The Ashkin-Teller model can be seen as two Ising models coupled together with four-spin interactions. Performing a dual transformation on one of the two Ising models and interpreting the result as a vertex model, one finds that the Ashkin-Teller model is equivalent to a *staggered* eight-vertex model [29], thus exhibiting two phase transitions.

Wu's analysis of spin models was not restricted to two dimensions. For instance, Barry and Wu have obtained exact results for a four-spin-interaction Ising model on the three-dimensional pyrochlore lattice [128], and Wu has also obtained various results for spin models on the Bethe lattice and Cayley trees [54, 56].

Wu also performed *real-space renormalization* studies for Ising models [129], but, not surprisingly, using some duality ideas, namely, the duality-decimation transformation of T. W.

Burkhardt. Wu had previously applied the duality-decimation transformation in order to solve the two-dimensional Ising model with nearest-neighbor, next-nearest-neighbor and four-spin interactions in a pure imaginary field [105] (see section III-4.2). In paper [129] Burkhardt's method, which combines a bond-moving and duality-decimation transformation, is modified, in order to preserve the free energy in the renormalization transformation.

IV-1. Generalized transmissivities for spin models

We will see below that a large number of F. Y. Wu's work correspond to graph expansions (see section (VII-1)). For spin models with edge interactions this requires the introduction of certain "transmissivity" variables. Thermal transmissivities are introduced when considering high-temperature expansions of an edge-interaction spin model or performing renormalization analyses. They are also the natural variables to use in the decimation of spins in a simple multiplicative way.

Introduce the edge Boltzmann weight $W(K_1, K_2, \cdots K_n; a, b)$, where $K_1, K_2, \cdots K_n$ denote a set of coupling constants describing the model, and a and b are two nearest-neighbor spin states which can take on q values. Let us assume that the decimation procedure yields a Boltzmann weight of the *same* form:

$$\sum_b W(K_1, K_2, \cdots K_n; a, b) \cdot W(K_1', K_2', \cdots K_n'; b, c) = \lambda \cdot W(K_1'', K_2'', \cdots K_n''; a, c). \quad (18)$$

Alternatively, one can build a $q \times q$ Boltzmann matrix $\mathbf{W}$ with entries $\mathbf{W}_{i,j} = W(K_1, K_2, \cdots K_n; i, j)$. In terms of such matrices relation (18) becomes $\mathbf{W} \cdot \mathbf{W}' = \mathbf{W}''$. The decimation procedures, and also the high-temperature expansions in such models, are greatly simplified by introducing a "transmissivity" function t_α, such that the matrix relation $\mathbf{W} \cdot \mathbf{W}' = \mathbf{W}''$ becomes one or more multiplicative relations of the form:

$$t_\alpha(\mathbf{W}) \cdot \mathbf{t}_\alpha(\mathbf{W}') = \mathbf{t}_\alpha(\mathbf{W}''), \qquad \alpha = \mathbf{1}, \cdots, \mathbf{r}.$$

The simplest example is the transmissivity variable for the q state standard scalar Potts model, for which one has $t = (e^K - 1)/(e^K + q - 1)$. This is the natural expansion variable for the high-temperature series of the model (see also the $f_{i,j}$'s in (45) and (46) introduced in section (VII-1) below). For the Ising model this reduces to the $\tanh(K)$ variable. F. Y. Wu *et al.* [147] underlined the fact that two quite different situations must be considered. If the family of Boltzmann matrices $\mathbf{W}$ is a set of *commuting* matrices, then they can be diagonalized simultaneously and the transmissivity variables are nothing but all the possible ratios of eigenvalues of the Boltzmann matrices $\mathbf{W}$. If, alternatively, the Boltzmann matrices $\mathbf{W}$ do not commute, then one must perform a simultaneous block-diagonalization of this family of Boltzmann matrices, and, therefore, some of the t_α's will be block matrices from which one can extract functions ϕ_α satisfying $\phi_\alpha(\mathbf{W}'') = \phi_\alpha(\mathbf{W}) \cdot \phi_\alpha(\mathbf{W}')$. One obvious choice for ϕ_α is the (ratio of) determinants of these blocks. A number of non-trivial non-commuting transmissivities are given in [147].

IV-2. Three-spin interactions: the Baxter-Wu model

Another important work in the history of exact solutions of lattice statistics is the Baxter-Wu model, which is an Ising model on the triangular lattice with *three-spin* interactions. This model was solved exactly by R. J. Baxter and F. Y. Wu in 1973 [45, 49].

The three spins surrounding every triangular face interact with a *three-body* interaction of strength $-J$, so that the Hamiltonian reads:

$$H = -J \cdot \sum \sigma_i \sigma_j \sigma_k. \tag{19}$$

Baxter and Wu found that the per-site partition function Z has a remarkably simple expression:

$$Z = \sqrt{6yt} \qquad \text{with:} \qquad t = \sinh(2\,|J|/kT), \tag{20}$$

and where y is the solution of the algebraic equation:

$$(y-1)^3\,(1+3y)\,(1+t^2)\cdot t = 2(1-t)^4 \cdot y^3. \tag{21}$$

The partition function has a singular part which behaves as $|t-1|^{4/3}$.

Some interesting duality properties of the Baxter-Wu model are very clearly detailed in [45], and used to convert the Baxter-Wu model into a *coloring* problem. This provides a very heuristic example showing that duality is *not* specific to edge-interaction spin models, but can also be introduced with many-body interactions. In the following we briefly describe how the Baxter-Wu model is transformed into a coloring problem.

First we introduce a $\mathbf{Z}_2$-Fourier transform with function $g(\lambda, \mu)$ which enables us to simply write the Kramers-Wannier duality for this three-spin model (λ and μ are Ising spins) as:

$$\begin{aligned} g(\lambda,\mu) &= +1 \quad \text{if:} \quad \lambda = +1, \\ g(\lambda,\mu) &= \mu \quad\;\; \text{if:} \quad \lambda = -1. \end{aligned} \tag{22}$$

Note that this function is symmetric in λ and μ, namely, $g(\lambda, \mu) = +1$ when $\mu = +1$ and $g(\lambda, \mu) = \lambda$ when $\mu = -1$.

Returning to the Baxter-Wu model, each spin σ_i of the triangular lattice belongs to six triangles around vertex i, which form a hexagon with the spin σ_i at the center. Let us now consider the close-packing of such hexagons. The spins σ_i now form a (triangular) sublattice of the initial triangular lattice.

Consider next the spins σ_i, and denote the edge connecting nearest-neighboring spins, σ_k and σ_l, sitting on the hexagon surrounding σ_i, by $< kl >$. Let us introduce Ising edge variables λ_r corresponding to the six edges $< kl >$ of the hexagon: $\lambda_r = \sigma_k \cdot \sigma_l$, $r = 1, \cdots 6$. The local Boltzmann weight of a hexagonal cell around a spin σ_i can be written as:

$$W_{\text{hex}} = \frac{1}{2} \cdot \Big(1 + \prod_{r=1}^{6} \lambda_r\Big) \cdot \exp\Big(K \cdot \sigma_i \cdot \sum_{r=1}^{6} \lambda_r\Big), \tag{23}$$

where the factor $(1 + \prod \lambda_r)$ takes into account the fact that the Ising edge variables λ_r are not independent, but are constrained by the condition $\prod \lambda_r = 1$. As usual this condition, associated with every hexagon, can be written by introducing a dummy variable μ_i also associated with every hexagon:

$$\sum_{\mu_i = \pm 1} \prod_{r=1}^{6} g(\lambda_r, \mu_i) = 1 + \prod_{r=1}^{6} \lambda_r,$$

enabling us to rewrite (23) as:

$$W_{\text{hex}} = \sum_{\mu_i=\pm 1} \prod_{r=1}^{6} g(\lambda_r, \mu_i) \cdot \exp\Big(K \cdot \sigma_i \cdot \sum_{r=1}^{6} \lambda_r\Big).$$

The partition function of the Baxter-Wu model is now seen as a summation over all the (initial) spins σ_i and the (dummy) spins μ_i of a triangular sublattice, and the edge Ising spins λ_r. Let us focus on one λ_r. The edge $r =< kl >$ belongs to a hexagon around spin σ_i and a neighboring hexagon around another spin, say, σ_j. The edge Ising spin λ_r thus occurs in the Boltzmann factors with a factor of

$$W_{\lambda_r} = e^{K\cdot(\sigma_i+\sigma_j)\cdot\lambda_r} \cdot g(\lambda_r, \mu_i) \cdot g(\lambda_r, \mu_j).$$

Summing over the edge Ising spin λ_r in the partition function and using relations (22), one thus obtains a factor:

$$\omega_{ij} = \sum_{\lambda_r=\pm 1} W_{\lambda_r} = e^{K(\sigma_i+\sigma_j)} + \mu_i\mu_j \cdot e^{-K(\sigma_i+\sigma_j)} \tag{24}$$

between two spins σ_i and σ_j on the sublattice. This can be interpreted as the edge weight associated with a coloring problem, and the Baxter-Wu model is transformed into a coloring problem.

IV-3. The Blume-Emery-Griffiths model

The Blume-Emery-Griffiths (BEG) model is a model that F. Y. Wu and his coworkers quite naturally considered [106, 116, 136, 148], since it reduces to an Ising model on the honeycomb lattice on a special manifold [106, 116].

The BEG model is defined by the Hamiltonian:

$$-\beta \cdot H = -J \cdot \sum_{<i,j>} S_i S_j - K \cdot \sum_{<i,j>} S_i^2, S_j^2 - \Delta \sum_i S_i^2 - H \cdot \sum_i S_i, \tag{25}$$

where the spins are classical spin-1 spins taking on the values $S_i = 0, \pm 1$.In the high-temperature expansion the nearest-neighbor Boltzmann factor assumes the form

$$\exp(J\, S_i\, S_j + K\, S_i^2\, S_j^2) = 1 + (e^K \sinh J) S_i S_j + (e^K \cosh J - 1) S_i^2 S_j^2. \tag{26}$$

It follows then in the subspace $K = -\ln(\cosh J)$, one has the simple relation

$$\exp(J S_i S_j + K S_i^2 S_j^2) = 1 + S_i\, S_j\, \tanh(J),$$

and the partition function of the BEG model assumes the simpler form:

$$Z_{BEG} = \sum_{S_i=0,\pm 1} \prod_{<i,j>} (1 + S_i S_j \tanh J) \prod_i \exp(-\Delta S_i^2 + H\, S_i).$$

Expanding the products over neighboring pairs, representing each term by a graph and making use of the identities

$$\sum_{S_i=0,\pm 1} S_i^n \cdot e^{-\Delta S_i^2} = \rho(n) \qquad \text{with:} \qquad \rho(2) = 2e^{-\Delta},$$
$$\rho(0) = 2e^{-\Delta} + 1, \qquad \rho(1) = 0, \tag{27}$$

one finds that one has eight possible configurations at each vertex of the honeycomb lattice, corresponding to the following Boltzmann weights of an (isotropic) eight-vertex model:

$$a = 1 + 2e^{-\Delta}\cosh(H), \qquad b = 2\sqrt{\tanh(J)}e^{-\Delta}\sinh(H),$$
$$c = 2\tanh(J)e^{-\Delta}\cosh(H), \quad d = 2(\tanh(J))^{3/2}e^{-\Delta}\sinh(H).$$

With these notations one deduces the identity of the partition function of the BEG model (25) with the eight-vertex model on a honeycomb lattice (see also sections (V-1) and (V-3) below):

$$Z_{BEG} = Z_{8v}(a, b, c, d).$$

Performing a weak-graph duality transformation on this eight-vertex model (see section (V) below) associated with the 2×2 (gauge) matrix [48]:

$$g_1 = g_2 = \frac{1}{\sqrt{2}} \cdot \begin{pmatrix} 1 & y \\ -y & 1 \end{pmatrix},$$

one finds that the partition function of the eight-vertex model $Z_{8v}(a, b, c, d)$ remains invariant under a weak-graph duality transformation [7, 116]:

$$\tilde{a} = (a + 3yb + 3y^2c + y^3d)/(1 + y^2)^{3/2}, \; \cdots \; Z_{8v}(a, b, c, d) = Z_{8v}(\tilde{a}, \tilde{b}, \tilde{c}, \tilde{d}). \tag{28}$$

The four parameters a, b, c, d or $\tilde{a}, \tilde{b}, \tilde{c}, \tilde{d}$ can be seen, as far as the calculation of the partition function is concerned, as four *homogeneous* parameters. Taking into account an irrelevant overall factor and the irrelevant gauge variable y from the weak-graph symmetry (28), one sees that the partition function of the eight-vertex model $Z_{8v}(a, b, c, d)$ basically depends on two variables instead of four. Not surprisingly, Wu found that $Z_{8v}(a, b, c, d)$ is equivalent to the partition function of an Ising model with nearest-neighbor interactions K_I and a magnetic field L:

$$\begin{aligned} Z_{\text{Ising}}(L, K_I) &= \sum_{\sigma} \prod_{<ij>} \exp(K_I \sigma_i \sigma_j) \prod_i \exp(L\sigma_i) \\ &= Z_{8v}(\tilde{a}, \tilde{b}, \tilde{c}, \tilde{d}) \cdot \left(\frac{2\cosh(L)\cosh^{3/2}(K_I)}{\tilde{a}} \right)^N, \end{aligned} \tag{29}$$

where N is the number of lattice sites. The explicit expressions of K_I and L in terms of the BEG parameters are complicated. But for $H = 0$, one has $L = 0$ and

$$\tanh(K_I) = \frac{2}{2 + e^{\Delta}} \cdot \tanh(J),$$

using which one determines the critical line $K_I = 1/\sqrt{3}$ in the $J > 0$ regime. The spontaneous magnetization of the BEG model for $J > 0$ and the phase boundary of the $J < 0$ BEG model can be similarly determined [116].

The proof of the equivalence of the honeycomb eight-vertex model with an Ising model in a field, as outlined in the above, is quite tedious. However, a more direct derivation has since been given by Wu [130].

The eight-vertex model on the honeycomb lattice can also be seen to be related to a lattice-gas grand partition function $\Xi^{Kag}(z, J, J_3)$ on the *Kagomé lattice* [126] (see also section (VI-3) below), where $-J$ is the nearest-neighbor interaction, $-J_3$ the triplet interactions existing among three sites surrounding a triangular face of the lattice, and z denotes the fugacity. Then one has the equivalence:

$$\begin{aligned} &\Xi^{Kag}(z, J, J_3) = Z_{8v}(a, b, c, d) \\ &a = 1, \quad b = \sqrt{z}, \quad c = z \cdot e^J, \quad d = z^{3/2} \cdot e^{3J+J_3}. \end{aligned} \tag{30}$$

From (29) and (30), F. Y. Wu and X. N. Wu were able to obtain results for the liquid and vapor densities, showing that an observed anomalous critical behavior occurs in the lattice gas only when there are nonzero triplet interactions [126]. This analysis has been extended to a lattice gas on the 3-12 lattice by J. L. Ting, S. C. Lin and F. Y. Wu [140].

Wu's tricks for the honeycomb BEG model are not limited to the weak-graph transformation for the eight-vertex model. Using a syzygy analysis of the invariants under the O(3) transformation L. H. Gwa and F. Y. Wu have obtained an expression for the critical variety of the honeycomb BEG model to an extremely high degree of accuracy [148] (see section (V-4) below).

IV-4. Other spin results: disorder points

Let us finally describe, among many results obtained by F. Y. Wu on spin models, one result concerning disorder points. Disorder solutions are particularly simple solutions corresponding to some "dimensional reduction" of the model, which provide simple exact results for models which are generically quite involved. While this yields severe constraints on the phase diagrams, the series expansion, and the analyticity properties of the model, it does lead to exact solutions of models which are otherwise nonintegrable.

For example, using a decimation approach, Wu [100] has deduced the disorder solution for the triangular Ising model in a nonzero magnetic field. Wu and K. Y. Lin [120] have used a checkerboard Ising lattice to illustrate that there may exist more than one disorder point in a given spin system. Along the same vein, N. C. Chao and Wu [101] have explored the validity of the decimation approach by considering the disorder solutions of a general checkerboard Ising model in a field.

V. Weak-graph dualities and Hilbert's syzygies

In a pioneering paper F. Y. Wu and Y. K. Wang [58] introduced a duality transformation for a general spin model which can have chiral interactions. This is the first time that a chiral spin model was explicitly considered. In terms of R matrices such as (12) these transformations are the tensor product of two similarities:

$$R \quad \longrightarrow \quad g_1 \quad g_2 \cdot R \cdot g_1^{-1} \quad g_2^{-1} \tag{31}$$

where g_1 and g_2 are two $q \times q$ matrices, R is a $q^2 \times q^2$ matrix ($q = 2$ for the sixteen-vertex model). This symmetry group is an $\mathrm{sl}(q) \times \mathrm{sl}(q)$ symmetry group. The high- and low-temperature duality (3) given in section (III-1) for the sixteen-vertex model is a particular case of such transformations, corresponding to g_1 and g_2 being two involutions:

$$g_1 = g_2 = \frac{1}{\sqrt{2}} \cdot \begin{pmatrix} 1 & i \\ -i & 1 \end{pmatrix}.$$

This duality relation [58, 131] is now well-known for vertex models. It corresponds to symmetries of the model and can be used, as will be seen in the next two subsections, to find good variables to express the critical manifolds of a lattice model, and, hopefully, to determine their exact expressions when algebraic.

V-1. Hilbert's syzygies, gauge-like dualities and critical manifolds

Let us give some hints as to how the gauge-like dualities enable us to deduce results on critical manifolds or varieties. The main idea is to construct algebraic invariants under these gauge transformations.

Hilbert has shown that all invariants of a linear transformation are algebraic and can be expressed in terms of a set of homogeneous polynomials, the *syzygies*. Considering the sixteen-vertex model, the transformation is O(2) and the fundamental invariants corresponding to the O(2) group have been constructed by J. H. H. Perk, F. Y. Wu and X. N. Wu in [131]. Likewise for 3-state vertex models the transformation is O(3) and the associated invariants have been constructed by L. H. Gwa and F. Y. Wu [146] (see section (V-4) below).

V-2. Hilbert's syzygies and the square lattice Ising model in a magnetic field

With an algebraic prejudice for critical manifolds, it is very tempting to conjecture closed algebraic formula for critical manifolds that will reproduce known exact results in various limits. For instance, closed-form expressions for the critical line of the square lattice antiferromagnetic Ising model in a magnetic field were proposed[15] by Müller-Hartmann and Zittartz. However, it has been shown that the expression is numerically incorrect.

X. N. Wu and F. Y. Wu [135] considered the square lattice antiferromagnetic Ising model in a magnetic field, which can be seen as a subcase of the sixteen-vertex model under the O(2) group. Introducing the variables

$$a = 1, \quad b = \sqrt{v}h, \quad c = v, \quad d = v^{3/2}h, \quad e = v^2$$

$$\text{with:} \qquad v = \tanh(J/kT), \qquad h = \tanh(H/kT),$$

the five fundamental Hilbert invariants of O(2) read:

$$I_1 = a + 2c + e \qquad I_2 = (a - 6c + e)^2 + 16(b - d)^2,$$

$$I_3 = (a - e)^2 + 4(b + d)^2, \qquad I_5 = a^2 d - be^2 - 3(a - e)(b + d)c,$$

$$I_4 = (a - 6c + e) \cdot ((a - e)^2 - 4(b + d)^2) + 4(a - e)(b^2 - d^2),$$

and the critical line proposed by Wu and Wu assumes the form

$$c_1 \cdot I_1^4 \, c_2 \cdot I_1^2 \, I_2 + c_3 \cdot I_1^2 \, I_3 + c_4 \cdot I_2^2 + c_5 \cdot I_3^2 + c_6 \cdot I_2 \, I_3 = 0.$$

[15] This is just one example in a very long list of incorrect algebraic conjectures for critical manifolds that one can find in the literature.

They then determined the c_i's using the various known results, including a constraint dictated by the known zero-field critical point, as well as the results of a finite-size analyses which they carried out. This lead to the values $c_1 = 1$, $c_2 = -0.044\ 338$, $c_3 = 0.362\ 73$, $c_4 = 0.000\ 4938$, $c_5 = 0.042\ 779$, $c_6 = -0.008\ 9149$. The resulting closed form expression for the critical line reproduces all known numerical data to a high degree of accuracy. For instance, the critical line yields for a small magnetic field: $T_c \simeq T_0 \cdot (1 - u \cdot (H/J)^2)$ with $u \simeq 0.038\ 022$. This is compared to the presumably exact value obtained by M. Kauffman: $u \simeq 0.038\ 0123\ 259 \cdots$

V-3. Hilbert's syzygies and the honeycomb lattice Ising model in a magnetic field

Similar analyses have been carried out for the honeycomb lattice by F. Y. Wu, X. N. Wu and H. W. J. Blöte [132]. For the corresponding honeycomb eight-vertex model we have

$$a = 1, \qquad b = \sqrt{v}h, \qquad c = v, \qquad d = v^{3/2}h$$

$$\text{with:} \qquad v = \tanh(J/kT), \qquad h = \tanh(H/kT).$$

Analogous to (32), we introduce the following Hilbert's syzygies:

$$P = a^2 + 3ac + 3bd + d^2, \qquad Q = b^2 - ac + c^2 - bd$$

$$P_2 = 2(a^4 + d^4) - 6(a^2\, c^2 + b^2\, d^2) + 12(a^2\, b^2 + c^2\, d^2) - 5a^2\, d^2$$

$$+27\, b^2\, c^2 + 36(ab + cd)bc + 18\, a\, b\, c\, d.$$

The critical line proposed by Wu, Wu and Blöte now reads [125]:

$$c_1 \cdot P_2 + c_2 \cdot P^2 + c_3 \cdot PQ + c_4 \cdot Q^2 = 0. \tag{32}$$

After an extensive search by mapping with all known exact results, they proposed the numbers: $c_1 = 1$, $c_2 = -(4 + 3\sqrt{3})/6$, $c_3 = -(1 - 9\sqrt{3})/8$, and $c_4 = -3(3 - \sqrt{3})/8$.

The initial slope of this critical frontier for small H is $-\ln(z_c)$ where z_c is the critical fugacity of the nearest-neighbor exclusion gas. Their expression leads to the value $z_c \simeq 7.851\ 780\ 04 \cdots$ which is in very good agreement with the value obtained from finite-size analysis, namely, $z_c \simeq 7.851\ 725\ 175(13)$. The critical line (32) is probably not the exact one but certainly a very accurate approximation.

Comment: Hilbert's invariant theory amounts to considering *linear* gauge-like symmetries of the model and the associated invariants. From the inversion relations one has further an infinite discrete set of birational *non-linear* symmetries, that one can couple with these continuous linear groups. In fact, all the above analyses can be revisited by combining the gauge transformation with the *infinite discrete symmetries* generated by the *inversion relations* of the sixteen- (or simply eight)-vertex models. This would lead us to consider a unique "superinvariant" which is, in fact, the *modular invariant* of the elliptic curves parametrizing the sixteen-vertex model.

V-4. Hilbert's syzygies and the honeycomb BEG model

To apply the syzygy consideration to the BEG model, which is a 3-state spin model, one needs to consider the O(3) gauge transformation and its associated invariants, but the construction of the O(3) invariants is very complicated. However, the day is saved, since there exists a mapping between O(3) and sl(2), and invariants for the latter have been worked out by mathematicians a long time ago. While the mathematics to decipher the old results is involved, L. H. Gwa and F. Y. Wu [146] have succeeded in carrying out such an analysis and deduced that there are 5 independent invariants for the O(3) group. They next applied the analysis to the isotropic BEG model on the honeycomb lattice [148], and found one of the invariants to vanish identically. The remaining 4 invariants are then used to determine the critical variety of the BEG model, as in the case of the O(2) gauge. The resulting closed-form expression for the critical variety agrees extremely well with numbers obtained from a finite-size analysis, which they also carried out.

VI. Critical manifolds and critical varieties

A problem solver like F. Y. Wu first tries to find the exact solution of a problem. He tries to "dig out" problems that can be solved. However since most of the problems one looks at cannot be solved exactly, one then tries to study models for which some exact results can be "salvaged". This could be the critical manifolds, which are submanifolds along which the models are Yang-Baxter integrable. In such cases the critical manifolds are, in fact, critical *varieties*. For other models the critical manifolds are *algebraic* varieties without hidden Yang-Baxter integrability [79, 161].

For two-dimensional lattice models the situation is more specific: one can have some "conformal prejudice" that critical manifolds should be submanifolds where the model has a two-dimensional conformal (infinite) symmetry yielding some integrability in the scaling limit. Therefore, as far as critical manifolds are concerned, it is crucial to understand the inter-relation between 1) algebraicity consequences of Yang-Baxter integrability, 2) conformal integrability consequences of a two dimensional criticality, and 3) self-duality.

Many criticality conditions have been obtained, or simply conjectured, in the literature of lattice models in statistical mechanics [47, 65, 72-74, 79, 80], and all these conjectures were *algebraic* [125]. A straightforward situation corresponds to the case where the model possesses a duality symmetry (see section (V)) for which it is always possible to give a *linear* representation of this duality transformation. One can sometimes find varieties which are globally invariant under this symmetry.

Let us, instead, consider the fixed points of the linear duality transformation, which belong to some algebraic variety (hyperplane). If the algebraic variety separates the phase diagram into two disconnected parts, and *if one assumes that the critical temperature is unique*, one can actually deduce that this algebraic variety is a critical variety [79]. Of course if the algebraic variety is only globally invariant (and not invariant *point-by-point* on the algebraic variety) one cannot draw any conclusion.

In fact, for most of the time one is not in a situation where a simple self-dual argument allows the determination of the critical points. A good example is the duality transformation (31) for the sixteen-vertex model for which self-dual arguments are insufficient. In fact there exists a "super-invariant" in this model after taking into account the gauge (weak-graph) duality symmetries (31) and the inversion relation symmetries[16]. But that is another story.

When the critical varieties are exact, *they are almost always related to some integrability of the model, the algebraicity thus being a consequence of the integrability*. A paradigm is the

[16] Remarkably this consideration even extends the $sl(2)$ $sl(2)$ weak-graph symmetry group to an $sl(2)$ $sl(2)$ $sl(2)$ $sl(2)$ symmetry group.

standard scalar Potts model (see section (VII-1)) which is integrable at criticality. However, we will give below, in the case of the two- and three-site interaction Potts model [161], an algebraic variety which is the critical condition [125] but is *unrelated* to any simple Yang-Baxter-like integrability.

Throughout the years F. Y. Wu has obtained numerous results on critical varieties for two- and three-dimensional spin models, and many related conjectures as well. One can only say that the seeking of critical manifolds and critical varieties is a fascinating subject matter by itself for specialists like Wu.

VI-1. Inversion relations, duality and critical varieties

The considerations of *inversion relations* has been shown to be a powerful tool for analyzing the phase diagram of lattice models and, particularly, for obtaining critical algebraic manifolds in the form of algebraic varieties (see (56) in section (VII-2)).

Let us consider the standard scalar q-state Potts model on an anisotropic triangular lattice with nearest-neighbor and three-spin interactions around up-pointing triangles [79, 161] as shown:

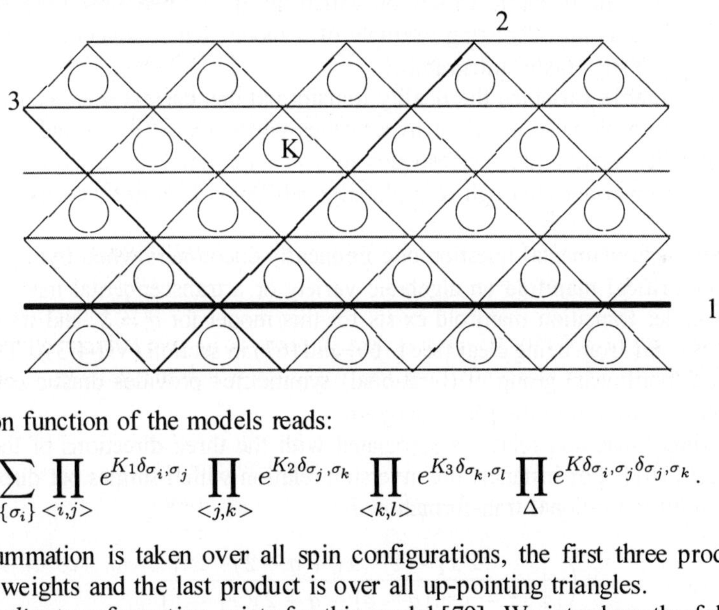

The partition function of the models reads:

$$\mathcal{Z} = \sum_{\{\sigma_i\}} \prod_{<i,j>} e^{K_1 \delta_{\sigma_i,\sigma_j}} \prod_{<j,k>} e^{K_2 \delta_{\sigma_j,\sigma_k}} \prod_{<k,l>} e^{K_3 \delta_{\sigma_k,\sigma_l}} \prod_{\Delta} e^{K \delta_{\sigma_i,\sigma_j} \delta_{\sigma_j,\sigma_k}}.$$

Here the summation is taken over all spin configurations, the first three products denote edge Boltzmann weights and the last product is over all up-pointing triangles.

A *duality* transformation exists for this model [79]. We introduce the following notation:

$$x = e^{K}, \qquad x_i = e^{K_i}, \qquad i = 1, 2, 3$$
$$y = x\, x_1\, x_2\, x_3 - (x_1 + x_2 + x_3) + 2. \tag{33}$$

With the notation (33) the duality

$$D: \begin{cases} x_i \quad \longrightarrow x_i^* = 1 + q\dfrac{x_i - 1}{y}, \qquad y \longrightarrow y^* = \dfrac{q^2}{y}, \\ x \quad \longrightarrow x^* = \dfrac{x_1 + x_2 + x_3 - 2 + q^2/y}{x_1\, x_2\, x_3}, \end{cases} \tag{34}$$

and the partition transforms as

$$Z(x_1, x_2, x_3, y) = (y/q)^N \cdot Z(x_1^*, x_2^*, x_3^*, y^*), \tag{35}$$

where N is the number of sites.

On the basis of this duality Baxter *et al.* proposed that the critical points are located on the *algebraic variety*:

$$x\, x_1\, x_2\, x_3 - (x_1 + x_2 + x_3) + 2 - q = 0, \tag{36}$$

which corresponds to the set of fixed points of D. The critical variety (36) is not only globally invariant under (34), it is also *point-by-point* invariant, namely, every point on the variety is invariant. In general when an algebraic variety is such that every point of the variety is invariant under a duality symmetry, it is possible to argue, subject to some continuity and uniqueness arguments, that the variety actually corresponds to the criticality variety. This has been done by Wu and Zia [125] for $q > 4$ in the ferromagnetic region. It is important to note that the critical variety (36) is *not* an algebraic variety on which the model becomes *Yang-Baxter (star-triangle) integrable.* This is an interesting example of a model where algebraic criticality does not automatically imply Yang-Baxter integrability.

Comment: In suitable variables the duality transformations can be seen as a *linear* transformation. There are two globally invariant hyperplanes under D: $y = +q$ and $y = -q$. The (ferromagnetic) criticality variety (36) corresponds to $y = +q$. The second hyperplane $y = -q$ is not a point-by-point invariant although it is globally self-dual. It is not a locus for critical or transition points.

This illustrates a fundamental question one frequently encounters when trying to analyze a lattice model: is the critical manifold an algebraic variety or a transcendental manifold? It will be seen that a first-order transition manifold exists for this model for $q = 3$, and its algebraic or transcendental status is far from being clear (see [166] and (67) in section (VII-4-3)). The existence of such a very large (nonlinear) group of (birational) symmetries provides drastic constraints on the critical manifold and therefore the phase diagram.

There exist three inversion relations associated with the three directions of the triangular lattice for this model [161]. For instance, the inversion relation which singles out direction 1 (see figure 1) is the (involutive) rational transformation I_1:

$$\begin{aligned} I_1 : (x, x_1, x_2, x_3) \longrightarrow \Big(&\frac{(x\, x_1 - 1)^2 (x_1 + q - 2)}{(x\, x_1^2 + x\, x_1 (q-3) - q + 2)(x_1 - 1)}, \\ 2 - q - x_1 + &\frac{x_1 (x-1)}{x_1\, x - 1}, \frac{x_1 - 1}{x_3 (x\, x_1 - 1)}, \frac{x_1 - 1}{x_2 (x\, x_1 - 1)} \Big). \end{aligned} \tag{37}$$

These three inversion relations generate a group of symmetries which is naturally represented in terms of birational transformations in a four dimensional space. This infinite discrete group of *birational* symmetries is generically a *very large one* (as large as a free group). The algebraic variety (36) is remarkable from an algebraic geometry viewpoint: it is invariant under this very large group generated by three involutions (37).

In this framework of a very large group of symmetries of the model, an amazing situation arises: the one for which q, the number of states of the Potts model, corresponds to *Tutte-Beraha numbers* $q = 2 + 2\cos(2\pi/N)$ where N is an integer. For these selected numbers of q, the group of birational transformations is generated by generators of finite order: it is seen as a Coxeter group generated by generators and relations between the generators. The elements of the group can be seen as the words one can build from an alphabet of three letters A, B and C with the constraints $A^{N+1} = A$, $B^{N+1} = B$, $C^{N+1} = C$. Since the generators A, B and C do not commute (nor does any power of A, B and C) the number of words of length L still grows exponentially with L (hyperbolic group). Among these values of q, two Tutte-Beraha numbers play a special role: $q = 1$ and $q = 3$. For these two values the hyperbolic Coxeter group degenerates[17] into a group isomorphic to $\mathbf{Z} \times \mathbf{Z}$.

For the standard scalar nearest-neighbor Potts model the Tutte-Beraha numbers correspond to the values of q for which the critical exponents of the model are *rational* (see (53) in section (VII-1)).

VI-2. The exact critical frontier of the Potts model on the 3-12 lattice

F. Y. Wu *et al.* considered a general 3-12 lattice with two *and three-site* interaction on the triangular cells [155]. This model has eleven coupling constants and includes the Kagomé lattice as a special case.

In a special parameter subspace of the model, condition (38) below, an exact critical frontier for this Potts model on a general 3-12 lattice Potts model was determined. The Kagomé lattice limit is unfortunately not compatible with the required condition (38).

The condition under which they obtained the exact critical frontier reads:

$$\begin{aligned} &x^2\, x_1^2\, x_2^2\, x_3^2 - x\, x_1\, x_2\, x_3 \cdot (x_1\, x_2 + x_2\, x_3 + x_1\, x_3 - 1) \\ &+(x_1 + x_2 + x_3 + q - 4) \cdot (x_1 x_2 + x_2 x_3 + x_1 x_3 + 3 - q) \\ &-q\, x_1\, x_2\, x_3 - (x_1^2 + x_2^2 + x_3^2) + q^2 - 6\, q + 10 = 0. \end{aligned} \tag{38}$$

This is nothing but the condition which corresponds to the star-triangle relation of the Potts model.

Comment: One can show that condition (38) *is actually invariant under the inversion relation* (37) of the previous section (VI-1), and therefore, since (38) is symmetric under the permutations of K_1, K_2 and K_3, under the three inversions generating the very large group of birational transformations previously mentioned in section (VI-1). More generally, introducing D_1, D_2 and D_3:

$$D_1 = x_1 + x_2 + x_3 - x\, x_1 x_2 x_3 + q - 2, \qquad D_3 = x\, x_1 x_2 x_3 - x_1\, x_2\, x_3,$$

$$D_2 = x_1 + x_2 + x_3 + x\, x_1 x_2 x_3 - 1 - (x_1 x_2 + x_2\, x_3 + x_1\, x_3),$$

one can show that the algebraic expression

$$\mathbf{I}_1(x_1, x_2, x_3, x) = \frac{D_1 \cdot D_2}{D_1 D_2 - q \cdot D_3} \tag{39}$$

is invariant under the three inversion relations and the large group of birational transformation they

[17] Up to semi-direct products by finite groups.

generate, the (star-triangle) condition (38) corresponding to $\mathbf{I}_1(x_1, x_2, x_3, x) = \infty$, namely[18] $D_1\, D_2 - q\, D_3 = 0$.

When $x = 1$, or $q = 1$ or 3, there are additional invariants of the three inversions (37). For instance, for $x = 1$, one can build an invariant from a covariant we give below (see (57)). For $q = 3$, introducing

$$D_5 = x_1 x_2 x_3 \cdot (x_1^2 x_2^2 x_3^2 x^2 - x_2^2 x x_1^2 - x_3^2 x x_1^2 - x_2^2 x_3^2 x + x_1^2 + x_2^2 + x_3^2 - 1),$$

one finds that the expression:

$$\mathbf{I}_2(x_1, x_2, x_3, x) = \frac{D_1^3 \cdot D_2}{3^5 \cdot D_5},$$

is invariant under the three inversions (37). One can try to find the manifold corresponding to the first order transition (see (VII-4-3) below) in the form $F(\mathbf{I}_1, \mathbf{I}_2) = 0$. It still remains an open question whether this variety is algebraic or transcendental. The $x = 1$ limit corresponds to $\mathbf{I}_1 = +1$. The condition $\mathbf{I}_2(x_1, x_2, x_3, x) = 1$ yields $x_1 = x_2 = x_3 = 0.215\,816$ (to be compared with 0.226 681 from (57) in section (VII-2) below), still different from 0.204 (see (66) in section (VII-4-3) below), which is believed to be the location of the first-order transition point.

VI-3. The embarrassing Kagomé critical manifold

At the end of the 80's there was a surge of interest in the Kagomé lattice coming from the theoretical study of high-T_c or strongly interacting fermions in two dimensions (the 2D Hubbard model, resonating valence bond (RVB), ground state of the Heisenberg model). The two-dimensional Gutzwiller product RVB ansatz strategy promoted by P. W. Anderson for describing strongly interacting fermions seemed to fail for regular lattices (square, triangular, ...). Thus, because of its ground state entropy and other specific properties, the Kagomé lattice seemed to be the "last chance" for the RVB approach.

Since one can obtain a critical frontier (38) for the general 3-12 lattice model, and since the 3-12 model includes the Kagomé lattice as a special case, it is tempting to try to obtain the critical frontier for the Potts Kagomé lattice.

The Kagomé Potts critical point was first conjectured by Wu [74] as

$$\begin{aligned} &y^6 - 6y^4 + 2(2-q)\cdot y^3 + 3(3-2\,q)\cdot y^2 - 6(q-1)\cdot(q-2)\,y \\ &\quad -(q-2)\,(q^2 - 4\,q + 2) = 0, \end{aligned} \tag{40}$$

which gives, for $q = 2$, the correct critical point $y^4 - 6y^2 - 3 = 0$ and for $q = 0$ gives (also correctly) $y = 1$. Furthermore, for large q, y behaves like $\sqrt{q}$, as it should. However in the percolation limit $q \to 1$, it gives a percolation threshold p_c for the Kagomé lattice of $p_c = 0.524\,43\cdots$, which compares to the best numerical estimate[19] obtained by R. M. Ziff and P. N. Suding, namely $p_c = 0.524\,405\,3\cdots$, with uncertainty in the last quoted digit. Wu's conjecture is thus wrong,

[18] For $x = 1$ (no three-spin interaction, $D_3 = 0$), condition (38) factorizes and one recovers the ferromagnetic critical condition (36) of the q-state Potts model on an anisotropic triangular lattice.

[19] R. M. Ziff and P. N. Suding, Determination of the bond percolation threshold for the Kagomé lattice, J. Phys. *A 30*, 5351 (1997) and cond-mat/9707110.

but by less than $5 \cdot 10^{-5}$. Some very long high-temperature series of I. Jensen, A. J. Guttmann and I. G.Enting on the q-state Potts model on the Kagomé lattice further confirm that the conjecture is wrong for arbitrary values of q. Nevertheless the Wu conjecture remains an extraordinary approximation.

It is a bit surprising that no exact result on integrability (along some algebraic subvariety) or exact expression for the critical variety is known for the standard scalar Potts model on the *Kagomé lattice*, as generally one expects that the integrability on one lattice, say the square lattice, implies integrability for most of the other Euclidian lattices. This is certainly not the case for the Kagomé lattice.

VII. Potts models

The Potts model encompasses a very large number of problems in statistical physics and lattice statistics. The Potts model, which is a generalization of the two-component Ising model to q components for arbitrary q, has been the subject matter of intense interest in many fields ranging from condensed matter to high-energy physics. It is also related to coloring problems in graph theory.

However, exact results for the Potts model have proven to be extremely elusive. Rigorous results are limited, and include essentially only a closed-form evaluation of its free energy for $q = 2$, the Ising model, and critical properties for the square, triangular and honeycomb lattices [70]. Much less is known about its correlation functions.

VII-1. Wu's review of the Potts model

F. Y. Wu's 1982 review of the Potts model is very well-known [89] (see also [98]). It is an exhaustive expository review of most of the results known about the Potts model up to 1981, a time when interest in the model began to mount. It has remained extremely valuable for anyone wishing to work on the standard scalar Potts model. In particular, it explains the $q \to 1$ limit of the percolation problem (see also [64]), the $q \to 1/2$ limit of the dilute spin glass problem, and the $q \to 0$ limit of the resistor network problem; the equivalences with the Whitney-Tutte polynomial [89] (see section (7.7) and also [57])) and many other related models are also detailed. For instance, the Blume-Capel and the Blume-Emery-Griffiths model (see (25) in (IV-3)) can also be seen as a Potts models. More generally, it is shown that any system of classical q-state spins, the Potts model included, can be formulated as a spin $(q-1)/2$ system.

However, Wu's review was not written in time to include discussions of the *inversion functional relations*. For the two- and three-dimensional anisotropic q-state Potts models, the partition functions satisfies, respectively, the functional relations:

$$Z(e^{K_1}, e^{K_2}) \cdot Z(2 - q - e^{K_1}, e^{-K_2}) = (e^{K_1} - 1) \cdot (1 - q - e^{K_1}), \tag{41}$$

$$Z_{cubic}(e^{K_1}, e^{K_2}, e^{K_3}) \cdot Z_{cubic}(2 - q - e^{K_1}, e^{-K_2}, e^{K_3}) = (e^{K_1} - 1) \cdot (1 - q - e^{K_1}). \tag{42}$$

There are also permutation symmetries like, in 3 dimensions, $Z_{cubic}(e^{K_1}, e^{K_2}, e^{K_3}) = Z_{cubic}(e^{K_3}, e^{K_1}, e^{K_2}) = Z_{cubic}(e^{K_3}, e^{K_2}, e^{K_1})$. Combining these relations one generates an infinite set of discrete symmetries which yield a canonical rational parametrization of the Potts model at and beyond[20] $T = T_c$, and shows clearly the role played by the Tutte-Beraha numbers. These infinite sets of discrete symmetries impose very severe constraints on the critical manifolds and the integrability (see sections (6), (VI-1)). An inversion relation study has subsequently been carried out by F. Y. Wu *et al.* [161].

Graph theory plays a central role in Wu's work on the Potts model. The Potts partition function can be written as [89]

$$Z \equiv Z_G(q, K) = \sum_{G' \subseteq G} (e^K - 1)^b q^n, \tag{43}$$

where $K = J/kT$, the summation is taken over all subgraphs $G' \subseteq G$, and b and n are, respectively, the number of edges and clusters, including isolated vertices, of G'. The duality relation of the Potts model can be obtained from a graph-theoretical viewpoint by using the Euler relation $c + N = b + n$, where c is the number of independent circuits in the subgraph G', and N is the total number of vertices in G. This leads to the duality relation

$$Z_G(q, K) = v^{|E|} \cdot q^{1-N_D} \cdot Z_D(q, K^*),$$

where D is the graph dual to G, and the dual variable K^* is given by:

$$(e^K - 1) \cdot (e^{K^*} - 1) = q. \tag{44}$$

The generalization of the duality to multisite interactions is also given in the review.

A consequence of (43) is that one finds the following connection with the *chromatic polynomial* $P_G(q)$ on G by taking the antiferromagnetic zero-temper-ature limit $K \to -\infty$:

$$Z_G(q, K = -\infty) = P_G(q).$$

The (high- and low-temperature) series expansions are described from a graph-theoretical viewpoint. For instance, the high-temperature expansions are written in the (Domb) form:

$$Z_G(q, K) = \sum_{\sigma_i = 0}^{q-1} \prod_{<ij>} \frac{q + v}{q} \cdot (1 + f_{ij}), \quad f_{ij} = \frac{v}{q + v} \cdot (q\delta(\sigma_i, \sigma_j) - 1). \tag{45}$$

The introduction of these f_{ij} variables comes from the fact that

$$\sum_{\sigma_j = 0}^{q-1} f_{ij} = 0, \tag{46}$$

and, consequently, all subgraphs with vertices of degree 1 give rise to zero contributions. The number of subgraphs that occur in the expansion is therefore greatly reduced.

The location of the critical points of the anisotropic Potts model on a square, triangular and honeycomb lattice were given in terms of the variables $x_r = (e^{K_r} - 1)/\sqrt{q}$ (see also (61) below).

---

[20] At $T = T_c$, one recovers the well-known rational parametrization of the model (occurring in (52) see below).

These expressions are invariably the various special cases of Wu's conjecture [89] for the critical point of the more general checkerboard lattice, namely,

$$\sqrt{q}+x_1+x_2+x_3+x_4 = x_1\,x_2\,x_3 + x_1\,x_2\,x_4 + x_1\,x_3\,x_4 + x_2\,x_3\,x_4 + \sqrt{q}\cdot x_1\,x_2\,x_3\,x_4. \tag{47}$$

Using the notation $a = e^{K_1}$, $b = e^{K_2}$, $c = e^{K_3}$, $d = e^{K_4}$, this critical algebraic variety (47) reads:

$$-(q-1)(q-3) + (a+b+c+d)(2-q) - (ab+ac+bc+ad+bd+cd) + abcd = 0. \tag{48}$$

The critical point of a mixed ferromagnetic-antiferromagnetic square Potts model considered by Kinzel, Selke and Wu [82]:

$$(e^{K_1}-1)\cdot(e^{K_2}+1) = -q, \tag{49}$$

was also given. While this expression coincides with the critical point for $q = 2$, it is incompatible with the inversion relations (41) for general q, and hence is not a critical variety. This is because the infinite discrete group of symmetries generated from the inversion relations of the square Potts model transforms (49) into an *infinite set* of other algebraic varieties, and hence cannot be critical.

Generally, critical manifolds need to be (globally) invariant under this infinite set of transformations (discrete symmetries). Actually, for the anisotropic square Potts model, for instance, one can show that, when q is not a Tutte-Beraha number[21], the *only algebraic varieties compatible with the inversion relation symmetries* (41) are given by the well-known ferromagnetic condition:

$$(e^{K_1}-1)\cdot(e^{K_2}-1) = q, \tag{50}$$

and the antiferromagnetic condition obtained by R. J. Baxter:

$$(e^{K_1}+1)\cdot(e^{K_2}+1) = 4-q, \tag{51}$$

(for which the model is exactly soluble). Note that these two varieties can be deduced from the conjecture (47) by taking $x_1 = x_3$ and $x_2 = x_4$. In this limit, the critical condition (47) factors into conditions (50) and (51). In fact, it has since been shown that the critical condition (47) corresponds to an integrability condition of the checkerboard Potts model.

At *criticality* the Potts model is *exactly solvable*. Let us give the example of the square lattice. The free energy of the isotropic Potts model at the *ferromagnetic* critical point $T = T_c$ reads:

$$f(q, T_c) = \frac{1}{2} + \theta + 2\cdot\sum_{n=1}^{\infty}\frac{e^{-n\theta}\cdot\tanh(n\theta)}{n}, \quad \text{for:} \quad q > 4$$

[21] When q is a Tutte-Beraha number the generically infinite discrete group Γ generated by the inversion relations on the model becomes a finite group, and many algebraic varieties can be invariant under such a finite group: a simple way to build such algebraic Γ-invariants amounts to performing summations over the group Γ.

$$f(q, T_c) = \ln(2) + 4 \cdot \ln\left(\frac{\Gamma(1/4)}{2\Gamma(3/4)}\right), \qquad \text{for:} \qquad q = 4 \tag{52}$$

$$f(q, T_c) = \frac{1}{2} + \int_{-\infty}^{+\infty} \frac{dx}{x} \tanh(\mu x) \frac{\sinh((\pi - \mu)x)}{\sinh(\pi x)}, \qquad \text{for:} \quad q < 4,$$

where the variables θ and μ correspond to the rational parametrization of the model at $T = T_c$, namely $\cosh(\theta) = \sqrt{q}/2$ or $\cos(\mu) = \sqrt{q}/2$.

The exact critical exponents of the standard scalar Potts model are also given in this review. These critical exponents, which are *rational* when q is the Tutte-Beraha numbers, are:

$$\alpha = \alpha' = \frac{2}{3} \cdot \frac{1 - 2u}{1 - u}, \quad \beta = \frac{1 + u}{12}, \quad \gamma = \gamma' = \frac{7 - 4u + u^2}{6 \cdot (1 - u)},$$

$$\delta = \frac{(3 - u) \cdot (5 - u)}{1 - u^2}, \quad \nu = \nu' = \frac{2 - u}{3 \cdot (1 - u)}, \quad \eta = \frac{1 - u^2}{2 \cdot (2 - u)},$$

where the parameter u is related to q by:

$$2\cos\left(\frac{\pi u}{2}\right) = \sqrt{q}, \qquad \text{or:} \qquad 2 + 2\cos(\pi u) = q. \tag{53}$$

These results played a key role in the emergence of the conformal theory.

VII-2. Comments on the checkerboard Potts model

The Wu conjecture (47) for the criticality condition of the q-state checkerboard has since been confirmed from an inversion relation analysis. To discuss the inversion relation we introduce variables u, v, w, z, t defined by:

$$\frac{e^{K_1} - 1}{e^{K_1} + q - 1} = t \cdot \frac{u - t}{1 - t^3 u}, \quad \frac{e^{K_2} - 1}{e^{K_2} + q - 1} = t \cdot \frac{v - t}{1 - t^3 v}, \cdots \tag{54}$$

$$\text{or:} \quad e^{K_1} = a = \frac{u - t^3}{t \cdot (1 - ut)}, \quad e^{K_2} = b = \frac{v - t^3}{t \cdot (1 - vt)}, \cdots$$

$$\text{with:} \qquad t + t^{-1} = \sqrt{q}.$$

In these variables the criticality condition (47) reads:

$$u\,v\,w\,z = 1. \tag{55}$$

Using the inversion trick[22], the partition function of the checkerboard model at *criticality* can be written in a multiplicative form

[22] With the well-suited variables (54) the inversion trick naturally yields exact formulae as products over an infinite discrete group, in this case Eulerian products (56).

$$Z(u,v,w,z)_{[uvwz=1]} = \frac{q}{t^2} \cdot \frac{F(u)\cdot F(1/u)}{1-tu} \cdot \frac{F(v)\cdot F(1/v)}{1-tv} \cdot \frac{F(w)\cdot F(1/w)}{1-tw} \cdot \frac{F(z)\cdot F(1/z)}{1-tz} \tag{56}$$

where: $$F(u) = \prod_{n=1}^{\infty} \frac{1-t^{4\,n-1}u}{1-t^{4\,n+1}u}.$$

This formula is reminiscent of (10) for the six-vertex model (see section (III-1)), as it should since it also follows from the Potts and six-vertex model correspondence (see section (VII-3) below).

Note that, in the anisotropic case $w = u$ and $z = v$, the critical condition (55) factorizes into $uv = +1$ and $uv = -1$ which are, respectively, the aforementioned ferromagnetic and antiferromagnetic criticalities (and integrability).

The checkerboard model reduces to a honeycomb model, and hence a triangular model by taking the dual, if one of the four interactions vanishes. Therefore it is useful to re-examine the triangular lattice limit of the checkerboard variety. P. Martin *et al.* have established two varieties for the triangular Potts model: a ferromagnetic variety which is precisely (36) with $x = 1$, and an antiferromagnetic variety:

$$(q-2)^2 - 2 + (a+b+c+abc)(q-2) + 2(ac+ab+bc) = 0. \tag{57}$$

These two algebraic varieties can be written, respectively, as $u\,v\,w = +t$ and $u\,v\,w = -t$, which can be deduced by taking, respectively, $d = 1$ and $d = -(q-2)/2$ in (48).

For $q = 3$ the antiferromagnetic algebraic variety yields an algebraic critical point very close to the first-order transition point for the three-state Potts model [91, 143].

Along the well-known ferromagnetic variety (44), which for the isotropic model reads $(e^K - 1)\cdot(e^{K^*} - 1) = q$, it is of interest to point out the existence of some "hidden" duality $K \to K^\dagger$, such that the antiferromagnetic condition (51) reads $K_2 = K_1^\dagger$. This hidden duality is the involution:

$$e^{K^\dagger} = -\frac{e^K + q - 3}{e^K + 1}. \tag{58}$$

Note that the dualities (44) and (58) commute, and their product gives the involution:

$$e^K \quad \longrightarrow \quad e^{K^{*\dagger}} = -\frac{(q-2)\cdot e^K + 2}{2e^K + (q-2)} \tag{59}$$

The ferromagnetic critical variety of the anisotropic triangular Potts model (36) now transforms into the anti-ferromagnetic critical variety (57) under (59). One also finds that the ferromagnetic critical variety of the anisotropic square-lattice Potts model (50), and its anti-ferromagnetic critical variety (51), are *both* invariant under transformations (44), (58) and thus (59). More generally, for the checkerboard model, one finds that (48) is invariant under both (44), (58), and thus (59). These results can be very simply seen in the variables u, v, w, z. For example, (59) is simply $u \to -u$. Involutions like the "hidden duality" (58), or like (59), do not yield simple functional equations on the partition function like the Kramers-Wannier duality does (see (44) and (44)).

They are not symmetries of the model, but, rather, "symmetries of the second kind": symmetries of the symmetries.

VII-3. Equivalence of the q-state Potts model with a six-vertex model

Recall that the partition function of the standard scalar q-state Potts model can readily be written as a Whitney-Tutte polynomial. In a further step, Temperley and Lieb used operator methods to show that, for the square lattice, the Whitney-Tutte polynomial is, in turn, equivalent to a staggered ice-type vertex model.

In a classic paper [57] by R. J. Baxter, S. B. Kelland and F. Y. Wu, this equivalence is *rederived* from a graphical approach (see (72) in section (VII-7)), which is easier than the algebraic method of Temperley and Lieb and is applicable to an *arbitrary* planar graph. In particular, the equivalence was extended to triangular or honeycomb Potts models and a staggered six-vertex model on the Kagomé lattice.

The equivalences are as follows (page 404 in [57]). For the square, triangular and honeycomb lattices, the equivalent ice-type vertex model has the Boltzmann weights:

$$(\omega_1, \omega_2, \cdots, \omega_6) = (1, 1, x_r, x_r, A_r, B_r), \tag{60}$$

where (A_r, B_r) are, respectively, for the square, triangular and honeycomb lattices:

$$\left(\frac{1}{s} + x_r \cdot s, \frac{x_r}{s} + s\right), \quad \left(\frac{1}{t} + x_r \cdot t^2, \frac{x_r}{t^2} + t\right), \quad \left(\frac{1}{t^2} + x_r \cdot t, \frac{x_r}{t} + t^2\right),$$

with:

$$s = e^{\theta}, \qquad t = e^{\theta/3}, \quad 2\cosh(\theta) = \sqrt{q}, \qquad x_r = \frac{e^{K_r} - 1}{\sqrt{q}}. \tag{61}$$

These mappings play an important role in the Lee-Yang theorem to be discussed latter (see section (VII-5-1)).

VII-4. Miscellaneous Potts model phase diagrams

VII-4-1. Potts model with competing interactions

Kinzel, Selke and Wu [82] have studied a square lattice Potts model with the competing interactions alluded to earlier in section (VII-1). A similar model in 3 dimensions with next-nearest-neighbor competing interactions has been studied by J. R. Banavar and Wu using mean-field theory and Monte-Carlo simulations [97]. A rich phase diagram was found, and they established positively that the behavior of the 4-state three dimensional Potts model is mean-field-like.

VII-4-2. First-order transition in the antiferromagnetic Potts model

F. Y. Wu *et al.* have analyzed specifically the three-state triangular Potts model and considered the (tricky) tricritical behavior of this model [166].

Recalling the results of section (VI-1), and in particular the special role played by $q = 3$ for the two- and three-site interaction Potts model (33), Monte-Carlo calculations of the $q = 3$ isotropic limit of the model have been performed [144]. These studies confirmed the existence of an antiferromagnetic critical point (in addition to the well-known ferromagnetic one), probably corresponding to a first-order transition occurring near the variety $x = 1$ isotropic region.

For the triangular Potts model with two-spin interactions, Monte-Carlo calculations [166] confirm the localization $a \simeq 0.204 \pm .003$ for a *first-order* transition point. This transition point is confirmed to be different from *the algebraic antiferromagnetic point localized* at $a = 0.22665$ derived below (see (67) below).

The antiferromagnetic transition point $a \simeq 0.204 \pm .003$ can be interpreted as belonging to some singular manifold in the parameter space of the model with two- and three-spin interactions. This singular manifold corresponds to a first-order transition frontier. Recalling discussions in section (VI-1), the question of the algebraic or transcendental status of this first-order transition frontier still remains open.

VII-4-3. Chiral Potts models

F. Y. Wu *et al.* have analyzed a particular two-dimensional chiral Potts model, namely, the 3-state chiral Potts, in order to understand the relation between the (higher genus) integrability and criticality conditions [143]. On the checkerboard model the higher genus integrability of the 3-state chiral Potts model is restricted to the following algebraic variety:

$$\begin{aligned} 3 \cdot (Q_1 P_2 P_3 P_4 + Q_2 P_1 P_3 P_4 + Q_3 P_1 P_2 P_4 + Q_4 P_1 P_2 P_3) \\ -(P_1 Q_2 Q_3 Q_4 + P_2 Q_1 Q_3 Q_4 + P_3 Q_1 Q_2 Q_4 + P_4 Q_1 Q_2 Q_3) = 0, \end{aligned} \tag{62}$$

where:

$$\begin{aligned} P_i = f_i - 3\,h_i, \qquad Q_i = f_i - 2\,g_i + 3\,h_i, \qquad h_i = a_i^2\, b_i^2\, c_i^2 \\ f_i = a_i\, b_i\, c_i \cdot (a_i^3 + b_i^3 + c_i^3), \qquad g_i = a_i^3\, b_i^3 + b_i^3\, c_i^3 + c_i^3\, a_i^3, \end{aligned} \tag{63}$$

where a_i, b_i, c_i denote the three possible values of the four edge Boltzmann weights $w_i(\sigma_k - \sigma_l)$ of the checkerboard lattice:

$$a_i = w_i(0), \quad b_i = w_i(1) = w_i(-2), \quad c_i = w_i(2) = w_i(-1), \quad i = 1, 2, 3, 4.$$

Recalling the conformal theory prejudice (criticality in two dimensions versus integrability), one can also wonder if an integrability condition like (62) could correspond to a critical subvariety of the phase diagram. Let us consider the standard scalar limit of this model. The higher genus integrability condition (62) reduces to the critical condition (47) or (48) of the standard scalar Potts model on the checkerboard lattice for $q = 3$:

$$abcd - (ab + ad + ac + bc + bd + cd) - (a + b + c + d) = 0, \tag{64}$$

together with another algebraic variety:

$$\begin{aligned} abcd + 2 \cdot (acd + bcd + abd + abc) + ab + bd + cd + ac + ad + bc \\ -(a + b + c + d) - 2 = 0. \end{aligned} \tag{65}$$

With the variables (54) taken for $q = 3$, the critical condition (64) reads $u\,v\,w\,z = +1$, and the algebraic condition (65) reads: $uvwz = -1$. Considering the similarity of (65), namely

$uvwz = -1$, with (64), namely $uvwz = +1$, it is tempting to imagine that (65) could also be in some domain of the parameter space a, b, c, d, a critical variety.

In the isotropic triangular and standard scalar limit the higher genus integrability condition (62) factorizes into two conditions. One is the ferromagnetic critical condition of the standard scalar Potts model (see also (36)) $2 - q - a - b - c + abc = 0$ (or $uvw = t$) with $q = 3$, the other one is:

$$(q-2)^2 - 2 + (a+b+c+abc)(q-2) + 2(ac+ab+bc) = 0, \quad \text{or:} \quad uvw = -t, \tag{66}$$

with $q = 3$. In the isotropic limit and for $q = 3$, these two algebraic varieties give, respectively

$$a^3 - 3a - 1 = 0 \qquad \text{and} \qquad a^3 + 6a^2 + 3a - 1 = 0, \tag{67}$$

yielding the ferromagnetic critical point, $a = 1.8793$ and *an antiferromagnetic* transition point at $a = 0.22665$. This antiferromagnetic point must be compared with the antiferromagnetic critical and *first-order* transition point $a \simeq 0.204 \pm 0.003$ obtained from series analysis by I. G. Enting and F. Y. Wu [91].

VII-5. Zeros of partition functions of Potts models

In 1952 Yang and Lee introduced the concept of considering the zeros of the grand partition function of statistical mechanical systems, a consideration that has since opened new avenues to the study of phase transitions. While Yang and Lee considered the zeros in the complex fugacity plane, or equivalently the complex magnetic field plane in the case of spin systems, Fisher in 1964 called attention to the relevance of the zeros of the canonical partition function in the complex temperature plane. Generally speaking, there exist several different kinds of exact results on lattice models in statistical mechanics. Ideally, one would like to obtain the exact, closed-form, expressions of thermodynamic quantities such as the per-site free energy, the surface tension, spontaneous magnetization, and correlation functions. A knowledge of these exact expressions leads to a complete description of the system including the phase boundary (critical frontier) and the location of the zeros of the partition function. However, exact evaluations of physical quantities are not always possible. In such cases one can sometimes determine the critical frontier from properties such as the duality and the inversion relations, or analyze the analyticity properties of the free energy by locating the zeros of the partition function. But other than in the case of some special one-dimensional model, exact results on the zeros have been confined mostly to the Ising model.

VII-5-1. Fugacity variable for checkerboard Potts model, staggering field, Lee-Yang theorem, duality

Let us consider a q-state Potts model on a square, or triangular, lattice, with no magnetic field. Since one does not have a magnetic field, one does not expect a Lee-Yang theorem to exist. Actually *this is not true*: there is a "hidden" field for the standard scalar q-state Potts model!

A. Hinterman, H. Kunz and F. Y. Wu [70] combined the equivalence of the Potts model with a staggered six-vertex model, together with the Lee-Yang circle theorem due to Suzuki and Fisher (see also [30]), to deduce the critical variety of the Potts model. In this consideration a fugacity variable z is associated with the staggering field that occurs in the aforementioned correspondence. In terms of the variables (A_r, B_r) given in (VII-3), the fugacity variable z reads, respectively, for the square and triangular lattices:

$$z^4 = \frac{A_1 A_2}{B_1 B_2}, \qquad z^6 = \frac{A_1 A_2 A_3}{B_1 B_2 B_3}. \tag{68}$$

A condition on the Suzuki-Fisher extension of the Lee-Yang theorem and for real temperatures requires that we must have $q > 4$.

All these results can be generalized to the q-state Potts model on the checkerboard lattice. Not surprisingly, (68) is generalized into:

$$z^8 = \frac{A_1 A_2 A_3 A_4}{B_1 B_2 B_3 B_4}.$$

Note that the duality (44) of the Potts model *has a very simple representation in terms of these fugacity variables* z : $z \to 1/z$.

The zeros of the partition function of the checkerboard model will later be seen to lie on $|z| = 1$, as they should from the Lee-Yang theorem which applies when $q > 4$. Remarkably, it was also seen later by other authors, that the $|z| = 1$ condition can be extended to $q < 4$, and that one then recovers the well-known Fisher's circles for the Ising model!!

The fugacity variable z is a fundamental variable. It corresponds to a crucial combination of variables encapsulating the action of the infinite discrete group generated by the inversion relations. The criticality conditions corresponding to $z = 1$ and $z = -1$ seem also to play some role (see (65) for $q = 3$ in section (VII-4-3)), but not a critical or transition point role. In terms of the variables u, v, w, z of section (VII-4-3), the fugacity variable z is simply the product $u\,v\,w\,z$.

VII-5-2. Zeros for the square lattice: A graph-theoretical viewpoint

Following F. Y. Wu's approach, let us consider the q-state Potts model on the square lattice from a graph-theoretical viewpoint with the partition function (43). Introducing the variable

$$x = (e^K - 1)/\sqrt{q}, \tag{69}$$

the partition function (43) can be written as a polynomial in x

$$Z \equiv P_G(q,x) = \sum_{b=0}^{E} c_b(q) x^b, \quad \text{where} \quad c_b(q) = q^{b/2} \sum_{G' \subseteq G} q^n,$$

where the second summation is taken over all $G' \subseteq G$ for a fixed b. Then, the duality relation (44) can be rewritten as a duality relation for the polynomial P_G [58]:

$$P_G(q,x) = q^{N-1-E/2}\, x^E \cdot P_D(q, x^{-1}). \tag{70}$$

In the case of the square lattice for which D is identical to G in the thermodynamic limit regardless of boundary conditions, (70) implies that the system is critical at $x_c = 1$. For finite self-dual lattices relation (70) gives an example of a self-dual polynomial[23] [197].

To describe the density of zeros on the Lee-Yang circle, we introduce an angle θ associated with the location of the zero on the unit circle. For small θ we have $g(\theta) = a|\theta|^{1-\alpha(q)}$, for $q \leq 4$, and $g(\theta) = \epsilon(q)$, when $q > 4$. This leads to the specific heat singularity $|t|^{-\alpha(q)}$, for $q \leq 4$, and a jump discontinuity of $\epsilon(q)$ in U for $q \leq 4$. This is the known critical behavior of the Potts model [89].

The zeros of the partition function of the q-state Potts model on the square lattice have been evaluated numerically [170]. On the basis of these numerical results, it was conjectured [170] that, for both finite self-dual lattices and for lattices with free or periodic boundary conditions in the thermodynamic limit, the zeros in the $\mathrm{Re}(x) > 0$ region of the complex x plane *are located on the unit circle* $|x| = 1$.

VII-5-3. Zeros for two- and three-spin interactions on triangular lattice

We now return to the q-state Potts model on the triangular lattice with two- and three-site interactions in alternate triangular faces [79] (section VI-1)). The partition function is:

$$Z(x, x_1, x_2, x_3) = \sum_G W(G), \qquad \text{where:}$$

$$W(G) = \prod_{\Delta} (1 + v\delta_{abc})(1 + v_1\delta_{bc})(1 + v_2\delta_{ca})(1 + v_3\delta_{ab}),$$

$$\text{and} \qquad v = e^K - 1, \qquad v_i = e^{K_i} - 1,$$

and the product is taken over all up-pointing triangles. It is convenient to represent terms in the expansion of the partition function by graphs G in which the up-pointing triangular faces are either occupied by a solid triangle with a fugacity v or unoccupied.

We next evaluate the weight $W(G)$ associated with the graph G. It is clear that each solid triangle contributes a factor v to $W(G)$, and each bond a factor v_i. In addition, by including the associated bond factors, each solid triangle contributes an additional factor $(1+v_1)(1+v_2)(1+v_3) = e^{K_1+K_2+K_3}$. Consider next the q dependence of $W(G)$. For the graph representing N isolated points, we have simply $W(G) = q^N$. For other graphs, each triangle reduces the factor q^N by q^2, and each bond by q. But whenever the triangles and bonds close up to form a circuit[24], this restores a factor q, due to the overlapping of one lattice site summation. Thus we have:

$$Z = q^N \sum_G \left[\frac{v}{q^2} e^{K_1+K_2+K_3}\right]^m \left[\frac{v_1}{q}\right]^{b_1} \left[\frac{v_2}{q}\right]^{b_2} \left[\frac{v_3}{q}\right]^{b_3} q^c, \tag{71}$$

where the summation is over the 2^N graphs G, m is the number of solid triangles, b_i is the number of bonds with weight v_i, and $c(G)$ is the number of independent circuits in G. Expression (71) generates the high-temperature expansion of the partition function.

In the case of pure three-site interactions, (71) reduces to

[23] A polynomial $P(x)$ in x is self-dual if it is proportional to $P(1/x)$. Self-dual polynomials occur naturally in lattice models in statistical physics and in restricted partitions of an integer in number theory (see section (VIII-3) below).

[24] Here, we use the term *circuit* in the topological sense that solid triangles can be regarded as stars having three branches, each of which can be connected to other triangles and bonds.

$$Z = q^N \sum_G (w/q)^{m(G)} q^{c(G)}, \quad \text{where} \quad w = (e^K - 1)/q.$$

For pure three-site interactions, the partition function is self-dual and the critical variety assumes the simple form $w = 1$. On the basis of a reciprocal symmetry and numerical results, Wu *et al.* [170] conjectured that zeros for the three-site Potts model (in up-pointing triangles) lie in the thermodynamic limit on the *unit circle* $|w| = 1$, as well as a line segment on the real negative axis.

VII-5-4. Density of Fisher zeros for the Ising model

One trademark of F. Y. Wu's research is that he often looks at old problems and finds new life or new solution that others have not previously seen. A good example of a new look at an old problem is the density of Fisher zeros for the Ising model, which is the $q = 2$ Potts model.

In 1964 Fisher pointed out that in the thermodynamic limit, the zeros of the Ising partition function for a square lattice lie on two circles, now known as the Fisher circles, in the complex $\tanh K$ plane, where K is the nearest-neighbor interaction. However, Fisher had not made the argument rigorous and, furthermore, no one had bothered to look into the distribution of the zeros on the circles, except at small angles which dictates the Ising critical behavior.

Both of these two deficiencies have been rectified by W. T. Lu and F. Y. Wu. First, by considering the zeros of the Fisher zeros for the Ising model on finite self-dual lattices, Lu and Wu [190] established rigorously that, indeed, the Fisher zeros approach two circles in the thermodynamic limit. In a subsequent paper published in 2000 [202], they deduced the close-form expression for the density of the Fisher zeros for many regular two-dimensional lattices, thus completing the story of the Fisher zeros some 25 years after it was first proposed!

VII-6. Duality relation for Potts correlation functions

VII-6-1. Correlation dualities

Duality considerations are not often applied to correlation functions [133], but F. Y. Wu initiated a new method for generating duality relations for correlation functions of the Potts model on planar graphs. In a pioneering paper [183], he obtained duality relations for 2- and 3-point correlation functions, for spins residing on the boundary of a lattice. The consideration was soon extended to n-point correlations [187, 189] and to the case where the spins reside on 2 or more faces in the interior of the lattice [206]. A graph-theoretical formulation of the results in terms of rooted Tutte polynomials (see section (VII-6-2) below) was also given [186, 187]. In addition, C. King and Wu [206] showed that, generally, it is linear combinations of correlation functions, not the individual correlations, that are related by dualities.

VII-6-2. Correlation functions as rooted Tutte Polynomials

As previously mentioned, the Potts partition function is also the Whitney-Tutte polynomial, or in short, the Tutte polynomial, considered in graph theory. In one further step, F. Y. Wu, C. King and W. L. Lu formulated the Potts correlation function as a rooted Tutte polynomial [193].

In graph theory a vertex is rooted if it is colored with a prescribed (fixed) color and a graph is rooted if it contains a rooted vertex. If one interprets the color of a site (vertex) as spin states, then the Potts correlation functions for which the spin states of given sites are fixed can naturally be formulated as rooted-Tutte polynomials. This is the basis of their graph-theoretical formulation of the Potts correlation function, from which duality relations of the Potts correlations become transparent and can be analyzed [193].

VII-7. Potts model and graph theory

F. Y. Wu has written several review papers dedicated to the analysis of the Potts model from a graph theoretical viewpoint [57, 71, 117, 122]. The most important one is the classic paper [57] written in collaboration with Baxter and Kelland alluded to above, which gives the graphical construction of the equivalence of the partition function of the Potts model with an ice-type model [57]. This derivation simplifies the algebraic method of Temperley and Lieb, which is based on the Temperley-Lieb algebra[25]:

$$U_{i,i+1}^2 = \sqrt{q}\cdot U_{i,i+1},$$

$$U_{i,i-1}\cdot U_{i,i+1}\cdot U_{i,i-1} = U_{i,i-1}, \qquad U_{i,i+1}\cdot U_{i,i-1}\cdot U_{i,i+1} = U_{i,i+1},$$

$$U_{i,i+1}\cdot U_{j,j+1} = U_{j,j+1}\cdot U_{i,i+1} \quad \text{if} \quad |i-j|>3,$$

and applies to an arbitrary planar graph. This then opens the door for analyzing the triangular and honeycomb Potts models.

Another important graphical analysis of a Potts model is the joint work with J. H. H. Perk [103, 104] on the non-intersecting string (NIS) model of Stroganov and Schultz, the close-packed loop model. The NIS model formulated by Perk and Wu in [103] turns out to be nothing but the bracket polynomial introduced by L. H. Kauffman in his state-model formulation of knot invariants. This fact offers a most natural approach to knot invariants from a statistical mechanical viewpoint [154].

VIII. Other miscellaneous topics

F. Y. Wu has worked on a diverse array of topics in mathematics and mathematical physics. In this section we present a random choice of topics that are not included above.

VIII-1. Topics in graph theory

F. Y. Wu is fond of graphs and has made many contributions to graph theory. A fine example is the aforementioned introduction of the rooted Tutte polynomial. Even when Wu does not obtain new results, he tries to provide simpler derivations, or find new consequences of known results. A good example is his work on random graphs [92].

Random graphs is a topic well-known to graph theorists after the work by Erdös and Renyi, who introduced the problem in 1960. In the simplest formulation each pair of points of a set of N can be connected (by a bond, say) with a probability α/N, where α is a constant. One then asks questions such as what is the mean cluster size and the probability $P(\alpha)$ that the set becomes fragmented, namely not connected, etc.

Using a Potts model formulation, Wu [93] reproduced the Erdös-Renyi result in just a few steps, showing that a transition occurs at $\alpha_c = 1$ and computed the critical exponents as well as the mean cluster size at criticality. This work has drawn considerable attention from graph

[25] A matrix representation of the $U_{i,i\pm1}$ is, for instance, the $q^n \times q^n$ matrices with entries $q^{-1/2}\prod_{j=1;j\neq i}^{n}\delta(\sigma_j,\sigma_j')$ and $q^{1/2}\delta(\sigma_i,\sigma_{i+1})\prod_{j=1}^{n}\delta(\sigma_j,\sigma_j')$.

theorists. Wu has also applied the result to evaluate the reliability probability of a communication network [92].

As another example of Wu's work in graph theory, one can mention his paper on the Temperley-Nagle identity for graph embeddings [69], where he provides a simple derivation of the Temperley-Nagle identity:

$$\sum_G x^v y^l = (1+x)^N \sum_L (y-1)^l \Big(\frac{x}{1+x}\Big)^v,$$

where G denotes section graphs of the original graph of v vertices and l lines, and L is a line set of G containing l lines covering v vertices. Using a weak-graph expansion he also deduced a sum-rule relation connecting the lattice constants of weak and strong embeddings.

VIII-2. The vicious neighbor problem

Consider N points randomly distributed in a bounded d-dimensional space. At a given instance, each point destroys his nearest neighbor (vicious neighbors) with a probability p. What is the probability $P_N(p)$ that a given point will survive in the $N \longrightarrow \infty$ limit?

The $d = 2$, $p = 1$ version of this problem was first posed by the Brandeis mathematician R. Abilock in American Mathematical Monthly in 1967, and remained unsolved for almost two decades. In 1986 the *Omni* magazine posted a prize for its correct solution, and R. Tao and Wu claimed the prize by publishing the solution for general d and p in 1987 [111].

The idea of their solution is very simple. In d dimensions a given point can be killed by at most a finite number n_d of other points. In two dimensions, for example, the number is $n_2 = 5$ (theoretically a point can also be killed by 6 other points, but the phase space for that to happen has a zero measure). Therefore, one computes the volumes of the phase space for a point to be killed by 1, 2, $\cdots$, n_d neighbors, and uses the inclusion-exclusion principle to write the probability in question as an alternate series in p, whose highest power is n_d. However, the evaluation of the volumes of the phase space is tedious, requiring special techniques.

For $d = 1$ the result is quite simple and one has $P_\infty(p) = 1 - p + p^2/2$. For $d = 2$ the result is:

$$P_\infty(p) = 1 - p + 0.316\ 3335\, p^2 - 0.032\ 9390\, p^3 + 0.000\ 6575\, p^4 - 0.000\ 0010\, p^5, \quad (72)$$

where the coefficient of each term is evaluated from integrals which can, in principle, be computed to any numerical accuracy. The coefficient of the last term, for example, is obtained by combining two 8-fold integrations.

For $p = 1$, (72) yields $P_\infty(1) = 0.284\ 051...$, a solution which claimed the *Omni* prize. Tao and Wu also carried out Monte Carlo simulations to obtain the solution for $d = 3, 4, 5$.

VIII-3. Counting partitions: from Potts to three-dimensional enumeration and beyond

F. Y. Wu *et al.* [172] considered a directed lattice animal problem on the d-dimensional hypercubic lattice, and established its equivalence first with the infinite-range Potts model and, in a second step, with the enumeration of $(d-1)$-dimensional restricted partitions of an integer. The directed compact lattice animal problem was solved exactly in two and three dimensions, using known results in number theory. They found that the number of lattice animals of n sites grows as:

$$\exp(c \cdot n^{(d-1)/d}).$$

Furthermore, the infinite-state Potts model solution leads to a conjectured limiting form for the generating function of restricted partitions for $d > 3$, which is a long-standing unsolved problem in number theory.

Let us denote by $A_n(L_1, L_2, \cdots L_d)$ the number of n-site animals that can grow on an $L_1 \times L_2 \times \cdots \times L_d$ lattice. F. Y. Wu *et al.* showed that A_n is precisely the number of $(d-1)$-dimensional restricted partitions of the integer n into non-negative parts to units of a hypercube of size $L_1 \times L_2 \times \cdots \times L_{d-1}$, with the size of each part being at most L_d.

Define the generating function

$$G(L_1, L_2, L_3, \cdots L_d; t) = 1 + \sum_{n=1}^{L_1 \cdot L_2 \cdot L_3 \cdots L_d} A_n(L_1, L_2, \cdots L_d) \cdot t^n,$$

which is of interest in number theory. F. Y. Wu *et al.* showed that G is precisely the partition function of a Potts model on the d-dimensional lattice in the infinite-state limit, provided one identifies t with x^d where $x = (e^K - 1)/q^{1/d}$. This then connects the Potts model with the theory of partitions in number theory.

For $d = 2$, the generating function corresponding to the square lattice (L_1, L_2) reads[26]:

$$G(L_1, L_2; t) = \frac{(t)_{L_1+L_2}}{(t)_{L_1} \cdot (t)_{L_2}}, \tag{73}$$

where $(t)_p = \prod_{q=1}^{p} (1 - t^q)$. $G(L_1, L_2)$ is a polynomial in t, also known as the Gaussian polynomial or the "q-coefficient". For $d = 3$, the generating function reads:

$$G(L_1, L_2, L_3; t) = \frac{[t]_{L_1+L_2+L_3} \cdot [t]_{L_1} \cdot [t]_{L_2} \cdot [t]_{L_3}}{[t]_{L_1+L_2} \cdot [t]_{L_2+L_3} \cdot [t]_{L_3+L_1}}, \tag{74}$$

where $[t]_L$ denotes:

$$[t]_L = \prod_{p=1}^{L-1} (t)_p, \quad L > 1, \qquad (t)_p = \prod_{q=1}^{p} (1 - t^q). \tag{75}$$

For $d = 4$, $G(L_1, L_2, L_3, L_4; t)$ is the generating function of restricted solid partitions of a positive integer into parts on a $L_1 \times L_2 \times L_3$ cubic lattice, with each part being no greater than L_4. The evaluation of a closed-form expression for G in this case has remained an unsolved problem for almost a century.

The expression which straightforwardly generalizes (73) and (74) would be:

$$G_{\text{straight}}(L_1, L_2, L_3, L_4; t) = \frac{N_G(t)}{D_G(t)}, \qquad \text{with:} \tag{76}$$

[26] Let us recall the classic analysis due to Rademacher, which yields the celebrated Hardy-Ramanujan asymptotic result: $A_n \simeq 1/(4n\sqrt{3}) \cdot \exp(\pi\sqrt{2n/3})$. One gets the following asymptotic behavior for $A_n(L_1, L_2)$ when $L_1 = L_2$: $A_n(L, L) \simeq (\sqrt{3}/(2\pi n)) \cdot 2^{2\sqrt{2n}}$.

$$N_G(t) = \{t\}_{L_1+L_2+L_3+L_4} \cdot \{t\}_{L_1+L_2} \cdot \{t\}_{L_1+L_3} \cdot \{t\}_{L_1+L_4}$$
$$\cdot \{t\}_{L_2+L_3} \cdot \{t\}_{L_2+L_4} \cdot \{t\}_{L_3+L_4},$$
$$D_G(t) = \{t\}_{L_1+L_2+L_3} \cdot \{t\}_{L_2+L_3+L_4} \cdot \{t\}_{L_1+L_3+L_4} \cdot \{t\}_{L_1+L_2+L_4}$$
$$\cdot \{t\}_{L_1} \cdot \{t\}_{L_2} \cdot \{t\}_{L_3} \cdot \{t\}_{L_4},$$

where:

$$\{t\}_L = \prod_{p=1}^{L-1} [t]_p \quad L > 2. \tag{77}$$

But the explicit expression of $G(2, 2, 2, L_4; t)$, obtained by Major P. A. MacMahon in 1916 is

$$G(2, 2, 2, L_4; t) = G_{\text{straight}}(2, 2, 2, L_4; t) + C(2, 2, 2, L_4; t), \tag{78}$$

where

$$G(2, 2, 2, L_4; t) = \sum_{i=0}^{4} g_i \cdot \frac{(t)_{L_4+8-i}}{(t)_8 \cdot (t)_{L_4-i}}, \qquad \text{with:}$$
$$g_0 = 1, \qquad g_1 = 2t^2 \cdot (1 + t + t^2 + t^3 + t^4) + t^4,$$
$$g_2 = t^5 \cdot (1 + 3t + 4t^2 + 8t^3 + 4t^4 + 3t^5 + t^6)$$
$$g_3 = 2t^{10} \cdot (1 + t + t^2 + t^3 + t^4) + t^{12}, \qquad g_4 = t^{16},$$

and

$$C(2,\ 2,\ 2,\ P;\ t) = -\Big(\frac{t^6 \cdot (t+1)^2 \cdot (t^4 - 2\,t^3 + t^2 - 2\,t + 1)}{t^2 + t + 1}\Big) \cdot \Big(\frac{(t)_{P+6}}{(t)_8 (t)_{P-2}}\Big).$$

H. Y. Huang and F. Y. Wu [179] decided to look into the zeros of the generating function $G(2, 2, 2, L_4; t)$ for various increasing values of L_4. They found that the zeros are not exactly on the unit circle, but seem to converge to the unit circle as L_4 increases. This indicates that a multiplicative correction $C_{\text{mult}}(2, 2, 2, L_4; t) = G(L_1, L_2, L_3, L_4; t)/G_{\text{straight}}(L_1, L_2, L_3, L_4; t)$, would not have any simple Eulerian product form as in (76) and (77):

$$C_{\text{mult}}(2, 2, 2, L_4; t) = \prod_{n=1} (1 - t^n)^{\alpha_n}, \tag{79}$$

where α_n are positive integers, since these product forms (79) would necessarily yield zeros on the unit circle. H. Y. Huang and F. Y. Wu conjectured however, on the basis on their numerical results, that the zeros tend to be on the unit circle in the limit, when any one of L_1, L_2, L_3, $L_4 \to \infty$.

VIII-3-1. Directed percolation and random walk problems

F. Y. Wu and H. E. Stanley [90] have considered a directed percolation problem on square and triangular lattices in which the occupation probability is unity along one spatial direction.

They formulated the problem as a random walk, and evaluated in closed-form the percolation probability, or the arriving probability of a walker. To this date this solution stands as the only exactly solved model of directed percolation.

In another random walk problem, Wu and H. Kunz [192] considered *restricted random walks* on graphs, which keep track of the number of immediate reversal steps, by using a transfer matrix formulation. A closed-form expression was obtained for the number of n-step walks with r immediate reversals for any graph. In the case of graphs of a uniform valence, they established a probabilistic meaning of the formulation, and deduced explicit expressions for the generating function in terms for the eigenvalues of the adjacency matrix.

IX. Knot theory

The connection between knot theory and statistical mechanics was probably first discovered by Jones. His derivation of the V. Jones polynomial reflects the resemblance to the von Neumann algebra when he uses with the Lieb-Temperley algebra occurring in the Potts model (see section (VII-7)). This direct connection came to light when L. Kauffman produced a simple derivation of the Jones polynomial using the very diagrammatic formulation of the non-intersecting string (NIS) model of J. H. H. Perk and F. Y. Wu [103, 104]. Soon thereafter Jones worked out a derivation of the Homfly polynomial using a vertex-model approach. The connection between knot theory and lattice statistical mechanics was further extended by Jones to include spin and IRF models.

F. Y. Wu has written several papers on the connection between knot theory and statistical mechanics [150, 151, 154], including a comprehensive review [150]. In hindsight, knot invariants arose naturally in statistical mechanics even before the connection with solvable models was discovered. In their joint paper [103], for example, J. H. H. Perk and F. Y. Wu described a version of an NIS model which is precisely the bracket polynomial of L. Kauffman. Similarly, the q-color NIS model studied by J.H.H. Perk and C. Schultz is a $q(q-1)$ vertex model which generates the Homfly polynomial. Here we briefly describe the latter connection.

The q-color NIS model has vertex weights ($\delta_{abcd} = 1$ if $a = b = c = d$ and zero otherwise):

$$w(a,b,c,d) = (W(u) - S(u) - T(u)) \cdot \delta_{abcd} + S(u) \cdot \delta_{ab}\delta_{cd} + T(u) \cdot \delta_{ac}\delta_{bd}, \tag{80}$$

where: $\quad W(u) = \sinh(u) = \sinh(\eta + u), \quad S(u) = \sinh(u),$

$$T(u) = \sinh(\eta - u), \qquad q = e^{\eta} + e^{-\eta},$$

and the *Homfly polynomial* is a *two variable* knot invariant polynomial, discovered after Jones' work, by Freyd *et al.* The Homfly polynomial knot invariant has since been re-derived and analyzed by Jones using the Hecke algebra of the braid group. It can also be constructed from the Perk-Schultz NIS model. Actually the partition function $Z(q, e^{\eta})$ of the NIS model is a knot invariant related to the Homfly polynomial $P(t, z)$:

$$P(t, z) = \frac{\sinh(\eta)}{\sinh(q\eta)} \cdot Z(q, e^{\eta}).$$

In the infinite rapidity limit this model leads to the Jones polynomial. The Boltzmann weight (80) of the non-intersecting string model becomes:

$$w(a,b,c,d) = -e^{\pm 2\eta}\delta_{a,b}\delta_{c,d} + e^{\pm\eta}\delta_{b,d} \quad \text{with:} \quad q = e^{\eta} + e^{-\eta}. \tag{81}$$

The Jones polynomial $V(t)$ is then obtained from the Homfly polynomial $P(t,z)$ by taking $z = \sqrt{t} - 1/\sqrt{t}$.

F. Y. Wu discussed many knot invariants in his review [154]: the Alexander-Conway polynomial, the Jones polynomial, the Homfly polynomial, the Kauffman polynomial and the Akutsu-Wadati polynomial, etc. The Alexander-Conway polynomial can be obtained from the Homfly polynomial by setting $t = 1$ in the Homfly polynomial $P(t,z)$. The Akutsu-Wadati polynomial is an example of a *new* knot invariant derived from exactly solvable models in statistical mechanics.

As our final example of F. Y. Wu's versatility, he and P. Pant and C. King [162] have obtained a new knot invariant using the exactly solvable chiral Potts model and a generalized Gaussian summation identity. Starting from a general formulation of link invariants using edge-interaction spin models, they establish the uniqueness of the invariant for self-dual models. They applied the formulation to the self-dual chiral Potts model, and obtain a *link invariant* in the form of a lattice sum defined by a matrix associated with the link diagram. A generalized Gaussian summation identity was then used to carry out this lattice sum, enabling them to cast the invariant into a tractable form. The resulting expression for the link invariant was characterized by the roots of unity and does not appear to belong to the usual quantum group family of invariants.

Finally, Pant and Wu [185] have derived a link invariant associated with the Izergin-Korepin model.

X. Conclusion

It would not be fair to summarize F. Y. Wu's contributions by a quick conclusion such as: he wrote several important monographs on vertex models, on the Potts model and on knot theory, obtained many important results, in particular the Lieb-Wu solution of the Hubbard model, the Fan-Wu free-fermion vertex model, the solution of the Baxter-Wu model, and many other results on dimers or free-fermion models, 3D dimers, d-dimensional free-fermion models, Potts models, Ising and vertex models, using a large set of tools including analytic calculations, expansions, series analysis, Monte-Carlo, ..., with a particular emphasis on graph-theoretical methods.

Most of the work of F. Y. Wu could be said to correspond to exact results in lattice statistical mechanics, or in mathematics, with particular emphasis on *graph theory* and enumerative combinatorics. We have tried to give here some hints as to the space of F. Y. Wu's very large "graph" of concepts, results, tools, models, with many "intellectual loops". We have not tried to provide an exhaustive description of F. Y. Wu's contributions but, rather, only to provide a few *comments* on some of his results, emphasizing the fruitful cross-fertilizations between the various domains of mathematical physics and mathematics, and also to show the motivation and relevance of these results, tools, concepts and methods.

Beyond a post-modern accountant's evaluation and from a research viewpoint, one must say that the important and numerous results F. Y. Wu has obtained are not due to publish-or-perish productivity pressure, but, on the contrary, are the natural consequence of the pleasure of a scientist who loves to play with concepts and mathematical objects (dimers, graphs with particular boundary conditions, dualities, Potts models, series expansions with transmissivities, ...) and who has a strong desire to reach ambitious goals, such as obtaining new results in three dimensions, new results for non-critical Potts models, or even for the Ising model in a magnetic field.

A scientist does not become as productive as F. Y. Wu in response to external pressure but, on the contrary, only by forces being in harmony with his deep personal motivations. This is the only way to be as efficient and productive as F. Y. Wu and, as the famous French mathematician Jean Dieudonné once wrote, to work efficiently, "pour l'honneur de l'esprit humain".

Acknowledgments

I would like to thank Professor Chin-Kun Hu for giving me the opportunity to participate in this conference in honor of Professor F-Y. Wu. I would also like to thank Fred for so many passionate discussions on lattice models throughout the years. I would like to thank A. J. Guttmann for many valuable comments on this paper, and J-C Anglès d'Auriac, S. Boukraa for careful readings of the manuscript.

References

*In honor of F.Y. Wu on the occasion of his 70th birthday. Statphys-Taiwan 2002. The second APCTP and sixth Taiwan International Symposium on Statistical Physics.

†E-mail: maillard@lpthe.jussieu.fr

[1] F. Y. Wu, On the discussions of 'free waveforms' (in Chinese), Naval Tech. **2**, 9-11 (1955).

[2] F. Y. Wu, Ground state of liquid He^4, in *Proceedings of the Midwest Conference on Theoretical Physics*, 182-196 (1961).

[3] F. Y. Wu and E. Feenberg, Ground state of liquid helium (Mass 4), Phys. Rev. **122**, 739-742 (1961).

[4] F. Y. Wu and E. Feenberg, Theory of Fermi liquids, Phys. Rev. **128**, 943-955 (1962).

[5] F. Y. Wu, T. S. Kuo and K. H. Sun, Four slow neutron converters, Nucleonics **20**, 61-62 (1962).

[6] F. Y. Wu, The ground state of liquid He^4, Prog. Theoret. Phys. (Kyoto) **28**, 568-569 (1962).

[7] F. Y. Wu, Cluster development in an N-body problem, J. Math. Phys. **4**, 1438-1443 (1963).

[8] F. Y. Wu, H. T. Tan and E. Feenberg, Necessary conditions on radial distributions functions, J. Math. Phys. **8**, 864-869 (1967).

[9] F. Y. Wu, Exactly soluble model of ferroelectric phase transition in two dimensions, Phys. Rev. Lett. **18**, 605-607 (1967).

[10] F. Y. Wu, Remarks on the modified KDP model, Phys. Rev. **168**, 539-543 (1968).

[11] E. H. Lieb and F. Y. Wu, Absence of Mott transition in an exact solution of the short-range one-band model in one dimension, Phys. Rev. Lett. **20**, 1445-1448 (1968).

[12] C. Fan and F. Y. Wu, Ising model with next-neighbor interactions: Some exact results and an approximate solution, Phys. Rev. **179**, 560-570 (1969).

[13] F. Y. Wu, Critical behavior of two-dimensional hydrogen-bonded antiferroelectrics, Phys. Rev. Lett. **22**, 1174-1176 (1969).

[14] F. Y. Wu, Exact solution of a model of an antiferroelectric transition, Phys. Rev. **183**, 604-607 (1969).

[15] F. Y. Wu and M. K. Chien, Convolution approximation for the n-particle distribution function, J. Math. Phys. **11**, 1912-1916 (1970).

[16] C. Fan and F. Y. Wu, General Lattice Model of Phase Transitions, Phys. Rev. **B2**, 723-733 (1970).

[17] M. F. Chiang and F. Y. Wu, One-dimensional binary alloy with Ising interactions, Phys. Lett. A **31**, 189-191 (1970).

[18] F. Y. Wu, Critical behavior of hydrogen-bonded ferroelectrics, Phys. Rev. Lett. **24**, 1476-1478 (1970).

[19] F. Y. Wu, C. W. Woo and H. W. Lai, Communications on a microscopic theory of helium submonolayers, J. Low Temp. Phys. **3**, 331-333 (1970).

[20] F. Y. Wu, H. W. Lai and C. W. Woo, Theory of helium submonolayers: Formulation of the problem and solution in the zero coverage limit, J. Low Temp. Phys. **3**, 463-490 (1970).

[21] F. Y. Wu, *Introduction to Crystal Physics* (in Chinese), translation of a popular science book of Doubleday science study series (Chung Hwa Book Co., Taipei 1970).

[22] F. Y. Wu, Density correlations in a many particle system, Phys. Lett. A **34**, 446-447 (1971).

[23] F. Y. Wu and D. M. Kaplan, On the eigenvalues of orbital angular momentum, Chin. J. Phys. **9**, 31-40 (1971).

[24] F. Y. Wu, Modified KDP Model in a staggered field, Phys. Rev. B **3**, 3895-3900 (1971).

[25] F. Y. Wu, H. T. Tan and C. W. Woo, Sum rules for a binary solution and effective interactions between He^3 quasi-particles in superfluid He^4, J. Low Temp. Phys. **5**, 261-265 (1971).

[26] F. Y. Wu, H. W. Lai and C. W. Woo, Theory of helium submonolayers II: Bandwidths and correlation effects, J. Low Temp. Phys. **5**, 499-518 (1971).

[27] F. Y. Wu and C. W. Woo, Physically adsorbed monolayers, Chin. J. Phys. **9**, 68-91 (1971).

[28] F. Y. Wu, Multiple density correlations in a many particle system, J. Math. Phys. **12**, 1923-1929 (1971).

[29] F. Y. Wu, Ising model with four-spin interactions, Phys. Rev. B **4**, 2312-2314 (1971).

[30] K. S. Chang and S. Y. Wang and F. Y. Wu, Circle theorem for the ice-rule ferroelectric models, Phys. Rev. A **4**, 2324-2327 (1971).

[31] E. H. Lieb and F. Y. Wu, Two-dimensional ferroelectric models, in *Phase Transitions and Critical Phenomena*, Vol. 1, Eds. C. Domb and M. S. Green (Academic Press, London, 1972) pp. 331-490.

[32] F. Y. Wu, Exact results on a general lattice statistical model, Solid State Commun. **10**, 115-117 (1972).

[33] F. Y. Wu, Solution of an Ising model with two- and four-spin interactions, Phys. Lett. A **38**, 77-78 (1972).

[34] F. Y. Wu, Phase transition in a sixteen-vertex lattice model, Phys. Rev. B **6**, 1810-1813 (1972).

[35] G. Keiser and F. Y. Wu, Electron gas at metallic densities, Phys. Rev. A **6**, 2369-1377 (1972).

[36] F. Y. Wu, Matrix elements of a many-boson Hamiltonian in a representation of correlated basis functions, J. Low Temp. Phys. **9**, 177-187 (1972).

[37] F. Y. Wu, Statistics of parallel dimers on an $m \times n$ lattice, Phys. Lett. A **43**, 21-22 (1973).

[38] H. J. Brascamp, H. Kunz and F. Y. Wu, Critical exponents of the modified F model, J. Phys. C **6**, L164-L166 (1973).

[39] K. S. Chang, S. Y. Wang and F. Y. Wu, Statistical mechanics of finite ice-rule ferroelectric models, Chin. J. Phys. **11**, 41-48 (1973).

[40] F. Y. Wu, Phase diagram of a decorated Ising system, Phys. Rev. B **8**, 4219-4222 (1973).

[41] F. Y. Wu, Modified KDP model on the Kagomé lattice, Prog. Theoret. Phys. (Kyoto) **49**, 2156-2157 (1973).

[42] G. Keiser and F. Y. Wu, Correlation energy of an electron gas in the quantum strong-field limit, Phys. Rev. A **8**, 2094-2098 (1973).

[43] H. J. Brascamp, H. Kunz and F. Y. Wu, Some rigorous results for the vertex model in statistical mechanics, J. Math. Phys. **14**, 1927-1932 (1973).

[44] F. Y. Wu, Phase transition in an Ising model with many-spin interactions, Phys. Lett. A **46**, 7-8 (1973).

[45] R. J. Baxter and F. Y. Wu, Exact solution of an Ising model with three-spin interactions on a triangular lattice, Phys. Rev. Lett. **31**, 1294-1297 (1973).

[46] F. Y. Wu, Phase transition in a vertex model in three dimensions, Phys. Rev. Lett. **32**, 460-463 (1974).

[47] F. Y. Wu and K. Y. Lin, Two phase transitions in the Ashkin-Teller model, J. Phys. C **7**, L181-L184 (1974).

[48] F. Y. Wu, Eight-vertex model on the honeycomb lattice, J. Math. Phys. **6**, 687-691 (1974).

[49] R. J. Baxter and F. Y. Wu, Ising model on a triangular lattice with three-spin interactions: I. The eigenvalue equations, Aust. J. Phys. **27**, 357-367 (1974).

[50] F. Y. Wu and K. Y. Lin, Staggered ice-rule model - The Pfaffian solution, Phys. Rev. B **12**, 419-428 (1975).

[51] F. Y. Wu, Spontaneous magnetization of the three-spin Ising model on the Union Jack lattice, J. Phys. C **8**, 2262-2266 (1975).

[52] C. S. Hsue, K. Y. Lin and F. Y. Wu, Staggered eight-vertex model, Phys. Rev. B **12**, 429-437 (1975).

[53] J. E. Sacco and F. Y. Wu, Thirty-two vertex model on a triangular lattice, J. Phys. A **8**, 1780-1787 (1975).

[54] Y. K. Wang and F. Y. Wu, Multi-component spin model on the Bethe lattice, IUPAP Conference on Statistical Physics (Akademiai Kiado, 1975), p. 142.

[55] M. L. C. Leung, B. Y. Tong and F. Y. Wu, Thermal denaturation and renaturation of DNA molecules, Phys. Lett. A **54**, 361-362 (1975).

[56] Y. K. Wang and F. Y. Wu, Multi-component spin model on the Cayley tree, J. Phys. A **9**, 593-604 (1976).

[57] R. J. Baxter, S. B. Kelland and F. Y. Wu, Equivalence of the Potts model or Whitney polynomial with the ice-type model: A new derivation, J. Phys. A **9**, 397-406 (1976).

[58] F. Y. Wu and Y. K. Wang, Duality transformation in a many component spin model, J. Math. Phys. **17**, 439-440 (1976).

[59] F. Y. Wu, Two phase transitions in triplet Ising models, J. Phys. C **10**, L23-L27 (1977).

[60] F. Y. Wu, Ashkin-Teller model as a vertex problem, J. Math. Phys. **18**, 611-613 (1977).

[61] K. G. Chen, H. H. Chen, C. S. Hsue and F. Y. Wu, Planar classical Heisenberg model with biquadratic interactions, Physica A **87**, 629-632 (1977).

[62] F. Y. Wu, Number of spanning trees on a lattice, J. Phys. A **10**, L113-L115 (1977).

[63] H. Kunz and F. Y. Wu, Site percolation as a Potts model, J. Phys. C **11**, L1-L4 (1978).

[64] F. Y. Wu, Percolation and the Potts model, J. Stat. Phys. **18**, 115-123 (1978).

[65] F. Y. Wu, Some exact results for lattice models in two dimensions, Ann. Israel Phys. Soc. **2**, 370-376 (1978).

[66] V. T. Rajan, C.-W. Woo and F. Y. Wu, Multiple density correlation for inhomogeneous systems, J. Math. Phys. **19**, 892-897 (1978).

[67] F. Y. Wu, Absence of phase transitions in tree-like percolation in two dimensions, Phys. Rev. B **18**, 516-517 (1978).

[68] F. Y. Wu, Phase diagram of a spin-one Ising system, Chin. J. Phys. **16**, 153-156 (1978).

[69] F. Y. Wu, On the Temperley-Nagle identity for graph embeddings, J. Phys. A **11**, L243 (1978).

[70] A. Hinterman, H. Kunz and F. Y. Wu, Exact results for the Potts model in two dimensions, J. Stat. Phys. **19**, 623-632 (1978).

[71] F. Y. Wu, Graph Theory in Statistical Physics, in *Studies in Foundations of Combinatorics*, Adv. in Math.: Supp. V.1, Ed. G-C. Rota, 151-166, (Academic Press, New York 1978).

[72] K. Y. Lin and F. Y. Wu, Phase diagram of the antiferromagnetic triangular Ising model with anisotropic interactions, Z. Phys. B **33**, 181-185 (1979).

[73] F. Y. Wu, Phase diagram of a five-state spin model, J. Phys. C **12**, L317-L320 (1979).

[74] F. Y. Wu, Critical point of planar Potts models, J. Phys. C **12**, 645-649 (1979).

[75] X. Sun and F. Y. Wu, Critical polarization of the modified F model, J. Phys. C **12**, L637-L641 (1979).

[76] F. Y. Wu and K. Y. Lin, On the triangular Potts model with two- and three-site interactions, J. Phys. A **14**, 629-636 (1980).

[77] F. Y. Wu, Exact results for a dilute Potts model, J. Stat. Phys. **23**, 773-782 (1980).

[78] F. Y. Wu, On the equivalence of the Ising model with a vertex model, J. Phys. A **13**, L303-L305 (1980).

[79] R. K. P. Zia and F. Y. Wu, Critical point of the triangular Potts model with two- and three-site interactions, J. Phys. A **14**, 721-727 (1981).

[80] X. Sun and F. Y. Wu, The critical isotherm of the modified F model, Physica A **106**, 292-300 (1981).

[81] F. Y. Wu, Dilute Potts model, duality and site-bond percolation, J. Phys. A **14**, L39-L44 (1981).

[82] W. Kinzel, W. Selke and F. Y. Wu, A Potts model with infinitely degenerate ground state, J. Phys. A **14**, L399-L404 (1981).

[83] S. Sarbach and F. Y. Wu, Exact results on the random Potts model, Z. Phys. B **44**, 309-316 (1981).

[84] F. Y. Wu and Z. R. Yang, Critical phenomena and phase transitions I: Introduction, Prog. in Physics (in Chinese) **1**, 100-124 (1981).

[85] F. Y. Wu and Z. R. Yang, Critical phenomena and phase transitions II: The Ising Model, Prog. in Physics (in Chinese) **1**, 314-364 (1981).

[86] F. Y. Wu and Z. R. Yang, Critical phenomena and phase transitions III: The vertex model, Prog. in Physics (in Chinese) **1**, 487-510 (1981).

[87] F. Y. Wu and Z. R. Yang, Critical phenomena and phase transitions IV: Percolation, Prog. in Physics (in Chinese) **1**, 511-525 (1981).

[88] F. Y. Wu and Z. R. Yang, Critical phenomena and phase transitions V: Model systems, Prog. in Physics (in Chinese) **1**, 525-541 (1981).

[89] F. Y. Wu, The Potts Model, Rev. Mod. Physics **54**, 235-268 (1982).

[90] F. Y. Wu and H. E. Stanley, Domany-Kinzel model of directed percolation: Formulation as a random-walk problem and some exact results, Phys. Rev. Lett. **48**, 775-777 (1982).

[91] I. G. Enting and F. Y. Wu, Triangular lattice Potts model, J. Stat. Phys. **28**, 351-378 (1982).

[92] F. Y. Wu, Random graphs and network communication, J. Phys. A **15**, L395-L398 (1982).

[93] F. Y. Wu, An infinite-range bond percolation, J. Appl. Phys. **53**, 7977 (1982).

[94] F. Y. Wu and H. E. Stanley, Universality in Potts models with two- and three-site interactions, Phys. Rev. B **26**, 6326-6329 (1982).

[95] F. Y. Wu and Z. R. Yang, The Slater model of $K(H_{1-x}D_x)_2PO_4$ in two dimensions, J. Phys. C **16**, L125-L129 (1983).

[96] F. Y. Wu and H. E. Stanley, Polychromatic Potts model: A new lattice-statistical problem and some exact results, J. Phys. A **16**, L751-755 (1983).

[97] J. R. Banavar and F. Y. Wu, Antiferromagnetic Potts model with competing interactions, Phys. Rev. B **29**, 1511-1513 (1984).

[98] F. Y. Wu, Potts model of ferromagnetism, J. Appl. Phys. **55**, 2421-2425 (1984).

[99] D. H. Lee, R. G. Caflish, J. D. Joannopoulos, and F. Y. Wu, Antiferromagnetic classical XY-model: A mean-field analysis, Phys. Rev. B **29**, 2680-2684 (1984).

[100] F. Y. Wu, Exact solution of a triangular Ising model in a nonzero magnetic field, J. Stat. Phys. **40**, 613-620 (1985).

[101] N. C. Chao and F. Y. Wu, Disorder solution of a General checkerboard Ising model in a field and validity of the decimation approach, J. Phys. A **18**, L603-L607 (1985).

[102] Z. R. Yang and F. Y. Wu, Exact solution of a dilute-bond Potts model on the decorated square lattice, Acta Physica Sinica (in Chinese), **34** 484-492 (1985).

[103] J. H. H. Perk and F. Y. Wu, Non-intersecting string model and graphical approach: Equivalence with a Potts model, J. Stat. Phys. **42**, 727-742 (1986).

[104] J. H. H. Perk and F. Y. Wu, Graphical approach to the non-intersecting string model: Star-triangle equation, inversion relation and exact solution, Physica A **138**, 100-124 (1986).

[105] F. Y. Wu, Two-dimensional Ising model with crossing and four-spin interactions and a magnetic field $i\pi kT/2$, J. Stat. Phys. **44**, 455-463 (1986).

[106] F. Y. Wu, On the Horiguchi's Solution of the Blume-Emery-Griffiths Model, Phys. Lett. A **116**, 245-247 (1986).

[107] F. Y. Wu, Thermodynamics of particle in a box, Ann. Report Inst. Phys. Academia Sinica **16**, 25-30 (1986).

[108] H. K. Sim, R. Tao, and F. Y. Wu, Ground-state energy of charged quantum liquids in two dimensions, Phys. Rev. B **34**, 7123-7128 (1986).

[109] W. Selke and F. Y. Wu, Potts models with competing interactions, J. Phys. A **20**, 703-711 (1987).

[110] Y. Chow and F. Y. Wu, Residual entropy and validity of the third law of thermodynamics in discrete spin systems, Phys. Rev. B **36**, 285-288 (1987).

[111] R. Tao and F. Y. Wu, The Vicious Neighbor Problem, J. Phys. A **20**, L299-L306 (1987).

[112] G. O. Zimmerman, A. K. Ibrahim, and F. Y. Wu, The effect of defects on a two-dimensional dipolar system on a honeycomb lattice, J. Appl. Phys. **61**, 4416-4418 (1987).

[113] F. Y. Wu, Book Review: Statistical mechanics of periodic frustrated Ising systems (R. Liebmann), J. Stat. Phys. **48**, 953 (1987).

[114] F. Y. Wu and K. Y. Lin, Ising model on the Union Jack lattice as a free-fermion model, J. Phys. A **20**, 5737-5740 (1987).

[115] G. O. Zimmerman, A. K. Ibrahim, and F. Y. Wu, A planar classical dipolar system on a honeycomb lattice, Phys. Rev. B **37**, 2059-2065 (1988).

[116] X. N. Wu and F. Y. Wu, The Blume-Emery-Griffiths Model on the honeycomb lattice, J. Stat. Phys. **50**, 41-55 (1988).

[117] F. Y. Wu, Potts model and graph theory, J. Stat. Phys. **52**, 99-112 (1988).

[118] K. Y. Lin and F. Y. Wu, Magnetization of the Ising model on the generalized checkerboard lattice, J. Stat. Phys. **52**, 669-677 (1988).

[119] K. Y. Lin and F. Y. Wu, Ising models in the magnetic field $i\pi/2$, Int. J. Mod. Phys. B **2**, 471-481 (1988).

[120] F. Y. Wu and K. Y. Lin, Spin model exhibiting multiple disorder points, Phys. Lett. A **130**, 335-337 (1988).

[121] H. Kunz and F. Y. Wu, Exact results for an $O(n)$ model in two dimensions, J. Phys. A **21**, L1141-L1144 (1988).

[122] F. Y. Wu, Potts model and graph theory, in *Progress in Statistical Mechanics*, Ed. C. K. Hu, (World Scientific, Singapore 1988).

[123] K. Y. Lin and F. Y. Wu, Three-spin correlation of the free-fermion model, J. Phys. A **22** 1121-1130 (1989).

[124] X. N. Wu and F. Y. Wu, Duality properties of a general vertex model, J. Phys. A **22**, L55-L60 (1989).

[125] F. Y. Wu, X. N. Wu and H. W. J. Blöte, Critical frontier of the antiferromagnetic Ising model in a nonzero magnetic field: The honeycomb lattice, Phys. Rev. Lett. **62**, 2273-2276 (1989).

[126] F. Y. Wu and X. N. Wu, Exact coexistence-curve Diameter of a lattice gas with pair and triplet interactions, Phys. Rev. Lett. **63**, 465 (1989).

[127] X. N. Wu and F. Y. Wu, Exact results for lattice models with pair and triplet interactions, J. Phys. A **22**, L1031-L1035 (1989).

[128] J. H. Barry and F. Y. Wu, Exact solutions for a four-spin-interaction Ising model on the three-dimensional pyrochlore lattice, Int. J. Mod. Phys. B **8**, 1247-1275 (1989).

[129] F. Lee, H. H. Chen and F. Y. Wu, Exact duality-decimation transformation and real-space renormalization for the Ising model on a square lattice, Phys. Rev. B **40**, 4871-4876 (1989).

[130] F. Y. Wu, Eight-vertex model and Ising model in a nonzero magnetic field on the honeycomb lattice, J. Phys. A **23**, 375-378 (1990).

[131] J. H. H. Perk, F. Y. Wu and X. N. Wu, Algebraic invariants of the symmetric gauge transformation, J. Phys. A **23**, L131-L135 (1990).

[132] H. W. J. Blöte, F. Y. Wu and X. N. Wu, Critical point of the honeycomb antiferromagnetic Ising model in a nonzero magnetic field: Finite-size analysis, Int. J. Mod. Phys. B **4**, 619-630 (1990).

[133] A. C. N. de Magalhães, J. W. Essam and F. Y. Wu, Duality relation for Potts multispin correlation functions, J. Phys. A **23**, 2651-2669 (1990).

[134] K. Y. Lin and F. Y. Wu, General eight-vertex model on the honeycomb lattice, Mod. Phys. Lett. B **4**, 311-316 (1990).

[135] X. N. Wu and F. Y. Wu, Critical line of the square-lattice Ising model, Phys. Lett. A **144**, 123-126 (1990).

[136] F. Y. Wu, Comment on 'Rigorous Solution of a Two-Dimensional Blume-Emery-Griffiths Model', Phys. Rev. Lett. **64**, 700 (1990).

[137] F. Y. Wu, Yang-Lee edge in the high-temperature limit, Chin. J. Phys. **28**, 335-338 (1990).

[138] F. Y. Wu, Ising Model, in *Encyclopedia of Physics*, Eds. R. G. Lerner and G. L. Trigg (Addison-Wesley, New York 1990).

[139] K. Y. Lin and F. Y. Wu, Unidirectional-convex polygons on the honeycomb lattice, J. Phys. A **23**, 5003-5010 (1990).

[140] J. L. Ting, S. C. Lin and F. Y. Wu, Exact analysis of a lattice gas on the 3 – 12 lattice with two- and three-site interactions, J. Stat. Phys. **62**, 35-43 (1991).

[141] D. L. Strout, D. A. Huckaby and F. Y. Wu, An exactly solvable model ternary solution with three-body interactions, Physica A **173**, 60-71 (1991).

[142] F. Y. Wu, Rigorous results on the Ising model on the triangular lattice with two- and three-spin interactions, Phys. Lett. A **153**, 73-75 (1991).

[143] J. C. Anglès d'Auriac, J.-M. Maillard, and F. Y. Wu, Three-state chiral Potts models in two dimensions: Integrability and symmetry, Physica A **177**, 114-122 (1991).

[144] J. C. Anglès d'Auriac, J.-M. Maillard, and F. Y. Wu, Three-state chiral Potts model on the triangular lattice: A Monte Carlo study, Physica A **179**, 496-506 (1991).

[145] K. Y. Lin and F. Y. Wu, Rigorous results on the anisotropic Ising model, Int. J. Mod. Phys. B **5**, 2125-2132 (1991).

[146] L. H. Gwa and F. Y. Wu, The $O(3)$ gauge transformations and three-state vertex models, J. Phys. A **24**, L503-L507 (1991).

[147] J.-M. Maillard, F. Y. Wu and C. K. Hu, Thermal transmissivity in discrete spin systems: Formulation and applications, J. Phys. A **25**, 2521-2531, (1992).

[148] L. H. Gwa and F. Y. Wu, Critical surface of the Blume-Emery-Griffiths model on the honeycomb lattice, Phys. Rev. B **43**, 13755-13777 (1991).

[149] C. K. Lee, A. S. T. Chiang and F. Y. Wu, Lattice model for the adsorption of benzene in silicalite I, J. Am. Int. Chem. Eng. **38**, 128-135 (1992).

[150] F. Y. Wu, Knot theory and statistical mechanics, in *AIP Conf. Proc.*, No. 248, Ed. C. K. Hu, 3-11 (1992).

[151] F. Y. Wu, Jones polynomial as a Potts model partition function, J. Knot Theory and Its Ramifications **1**, 47-57 (1992).

[152] F. Y. Wu, Transformation properties of a spin-1 system in statistical physics, Chin. J. Phys. **30**, 157-163 (1992).

[153] F. Y. Wu, Jones polynomial and the Potts model, Helv. Phys. Acta **65**, 469-470 (1992).

[154] F. Y. Wu, Knot theory and statistical mechanics, Rev. Mod. Phys. **64**, 1099-1131 (1992).

[155] J.-M. Maillard, G. Rollet and F. Y. Wu, An exact critical frontier for the Potts model on the 3 – 12 lattice, J. Phys. A **26**, L495-L499 (1993).

[156] F. Y. Wu, The Yang-Baxter equation in knot theory, Int. J. Mod. Phys. B **7**, 3737-3750 (1993).

[157] F. Y. Wu, Knot Invariants and Statistical Mechanics: A Physicist's Perspective, in *Braid Group, Knot Theory and Statistical Mechanics*, Eds. M. L. Ge and C. N. Yang (World Scientific, Singapore 1993).

[158] F. Y. Wu and H. Y. Huang, Exact solution of a vertex model in d dimensions, Lett. Math. Phys. **29**, 205-213 (1993).

[159] J. C. Anglès d'Auriac, J.-M. Maillard, G. Rollet, and F. Y. Wu, Zeros of the partition function of the triangular Potts model: Conjectured distribution, Physica A **206**, 441-453 (1994).

[160] H. Y. Huang, and F. Y. Wu, Exact solution of a model of type-II superconductors, Physica A **205**, 31-40 (1994).

[161] J.-M. Maillard, G. Rollet, and F. Y. Wu, Inversion relations and symmetry groups for Potts models on the triangular lattice, J. Phys. A **27**, 3373-3379 (1994).

[162] F. Y. Wu, P. Pant, and C. King, A new knot invariant from the chiral Potts model, Phys. Rev. Lett. **72**, 3937-9340 (1994).

[163] C. K. Hu, J. A. Chen, and F. Y. Wu, Histogram Monte-Carlo renormalization group approach to the Potts model on the Kagomé lattice, Mod. Phys. Lett. B **8**, 455-459 (1994).

[164] F. Y. Wu, P. Pant and C. King, Knot invariant and the chiral Potts model, J. Stat. Phys. **78**, 1253-1276 (1995).

[165] F. Y. Wu and H. Y. Huang, Soluble free-fermion model in d dimensions, Phys. Rev. E **51**, 889-895 (1995).

[166] H. Meyer, J. C. Anglès d'Auriac, J.-M. Maillard, G. Rollet, and F. Y. Wu, Tricritical behavior of the three-state triangular Potts model, Phys. Lett. A **201**, 252-256 (1995).

[167] F. Y. Wu, Statistical mechanics and the knot theory, AIP Conf. Proc. No. **342**, Ed. A. Zepeda, 750-756 (1995).

[168] G. Rollet and F. Y. Wu, Gauge transformation and vertex models: A polynomial formulation, Int. J. Mod. Phys. B **9**, 3209-3217 (1995).

[169] C. K. Hu, C. N. Chen, and F. Y. Wu, Histogram Monte-Carlo renormalization group: Applications to the site percolation, J. Stat. Phys. **82**, 1199-1206 (1996).

[170] C. N. Chen, C. K. Hu, and F. Y. Wu, Partition function zeros of the square lattice Potts model, Phys. Rev. Lett. **76**, 169-172 (1996).

[171] *Statistical Models, Yang-Baxter Equation and Related Topics and Symmetry, Statistical Models and Applications*, Eds. M. L. Ge and F. Y. Wu, Advances in Statistical Mechanics, Vol. 11 (World Scientific, Singapore 1996).

[172] F. Y. Wu *et al.*, Directed compact lattice animals, restricted partitions of numbers, and the infinite-state Potts model, Phys. Rev. Lett. **76**, 173-176 (1996).

[173] H. Y. Huang, F. Y. Wu, H. Kunz, and D. Kim, Interacting dimers on the honeycomb lattice: An exact solution of the five-vertex model, Physica A **228**, 1-32 (1996).

[174] F. Y. Wu, Soluble free-fermion models in d dimensions, in *Statistical Models, Yang-Baxter Equation and Related Topics, and Symmetry, Statistical Mechanical Models and Applications*, Eds. M. L. Ge and F. Y. Wu, 334-339, (World Scientific, Singapore 1996).

[175] A. S. T. Chiang, C. K. Lee, and F. Y. Wu, Theory of adsorbed solutions: analysis of one-dimensional systems, J. Am. Chem. Eng. **42**, 2155-2161 (1996).

[176] F. Y. Wu, The infinite-state Potts model and restricted multidimensional partitions of an integer, J. Comp. and Math. Comp. Modeling **26**, 269-274 (1997).

[177] H. Y. Huang, V. Popkov, and F. Y. Wu, Exact solution of a three-dimensional dimer system, Phys. Rev. Lett. **78**, 409-413 (1997).

[178] C. King and F. Y. Wu, Star-star relation for the Potts model, Int. J. Mod. Phys. B **11**, 51-56 (1997).

[179] H. Y. Huang and F. Y. Wu, The infinite-state Potts model and solid partitions of an integer, Int. J. Mod. Phys. B **11**, 121-126 (1997).

[180] *Exactly Soluble Models in Statistical Mechanics, Historical Perspectives and Current Status*, Eds. C. King and F. Y. Wu, Advances in Statistical Mechanics, Vol. 13 (World Scientific, Singapore 1997).

[181] P. Pant, F. Y. Wu, and J. H. Barry, Exact results for a model ternary polymer mixture, Int. J. Mod. Phys. B **11**, 169-174 (1997).

[182] J. H. Barry, P. Pant, and F. Y. Wu, A lattice-statistical model for ternary polymer mixtures: Exact phase diagrams, Physica A **238**, 149-162 (1997).

[183] F. Y. Wu, Duality relations for Potts correlation functions, Phys. Lett. A **228**, 43-47 (1997).

[184] V. Popkov, D. Kim, H. Y. Huang, and F. Y. Wu, Lattice statistics in three dimensions: Exact solution of layered dimer and layered domain wall models, Phys. Rev. E **56**, 3999-4008 (1997).

[185] P. Pant and F. Y. Wu, Link invariant of the Izergin-Korepin model, J. Phys. A **30**, 7775-7782 (1997)

[186] F. Y. Wu and H. Y. Huang, New sum rule identities and duality relation for the Potts n-point correlation function, Phys. Rev. Lett. **79**, 4954-4957 (1997).

[187] F. Y. Wu and W. T. Lu, Correlation duality relation of the (N_α, N_β) model, Chin. J. Phys. **35**, 768-773 (1997).

[188] F. Y. Wu, The exact solution of a class of three-dimensional lattice-statistical models, in *Proceedings of the 7th Asia Pacific Physics Conference*, Ed. H. Chen, 20-28 (Science Press, Beijing 1998).

[189] W. T. Lu and F. Y. Wu, On the duality relation for correlation functions of the Potts model, J. Phys. A **31**, 2823-2836 (1998).

[190] W. T. Lu and F. Y. Wu, Partition function zeros of a self-dual Ising model, Physica A **258**, 157-170 (1998).

[191] C. A. Chen, C. K. Hu and F. Y. Wu, Critical point of the Kagomé Potts model, J. Phys. A **31**, 7855-7864 (1998).

[192] F. Y. Wu and H. Kunz, Restricted random walks on a graph, Ann. Combinatorics **3**, 475-481 (1999).

[193] F. Y. Wu, C. King and W. T. Lu, On the rooted Tutte polynomial, Ann. Inst. Fourier **49**, 101-112 (1999).

[194] W. T. Lu and F. Y. Wu, Dimer statistics on a Möbius strip and the Klein bottle, Phys. Lett. A **259**, 108-114 (1999).

[195] F. Y. Wu, Exactly solvable models in statistical mechanics, Notices of the AMS, 1353-1354 (1999).

[196] F. Y. Wu, Two solvable lattice models in three dimensions, in *Statistical Mechanics in the Eve of the 21st century*, Eds. M. T. Batchelor and T. Wille, 336-350 (World Scientific, Singapore 1999).

[197] F. Y. Wu, Self-dual polynomials in statistical physics and number theory, in *Yang-Baxter Systems, Nonlinear Models and Their Applications*, Eds. B. K. Chung, Q-Han Park and C. Rim, 77-82 (World Scientific, Singapore 2000).

[198] W. J. Tzeng and F. Y. Wu, Spanning trees on hypercubic lattices and non-orientable surfaces, Appl. Math. Lett. **13**, 19-25 (2000).

[199] N. Sh. Izamailian, C. K. Hu and F. Y. Wu, The 6-vertex model of hydrogen-bonded crystals with bond defects, J. Phys. A **33**, 2185-2193 (2000).

[200] R. Shrock and F. Y. Wu, Spanning trees on graphs and lattices in d dimensions, J. Phys. A **33**, 3881-3902 (2000).

[201] W. Guo, H. W. J. Blöte and F. Y. Wu, Phase transitions in the $n > 2$ honeycomb $O(n)$ model, Phys. Rev. Lett. **85**, 3874-3877 (2000).

[202] W. T. Lu and F. Y. Wu, Density of the Fisher zeros for the Ising model J. Stat. Phys. **102**, 953-970 (2000).

[203] W. T. Lu and F. Y. Wu, Ising model for non-orientable surfaces, Phys. Rev. E **63**, 026107 (2001).

[204] F. Y. Wu and J. Wang, Zeros of the Jones polynomial, Physica A **296**, 483-494 (2001).

[205] W. T. Lu and F. Y. Wu, Closed-packed dimers on non-orientable surfaces, Phys. Lett. A **293**, 235-246 (2002), cond-mat/0110035.

[206] C. King and F. Y. Wu, New correlation relations for the planar Potts model, J. Stat. Phys. to appear (2002).

[207] W. J. Tzeng and F. Y. Wu, Dimers on a simple-quartic net with a vacancy, J. Stat. Phys. to appear (2002), cond-mat/0203149.

[208] F. Y. Wu, Dimers and spanning trees: Some recent results, Int. J. Mod. Phys. B, to appear (2002).

Fa Yueh Wu: Vita

Address

Department of Physics
Northeastern University
Boston, Massachusetts 02115, USA
fywu@neu.edu

Personal Information

Born in China, January 5, 1932
Ph.D. Washington University (St. Louis), 1963
M.S. National Tsing Hua University, Taiwan, 1959
B.S. Naval College of Engineering, Taiwan, 1954

Positions Held

1992-2006	Matthews Distinguished University Professor, NU
1989-1992	University Distinguished Professor, NU
1975-1989	Professor, Northeastern University
1969-1975	Associate Professor, Northeastern University
1967-1969	Assistant Professor, Northeastern University
1963-1967	Assistant Professor, Virginia Polytechnic Institute

Visiting and Other Positions Held

2005	Tsing Hua University, Beijing
2002	University of California, Berkeley
1999	National Center for Theoretical Physics, Taiwan
1996, 1991	University of Paris VI
1995, 1990	Institute of Physics, Academia Sinica, Taiwan
1995, 1990, 1973	Australian National University
1994	University of Amsterdam
1991	Chair Lecturer, National Research Council, Taiwan
1991, 1988, 1985, 1978, 1975	École Polytechnic Fédérale, Laussane, Switzerland
1988, 1974	National Tsing Hua University, Taiwan
1987	Brazilian Center of Theoretical Physics
1987	University of Washington
1984	National Taiwan University, Taiwan
1983-1984	Program Director, National Science Foundation
1981	Institute of Nuclear Energy, Jülich, West Germany
1980	Lorentz Institute and Delft University, Holland

1973	Institut des Hautes Études Scientifiques, Paris
1968	Institute for Theoretical Physics, Stony Brook

Affiliations and Honors

Fulbright-Hays Senior Research Fellow, 1973
Fellow, American Physical Society, 1975
Honorary Professor, Beijing Normal University, 1979
Permanent member, Chinese Physical Society, 1982
Honorary Professor, Southwest Normal University, China, 1985
Honorary Guest Professor, Nankai University, China, 1992
Outstanding Alumnus, National Tsing Hua University, Taiwan, 2003

List of Publications 2002-2009 (For 1955-2001 publications see pp. 626-634)

1 C. King and F. Y. Wu, New correlation duality relations for the planar Potts model J. Stat. Phys. **107**, 919-940 (2002).

2 W. T. Lu and F. Y. Wu, Close-packed dimers on nonorientable surfaces *Phys. Lett. A* **293** 235-246 (2002); Erratum, *ibid.* **298**, 293 (2003).

3 W. J. Tzeng and F. Y. Wu, Dimers on a simple-quartic net with a vacancy, J. Stat. Phys. **116**, 67-68 (2003).

4 F. Y. Wu, Dimers and spanning trees: Some recent results, Int. J. Mod. Phys. B **16**, 1951-1961 (2003).

5 W. T. Lu and F. Y. Wu, Generalized Fibonacci numbers and dimer statistics, Mod. Phys. Lett. B **16**, 1177-1182 (2003); Erratum, *ibid.* **17**, 789 (2003).

6 E. H. Lieb and F. Y. Wu, The one-dimensional Hubbard model: A reminiscence, Physica A **321**, 1-27 (2003).

7 D. H. Lee and F. Y. Wu, Duality relation for frustrated spin systems, Phys. Rev. E **67**, 026111 (2003).

8 F. Y. Wu and H. Kunz, The odd eight-vertex model, J. Stat. Phys. **116**, 67-78 2004).

9 F. Y. Wu, Theory of Resistor Network: The Two-Point Resistance, J. Phys. A **37**, 6653-6673 (2004).

10 W. T. Lu and F. Y Wu, Soluble kagome Ising model in a magnetic field, Phys. Rev. E **71**, 042160 (2005).

11 L. M. Gasser and F.Y. Wu, On the entropy of spanning trees on a large triangular lattice, Ramanujan Journal **10**, 205-214 (2005).

12 L. C. Chen and F. Y. Wu, Random cluster model and a new integration identity, J. Phys. A **38**, 6271-6276 (2005).

13 A. Widom and F. Y. Wu, Book review: "Lectures on the Kinetic Theory of Gases, Non-equilibrium Thermodynamics and Statistical Theories", by Ta-You Wu, J. Stat. Phys. **119**, 945-948 (2005).

14 F. Y. Wu, Dimers on two-dimensional lattices, Int. J. Mod. Phys. B **20**, 5357-5371 (2006).

15 W. Guo, X. Qian, H. W. J. Blöte and F. Y. Wu, Critical line of an n-component cubic model, Phys. Rev. E **73**, 026104 (2006).

16 W. J. Tzeng and F. Y. Wu, Theory of impedance networks: The two-point impedance and *LC* resonances, J. Phys. A **39**, 8579-8591 (2006).

17 L. C. Chen and F. Y. Wu, Directed percolation in two dimensions: An exact solution, in *Differential Geometry and Physics, Nankai Tracts in Mathematics*, Vol. **10**, Eds. M. L. Ge and W. Zhang (World Scientific, Singapore 2006) pp. 160-168.

18 F. Y. Wu, New critical frontiers for the Potts and percolation models, Phys. Rev. Lett. **96**, 090602 (2006).

19 F. Y. Wu, The Pfaffian solution of a dimer-monomer problem: Single monomer on the boundary, Phys. Rev. E **74**, 020104(R) (2006); Erratum, *ibid* **74**, 039907 (2006).

20 F. Wang and F. Y. Wu, Exact solution of closed packed dimers on the kagome lattice, Phys. Rev. E **75**, 040105(R) (2007).

21 F. Y. Wu and F. Wang, Dimers on the kagome lattice I: Finite lattices, Physica A **387**, 4148-4156 (2008).

22 F. Wang and F. Y. Wu, Dimers on the kagome lattice II: Correlations and the Grassmannian approach, Physica A **387**, 4157-4162 (2008).

23 F. Y. Wu, Professor C. N. Yang and statistical mechanics, Int. J. Mod. Phys. B **22**, 1899-1909 (2008).

24 F. Y. Wu, B. M. McCoy, M. E. Fisher and L. Chayes, On a recent conjectured solution of the three-dimensional Ising model, Phil. Mag. **88**, 3093-3095 (2008).

25 F. Y. Wu, B. M. McCoy, M. E. Fisher and L. Chayes, Rejoinder to the response to 'Comment on a recent conjectured solution of the three-dmensional Ising model', Phil. Mag. **88**, 3103 (2008).

26 J. W. Essam and F. Y. Wu, The exact corner-to-corner resistance of an $M \times N$ resistor network: Asymptotic expansion, J. Phys. A **42**, 025205 (2009).

Home page

For a more detailed vita, see
http://www.physics.neu.edu/wu.html/

Index of Names in the Commentaries